Viva in
Anatomy, Physiology and Biochemistry

Viva in
Anatomy, Physiology and Biochemistry

Second Edition

Compiled by
Anjula Vij MBBS
USA

The Health Sciences Publisher
New Delhi | London | Philadelphia | Panama

Jaypee Brothers Medical Publishers (P) Ltd.

Headquarters
Jaypee Brothers Medical Publishers (P) Ltd.
4838/24, Ansari Road, Daryaganj
New Delhi 110 002, India
Phone: +91-11-43574357
Fax: +91-11-43574314
E-mail: jaypee@jaypeebrothers.com

Overseas Offices

J.P. Medical Ltd.
83, Victoria Street, London
SW1H 0HW (UK)
Phone: +44-20 3170 8910
Fax: +44(0)20 3008 6180
E-mail: info@jpmedpub.com

Jaypee-Highlights Medical Publishers Inc.
City of Knowledge, Bld. 237, Clayton
Panama City, Panama
Phone: +1 507-301-0496
Fax: +1 507-301-0499
E-mail: cservice@jphmedical.com

Jaypee Medical Inc.
The Bourse
111, South Independence Mall East
Suite 835, Philadelphia, PA 19106, USA
Phone: +1 267-519-9789
E-mail: jpmed.us@gmail.com

Jaypee Brothers Medical Publishers (P) Ltd.
17/1-B, Babar Road, Block-B, Shaymali
Mohammadpur, Dhaka-1207
Bangladesh
Mobile: +08801912003485
E-mail: jaypeedhaka@gmail.com

Jaypee Brothers Medical Publishers (P) Ltd.
Bhotahity, Kathmandu, Nepal
Phone: +977-9741283608
E-mail: kathmandu@jaypeebrothers.com

Website: www.jaypeebrothers.com
Website: www.jaypeedigital.com

Inquiries for bulk sales may be solicited at: jaypee@jaypeebrothers.com

Viva in Anatomy, Physiology and Biochemistry

First Edition: 2010

Second Edition: **2015**

ISBN: 978-93-5152-599-8

Printed at Sanat Printers, Kundli

Preface to the Second Edition

Practical examinations form an important component of the professional examinations during the MBBS. It is as important to get through the theory papers with flying colors as that through the practical papers. Since the students of first professional examination are relatively new to the concept of extensive viva voce examinations, it is important for them to get familiar with the kinds of questions they might have to face before examiners.

The second edition is inspired by the success of the first edition. This book represents a unique combination of important viva questions and answers of all the three subjects (Anatomy, Physiology and Biochemistry) taught in the first professional examinations. Its unique presentation in the form of three-column format adequately equipped with appropriate illustrations would make it an interesting reading for the students. Three new segments, i.e. Embryology, Histology and Genitourinary system are added to the chapter of "General Anatomy". In addition to this, several new diagrams (more than 50), color plates (21) and appendix have been added to the new edition.

The students must, however, remember that the book is in no way a replacement for standard textbooks in anatomy, physiology and biochemistry. Nothing can be a replacement for a standard textbook in a particular subject, which would help clarify the various concepts and fundamentals. The student must remember that "the mouth speaks only what the mind knows", so nothing can be replacement for a sound and effective examination preparation. Strong foundation in these three basic subjects goes a long way in the development of an undergraduate student into a full-fledged doctor. Thus, the students must try to grasp all the important concepts before they start reading this book. This book is meant only for the aid and assistance to the first professional examination and for removing all the fears from the students' mind.

Making a book is a mammoth task, involving the inputs of several people. I would like to acknowledge various doctors, from whose books images have been used. Last but not least, I would like to thank the medical editors, Ms. Dikshika Joshi and Ms. Alpana Priya for helping me in revising the book. I am also especially grateful to the entire staff of Jaypee Brothers for bringing out this fantastic book.

Anjula Vij

Preface to the First Edition

Practical examinations form an important component of the professional examinations during the MBBS. It is as important to get through the theory papers with flying colors as that to the practical papers. Since the students of first professional examination are relatively new to the concept of extensive viva voce examinations, it is important for them to get familiar with the kinds of questions they might have to face before the examiners.

This book presents a unique combination of important viva questions and answers of all the three subjects (Anatomy, Physiology and Biochemistry) taught in the first professional examinations. Its unique presentation in the form of three-column format, adequately equipped with appropriate illustrations would make it an interesting reading for the students. The students must, however, remember that the book is in no way a replacement for standard textbooks in anatomy, physiology and biochemistry. Nothing can be a replacement for a standard textbook in a particular subject, which would help clarify the various concepts and fundamentals. The students must remember that "the mouth speaks only those what the mind knows", so nothing can be replacement for a sound and effective examination preparation. Strong foundation in these three basic subjects goes a long way in the development of an undergraduate student into a full-fledged doctor. Thus, the students must try to grasp all the important concepts before they start reading this book. This book is meant only for the aid and assistance to the first professional examination and for removing all the fears from the students' mind.

Anjula Vij

Contents

Section 1: ANATOMY

Section 2: PHYSIOLOGY

Section 3: BIOCHEMISTRY

Section 1

ANATOMY

CHAPTERS

Chapter 1

General Anatomy

Embryology

Q.1 When does the blastocyst get implanted in the uterine wall?
The implantation starts at 7½ days and is completed by the 10th day (Fig. 1.1).

Q.2 What is an organizer?
Organizers are those which are concerned with early histogenesis and development.

Q.3 Give an example of a good organizer.
Primitive streak is the best organizer in the early stage.

Q.4 What is twinning?
The birth of two human infants at a time is twinning.

Q.5 What is ectopia cardia?
This is a condition where there is a developmental defect of the sternum and pericardium.

Q.6 What are the types of twinning present?
There are monozygotic and dizygotic twins. The monozygotic or monoovulatory twins are due to the fertilization of a single ovum by a single sperm. They are defined as true or identical twins.

The dizygotic twins are a result of the fertilization of two ova produced during the same ovulatory phase by two sperms. The dizygotic twins unlike monozygotic have different genetic constitution and could be of different sex. They are also called fraternal or unlike twins and are usually dischorial (Figs 1.2 and 1.3).

Q.7 What is superfecundation?
It is the fertilization of two ova produced from the same ovulatory phase by two sperms from two separate sexual encounters.

Q.8 What is superfetation?
It is the fertilization of two ova produced during separate ovulation cycles.

Q.9 What are the arterial arches left over in the humans?
There are three arterial arches which are left over which are 3rd, 4th and 6th arches. The common and the internal carotid arteries are from 3rd arch. The right subclavian is from right fourth arch. The arch of aorta is from the left 4th arch. The pulmonary arteries present on both sides are from 6th arch.

Q.10 Why is the left innominate vein longer than the right innominate vein?
The left innominate vein is longer than the light one because it is formed by the communication between the anterior cardinal veins and the functional part of the left cardinal vein.

Q.11 Why is the left branch of portal vein longer than the right branch?
This is because it is formed by the fusion of the left vitelline vein and anastomoses present between the two vitelline veins.

Q.12 Why does the thoracic duct opens at the junction of internal jugular and subclavian veins?
The thoracic duct begins at the cisterna chyli, passes obliquely from 7th to 5th thoracic vertebra and goes upwards to the left side. During development of the thoracic ducts, there is a communication between the ducts, passing obliquely from the right 7th thoracic to the left 5th thoracic vertebra. The right part of the duct above the communication and the left part below the communication disappear. The thoracic duct finally opens at the junction of internal jugular and subclavian veins.

Q.13 What is supernumerary cusps of the cardiac valves?
This is a defect due to the atypical division of the cusps of the primordia.

Q.14 What is double aorta?
This is due to the non-disappearance of the right dorsal aorta.

Q.15 Why does the diaphragm get its nerve supply from the phrenic nerve?
The diaphragm during development is in the cervical region which is supplied by the cervical 3rd, 4th and 5th segment nerves, which is the phrenic nerve. With the descent of the diaphragm the phrenic nerve also descends and this phenomenon is known as "neurobiotaxis".

Q.16 Why is the left renal vein longer than the right renal vein?
This is because it is formed by the fusion of the subcardinal anastomotic channel and the primitive left renal vein.

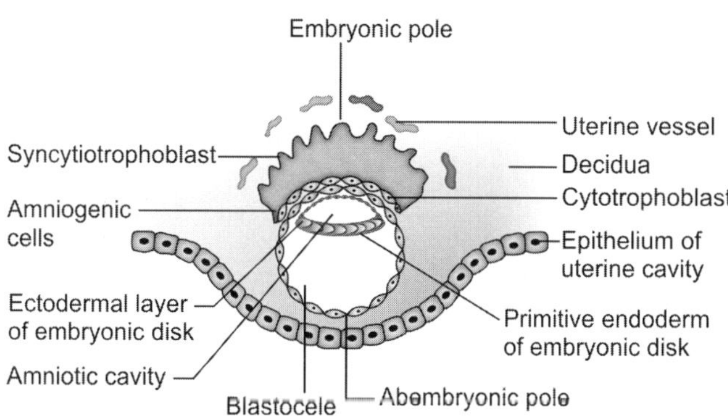

Fig. 1.1: Implantation of blastocyst

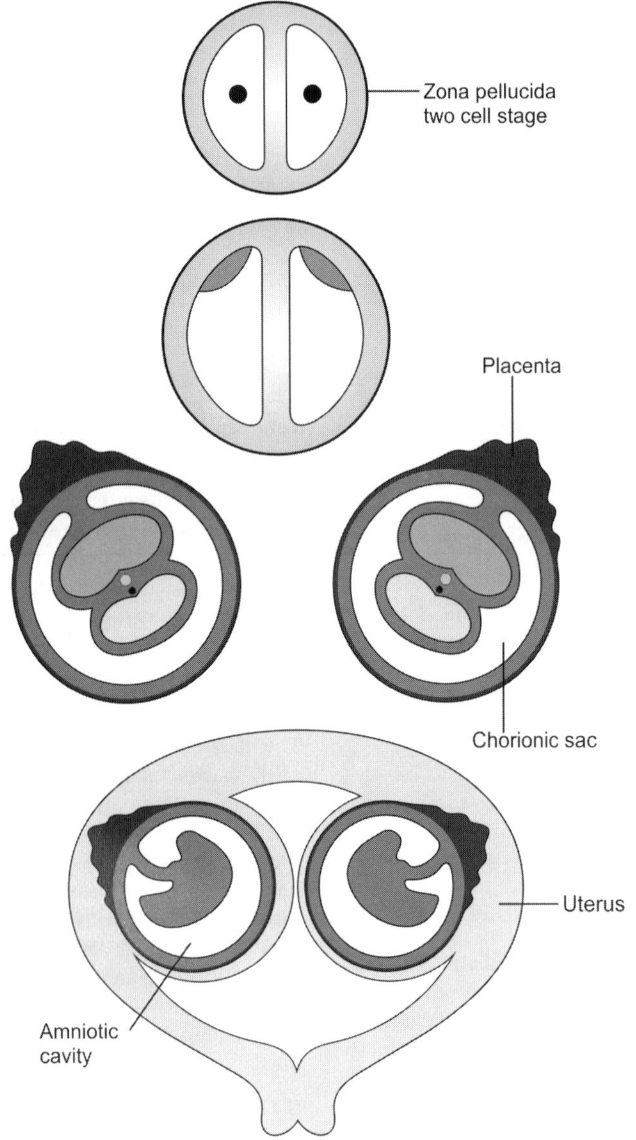

Fig. 1.2: Monozygotic twins with separate amniotic, chorionic sacs and separate placentas

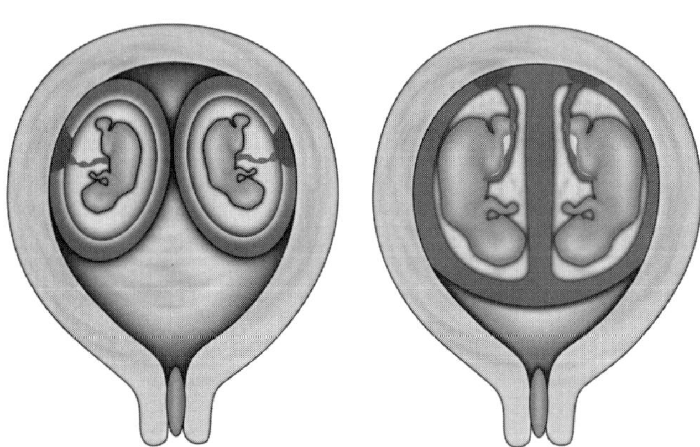

Fig. 1.3: Dizygotic twins having its own amnion, chorion and placenta, however placentas get fused

Q.17 Why is the submandibular salivary gland fixed anteriorly, whereas, the posterior part is free?

The submandibular gland is fixed anteriorly as it develops in the floor of the pharynx and the starting point is at the frenulum. The posterior part is free because it is not in the floor of the mouth cavity.

Q.18 What is coloboma of the iris?

It is due to the non-fusion of the fetal fissure, which is the inferomedial part of the eye (Fig. 1.4).

Q.19 What is the reason for the oblique facial cleft?

This is due to the non-fusion of the frontonasal process with the maxillary process.

Q.20 Why are there three lobes in the right lung and two lobes in the left lung?

The lung is an endodermal derivative. The right lung bud gives rise to two monopodial diverticulae, whereas the left bud gives rise to only one monopodial diverticula and hence there are three lobes in the right lung and two lobes in the left lung.

Q.21 Why there is more growth of the dorsal part of the stomach than the ventral part?

The dorsal part is mainly supplied by the dorsal aorta, hence the growth is more on the dorsal aspect of the stomach.

Q.22 What is the cause of sinus in the middle of neck?

The sinus in the midline of the neck may represent the remnant of thyroglossal duct.

Q.23 What is the cause of lateral sinuses in the neck?

It is due to remnant second and third pouch, which are located in front of the sternocleidomastoid muscle.

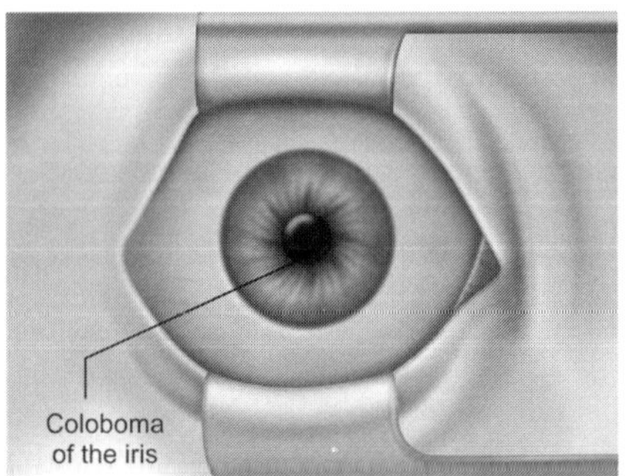

Fig. 1.4: Coloboma of the iris

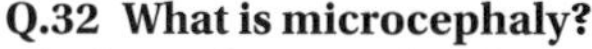

Q.24 What is hiatus hernia?
It is due to intrathoracic herniation of the cardia with a part of the stomach and peritoneal covering over it. It occurs through a weakness in the esophageal opening of the diaphragm (Fig. 1.5).

Q.25 What is cleidocranial dysostosis?
This is a condition, where there is partial failure of ossification of membrane bones and this affects the vault of the skull primarily. This may also lead to the non-fusion of the fontanelles.

Q.26 What is plagiocephaly?
Plagiocephaly is a condition developed due to fusions and asymmetrical conditions.

Q.27 What is cranioschisis?
The membranous bones of the skull fail to ossify giving rise to a wide open vault of the skull which is known as cranioschisis. This is associated with anencephaly, where the brain fails to develop.

Q.28 What is hypertelorism?
This occurs when there is overgrowth of the lesser wing of the sphenoid bone. In this condition, the eyes are separated abnormally and this is seen in Down syndrome.

Q.29 What is amastia?
The absence of one or both breasts is termed as amastia.

Q.30 What is a crater nipple?
The "crater nipple" occurs due to the failure of mesodermal elevation at the epidermis. There is congenital retraction of the nipples and downward growth of the lactiferous duct.

Q.31 What is Meckel's cartilage?
This is the mandibular arch (I arch) cartilage. It is represented in the mandibles as the genital tubercles, anterior part of the coronoid process and a thickened area between the head of the mandible and the lingula.

Q.32 What is microcephaly?
The brain does not develop properly. It is characterized with the presence of a small brain and cranial cavity.

Q.33 What is horseshoe kidney?
This is a condition, where the caudal ends of the metanephros fuse to form a "u" shaped kidney. It is located in front of the abdominal aorta. The ureters are in the normal position (Fig. 1.6).

Q.34 What is ductus venosus?
It is a communication between the left umbilical vein and the right hepatocardiac channel. This becomes ligamentum venosum after birth.

Q.35 How would you differentiate between embryological and pathological Meckel's diverticulum?
The embryological diverticulum has all the layer of the intestine, whereas, the pathological may not have all the layers as it is diseased.

Q.36 How can you differentiate exomphalos and congenital umbilical hernia?
Exomphalos has no umbilical sac. The congenital umbilical hernia may be paraumbilical or may above or below the umbilicus and usually has a sac (Figs 1.7 and 1.8).

Q.37 What is the cause for the development of double ureters?
The ureter develops from the ureteric bud of the mesonephric duct, if it starts to divide before reaching the metanephric cap, it gives rise to double ureters.

Q.38 What is a weeping umbilicus?
It is due to the externalization of the epithelium of the vitellointestinal duct which is red in color.

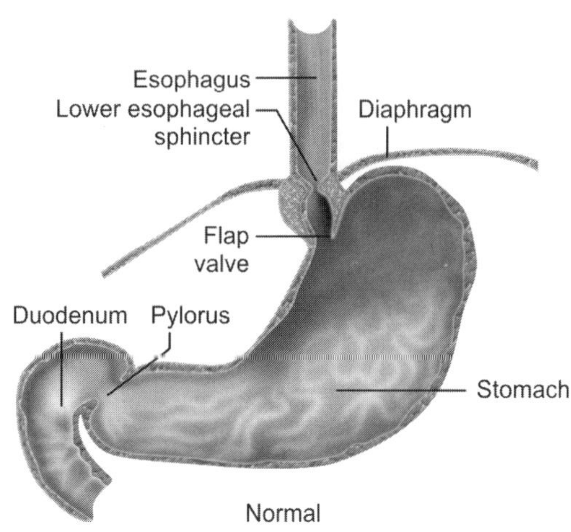

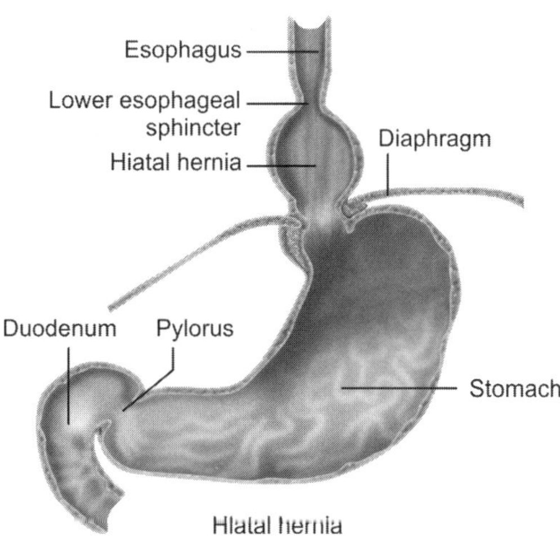

Fig. 1.5: Hiatus hernia

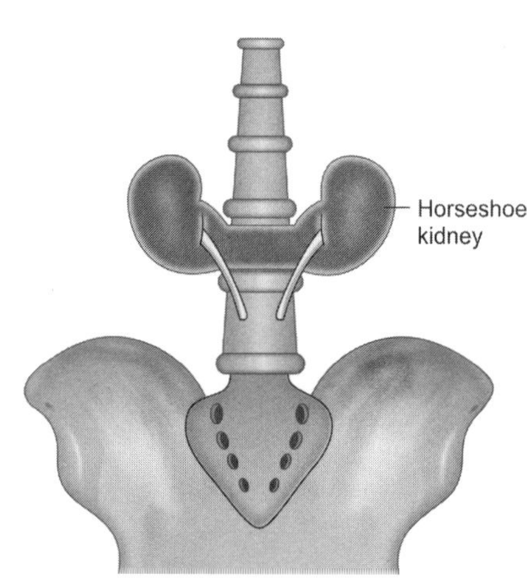

Fig. 1.6: Horseshoe kidney

Q.39 What is urachus?

It is allantoic diverticulum from the yolk sac. This will form a median umbilical ligament in the adult.

Q.40 What are the components of placenta?

Placenta is defined as having two components with the maternal component being decidua basalis and the fetal component being chorion frondosum (Fig. 1.9).

Q.41 What is chorion frondosum?

It is tertiary chorionic villi which are composed of umbilical vessels, extra embryonic mesoderm and trophoblasts (Fig. 1.9).

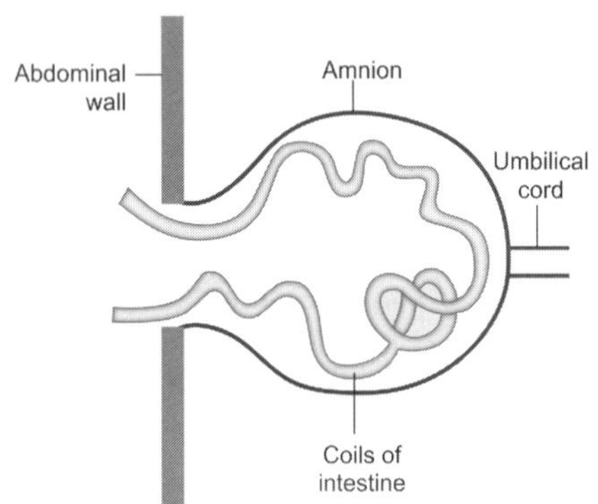

Fig. 1.7: Exomphalos

Q.42 What is a true hermaphrodite?

This occurs when there is an ovary on one side and the testis on the other side. Sometimes both could be either side and this is known as ovotestis. This usually has a genetic predilection with nearly 53–54% having normal chromosomal build up whereas the remainder may have mosaic patterns such as 44 XXY, 44 XX, or 44 XY.

Q.43 What is a pseudohermaphrodite?

This is a condition when a person externally may appear to be a male or female but characterized by the presence of an opposite phenotype.

Q.44 What is Down syndrome?

This is also called "trisomy 21". It is quite a common condition, where the chromosome number is 47. It is due to meiotic nondisjunction with translocation occurring at the level of the long arm of chromosome 21. The extra chromosome is maternal in origin. The clinical features are mental retardation, cardiac malformations, presence of epicanthic folds, recurrent infections, wide gap between great toe and second toe, small mouth, large tongue.

Q.45 What is Edward's syndrome?

This is trisomy 18, 47X. It commonly affects the females. The clinical features are mental retardation, prominent occiput, overlapping fingers, micrognathia, developmental defects of heart and kidney and presence of rocker-bottom feet.

Q.46 What is Patau's syndrome?

This is trisomy 13. The clinical features are microcephaly, mental retardation, microphthalmia, harelip, cleft palate, cardiac abnormalities, polydactyly and short fingers, umbilical hernia and developmental defects of the kidney.

Q.47 What is Klinefelter's syndrome?

This is seen in males characterized with the presence of an extra sex-chromosome. The extra "x" chromosome may be from paternal or maternal side. The clinical features are hypogonadism, elongated body due to increased length between pubis and sole, small penis, azoospermia, sterility, gynecomastia, testis are reduced in size, reduction in facial and body hair, feminine habitus and a high pitched voice (Fig. 1.10).

Q.48 What is Turner's syndrome?

The incidence is one in 3000. Females are affected. The karyotype is 45 XO. There is primary hypogonadism in females thought to be due to the absence of one chromosome. The clinical features are short stature due to growth retardation, webbing of the neck, increase in the carrying angle of the lower extremity known as cubitus valgus, low posterior hair line, far placed nipples, cardiac defects, arched palate, lymphedema of hands and feet, absence of secondary sexual characters, infantile genitalia, inadequate breast development, and streak-ovaries.

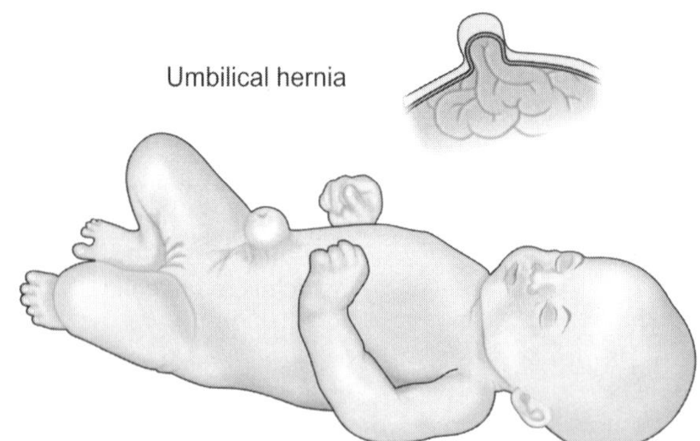

Fig. 1.8: Congenital umbilical hernia

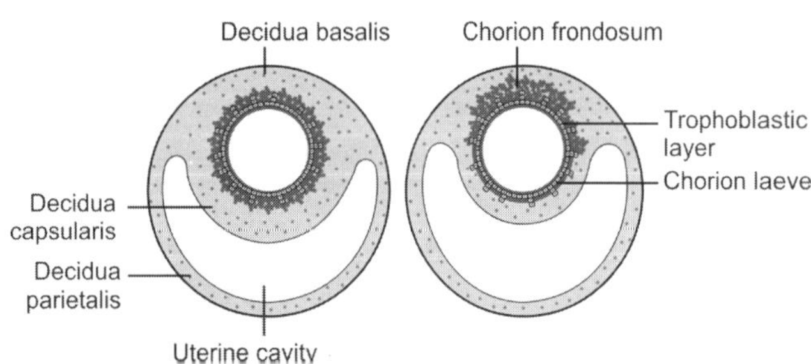

Fig. 1.9: Formation of chorion frondosum and chorion laeve

Note: Three parts of decidua—1. Decidua basalis, 2. Decidua capsularis and 3. Decidua parietalis

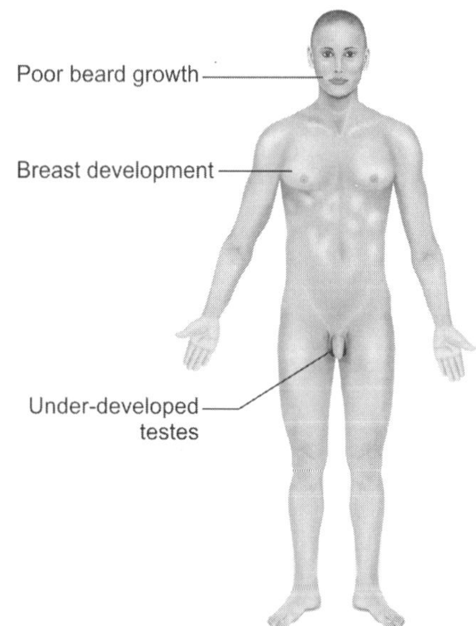

Fig. 1.10: Klinefelter's syndrome

PLATE-1

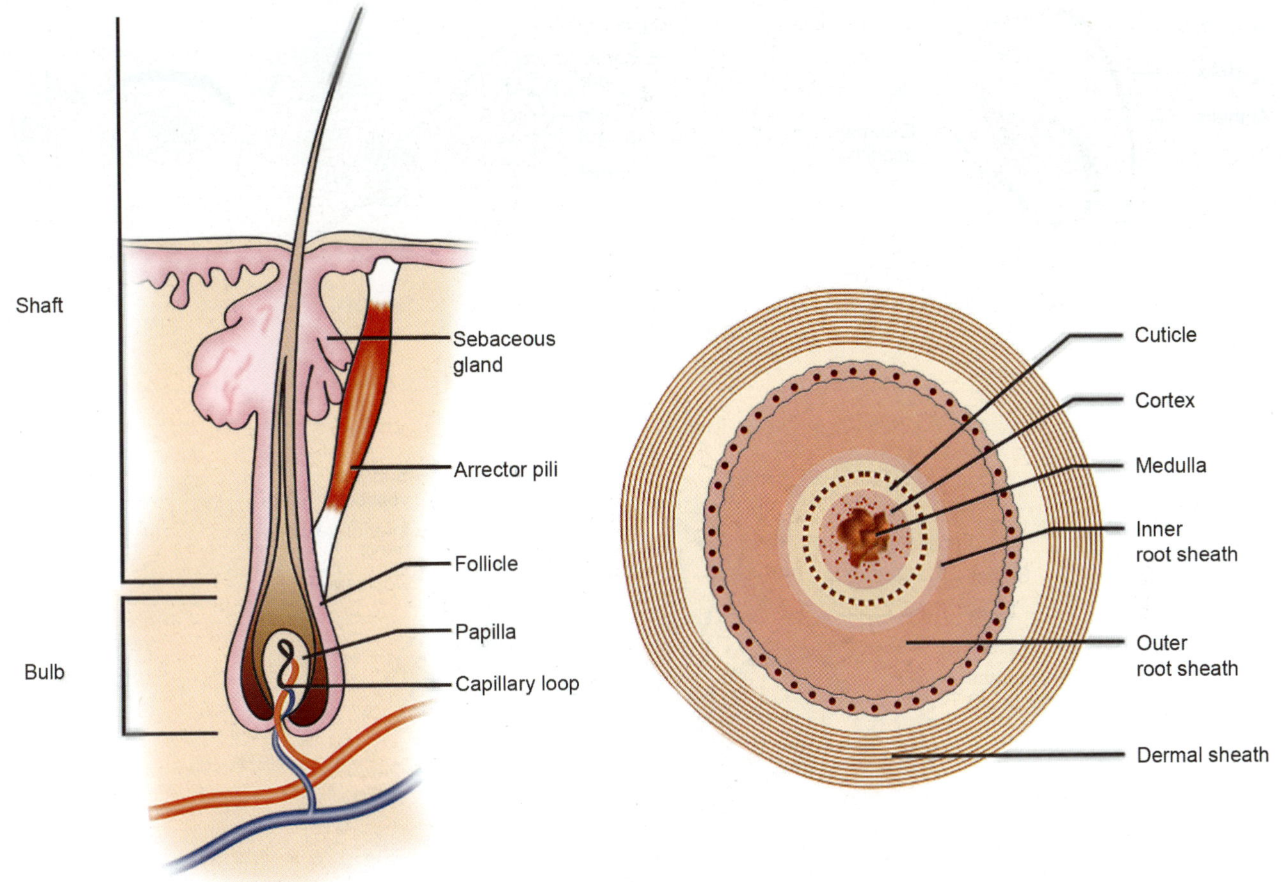

Hair in longitudinal and cross-section

PLATE-2

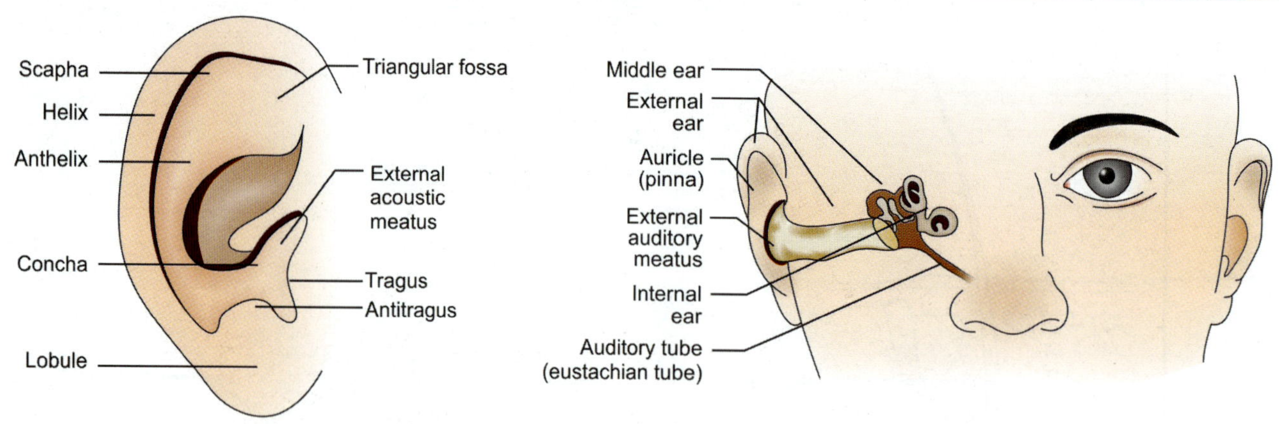

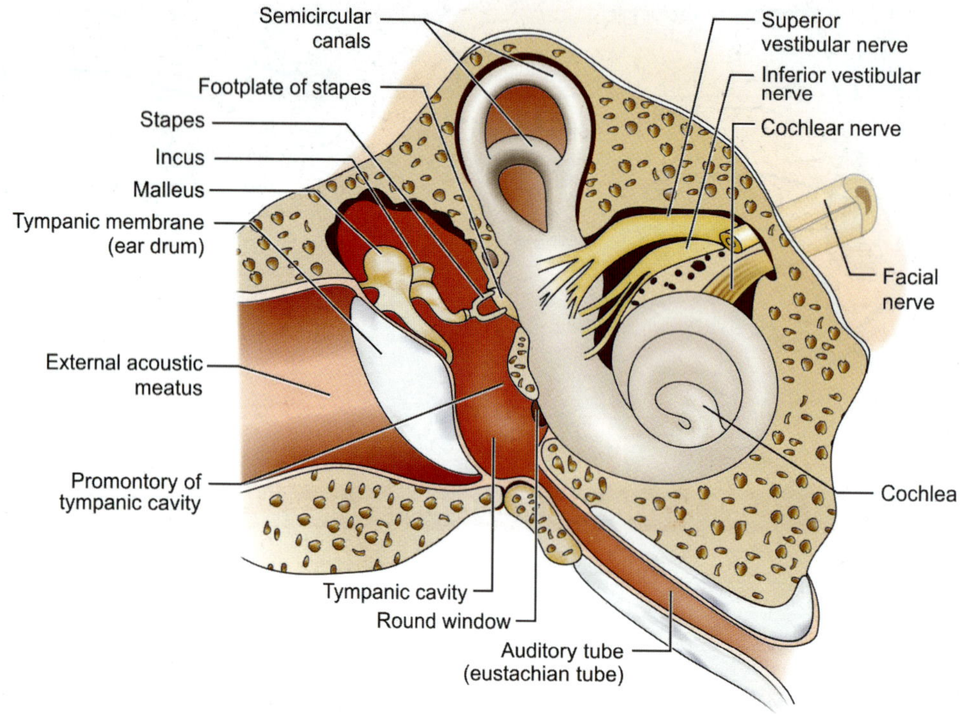

Structures of the ear

PLATE-3

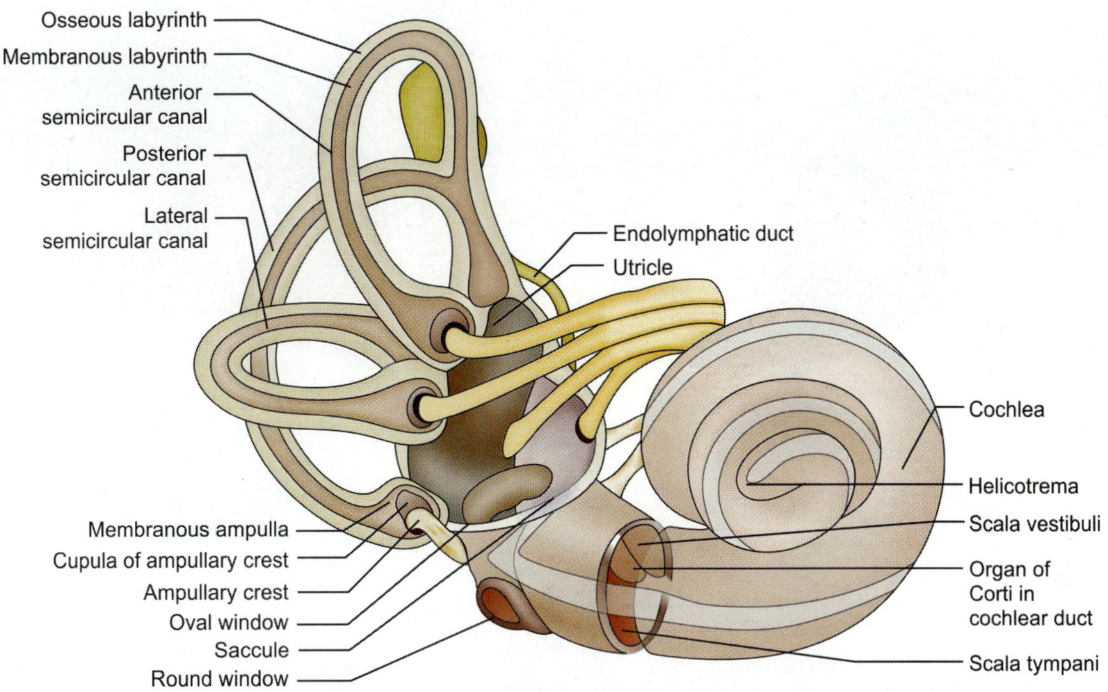

Structure of the ear

PLATE-4

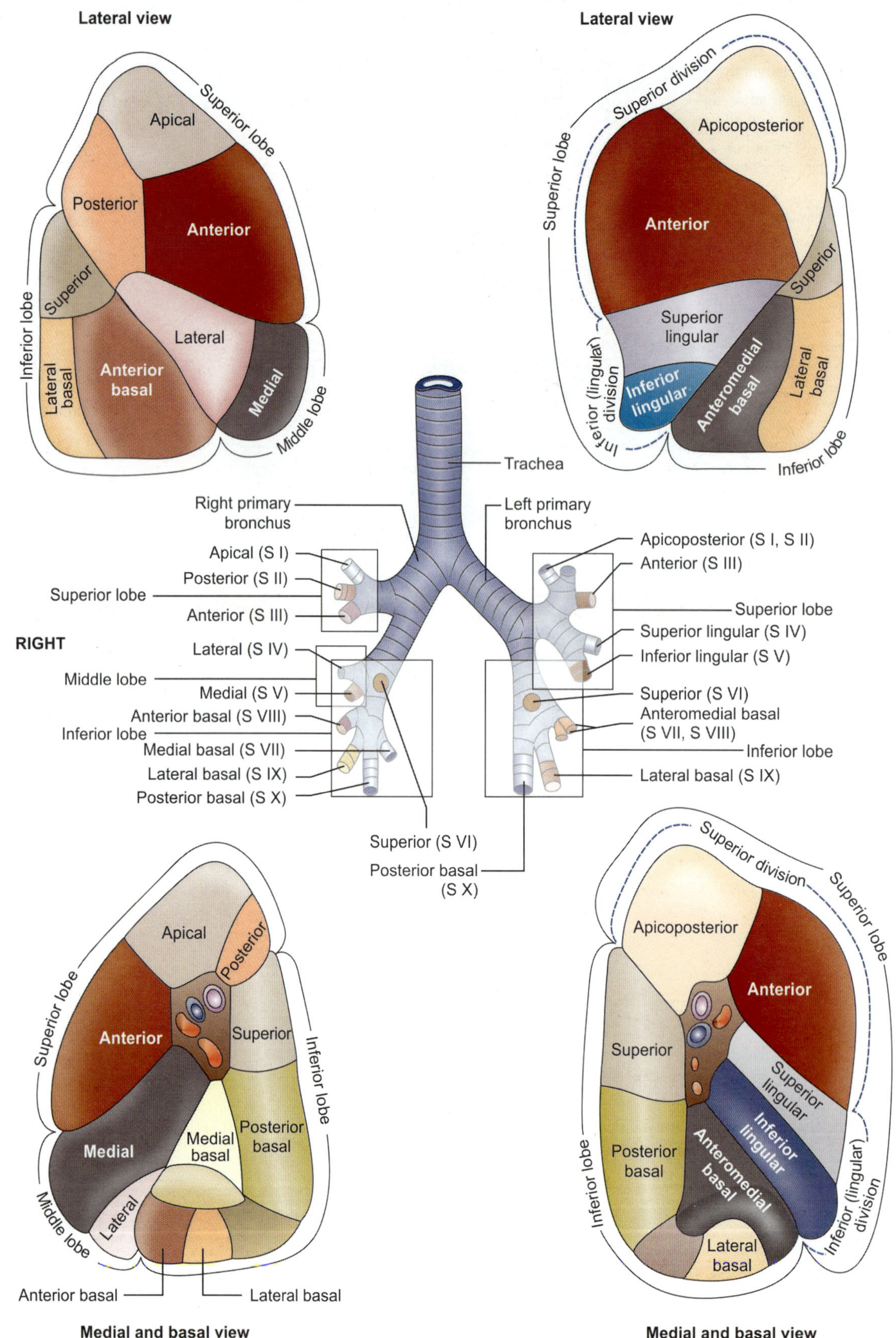

Pulmonary segments

PLATE-5

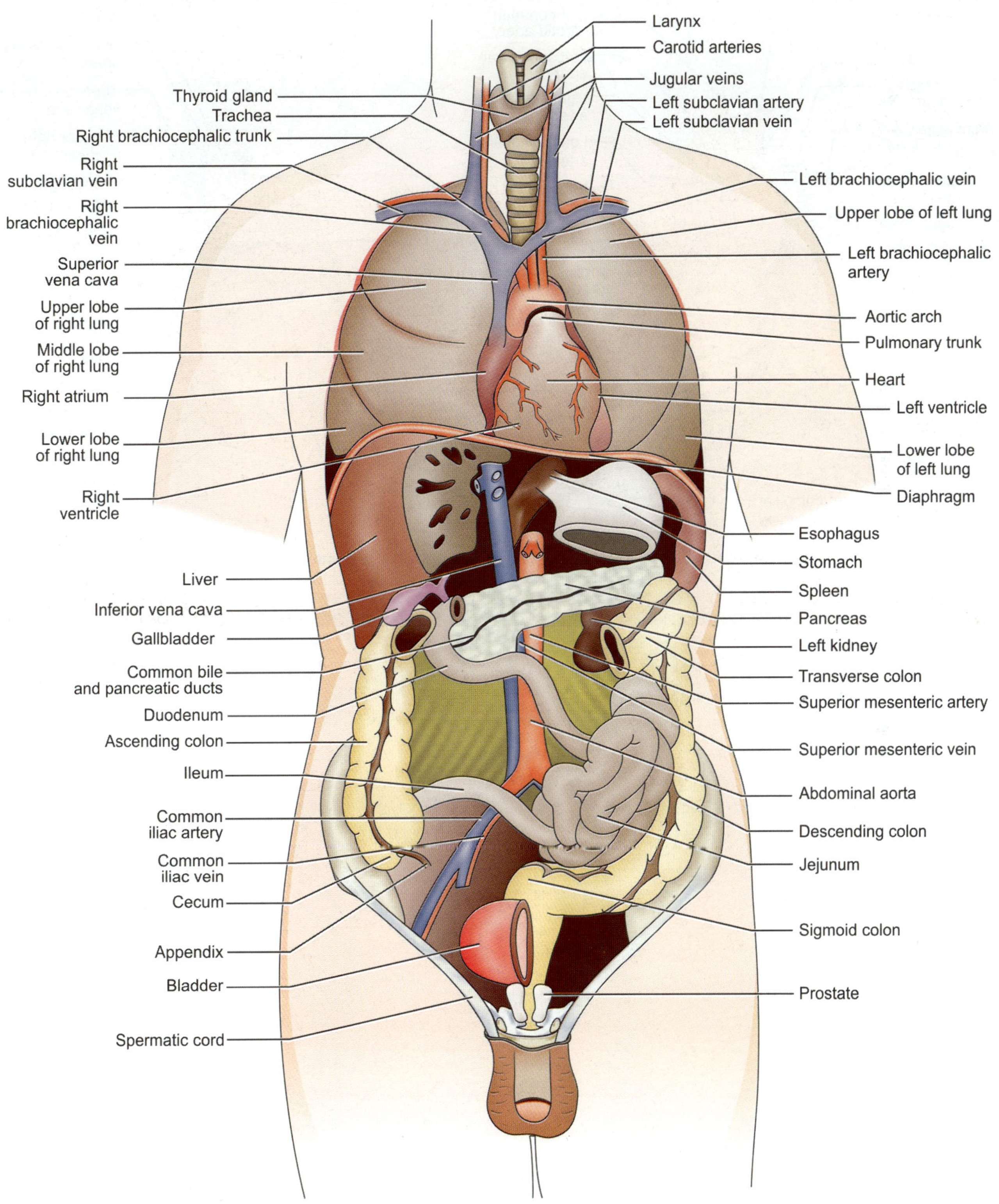

Thoracic and abdominal viscera

PLATE-6

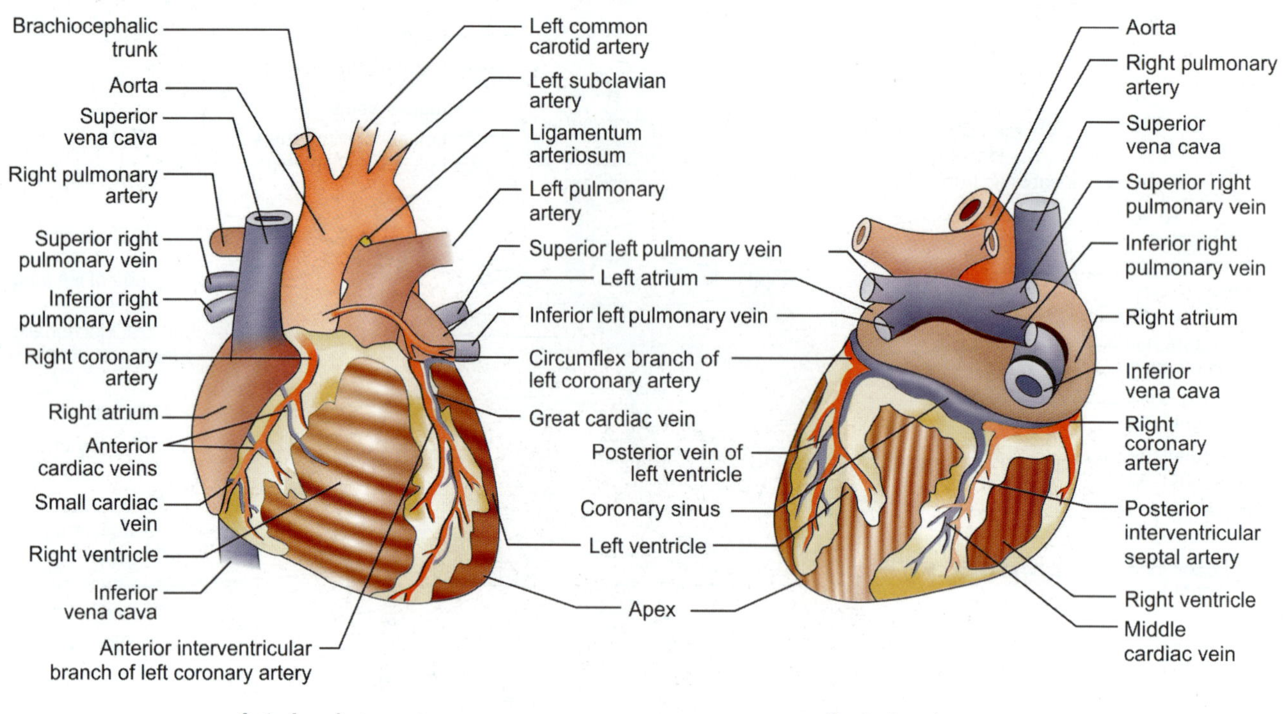

Structures of heart

PLATE-7

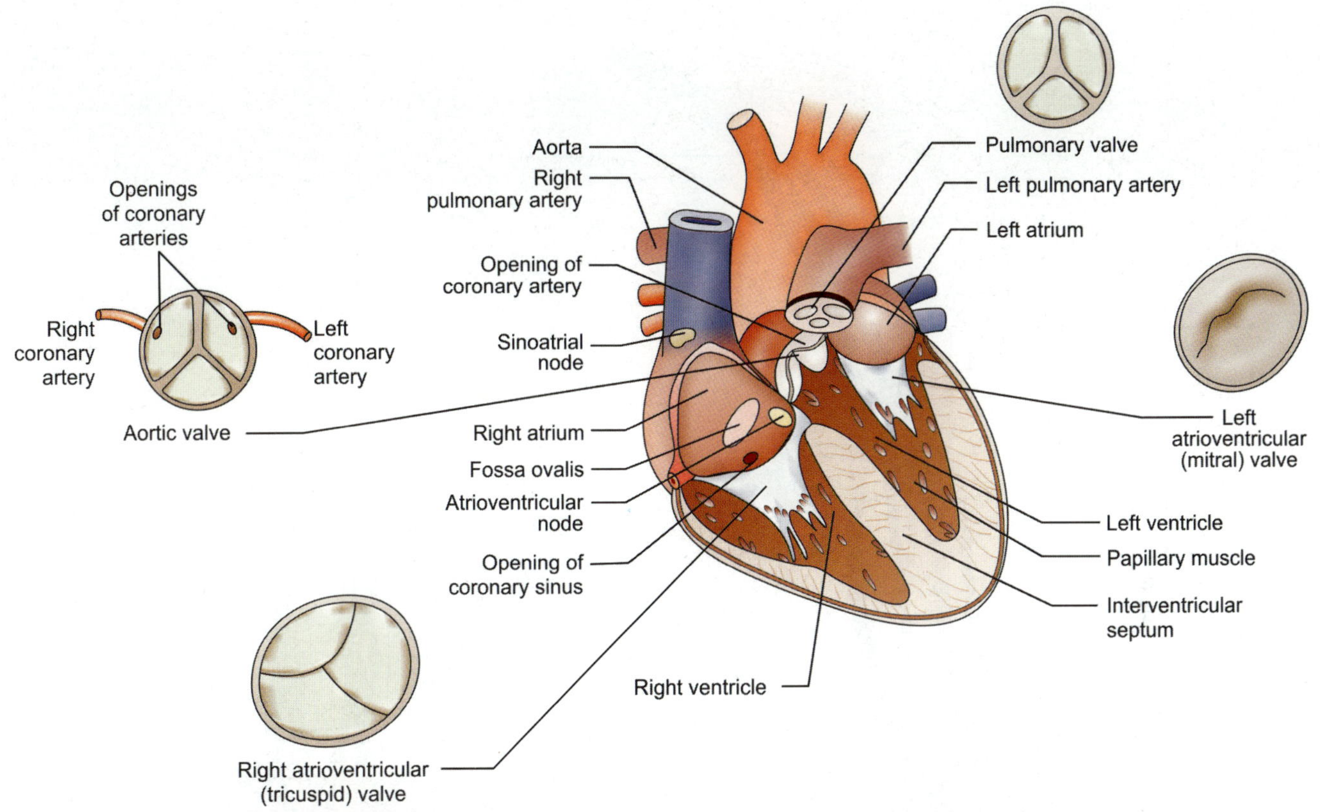

Structure of heart

PLATE-8

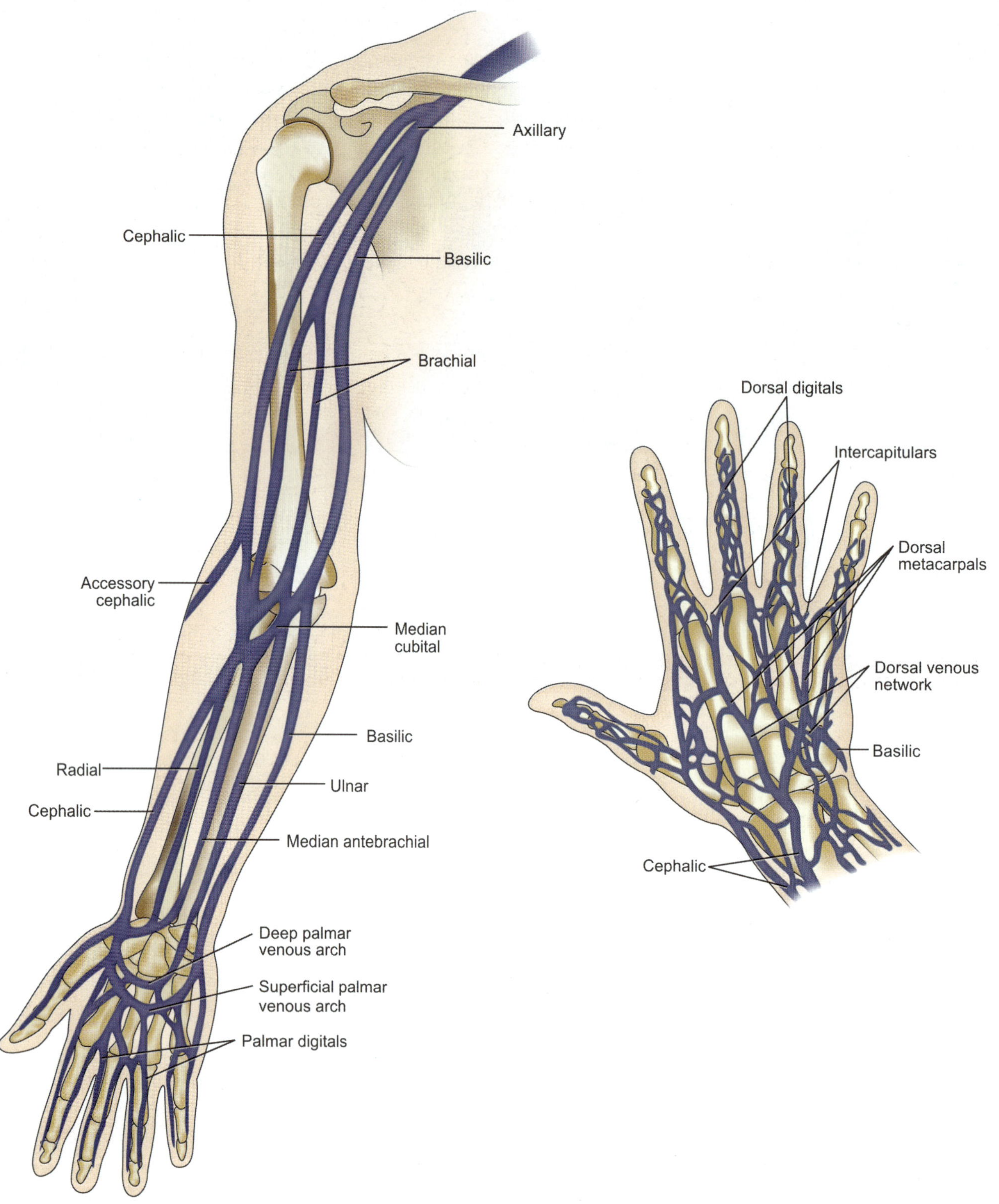

Veins of upper limbs

PLATE-9

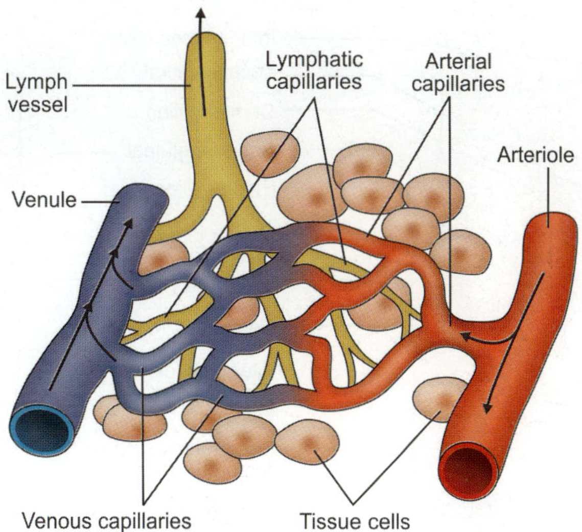

Capillary bed

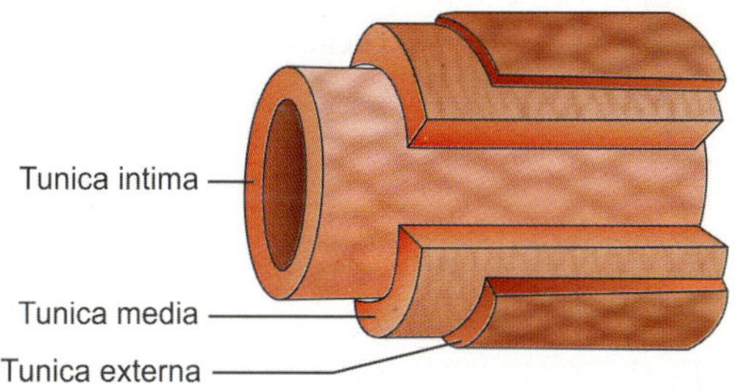

The three coats of a vein

Veins of upper limbs

PLATE-10

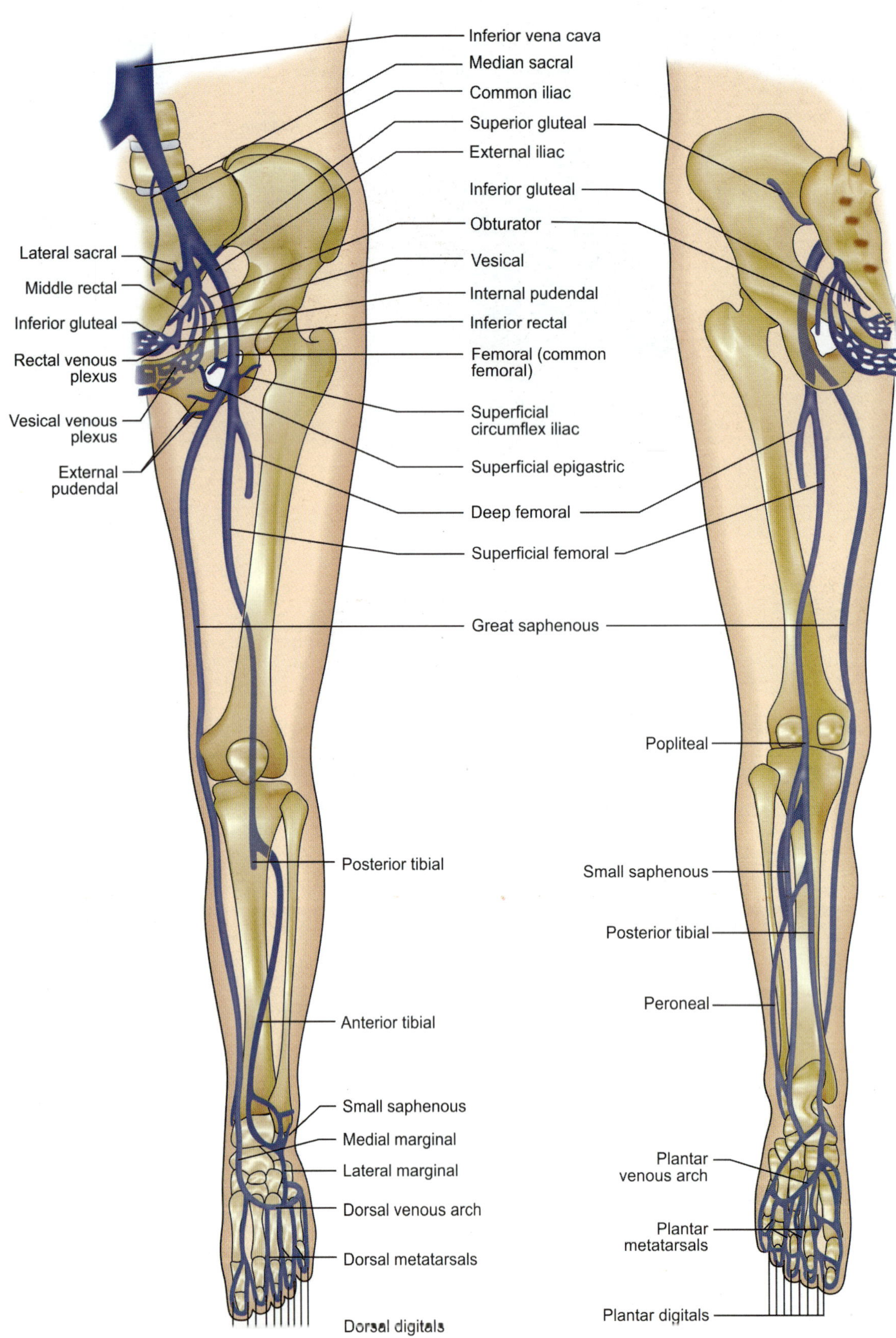

Veins of lower limbs

PLATE-11

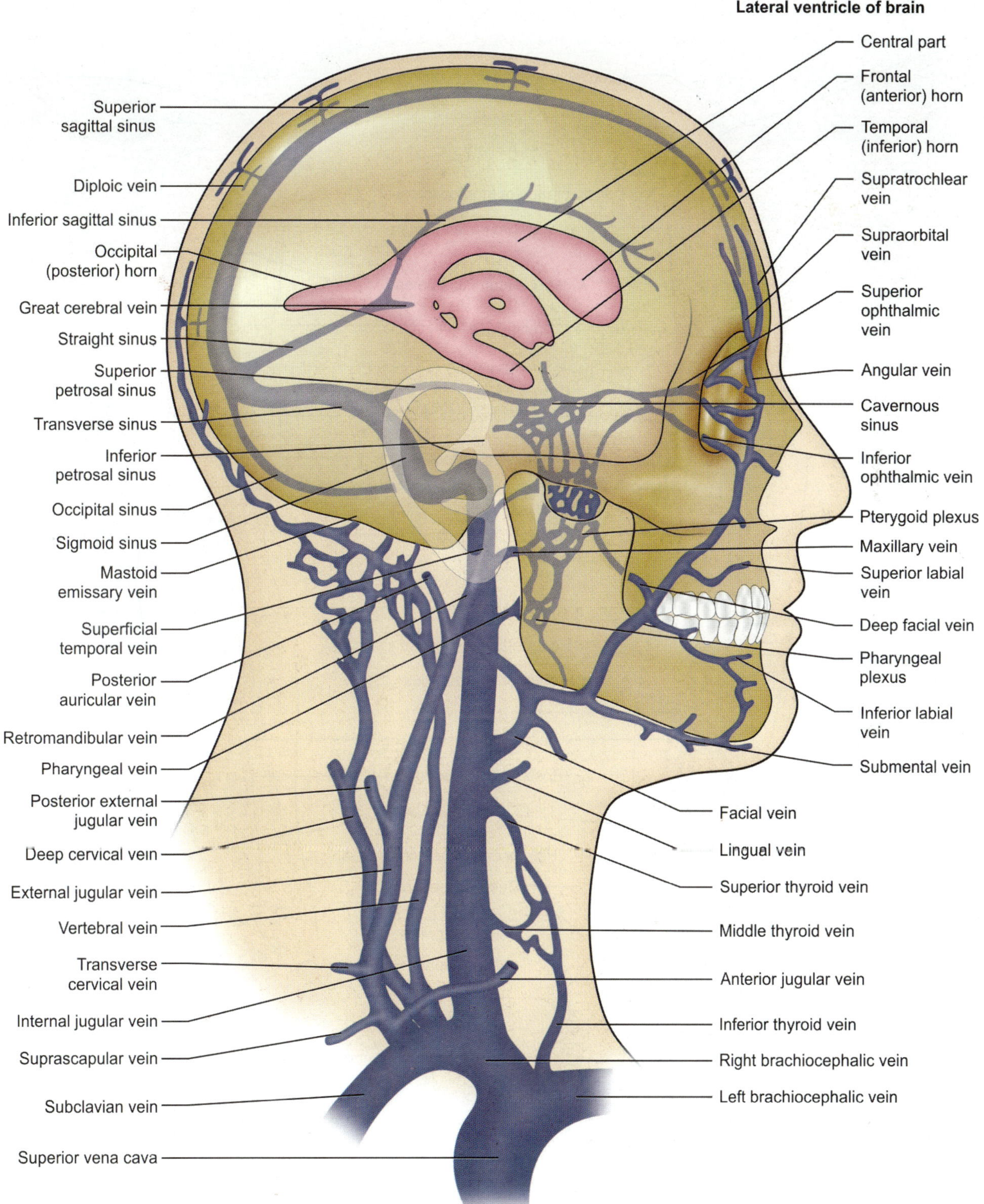

Veins of head and neck

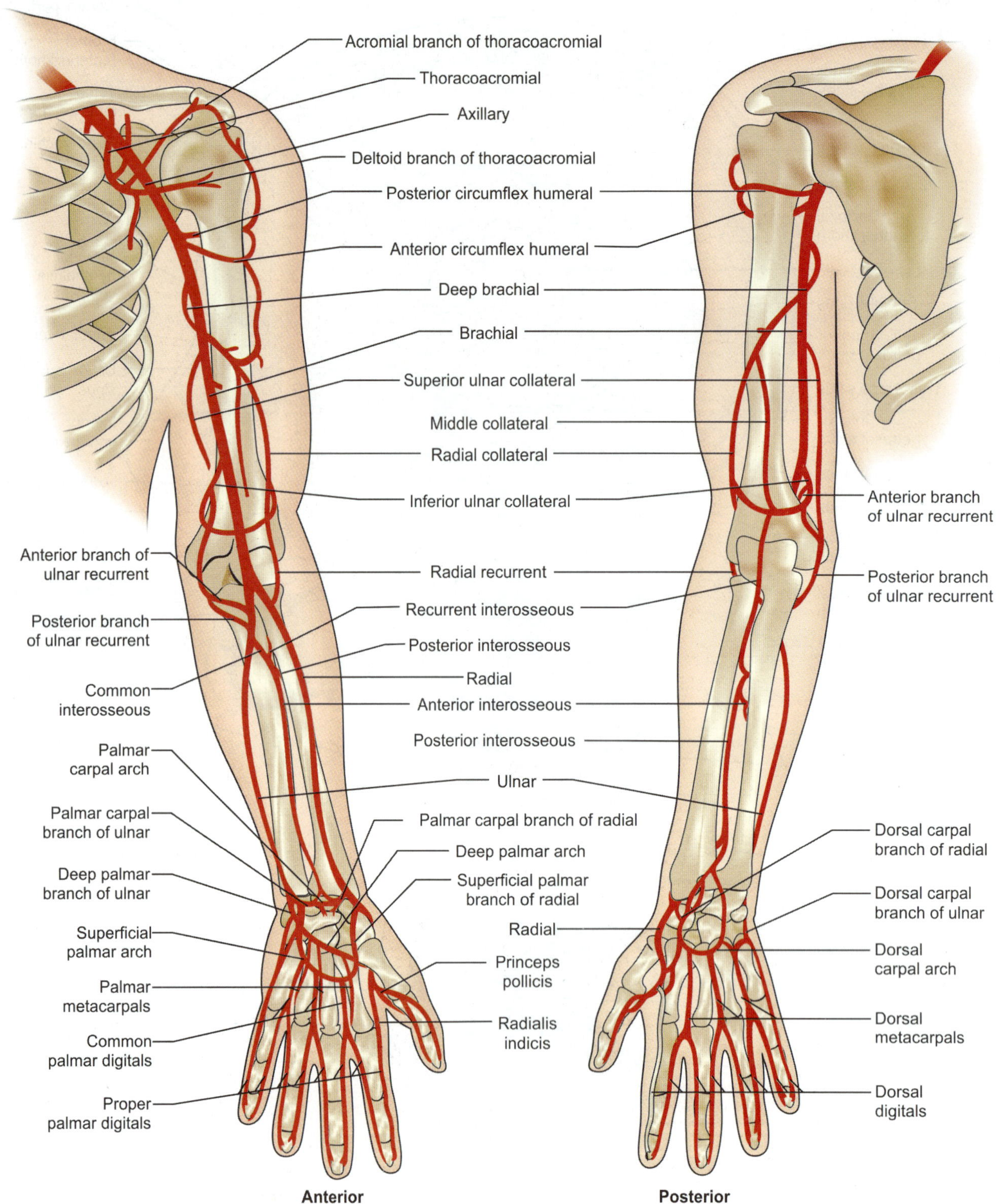

Arteries of upper limbs

PLATE-13

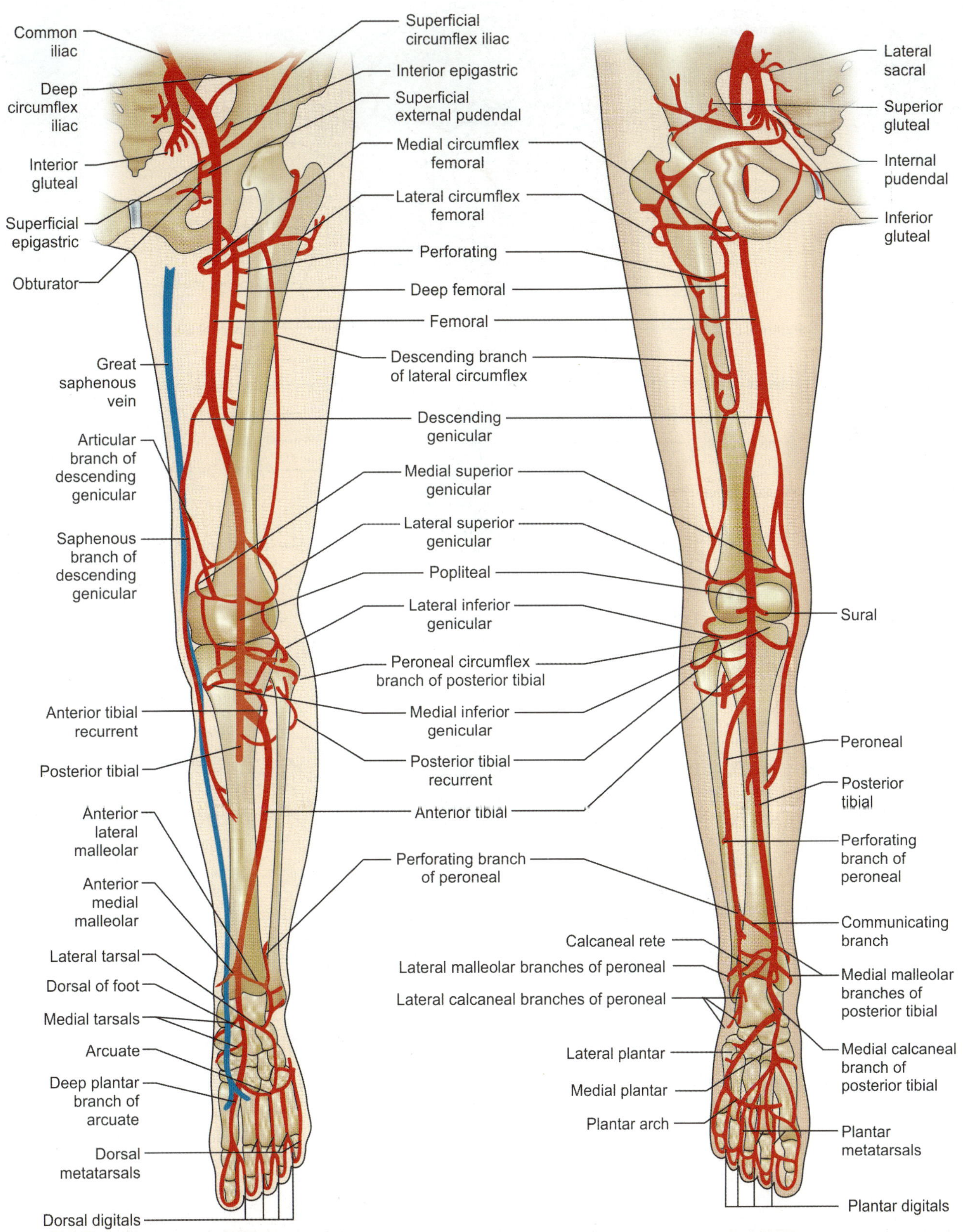

Arteries of lower limbs

PLATE-14

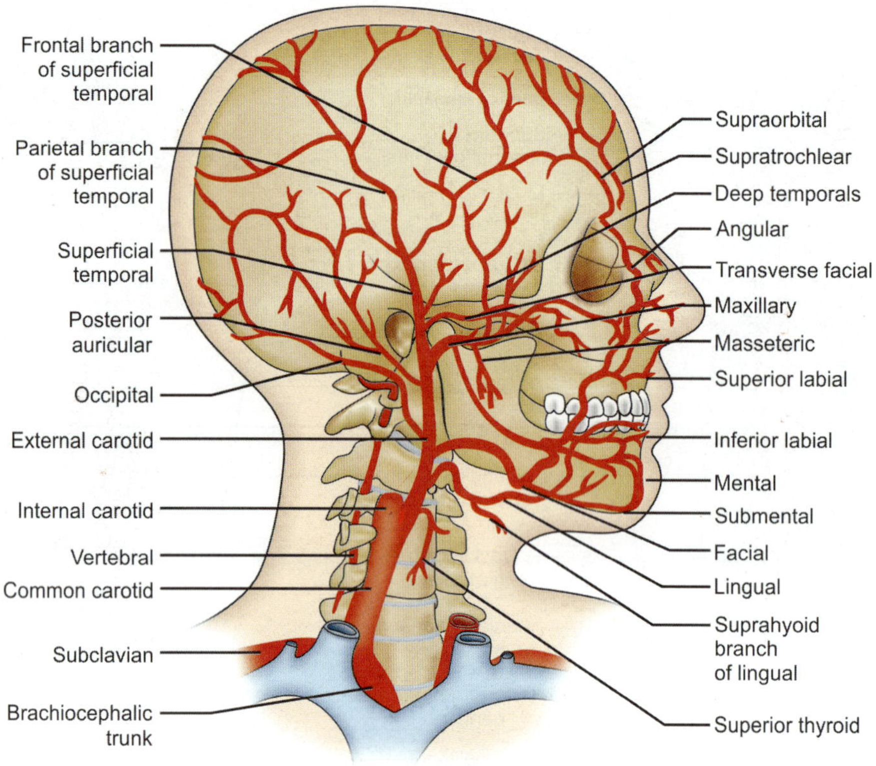

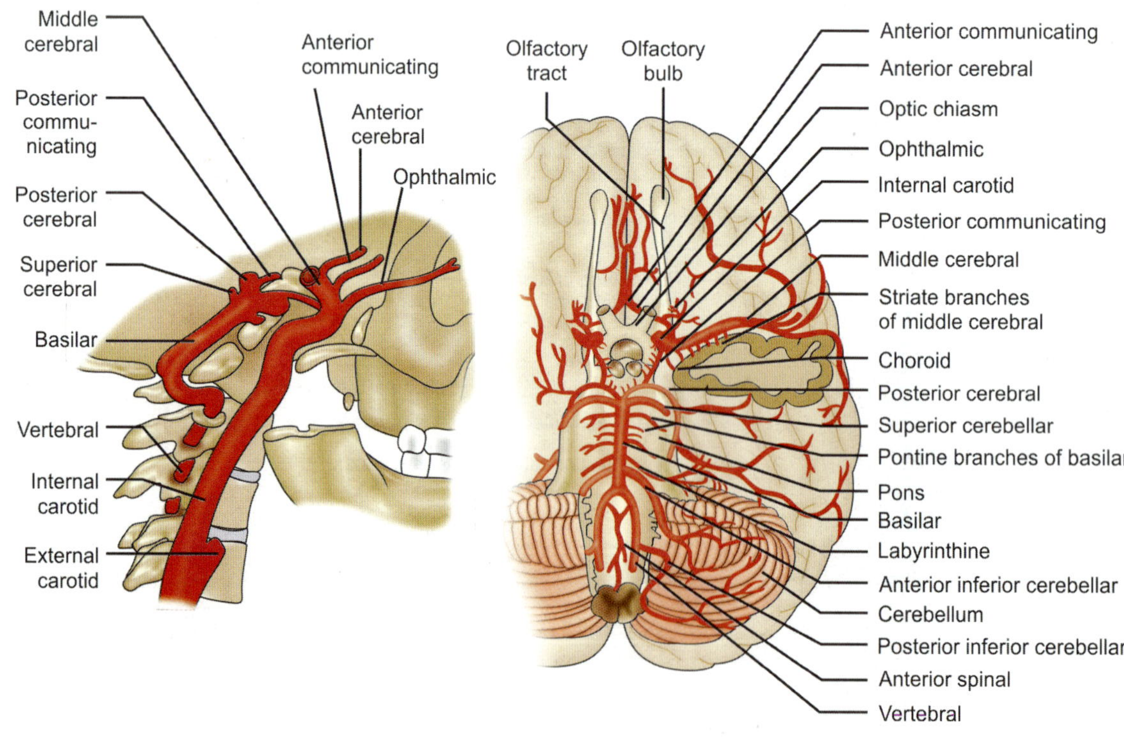

Arteries of head, neck and base of brain

PLATE-15

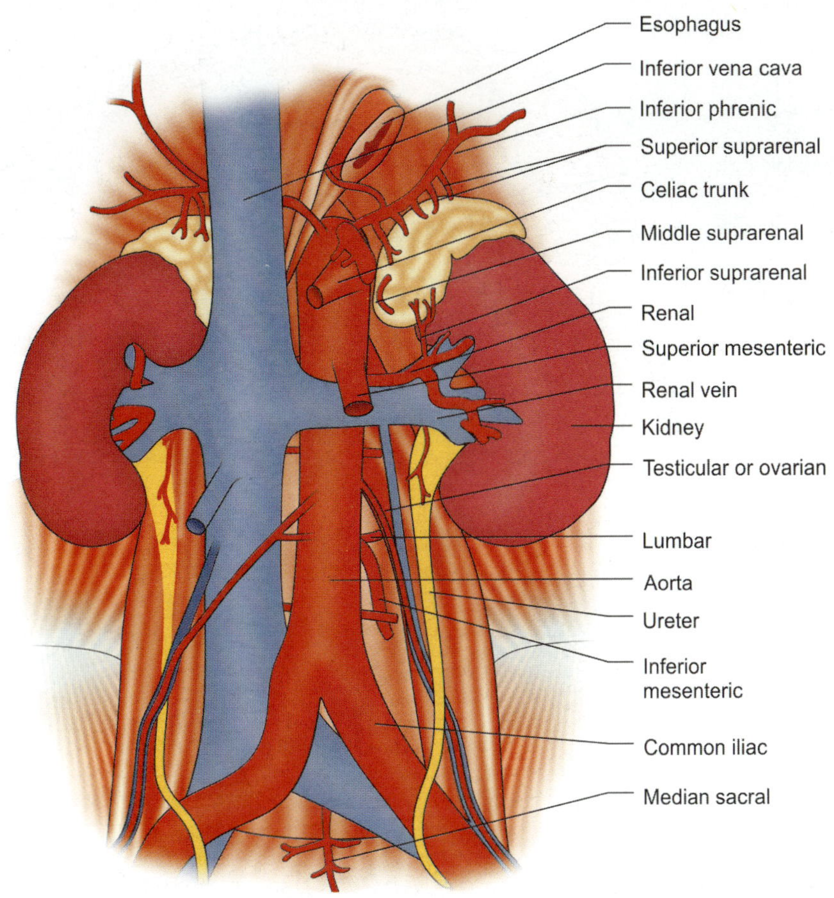

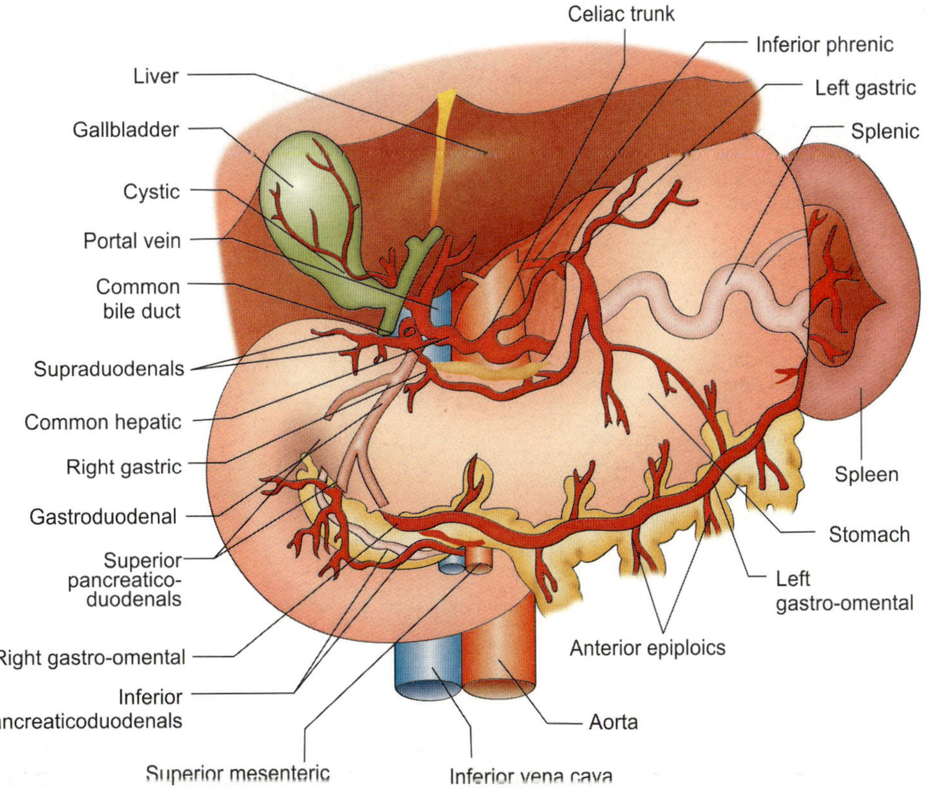

Arteries of abdomen and pelvis

PLATE-16

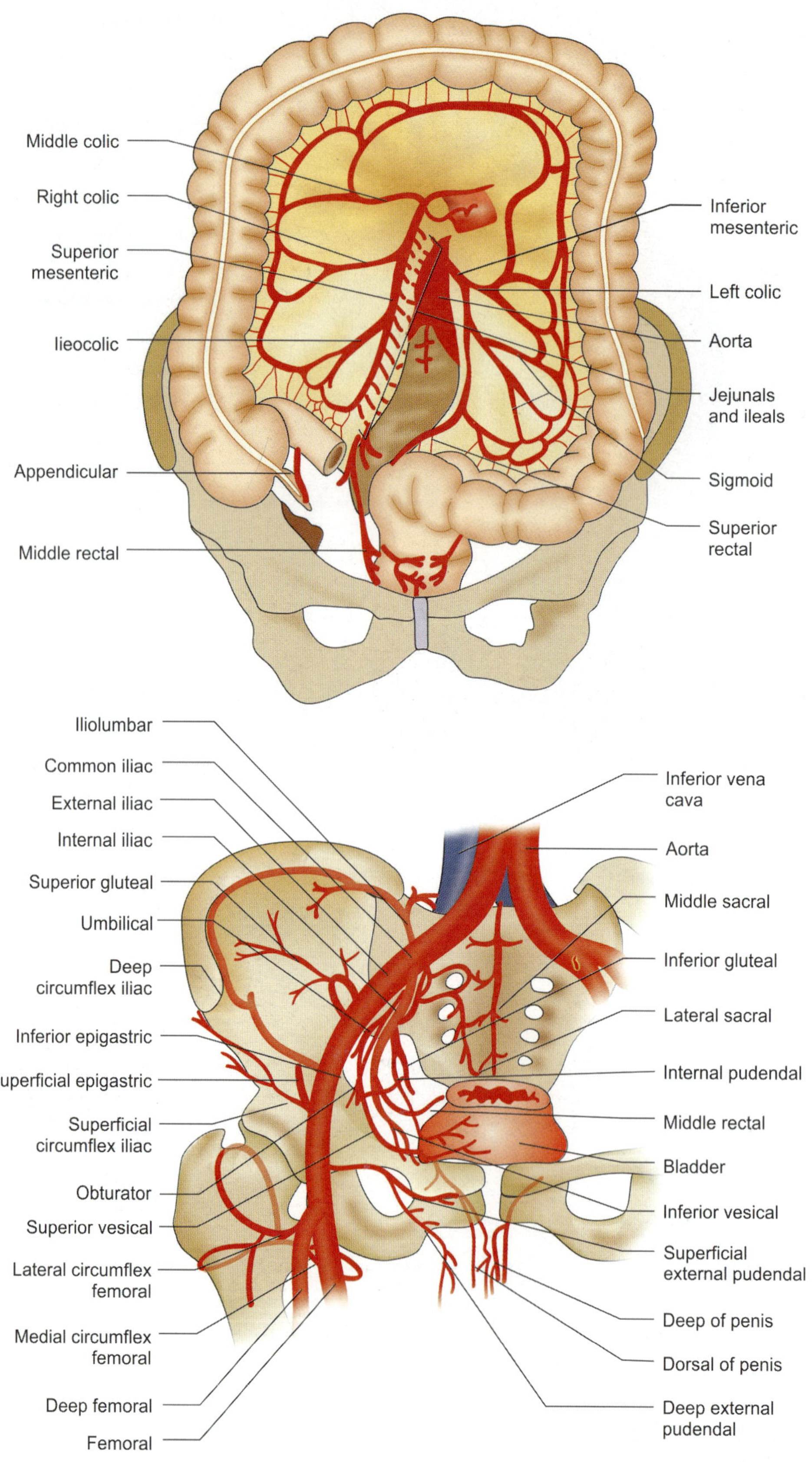

Arteries of abdomen and pelvis

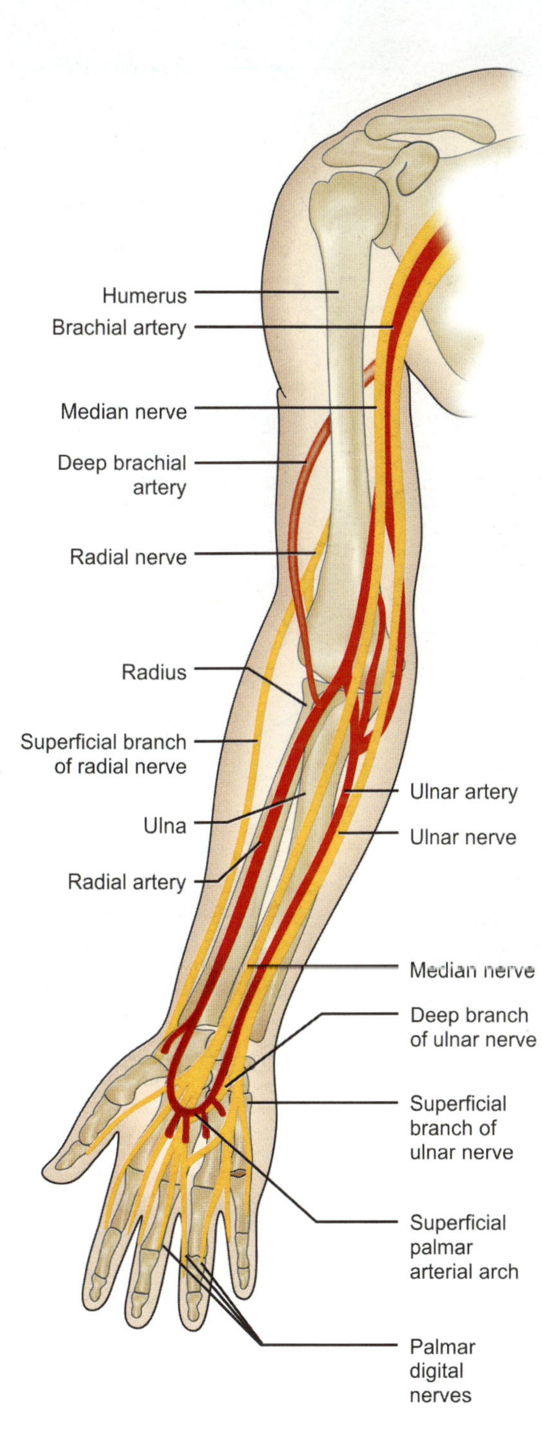

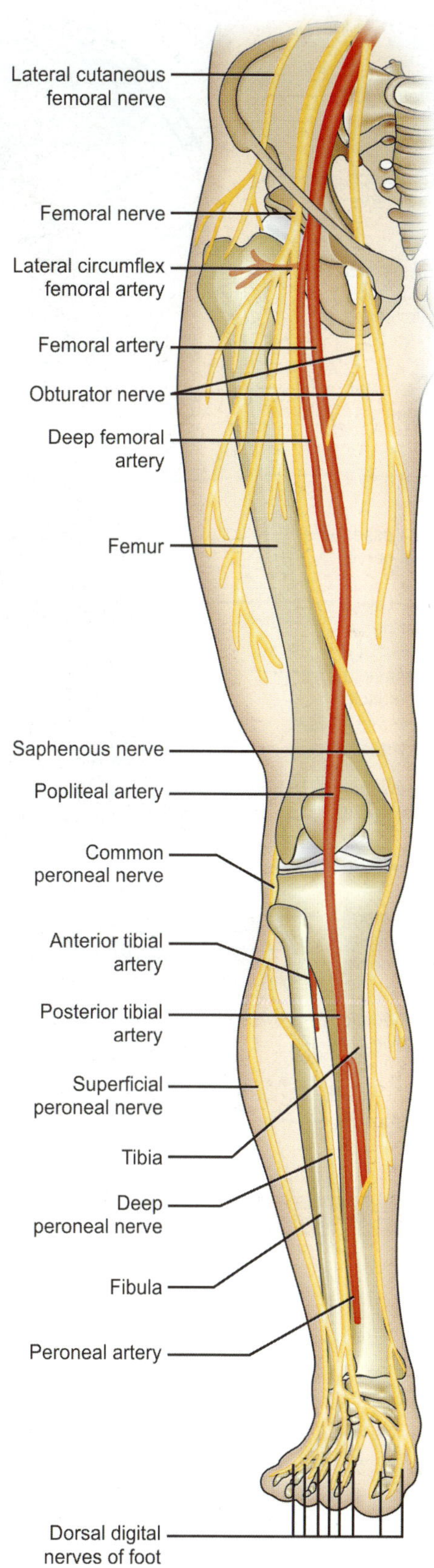

Nerves of limbs

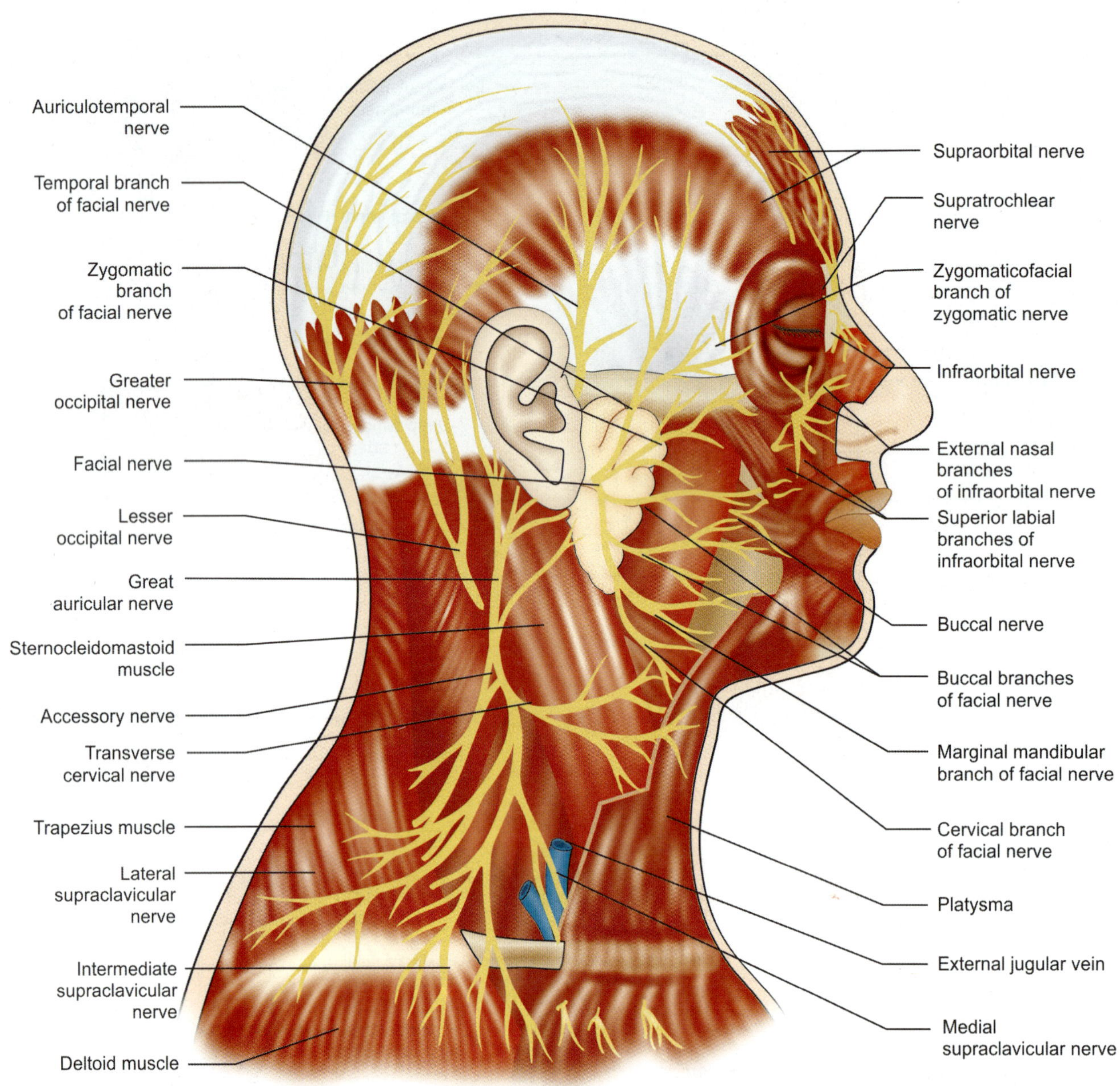

Superficial nerves of head and neck

PLATE-19

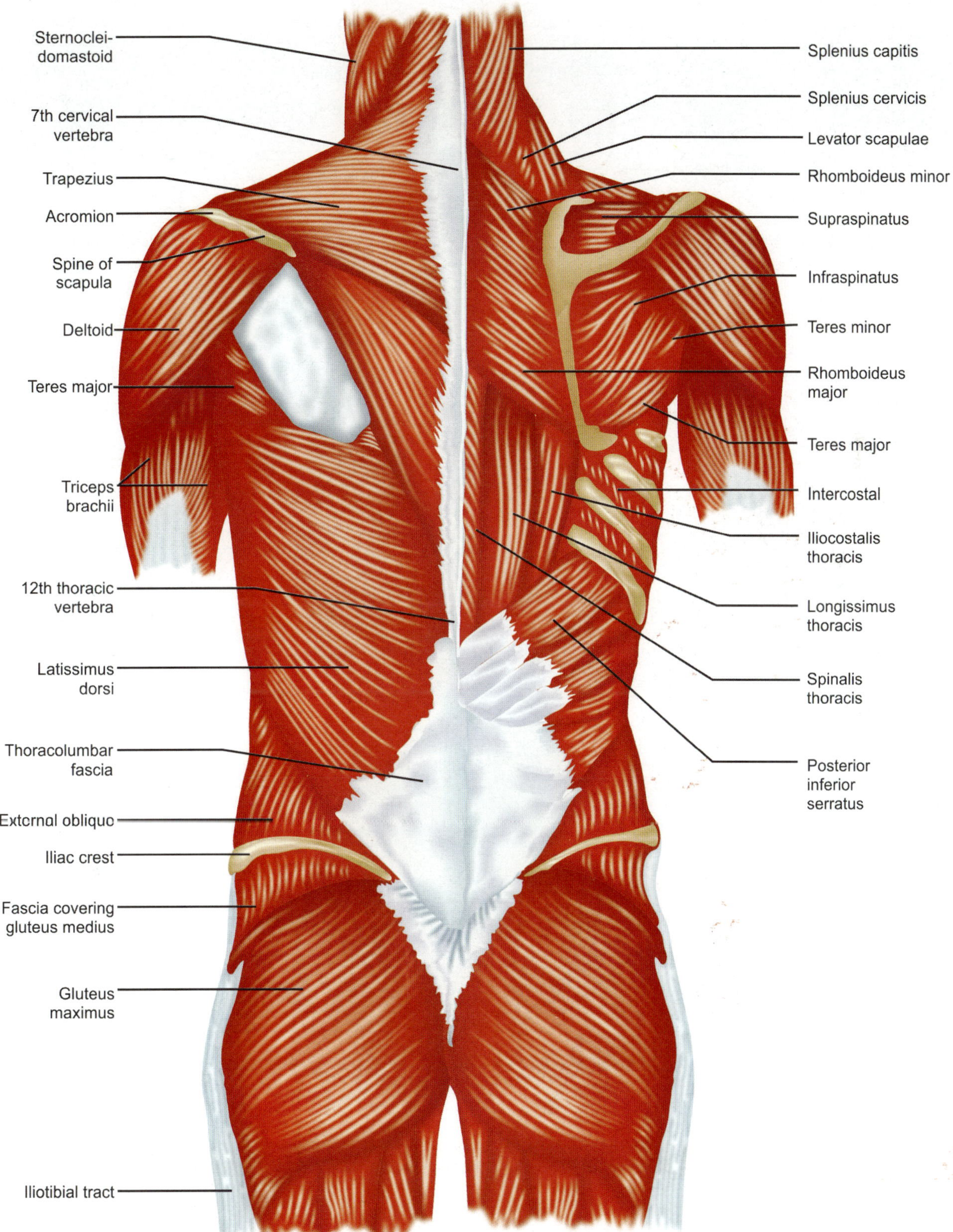

Muscles of trunk (Posterior view)

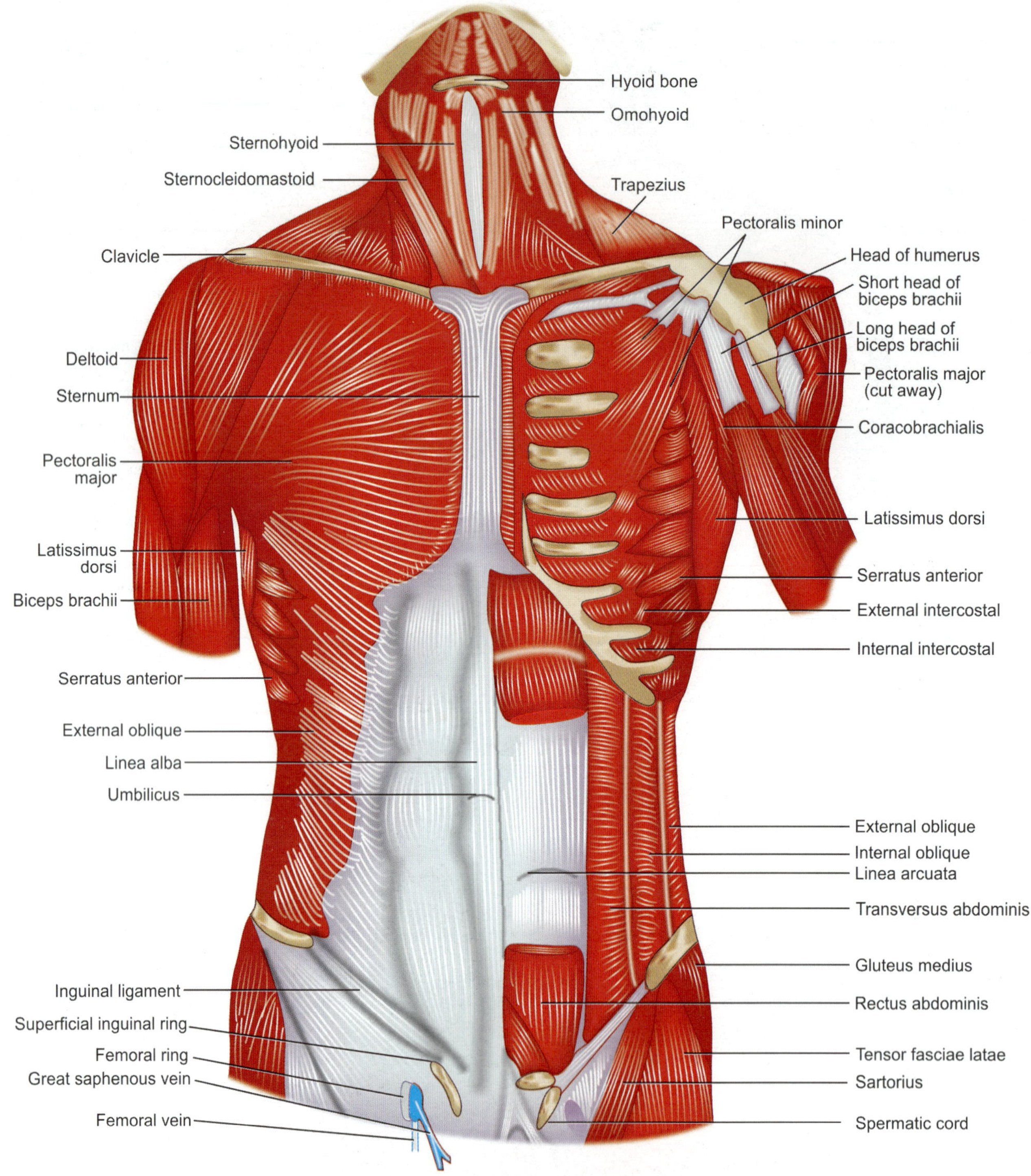

Muscles of trunk (Anterior view)

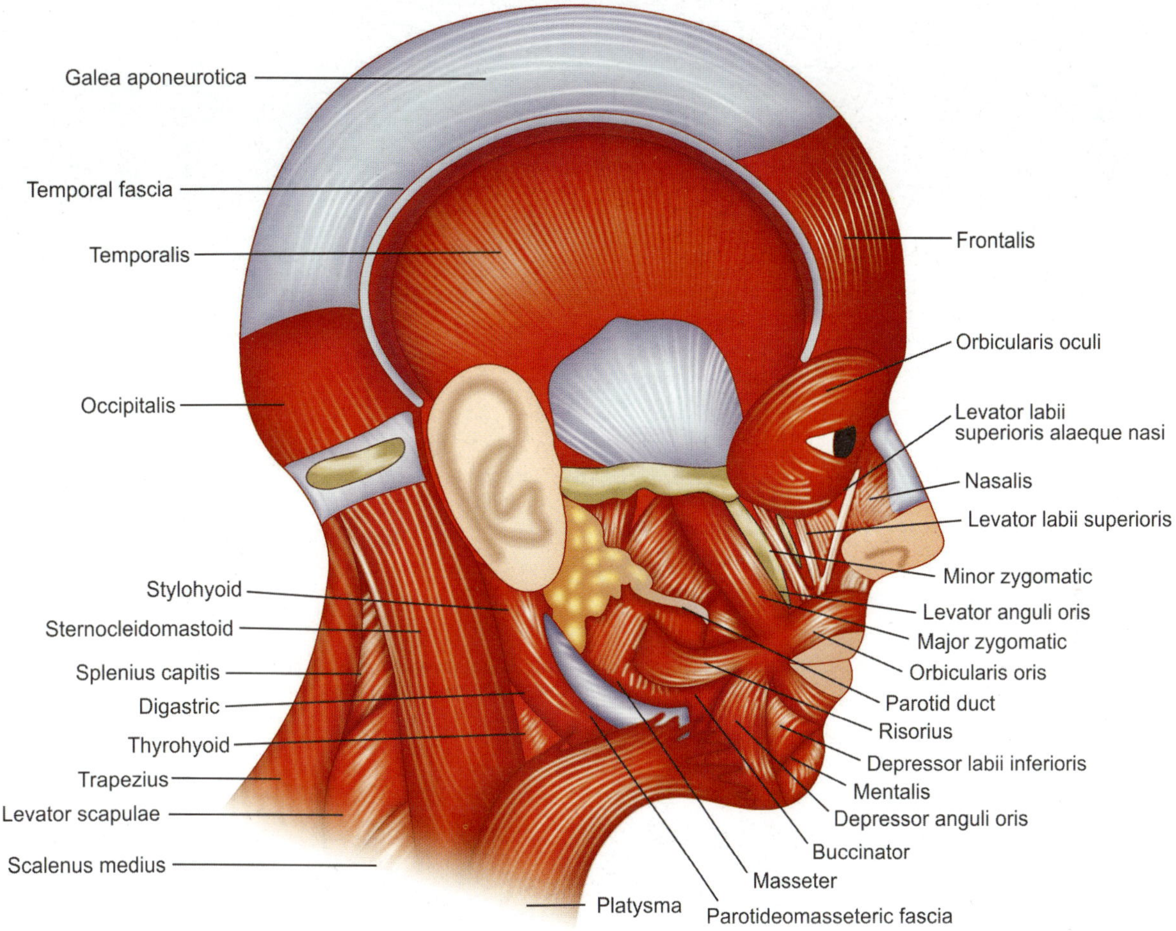

PLATE-21

Galea aponeurotica
Temporal fascia
Temporalis
Occipitalis
Frontalis
Orbicularis oculi
Levator labii superioris alaeque nasi
Nasalis
Levator labii superioris
Minor zygomatic
Levator anguli oris
Major zygomatic
Orbicularis oris
Parotid duct
Risorius
Depressor labii inferioris
Mentalis
Depressor anguli oris
Buccinator
Masseter
Parotideomasseteric fascia
Platysma
Stylohyoid
Sternocleidomastoid
Splenius capitis
Digastric
Thyrohyoid
Trapezius
Levator scapulae
Scalenus medius

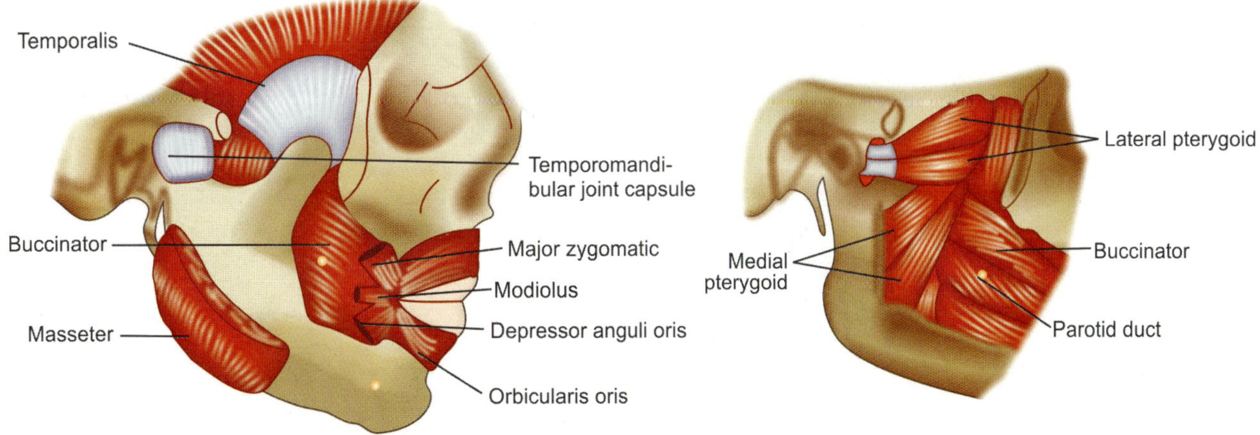

Muscles of the head and neck

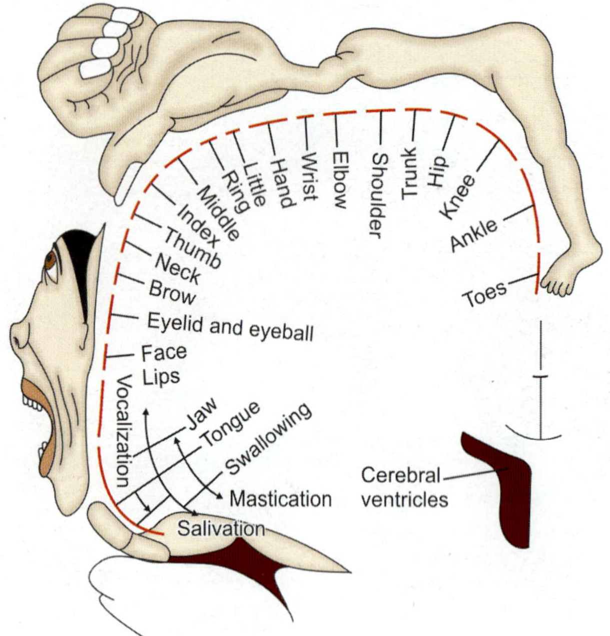

Motor homunculus demonstrating the relative somatotopical representation in the principal motor area of the brain

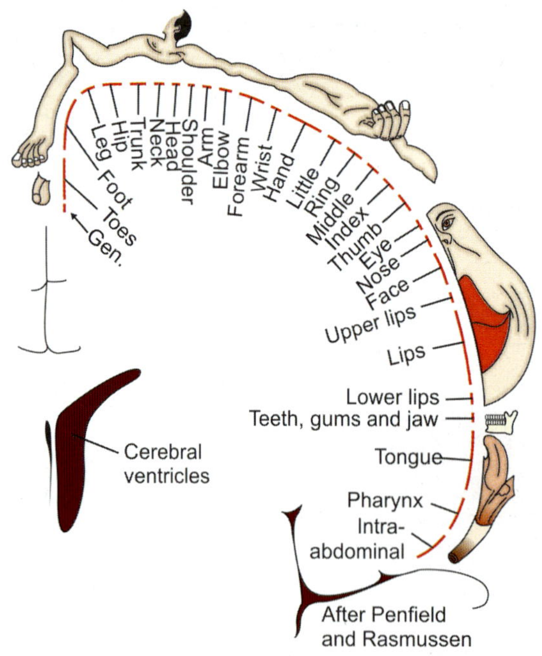

Sensory homunculus demonstrating the relative somatotopical representation in the somesthetic cortex of the brain

Motor-sensory homunculus

PLATE-23

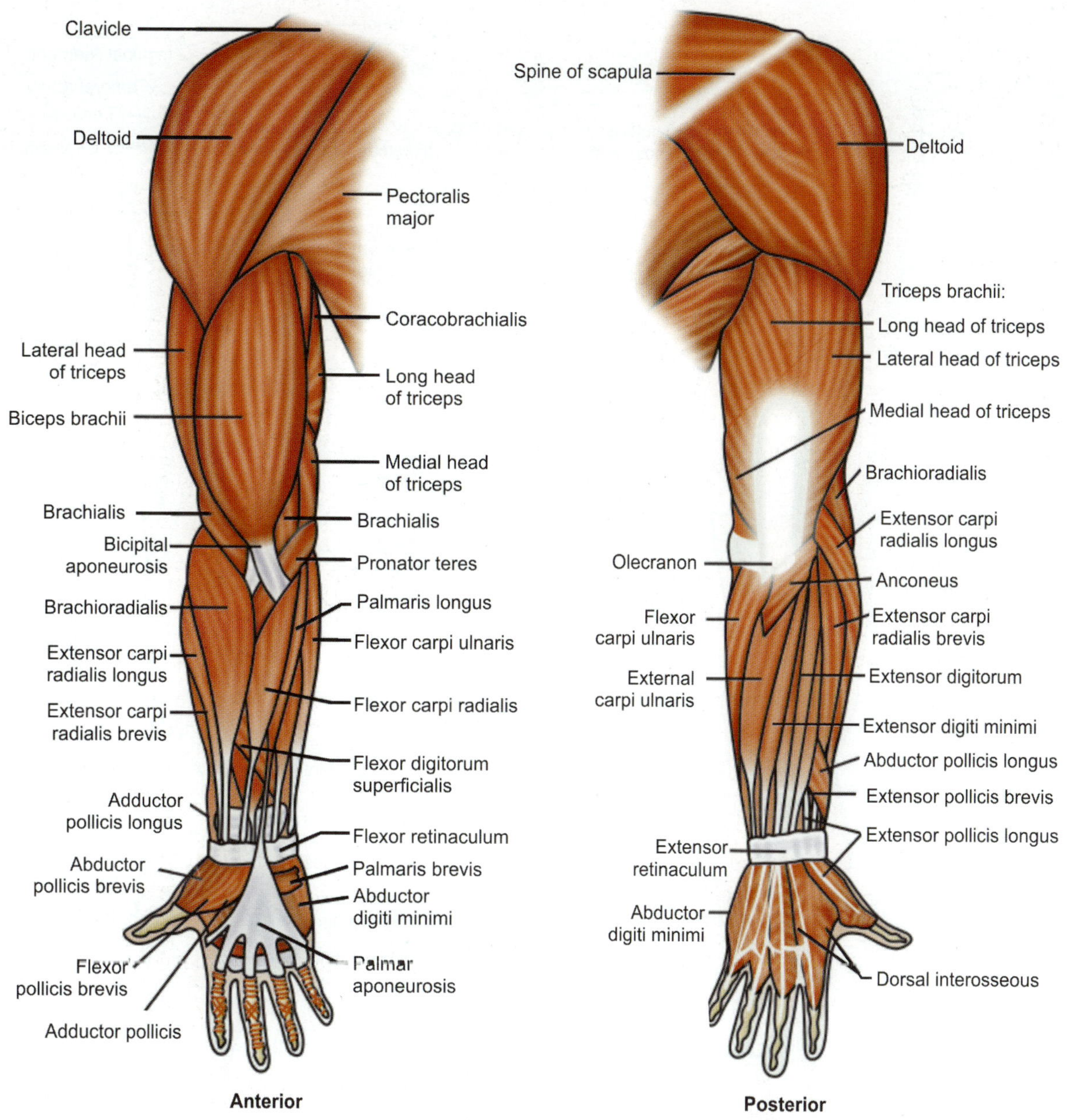

Superficial muscles of the upper limb

PLATE-24

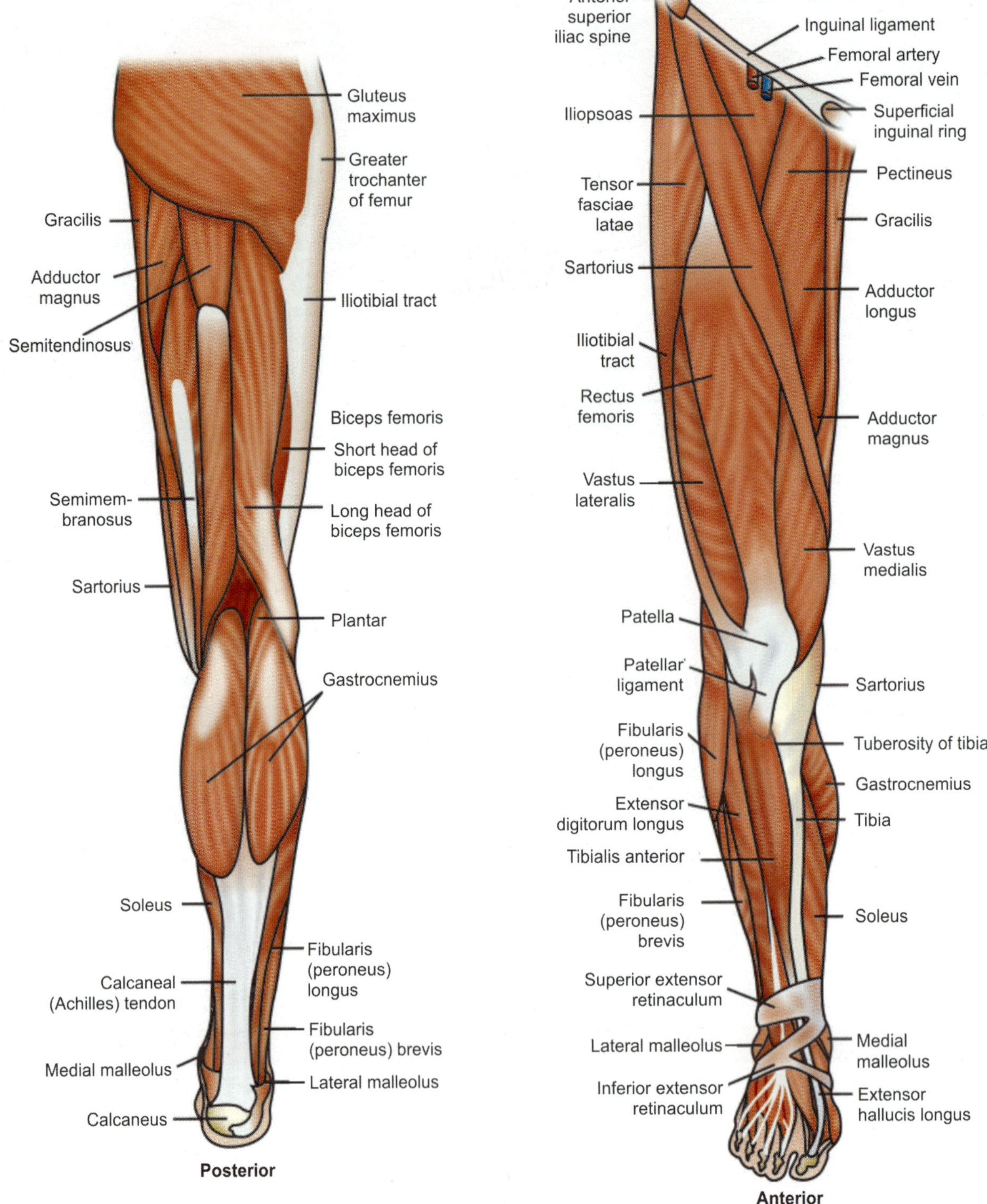

Superficial muscles of the lower limb

Q.49 What are the anomalies of the thyroid gland?
The defects are as outlined below:
a. accessory thyroid
b. thyroid crypt sinus
c. ectopic thyroid

Accessory thyroid can occur along the path of the thyroglossal duct and it is in association with the normal thyroid gland. It is usually seen at the foramen cecum. The foramen cecum never opens on the skin. Ectopic thyroid is very rare and if it is present, it is below the hyoid bone.

Q.50 How is a tooth developed?
The enamel of the tooth develops from ameloblast and the dentin develops from odontoblast. The enamel is an ectodermal derivative whereas the dentin is mesodermal in origin.

Q.51 How does the lacrimal gland develop?
This develops by 25 mm stage as a small epithelial bud from the ectoderm.

Q.52 What is "x-zone" in the suprarenal gland?
The primitive cortex is the x-zone.

Q.53 How does the suprarenal gland develop?
It develops in two parts. The cortex develops from the mesoderm and medulla develops from the sympathetic cells.

Q.54 Why is the suprarenal relatively larger than the size of kidney in the fetal life?
This is due to big acidophilic cells in the cortex which become basophilic in the postnatal life.

Q.55 What is Gartner's duct?
The mesonephric ducts in females will disappear normally. If they persist in the postnatal life then this condition is known as Gartner's duct, which is attached to the labia majora.

Q.56 Name the three parts of urogenital sinus?
1. Vesicoureteral part
2. Pelvic part
3. Phallic or penile part.

Q.57 What is harelip and its cause?
There is non-fusion in the upper lip either on one or both sides. This is a developmental defect, where the frontonasal process does not fuse with the maxillary process (unilateral harelip or bilateral harelip) (Fig. 1.11).

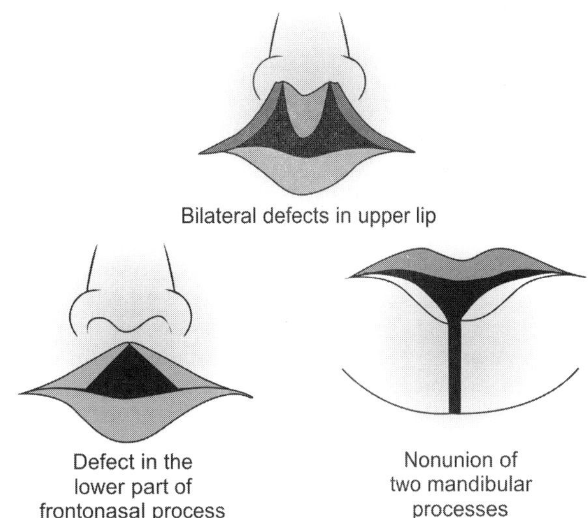

Fig. 1.11: Types of harelip

Histology

Q.58 What is podocyte?
The Bowman's capsule's visceral layer cells are large with foot like processes, whose cytoplasm projects into the capsular space. These are podocytes. The space in between the processes is 0.02–0.04 mm.

Q.59 What are the different types of cells in the tubules of the kidney?
The proximal convoluted tubule is formed of a single layer of low columnar cells with rounded nucleus and granular cytoplasm, stained deeply by eosin. The straight proximal or medullary portion more or less looks like the cortical portion. The loop of Henle has flattened epithelial cells with nuclei bulging into the lumen with abundance of cytoplasm. The thick ascending loop is lined by cuboidal cells.

Q.60 What is spermiogenesis?
The process of transformation of spermatids to spermatozoa is called spermiogenesis.

Q.61 Why does the esophagus have stratified squamous epithelium?
The reason for the above is to enable the wear and tear secondary to the swallowing of hard food.

Q.62 Why does the transitional epithelium line the bladder?
This is to help to withstand the action of chemicals in the bladder.

Q.63 What is duct of Bellini?
It is the main collecting duct which is located in the medulla of the kidney into which a number of the collecting tubules drain urine.

Q.64 What is the space of Disse?
This space is in between the Kupffer cells of liver (which lines the sinuses) and hepatic cells, and it is traversed by reticular fibers.

Q.65 What are centroacinar cells?
These cells are noticed in the lumen of the pancreatic acini.

Q.66 What are interstitial cells?
The interstitial cells are those which are in the connective tissue of the testis, in between the seminiferous tubules, and they are also known as Leydig cells.

Q.67 What is the purpose of simple squamous epithelial lining in certain areas of the body?
The main purpose of this epithelium is for passive transport of gases or fluids. It lines the pleura, pericardium and the peritoneum.

Q.68 What is the difference between the stratified squamous epithelium and transitional epithelium?
The transitional epithelium has similar cells whereas the stratified epithelium has cells of different shapes and morphology.

Q.69 What cellular change occurs during the distended and contracted phases of the bladder?
During the contracted phase the cells look like cuboidal or columnar cells and numerous layers of epithelium are present. During the distended state the layers are reduced and the superficial cells are flattened.

Q.70 Is there a basement membrane in transitional epithelium?
The basement membrane is well developed and makes an even contour.

Q.71 What are rectal columns or columns of Morgagni?
These are seen in the rectum, formed of submucosa and mucosa. The mucosa is formed by columnar epithelium, lamina propria, muscularis mucosae and lymphocytes.

Q.72 What are concretions?
They are the calcified organic material found in the alveoli of the prostate gland.

Q.73 What is Bowman's membrane?
This is found in the cornea. It is a homogenous membrane on which the epithelium of the cornea rests and it stops at the limbus.

Q.74 What is Descemet's membrane?
This is also found in the cornea, and is also called as the posterior elastic lamina. It is a transparent elastic membrane, and on it rests the endothelium of the cornea.

Q.75 What are rods and cones?
These are specialized cells, homologous with the ependymal cells of the ventricular cavity which are present in the outer nuclear layer of the retina. They are photosensitive receptor cells.

Q.76 Where do you get the pseudostratified columnar ciliated epithelium?
The epithelium is present in the trachea and bronchi. It is characterized by the presence of nuclei which are arranged at different layers.

Q.77 What is stratified squamous epithelium and where do you find it?
It is seen in the esophagus, where it is a non-cornified and the skin where it is a cornified epithelium. The cells are arranged in a different fashion and are mostly squamous type of cells. The intermediate cells are polyhedral with rounded and oval nuclei. In these cells the mitotic division can also be seen sometimes. The deeper layer of cells can be cuboidal or columnar. The cytoplasm of the deeper cells is granular and is rich in chondroitin, with oval nuclei. The basement membrane is rarely visible.

Q.78 What is the structural unit of bone?
Osteon or "haversian system" is called the structural unit of a bone, each system consists of a haversian canal surrounded by concentric lamellae.

Q.79 What is a cement line?
This is a refractive line forming the boundary of each osteon.

Q.80 What type of collagen does hyaline cartilage consist of?
Hyaline cartilage consists primarily of type II collagen.

Q.81 What is the difference between the cardiac and striated muscle?
The striated muscles have the nucleus in the periphery and the striations are better visible, whereas in the cardiac muscle, the nucleus is centrally positioned and the fibers are branched.

Q.82 What is the structure of the gallbladder?
The gallbladder has mucous membrane which is thrown into folds and has numerous mucous cells. The tall columnar cells lines the lamina propria.

Q.83 What is the structure of pancreas?
The lobules of the pancreas are covered by a loose connective tissue and there is no definite capsule present. The alveoli are lined by pyramidal cells having a basement membrane. The inner part of the cell have secretory granules, whereas, the outer part have staining capacity to basophilic dye. The islets of Langerhans are present between the alveoli which are responsible for producing insulin. They vary in number from 1/2 million to 1 million. They are highly vascular having "A", "B" and "D" cells.

Q.84 What is respiratory bronchi's structure?
Respiratory bronchi have cuboidal epithelium, goblet cells and these cells lack cilia.

Q.85 What is the histology of trachea?
It is formed by the respiratory epithelium or pseudostratified columnar ciliated epithelium with goblet cells. The lamina propria has glands which form the secretions that are released into the epithelium. The longitudinal fibers present on the external surface with the transverse fibers form a very definite layer known as trachealis muscle.

Q.86 What is the structural differences between the interlobular, intralobular and respiratory bronchiole?
a. Interlobular bronchus have respiratory type of epithelium, smooth muscle fibers (reiseeineres muscle) and plates of cartilage.
b. Intralobular bronchus has a stellate lumen, columnar ciliated epithelium and goblet cells.
c. Respiratory bronchi have cuboidal epithelium, goblet cells but lack cilia.

Q.87 How does the trachea extend longitudinally during inspiration?
This is due to the yellow elastic and white collagenous fibers which are enclosed by fibrous membrane and extend during respiration.

Q.88 What is the histology of the urinary bladder?
The bladder has four layers which are:
a. The outer layer which is the serous layer, formed by the peritoneum covering the superior surface of the bladder
b. The muscular layer has three layers namely: external, middle and internal layers which are arranged like bundles
c. The submucous layer with areolar cells which have elastic fibers
d. The mucous layer, formed of transitional epithelium.

Q.89 What is the histology of urethra?
a. The prostatic part is lined by transitional epithelium
b. The membranous and spongy part is lined by the stratified columnar epithelium
c. Fossa navicularis is lined by stratified squamous epithelium.

Q.90 What are vaginal rugae?
The mucous coat of the vagina is corrugated by elevations and they are known as the rugae.

Q.91 What is the difference between the normal and lactating mammary gland?
The normal mammary gland is formed by the ducts and fibroareolar tissue. The alveoli are small and few in number, whereas, during the pregnancy the alveoli are enlarged and distended and there are increased number of ducts.

Q.92 What are the layers forming the vagina?
The layers forming the vagina are mucous and the muscular coats. The mucous coats have folds or rugae whereas the muscular coats have plain muscle fibers, disposed longitudinally. The lower ends of the vagina have circular fibers bundles of striated muscle. The vaginal rugae are covered by stratified squamous epithelium.

Q.93 What is the structure of the thyroid gland?
The thyroid gland has lobules. The acini are lined by cuboidal cells and have colloid material in the center.

Q.94 How do you differentiate the parathyroid tissue?
It is differentiated by the presence of epithelial cells interconnected with trabeculae which are less vascular.

Q.95 How does the thymus looks under the microscope?
The lobules have a cortex and medulla. The cortex has lymphocytes embedded in the fibrous reticulum but has no lymph vessels. The medulla is differentiated by the presence of reticular cells and lymphocytes. The most important structure present in the medulla is known as the Hassall's corpuscles, these are concentric epithelioid cells varying in size between 25 and 75 mm in diameter.

Q.96 What are the types of cells seen in the pituitary gland?
There are two lobes present in the pituitary gland, the anterior and posterior lobes. The anterior lobe is lined by the chromophobes and chromophils. The chromophil cells could be acidophils or basophils. The basophils form 1/3 of the total number of cells whereas the acidophils form 1/10 of the total cells. The pars posterior has neuroglial cells and pituicytes which produce oxytocin which causes uterine contraction.

Q.97 How would you identify the vas deferens?
The vas deferens has a small irregular lumen with ciliated columnar epithelium or pseudostratified epithelium. It is highly muscular having outer longitudinal and inner circular fibers. The lamina propria surrounds the lumen.

Q.98 How will you identify the prostate gland?
The presence of fibromuscular stroma that is the connective tissue in between the alveoli is the characteristic histologic feature of the prostate gland. The alveoli are lined by columnar epithelium and the ducts are stained darkly. The alveoli have calcified organic material called corpora amylacea present in them.

Q.99 What is the difference between the testis and the epididymis?
The testis and epididymis both have basement membranes. The testis is lined by specialized epithelial cells, whereas, the epididymis is lined by ciliated columnar cells, and clusters of spermatozoa are also present. The testis is enclosed by a thick connective tissue known as "tunica albuginea".

Q.100 What are the different layers of the ovary?
The outermost layers are formed by the theca externa and theca interna. The ovarian follicles are surrounded by zona pellucida and externally have follicular cells known as corona radiata. The maturing sex cells and corona radiata are known as cumulus oophorus and in case they are stained deeply, they are known as "call-exnervacuoles". The nucleus lies inside the zona pellucida.

Q.101 What constitutes the lacrimal gland?
This is formed by the acini with serous cells. Depending on the function, the size of the lumen changes. The ducts have a double layer of prismatic cells.

Q.102 What is the histology of the iris?
It is a mass of stroma formed by elastic fibers along with the pigmented cells, plain muscle fibers and blood vessels.

Q.103 What is the structure of the eyelid?
It has a firm plate of condensed fibrous tissue known as tarsus. The skin is lined by stratified squamous epithelium and has a muscular layer internally.

Musculoskeletal System (Fig. 1.12)

Osteology

Q.104 What are the subdivisions of skeleton?
The skeleton is divided into:

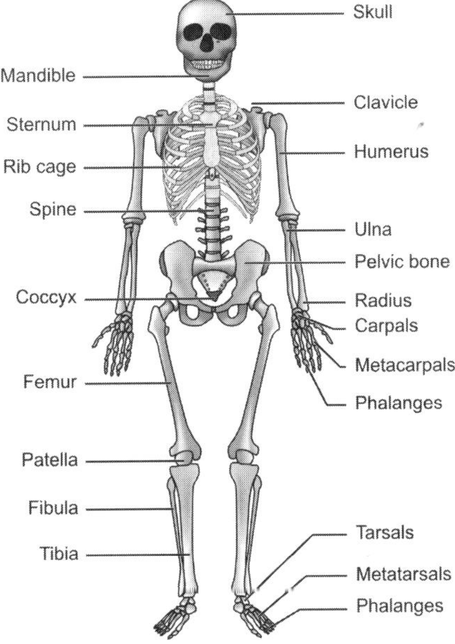

Fig. 1.12: Human anatomy of skeleton

- *Axial skeleton:* It is central bony framework, e.g. skull, vertebral column and thoracic cage.
- *Appendicular skeleton:* Formed by peripheral bones of the limbs.

Q.105 How the bones are classified according to shape?
- *Long bones:* Characterized by elongated tubular shaft, having a central medullary cavity and expanded articular ends (epiphyses), e.g. humerus, radius, femur, etc.
- *Smaller long bones:* They have only one epiphysis, e.g. metacarpals, metatarsal.
- *Short bones:* They are cuboid, cuneiform, scaphoid or trapezoid in shape, e.g. carpal and tarsal bones.
- *Flat bones:* Like shallow plates, e.g. ribs, scapula and bones of cranial vault.
- *Irregular bones:* Includes those bones which cannot be assigned to any of above groups, e.g. hip bone, vertebrae, etc.
- *Pneumatic bones:* They contain air spaces and are lined by mucous membrane, e.g. maxilla.
- *Accessory bones:* These are sometimes present in relation to limbs and skull bones.

Q.106 What are the functions of the bones?
- Provide shape and size to body
- Provide attachment to muscles, ligaments and tendons
- Protect vital organs
- Resist compression and tension stresses due to collagen tissue in bones
- Act as store house for calcium and phosphorus
- Act as a system of levers for movements by muscles
- Ear ossicles help in audition
- Bone marrow has blood forming function
- Reticuloendothelial cells of marrow are phagocytic and have a role in immune reactions
- Air sinuses in skull provide resonance to the voice.

Q.107 What are sites where red bone marrow is present in adults?
Proximal ends of femur and humerus, ribs, sternum, skull, vertebrae and hip bone.

Q.108 What is anthropometry?
It is the study of variation in dimensions and bodily proportions of various bones in different races and with age and sex in a single race.

Q.109 What are the parts of long bone?
- *Epiphysis:* Ends of a long bone which ossifies from secondary centers.

- *Diaphysis:* Shaft of a long bone which ossifies from a primary center. It consists of an outer cortex of compact bone and inner medullary cavity filled with bone marrow.
- *Metaphysis:* The epiphysial ends of diaphysis. It is the zone of active growth of bone.
- *Epiphysial plate of cartilage:* It separates metaphysis and epiphysis. Proliferation of this cartilaginous plate leads to lengthwise growth of bone.

Q.110 What are the different types of epiphyses?

- *Pressure epiphysis:* Articular end takes part in transmission of weight, e.g. head of femur, lower end of radius, and medial end of clavicle.
- *Traction epiphysis:* Non-articular. One or more tendon is attached to it which exerts a traction on it, e.g. trochanters of femur.
- *Atavistic epiphysis:* Phylogenetically represents a separate bone which in man has become fused to another bone, e.g. coracoid process of scapula.
- *Aberrant epiphysis:* These are not always present, e.g. epiphysis at head of first metacarpal.

Q.111 How the bones are classified according to their structure?

- *Compact bone:* Dense and is developed in cortex of long bones. It is able to resist mechanical pressure.
- *Cancellous bone* (spongy)*:* Consists of meshwork of trabeculae (lamellae) within which are intercommunicating spaces, e.g. vertebral bodies, ribs, and sternum.

Q.112 What are Sharpey's fibers?

These are the transverse fibers which hold the lamellae of the compact bone together and periosteum to the underlying bone.

Q.113 What are the different types of lamellae in a bone?

- *Circumferential lamellae:* Lie parallel to bony surface
- *Osteonic lamellae:* Concentric lamellae found around vascular canals of bone.
- *Interstitial lamellae:* Lie in space between osteons, i.e. vascular canals.

Q.114 How the bones are classified according to their developmental origin?

- *Intramembranous (dermal) bone:* Develops from direct transformation of condensed mesenchyme, e.g. bones of skull.
- *Intracartilaginous (endochondral) bone:* Replaces a preformed cartilage model, e.g. bones of limb and thoracic cage.
- *Membranocartilaginous bone:* Develops partly in membrane and partly in cartilage, e.g. clavicle and mandible.

Q.115 What is Wolff's law?

The mechanical stresses are directly proportional to the bone formation.

Q.116 What are centers of ossification?

These are certain constant points in a bone where the mineralization of connective tissue begins and the process of transformation spreads, until whole skeletal element is ossified.

Q.117 What is 'law of ossification' for a long bone?

Where a bone has an epiphysis at either end, the epiphysis which is first to appear is last to join and the epiphysis which is last to appear is the first to join except fibula.

Q.118 What is the arterial supply of a long bone?

The arterial supply of a long bone is derived from four sources:

- *Nutrient artery:* It enters the shaft through nutrient foramen and runs obliquely in cortex and divides into ascending and descending branches in medullary cavity. Each branch inturn divides and redivides into parallel vessels, which run in metaphysis.
  - These terminate by anastomizing with epiphyseal, metaphyseal and periosteal arteries.
  - It supplies medullary cavity and inner 2/3 of cortex.
  - The nutrient foramen is directed opposite to the growing end of bone.
- *Juxta-epiphyseal (metaphyseal) arteries of Lexer:* These are derived from anastomosis around the joint. They pierce the metaphysis along line of attachment of joint capsule.
- *Epiphyseal arteries:* Derived from periarticular vascular arcades found on non-articular bony surface.
- *Periosteal arteries:* These ramify beneath periosteum and supply outer 1/3 of cortex.

Q.119 What are 'sesamoid bones'? What are their characteristic features?

- These are bone nodules found embedded in tendons where they lie close to articular surface or turn around a bony surface and joint capsules.
- These have no periosteum.
- They are not always completely ossified and consist of fibrous tissue, cartilage and bone in varying proportion, e.g. in tendon of adductor pollicis and flexor pollicis brevis and in 70% cases in tendons anterior to metacarpophalangeal joint; patella; in tendon of flexor hallucis brevis, peroneus longus and tibialis posterior.
- They ossify after birth.
- They have no Haversian system.

Q.120 What are the functions of sesamoid bones?

- Alter the direction of pull of muscle or improve the pull of the muscles.
- To minimize friction.
- To modify pressure.
- Aids in maintaining the local circulation.
- Provide additional articular surface to a joint.

Cartilage

Q.121 What is cartilage and what are its characteristic features?

- It is a type of connective tissue, which has gel like ground substance known as matrix in which are embedded cartilage cells (chondrocytes).
- The matrix is made up of mucopolysaccharide and contains elastic or collagen fibers.
- The cartilage is firm in consistency and has elasticity.
- It has no lymphatics or blood supply.
- It may become calcified.

Q.122 What are the different types of cartilage and their distribution?

- *Hyaline cartilage:* No fibers seen in matrix. Does not regenerate because chondrocytes cannot redivide. Present at articular surface of synovial joint bones, costal cartilage, bronchial cartilage.
- *Fibrocartilage:* Collagen fibers present in matrix. Present in intervertebral disk, disks in joints, and on the articular surfaces of clavicle and mandible.
- *Elastic cartilage:* Elastic fibers present in cartilage, e.g. auditory tube, pinna, and epiglottis.

Q.123 Name the cartilages which calcify.

- Hyaline cartilage
- Fibrocartilage

Q.124 How the different cartilages obtain their nourishment?

Fibrocartilage is supplied by blood vessels but hyaline and elastic cartilage have no capillaries and their cells are being nourished by diffusion of lymph.

Arthrology

Q.125 How the joints are classified according to their structure?

- *Fibrous joint:* Bones are joined together by fibrous tissue. These joints are immobile or permit only slight movement.
- *Cartilaginous joint:* Bones are joined together by cartilage.
- *Synovial joint:* Articular surfaces of bone are covered by articular (hyaline) cartilage and between articular surface is joint cavity, containing synovial fluid. These joints permit maximum degree of movement.

Q.126 What are the different types of fibrous joints?

- *Sutures:* Found in skull and are immobile. Sutural ligament is present between two bones, which is attached on outside to pericranium and endocranium (outer layer of dura mater) on inside.
- *Syndesmosis:* Bones are connected by interosseous ligament, e.g. inferior tibiofibular joint.
- *Gomphosis:* Peg and socket type of joint, e.g. tooth in its socket.

Q.127 What are the characteristic features of synovial joint?

- Bony articular surfaces are covered with hyaline cartilage. It is insensitive to pain.
- Articular bones are connected by a fibrous capsule. The capsule has poor blood supply and heals very slowly. It is sensitive pain and stretch.
- Inner surface of capsule and all intra-articular structures which are not covered with cartilage are covered by synovial membrane, which secretes synovial fluid. It is highly vascular.

Q.128 What is the characteristic feature of synovial fluid?

It is presence of large amounts of mucopolysaccharide (hyaluronic acid) which gives it characteristic viscosity and it does not clot.

Q.129 What are the functions of synovial fluid?

1. Lubrication of joint
2. Nourishes the articular cartilage.

Q.130 What are the different types of synovial joint?

- *Arthrodial (plane):* Flat surfaces are in contact. Only gliding movement is possible, e.g. intercarpal joints, intertarsal joints.
- *Hinge:* Movements take place around a transverse axis, e.g. elbow joint between humerus and ulna.
- *Pivot:* A bony pivot like process moves within a ring. So movements are possible only around longitudinal axis through center of pivot, e.g. upper radioulnar joint and median atlantoaxial joint.
- *Condylar:* Two convex condyles (articular surface) moves within two concavities on opposite side. Movements occur mainly in transverse axis but partly in vertical axis (rotation), e.g. knee joint, temporomandibular joint, interphalangeal joints.
- *Ellipsoid:* Formed by a oval convex surface and an elliptical concavity, e.g. radiocarpal joint (wrist joint), metacarpophalangeal joint. Movements possible are flexion, extension, abduction, adduction and circumduction. No rotation occurs around central axis.
- *Saddle:* Articular surfaces are both concavoconvex. Movements permitted are same as in condylar type with some rotational movements, e.g. carpometacarpal joint of thumb, ankle joint.
- *Ball and socket:* Articular surfaces are globular head which fit into a cup like cavity. Movements are possible in every direction around a common center, e.g. hip joint, shoulder joint.

Q.131 What is a compound joint?

When more than two bone ends are enclosed with in a single capsule, the joint is known as compound, e.g. elbow joint has humeroulnar, humeroradial and superior radioulnar joint.

Q.132 What is a complex joint?

The joint cavity is divided completely or incompletely into two parts by intra-articular disk or fibrocartilage, e.g. temporomandibular joint, sternoclavicular joint and knee joint.

Q.133 How the joints are divided according to axis of movements?

- *Multiaxial* : Ball and socket joints
- *Biaxial* : Ellipsoid and saddle joints
- *Uniaxial* : Hinge and pivot joints

Q.134 What are fatty pads? What is their importance?

These are found in some synovial joints, occupying spaces where bony surfaces are incongruous and are covered by synovial membrane, e.g.

- Hip joint (Haversian fat pad)
- Talocalcaneonavicular joint
- Infrapatellar fold and
- Alar folds of knee joints.

Q.135 What are different types of cartilaginous joints?

- *Primary (synchondroses):* The related bones are united by hyaline cartilage. They are immovable and the cartilage is replaced by bone with age, e.g. costochondral joints, joint between epiphysis and diaphysis of a growing long bone, between sphenoid and temporal bones.
- *Secondary (symphysis):* These joints occur in median plane. The bone ends are covered by hyaline cartilage and are connected by a disc of fibrocartilage, e.g. manubriosternal joints, symphysis pubis, intervertebral joint between vertebral bodies. These do not disappear with age. Slight movement is possible.

Q.136 Why symphysis menti, joining two halves of mandible is not a true symphysis?

Because it disappears with age.

Q.137 What are the different intra-articular structures present in joints?

Cartilaginous structures:

- *Articular disk*
  - *Complete:* Mandibular joint and strenoclavicular joint
  - *Incomplete:* Acromioclavicular joint.
- *Articular menisci:* Semilunar cartilages of knee joint.
- *Labrum glenoidale*: Glenoid cavity of scapula and acetabulum.
- *Ligaments traversing joints:* Bind articular surfaces, e.g. ligamentum teres of hip joint, cruciate ligaments of knee joint.

Muscle tendons: These arise inside capsule of joint, e.g.

- At shoulder joint, long head of biceps
- At knee joint, tendon of popliteus.

Q.138 What are the functions of intra-articular disks?

- They act as a buffer and absorb shock.
- They strengthen the joint
- They make the articulation between bony surfaces smooth and harmonious.

Q.139 What is Hilton's law?

- A joint is supplied by the same nerves which supply the muscles crossing the joint and skin over the joint.
- Therefore, in joint diseases, irritation of nerves cause reflex spasm of muscles and referred pain to the overlying skin.

Myology (Figs 1.13 to 1.15)

Q.140 What are the distinguishing features of different types of muscle?

Features	*Smooth muscle*	*Skeletal muscle*	*Cardiac muscle*
Location	Found in viscera and blood	Found attached to skeleton vessels	Found in myocardium of heart
Nerve supply	Autonomic nerves, so they are involuntary	Somatic nerves, so they are involuntary	Autonomic nerves, so they are involuntary
Muscle fiber	Has no cross striations Each fiber is elongated, spindle shaped	Has cross striations Cylindrical cell	Has cross striations Muscle fiber show branches and anastomoses with neighbouring fibers
	Single central nucleus	Multiple peripheral nuclei	Single central nucleus
Rhythmicity	Present	Absent	Present
Automaticity	Present	Absent	Present

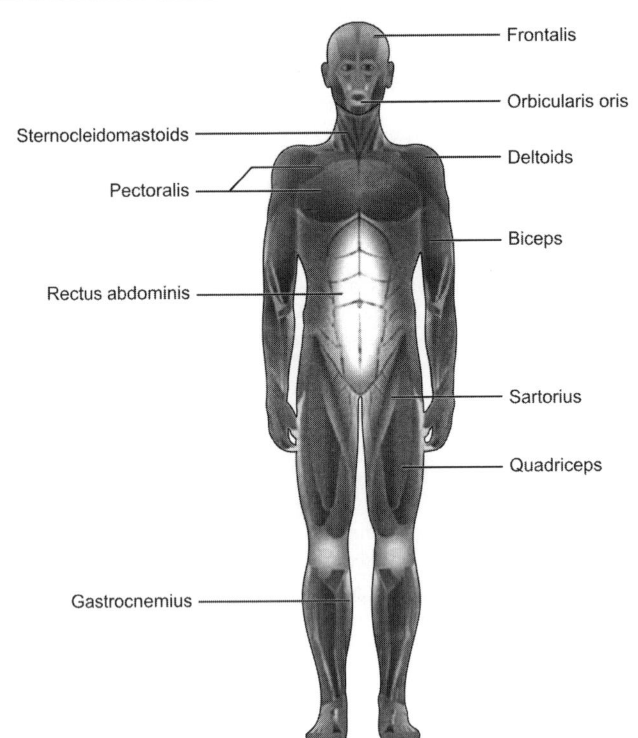

Fig. 1.13: Human anatomy—front view of muscle

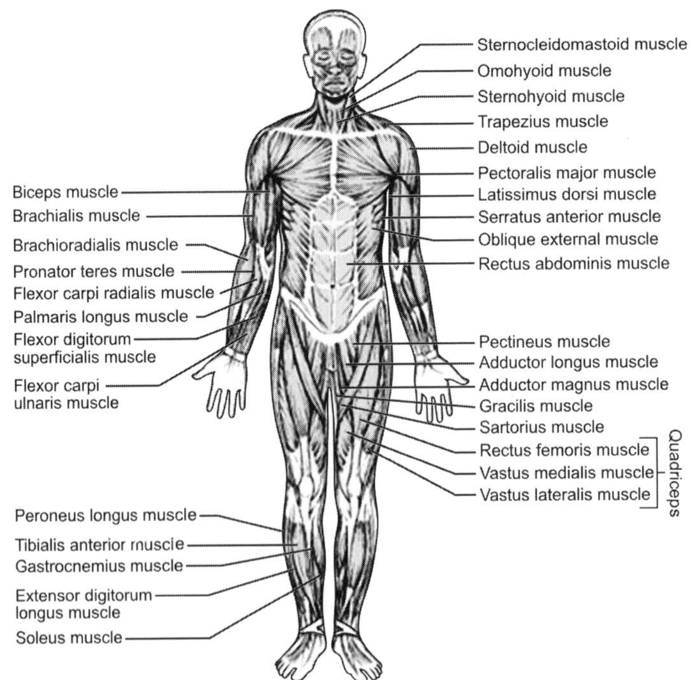

Fig. 1.14: The muscle front view

Q.141 How the skeletal muscles are classified according to direction of muscle fibers?

1. When the fasciculi (groups of muscle fibers) are parallel to line of pull, e.g.
   - *Strap like:* Sternohyoid, sartorius.
   - *Fusiform:* Biceps
   - *Quadrilateral:* Thyrohyoid.
2. When the fasciculi are oblique to line of pull ,e.g.
   - *Triangular:* Temporalis, adductor longus.
   - *Pennate (Feather like):*
     - *Unipennate:* Extensor digitorum longus, flexor pollicis longus.
     - *Bipennate:* Rectus femoris
     - *Multipennate:* Deltoid, subscapularis
     - *Circumpennate:* Tibialis anterior.
3. When muscle fibers are arranged in a twisted manner, e.g. trapezius, pectoralis major.

Q.142 What is the nerve supply of skeletal muscle?

Supplied by somatic nerves.

1. *Motor fibers:* Has
   - *Alpha efferents:* Myelinated anterior horn motor neurons, supply muscle fibers.
   - *Gamma efferents:* Myelinated fibers, supply muscle spindle (sensory end organ of skeletal muscle).
   - *Autonomic efferents:* Non myelinated, supply smooth muscle fibers of blood vessels.
2. *Sensory fibers:*
   - *Myelinated fibers:* Distributed to muscle spindle, tendon and fascia of the muscle.
   - *Non-myelinated fibers:* Distribution not known.

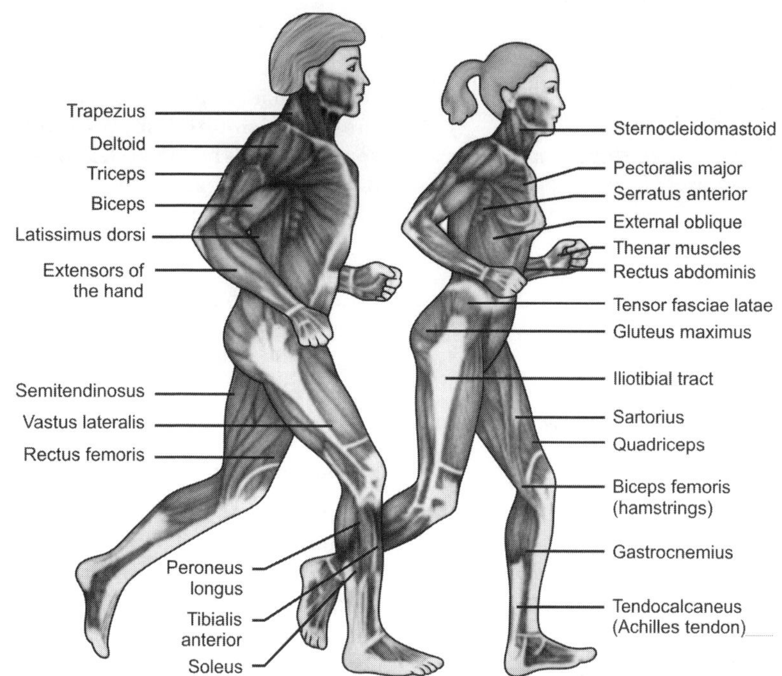

Fig. 1.15: The muscle side view

Q.143 What is a 'motor unit'?
It is a functional subdivision of muscle. It includes a single alpha motor neuron together with muscle fibers which it innervates.

Q.144 What is myotome?
A myotome is amount of muscle supplied by one segment of the spinal cord and muscles sharing a common primary action on a joint irrespective of their anatomical situation are supplied by the same segments.

Q.145 What are the features of muscles which receive 'double innervation'?
Generally they are flexor muscles that receive nerve supply from the extensor compartment. These muscles develop in the extensor compartment of fetal limb but for functional reasons, come to lie in the flexor compartment of the adult limb, bringing its nerve supply with, e.g.

- Lateral portion of brachialis (supplied by radial nerve).
- Short head of biceps femoris (by personal part of sciatic nerve).
- Brachioradialis (by radial nerve).

Q.146 What are bursae and where they are found?
Bursae are sacs of synovial membrane supported by dense irregular connective tissue.

They are found at the places where structures which move relative to each other are in tight apposition, e.g.

- Between skin and bone *(Subcutaneous bursae)*
- Between muscle and bone, tendon or ligament *(Submuscular)*.
- Between fascia and bone *(Subfascial)*
- Between ligaments *(Interligamentous)*
- *Adventitious bursae:* Normally not present but develop over bony situations which are subject to much friction or pressure, e.g.
  1. *Tailor's ankle:* Above lateral malleolus a bursa appears in tailors, who sit in cross legged position, thus bringing this area in contact with table.
  2. *Porter's shoulder:* In porters, between upper surface of clavicle and skin.
  3. *Weaver's bottom:* Between gluteus maximus and ischial tuberosity.

Q.147 What is 'aponeuroses'?
These are flat sheets of densely arranged collagen fibers associated with the attachment of muscle.

Q.148 What are the different types of muscles according to their action?

- *Prime movers:* These are active in initiation and maintenance of a particular movement.
- *Antagonists:* Muscles which oppose prime movers or initiate and maintain its converse.
- *Fixators:* Stabilize the position of a joint to provide a fixed base on which other muscles can act.
- *Synergists:* These help the prime movers in bringing the movement. They eliminate the undesired actions when prime movers cross more than one joint.

Circulatory System

Q.149 What is the difference between arteries and veins?

Features	*Arteries*	*Veins*
Thickness	Thick walled	Thin walled
Valves	Absent	Present
Lumen	Narrow	Larger
Fibromuscular tissue	More	Less
Elasticity	More	Less

Arteries carry oxygenated blood except pulmonary artery and veins carry deoxygenated blood except pulmonary veins.

Q.150 What are the differences between capillaries and sinusoids?

Features	*Capillary*	*Sinusoid*
1. Lumen	Smaller, regular	Larger (up to 30 m) irregular
2. Structure	*Endothelial lining:* Continuous	May be incomplete; some phagocytic cells are present.
	Basal lamina: Thicker and surround endothelial cells	Thinner
	Adventitial support: Present	Absent
3. Location	Connect metaarterioles and venules	Connect arteriole with venule or venule with venule

Q.151 Name the structures where fenestrated capillaries are present.

- Pancreas
- Intestine
- Renal glomeruli
- Endocrine glands

Q.152 Name the structures where continuous capillaries are present.
- Skin
- Muscles
- Fascia
- Brain

Q.153 What is the nerve supply of an artery?
The arteries are supplied by sympathetic nerves via nervi vasorum. These are vasoconstrictor. Few myelinated sympathetic fibers are also present, which carry pain sensation.

Q.154 Name the sites where sinusoids are present.
- Suprarenal gland
- Carotid body
- Liver
- Spleen.

Q.155 What are 'anastomosis'?
Arteries do not end always in capillaries, they unite with one another forming anastomosis.

Q.156 What are the different types of anastomosis?
- *Actual:* Arteries meet end to end, e.g. labial branches of facial arteries, intercostal arteries, uterine and ovarian arteries, arterial arcades in mesentery, arteries of greater and lesser curvatures of stomach.
- *Potential:* Anastomosis is by terminal arterioles and given sufficient time the arterioles can dilate to take sufficient blood, e.g. coronary arteries, cortical arteries of cerebral hemispheres, anastomoses around joints of extremities.

Q.157 What are the functions of anastomosis?
- Equalization of pressure over territories which they connect.
- Provide collateral circulation when a vessel is interrupted.

Q.158 What are 'end arteries'? What is their importance?
These are arteries which have no anastomoses with their neighbors, e.g. central artery of retina, arteries of spleen, liver, kidneys, metaphyses of long bones, medullary branches of the central nervous system, coronary arteries.

Importance: If an end artery is occluded, necrosis (death) of tissue takes place in area supplied by the vessel.

Q.159 What are 'arteriovenous shunts'?
These are vessels of communication between arteries and veins and when open, they bypass the capillaries, e.g. in skin of nose, lips and external ear, mucous membrane of alimentary canal, thyroid gland, palmar skin.

Q.160 What are the functions of arteriovenous shunts?
- Regulate the regional blood flow
- Regulate blood pressure
- Pressor reception
- Regulation of the temperature.

Lymphatic System

Q.161 What are the components of lymphatic system?
- *Lymph vessels:* Formed by lymph capillaries.
- *Peripheral lymphoid tissue:* Spleen, epitheliolymphoid system, lymph nodes and lymph nodules.
- *Central lymphoid tissue:* Bone marrow and thymus.
- *Lymphocytes:* Circulating in vessels.

Q.162 How the lymph capillaries differ from blood capillaries?
Lymph capillaries have
- Bigger lumen
- Lumen is less regular
- Permeable to bigger molecules
- Form pathways for absorption of colloid from tissue spaces

Q.163 Name the sites were lymph capillaries are absent.
- Epidermis
- Hair
- Nails
- Cornea
- Articular cartilage
- Splenic pulp
- Spinal cord
- Brain and
- Bone marrow

Q.164 What is the structure of lymph trunk?
It consists of three coats:
- *Tunica adventitia:* Composed mainly of fibrous tissue and some smooth muscle fibers.
- *Tunica media:* Consists of smooth muscle cells, fibers of which are arranged circularly and separated from one another by fibrous tissue.
- *Tunica intima:* Consist of endothelial cells and fibrous tissue.

  They possess more number of valves than small veins. The valve consists of reduplicated endothelium and lumen of lymph vessel immediately proximal to valve is expanded into a *sinus,* which gives the vessel a *beaded appearance.*

Q.165 What are the factors which favor the propulsion of lymph from tissue spaces towards lymph nodes and venous blood stream?
- In tissue spaces, filtration pressure generated by filtration of fluid from blood capillaries.
- Concentration of surrounding muscles compressing lymph vessels.
- Pulsation of artery near lymph vessels.
- Respiratory movements.
- Negative pressure in brachiocephalic veins.
- Contraction of smooth muscle of vessel.

Q.166 What is the function of lymph capillaries?
Lymph capillaries are concerned with the absorption of fluid from tissue spaces.

Q.167 What is structure of lymph node?
Grossly: These are oval bodies situated in the course of lymph vessels. The blood vessels enter and leave node at the hilus. A lymph node has a *cortex* into which afferent vessels drain and a *medulla* from which efferent vessels arise.
Microscopic:
- *Capsule and trabeculae:* Composed of collagen fibers, fibroblasts and elastic fibers.
- *Reticulum:* Fibrocellular and forms a meshwork within spaces outlined by capsule and trabeculae. In medulla, fewer cells in loose reticulum are present. Such parts allow rapid passage of lymph and are termed *lymph sinuses.* Reticular fibers are thin collagen fibers, ensheathed by fixed macrophages in an amorphous matrix. Reticular cells lining lymph sinuses are termed as *littoral cells.* Majority of cells are lymphocytes with some macrophages. In cortex, cells are densely packed to form *lymphatic follicles.* The central part of follicle has a *germinal center,* which consists of large cells. In medulla, cells are loosely packed.

  The outer part of cortex has B-lymphocytes and inner part has T-lymphocytes. The medulla has mature B-lymphocytes, plasma cells and macrophages.

Q.168 What is epitheliolymphoid system and where it is found?
- These are collections of lymphoid tissue found under the epithelium.
- These are found in Peyer's patches in intestine, appendix, pharyngeal tonsil, palatine and lingual tonsil.

Q.169 What are the different type of cells in reticuloendothelial system?

These cells are concerned with phagocytosis.

- Pericytes (Rouget cells) in capillaries
- Dust cells in lungs
- Macrophages in connective tissue, bone marrow and suprarenal gland
- Reticular cells in spleen and lymphoid tissue
- Monocytes in blood
- Kupffer cells in liver
- Microglia in nervous system.

Q.170 What are the functions of a lymph node?

- Act as a filter for lymph. Thus, foreign particles are prevented from entering the bloodstream.
- Macrophages in sinuses engulf foreign particles.
- Trapping of antigens by phagocytes
- Mature B and T lymphocytes are produced in the lymph node.
- Provides interaction between antigen laden phagocytes and lymphoid tissue with mounting of both cellular and humoral immune response.
- Provides portal of entry for blood borne lymphocytes back into lymphatic channels.

Nervous System

Q.171 What are the subdivisions of nervous system?

- *Central nervous system:* Includes brain and spinal cord.
- *Peripheral nervous system:* Divides into:
  1. *Cerebrospinal nervous system:* Includes 12 pairs of cranial nerves and 31 pairs of spinal nerves.
  2. *Peripheral autonomic nervous system:* Includes sympathetic and parasympathetic nervous system.

Q.172 What are the cell types forming nervous tissue?

- *Neurons (Nerve cells):* Excitable cells.
- *Neuroglia:* Non excitable cells, forming connective tissue of the nervous system.

Q.173 What is the function of neurons?

Reception, transmission, integration and transformation of impulses.

Q.174 What is the histological structure of a neuron?

Each neuron consists of:

- *Cell body (Perikaryon):* Mass of cytoplasm with a diploid nucleus and bound by a membrane. The cytoplasm contains basophilic Nissl bodies. Nissl body is made of ribonucleic acid and is concerned with the protein synthesis.
- *Neurites:* Extensions from periphery of cell body.

  They are of two types:
  1. *Dendrites:* Conduct impulses towards cell body. May branch to form a dendritic tree.
  2. *Axon:* Conduct impulses away from cell body. Begins at *axon hillock* and terminate by dividing into axon terminals *(telodendria)*.

Q.175 What are the different types of neurons?

- *Unipolar:* Single extension from cell body, e.g. mesencephalic nucleus of fifth cranial nerve.
- *Bipolar:* Extension at each end of the cell body, e.g. retinal bipolar cells, olfactory neuroepithelium and ganglion of 8th cranial nerve.
- *Multipolar:* Several extensions from cell body, e.g. most cells of brain and spinal cord.
- *Pseudounipolar:* Usually have one process arising are pole of cell body but actually two extensions emerge at same pole, e.g. dorsal root ganglion of spinal cord.

Q.176 What are Amacrine cells?

These are small neurons present in retina, olfactory bulb which lack an obvious axon and permit conduction in either direction.

Q.177 What are the different types of neurons in brain?

- *Stellate cells:* Dendrites extend in all directions from cell body.
- *Pyramidal cells:* Cell body is conical in shape and dendrites extend from angles of cone or pyramid.
- *Fusiform cells:* Spindle shaped dendrites emerge at both ends.
- *Glomerular cells:* Dendrites at their tip are highly coiled.

Q.178 How transmission occurs across the synapse?

Due to the release of transmitters (neurochemicals) released into the synaptic cleft by presynaptic process, which cause the stimulation or inhibition of postsynaptic process.

Q.179 What are the different types of synapse?

1. *Excitatory synapse:* Neurotransmitters released causes stimulation of postsynaptic neuron.
2. *Inhibitory synapse:* Neurotransmitter released causes inhibition of postsynaptic neuron.
3. *Reciprocal synapses:* Transmission between two processes occurs in either direction by staggered synaptic zones on each side of synaptic cleft.

Q.180 What are different types of neuroglial cells in brain and spinal cord?

- *Macroglia:* Larger cells, develop from neural plate. They are of following types:
  - *Astrocytes:* Have a small cell body with dendrite like extensions.
  - *Oligodendrocytes:* They have fewer cell processes.
  - *Pituicytes:* In posterior pituitary
  - *Muller cells:* In retina.
  - *Ependymal cells:* Line the ventricles of brain and central canal of spinal cord
  - *Bergmann cells:* In cerebellum.
- *Microglia:* Smallest glial cells. They have fine dendritic processes and flattened outlines. Develop from mesodermal tissue surrounding nervous system.

Q.181 What are the functions of glial cells?

- Act as mechanical support for nervous system.
- Act as insulators, separating neurons and their processes from each other. They prevent impulses from spreading in unwanted directions due to their non conducting nature.
- Act defensively by phagocytosing foreign material and cell debris.
- Help in regulating biochemical environment of neurons.
- Oligodendrocytes form myelin sheath in central nervous system.
- Ependymal cells take part in secretion, transport and uptake of cerebrospinal fluid.
- By proliferation, glial cells repair, by filling the gaps left by dead or degenerating neurons.
- They take up, store and metabolise the neurotransmitters.

Q.182 What are the different types of fibers in peripheral nerve?

Depending on diameter and rate of impulse conduction fibers in peripheral nerve are of three types:

- *Type A:* Subdivided into:
  1. Sensory (Afferent) fibers
  2. Motor (Efferent) fibers
- *Type B:* Preganglionic autonomic fibers
- *Type C:* Nonmyelinated, postganglionic autonomic fibers

Q.183 What are the factors on which conduction in myelinated fibers depend?

- Diameter of axon
- Thickness of myelin sheath
- Internodal distance between nodes of Ranvier

- Area and character of axonal membrane.

Q.184 What are the non-nervous cells present in peripheral nervous system?

- *Capsular cells:* Present around cell body of sensory and autonomic ganglia.
- *Schwann cells:* Present around axons of peripheral nerves and form myelin sheath.

Q.185 What is the composition of myelin?

Myelin contains lipid and basic proteins but has less proteins than cell membrane.

Q.186 What are 'incisures of Schmidt-Lanterman'?

Ans. These are oblique clefts in the myelin and provide conduction channels for metabolities into depths of myelin sheath and axon.

Q.187 What is the characteristic feature of distribution of sympathetic and parasympathetic nervous system?

All parts of body, whether somatic or visceral, receive a sympathetic supply.

But the parasympathetic supply has no somatic distribution but is wholly visceral, but does not innervate all viscera (e.g. suprarenal glands and gonads, which have only a sympathetic supply).

Q.188 What is the origin of autonomic nervous system outflow?

Sympathetic outflow emerges at T_1 to L_2 segments of spinal cord.

Parasympathetic outflow emerges from brain via 3rd, 7th, 9th and 10th cranial nerves and from S_{2-4} segments of spinal cord.

Q.189 To which gland the secretomotor nerves are sympathetic?

Sweat glands.

Q.190 Which cranial nerves contribute to the cranial parasympathetic outflow?

Preganglionic fibers from third, seventh, ninth and tenth cranial nerves.

Q.191 What is the neurotransmitter of autonomic nervous system?

- Noradrenaline is neurotransmitter of sympathetic system except at nerves ending for sweat gland and blood vessels of muscles, where neurotransmitter is acetylcholine.
- Acetylcholine is neurotransmitter of parasympathetic system.

Genitourinary System

Q.192 What is the gubernaculum?

It is a continuous mesenchymal condensation extending from the gonadal anlage to the sexually undifferentiated labioscrotal fold. This guides the descent of the testis into the scrotum in the male and the descent of the ovary into the deep pelvis in the female.

Q.193 What represents the gubernaculum in the female?

It is represented by the round ligament of the uterus and the ligament of the ovary.

Q.194 What are the contents of the deep perineal pouch?

The deep perineal pouch consists of the urogenital diaphragm, that is, the deep transverse perineal muscle and it is associated inferior and superior fascias, the external urethral sphincter, and the membranous urethra in the male.

Q.195 Why is the nerve responsible for back pain during menstruation?

This is because the visceral afferent fibers that mediate pain from the fundus and the uterine body travel along the sympathetic nerve pathways mediated via the hypogastric nerve and the lumbar splanchnics to reach the upper lumbar levels (L_1-L_2) of the spinal cord.

Q.196 What is the normal posture of the uterus?

The normal posture of the uterus is that it is anteverted and anteflexed.

Q.197 What is the external urethral sphincter in a female consist of?

The external urethral sphincter in the female is a modified portion of the deep transverse perineal muscle of the urogenital diaphragm.

Q.198 What is the pelvic diaphragm? And what is its function?

The pelvic diaphragm comprises the levator ani and the coccygeus muscle and their associated fascia, and is shaped like an inverted cone and forms the floor of the abdominopelvic cavity. Its function is to suspend and support the pelvic organs as well as to provide the principal component for continence.

Q.199 What does a posterolateral episiotomy incise?

A posterolateral episiotomy involves incising the skin over the ischioanal fossa, the bulbospongiosus muscle, the superficial transverse perineal muscle and associated transverse neurovascular bundle, a portion of the deep transverse perineal muscle and associated fascial layers (urogenital diaphragm) and frequently the medial fibers of the pubovaginalis and puborectalis portions of the levator ani.

Q.200 What is the autonomic innervation of the penis and its functions?

The sympathetic innervation of the penis is through the pelvic plexus and is responsible for ejaculation. The parasympathetic innervation is responsible for erection.

Q.201 How is the uterus formed?

The uterus is formed by the fusion of two paramesonephric tubes.

Chapter 2

Upper Limb

Bones of Upper Limb (Fig. 2.1)

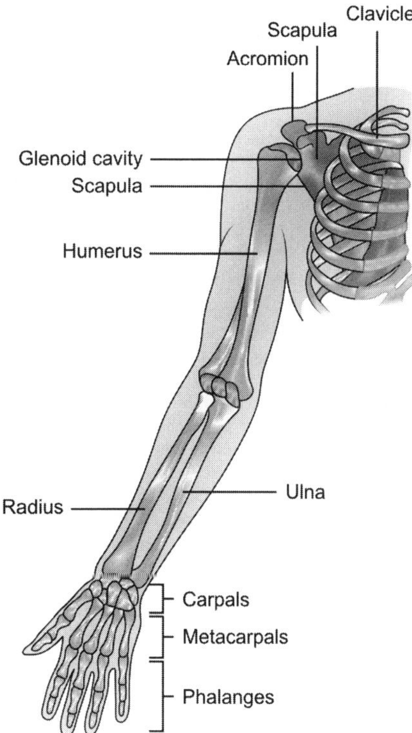

Fig. 2.1: Skeleton of hand

Clavicle (Figs 2.2A and B)

Q.1 What are the characteristic features of clavicle?

- It is a long bone which lies horizontally in body.
- It has no medullary cavity.
- It is subcutaneous throughout.
- It is the first bone to ossify in body of fetus.
- Only long bone that ossifies in membrane except sternal and acromial end.
- It is the only long bone which ossifies from two primary centers.
- It is the most commonly fractured bone in body.

Q.2 How will you determine the side to which clavicle belongs?

- It has two ends, lateral and medial. Lateral end is flat and medial end is large and quadrilateral.
- Shaft is convex forwards in medial 2/3 and concave forwards in lateral 1/3.
- Inferior surface is grooved longitudinally in middle 1/3.

Q.3 What is the nutrient artery supplying clavicle?

Nutrient branch from suprascapular artery.

Q.4 What are the muscles attached to the medial part of clavicle?

- Clavicular part of pectoralis major: From anterior surface of medial half.
- Clavicular head of sternocleidomastoid: From upper surface of medial part.
- Lateral part of sternohyoid: Posterior surface of medial end.

Q.5 What are the functions of clavicle?

- Transmits force from upper limb to axial skeleton.
- Provides attachment to muscles.
- Acts as a strut holding arm free from trunk.

Q.6 Name the structures attached to the edges of groove for subclavius.

Clavipectoral fascia.

Q.7 To which structure the medial end of clavicle articulates?

- Manubrium sterni
- First costal cartilage

Q.8 At which site the clavicle fracture occurs commonly?

At junction of middle and outer third, which is the weakest point.

Scapula

Q.9 What is the extent and position of scapula?

- It lies on posterolateral aspect of chest wall.
- It extends from II to VII rib.

Q.10 How will you determine side to which scapula belongs?

- Lateral angle is large and has a glenoid cavity.
- Lateral thickest border runs from glenoid cavity above to the inferior angle below.
- Dorsal surface is convex and is divided into supraspinatus and infraspinatus fossa by triangular spine.
- Costal surface is concave.

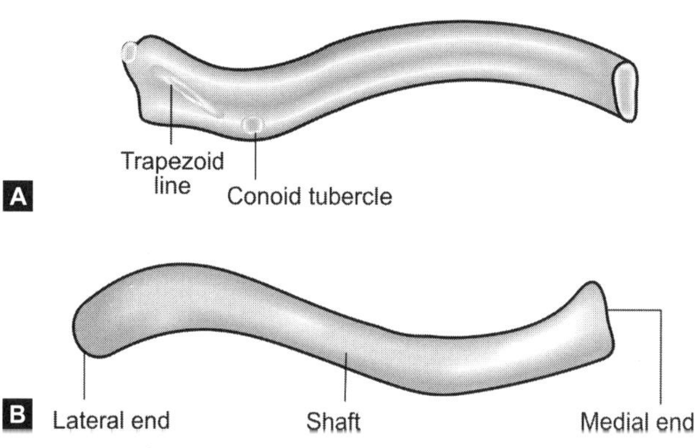

Figs 2.2A and B: Right clavicle (A) Interior aspect, (B) Superior aspect

Q.11 Name the structures passing above and below suprascapular notch.
- The suprascapular notch is converted into a foramen by the suprascapular ligament.
- The suprascapular artery passes above the ligament and suprascapular nerve below the ligament.

Q.12 Name the structures attached to acromion.
- *Trapezius:* Inserted on its medial border
- *Deltoid:* Originates from lateral margin, tip and upper surface.
- *Coracoacromial ligament:* Attached to apex of acromion.

Q.13 What are the structures attached to coracoid process?
Muscles:
- *Short head of biceps:* Origin from tip of coracoid process.
- *Coracobrachialis:* Origin from tip of coracoid process.
- *Pectoralis minor:* Insertion on medial border and superior surface.
Ligaments:
- *Coracoacromial ligament:* To lateral border.
- *Coracoclavicular ligament: Conoid part* to knuckle of process. *Trapezoid part* to ridge between pectoralis minor and coraco-acromial ligament.
- *Coracohumeral ligament:* To root of coracoid process.

Q.14 Name the muscles inserting on the medial border of scapula?
On costal surface: Insertion of two digitations of serratus anterior.
On dorsal surface: Insertion of
- *Levator scapulae:* Above root of spine
- *Rhomboideus minor:* Opposite the root
- *Rhomboideus major:* Below the root (Figs 2.3A and B).

Q.15 Which muscle originates from supraglenoid tubercle?
Long head of biceps.

Q.16 Which muscle originates from infraglenoid tubercle?
Long head of triceps.

Q.17 How many ossification centers are present in scapula?
Eight.

Humerus

Q.18 How will you determine the side to which the humerus belongs?
- Upper end is rounded and forms the head. Lower end is flattened from before backwards.
- Head is directed medially and backwards.
- Lesser tubercle projects from front of upper end.
- The anterior aspect of upper end shows a vertical groove called intertubercular sulcus (Figs 2.4A and B).

Q.19 What is the anatomical position of the humerus in body?
- Head is directed medially, upwards and backwards.
- Medial epicondyle is directed medially and slightly backwards.

Q.20 What are contents of intertubercular sulcus (bicipital groove)?
- Tendon of long head of biceps.
- Synovial sheath of the tendon.
- Ascending branch of anterior circumflex humeral artery.

Q.21 What are the structures attached to intertubercular sulcus?
- *To medial lip:* Teres major, insertion.
- *To floor:* Latissimus dorsi, insertion.
- *To lateral lip:* Pectoralis major, insertion.

Q.22 How the tendon of pectoralis major is inserted?
It is inserted by a bilaminar tendon on the lateral lip of bicipital groove, the two laminae are continuous with each other inferiorly.

Q.23 What are the muscles inserting on greater tubercle?
- *Supraspinatus:* On upper impression.
- *Infraspinatus:* On middle impression.
- *Teres minor:* On lower impression (Figs 2.5A and B).

Q.24 Which muscle is inserted into lesser tubercle?
Subscapularis

Q.25 What is the 'anatomical neck' and 'surgical neck' of humerus?
- *Anatomical neck*: Surrounds the margin of head. Provides attachment to capsular ligament which is deficient inferiorly on medial side.
- *Surgical neck:* Lies at upper end of shaft, below the epiphyseal line. It is a common site for fracture.

Q.26 What are the structures related to surgical neck of humerus?
- Axillary nerve
- Anterior and posterior circumflex humeral vessels.

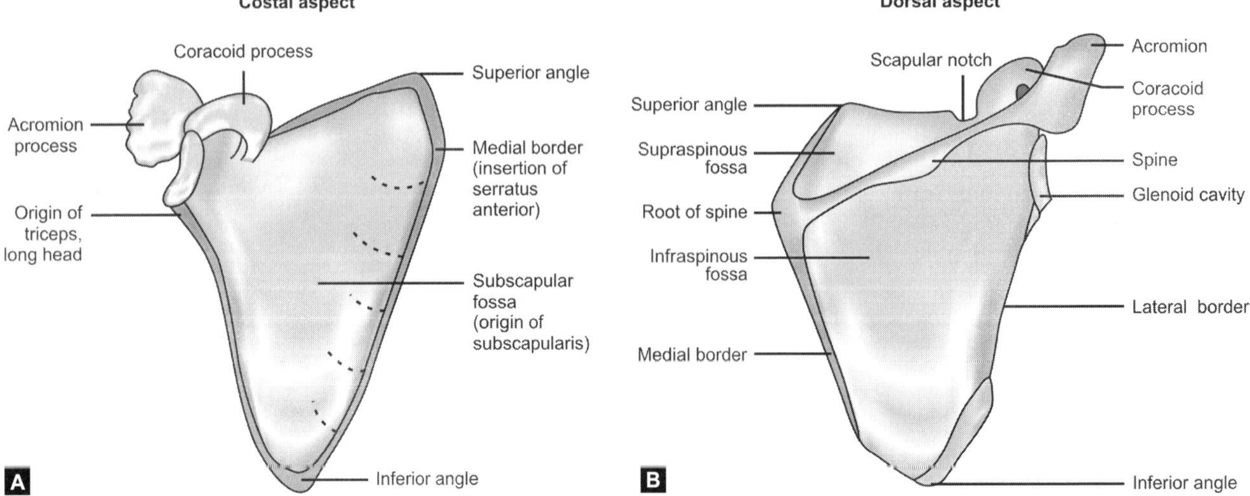

Figs 2.3A and B: (A) Right scapula, showing attachments, seen from the front, (B) Right scapula, showing attachments, seen from behind

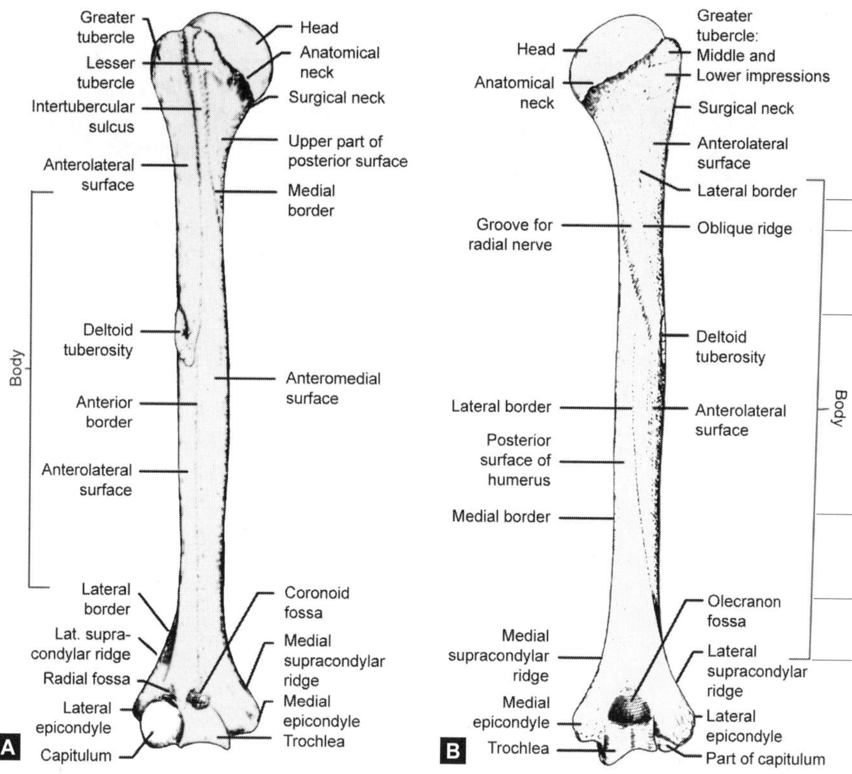

Figs 2.4A and B: (A) Right humerus seen from the front, (B) Right humerus seen from behind

Q.27 Name the structures related to radial groove of shaft of humerus.
- Radial nerve and
- Profunda brachii vessels.

Q.28 Which muscle originates from medial supracondylar ridge?
Pronator teres, from lower end of medial supracondylar ridge.

Q.29 Which muscle originates from lateral supracondylar ridge?
- Brachioradialis: From upper two-thirds.
- Extensor carpi radialis longus: From lower one-third.

Q.30 Which muscles arise from lateral epicondyle of humerus (common extensor origin)?
- Extensor carpi radialis brevis
- Extensor digitorum
- Extensor digiti minimi
- Extensor carpi ulnaris

Q.31 Which muscles also arise from lateral epicondyle of humerus except from common extensor origin?
- Anconeus
- Supinator

Q.32 What is angle of 'humeral torsion'?
- It is an angle formed by superimposition of long axes of upper and low articular surfaces of humerus.
- It is about 164 degrees.
- It is greater in man and in adults.

Q.33 What are fossae present in lower end of humerus?
- *Olecranon fossa:* Posteriorly, just above trochlea. Accommodates olecranon process in extended elbow.
- *Radial fossa:* Anteriorly, above capitulum. Accommodates radial head in flexed elbow.
- *Coronoid fossa:* Anteriorly, above trochlea. Accommodates coronoid process of ulna in flexed elbow.

Q.34 Why the fracture of humerus at junction of upper and middle third leads to delayed union?
Because of poor blood supply at the junction.

Q.35 Which nerve is most commonly involved in the supracondylar fracture of the humerus?
Median nerve.

Q.36 What are the ossification centers for lower end of humerus and at what age they appear?
- Capitulum: First year
- Medial epicondyle: Fifth year
- Medial part of trochlea: Ninth year
- Lateral epicondyle: Twelfth year
  Medial epicondyle fuses with shaft at 20th year and others fuse together to form a single epiphysis, which fuses with shaft at 15 years of age.

Radius and Ulna

Q.37 How will you determine the side to which the radius belongs?
- *Upper end:* Narrow and has a disk shaped head
- *Lower end:* Wide, smooth anteriorly and ridges and grooves on posterior aspect.
- *Shaft:* Convex laterally and concave forwards in lower part. Medial border is sharp (Figs 2.6A and B).

Q.38 Where the biceps brachii is inserted?
Into rough posterior part of radial tuberosity (Figs 2.7A and B).

Q.39 What structures are related to posterior aspect of lower end of radius?
- The groove behind medial part of lower end lodges tendons of extensor digitorum and extensor indicis.
- The oblique groove medial to dorsal tubercle lodges tendon of extensor pollicis longus.
- The groove lateral to dorsal tubercle lodges tendons of extensor carpi radialis longus and brevis.
- The lateral aspect is crossed by tendons of abductor pollicis longus and extensor pollicis brevis.

Q.40 How will you determine the side to which ulna belongs?
- Upper end is hook like, with its concavity directed forwards.
- Lateral border of shaft is sharp (Figs 2.8A and B).

Q.41 Name the structures attached to medial surface of olecranon process.
Upper part:
- Origin of ulnar head of flexor carpi ulnaris.
- Posterior and oblique bands of ulnar collateral ligaments.

Lower part: Upper fibres of flexor digitorum profundus.

Q.42 Name the structures attached to anterior surface of coronoid process.
Whole of surface: Insertion to brachialis (Figs 2.9A and B).
Medial margin:
- Ulnar head of pronator teres,
- Ulnar head of flexor pollicis longus,

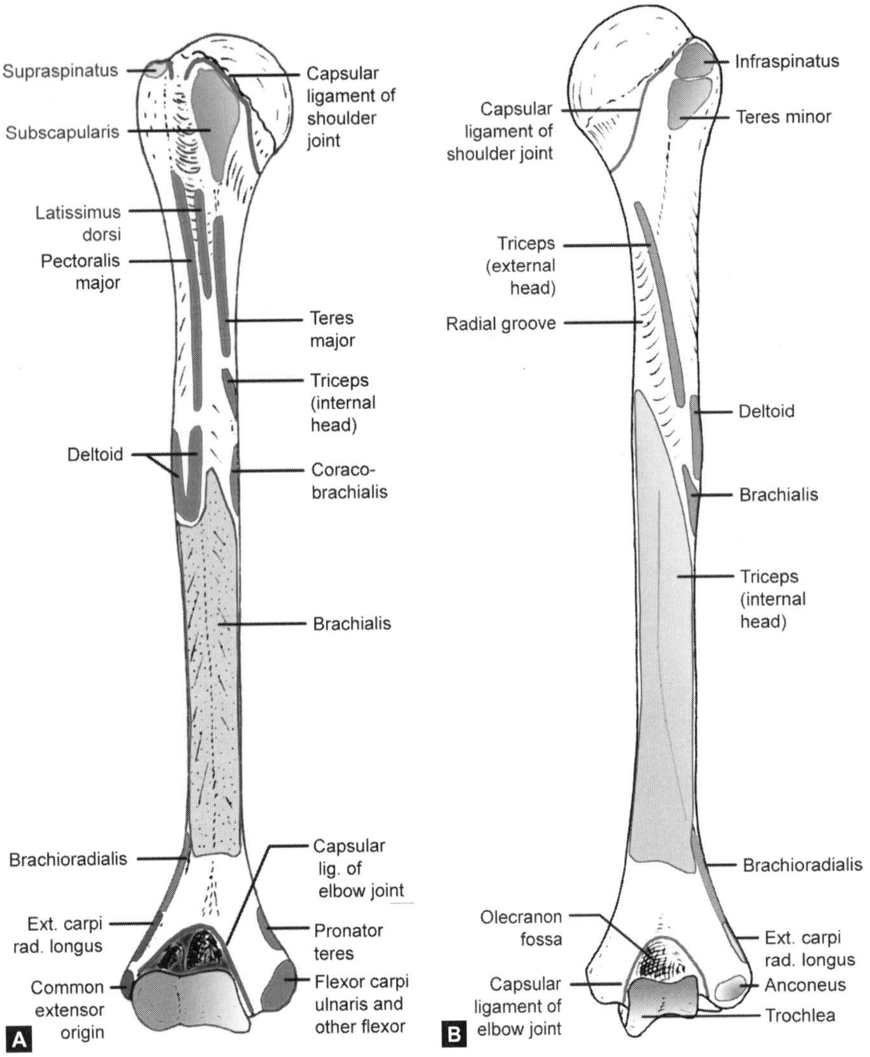

Figs 2.5A and B: (A) Right humerus, showing attachments, seen from the front, (B) Right humerus, showing attachments, seen from behind

- Humeroulnar head of flexor digitorum superficialis,
- Anterior and oblique bands of ulnar collateral ligament.

Q.43 What are the flexor muscles in front of the forearm?

The flexor muscles infront of the forearm can be divided into three layers (Fig. 2.10):

- *Deep layer*
  - Flexor digitorum profundus
  - Flexor pollicis longus
- *Intermediate layer*
  - Flexor digitorum superficialis
- *Superficial layer*
  - Pronator teres
  - Flexor carpi radialis
  - Flexor carpi ulnaris

Q.44 Name the muscles arising from aponeurosis of posterior border of shaft of ulna.

- Flexor digitorum profundus,
- Flexor carpi ulnaris and
- Extensor carpi ulnaris.

Q.45 What are the muscles arising from the lateral part of the posterior surface?

From above downwords:

- Abductor pollicis longus
- Extensor pollicis longus
- Extensor indicis

Q.46 What is Colles' fracture?

In adults, fall on the out stretched hand causes fracture of radius about one inch proximal to wrist joint and distal fragment is impacted dorsally and laterally.

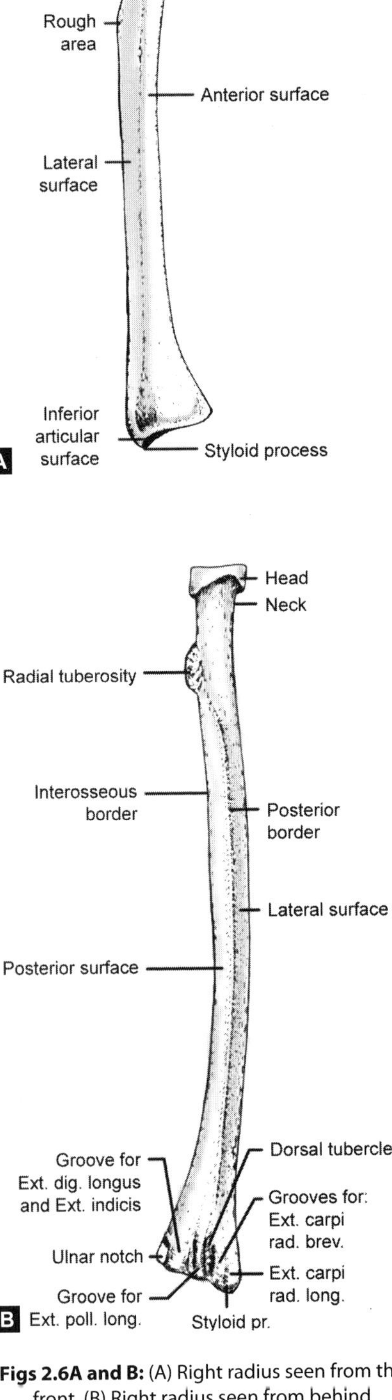

Figs 2.6A and B: (A) Right radius seen from the front, (B) Right radius seen from behind

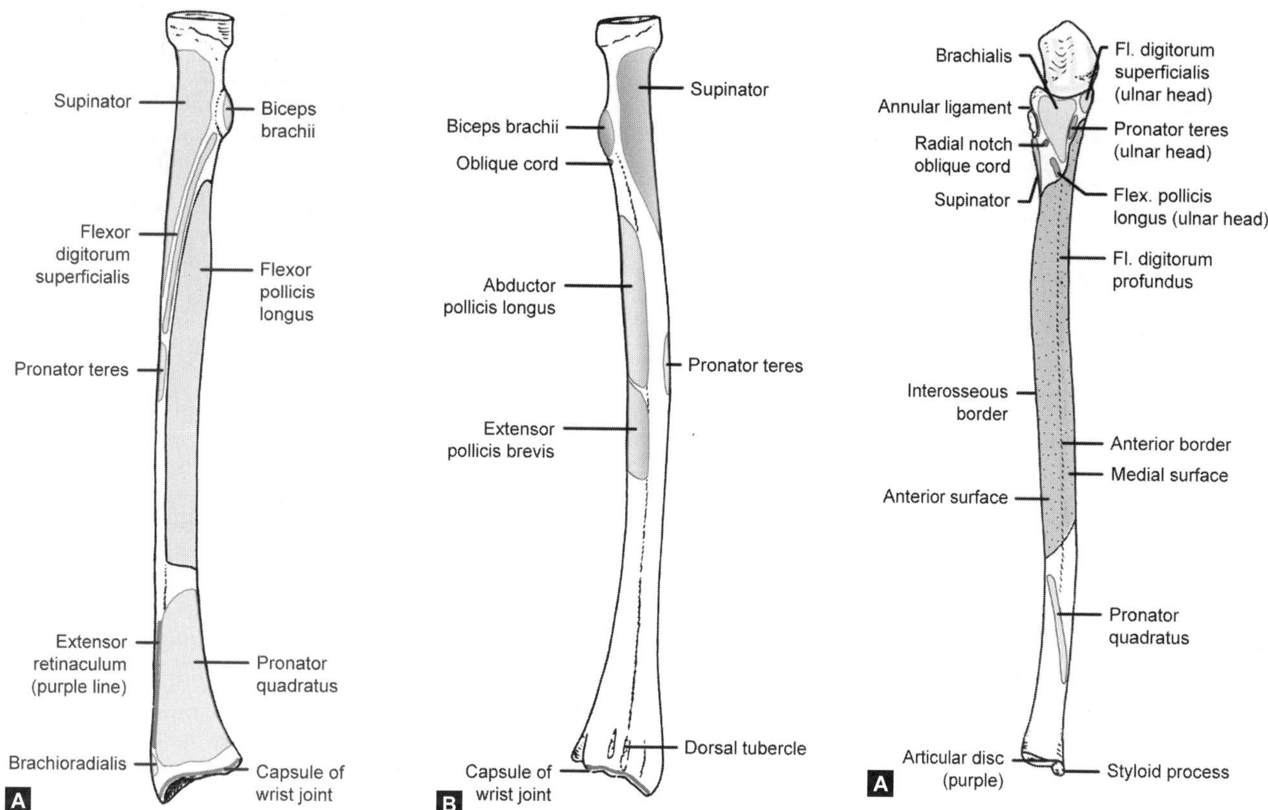

Figs 2.7A and B: (A) Right radius, showing attachments seen from the front, (B) Right radius, showing attachments seen from behind

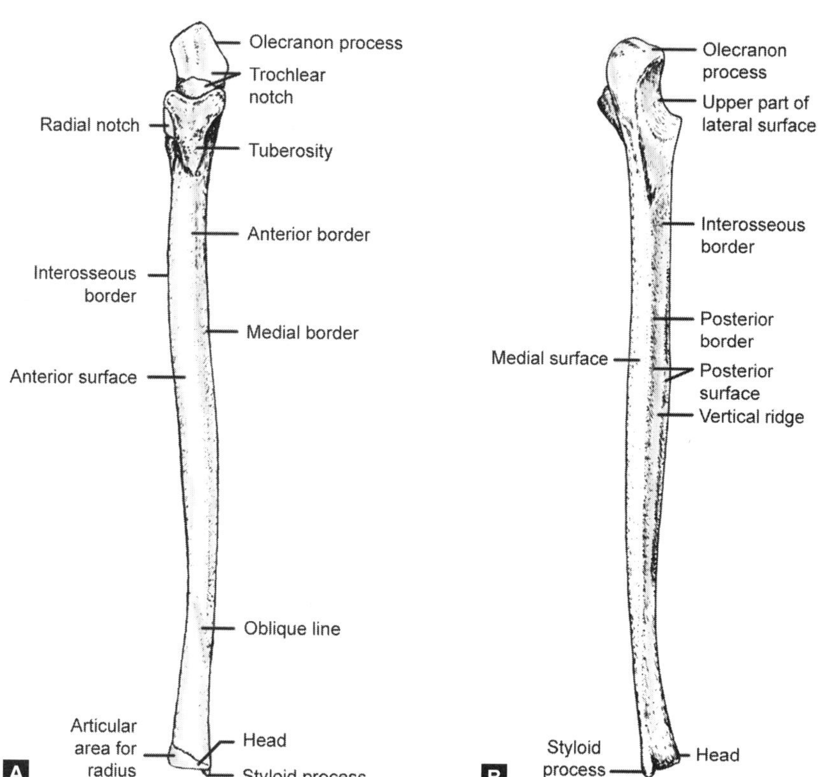

Figs 2.8A and B: (A) Right ulna seen from the front, (B) Right ulna seen from behind

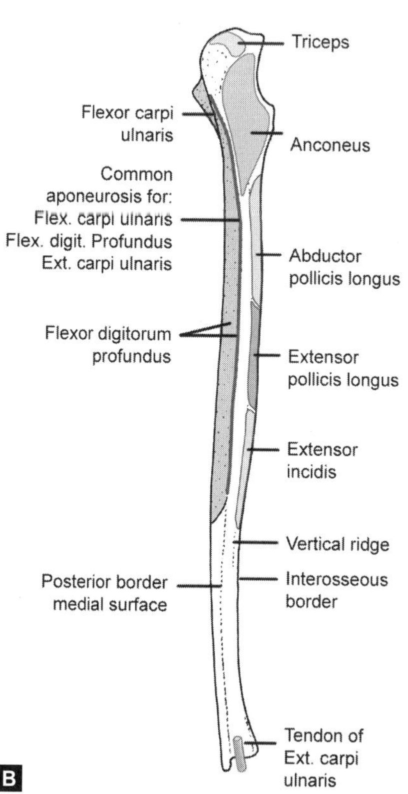

Fig. 2.9A and B: (A) Right ulna, showing attachments, seen from the front, (B) Right ulna, showing attachments, seen from behind

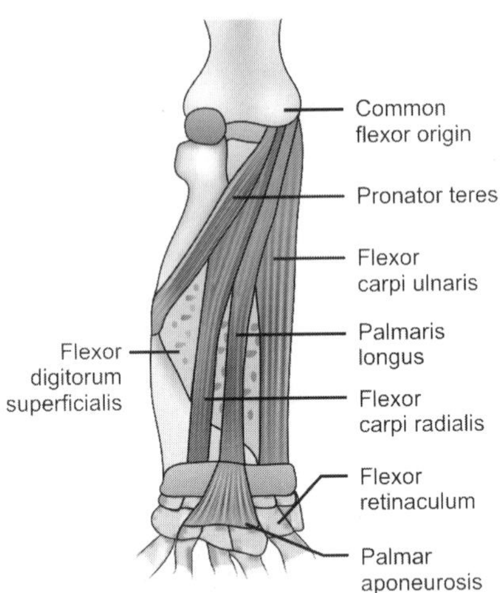

Fig. 2.10: Superficial muscles in the front of the forearm

Q.47 What is 'student's elbow or 'miners' elbow?

Inflammation of subcutaneous bursa present over olecranon due to repeated trauma. Commonly seen in miners and students.

Bones of Hand

Q.48 Name the carpal bones.

See Figures 2.11 and 2.12.

Proximal row: Contains from lateral to medial side

- Scaphoid,
- Lunate,
- Triquetral and
- Pisiform.

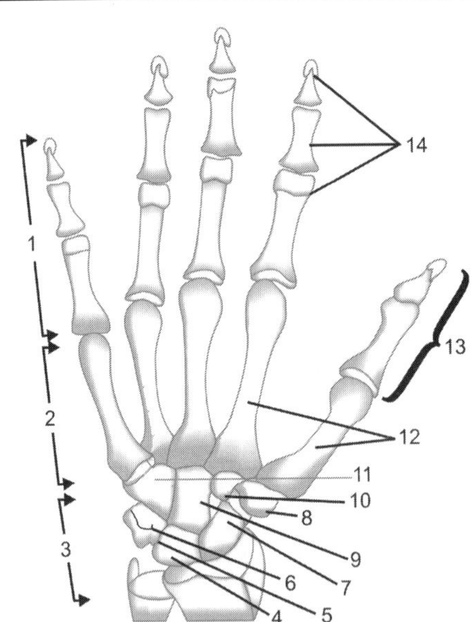

Fig. 2.11: Bones of the hand: (1) Digit; (2) Metacarpus; (3) Carpus; (4) Lunate; (5) Pisiform; (6) Triquetrum; (7) Navicular; (8) Greater multiangular; (9) Capitulum; (10) Lesser multiangular; (11) Hamate; (12) Metacarpals; (13) Thumb digit; (14) Phalanges

Distal row: Contains from lateral to medial side

- Trapezium,
- Trapezoid.
- Capitate and
- Hammate.

Q.49 Name the structures attached to tubercle of scaphoid.

- Abductor pollicis brevis and
- Flexor retinaculum.

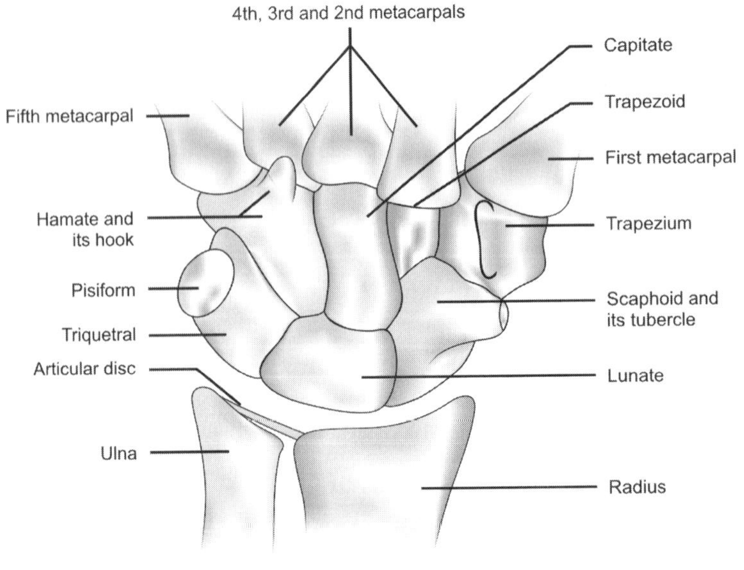

Fig. 2.12: Right carpus, seen from the front

Q.50 Name the structures attached to pisiform.

- Flexor carpi ulnaris,
- Abductor digiti minimi,
- Flexor retinaculum and
- Extensor retinaculum.

Q.51 Name the site of insertion of various muscles of thumb.

- *Flexor pollicis longus:* Volar surface of base of distal phalanx
- *Extensor pollicis longus:* Base of distal phalanx of thumb
- *Extensor pollicis brevis:* Dorsal surface of base of proximal phalanx
- *Abductor pollicis longus:* Base of first metacarpal
- *Flexor pollicis brevis and abductor pollicis brevis:* Base of proximal phalanx of thumb
- *Opponens pollicis:* Shaft of first metacarpal
- *First palmar interossei:* Base of proximal phalanx.

Q.52 Name the muscles attached to middle phalanx.

- *Flexor digitorum superficialis:* Inserted on sides of shaft.
- *Extensor digitorum:* Central slip inserted on dorsal surface of base.

Q.53 Name the structures attached to base of proximal phalanx.

Insertion of lumbricals and interossei.

- In thumb, insertion of
  - *On lateral side:* Abductor pollicis brevis and flexor pollicis brevis.
  - *On medial side:* First palmar interosseous.
  - *On dorsal surface:* Extensor pollicis brevis.
- In little finger,

On medial side: Insertion of abductor digiti minimi and flexor digiti minimi.

Q.54 What are the sesamoid bones found in upper limb?

- *Pisiform:* Sesamoid bone within tendon of flexor carpi ulnaris.
- Two sesamoid bones on head of first metacarpal bone.
- In capsule of interphalangeal joint of thumb.
- On ulnar side of capsule of metacarpophalangeal joint of little finger.

Joints of Upper Limb

Shoulder Joint

Q.55 What type of joint shoulder joint is?

Ball and socket variety of synovial joint.

Q.56 What are the articular surfaces of shoulder joint?

Glenoid cavity of scapula and head of humerus.

Q.57 Name the ligaments of shoulder joint?

- Capsular ligament.
- Coracohumeral ligament.
- Transverse humeral ligament.
- Glenoid labrum (Figs 2.13A and B).

Q.58 What is the arterial supply of shoulder joint?

- Anterior circumflex humeral artery.
- Posterior circumflex humeral artery.
- Suprascapular artery and
- Subscapular artery.

Q.59 What are the movements possible at shoulder joint? Name also the main muscle producing these movement.

1. *Flexion:*
   - Clavicular head of pectoralis major
   - Anterior fibers of deltoid.
   - Coracobrachialis
   - Short head of biceps
2. *Extension:*
   - Sternocostal head to pectoralis major
   - Posterior fibers of deltoid
   - Latissimus dorsi
   - Teres major.
3. *Adduction:*
   - Pectoralis major
   - Latissimus dorsi
   - Subscapularis
   - Teres major.
4. *Abduction:*
   - Middle fibers of deltoid
   - Supraspinatus
5. *Medial rotation:*
   - Pectoralis major
   - Anterior fibers of deltoid
   - Latissimus dorsi
   - Teres major
   - Subscapularis.
6. *Lateral rotation:*
   - Posterior fibers of deltoid
   - Infraspinatus
   - Teres minor.
7. *Circumduction:* Combination of different movements.

Q.60 Name the bursa around shoulder joint.

- Subacromial bursa
- Subscapularis bursa
- Infraspinatus bursa
- Bursa related to muscles around shoulder joint, e.g. teres major, long head of triceps, coracobrachialis (Fig. 2.14).

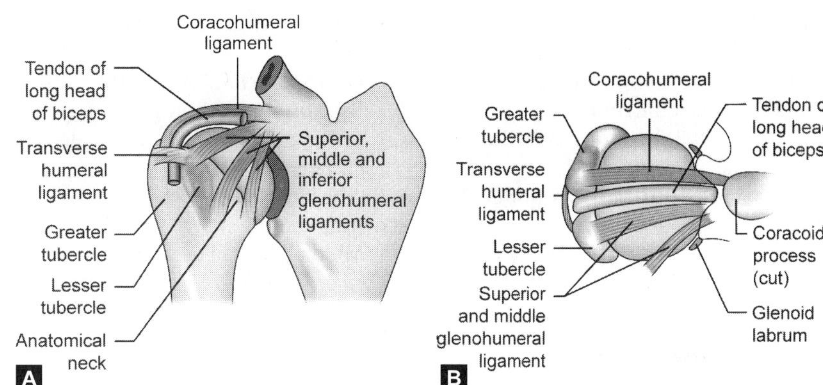

Figs 2.13A and B: Some ligaments of the shoulder joint. The scapula and humerus are viewed from the front in (A), and from above in (B)

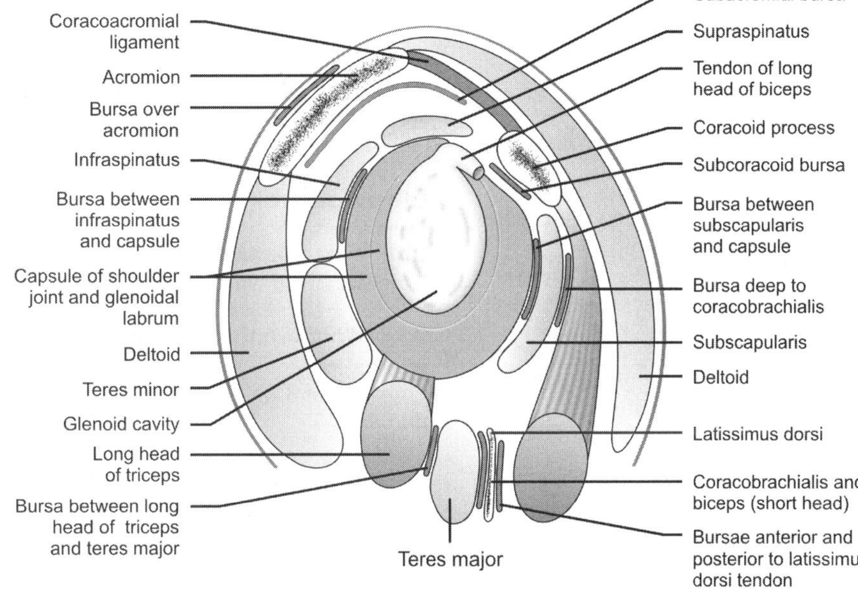

Fig. 2.14: Schematic diagram to show muscles and bursae around the shoulder joint

Q.61 a) Why shoulder joint is a weak joint?
b) How its stability is increased?

a)
- The glenoid cavity is shallow and small.
- Head of humerus is larger than glenoid cavity.

b)
- By musculotendinous cuff of shoulder.
- Coracoacromial arch.
- Glenoidal labrum, which deepens the glenoid cavity and articular cartilage lining it.
- Long muscles of shoulder, e.g. deltoid, long head of triceps, latissimus dorsi and teres major.

Q.62 What is 'rotator cuff' or 'musculotendinous cuff'?

It is a fibrous sheath of tendons of short muscles of shoulder which cover all except inferior aspects of shoulder joint. The muscles are supraspinatus (superiorly) subscapularis (anteriorly), infraspinatus and teres minor (posteriorly).

The cuff gives strength to the capsule of shoulder joint.

Q.63 Which tendon is most commonly injured in rotator cuff lesions?

Supraspinatus.

Q.64 Why the dislocation of the shoulder joint occurs inferiorly?

Because the inferior aspect is unprotected by musculotendinous cuff.

Q.65 What is the clinical importance of shoulder tip pain?

- Irritation of undersurface of diaphragm from surrounding pathology causes *referred pain* in the shoulder, because

the phrenic nerve and supraclavicular nerves have similar root values ($C_{3,4}$).
- Pain in the left shoulder tip due to irritation by splenic rupture.
- Pain in the right shoulder, due to subphrenic abscess.
- Acute pancreatitis and gas under the diaphragm due to perforation of peptic ulcer causes referred pain in the either of the shoulder tip.

Shoulder Girdle (Fig. 2.15)

Q.66 What are the joints of shoulder girdle?
- Sternoclavicular joint
- Acromioclavicular joint.

Q.67 What type of joints are joints of shoulder girdle?
- *Sternoclavicular joint:* Saddle variety of synovial joint.
- *Acromioclavicular joint:* Plane variety of synovial joint.

Q.68 What is the characteristic feature of acromioclavicular joint?
It is partially divided by an incomplete fibrocartilage articular disc, which is perforated in the center.

Q.69 Name the ligaments forming acromioclavicular joint?
- Coracoclavicular ligament: Main ligament
- Coracoacromial ligament

Q.70 What are the movements produced at the shoulder girdle?
1. *Elevation of scapula:*
   - By upper fibers of trapezius and
   - Levator scapulae. For example, shurgging of shoulders.
2. *Depression of scapula:* By
   - Lower fibers of serratus anterior
   - Pectoralis minor
   - Levator scapulae and rhomboids also assist.
3. *Protraction of the scapula:* By
   - Serratus anterior and
   - Pectoralis minor. For example, punching movements.
4. *Retraction of scapula:* By
   - Rhomboids and
   - Middle fibers of trapezius.
5. *Forward rotation of scapula around chest wall:* In overhead abduction of shoulder by:
   - Upper fibers of trapezius and
   - Lower fibers of serratus anterior.
6. *Backward rotation of scapula:* By
   - Levator scapulae and
   - Rhomboids.

Q.71 What is the function of shoulder girdle?
It suspends the upper limb to axial skeleton.

Elbow Joint

Q.72 What type of joint elbow joint is?
Hinge variety of synovial joint.

Q.73 What are the surfaces of elbow joint?
From above: Capitulum and trochlea of humerus.
From below:
- Capitulum articulates with upper surface of head of radius
- Trochlear notch of ulna articulates with trochlea of humerus (Figs 2.16A to C).

Q.74 Name the ligaments of elbow joint.
- Capsular ligament.
- Anterior ligament.
- Posterior ligament.
- Ulnar collateral ligament.
- Radial collateral ligament.

Fig. 2.15: Joints of upper limb: Sternoclavicular and acromioclavicular joints

Q.75 What are the movements of elbow joint? Name the muscles producing these movements.
Flexion: By
- Brachialis,
- Biceps and
- Brachioradialis.

Extension: By
- Triceps and
- Anconeus.

Q.76 How will you clinically test for dislocation of elbow joint?
Normally, in semiflexed position, olecranon and two humeral epicondyles form a equilateral triangle. In dislocation of elbow, this relationship is disturbed.

Q.77 What is 'tennis elbow'?
It is due to the partial tear of the common origin of the superficial extensor muscles of forearm.

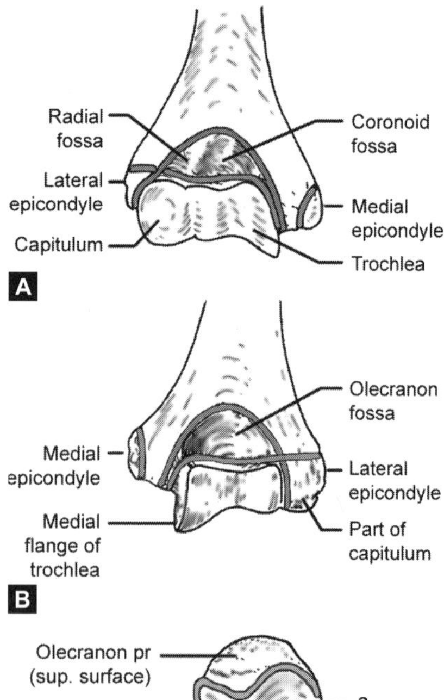

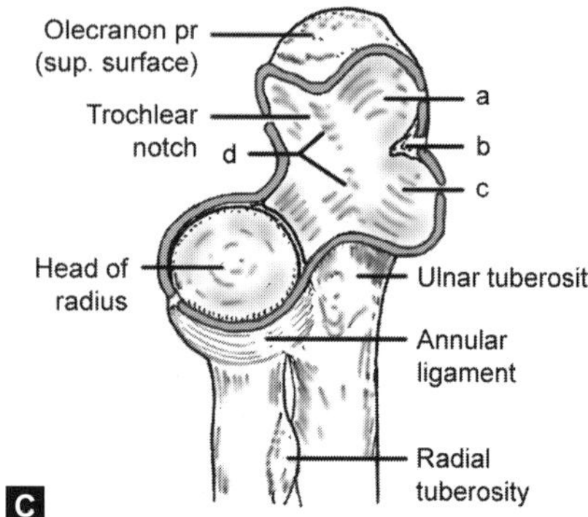

Figs 2.16A to C: (A and B) Attachment of the capsular ligament (dark line) of the elbow joint to the humerus. A. Anterior aspect. B. Posterior aspect, (C) Lower articular surfaces of elbow joint, and capsular attachment. The radius and ulna are viewed from the anterosuperior aspect

Q.78 What is 'golfer's elbow'?
It is due to partial tear of the common origin of the superficial flexor muscles of forearm.

Cubital Fossa

Q.79 What is cubital fossa?
It is a triangular hollow in front of elbow.

Q.80 What are the boundaries of cubital fossa?
- *Laterally:* Medial border of brachioradialis.
- *Medially:* Lateral border of pronator teres.
- *Base:* By an imaginary line joining two epicondyles of the humerus.
- *Apex:* By meeting point of lateral and medial boundaries.
- *Floor:* By
  - Brachialis and
  - Supinator.
- *Roof:*
  - Skin
  - Superficial fascia
  - Deep fascia and
  - Bicipital aponeurosis (Fig. 2.17).

Q.81 What are the contents of cubital fossa?
- Median nerve.
- Termination of brachial artery.
- Tendon of biceps with bicipital aponeurosis.
- Radial nerve.

Q.82 Name the structures lying in superficial fascia of cubital fossa.
- Median cubital vein
- Lateral cutaneous nerve of forearm
- Medial cutaneous nerve of forearm.

Q.83 What is carrying angle?
It is the angle between long axis of arm with long axis of forearm, when forearm is extended and supinated. It disappears in full flexion of elbow and in pronation.

It is about 170°.

Q.84 What are the factors responsible for carrying angle?
- Medial flange of trochlea is larger than lateral flange and projects downward to a lower level. As a result lower edge of trochlea passes downwards and medially.
- Superior articular surface of coronoid process of ulna is oblique.

Q.85 What is the importance and sex differences in carrying angle?
Importance:
- It allows the arm to swing clearly away from the body.
- The forearm comes in line with long axis of arm in midprone position in which the hand is mostly used.

Sex differences: Greater in females because of wider pelvis.

Radioulnar Joints

Q.86 What type of joint radioulnar joints are?
- *Superior radioulnar joint:* Pivot type of synovial joint.
- *Inferior radioulnar joint:* Pivot type of synovial joint.
- *Middle radioulnar joint:* Syndesmoses type of fibrous joint (Figs 2.18 A, B and 2.19A to C).

Q.87 What are the functions of interosseous membrane of middle radioulnar joint?
- Attachment to muscles,
- Transmits weight of hand from radius to ulna.

Q.88 What is pronation and supination?
- These are rotatory movements of forearm with hand around a vertical axis in semiflexed position.
- In pronation, palm faces downwards
- In supination, palm faces upwards.

Q.89 What is the axis of pronation and supination?
Vertical axis passing superiorly, through center of head of radius and inferiorly, through apex of articular disc when ulna is fixed or through any fixed finger when ulna is free to move.

Q.90 Name the muscles producing pronation and supination.
- *Pronation:*
  Principal muscles: Pronator teres, Pronator quadratus.
  Accessory muscles: Flexor carpi radialis, palmaris longus.
- *Supination:* Supinator and biceps brachii.

Wrist Joint

Q.91 What type of joint wrist joint is?
Ellipsoid variety of synovial joint.

Q.92 What are the articular surfaces of wrist joint?
From above:
- *Radius:* Inferior surface of lower end,
- Triangular articular disc of inferior radioulnar joint.

From below: Carpal bones: Scaphoid, lunate and triquetral (Fig. 2.20).

Q.93 Name of the ligaments of wrist joint.
- Capsular ligament
- Anterior radiocarpal and ulnocarpal
- Posterior radiocarpal ligament
- Radial collateral ligament
- Ulnar collateral ligament.

Q.94 At which joint, movements of wrist take place?
- *Radiocarpal joint:* Mainly extension and adduction.
- *Midcarpal joint:* Mainly flexion and abduction.

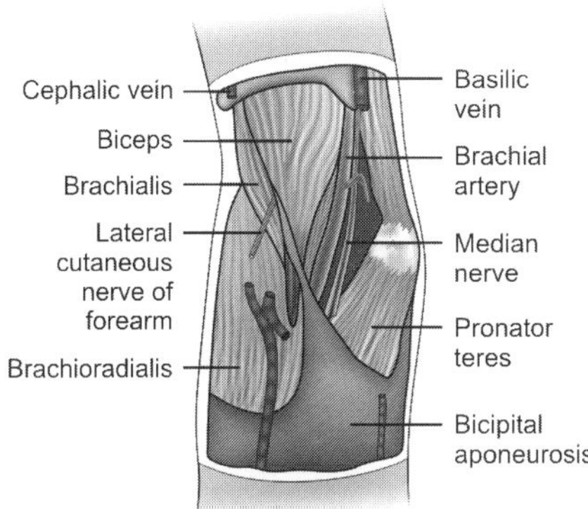

Fig. 2.17: Boundaries of the cubital fossa

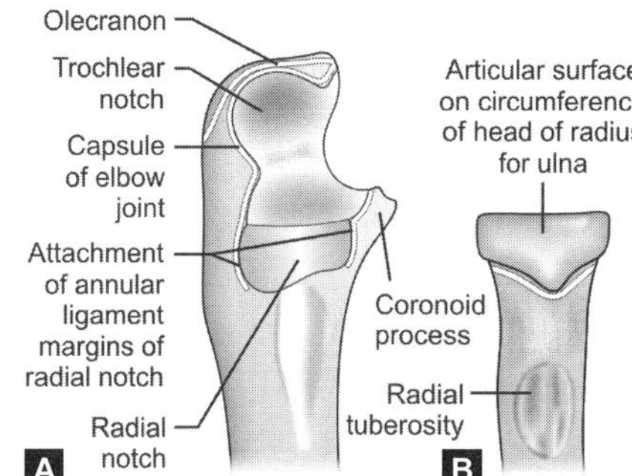

Figs 2.18A and B: Articular surfaces of the superior radioulnar joint: (A) Upper end of ulna, lateral aspect, (B) Upper end of radius, medial aspect

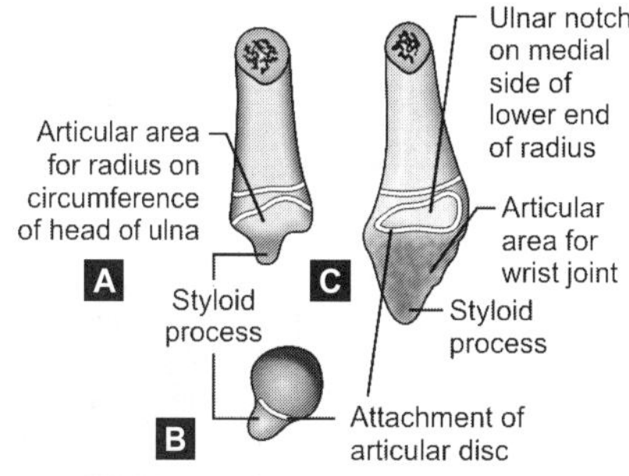

Figs 2.19A to C: Articular surfaces and capsular attachments of inferior radioulnar joint: (A) Lower end of ulna, lateral aspect, (B) Lower end of ulna, inferior aspect, (C) Lower end of radius, medial aspect

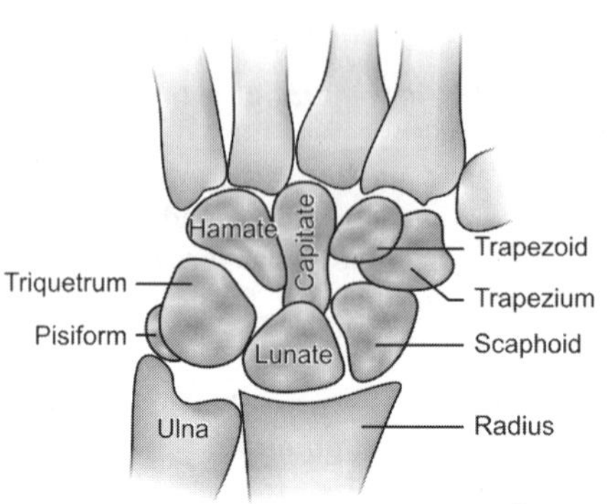

Fig. 2.20: Schematic coronal section through the wrist to show the formation of the articular surfaces of the inferior radioulnar, wrist and midcarpal joints

Q.95 Name the muscles producing abduction and adduction at wrist joint.

Abduction (Fig. 2.21):
- Flexor carpi radialis,
- Extensor carpi radialis longus and brevis,
- Extensor pollicis brevis and
- Abductor pollicis longus.

Adduction:
- Flexor carpi ulnaris and
- Extensor carpi ulnaris.

Q.96 Why the range of adduction is greater than abduction?

Because of longer styloid process of radius, which limits the abduction.

Q.97 What are the boundaries of 'anatomical snuffbox'?

It is depression on lateral side of wrist, when the thumb is extended.

Anterior:
- Abductor pollicis longus and
- Extensor pollicis brevis.

Posterior: Extensor pollicis longus.

Pulsations of the radial artery can be felt in the floor of the depression against the scaphoid and trapezium and in the proximal part, styloid process of the radius and base of the thumb metacarpal distally.

Joints of Hand

Q.98 What type of joint first carpometacarpal joint is?

Saddle variety of synovial joint.

Q.99 What is characteristic of movements of the carpometacarpal joint of thumb?

The thumb is rotated by 90° on its long axis, relative to other digits. As a result, ventral surface faces medially and dorsal surface laterally. Therefore, flexion and extension take place in plane parallel to palm while in other digits, it takes place in planes at right angles to palm.

Q.100 Why the movements at first carpometacarpal joint are freer than the other corresponding joints?

Because this has a separate joint cavity.

Q.101 Why the abduction and adduction are not possible at metacarpophalangeal joint when fingers are flexed?

- Because each metacarpal head is flattened anteriorly and when base of proximal phalanx moves on this flattened surface abduction and adduction become impossible.
- The collateral ligament becomes taut in flexion and prevent sideways movement.

Q.102 What are the attachments of flexor retinaculum?

Medial: Hook of hamate and pisiform.
Lateral: Tubercle of trapezium and tubercle of scaphoid (Fig. 2.22).

Q.103 Name the structures passing superficial to the flexor retinaculum.

- Tendon of palmaris longus.
- Palmar cutaneous branch of median nerve.
- Palmar cutaneous branch of ulnar nerve,
- Ulnar nerve and
- Ulnar vessels.

Q.104 Name the structures passing deep to flexor retinaculum.

- Median nerve
- Tendons of flexor digitorum sublimis
- Tendons of flexor digitorum profundus
- Tendon of flexor pollicis longus
- Ulnar bursa
- Radial bursa

Q.105 Name the structures piercing flexor retinaculum.

- Flexor carpi radialis and
- Flexor carpi ulnaris.

Q.106 Name the structures passing deep to extensor retinaculum.

The structures deep to extensor retinaculum lie in 6 compartments formed by septa passing from retinaculum to posterior surface of radius.

The structures from lateral to medial side in each compartment (Fig. 2.23) are:

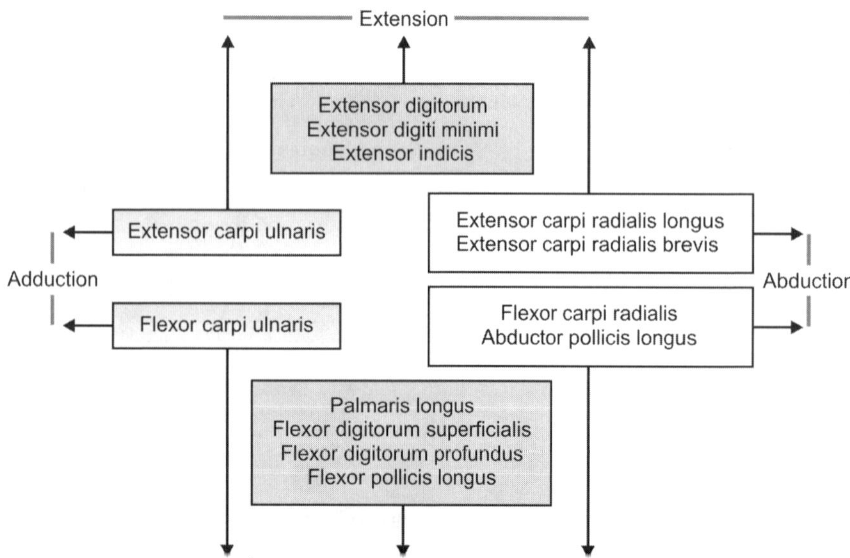

Fig. 2.21: Scheme to show the muscles responsible for movements at the wrist joint

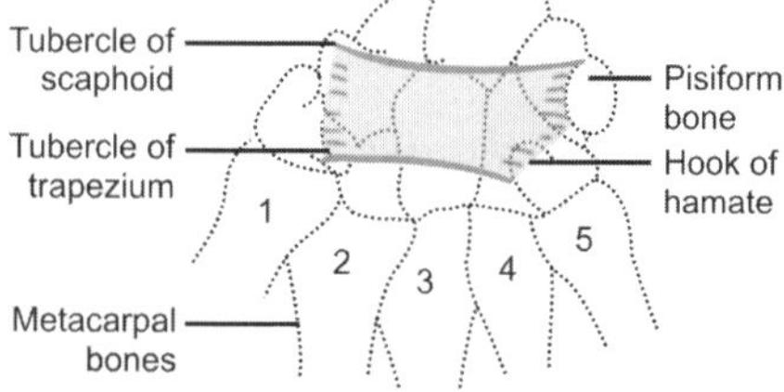

Fig. 2.22: Attachments of the flexor retinaculum

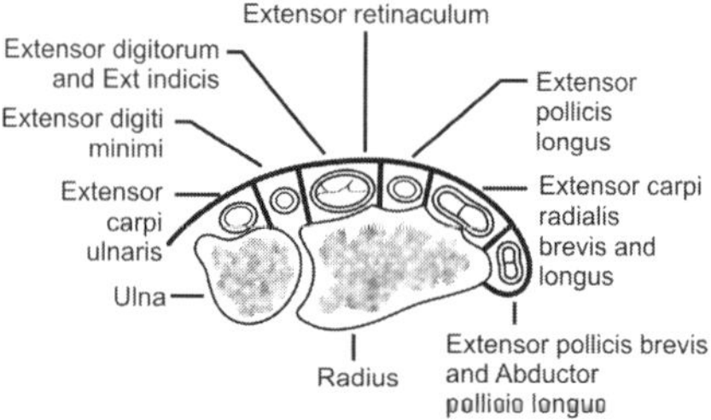

Fig. 2.23: Tendons passing under cover of the extensor retinaculum

- Abductor pollicis longus and extensor pollicis brevis.
- Extensor carpi radialis longus and extensor carpi radialis brevis.
- Extensor pollicis longus.
- Extensor digitorum
  - Extensor indicis
  - Posterior interosseous nerve and anterior interosseous artery.
- Extensor digiti minimi.
- Extensor carpi ulnaris.

Q.107 What is 'palmar aponeurosis'?
It is central part of deep fascia of palm. It improves the grip by fixing the skin of palm. Digital nerves, vessels and tendons pass deep to it, so it protects these.

Q.108 How fibrous flexor sheaths of fingers are formed? What is their importance?
These are made up of deep fascia of the fingers, which is thickened and arched to be attached to side of the phalanges and across the base of distal phalanx.

It forms a fascial tunnels which contains long flexor tendons enclosed in digital synovial sheath and it holds the tendons in position during flexion of the digits.

Q.109 What are the muscles forming the thenar eminence?
- Abductor pollicis brevis
- Flexor pollicis brevis
- Opponens pollicis
- Adductor pollicis.

Q.110 Name the muscle lying deepest at the thenar eminence?
Adductor pollicis.

Q.111 Name the muscles forming the hypothenar eminence?
- Abductor digiti minimi
- Flexor digiti minimi
- Opponens digiti minimi.

Q.112 What is Dupuytren's contracture?
- It is thickening and contraction of ulnar side of palmar aponeurosis.
- This usually affects ring finger in which proximal and middle phalanx are flexed and cannot be straightened.

Q.113 What are the differences between lumbricals of hand?
Four lumbricals in hand

	Medial two	*Lateral two*
1. Structure	Bipennate	Unipennate
2. Nerve supply	Median nerve	Ulnar nerve

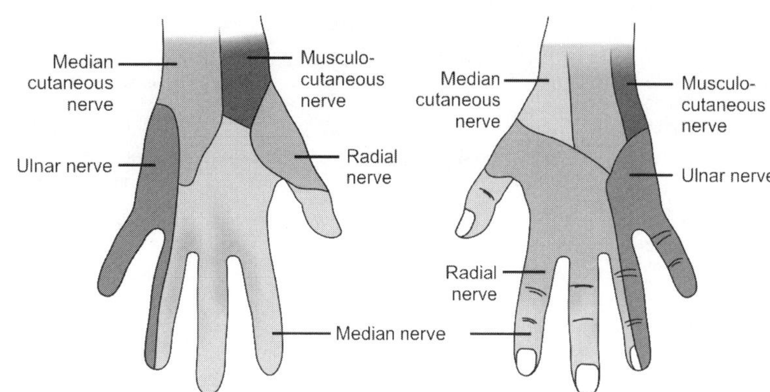

Fig. 2.24: Nerve supply of the hand

Q.114 Which digit does not have palmar interossei?
Third digit.

Q.115 Which digit does not have dorsal interossei?
First and fifth.

Q.116 What are the functions of lumbricals and interossei?
- Lumbricals and interossei together bring about
  - Flexion at metacarpophalangeal joint and
  - Extension at interphalangeal joints.
- Lumbricals alone are weak flexors of metacarpophalangeal joint.
- Palmar interossei are adductors of fingers.
- Dorsal interossei are abductors of fingers.

Q.117 Describe the nerve supply of the hands.
The nerve supply of the back and front of the hands is shown in the Figure 2.24.

Axilla

Q.118 What is the shape of axilla?
It is a four sided pyramidal shaped space, situated between upper part of arm and chest wall (Fig. 2.25).

Q.119 What is the direction of apex of axilla?
It is direct upwards and medially towards the root of neck.

Q.120 What is 'cervicoaxillay canal'?
It is a triangular interval bound by:
Arteriorly: Posterior surface of clavicle.
Posteriorly: Superior border of scapula.
Medially: Outer border of first rib.

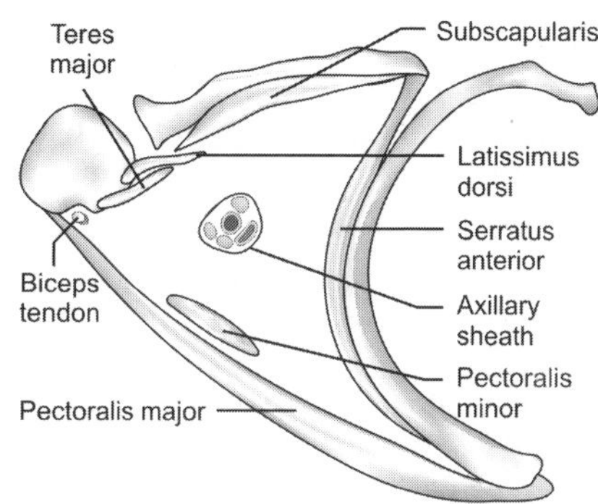

Fig. 2.25: Transverse section through the axilla to show its walls

It corresponds to apex of axilla and through it axillary vessels and brachial plexus enter the axilla from the neck.

Q.121 What are the contents of the axilla?
- Axillary artery and its branches.
- Axillary vein and its tributaries.
- Infraclavicular part of the brachial plexus.
- Axillary lymph nodes and lymphatics.
- Long thoracic and intercostobrachial nerves.
- Axillary fat and areolar tissue.

Q.122 What are the boundaries of axilla?
- *Apex:* Truncated.
- *Base:* Skin and axillary fascia.
- *Anterior wall:* Pectoralis major, pectoralis minor and clavipectoral fascia.
- *Posterior wall:* Subscapularis above, teres major and latissimus dorsi below.
- *Medial wall:* Upper four ribs with intercostal muscles, upper part of serratus anterior.
- *Lateral wall:* Upper part of shaft of humerus, coracobrachialis, short head of biceps muscle (Fig. 2.26).

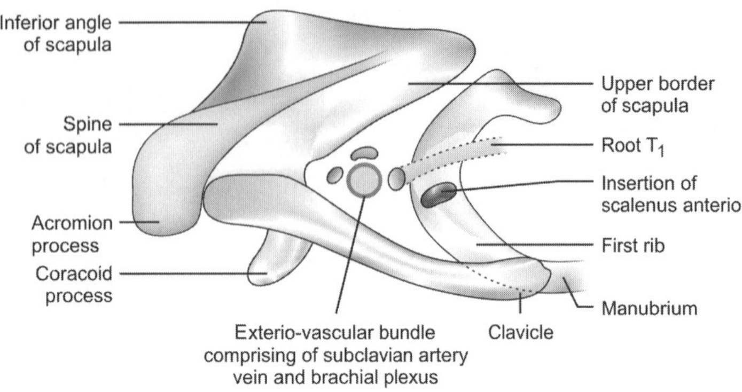

Fig. 2.26: Cervicoaxillary canal viewed from above

Arteries of Upper Limb

Q.123 What is the extent of axillary artery?

It extends from outer border of first rib to lower border of teres major muscle. It is a continuation of subclavian artery and it continues as brachial artery.

Q.124 What are the parts of axillary artery?

Pectoralis minor muscle crosses it and divides it into three parts:

First Part: Proximal to muscle.
Second Part: Posterior to muscle.
Third Part: Distal to muscle.

Q.125 What are the relations of various nerves with the axillary artery?

First part:
- *Anterior:*
  - Supraclavicular nerves
  - Lateral pectoral nerve
  - Loop of communication between lateral and medial pectoral nerve.
- *Posterior:*
  - Nerve to serratus anterior
  - Medial cord of brachial plexus
- *Lateral:* Lateral and posterior cords of brachial plexus.

Second part:
- *Posterior:* Posterior cord of brachial plexus
- *Medial:*
  - Medial cord of brachial plexus
  - Medial pectoral nerve

Lateral: Lateral cord of brachial plexus.

Third part:
- *Anterior:* Medial root of median nerve
- *Posterior:* Axillary nerve
- *Lateral:*
  - Musculocutaneous nerve and lateral root of median nerve in upper part.
  - Trunk of median nerve in lower part.
- *Medial:*
  - Medial cutaneous nerve of forearm
  - Ulnar nerve
  - Medial cutaneous nerve of arm

Q.126 What is the relation of various muscles with axillary artery?

Anteriorly:
- Pectoralis major to whole artery except lowermost part
- Pectoralis minor to second part
- Clavipectoral fascia to first part (Fig. 2.27).

Posteriorly:
- Intercostal muscles of first space and serratus anterior to first part
- Subscapularis to second and upper portion of third part. Teres major and tendon of latissimus dorsi to lower portion of third part.

Laterally:
- Coracobrachialis to second and third part

Q.127 Which veins cross the axillary artery?

Cephalic vein and thoracoacromial vein, the tributaries of axillary vein cross the first part.

Q.128 Name the branches of axillary artery.

From first part: Superior thoracic artery

From second part:
- Thoracoacromial artery
- Lateral thoracic artery.

From third part:
- Subscapular artery
- Anterior circumflex humeral artery
- Posterior circumflex humeral artery (Fig. 2.28).

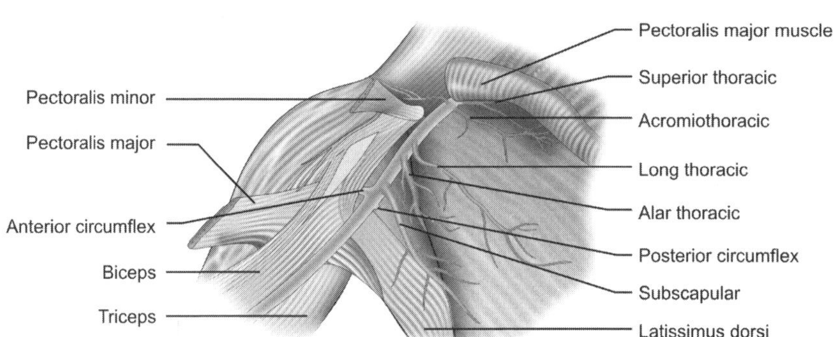

Fig. 2.27: Muscles related to the axillary artery

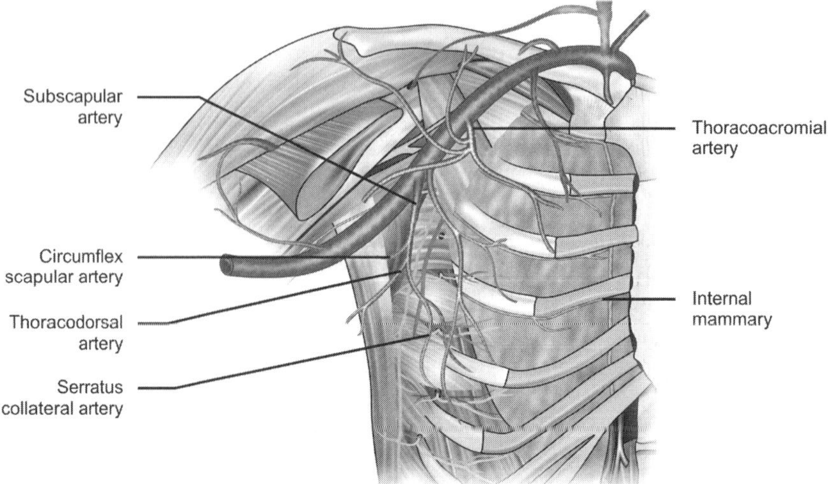

Fig. 2.28: Branches of axillary artery

Q.129 What is the extent of brachial artery?

It extends from the lower border of teres major muscle to elbow at the level of neck of radius just medial to tendon of biceps.

Q.130 What are the nerves related to brachial artery in its course?

1. *In upper part of arm:*
   - Anteriorly to medial cutaneous nerve of forearm
   - Medially, to ulnar nerve
   - Laterally, to median nerve.
2. *In middle of arm:* Crossed by median nerve from lateral to medial side.
3. *In lower part of arm:* Medially, median nerve.
   - Posteriorly, it is related to radial nerve, only in the upper most part.
4. *In elbow:*
   - *Laterally:* Radial nerve
   - *Medially:* Median nerve

Q.131 Name the branches of brachial artery.

- Muscular branches
- Profunda brachii artery
- Superior ulnar collateral
- Inferior ulnar collateral
- Nutrient artery to humerus
- *Terminal branches:* Radial and ulnar (Fig. 2.29).

Q.132 What is the clinical importance of brachial artery?

Brachial pulsations are auscultated in front of elbow just medial to tendon of biceps while recording the blood pressure.

Q.133 Name the branches of profunda brachii artery.

- Anterior descending
- Posterior descending
- Ascending branch

Q.134 Name the branches of radial artery.

- Muscular branches
- Radial recurrent branch
- Palmar carpal branch
- Superficial palmar branch
- Dorsal carpal branch
- First dorsal metacarpal artery
- Princeps pollicis artery
- Radialis indicis artery

Q.135 Where the pulsations of radial artery are felt?

At the wrist against the anterior surface of lower end of radius.

Q.136 Name the branches of ulnar artery.

- Muscular branches
- Anterior and posterior ulnar recurrent branches
- Palmar and dorsal carpal branches
- Common interosseous artery
- Superficial and deep palmar branch.

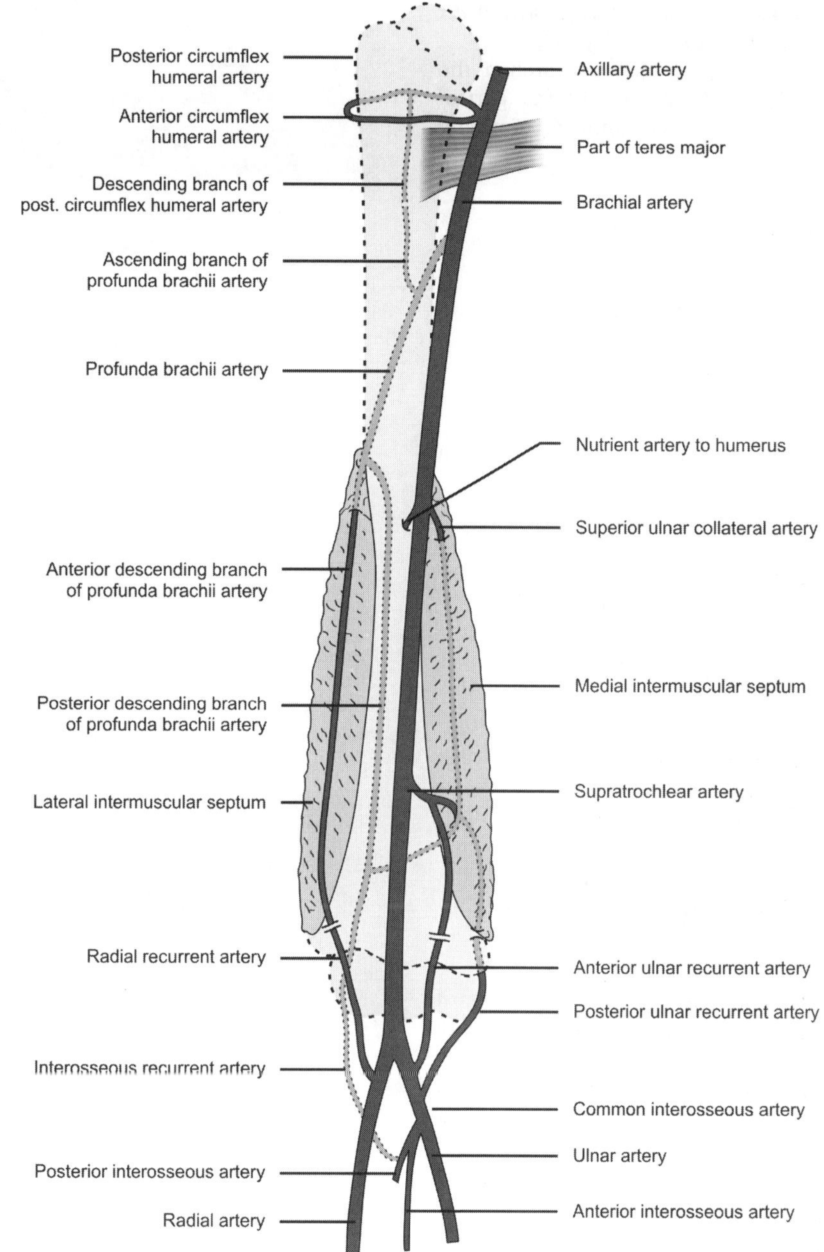

Fig. 2.29: Scheme to show the arteries of the arm and various anastomoses in the region

Q.137 Anterior interosseous artery lies between which muscles?

Flexor digitorum profundus and flexor pollicis longus.

Q.138 Name the structure separating the ulnar artery from median nerve at elbow.

Ulnar head of pronator teres.

Q.139 What are the relations of radial artery in the forearm.

Radial artery lies between flexor carpi radialis and brachioradialis.

Q.140 What is clinical importance of Allen's test?

This test is performed to test the patency of the radial or ulnar artery.

Q.141 Name the arteries forming an anastomosis around the scapula.

- Suprascapular artery branch of cervical artery
- Deep branch of transverse thyrocervical joint
- Circumflex scapular artery, branch of subscapular artery.

Q.142 How the circulation is carried out when axillary artery is ligated.

By the collateral channels around the scapula, which connect the first part of subclavian artery with third part of axillary artery.

Venous Drainage of Upper Limb

Q.143 What are the main superficial veins of upper limb?

- *Cephalic (preaxial) vein:* Begins from lateral end of dorsal venous arch and drains into axillary vein.
- *Basilic (Postaxial) veins:* Begins from medial end of the dorsal venous arch and above the lower border of teres major continues as the axillary vein
- *Median cubital vein:* Large communicating vein which shunts blood from cephalic to basilic in 70%.
- *Median vein of forearm:* Drains the palmar venous arch and ends in basilic or median cubital veins (Fig. 2.30).

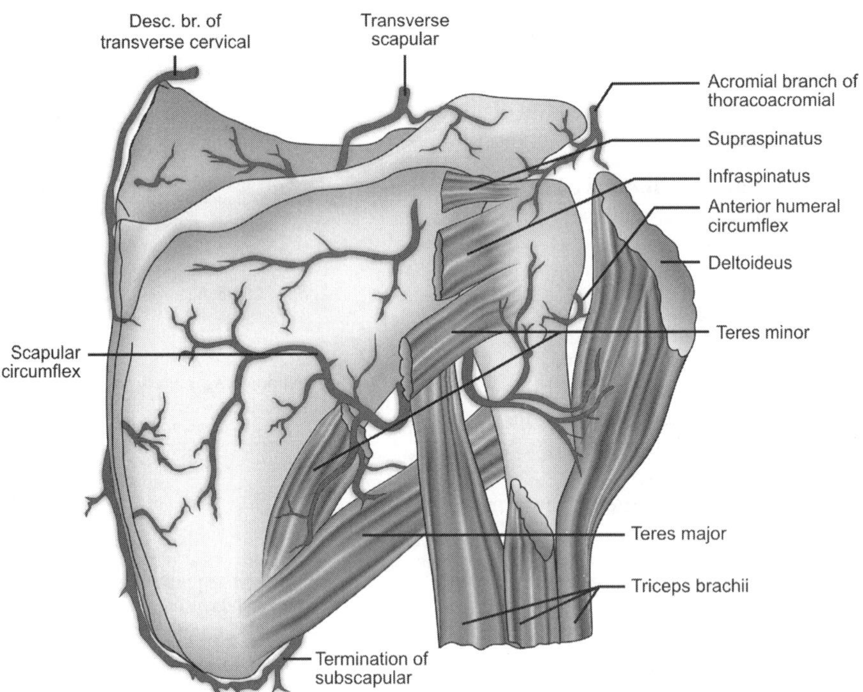

Fig. 2.30: Scheme to show the anastomoses around the scapula, as seen from the front. Arteries on the dorsal side of the scapula are shown in interrupted line

Q.144 Name the structures lying between axillary artery and vein.

The vein lies anteromedial to the artery. The two are separated by:

- Medial cord of brachial plexus
- Medial pectoral nerve
- Ulnar nerve
- Medial cutaneous nerve of forearm

Q.145 How the venous drainage of upper limb is maintained in axillary vein obstruction?

By communication between upper part of cephalic vein with the external jugular vein in neck.

Q.146 What is the clinical importance of median cubital vein?

It is connected to the deep veins of upper limb through a perforator which pierces the bicipital aponeurosis, which fixes it. So it does not slip away when intravenous injections are given.

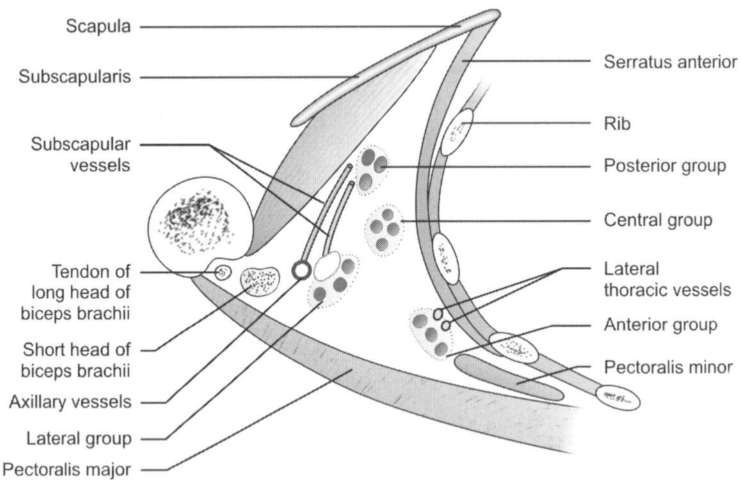

Fig. 2.31: Schematic transverse section through the axilla to show the axillary lymph nodes

Lymphatic Drainage of Upper Limb

Q.147 Which is main lymph node of upper limb?

Lateral group of axillary nodes (Fig. 2.31).

Q.148 What is the area of lymphatic drainage of axillary lymph nodes?

- *Anterior group:* Drains skin and muscles of anterior and lateral walls of trunk up to level of umbilicus and part of breast.
- *Posterior group:* Drains skin and muscles of back of trunk from iliac crest to lower part of neck.
- *Lateral group:* Upper limb.

Q.149 How the axillary lymph nodes drain?

Anterior, posterior and lateral group drain into central group which in turn drains into apical group. Subclavian lymphatic trunk from apical group drains into thoracic duct on left side and to right lymphatic duct on right side.

Breast

Q.150 What is breast?

It is *modified gland of apocrine type,* which is present in both the sexes, but is rudimentary in males and well developed in females after puberty.

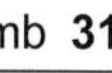

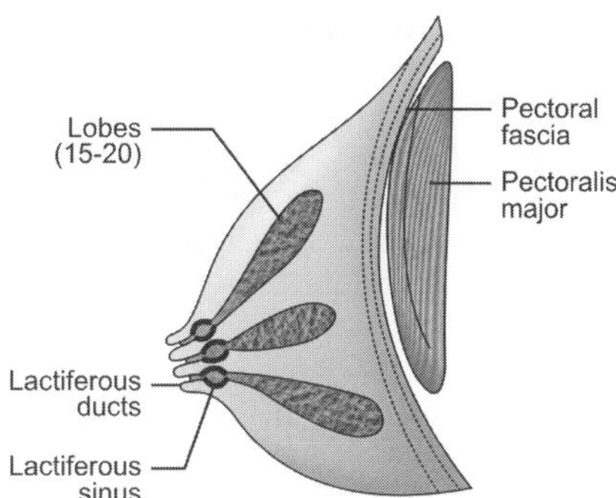

Fig. 2.32: Schematic vertical section through the breast

It forms as important accessory organ of female reproductive system (Fig. 2.32).

Q.151 What is extent of female breast?

- *Superiorly:* 2nd rib.
- *Inferiorly:* 6th rib.
- *Medially:* Lateral border of sternum.
- *Laterally:* Midaxillary line.
- The superolateral part of gland is prolonged upwards and laterally, pierces the deep fascia at anterior fold of axilla and lies in the axilla at the level of third rib. This process of gland is known as '*axillary tail of Spence*' and the opening in deep fascia is known as '*Foramen of Langer*'.

Q.152 What is situation of breast?

Breast lies in the superficial fascia of pectoral region except for axillary tail which pierces the deep fascia through foramen of Langer and lies in axilla.

Q.153 What is shape of breast?

- In young adult female, it is hemispherical.
- In later life, it is usually pendulous.

Q.154 What are deep relations of breast?

1. Retromammary space of loose areolar tissue. According to former concept of free flow of lymphatics, it was known as *Lake of Marcille.*
2. Pectoral fascia
3. Pectoralis major, serratus anterior and external oblique.

Q.155 What is the structure of breast?

- *Glandular tissue:* This consists of 15-20 lobes. Each lobe consists of several lobules and each lobule consists of a cluster of *alveoli* which open into the smallest branches of lactiferous ducts. These branches unite to form larger branches of duct.
  Each lactiferous duct, drains a lobe of gland and opens at nipple. At the bottom of the nipple each duct is dilated to form a *sinus*. The ducts are arranged radially around the nipple. The glandular tissue is the functional portion of the breast and secretes milk.
- *Fibrous tissue stroma*: This consists of numerous septa connecting the lobules and supporting them. These septa link the pectoral fascia to the skin of the breast. These are known as '*suspensory ligaments of Cooper*'.
- *Adipose tissue:* This fills the interalveolar and interductular intervals and accounts for the smooth contour and most of the bulk of breast.
- *Skin:*
  - *Nipple:* Cylindrical or conical projection directed superolaterally. It lies at the level of 4th intercostal space in nulliparous females.
  - *Areola:* Pigmented area around nipple. Rose pink in virgins and dark brown or black after pregnancy. The nipple and the subareolar tissue contain smooth muscle but lack the fat.
  - *Montgomery's tubercles:* These are sebaceous glands underlying the areolar skin and are called areolar glands. They enlarge during pregnancy and lactation and form raised tubercles. Oily secretions of these glands lubricates nipple and areola and prevent them from cracking during lactation.

Q.156 How does the structure of male breast differs from the female breast?

The male breast is rudimentary. It consists of small ducts without alveoli. There is little supporting fibrous tissue and fat.

Q.157 What is retromammary space and what is its clinical significance?

It is a space which lies between the deep aspect of the breast and the fascia covering the pectoralis major. It contains loose areolar tissue and allows the breast some degree of movement on pectoral fascia.

Fixity of the breast to the pectoral fascia and the muscle may occur, by invasion, in advanced carcinoma of breast. This is of great significance in clinical staging of breast carcinoma.

Q.158 What is the clinical significance of retraction of nipple?

- *Retraction occurring at pregnancy:* It is due to a developmental abnormality. The nipple, for some unknown reason, does not develop with breast.
- Recent retraction of nipple may be due to the fibrous contraction of the lactiferous ducts in breast carcinoma or chronic abscess.

Q.159 What is the clinical significance of the dimpling of skin over breast?

This is due to contraction of ligaments of Cooper. It can occur in chronic infection, after trauma or the breast carcinoma infiltrating the ligaments.

Q.160 What is arterial supply of breast?

Breast is supplied by:

1. Internal thoracic artery, through its perforating branches in 2nd–6th intercostal space.
2. Lateral thoracic
3. Superior thoracic and Acromiothoracic (Branches of axillary artery)
4. Lateral branches of posterior intercostal arteries

Q.161 What is the venous drainage of breast?

Veins converge towards the base of nipple where they form an anastomotic venous circle, from where veins run in superficial and deep sets.

- Superficial veins drain into internal thoracic and superficial veins of lower neck.
- Deep veins drain into internal thoracic, axillary and posterior intercostal veins.

Q.162 What is the importance of knowing venous drainage of breast?

1. Veins indicate lymphatic pathways because the lymphatics run with the veins.
2. Carcinoma of breast can spread through veins.
3. As the posterior intercostal veins communicate with the vertebral plexus of veins through which the malignancy can spread to bones and nervous system.

Q.163 What is the nerve supply of breast?

Breast is supplied by anterior and lateral cutaneous branches of 4th to 6th intercostal nerves. Nerves do not control the secretion of milk. The nerves supplying the glandular tissue are sympathetic.

Q.164 What is lymphatic drainage of breast?

See Figures 2.33A and B.

- *Superficial lymphatics:* Draining the skin over breast except for nipple and areola. They pass into axillary, internal mammary, supraclavicular and cephalic lymph nodes.
  Superficial lymphatics of one side communicate with those of the other side so unilateral malignancy can become bilateral.
- *Deep lymphatics:* Drain the parenchyma of breast, nipple and areola.

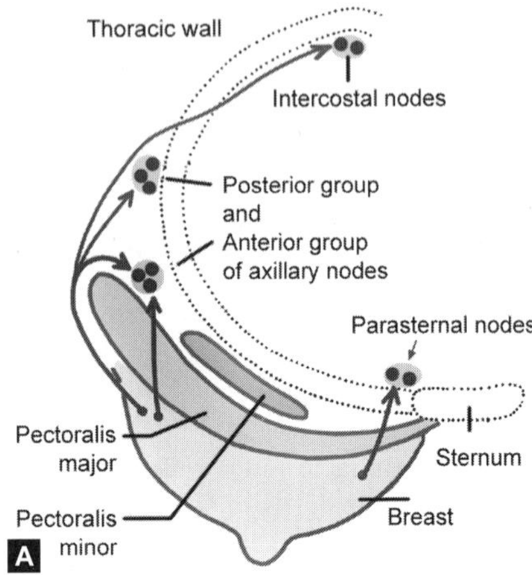

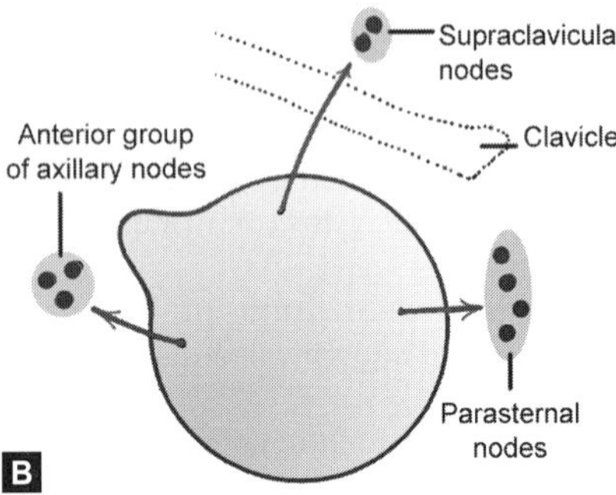

Figs 2.33A and B: Scheme to show some routes followed by lymphatic vessels draining the breast, (B) Lymphatic drainage of the skin of the breast (excluding that over the areola and nipple)

- 75% of lymph drains into axillary lymph nodes mainly anterior group.
- 20% drains into internal mammary group, which drain the lymph not only from the medial quadrant, but also from lateral quadrant.
- 5% drain into posterior intercostal nodes.
- The *subareolar plexus of Sappy* situated beneath the areola drains the nipple and areola and communicates with the lymphatics of parenchyma.
- Lymphatics from the deep surface pass through pectoralis major and clavipectoral fascia, to reach apical group of lymph nodes.
- Lymphatics from the lower and inner quadrants may communicate with subdiaphragmatic and subperitoneal lymph plexus, after crossing the costal margin and then piercing the anterior abdominal wall through upper part of linea alba. Thus, the carcinoma of breast may spread to the liver and can gravitate through the peritoneal cavity to lie on the pelvic organs, e.g. on ovary, when the condition is known as *Krukenberg's tumor*.

Q.165 What is the lymphatic drainage of axillary tail?
It drains into the scapular (anterior) axillary group.

Q.166 What is peau d'orange skin?
Peau d'orange is due to cutaneous lymphatic edema. Where the infiltrated skin is tethered by the sweat ducts, it cannot swell. The characteristic appearance is like that of orange peel. It is a classical physical sign of advanced carcinoma of breast. It is also seen over an abscess, particularly chronic abscess of the breast.

Q.167 How the carcinoma breast spreads to the vertebrae?
By spread through the veins.

Q.168 What is cancer of cuirass?
In it, there is persistent, non-pitting edema of the arm and the affected side of the thoracic wall is studded with carcinomatous nodules and the skin is so infiltrated that it is like the coat of armour. The condition appears in cases where local recurrence after surgery of breast occurs.

Q.169 How does the breast develop?
The breast develops as an in vagination of ectoderm of the ventral wall of the body. In the 6th week of intrauterine life, two longitudinal ectodermal thickening develop, one on each side called *mammary ridge* or *milk ridge*. This ridge extends from the axilla to the groin, but in the human embryo it persists only in the pectoral region. Ingrowths from the milk ridge gives rise to the glandular tissue, the ducts and alveoli of breasts. The connective tissue supporting the glandular tissue is derived from the surrounding mesenchyme.

Q.170 What is polymastia (Supernumerary breast)?
This is congenital anomaly in human in which there are more than one breast on one or both sides. This is due to the persistence of the milk ridge which normally disappears except in the pectoral region.

Q.171 What is polythelia?
This is the presence of supernumerary nipples.

Q.172 What type of incisions is given to drain a breast abscess?
By a radial incision to avoid cutting across a number of lactiferous ducts.

Q.173 What is the extent of clavipectoral fascia?
Vertically:
Superiorly, splits to enclose subclavius muscle and is attached to the clavicle.
Inferiorly, splits to enclose pectoralis minor and continues as suspensory ligament.
Horizontally:
Medially, attached to first rib, costoclavicular ligament and fascia covering the two intercostal spaces.
Laterally, attached to coacoid process and blends with the coracoclavicular ligament.

Q.174 Name the structures piercing clavipectoral fascia.
- Lateral pectoral nerve
- Cephalic vein
- Thoracoacromial vessels
- Lymphatics.

Brachial Plexus

Q.175 What is brachial plexus?
Brachial plexus (Fig. 2.34) is formed by the union of the ventral rami of lower four cervical nerves ($C_{5,6,7,8}$) and the greater part of the ventral ramus of the first thoracic nerve (T_1). The fourth cervical nerve usually gives a branch to the fifth cervical and the first thoracic nerve frequently receives one from the second thoracic nerve.

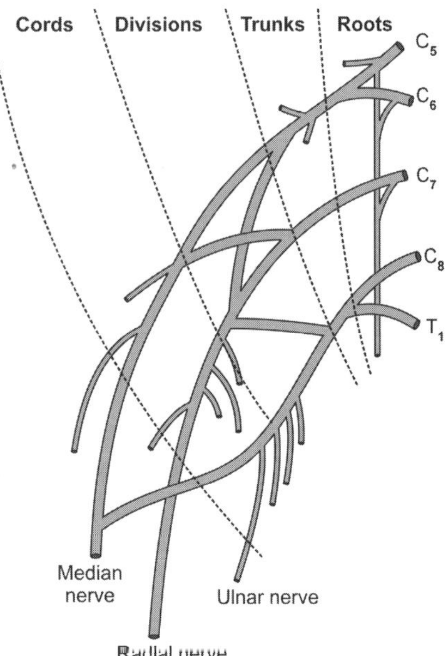

Fig. 2.34: Scheme to show the formation and branches of the brachial plexus

Q.176 What is prefixed and postfixed type of plexus?
When the branch from C_4 is large, the branch from T_2 is frequently absent and the branch of T_1 is reduced in size. This is prefixed type of plexus. On the other hand, the branch form C_4 may be very small or entirely absent. In the event, the contribution of C_5 is reduced in size but that of T_1 is larger and T_2 is always present. That constitutes postfixed type of plexus.

Q.177 How the brachial plexus forms trunks?
The C_5 and C_6 join to form upper trunk, C_7 forms the middle trunk and C_8 and T_1 join to form the lower trunk.

Each trunk divides into ventral and dorsal division, which ultimately supply anterior and posterior aspect of upper limb.

Q.178 How the cords of brachial plexus are formed?
Lateral cord is formed by union of ventral division of the upper and middle trunks. The medial cord is formed by the ventral divisions of lower trunk. Posterior cord is formed by union of dorsal divisions of all the three trunks.

Q.179 What are the branches of roots of brachial plexus?
Long thoracic nerve (Nerve to serratus anterior) $C_{5,6,7}$.
- Dorsal scapular nerve (Nerve to rhomboids) C_5.

Q.180 What are the branches of trunks of brachial plexus?
- Suprascapular nerve ($C_{5,6}$).
- Nerve to subclavius ($C_{5,6}$).

Q.181 What are the branches of lateral cord of brachial plexus?
- Lateral pectoral nerve
- Lateral root of median nerve
- Musculocutaneous nerve.

Q.182 What is the main nerve supply of pectoralis major?
Lateral pectoral nerve.

Q.183 What are the branches of medial cord of brachial plexus?
- Medial root of median nerve
- Medial pectoral nerve
- Ulnar nerve
- Medial cutaneous nerve of arm
- Medial cutaneous nerve of forearm.

Q.184 What are the branches of posterior cord of brachial plexus?
- Radial nerve
- Axillary nerve
- Thoracodorsal nerve (Nerve to latissimus dorsi)
- Upper subscapular nerve
- Lower subscapsular nerve.

Q.185 What is main nerve supply of pectoralis minor?
Medial pectoral nerve.

Q.186 What is the distribution of suprascapular nerve?

Muscular:	Supraspinatus
	Infraspinatus
Articular:	Shoulder joint
	Acromioclavicular joint

Q.187 What is Erb's point?
It is the junction of the ventral primary rami of C_5 and C_6 forming the upper trunk of the brachial plexus. The trunk being short, the suprascapular nerve and nerve to subclavius which arise directly from it and the anterior and posterior divisions of trunk all lie close to the Erb's point and may be involved in any injury at this point.

Q.188 What is Erb's paralysis?
It is the paralysis resulting from a lesion of the upper trunk at the Erb's point. It is caused by the forcible downward traction of the shoulder with lateral displacement of the head to the other side.

In this lesion, the area of distribution of C_5 and C_6 is affected and most commonly the muscles supplied by C_5 are involved. The deltoid, biceps, brachialis, brachioradialis and sometimes supraspinatus, infraspinatus and supinator are paralyzed. The affected limb assumes a characteristic '*waiter's tip' position.*
- The arm hangs simply by the side due to paralysis of the deltoid and supraspinatus and is rotated medially, due to paralysis of infraspinatus.
- The elbow is extended due to paralysis of the biceps and brachialis and imposed action of the extensors of elbow.
- The forearm is pronated due to paralysis of the biceps and supinator.
- The wrist is slightly flexed due to weak wrist extensors.

Q.189 What is Klumpke's paralysis?
This is a paralysis resulting from the lesion of the lower trunk (C_8 and T_1 nerve roots). It is caused by forceful upward traction of the arm. The area of distribution mainly of T_1 is involved, i.e. all the intrinsic muscles of the hand are affected and flexors of wrist are affected due to C_8 root involvement.
- The hand assumes a characteristic deformity described as *claw hand.* In this, the metacarpophalangeal joints are hyperextended due to unopposed action of the long extensors as the lumbricals and interossei are paralyzed while the interphalangeal joints are flexed due to unopposed actions of the long flexors of the fingers.
- There is sensory loss along the ulnar side of the hand, forearm and arm.
- There may also be Horner's syndrome characterized by moisis, ptosis, anhydrosis and anophthalmos.

Q.190 What is the clinical importance of postfixed type of brachial plexus?
The T_2 root has to curve up over the first rib to form the brachial plexus. This results in greater pressure on T_2 nerve root in post fixed brachial plexus as compared to normal. Hence symptoms associated cervical rib can be present in absence of such a rib.

Q.191 What is the cause of the referred pain to the skin over shoulder?
This is due to inflammation of the diaphragmatic pleura or peritoneum, (usually occurring in cholecystitis and splenic infarction) which has the same segmental nerve supply (C_4) via phrenic nerve as the shoulder skin via the lateral supraclavicular nerve.

Q.192 What is 'crutch paralysis'?
It is due to the damage to the brachial plexus in the axilla from the pressure of crutch. In these, the radial nerve is frequently implicated and ulnar nerve suffers next in frequency.

Q.193 What is 'Saturday night palsy'?
This is radial nerve palsy due to the prolonged pressure on the nerve in the spiral groove of the humerus. This occurs when a drunkard falls into sleep (on Saturday night!) with his arm hanging over the back of chair. In the morning, he is suffering from the wrist drop which is temporary.

Q.194 What is 'Winging of scapula'?
This is clinical condition in which the inferior angle and the medial border of the scapula becomes unduly prominent. It occurs in the paralysis of long thoracic nerve (nerve of Bell) which supplies serratus anterior. It can be demonstrated by asking the patient to push against the wall with outstretched hands. The scapula on affected side becomes winged due to unopposed action of the rhomboids and levator scapulae, while the paralyzed serratus anterior is not contracting.

Q.195 Name the clinical conditions in which the axillary nerve is likely to be injured.

- Fracture of surgical neck of humerus
- Dislocation of shoulder joint.

Q.196 What will be the effects of complete damage of the axillary nerve?

1. There is paralysis of the deltoid and teres minor. Paralysis of deltoid causes inability to abduct the shoulder joint while the wasting of the muscles causes undue prominence of the acromion. Paralysis of teres minor is not easily demonstrated clinically.
2. There will be sensory loss over the lower part of deltoid.

Radial Nerve

Q.197 What is origin of radial nerve?

It is a branch of posterior cord of brachial plexus with a root value of $C_{5,6,7,8}$ T_1.

Q.198 What are the branches and distribution of radial nerve?

- *Muscular branches*: To
  - Triceps
  -. Anconeus
  - Brachialis, only lateral part
  - Brachioradialis and
  - Extensor carpi radialis longus
- *Cutaneous branches:*
  - Lower lateral cutaneous nerve of arm
  - Posterior cutaneous nerve of forearm
  - Posterior cutaneous nerve of arm
  - Dorsal digital branches from superficial terminal branch
- *Articular branches:* To elbow and wrist joint.

Q.199 What are the structures supplied by posterior interosseous nerve?

Muscular branches:

- Extensor carpi radialis brevis
- Supinator
- Extensor digitorum
- Extensor digiti minimi
- Extensor carpi ulnaris
- Extensor pollicis longus
- Extensor indicis
- Abductor pollicis longus
- Extensor pollicis brevis

Q.200 What is the commonest site of radial nerve injury? What are the common causes of lesion?

In the region of radial (spiral) groove of humerus.

The common causes of injury are:

- Fracture of shaft of humerus
- Intramuscular injections in arm

Q.201 What are the effects of the lesion of radial nerve in the spiral groove?

- Triceps is not paralyzed since the branches supplying arise from the radial nerve more proximally.
- There is *wrist* drop, i.e. hand is flexed at wrist and it lies flaccid due to the paralysis of the extensors of the wrist. The fingers are also flexed and when an attempt is made to extend them, the last two phalanges only will be extended, through the action of lumbricals and interossei.
- Supination is completely lost when the forearm is extended on the arm, but is possible to a certain extent if the forearm is flexed to allow effective action of the biceps.

Q.202 What will be the effect of cutting the radial nerve just below the elbow?

Sensory loss: It is marked on the lateral part of the dorsum of the hand.

Motor loss:

- Wrist drop
- Loss of power of supination
- Extension of elbow is retained because of the intact triceps
- Flexion of elbow in normal position will also be retained because of intact biceps brachii and brachioradialis.

Q.203 What is the 'high' and 'low' radial nerve palsy?

When radial nerve is damaged above the origin of nerve to brachioradialis which arises above the elbow joint then it is called 'high' radial nerve palsy.

When radial nerve is damaged below the origin of nerve to brachioradialis then it is called 'low' radial nerve palsy. The brachioradialis is not paralyzed and cause flexion of the elbow joint in the midprone position which when elbow is flexed against resistance becomes prominent, can be felt superficially on lateral aspect of the forearm.

Musculocutaneous Nerve

Q.204 What is the origin of musculocutaneous nerve?

It is a branch of lateral cord of brachial plexus, arising at the lower border of pectoralis minor ($C_{5,6,7}$).

Q.205 What are the branches of musculocutaneous nerve?

Muscular: To

- Coracobrachialis
- Biceps and
- Brachialis

Cutaneous: It continues as lateral cutaneous nerve of forearm and supplies skin of lateral side of forearm.

Articular: To elbow joint

Communicating: To radial nerve, posterior cutaneous nerve of forearm and palmar cutaneous branch of median nerve.

Q.206 What will be the effect of lesion of the musculocutaneous nerve.

Motor loss: There will be paralysis of the biceps, coracobrachialis and the medial part of brachialis, the lateral part being supplied by the radial nerve.

Effects:

- Flexion of the elbow joint will be weak but is still possible,
  - With the forearm supinated due to the action of the unaffected lateral part of brachialis and superficial flexors and
  - In the prone or midprone position by the brachioradialis and extensor carpi radialis longus.
- There will be very weak supination with the elbow flexed at 90°.
- There will be loss of biceps jerk.
- There will be muscles atrophy.

Sensory loss: Will be present over the lateral half of the forearm but the area will be less due to overlapping of the intact adjacent cutaneous nerves.

Median Nerve

Q.207 What is the origin of median nerve?

Median nerve is formed by the union of a medial root (C_8, T_1) from medial cord and lateral root ($C_{5,6,7}$) from lateral cord.

Q.208 Name the branches and structures supplied by median nerve?

In arm:

- Muscular branch to pronator teres
- Vascular branches to brachial artery

In forearm:

- Muscular branches to flexor carpi radialis, palmaris longus, flexor digitorum superficialis.
- *Anterior interosseous branch* to flexor pollicis longus, lateral half of flexor digitorum profundus, pronator quadratus and to distal radioulnar and wrist joints.
- Palmar cutaneous branch, to skin over thenar eminence and middle of the palm.
- Articular branch to elbow and proximal radioulnar joints.

- Vascular branches to radial and ulnar arteries.
- Communicating branch to ulnar nerve.

In hand:

- Muscular branches to abductor pollicis brevis, flexor pollicis brevis, opponens pollicis and first and second lumbrical
- Cutaneous branches to skin of lateral 3½ digits (palmar digital branches).

Q.209 What will be the effect of a lesion of the median nerve at the wrist?

- *Motor loss:* There will be paralysis of the thenar muscles and the 1st and 2nd lumbricals.

  Effect:
  - There will be loss of opposition of the thumb due to paralysis of the opponens pollicis
  - Abduction of the thumb will not be greatly affected due to intact abductor pollicis longus which is supplied by the radial nerve
  - Paralysis and wasting of the thenar muscles and unopposed extension by extensor pollicis longus and adduction by adductor pollicis will be rise to *'ape-thumb' deformity.*
- *Sensory loss:* There will be loss of sensation over the thumb, adjacent 3½ fingers and the radial two thirds of the palm.

Effect: The sensory loss will prevent the accurate and delicate adjustments which the hand makes in response to tactile stimuli.

Q.210 What is 'carpal tunnel' syndrome?
This is a neuropathy resulting from compression of median nerve as it passes beneath the flexor retinaculum through carpal tunnel (Fig. 2.35).
It causes:

- *Motor loss:* Progressive weakness and wasting of thenar muscle
- *Sensory loss:* In lateral 3½ digits.

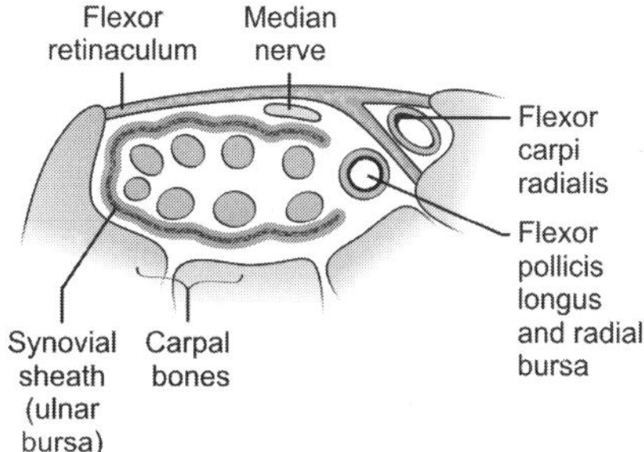

Fig. 2.35: Structures passing through the carpal tunnel

Ulnar Nerve

Q.211 What is the origin of ulnar nerve?
It arises from medial cord of brachial plexus C_8, T_1.

Q.212 What are the branches of ulnar nerve?
In forearm:

- *Muscular:* To flexor carpi ulnaris and medial half of flexor digitorum profundus.
- Palmar and dorsal cutaneous branches.

In hand:

- *Muscular:*
  - *By deep terminal branch:* Abductor digiti minimi, flexor digiti minimi, opponens digiti minimi, medial two lumbricals, palmar and dorsal interossei and adductor pollicis.
  - Palmaris brevis by palmar cutaneous or *superficial terminal branch.*
- *Articular:* To elbow joint.
- *Skin:* Medial 1½ fingers by palmar digital branches.

Q.213 What will be the effect of a lesion of the ulnar nerve at the wrist?

1. *Motor loss:* There will be paralysis of all the intrinsic muscles of the hand (except those supplied by the median nerve), i.e. all interossei, 3rd and 4th lumbricals, hypothenar muscles and adductor pollicis.

   Effect:
   - There will be *'main-en-griffe'* deformity or *'clawing'* of the ring and little fingers. These fingers are hyperextended at the metacarpophalangeal joints (due to the unopposed action of the extensor digitorum as the 3rd and 4th lumbricals and all the interossei are paralyzed) and flexed at the interphalangeal joints (due to the unopposed action of the long flexors).
   - Abduction of 2nd to 5th fingers will be weak due to paralysis of the dorsal interossei and abductor digiti minimi.
   - There will be loss of power of adduction of the fingers due to paralysis of the palmar interossei.
   - There will be loss of power of adduction of the thumb due to paralysis of the adductor pollicis.
   - There will be wasting of the hypothenar eminence in long-standing injuries.
   - There will be hollowing between the metacarpal bones, clearly apparent on the dorsum, due to atrophy of the interossei muscles in long-standing injuries.
2. *Sensory loss:*
   - There will be sensory loss on the medial side of the palm and the palmar surfaces of the little and the medial half of the ring fingers and on the dorsal aspect of the distal and middle phalanges of these fingers
   - There will be no sensory loss over the dorsum of the hand as the dorsal cutaneous branch of the ulnar nerve will escape the injury. If the nerve is damaged proximal to the origin of this branch, then there will also be sensory loss over the dorsum of the hand.

Q.214 What will be the effect of a lesion of the ulnar nerve at the elbow?

1. *Motor loss:*
   - Same as when the nerve is damaged at the wrist.
   - There will also be paralysis of the medial half of the flexor digitorum profundus supplying the little and ring fingers and of the flexor carpi ulnaris.

   Effect:
   - Same as when the nerve is damaged the wrist. 'Clawing' of the ring and little fingers will be less marked as their distal phalanges are not flexed due to paralysis of only the medial half of the flexor digitorum profundus.
   - Loss of power in the flexor carpi ulnaris will result in weak flexion with radial deviation of the wrist.
2. Sensory loss will be present over the ulnar 1½ fingers and the hand.
3. Vasomotor and trophic changes will be present in the skin over the hypothenar eminence and little finger which will appear cold and dry and at times discolored. The nail of the little finger may be deformed.

Scapular Spaces

Q.215 What are the boundaries and contents of quadrangular space?
Boundaries (Fig. 2.36)*:*

- Superior:
  - Subscapularis
  - Capsule of shoulder joint
  - Teres minor
- Inferior: Teres major
- Medial: Long head of triceps
- Lateral: Surgical neck of humerus

Contents:

- Axillary nerve
- Posterior circumflex humeral vessels.

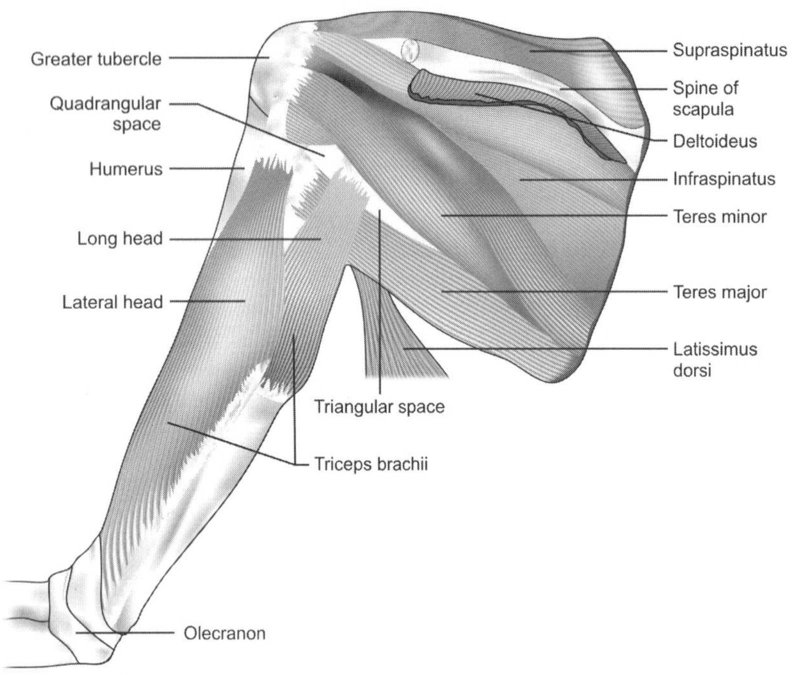

Fig. 2.36: The triangular space and the quadrangular space of the scapular region

Q.216 What are the boundaries and contents of upper and lower triangular space?

Upper triangular space:

Boundaries:
- Superior: Teres minor
  Subscapularis
- Lateral: Long head of triceps
- Inferior: Teres major.

Contents: Circumflex scapular artery

Lower triangular space:

Boundaries:
- Superior: Teres major
- Medial: Long head of triceps
- Lateral: Medial border of humerus

Contents:
- Profunda brachii vessels
- Radial nerve

Q.217 What are the boundaries of Triangle of auscultation?
- *Medial:* Lateral border of trapezius
- *Lateral:* Medial border of scapula
- *Below:* Upper border of latissimus dorsi
- *Floor:*
  - 7th rib and 6th and 7th intercostal spaces.
  - Rhomboideus major and latissimus dorsi.

Q.218 What is the clinical importance of 'triangle of auscultation'?

It is the only part of the back which is not covered with muscles and breath sounds are better heard there.

Spaces of the Hand

Q.219 Name the spaces of hand.
- *Palmar spaces:*
  - Superficial pulp spaces of the fingers
  - Synovial tendon sheats of 2nd, 3rd and 4th fingers
  - Ulnar bursa
  - Radial bursa
  - Midpalmar space
  - Thenar space.
- *Dorsal spaces:*
  - Dorsal subaponeurotic space
  - Dorsal subcutaneous space.
- Forearm space of parona

Q.220 What are the characteristic features of the pulp space of fingers?
- Front of distal phalanx is covered with subcutaneous fat.
- Dense fibrous processes bind the skin to the periosteum and divide fat into compartments.

Q.221 Why the infections of pulp space of fingers are painful?

Because it cannot expand due to fibrous processes attaching skin to periosteum and thus little swelling causes much increase in tension.

Q.222 Why the infections of pulp space of fingers cause necrosis of distal 4/5 of terminal phalanx?

Because of distal 4/5 receives its blood supply from arteries which transverse fibrous processes and increase in tension due to infection causes their occlusion and proximal 1/5 escapes necrosis because it receives its blood supply by vessels which do not traverse fibrous processes.

Q.223 Why the infections of little finger and thumb are more dangerous?

Because the synovial sheath of little finger is continuous with ulnar bursa and that of thumb with radial bursa so, the infections of these can spread of the forearm space of parona.

Q.224 What is the position of dorsal spaces?

The subcutaneous space lies deep to skin and subaponeurotic space deep to extensor tendons on dorsal aspect of hand.

Q.225 What is 'forearm space of parona' and its clinical importance?

It is space between long flexor tendons and pronator quadratus. Proximally, upward extent is limited by origin of flexor digitorum superficialis and inferiorly, it extends up to upper border of flexor retinaculum.

The proximal parts of flexor tendons synovial sheath protrude into it.

Clinical importance: It may be infected by the extension of synovial sheath infections from ulnar or radial bursa, leading to hour glass swelling.

Q.226 What is position and clinical importance of midpalmar and thenar space?

These are potential spaces deep to palmar aponeurosis and flexor tendons.
- *Midpalmar space:* Situated under inner half of hollow of palm
- *Thenar space:* Situated under outer half of hollow of palm
- *Clinical importance:* They become infected in wounds of palm and synovial sheath infections. They frequently communicate with each other, so infection can pass from one to the other.

Q.227 Why the collection of fluid is more on dorsal surface of hand in infections of palmar aspect of fingers?

Skin on dorsum of fingers and hand is loose, therefore fluid readily collects beneath it. But on the palm of hand, there is little subcutaneous tissue and skin is adherent to underlying palmar fascia.

Chapter 3

Lower Limb

Bones of Lower Limb

Hip Bone (Fig. 3.1)

Q.1 What are the different parts of a hip bone?

The hip bone is made up of three parts, the *ilium* superiorly, *ischium* postero-inferiorly and *pubis* antero-inferiorly. The three parts join to form a cup-shaped hollow articular surface, the *acetabulum.*

Q.2 How will you determine to which side the hip bone belongs?

In a hip bone, the acetabulum is directed laterally and the flat ilium forms upper part of bone, lying above the acetabulum. The obturator foramen lies below the acetabulum.

Q.3 What is the normal anatomical position of the hip bone in the body?

- Pubic tubercle and anterior superior iliac spine lie in the same vertical plane.
- The pelvic surface of the body of pubis is directed backwards and upwards.
- The ischial spine and upper border of symphysis pubis lie in same horizontal plane and
- Symphysis pubis lies in the median plane.

Q.4 What is the level through which the highest point of the iliac crest passes (intercristal plane)?

The intercristal plane passes at the level of interval between the spines of L_3 and L_4 vertebrae.

Q.5 What is the clinical importance of intercristal plane?

In clinical practice, lumbar puncture is done between the L_3 and L_4 vertebrae.

Q.6 What are the structures attached to the anterior superior iliac spine?

It provides:

- Attachment to the lateral end of inguinal ligament and
- Origin of sartorius.

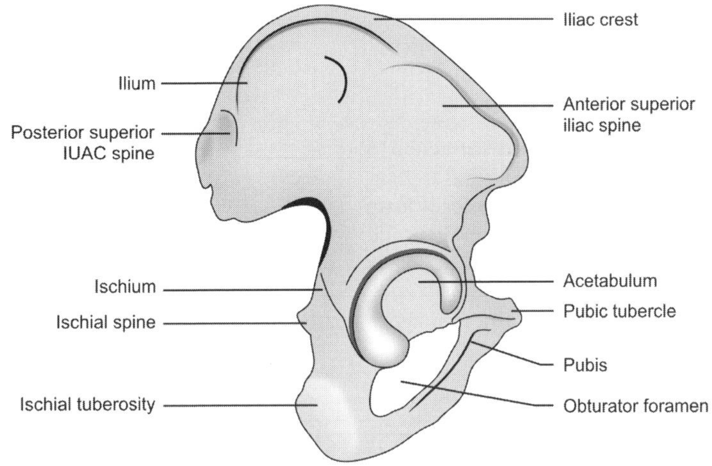

Fig. 3.1: Hip bone

Q.7 Name the structures attached to the iliac crest.

Anterior 2/3 of iliac crest has:

- *Outer lip* which provides
  - Attachment of fascia lata,
  - Origin of tensor fasciae lata,
  - Insertion to external oblique muscle and
  - Origin to latissimus dorsi just behind the highest point.
- *Intermediate area* provides origin to internal oblique muscle.
- *Inner lip* provides
  - Origin to transversus abdominis,
  - Attachment to fascia iliaca and fascia transversalis,
  - Origin to quadratus lumborum in posterior 1/3 and
  - Attachment to thoracolumbar fascia.

Posterior 1/3 segment of iliac crest has:

- *Lateral slope:* Origin of gluteus maximus.
- *Medial slope:* Origin of erector spinae.
- *Medial margin:* Interosseous and dorsal sacroiliac ligaments.

Q.8 Name the structures attached to anterior inferior iliac spine.

Anterior inferior iliac spine gives:

- Origin to straight head of rectus femoris in superior half and
- Attachment to iliofemoral ligament in inferior half.

Q.9 Name the structures attached to posterior border of ilium.

It provides:

- Attachment to upper fibers of sacro-tuberous ligament and
- Origin to fibers of piriformis.

Q.10 What are the structures attached to gluteal surface of ilium?

- Gluteus medius arises between anterior and posterior gluteal lines.
- Gluteus minimus arises between anterior and inferior gluteal line.
- Gluteus maximus (upper fibers) arise behind the posterior gluteal line.
- Below inferior gluteal line reflected head of rectus femoris arises.

Q.11 Name the structures attached to the pubic tubercle.

- Medial end of inguinal ligament.
- Ascending loops of cremaster muscle.

Q.12 Name the structures attached to the crest of pubis.

- Lateral head of rectus abdominis (origin)
- Pyramidalis (origin).

  Medial head of rectus abdominis arises from anterior pubic ligament.

Q.13 What are the structures attached to pectineal line?

The structures attached to pectineal line are:

- Conjoint tendon and lacunar ligament at medial end.
- Pectineal ligament lateral to lacunar ligament.
- Origin of pectineus muscle and fascia covering it, from the whole length.
- Insertion of psoas minor.

Q.14 Name the structures attached to ischial spine.

The structures attached to ischial spine are:

- Sacrospinous ligament
- Origin of coccygeus and levator ani.
- Origin of superior gemellus

Q.15 What are the structures attached to ischial tuberosity?

From upper area of ischial tuberosity arise semimembranous superolaterally and semitendinosus and long head of biceps femoris superomedially.

From lower lateral area adductor magnus arise.

Q.16 What are the nerves related to hip bone?

- Sciatic nerve related to lower margin of greater sciatic notch.
- Obturator nerve in the obturator canal.
- Nerve to obturator internus crosses the base of ischial spine.
- Pudendal nerve crosses base of ischial spine.
- Nerve to quadratus femoris runs on ischium as it crosses the greater sciatic notch.

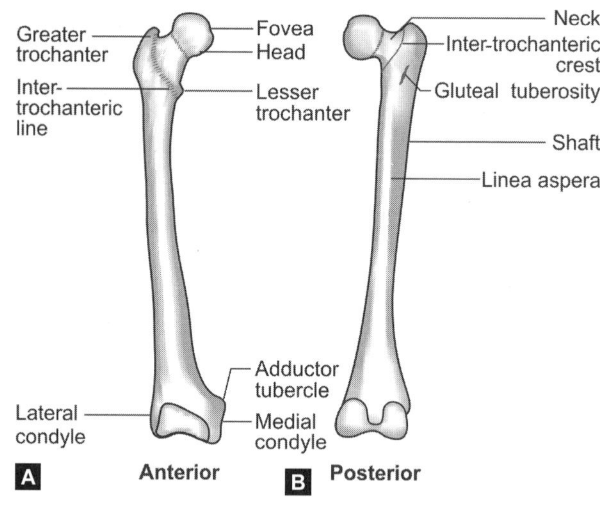

Figs 3.2A and B: Right femur

Femur (Figs 3.2A and B)

Q.17 What is the normal anatomical position of the femur in the body?

The head of femur is directed medially, upwards and slightly forwards and the shaft is obliquely downwards and medially, so that the two condyles at lower surface lie in same the horizontal plane.

Q.18 What is the arterial supply of the head of femur?

The medial part near fovea, supplied by medial epiphyseal arteries derived from ascending branch of medial circumflex femoral artery and posterior division of obturator artery.

The lateral part of head is supplied by lateral epiphyseal arteries derived from lateral circumflex femoral artery.

Q.19 What is the nutrient artery of the femur?

It is derived from second perforating artery.

Q.20 What is angle of anteversion?

The angle of anteversion (angle of femoral torsion) is the angle between the transverse axes of upper and lower ends of femur. It is about 15 degrees.

Q.21 Name the structures attached to intertrochanteric line of femur.

The following structures are attached to intertrochanteric line:

- Capsular ligament of hip joint
- Iliofemoral ligament
- Upper fibers of vastus lateralis and vastus medialis.

Q.22 Which muscle is inserted into trochanteric fossa?

Obturator externus.

Q.23 Which muscle is inserted in gluteal tuberosity?

Deep fibers of gluteus maximus.

Q.24 What is the origin of popliteus muscle?

From anterior part of groove on lateral aspect of lateral condyle of femur.

Q.25 What is the importance of ossification center for lower end of femur?

The ossification center for lower end of femur appears at end of 9th month of intrauterine life (the day of birth). It is of medicolegal importance in cases of newly born child found dead to decide whether it was viable or not.

Q.26 What is characteristic of primary ossification center of femur?

It is the second long bone in body to start ossifying.

Q.27 Why the fractures of neck of femur, leads to the necrosis of the head?

Because it will interrupt the blood supply to the head which is derived from:

- Vessels travelling up from diaphysis
- Vessels in the retinacula of the hip capsule.

Q.28 Why the intracapsular fracture of the neck of the femur are more dangerous than extracapsular fracture?

The intracapsular fracture interrupts the blood supply, to the femoral head resulting in necrosis whereas in the extracapsular fracture, the blood supply to the head remains unaffected and so there is no danger of avascular necrosis.

Q.29 What is coxa vara?

In this condition the angle between the femoral neck and shaft is decreased i.e., less than 160°. This results from adduction fractures.

Q.30 What is coxa valga?

Increase in the angle between femoral neck and shaft due to abduction fractures.

Q.31 At which level fracture of shaft of femur is dangerous?

Fracture of the lower end of femur are dangerous because proximal edge of the distal fragment is tilted backwards by the gastrocnemius, which tears the popliteal artery which lies directly behind it.

Tibia and Fibula (Figs 3.3A and B)

Q.32 Name the structures attached to the intercondylar area of tibia.

From before backwards, it provides attachment to:
- Anterior horn of medial meniscus
- Anterior cruciate ligament
- Anterior horn of lateral meniscus
- Posterior horn of lateral meniscus
- Posterior horn of medial meniscus
- Posterior cruciate ligament.

Q.33 What are the structures related to anterior surface of lower end of tibia?

From medial to lateral side it is related to tibialis anterior, extensor hallucis longus, anterior tibial vessels, deep peroneal nerve and extensor digitorum longus.

Q.34 What are the structures related to the posterior surface of lower end of tibia?

From medial to lateral side it is realted to tibialis posterior, flexor digitorum longus, posterior tibial artery, tibial nerve and flexor hallucis longus.

Q.35 What is the arterial supply of tibia?

Nutrient artery to tibia is a branch of the posterior tibial artery. It is the largest nutrient artery in the body.

Q.36 Although the tibia is one of the commonest sites of acute osteomyelitis but knee joint is not involved. Explain?

The knee joint is not involved because the capsule is attached near articular margins of tibia, proximal to epiphyseal line.

Q.37 The fracture of tibia is slow healing. Why?

The tibia is commonly fractured at the junction of upper 2/3 and lower 1/3 of its shaft, where it is most slender and this site is poorly supplied by blood vessels.

Q.38 How will you determine the side to which the fibula belongs?

The head is slightly expanded in all directions and lateral malleolus is expanded anteroposteriorly and is flattened from side to side. The medial side of lower end bears a triangular articular facet anteriorly and malleolar fossa posteriorly.

Q.39 Which structure lies between two heads of origin of peroneus longus?

Common peroneal nerve.

Q.40 Name the structures attached to malleolar fossa.

Malleolar fossa provides attachment to posterior talofibular and posterior tibio-fibular ligament.

Q.41 Fibula violates the general rule of ossification. Explain.

Normally in a long bone, growing end of a long bone ossifies first and unites with the shaft last while the non-growing end ossifies last and fuses with the shaft first. But in fibula, the ossification center for nongrowing end, i.e. lower end appears first but does not fuse last. This occurs because:
- The upper epiphysis (fuses last) is the growing end of the bone and
- Center for lower end appears first because it is a pressure epiphysis.

Q.42 What are the functions of fibula? (Fig. 3.4)
- It provides origin to muscles.
- It acts as a pulley for the tendons of peroneus longus and brevis.
- It forms a part of the ankle joint.
- It helps to increase the stability of ankle joint by lateral malleolus and ligaments attached to it.

Patella

Q.43 What is the function of patella?

The patella improves the leverage of the quadriceps femoris by increasing the angulation of the line of pull on the leg.

Q.44 How the stability of the patella is increased?

Due to outward angulation between long axes of thigh and leg the patella has a tendency to dislocate outwards. This is prevented by:
- *Muscular factor:* Insertion of vastus medialis on medial border of patella extends of a lower level than that of the vastus lateralis laterally.
- *Bony factor:* Lateral edge of patellar articular surface of femur is deeper than medial edge.

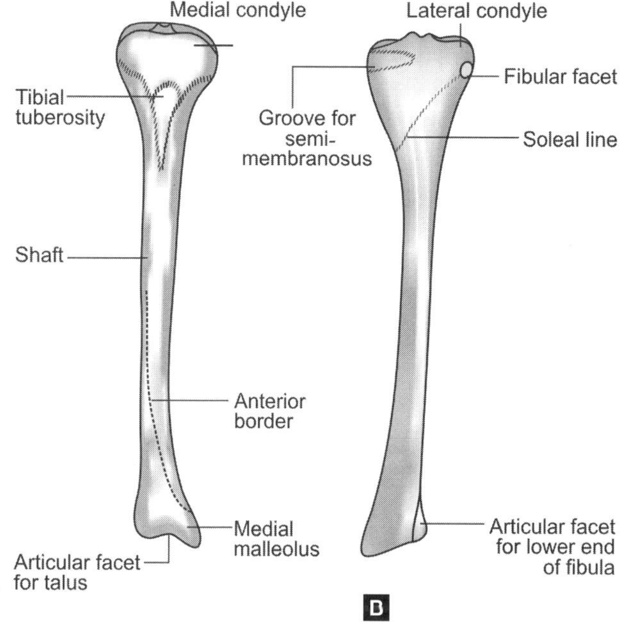

Figs 3.3A and B: Tibia (right) (A) Anterior view, (B) Posterior view

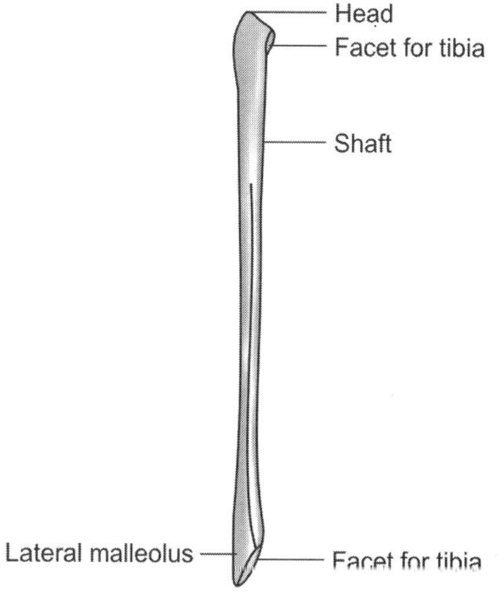

Fig. 3.4: Right fibula

Q.45 What are the different sesamoid bones present in the lower limb?

The following sesamoid bones are present in the lower limb:
- Patella, articulates with femur.
- Two small sesamoid bones in the tendons of flexor hallucis brevis, articulate with the head of the first metatarsal bone.
- One in the peroneus longus tendon, articulates with cuboid.
- Others may be present in the tendons of tibialis anterior, lateral head of gastrocnemius and gluteus maximus.

Q.46 What is 'fabella'?

It is a small, rounded sesamoid bone present in the lateral head of gastrocnemius. It articulates with the posterior surface of the lateral condyle of femur.

Bones of Foot (Fig. 3.5)

Q.47 Name the tarsal bone of foot.
- *Proximal row:* Talus, calcaneum,
- *Distal row:* Cuboid and medial, intermediate and lateral cuneiform.
  Navicular is interposed between the two rows

Q.48 Name the structures attached to medial tubercle of calcaneum.

Medially:
- Origin of abductor hallucis
- Flexor retinaculum.

Anteriorly:
- Origin of flexor digitorum brevis
- Plantar aponeurosis

Q.49 Name the structures attached to lateral tubercle of calcaneum.

Origin of abductor digiti minimi.

Q.50 What are the structures attached to sustentaculum tali?

To its medial margin are attached
- Spring ligament, anteriorly
- Slip from tibialis posterior, in middle
- Superficial fibers of deltoid ligament, along its whole length and
- Medial talocalcaneal ligament, posteriorly.

Q.51 What is the structure attached to tuberosity of navicular bone?

Insertion for tibialis posterior.

Q.52 Name the structures related to plantar groove of cuboid.
- Through groove pass tendon of peroneus longus.
- To posterior ridge, deep fibers of long plantar ligament.

Q.53 What are the differences between metacarpal and metatarsal?

Metacarpal	*Metatarsal*
Head and shaft: Prismoid	Flattened from side to side
Shaft: Uniform thickness	Tapers distally
Dorsal surface of shaft	Elongated, flat triangular area Uniformly convex
Base: Irregular	Cuts sharply and obliquely

Q.54 What are the "accessory bones"?

These are separate small pieces of bone which have not fused with the main bone e.g.,
- Os trigonum (posterior tubercle of talus) and
- Os vesalianum (tuberosity of fifth metatarsal).

Q.55 What is 'bunion'?

It is inflamed adventitial bursa over the head of first metatarsal bone.

Joints of Lower Limb

Hip Joint (Fig. 3.6)

Q.56 What is the type of hip joint?

Hip joint is a ball and socket type of synovial joint.

Q.57 What are the factors which increase the stability of the hip joint?

The stability of hip joint is increased by the following factors:
- Depth of acetabulum with a narrow mouth, made by acetabular labrum.
- Tension and strength of ligaments.
- Strength of the surrounding muscles.
- Length and obliquity of neck of femur.

The wide range of mobility depends upon the neck of femur which is narrower than the equatorial diameter of the head.

Q.58 What is the attachment of ligament of the head of femur?

It is attached laterally to fovea on head of femur and medially to two ends of acetabular notch and to transverse ligament.

Q.59 What are the ligaments strengthening the capsule of hip joint?
- Iliofemoral ligament: Strongest, Y-shaped ligament.
- Pubofemoral ligament
- Ischiofemoral ligament (Figs 3.7A and B).

Q.60 What are the relations of the hip joint?

The relations of the hip joint are:
- *Anteriorly:* Lateral fibers are pectineus, iliopsoas, straight head of rectus femoris.
- *Posteriorly:* Quadratus femoris covering obturator externus and ascending branch of medial circumflex femoral artery, the piriformis, obturator internus with two gemelli separate the sciatic nerve from the nerve to quadratus femoris.

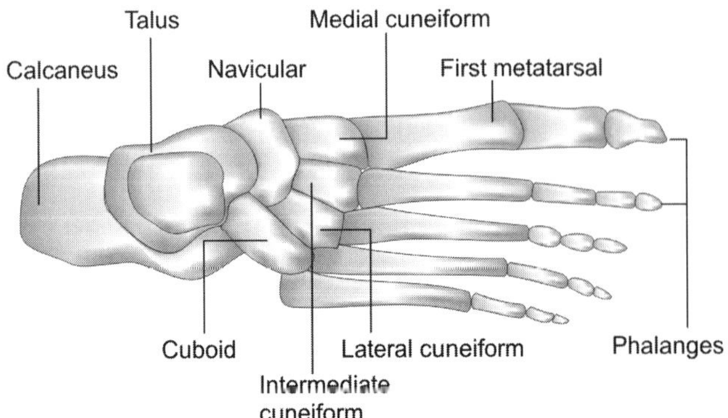

Fig. 3.5: Skeleton of the foot (viewed from above)

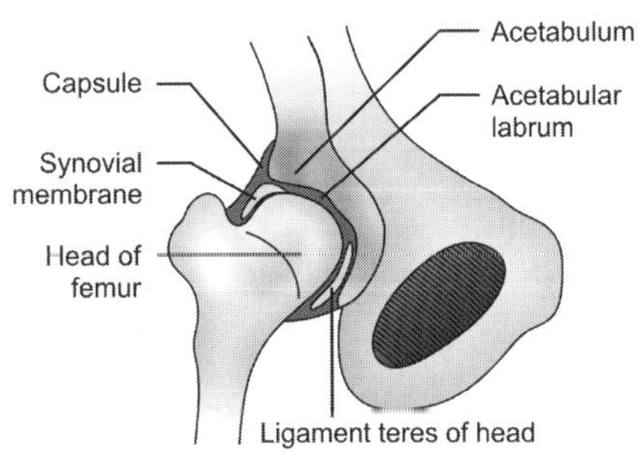

Fig. 3.6: Hip joint

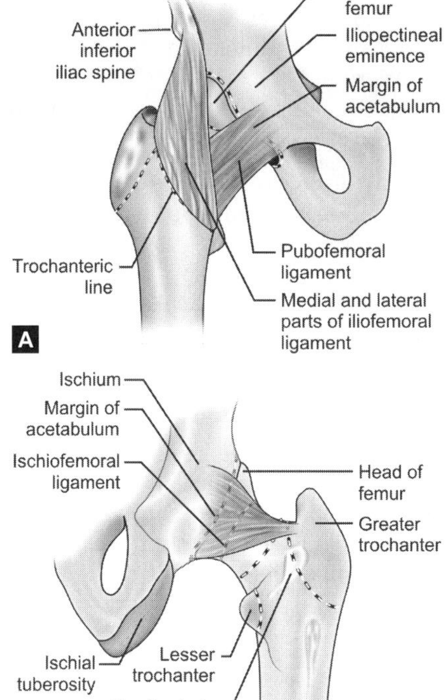

Figs 3.7A and B: Hip joint. (A) anterior aspect, (B) Posterior aspect. The capsular attachments (blue) and epiphyseal lines (magenta) are shown

- *Superior:* Reflected head of rectus femoris covered by gluteus minimus.
- *Inferior:* Lateral fibers of pectineus and obturator externus.

Q.61 What is the blood supply to the hip joint?

The hip joint is supplied by the medial circumflex femoral and the lateral circumflex femoral vessels (Fig. 3.8). There also may be contribution by the acetabular branch of femoral artery.

Q.62 What is the axis of different movements of hip joint?

- For rotation, vertical axis passing through the center of head of femur and its lateral condyle.
- Extension and flexion, occur around a transverse axis.
- Adduction and abduction, occur around an anteroposterior axis.

Q.63 What is the range of movements at the hip joint?

Flexion is limited by contact of thigh with anterior abdominal wall.

Adduction is limited by contact with opposite limb.

Range of other movements: Lateral rotation 60°, Medial rotation 25°, Abduction 50° and extension 15°.

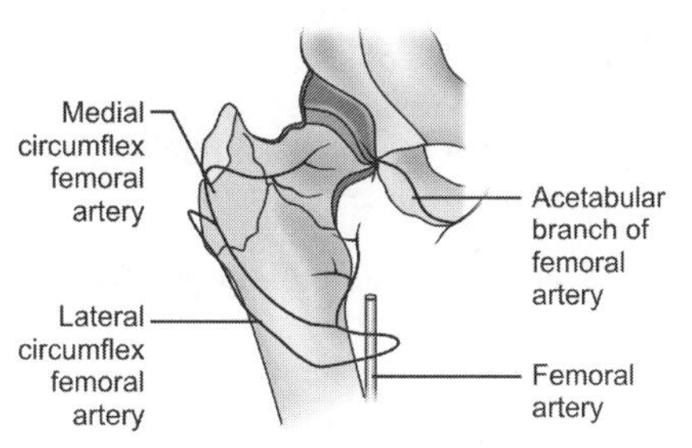

Fig. 3.8: Blood supply of hip joint

Q.64 What are the nerves supplying the hip joint?

The hip joint is supplied by:

- Femoral nerve, through nerve to rectus femoris,
- Anterior division of obturator nerve,
- Accessory obturator nerve,
- Nerve to quadratus femoris and
- Superior gluteal nerve.

Q.65 What are the different muscles producing extension of the hip joint?

Gluteus maximus and hamstrings.

Q.66 Which muscles produce abduction of the hip joint?

Chief muscles: Gluteus medius and minimus.

Accessory muscles: Tensor fasciae latae and sartorius.

Q.67 What is Trendelenburg test?

This test is employed for testing the stability of the hip joint. A positive test indicates a defect in osseomuscular stability especially abductors of hip joint and the patient has a "lurching" gait. If the patient is asked to stand on one leg. If the abductors of thigh are paralyzed on that side, they will be unable to sustain the pelvis against the body weight and pelvis tilts downwards on unsupported side.

Q.68 Name the adductors of hip joint.

- Adductor longus
- Adductor brevis
- Adductor magnus
- Gracilis
- Pectineus.

Q.69 Name the medial rotators of hip joint.

Gluteus medius and minimus:

- Tensor fasciae latae
- Adductor longus, brevis and magnus.

Q.70 Name the flexors of hip joint.

Mainly: Psoas major, iliacus, rectus femoris

Accessory muscles: Adductors are also flexors of hip joint.

Q.71 What is the cause of Weaver's bottom?

Inflammation of bursa over ischial tuberosity.

Q.72 In which injury of the hip joint sciatic nerve is likely to be damaged?

It is likely to be injured in the posterior dislocation of the hip joint associated with fracture of the posterior lip of the acetabulum, to which the nerve is closely related.

Knee Joint

Q.73 What is the function of anterior and posterior cruciate ligament?

- Anterior cruciate ligament: Prevents hyperextension of knee joint.
- Posterior cruciate ligament: Prevents hyperflexion of knee joint.

Q.74 What is compartment syndrome?

It is an increase in fluid pressure (> 30 mm) within an osseofascial compartment and leads to muscle and nerve damage. Usually occurs in anterior compartment of thigh as a result of crush injury can also occur in anterior compartment of leg due to fracture of the tibia.

Q.75 What is Legg-Perthes disease?

It is characterized by idiopathic avascular necrosis of the head of femur. Caucasian boys are more commonly affected and it is usually characterized by unilateral hip pain external rotation (slight) and a limp.

Q.76 What type of joint is the knee joint?

Compound synovial joint, having

- *Condylar synovial joint:* Between the condyles of femur and tibia.
- *Saddle synovial joint:* Between femur and patella (Fig. 3.9).

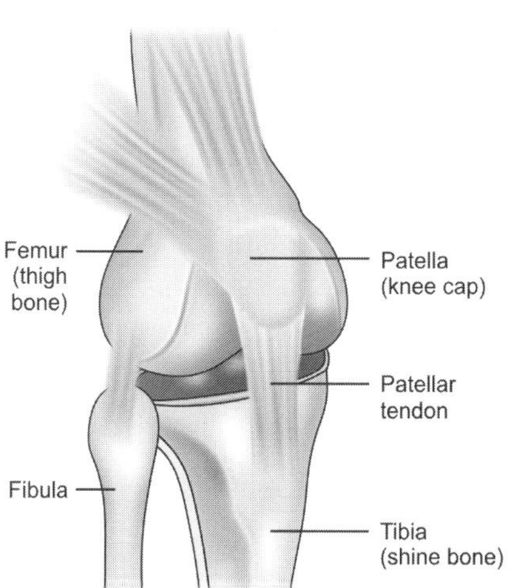

Fig. 3.9: The knee joint

Q.77 What are the articular surfaces in knee joint?

Condyles of femur:
- Condyles of tibia and
- Patella.

Q.78 What are the bony landmarks in the region of the knee?

1. *Adductor tubercle:* is felt just above the medial condyle of the femur.
2. *Head of the fibula:* is felt at posterolateral aspect of the knee. Lies 1.5 cm below the level of the knee joint.
3. *Tibial tubercle:* is felt in front of the knee in upper part of the tibia.
4. *Tibial condyles:* are felt on each side of the lower part of the ligamentum patellae.

Q.79 Name the ligaments of knee joint.

- Fibrous capsule
- Ligamentum patellae
- *Collateral ligaments:* Tibial and fibular
- *Popliteal ligaments:* Oblique and arcuate
- *Cruciate ligaments:* Anterior and posterior
- *Meniscus:* Medial and lateral
- Transverse ligament (Figs 3.10A and B).

Q.80 What is coronary ligament?

It is the part of fibrous capsule lying between the menisci and tibia.

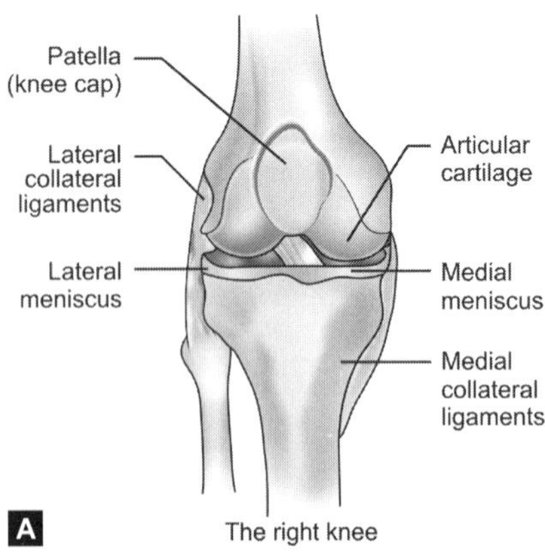

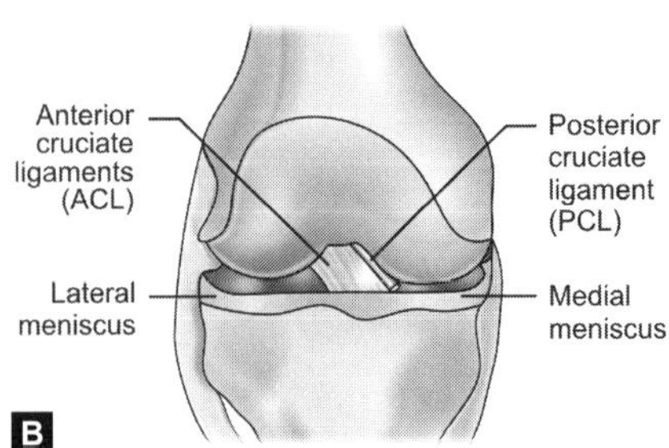

Figs 3.10A and B: Ligament of the knee joint

Q.81 What are the openings in the fibrous capsule of knee joint?

1. For suprapatellar bursa and
2. For the exit of tendon of popliteus with its synovial bursa.

Q.82 What are the attachments of arcuate popliteal ligament?

It passes from the head of fibula to the posterior margin of the intercondylar area of tibia.

Q.83 What are the attachments of oblique popliteal ligament?

It arises as an expansion from the tendon of semi-membranous. It blends with the posterior surface of fibrous capsule. It is attached to the intercondylar line and lateral condyle of femur and posterior aspect of medial condyle of tibia.

Q.84 Name the structures piercing oblique popliteal ligament.

1. Posterior division of obturator nerve and
2. Middle genicular nerve and vessels.

Q.85 What are menisci and what are their functions?

These are two fibrocartilaginous structures, semilunar in shape, which make the tibial articular surface deeper and divide the joint cavity partially into upper and lower compartment (Fig. 3.11).

Functions:
- They act as shock absorbers.
- They make the articular surfaces more congruent. They can adapt to varying curvatures of different parts of femoral condyles.

Q.86 What is the arterial supply of knee joint?

- Genicular branches of popliteal artery,
- Descending genicular branch of femoral artery,
- Descending branch of lateral circumflex femoral artery.
- Recurrent branches of anterior tibial artery and
- Circumflex fibular branch of posterior tibial artery.

Q.87 Name the arteries forming the anastomosis around the knee joint.

Medially:
- Descending genicular
- Superior medial genicular
- Inferior medial genicular

Laterally:
- Descending branch of lateral circumflex femoral
- Superior lateral genicular
- Inferior lateral genicular
- Anterior lateral recurrent
- Posterior lateral recurrent
- Circumflex fibular (Fig. 3.12).

Q.88 What is the nerve supply of knee joint?

1. Femoral nerve,
2. Genicular branches of tibial and common peroneal nerves and
3. Posterior division of obturator nerve.

Q.89 What are the movements possible at knee joint?

- Flexion
- Extension
- Medial and lateral rotation.

Q.90 What is 'conjunct' and 'adjunct' rotation?

- *Conjunct rotation:* Rotation of knee joint combined with flexion and extension.
- *Adjunct rotation:* Rotation of knee joint occurring independently in a partially flexed knee.

Q.91 What are the changes in the axis of movement of the knee joint with flexion and extension?

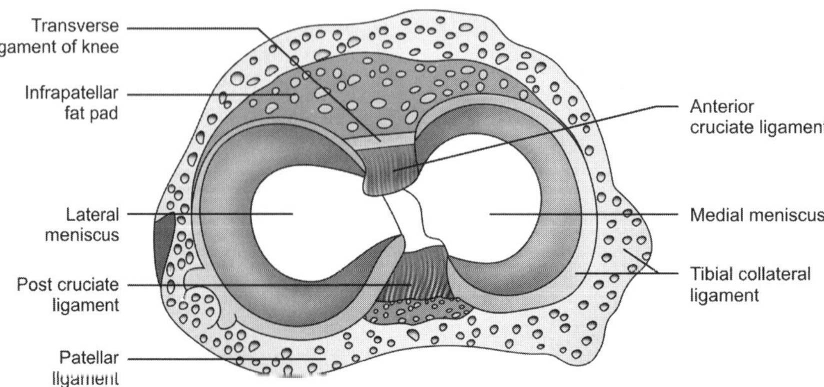

Fig. 3.11: Knee menisci

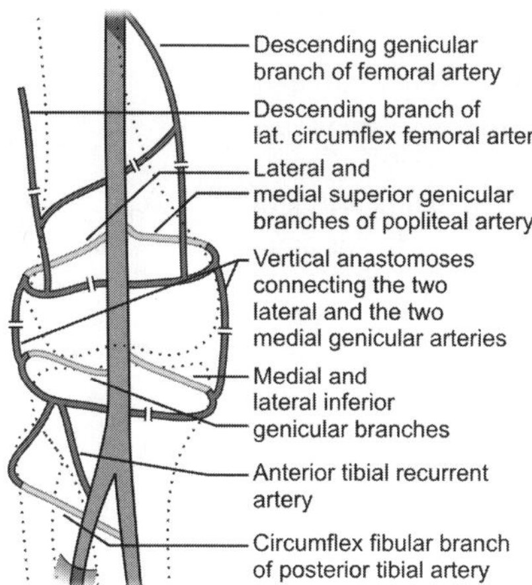

Fig. 3.12: Anastomoses around the knee joint

The flexion and extension of the knee joint takes place on a transverse axis which shifts along with the movements. Because of the spiral profiles of the femoral condyles, the axis shifts upwards and forwards during extension and backwards and downwards during flexion.

Q.92 What are the locking and unlocking movements of the knee joint?

In full extension from the position of flexion the last 30° of extension is accompanied by medial rotation of the femur on the tibia or lateral rotation of the tibia on the femur depending on whether the tibia or the femur is fixed. This is conjunct rotation and occurs passively as a part of the extension movement, is described as 'locking' of the knee joint.

From the position of full extension, the beginning of flexion is accompanied by lateral rotation of the femur or medial rotation of the tibia depending on whether the tibia or the femur is fixed. This rotation is called 'unlocking' of the knee joint. The contraction of popliteus is responsible for this unlocking movement.

Q.93 Name the intra-articular structures of the knee joint.

- *Cruciate ligaments:* Anterior and posterior
- *Menisci:* Medial and lateral
- Infrapatellar pad of fat
- Synovial membrane
- Origin of popliteus.

Q.94 Name the bursa around knee joint?

Anteriorly:
- Subcutaneous prepatellar bursa
- Subcutaneous infrapatellar bursa
- Deep infrapatellar bursa and
- Suprapatellar bursa.

Medially:
- Bursa deep to medial head of gastrocnemius.
- Bursa deep to tibial collateral ligament.
- Semimembranosus bursa
- Anserine bursa and
- Occasionally, bursa between tendons of semitendinosus and semimembranosus.

Laterally:
- Bursa deep to lateral head of gastrocnemius
- Bursa between fibular collateral ligament and tendon of popliteus
- Bursa between fibular collateral ligament and biceps femoris and
- Bursa between tendon of popliteus and lateral condyle of tibia.

Q.95 Name the bursa communicating with the knee joint.

- Suprapatellar bursa
- Popliteal bursa
- Bursa deep to medial head of gastrocnemius (Figs 3.13A and B).

Q.96 What is anserine bursa?

It is bursa with several diverticula which separate the tendons of sartorius, gracilis and semitendinosus from bony surface of tibia.

Q.97 Name the different muscles producing movements of knee joint.

	Principal muscles	*Accessory muscles*
Extension	Quadriceps femoris	Tensor fasciae lata
Flexion	Semitendinosus, Biceps femoris, Semimembranosus.	Sartorius, Gracilis, Popliteus, Gastrocnemius
Medial rotation	Semitendinosus, Semimembranosus	Sartorius, Gracilis
Lateral rotation	Biceps femoris	

Q.98 Name ligaments which become taut in full extension and flexion of the knee joint.

- *In full extension*:
  - Anterior cruciate ligament.
  - Tibial and fibular collateral ligament.
  - Oblique popliteal ligament.
- *In full flexion:*
  - Posterior cruciate ligament.

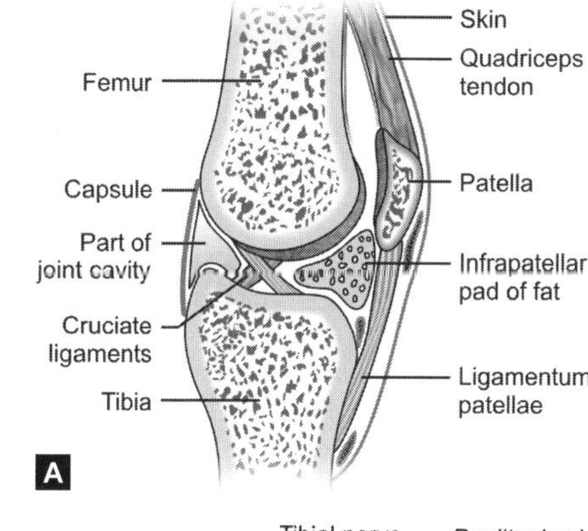

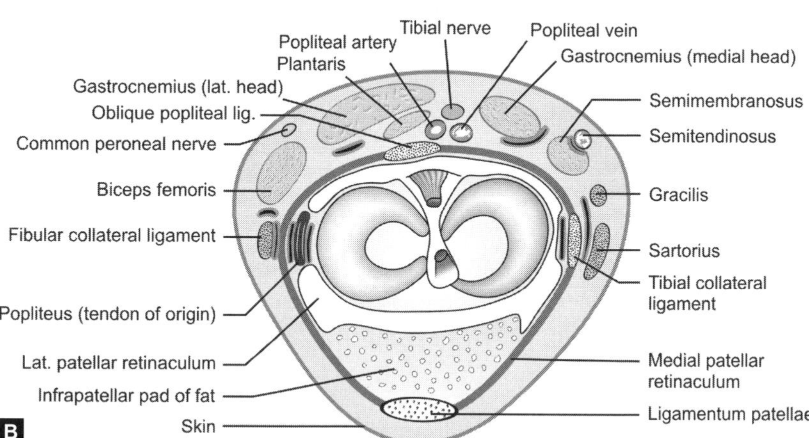

Figs 3.13A and B: (A) Schematic sagittal section through the knee joint to show some bursae related to the joint, (B) Schematic transverse section to show some bursae around the knee joint

Q.99 What could cause a tear of the menisci (semilunar cartilages) of the knee joint?

The menisci are usually torn by a twisting force with knee flexed. When the flexed knee is forcibly abducted and externally rotated, the medial meniscus is trapped between the medial condyles of the femur and tibia and is torn.

A severe adduction and internal rotation of the flexed knee may result in a tear of the lateral meniscus. But this injury is less common.

Q.100 Why the tears of medial meniscus are more frequent than that of lateral meniscus?

Because the medial meniscus is more firmly attached to the upper surface of the tibia, capsule and the tibial collateral ligament and therefore, is less able to adapt itself to sudden changes of position. The lateral meniscus on the other hand, is drawn backwards and downwards on the groove on the posterior aspect of the lateral tibial condyle by the medial fibers of popliteus. This prevents, the lateral meniscus from being impacted between the articular surfaces of the femur and the tibia during movements of the knee joint.

Q.101 Why in tear of medial meniscus there is locking of the knee before it is fully extended?

Because the torn segment of the cartilage is displaced and lodges between the femoral and tibial condyles and prevents full extension of the knee.

Q.102 Why the pain of hip joint is referred to the knee?

Because of the common nerve supply of the two joints.

Ankle Joint

Q.103 What type of joint is ankle joint?

Hinge variety of synovial joint.

Q.104 What are the articular surface of ankle joint?

- *From above:*
  - Lower end of tibia with medial malleolus
  - Lateral malleolus and
  - Inferior transverse tibiofibular ligament.
- *From below:* Body of talus.

Q.105 Name the ligaments of ankle joint.

- Fibrous capsule
- *Lateral ligament:* Consists of
  - Anterior talofibular ligament,
  - Posterior talofibular ligament and
  - Calcaneofibular ligament.
- *Medial (deltoid) ligament:* It has
  1. *Superficial part:* Consists of
     - Anterior fibers (tibionavicular)
     - Middle fibers (tibiocalcanean) and
     - Posterior fibers (Posterior tibiotalar).
  2. *Deep part* (Anterior tibiotalar) (Figs 3.14A and B).

Q.106 Name the tendons crossing the deltoid ligament.

- Tibialis posterior and
- Flexor digitorum longus.

Q.107 Name the structures related to ankle joint.

Anteriorly: From medial to lateral side:

- Tibialis anterior
- Extensor hallucis longus
- Anterior tibial vessels
- Deep peroneal nerve
- Extensor digitorum longus and
- Peroneus tertius.

Posteriorly: From medial to lateral side

- Tibialis posterior
- Flexor digitorum longus
- Posterior tibial vessels
- Tibial nerve
- Flexor hallucis longus
- Peroneus brevis and
- Peroneus longus.

Q.108 What are the movements produced at ankle joint?

- Dorsiflexion
- Plantar flexion
- *Accessory movements:* With plantar flexion, slight amount of side to side gliding, abduction and adduction are permitted.

Q.109 What is the axis of movements of the ankle joint?

It is represented by a transverse line drawn across the front of the ankle about 1.25 cm above the tip of the medial malleolus.

Q.110 What is the close-pack position of the ankle joint?

Dorsiflexion is the close-pack position of the joint in which the wider front part of the talus articulates with the mortise formed by the malleoli and lower end of the tibia. In this position, there is maximal congruence of the joint surface and tension of the ligaments.

Q.111 Name the muscles producing movements at ankle joint.

Dorsiflexion:

- *Main muscle:* Tibialis anterior.
- *Accessory muscles:*
  - Externsor digitorum longus,
  - Extensor hallucis longus and
  - Peroneus tertius.

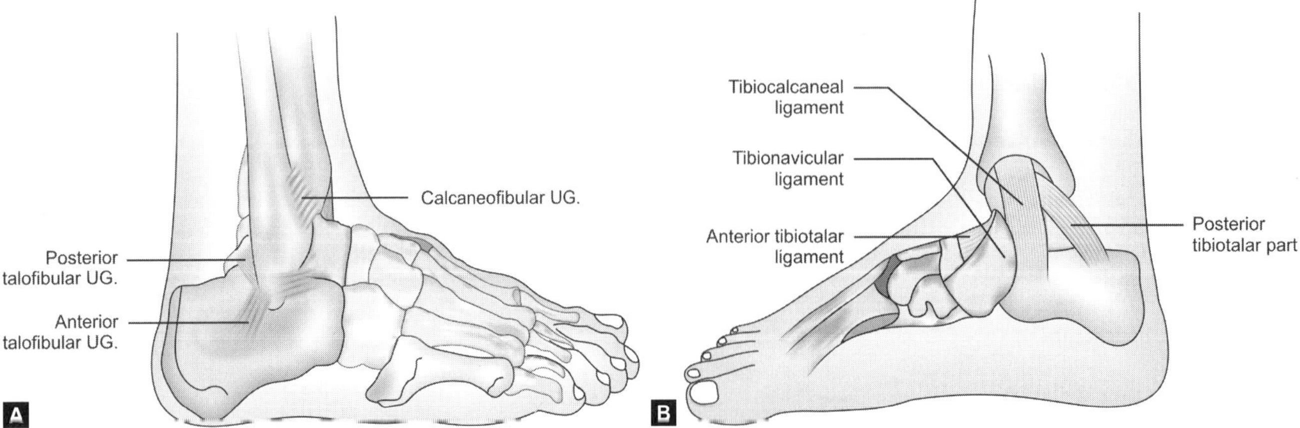

Figs 3.14A and B: (A) Ankle joint: lateral collateral ligament of ankle, (B) Ankle joint: medial ligament of ankle

Plantar flexion:
- *Main muscles:*
  - Gastrocnemius and
  - Soleus
- *Accessory muscles:*
  - Flexor digitorum longus,
  - Flexor hallucis longus,
  - Tibialis posterior and
  - Plantaris.

Q.112 What is most frequent fracture at the ankle joint?
Pott's fracture, usually produced by an abduction external rotation injury.

Tibiofibular Joints

Q.113 What type of joints are tibiofibular joints?
- *Superior tibiofibular joint:* Plane synovial joint.
- *Lower tibiofibular joint:* Syndesmosis type of fibrous joint.

Q.114 Name the structures passing through interosseous membrane of tibiofibular joint.
- Anterior tibial vessels
- Perforating branch of peroneal artery.

Joints of Foot

Q.115 What do you understand by inversion and eversion of foot?
Inversion: Movement in which medial border of foot is elevated and sole faces medially and inwards.
Eversion: Movement in which lateral border of foot is elevated and sole faces laterally and outwards.

Q.116 Name the joints at which inversion and eversion takes place.
- Subtalar (talocalcaneal) joint and
- Talocalcaneonavicular joint.

Q.117 What is the axis of the inversion and eversion?
Oblique axis which runs forwards, upwards and medially.

It passes between back of calcaneum, sinus tarsi and superomedial aspect of neck of talus.

Q.118 Name the evertors of foot.
- Mainly by, peroneus brevis and longus.
- Also by, peroneus tertius.

Q.119 Name the invertors of foot.
Principal muscles:
- Tibialis anterior and
- Tibialis posterior.

Accessory muscles:
- Flexor hallucis longus and
- Flexor digitorum longus

Q.120 Why are the movements of inversion and eversion required in man?
Inversion and eversion of the foot are essential for walking on rough, uneven or sloping surfaces.

Arches of Foot

Q.121 Classify the arches of foot.
- Longitudinal arches: Two
  - Medial and
  - Lateral
- Transverse arch (Figs 3.15A to C).

Q.122 How the arches of foot are maintained?
- By the configuration of articulating bones forming the arch.
- By the ligaments and muscles binding the adjacent bones and ends of an arch.
- By tendons of muscle which act as sling and thus help to suspend the arch from above.
- Plantar aponeurosis by connecting anterior and posterior ends of longitudinal arches like a tie beam.

Q.123 What are the functions of arches of foot?
- Rigid support for the weight of body in standing position.
- As mobile spring board during walking and running.

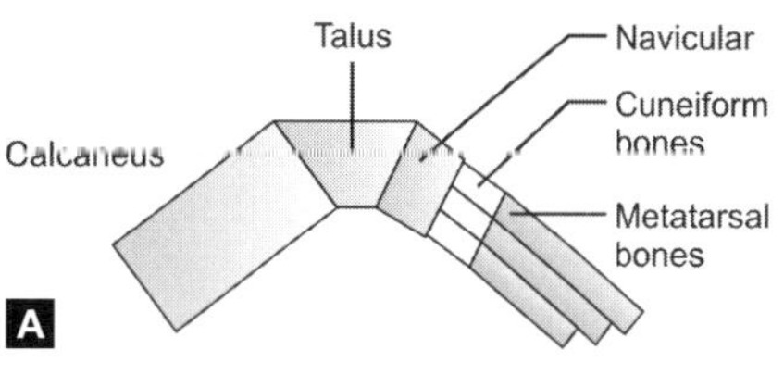

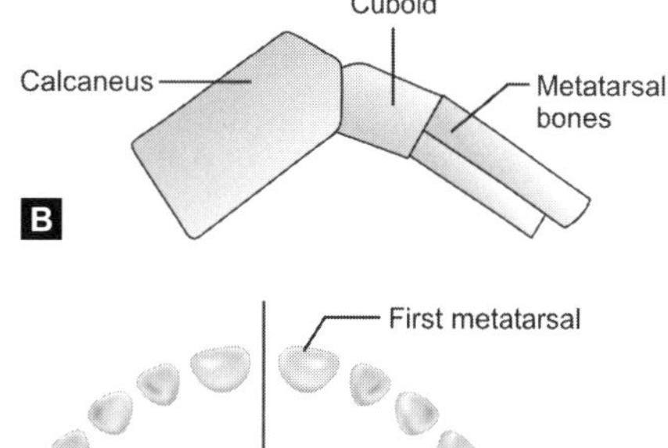

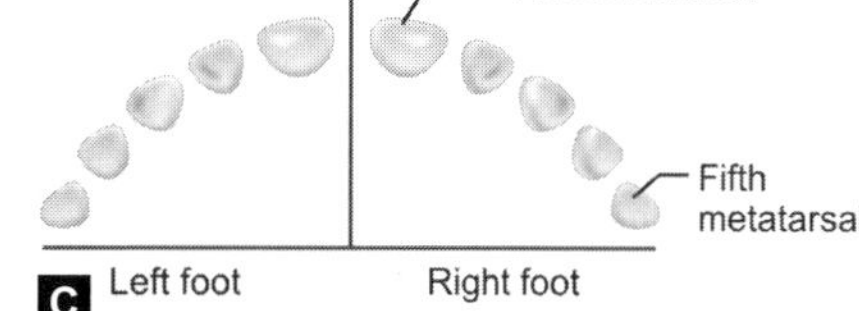

Figs 3.15A to C: (A) Scheme to show constitution of the medial longitudinal arch of the foot, (B) Scheme to show constitution of the lateral longitudinal arch of the foot, (C) Scheme to show the transverse arch formed by the two feet. Note that each foot forms half of the arch

- As shock absorbers in jumping.
- Protects the soft tissues of sole of foot.

Q.124 How the medial longitudinal arch is formed?
By calcaneum, talus, three cuneiforms and three medial metatarsals. The summit of arch is formed by talus.

Q.125 How the lateral longitudinal arch is formed?
By the calcaneum, cuboid and lateral two metatarsals.

Q.126 How the transverse arch is formed?
By the bases of the five metatarsals and the adjacent cuboid and cuneiforms of both feet.

Q.127 What are the attachments of spring ligament?
It passes from anterior margin of sustentaculum tali of calcaneus to plantar surface of navicular bone.

Q.128 What are the attachments of long plantar ligament?
It is attached posteriorly to plantar surface of calcaneus in front of lateral and medial tubercles and anteriorly to plantar surface of cuboid distal to groove for peroneus longus.

Q.129 Which structures maintain the medial longitudinal arch?
The bony configuration do not contribute to the maintenance of this arch.
- *Ligaments:*
  - The medial part of the plantar aponeurosis acts as a tie beam.
  - The plantar calcaneonavicular ('spring') ligament supports head of talus and forms intersegmental ties (connect adjacent bones).
- *Muscles:*
  - Medial half of the flexor digitorum brevis and abductor hallucis act as tie beams (connect ends of arch).
  - Tibialis anterior, tibialis posterior and flexor hallucis longus act by forming sling and suspend the arch.

Q.130 How the lateral longitudinal arch of the food is maintained?
- *Ligaments:*
  - The short plantar ligament, long plantar ligament and dorsal ligaments form intersegmental ties.
  - Lateral part of the plantar aponeurosis acts as a tie beam.
- *Muscles:*
  - The peroneus longus and peroneus brevis muscles form the slings.

- Lateral half of the flexor digitorum brevis and abductor digiti minimi act as tie beam.

Q.131 How the transverse arch of the foot is maintained?

Tarsal and metatarsal bones contribute in maintaining the concavity of arch.

- *Ligaments*
  - Ligaments that bind together the cuneiforms and the bases of the metatarsals form intersegmental ties.
  - Superficial and deep transverse metatarsal ligaments act as tie beams.
- *Muscles:*
  - The peroneus longus and tibialis posterior form slings.
  - Abductor hallucis acts as tie beam.

Q.132 What are the deformities of the foot resulting from defects of the longitudinal arches of the foot?

1. *Pes planus (flat foot):* Due to flattening of the longitudinal arch, in particular the medial arch.
2. *Pes cavus (high arched foot):* The congenital form is probably due to shortness of the plantar fascia (aponeurosis). The acquired form can be due to contracture of the intrinsic muscles of the foot.

Q.133 What is the 'talipes deformity' of the foot?

In talipes the foot no longer lies in the plantigrade position. The person walks either on the heels or on the toes. When he walks on the heel the condition is known as *talipes calcaneus* while walking on the toes is known as *talipes equinus.* In both these conditions the foot may be inverted (*varus*) or everted (*valgus*).

Q.134 What is hallux valugs?

In hallux valugs, there is lateral deviation of the great toe at the metatarsophalangeal joint. More common in women than men.

Q.135 What is 'hammer toe'?

The affected toe is hyperextended at metatarsophalangeal and distal interphalangeal joint and flexed at proximal interphalangeal joint.

Thigh

Q.136 What is midinguinal point and what is its importance?

Midinguinal point is a point midway between anterior superior iliac spine and the pubic symphysis. It is an important land-mark. The femoral artery and head of femur lie beneath the midinguinal point.

Q.137 What is Holden's line and what is its importance?

The deep layer of superficial fascia is firmly attached to the deep fascia of thigh along a horizontal line a little lateral to pubic tubercle and extends for about 8 cm laterally. This line of firm attachment is called Holden's line.

Clinical importance: The extravasation of urine between these two layers cannot extend into thigh because of the firm attachment.

Q.138 How is patellar plexus formed?

It is a plexus of nerves in front of patella and upper end of tibia. It is formed by

- Anterior division of lateral and medial cutaneous nerve of thigh
- Intermediate cutaneous nerve of thigh and
- Infrapatellar branch of saphenous nerve.

Q.139 What is housemaid's knee?

Chronic enlargement of prepatellar bursa is known as housemaid's knee because it commonly occurs in housemaid's who have to kneel regularly for sweeping the floor.

Q.140 What is miner's beat knee?

It is acute suppurative prepatellar bursitis in miners.

Q.141 What is clergyman's knee?

It is enlargement of subcutaneous infra-patellar bursa in clergyman.

Q.142 What is iliotibial tract and what are its functions?

The thickening of fascia lata on the lateral side of the thigh is called the iliotibial tract.

Functions:

1. Iliotibial tract stabilizes knee both in extension and partial flexion, i.e., during walking and running.
2. In leaning forwards with slightly flexed knees, it is the only antigravity force to support the knee.

Q.143 What are the modifications of deep fascia of thigh?

- *Saphenous opening*: Oval gap 4 cm below and lateral to pubic tubercle. Upper, lateral and lower margins form a crescentic sharp edge and medially deep part of fascia passes behind the femoral sheath.
- *Cribriform fascia:* Cover the saphenous opening and is pierced by great saphenous vein, two superficial arteries and lymphatics.
- *Iliotibial tract:* Receives insertion of ¾ of gluteus maximus and tensor fasciae latae.

Femoral Triangle

Q.144 Why femoral triangle is known as Scarpa's triangle?

Because it was first described by Antonio Scarpa (1747-1832) in Italy.

Q.145 What are the boundaries of femoral triangle?

It is bounded by (Figs 3.16 to 3.18)

- *Laterally:* Medial border of sartorius.
- *Medially:* Medial border of adductor longus.
- *Base:* Inguinal ligament.
- *Apex:* Directed downwards and is formed by meeting of medial and lateral boundaries.
- *Roof:*
  - Skin,
  - Superficial fascia and
  - Deep fascia.
- *Floor:*
  - Laterally by iliacus and psoas major.
  - Medially by adductor longus and pectineus.

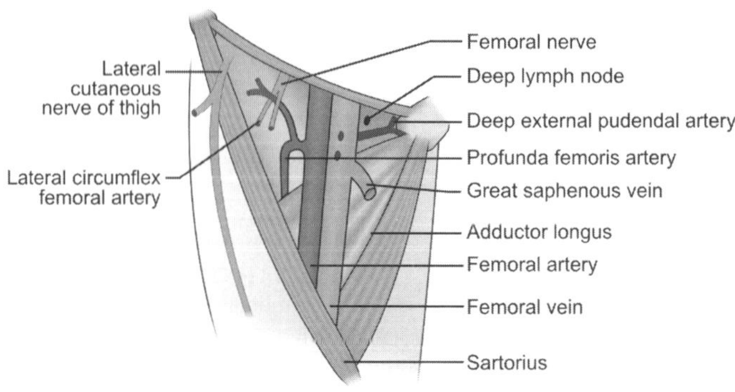

Fig. 3.16: Femoral triangle and its contents

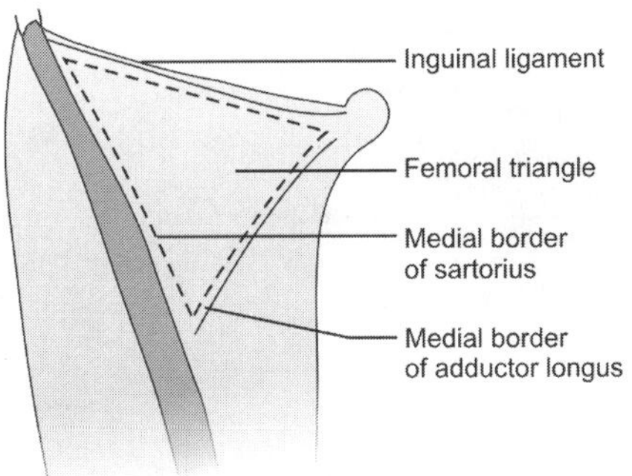

Fig. 3.17: Boundaries of femoral triangle

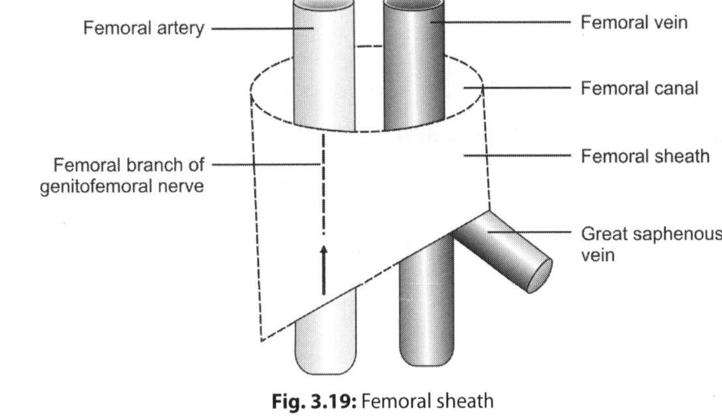

Fig. 3.19: Femoral sheath

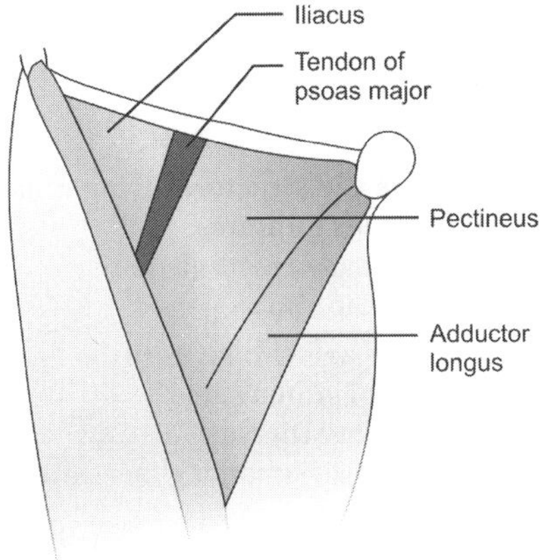

Fig. 3.18: Floor structure of femoral triangle

Q.146 What are the contents of femoral triangle?
- Femoral artery
- Branches of femoral artery:
  - Deep branches: Profunda femoris, deep external pudendal, descending genicular, saphenous and muscular.
  - Superficial branches: Superficial external pudendal, superficial epigastric and superficial circumflex iliac.
- Femoral vein (medial to artery) and its tributaries
- Femoral sheath
- Femoral nerve (lateral to artery)
- Nerve to pectineus
- Femoral branch of genitofemoral nerve
- Lateral cutaneous nerve of thigh and
- Deep inguinal lymph nodes.

Q.147 What is femoral sheath?
It is a funnel shaped fascial sleeve enclosing the upper 1½ inches of the femoral vessels (Fig. 3.19).

Q.148 How is femoral sheath formed?
It is formed by the downward extension of the abdominal fasciae. The anterior wall is formed by fascia transversalis and posterior wall by fascia iliaca.

Q.149 What are the relations of femoral sheath?
Anterior:
- Skin
- Superficial fascia and
- Deep fascia with saphenous opening and great saphenous vein.

Posterior:
- Iliopectineal fascia
- Pectineus and
- Iliopsoas.

Lateral:
- Femoral nerve and
- Iliacus.

Medial:
- Lacunar ligament
- Pectineus and
- Pubic bones.

Q.150 What are the parts of femoral sheath?
The cavity within femoral sheath is divisible in three parts. Lateral part contains femoral artery and femoral branch of genitofemoral nerve. Middle part contains femoral vein and medial part is called femoral canal.

Q.151 What is femoral canal?
It is the medial compartment of the femoral sheath. It is conical and ½ inch wide at base and ½ inch long.

Q.152 What is femoral ring?
The base or upper end of the femoral canal is called the femoral ring.

The femoral ring is filled by condensed extraperitoneal tissue, the *femoral septum*, containing a lymph node and covered by parietal peritoneum.

Q.153 What are the boundaries of the femoral ring?
- *Anterior*: Inguinal ligament.
- *Posterior:* Pectineus and its fascia.
- *Lateral:* Septum separating it from the femoral vein.
- *Medial:* Concave margin of lacunar ligament.

Q.154 What are the contents of femoral canal?
- Lymph node (of Cloquet or of Rosenmuller).
- Lymphatics.
- Areolar tissue.

Q.155 What are the functions of femoral canal?
- It serves as a dead space for expansion of the femoral vein.
- It allows a lymphatic pathway from the lower limb to the external iliac lymph nodes.

Q.156 What structure is drained by lymph node of femoral canal?
Glans penis in the male and clitoris in female.

Q.157 What is the clinical importance of the femoral canal?
The femoral canal is a potential point of weakness in the lower abdominal wall through which a viscus (intestines or urinary bladder) may protrude and give rise to a femoral hernia.

Q.158 Why is a femoral hernia commoner in females?
Because the femoral canal is larger in the females due to the greater width of the pelvis and smaller size of the femoral vessels. In the females, there is a rise in intra-abdominal pressure due to pregnancy predisposing to femoral hernia.

Q.159 Why is strangulation more common in femoral hernia?
Because the neck of the femoral canal is narrow.

Q.160 What is the risk of enlarging the opening of the femoral canal in releasing the strangulation of a femoral hernia?
In order to enlarge the opening of the femoral canal the sharp lateral edge of the lacunar (Gimbernat's) ligament may

require incision. An abnormal obturator artery may occasionally be present, which passes behind the lacunar ligament and is then in danger of being cut.

Q.161 What are the coverings of femoral hernia?

From within outwards:

- Peritoneum
- Femoral septum
- Femoral sheath
- Cribriform fascia
- Superficial fascia
- Skin.

Adductor Canal

Q.162 What are the boundaries of the adductor canal?

Posteriorly:

- Adductor longus above and
- Adductor magnus below.

Anteriorly: Vastus medialis.

Medially: Sartorius which lies on a fascial sheet extending across the anterior and posterior walls.

Q.163 What is the extent of the adductor canal?

It extends from the apex of the femoral triangle to the tendinous opening in the adductor magnus.

Q.164 What are the contents of the adductor canal?

- Femoral artery
- Femoral vein
- Descending genicular branch of the femoral artery
- Saphenous nerve
- Nerve to vastus medialis
- Obturator nerve.

Popliteal Fossa

Q.165 What are the boundaries of the popliteal fossa?

Superolaterally: Biceps femoris tendon.
Superomedially: Semimembranosus and semitendinosus.
Inferomedially: Medial head of gastrocnemius.
Inferolaterally: Lateral head of gastrocnemius and plantaris.

See Figure 3.20.

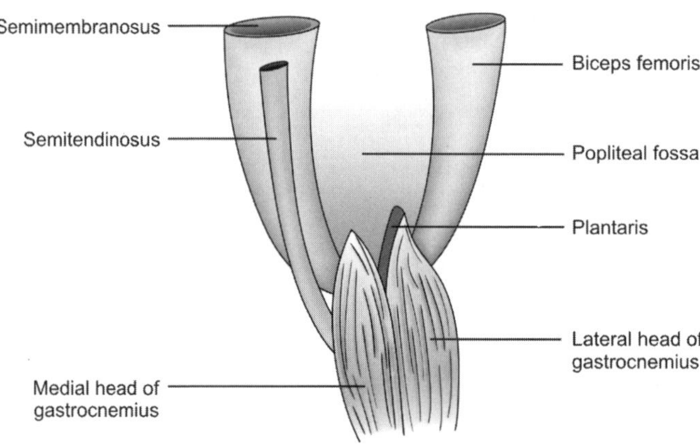

Fig. 3.20: Boundaries of the popliteal fossa

Q.166 Which structures form the floor of the popliteal fossa?

From above downwards:

- The popliteal surface of the femur
- The capsule of the knee joint
- Popliteal fascia.

Q.167 What is the relationship between the tibial nerve and popliteal vessels in the popliteal fossa?

From superficial to deep lie, the tibial nerve, popliteal vein and popliteal artery. The popliteal artery is crossed by the popliteal vein and tibial nerve posteriorly from the lateral to medial side.

Q.168 What are the contents of popliteal fossa?

- Popliteal artery and its branches
- Popliteal vein and its tributaries
- Tibial nerve and its branches
- Common peroneal nerve and its branches
- Genicular branch of obturator nerve
- Posterior cutaneous nerve of thigh
- Popliteal lymph nodes
- Fat.

Gluteal Region

Q.169 Name the structures passing through greater sciatic foramen.

- Piriformis
- Structures passing above piriformis
  - Superior gluteal nerve
  - Superior gluteal vessels
- Structures passing below piriformis
  - Inferior gluteal vessels
  - Internal pudendal vessels
  - Inferior gluteal nerve
  - Sciatic nerve
  - Posterior cutaneous nerve of thigh
  - Nerve to quadratus femoris
  - Pudendal nerve
  - Nerve to obturator internus.

Q.170 Name the structures passing through lesser sciatic foramen.

- Tendon of obturator internus
- Internal pudendal vessels
- Pudendal nerve
- Nerve to obturator internus.

Q.171 Name the structures lying under cover of gluteus minimus.

- Reflected head of rectus femoris
- Capsule of hip joint.

Q.172 What are the structures lying under cover of gluteus medius?

- Superior gluteal nerve
- Deep branch of superior gluteal artery
- Gluteus minimus
- Trochanteric bursa of gluteus medius.

Q.173 Name the structures lying under the cover of gluteus maximus.

- *Ligaments*
  - Sacrotuberous
  - Sacrospinous and
  - Ischiofemoral
- *Bones and joints*
  - Ilium
  - Ischium with ischial tuberosity
  - Upper end of femur with greater trochanter
  - Sacrum
  - Coccyx
  - Hip joint
  - Sacroiliac joint.
- *Bursae*
  - Trochanteric bursa of gluteus maximus
  - Bursa over ischial tuberosity and
  - Bursa between gluteus maximus and vastus lateralis.
- *Muscles*
  - Gluteus medius
  - Gluteus minimus
  - Reflected head of rectus femoris
  - Piriformis
  - Obturator internus
  - Superior and inferior gemelli
  - Quadratus femoris
  - Obturator externus
  - Origin of hamstrings
  - Insertion of adductor magnus.

- *Vessels*
  - Superior gluteal vessels
  - Inferior gluteal vessels
  - Internal pudendal vessels
  - Ascending branch of medial circumflex femoral artery
  - Trochanteric anastomosis
  - Cruciate anastomosis
  - First perforating artery.
- *Nerves*
  - Superior gluteal ($L_{4,5}$ S_1)
  - Inferior gluteal (L_5, $S_{1,2}$)
  - Sciatic ($L_{4,5}$ $S_{1,2,3}$)
  - Posterior cutaneous nerve of thigh ($S_{1,2,3}$)
  - Nerve to quadratus femoris ($L_{4,5}$ S_1)
  - Pudendal nerve ($S_{2,3,4}$)
  - Nerve to obturator internus (L_5, $S_{1,2}$)
  - Perforating cutaneous nerve ($S_{2,3}$).

Q.174 What is waddling gait?

Results from bilateral paralysis of gluteus medius and minimus so that the patient walks with swaying to clear the feet off the ground. When unilateral then it is known as *lurching gait.*

Leg and Foot

Q.175 Name the bony prominences felt in the leg and foot.

- Medial and lateral condyles of tibia.
- *Tibial tuberosity:* In front of upper part of tibia, 2.5 cm below the line passing between tibia condyles.
- *Head of fibula:* Posterolaterally at level of tibial tuberosity.
- Anterior border and medial surface of tibia.
- *Medial malleolus of tibia:* On medial side of ankle.
- Lateral malleolus of fibula.
- *Peroneal trochlea:* About a finger breadth below lateral malleolus.
- *Sustentaculum tali:* About a finger breadth below medial malleolus.
- *Tuberosity of navicular bone:* 2.5 to 3.5 cm anteroinferior to medial malleolus.
- *Tuberosity of base of fifth metatarsal:* On lateral border of foot.

Q.176 What are the parts of deep fascia of leg?

- *Intermuscular septa:*
  - *Anterior and posterior intermuscular septa:* Divide leg into three compartments anterior, lateral and posterior.
  - *Superficial transverse fascial septum:* Separates superficial and deep muscles of back of leg. Also forms flexor retinacula.
  - *Deep transverse fascial septum:* Separates tibialis posterior from long flexors of toes.
- *Retinacula:*
  - *Extensor retinacula:* Superior and inferior.
  - *Peroneal retinacula:* Superior and inferior.

Q.177 What are the attachment of inferior extensor retinacula?

It is a Y-shaped retinacula.

- *Stem:* Attached to anterior and articular part of superior surface of calcaneum.
- *Upper band:* Attached to anterior border of medial malleolus.
- *Lower band:* Attached to plantar aponeurosis.

Q.178 Name the structures passing deep to inferior extensor retinacula.

- Tibialis anterior
- Extensor hallucis longus
- Deep peroneal nerve
- Anterior tibial vessels.

Q.179 Name the muscles of posterior compartment of leg.

Superficial muscles:
- Gastrocnemius,
- Soleus and
- Plantaris.

Deep muscles:
- Popliteus,
- Flexor digitorum longus,
- Flexor hallucis longus and
- Tibialis posterior.

Q.180 Name the structures passing under the flexor retinaculum.

From medial to lateral and above downwards are:
- Tibialis posterior tendon
- Flexor digitorum longus tendon
- Posterior tibial vessels
- Tibial nerve
- Flexor hallucis longus tendon.

Q.181 What is tendocalcaneus?

It is a long tendon, receiving the insertion of fibers of soleus, gastrocnemius, both medial and lateral head.

Q.182 What is the insertion of tibialis anterior?

Tibialis anterior is inserted into medial side of medial cuneiform and base of first metatarsal.

Q.183 Where is peroneus longus inserted?

It is inserted into lateral side of medial cuneiform and base of first metatarsal.

Q.184 Name the muscles found in different layers of sole of foot.

From without inwards:

First layer:
- Flexor digitorum brevis
- Abductor hallucis
- Abductor digiti minimi.

Second layer:
- Flexor digitorum accessorius
- Lumbricals: Four in number

Third layer:
- Flexor hallucis brevis
- Flexor digiti minimi brevis
- Adductor hallucis

Fourth layer: Three plantar and four dorsal interossei.

Q.185 What is plantar aponeurosis and what are its functions?

It is the thickened central part of the deep fascia of sole.

Functions:
- Provides attachment to skin of sole.
- Gives origin to muscles of first layer of sole.
- Protects the digital vessels and nerves and deeper muscles.
- Helps in maintaining the longitudinal arch of the foot.

Q.186 What are the functions of interossei of sole?

- *Dorsal interossei:* Abductors of the toes.
- *Plantar interossei:* Adductors of the toes.

Arterial Supply of Lower Limb (Fig. 3.21)

Q.187 What are the branches of femoral artery.

- The branches of femoral artery (Fig. 3.22) can be either superficial or deep. The superficial branches include:
  - Superficial circumflex iliac artery
  - Superficial epigastric artery
  - Superficial external pudendal artery
- The deep branches include:
  - Deep external pudendal artery
  - Profunda femoris artery
  - Descending genicular artery

Q.188 What is the extent of femoral artery?

It begins at mid inguinal point and ends at medial side of middle and lower one-third of thigh by passing through an aperture in adductor magnus muscle to reach back of thigh and become popliteal artery.

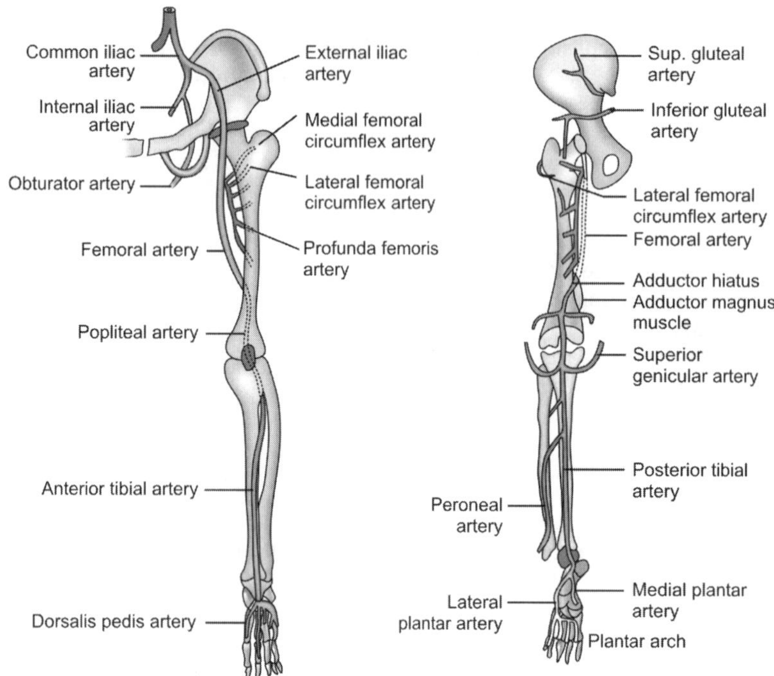

Fig. 3.21: Arterial supply of lower limb

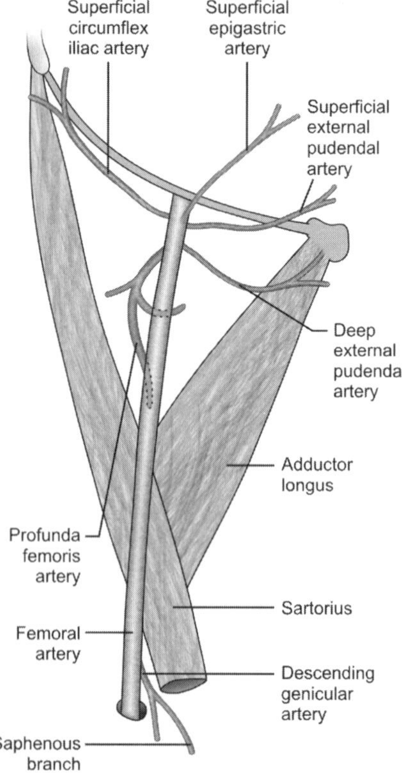

Fig. 3.22: Branches of femoral artery

Q.189 Name the branches of profunda femoris.

- Lateral circumflex femoral artery
- Medial circumflex femoral artery

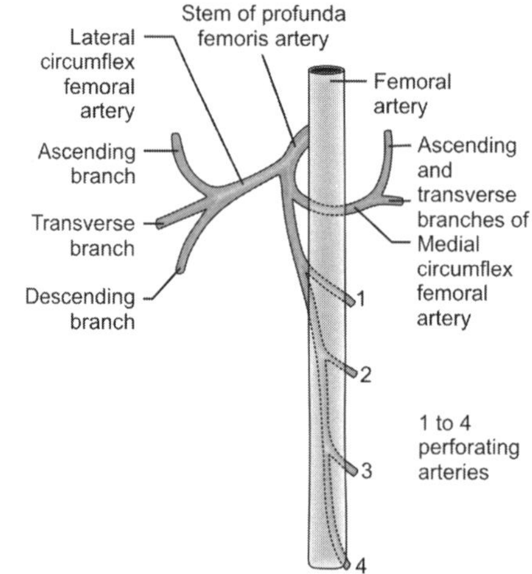

Fig. 3.23: Branches of the profunda femoris artery

- Perforating arteries
- Muscular branches (Fig. 3.23).

Q.190 Name the arteries forming the cruciate anastomosis.

- Inferior gluteal artery
- First perforating artery
- Transverse branch of medial circumflex femoral artery
- Transverse branch of lateral circumflex femoral artery.

Q.191 Name the arteries forming the trochanteric anastomosis.

- Descending branch of superior gluteal artery
- Ascending branch of medial circumflex femoral artery
- Ascending branch of lateral circumflex femoral artery.

Q.192 How the circulation is maintained in case of blockage of femoral artery?

In blockage in proximal part, circulation is maintained through cruciate and trochanteric anastomosis. When blockage is in lower thigh then circulation is maintained through perforating branches of profunda femoris artery and its anastomoses with branches of the popliteal artery.

Q.193 Name the branches of popliteal artery.

- Cutaneous branches
- *Superior muscular branches:* To adductor magnus and hamstrings
- *Sural arteries:* To gastrocnemius, soleus and plantaris.
- *Superior genicular arteries:* Medial and lateral
- Middle genicular artery
- *Inferior genicular arteries:* Medial and lateral
- *Terminal branches:* Anterior and posterior tibial (Fig. 3.24).

Q.194 What are the relations of anterior tibial artery in anterior compartment of leg?

- *Relation to muscles:* In upper 1/3, lies between tibialis anterior and extensor digitorum longus. In middle 1/3, lies between tibialis anterior and extensor hallucis longus. In lower 1/3, lies between extensor hallucis longus and extensor digitorum longus.

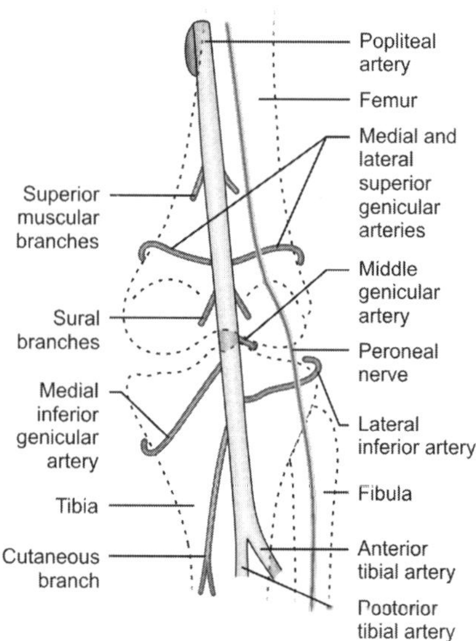

Fig. 3.24: Branches of popliteal artery

- *To veins:* Artery is accompanied by two venae comites.
- *To nerve:* Deep peroneal nerve is lateral to it in upper 1/3 and lower 1/3 and anterior to it in middle 1/3.

Q.195 What are the branches of anterior tibial artery?
- Muscular branches.
- *Recurrent branches:* Anterior and posterior tibial
- *Malleolar branches:* Anterior medial and anterior lateral.

Q.196 How dorsalis pedis artery is formed?
It is the continuation of anterior tibial artery in front of ankle between the two malleoli.

Q.197 Name the branches of dorsalis pedis artery.
- Lateral tarsal artery
- Medial tarsal artery
- Arcuate artery
- First dorsal metatarsal artery.

Q.198 Where the pulsations of dorsalis pedis artery are felt?
Between the tendon of extensor hallucis longus and first tendon of extensor digitorum longus on dorsum of foot about 5 cm distal to medial and lateral malleoli, over intermediate cuneiform bone.

Q.199 Name the branches of posterior tibial artery.
- Peroneal: Largest branch
- Muscular
- Nutrient artery to tibia
- *Anastomotic branches:*
  - Circumflex fibular
  - Communicating branch to peroneal
  - Malleolar
  - Calcaneal
- *Terminal branches:* Medial and lateral plantar.

Q.200 To which bone peroneal artery gives a nutrient artery?
Fibula

Q.201 Which artery forms the plantar arch?
Lateral plantar artery

Q.202 How the lateral plantar artery terminates?
It ends by joining termination of dorsalis pedis artery in interval between bases of first and second metatarsal bone.

Also see page 46 Femoral triangle and page 48 Popliteal fossa.

Venous Drainage

Q.203 What are the different factors which facilitate the return of venous blood to heart?
- *Local factors:*
  - Veins of lower limb are larger than veins of other parts of body. They also have greater number of valves, which prevent the back flow of blood.
  - Muscular contraction, compresses the deep veins and drives the blood upwards.
  - Muscular compression of veins is made more effective by tight deep fascia.
- *General factors:*
  - The valves which maintain a unidirectional flow.
  - Negative intrathoracic pressure, which pulls the column of blood up and it is made more negative during inspiration.
  - Vis-a-tergo (compulsion from behind) produced by arterial pressure and over flow from capillary bed.

Q.204 What are the main superficial veins of lower limb?
- *Great saphenous vein:* Continuation of medial marginal vein of foot. It ascends into thigh and after passing through saphenous opening in deep fascia ends in femoral vein. It receives superficial epigastric, superficial circumflex iliac, external pudendal, anterior vein of leg and posterior arch veins.
- *Anterior cutaneous vein of thigh:* Drains front of lower part of thigh and it drains into great saphenous vein.
- *Short saphenous vein:* Continuation of lateral marginal vein of foot and ends in popliteal vein above knee joint.
- *Perforating veins:* These are the veins connecting superficial veins with the deep veins after perforating the deep fascia. They permit only unidirectional flow of blood, from superficial to deep veins by means of valves. These are present both in thigh and leg, but a number of these are present in lower one-third of leg.

Q.205 What is 'calf pump' or 'peripheral heart'?
In upright position, venous return from lower limb depends largely on the contraction of calf muscles, these are known as calf pump, the soleus is called "peripheral heart" for same reason.

Q.206 What are varicose veins?
If the valves in veins become incompetent, the pressure during muscular contraction is transmitted from deep veins to the superficial veins and hence, leakage of blood. This causes dilatation of the superficial veins, known as varicose veins. Later on gradual degeneration occurs, leading to "varicose ulcers".

Q.207 What is the clinical importance of sural sinuses?
Sural sinuses are the common site for thrombosis and commonly leads to pulmonary embolism due to the detachment of thrombus.

Lymphatic Drainage of Lower Limb

Q.208 What is the lymphatic drainage of various inguinal lymph nodes?
- *Upper lateral superficial group:* Drains skin of anterior abdominal wall below umbilicus.
- *Upper medial superficial group:* Drains skin of anterior abdominal wall below umbilicus, external genitalia except glans penis or clitoris, lower part of anal canal and lower part of vagina and some lymphatics from inguinal canal.
- *Lower superficial inguinal group:* Drains superficial lymphatics of lower limb except from back of leg.
- *Deep inguinal group:* Drains deep lymphatics of thigh, glans penis or clitoris and popliteal lymph nodes.
- *Popliteal lymph nodes:* Drains deep lymphatics of foot and leg and superficial lymphatics of back of leg.

All lymphatics from inguinal nodes drain into external iliac lymph nodes.

Nerves of Lower Limb

Lumbar Plexus (Fig. 3.25)

Q.209 How lumbar plexus is formed?
By the ventral rami L_{1-3} and greater part of ventral ramus of L_4. The first lumbar nerve also receives a branch from T_{12} nerve.

Q.210 What is the distribution of obturator nerve?
- *Anterior branch* supplies:
  - *Muscular branches*: To adductor longus, gracilis, obturator externus and occasionally adductor brevis and pectineus.

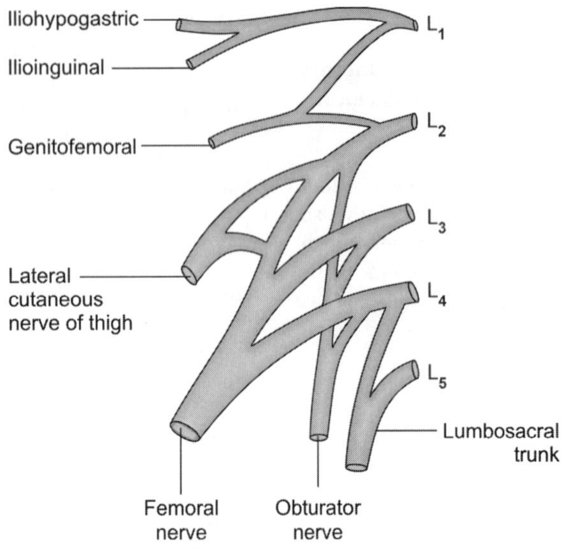

Fig. 3.25: Lumbar plexus

- *Articular*: To hip joint.
- *Cutaneous*: To subsartorial plexus
-. *Vascular branches*: To femoral artery
- *Posterior branch* supplies:
  - *Muscular branches:* To obturator externus, adductor magnus and adductor brevis.
  - *Articular:* To knee joint.

Q.211 Name the branches of femoral nerve.

- *Anterior division* supplies:
  - Nerve to pectineus
  - Intermediate cutaneous nerve of thigh
  - Medial cutaneous nerve of thigh
  - Nerve to sartorius
  - Nerve to iliacus
- *Posterior division* supplies:
  - Saphenous nerve
  - Muscular branches to quadriceps femoris
  - Vascular branches to femoral artery
  - Articular branches to hip and knee joint

Q.212 Name the nerves forming the subsartorial plexus.

- Medial cutaneous nerve of thigh
- Saphenous nerve
- Cutaneous branch of anterior division of obturator nerve.

Q.213 Name the nerves forming the patellar plexus.

- Saphenous nerve
- Medial, intermediate and lateral cutaneous nerve of thigh
- Saphenous nerve and its infrapatellar branch

Q.214 What is 'meralgia paresthetica'?

It is a clinical condition characterized by pain, tingling, numbness or anesthesia in the area of distribution of the lateral cutaneous nerve of the thigh. This nerve (a branch of the lumbar plexus) usually enters the thigh, passing deep to the inguinal ligament. Occasionally, the nerve pierces the ligament and may then be compressed by it with resultant irritation of the nerve.

Q.215 How can the pain of the adductor spasm be relieved?

By division of the obturator nerve.

Q.216 Why does a patient sometimes complain of pain in the knee when the disease is actually in the hip joint?

This is referred pain because both the hip and knee joints are supplied by the same nerves, i.e. the femoral and obturator nerves.

Sacral Plexus

Q.217 What are the branches of sacral plexus?

- Sciatic nerve ($L_{4,5}$ $S_{1\text{-}3}$)
- Superior gluteal nerve (Posterior division of $L_{4,5}$ S_1)
- Inferior gluteal nerve (Posterior division of L_5, $S_{1,2}$)
- Perforating cutaneous nerve (Posterior division of $S_{2,3}$)
- Nerve to piriformis (Posterior division of $S_{1,2}$)
- Pudendal nerve (Anterior division of $S_{1\text{-}3}$)
- Posterior cutaneous nerve of thigh (Anterior division of $S_{1,2}$ and posterior division of $S_{2,3}$)
- Nerve to obturator internus (Anterior division of L_5, $S_{1,2}$)
- Nerve to quadratus femoris (Anterior division fo $L_{4,5}$ S_1)
- Nerve to levator ani and coccygeus and sphincter ani externus from S_4 branches
- Pelvic splanchnic nerve from $S_{2\text{-}4}$

Q.218 How the sciatic nerve is formed? What are its branches?

The sciatic nerve is the continuation of the sacral plexus and derives its fibers from the $L_{4,5}$, $S_{1,2,3}$. It is the largest nerve in the body. The main trunk of the sciatic nerve is the nerve of the flexor compartment of the thigh.

Branches:

- *Articular:* To hip joint.
- *Muscular:* To biceps femoris, semitendinosus, semimembranosus and ischial head of adductor magnus.
- Terminal:
  - The *tibial nerve* is the nerve of the flexor compartments of the thigh (through the parent trunk), leg and sole of the foot. It receives fibers from the anterior divisions of $L_{4,5}$ $S_{1,2}$ and S_3 (which does not divide into anterior and posterior division)
  - The *common peroneal nerve* is the nerve of the extensor and peroneal compartments of the leg and dorsum of the foot. It is derived from the posterior divisions of $L_{4,5}$ $S_{1,2}$.

Q.219 Give the surface marking of the sciatic nerve.

- The sciatic nerve is represented by a thick line (2 cm broad) joining the following three points.
- The *first point* is taken 2.5 cm lateral to the mid-point of a line joining the posterior superior iliac spine (marked by a dimple lateral to the natal cleft) and the ischial tuberosity.
- The *second point* is taken at the mid-point between the greater trochantar of the femur and the ischial tuberosity.
- The *third point* is taken at the mid-point of a transverse line drawn at the junction of the middle and lower 2/3 of the back of the thigh, i.e. apex of the popliteal fossa.

Q.220 What will be the effect of a complete lesion of the sciatic nerve in the gluteal region?

- *Motor loss:*
  - Loss of flexion of the knee due to paralysis of the hamstring muscles, but some weak movement is possible due to the action of the sartorius (femoral nerve) and gracilis (obturator nerve).

- Loss of all movements below the knee due to paralysis of all the muscles of the leg and foot. There will be a *'foot drop'* deformity.
- Loss of achilles jerk and plantar reflex.

- *Sensory loss:* On the outer side of the leg and almost the entire foot.

Q.221 What is 'sciatica' and what is its common cause?

Sciatica is the term applied when pain is felt along the course and distribution of the sciatic nerve, i.e., in the buttock, posterior aspect of the thigh and leg and lateral aspect of the leg and foot. This is due to irritation of one or more of the roots of the sciatic nerve and commonly occurs due to a prolapsed intervertebral disc in the lumbar region.

Q.222 At what site intramuscular injections are given in gluteal region?

The injections are given in upper and outer quadrant of the gluteal region to avoid injury to the sciatic nerve.

Q.223 What is the site for local anesthetic to be injected for sciatica to relieve the pain?

The site of injection is midway between the greater trochanter of the femur and the ischial tuberosity.

Q.224 What are the branches of common peroneal nerve?

- Lateral cutaneous nerve of calf
- Communicating branch to sural nerve
- Terminal branches: Deep and superficial peroneal nerve.

Q.225 Where is the common peroneal (lateral popliteal) nerve commonly injured and what are the common causes of the injury?

The nerve is commonly injured where it winds round the neck of the fibula. It may be damaged at this site by the pressure of a tight bandage of plaster cast, in severe adduction injury to the knee or from direct trauma.

Q.226 What will be the effects of a complete section of the common peroneal (lateral popliteal) nerve at the level of the neck of the fibula?

- *Motor loss:*
  - Inability to extend the foot or toes due to paralysis of the ankle and foot extensors (tibialis anterior, extensor hallucis longus, extensor digitorum longus, peroneus tertius and extensor digitorum brevis). This results in "foot drop" which is characteristic of the common peroneal nerve injury.
  - Inability to evert the foot due to paralysis of the peroneal muscles.
  - Paralysis of the extensor and evertor muscles of the foot causes the foot to assume a position of *equino-varus* (equinus: plantar flexion, varus: inversion), results in a slapping or high steppage gait (the patient-raises the knee high and the foot hangs flexed and inverted).
- *Sensory loss:* Over the anterior and lateral aspects of the leg and foot. The lateral border of the foot and the lateral side of the little toe are unaffected since they are supplied by the sural branch of the tibial nerve.

Q.227 What are the structures supplied by deep peroneal nerve?

- *Muscular branches: To*
  - Tibialis anterior
  - Extensor hallucis longus
  - Extensor digitorum longus
  - Peroneus tertius and
  - Extensor digitorum brevis
- *Cutaneous branches:* To adjacent sides of first and second toes on dorsum of foot.
- *Articular branches:* To ankle joint, tarsal and metatarsal joints.

Q.228 What is the effect of lesion of deep peroneal nerve?

- *Sensory loss:* Adjacent sides' of I and II toe.
- *Motor loss:* Paralysis of muscles supplied by it. So over activity of peroneal and flexor muscles leads to *talipes equinovalgus.*

Q.229 Name the branches of superficial peroneal nerve.

- *Muscular branches:* To peroneus longus and peroneus brevis.
- *Cutaneous branches:* To lower 1/3 of lateral side of leg and dorsum of foot supplying medial side of I toe, lateral side of II toe and III, IV, V toes.
- *Communicating branches:* To sural, deep peroneal and saphenous nerve.

Q.230 What will occur if nerve supply to peroneal muscles is cut off?

Talipes varus.

Q.231 What is the distribution of tibial nerve?

- *Muscular branches* to gastrocnemius plantaris, soleus, popliteus, tibialis posterior, flexor digitorum longus, flexor hallucis longus.
- *Cutaneous branches:*
  - Sural nerve
  - Medial calcaneal branch
- *Articular branches:* To knee and ankle joint
- *Terminal branches:* Medial and lateral plantar nerves

Q.232 What is the distribution of medial plantar nerve?

- *Cutaneous branches:*
  - From trunk, skin to medial part of sole
  - Skin on medial side of great toe
  - Three plantar digital nerves to medial 3½ digits
- *Muscular branches:*
  - From trunk to abductor hallucis and flexor digitorum brevis.
  - From digital nerve to great toe to flexor hallucis brevis
  - From first plantar digital nerve to first lumbrical
- *Articular branches:*
  - Tarsal and tarsometatarsal joints from main trunk
  - Metatarsophalangeal and inter-phalangeal joints from digital nerves.

Q.233 What is the distribution of lateral plantar nerve?

- *Cutaneous branches:*
  - From trunk to skin of lateral part of sole
  - Digital branches to lateral 1½ toes.
- *Muscular branches:*
  - From trunk to flexor digitorum accessorius and abductor digiti minimi.
  - Digital branch to lateral side of fifth toe supplies flexor digiti minimi, 3rd plantar and 4th dorsal interossei
  - Deep branch to adductor hallucis, 2nd, 3rd and 4th lumbricals, all interossei except above.

Q.234 Where is the tibial (medial popliteal) nerve commonly injured and what are the common causes of the injury?

The tibial nerve may be damaged in or below the popliteal fossa by automobile accident, fractures of leg or by gunshot or stab wounds. The frequency of injuries to the tibial nerve is far less than the common peroneal nerve because of its deeper position and more protected course.

Q.235 What will be the effects of a complete section of the tibial (medial popliteal) nerve in the popliteal fossa?

- *Motor loss:*
  - Inability to fully flex the ankle joint due to paralysis of the gastrocnemius and soleus. A small degree of flexion is possible by the peroneus longus (which is supplied by the superficial peroneal nerve).

- Inability to invert the foot against resistance due to paralysis of the tibialis posterior.
- The foot assumes the position of a *calcaneo-valgus* (calcaneus: dorsiflexion, valgus: eversion) by the unopposed action of the extensors and evertors. The patient cannot stand on tip-toe. Walking is difficult due to difficulty in taking off.
- Inability to flex the toes due to paralysis of both the long and short flexors of the toes.
- Ankle jerk is absent.

• *Sensory loss* over the sole (except the inner border).

• *Vasomotor and trophic changes* are common. The foot becomes edematous, discolored and cold. Trophic ulcers are almost inevitable.

Q.236 What is the cutaneous nerve supply of back of leg?

• *Saphenous nerve* ($L_{3,4}$): Branch of posterior division of femoral nerve. Supplies skin of medial area of leg and medial border of foot upto ball of I toe.
• *Posterior division of medial cutaneous nerve of thigh* ($L_{2,3}$): Supplies upper most part of medial area of calf.
• *Posterior cutaneous nerve of thigh* ($S_{1,2,3}$): Supplies upper ½ of central area of calf.
• *Sural nerve* (L_5,$S_{1,2}$): Branch of tibial nerve. Supplies lower ½ of central area and lower 1/3 of lateral area of calf and lateral border of foot.
• *Lateral cutaneous nerve of calf* ($L_{4,5}$ S_1): Branch of common peroneal nerve. Supplies skin of upper 2/3 of lateral area of leg.
• *Peroneal (sural) communicating nerve* ($L_5S_{1,2}$): Branch of common peroneal nerve. Supplies skin of lateral area of calf.
• *Medial calcanean branches* ($S_{1,2}$): Supplies skin of heel and medial side of sole of foot.

DO YOU KNOW?

• Artery of ligamentum teres branch of obturator artery is important in children as it supplies the head of femur proximal to epiphyseal growth plate. Once this growth plate closed as in adults this artery is of no significance.
• Femoral neck fracture most commonly occurs in elderly woman who have osteoporosis. As a result, the lower limb is externally rotated and shorter than the uninjured limb.
• Femoral artery is commonly used for percutaneous arterial catheterization because it is easily palpated and also hemostatis can achieved easily by applying pressure even the head of femur.
• Common peroneal nerve usually get lesion in the lower limb. It is the most common nerve to be injured.
• In diabetic patients, the anterior tibial artery, posterior tibial artery and peroneal artery are susceptible to chronic occlusion.

Chapter 4

Thorax

Thoracic Cage

Q.1 How thoracic cage is formed?
Anteriorly: Sternum

Posteriorly: Twelve thoracic vertebrae and intervertebral discs

One each side: Twelve ribs with their cartilages.

Q.2 What variations occur in thorax with age?
- In adults, in transverse section thorax is reniform, with a greater transverse diameter than anteroposterior. In infants, circular in transverse section.
- In adults, ribs are oblique. In infants, ribs are horizontal.

Q.3 What are the boundaries of 'thoracic inlet'?
- *Anteriorly:* Upper border of manubrium sterni.
- *Posteriorly:* Upper border of body of T_1 vertebra.
- *One each side:* First rib with its cartilage.

Q.4 What is the direction of plane of inlet of thorax?
Downwards and forwards with a obliquity of about 45 degrees. The upper border of manubrium sterni lies at level of upper border of T_3 vertebra.

Q.5 What is Sibson's fascia and what are its attachments?
It is a triangular membrane at thoracic inlet (diaphragm of inlet of thorax).
Attachments:
- *Apex:* Tip of transverse process of C_7 vertebra.
- *Base:* Laser border of first rib and its cartilage.
- *Inferior surface:* Fused with cervical pleura.

Q.6 Name the structures passing through thoracic inlet.
See Figure 4.1.

Muscles:
- Sternohyoid
- Sternothyroid and
- Longus colli.

Arteries:
- Right and left internal thoracic arteries
- Brachiocephalic artery
- Left common carotid
- Left subclavian and
- Right and left superior intercostal arteries.

Nerves:
- Left recurrent laryngeal nerve
- Right and left phrenic nerve
- Right and left vagus nerve
- Right and left first thoracic nerve and
- Right and left sympathetic chain.

Veins:
- Right and left brachiocephalic vein
- Right and left posterior intercostal vein and
- Inferior thyroid veins.

Others:
- Thymus
- Trachea
- Esophagus
- Anterior longitudinal ligament
- Right and left pleura and
- Apex of right and left lung.

Q.7 What are the boundaries of outlet of thorax?
- *Anteriorly:* Infrasternal (subcostal) angle between two costal margins.
- *Posteriorly:* Inferior surface of body of 12th thoracic vertebra.
- *On each side:*
  - Costal margin formed by 7th, 8th, 9th and 10th ribs and
  - 11th and 12th ribs.

Ribs

Q.8 What are 'true ribs'?
First seven ribs connected through costal cartilages to sternum are called true ribs.

Q.9 What are 'false ribs'?
- Last five ribs are known as false ribs.
- Cartilages of 8th, 9th and 10th ribs are joined to each other and form *costal margin.*
- Anterior ends of 11th and 12th ribs are free and are called *'floating ribs.'*

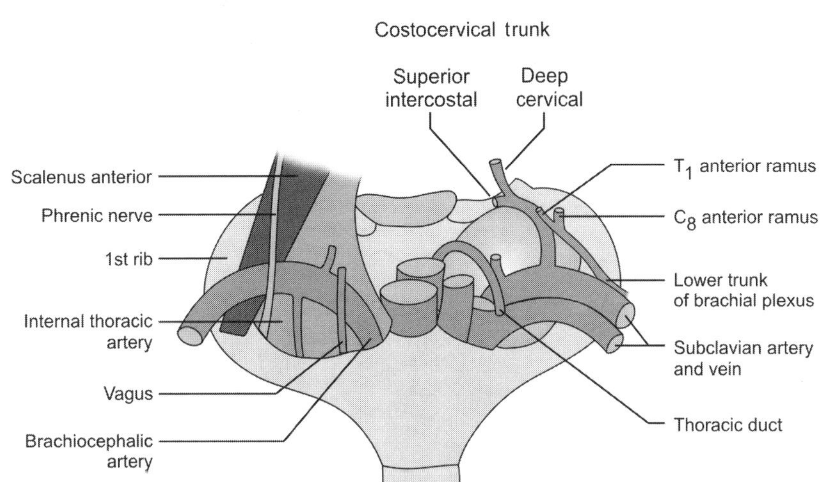

Fig. 4.1: Thoracic inlet

Q.10 What are typical and atypical ribs?
First two and last three ribs are called atypical because they present special features. The 3rd to 9th ribs are called typical because they have common features.

Q.11 What are the features of a typical rib?
Each typical rib has
- *Anterior end:* Oval and articulates with costal cartilage.
- *Posterior end:* It is made up of:
  - *Head:* Has two articular facets for articulation with vertebrae.
  - *Neck:* Has anterior and posterior surfaces and superior and inferior borders.
  - *Tubercle:* Medial part is articular.
- *Shaft:* Has outer and inner surfaces and upper and lower borders.

Q.12 What are the relations of head of typical rib?
- Sympathetic chain and
- Costal pleura.

Q.13 What is costal groove? Name the structures attached and lying with in the costal groove?
Costal groove is a depression present between inferior border and ridge on inner surface.

Attachments: Origin of internal intercostal muscle from floor of groove.

Contents: From above downwards:
- Intercostal vein
- Intercostal artery and
- Intercostal nerve.

Q.14 What are the special features of 1st rib?
- It is flattened from above downwards.
- It is shortest, broadest and most curved rib.
- Twist at angle of shaft is absent.
- Head has one articular facet.
- It has no costal groove (Fig. 4.2).

Q.15 Name the structures related to neck of first rib.
Anteriorly from medial to lateral side:
- Sympathetic trunk with cervicothoracic ganglion
- First posterior intercostal vein
- Superior intercostal artery and
- First thoracic nerve.

Superiorly:
- Deep cervical vessels and
- C_8 nerve.

Q.16 What are the structures related to grooves on superior surface of first rib?
- *Anterior groove:* Subclavian vein.
- *Posterior groove:* Subclavian artery and lower trunk of brachial plexus.

Q.17 What is the difference in ossification of 1st and other typical ribs?
The typical ribs have three secondary centers, one for head and two for tubercle.

The first rib has two secondary centers, one for head and one for tubercle.

Q.18 What are the special features of 2nd, 10th, 11th and 12th ribs?
Second rib:
- Sharply curved like first rib
- Shaft has no twist
- Outer surface of shaft has a rough tubercle
- Inner surface faces more downwards.

Tenth rib:
- Shorter than typical rib
- Single facet on head.

Eleventh rib:
- Short
- Neck and tubercle absent
- Anterior end pointed
- Angle slight
- Groove shallow.

Twelfth rib:
- Short
- Neck and tubercle absent
- Anterior end pointed
- Angle absent
- Groove absent

Q.19 In which rib both superior and inferior costotransverse ligaments are absent?
12th rib.

Q.20 What is the characteristic of costovertebral joint of 1st, 10th, 11th and 12th ribs?
These ribs articulate only with corresponding vertebrae.

Q.21 What is the commonest site of rib fracture in adults?
The rib is fractured at the angle, which is its weakest point.

Q.22 Why the fracture of ribs is rare in children?
Because in the children the chest wall is highly elastic.

Q.23 What is 'stove in chest'?
This is produced in severe crush injuries in which multiple rib fractures are produced along with permanent indentation of chest wall.

Q.24 What is 'flail chest'?
This results from more severe injury to chest. Multiple rib fractures at two or more sites result in unstable chest wall in which flail area is sucked inwards during inspiration and pushed out in expiration.

Q.25 What is cervical rib?
- Present in 0.5% of cases.
- It is attached to the transverse process of 7th cervical vertebra and its distal extremity is free or articulates with first thoracic rib.

Q.26 What is the clinical importance of a cervical rib?
It may press on the lower trunk of brachial plexus producing paraesthesia along ulnar border of forearm and wasting of small muscles of hand. Less commonly, vascular changes are produced due to pressure on subclavian artery.

Q.27 What is the histological structure of costal cartilage?
It is made up of hyaline cartilage.

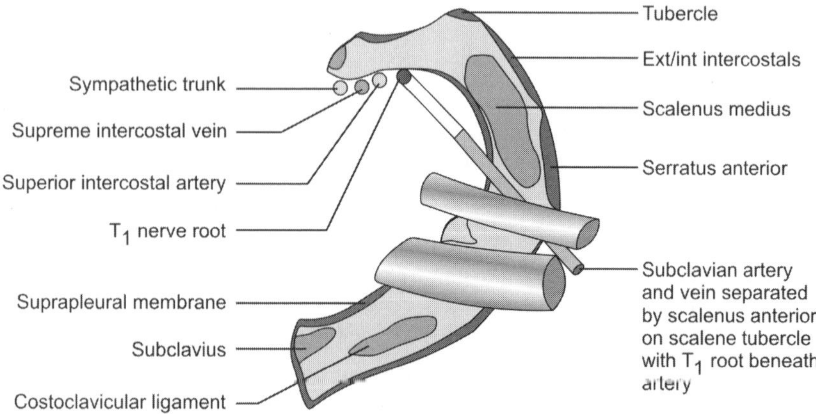

Fig. 4.2: The structures related to the first rib

Q.28 What type or joints costal cartilage form?

- *With the rib and manubrium:* Primary cartilaginous joint:
- *With the sternum:* Synovial joint.
- 8th to 10th cartilages medially are connected to each other.

Sternum

Q.29 What are the parts of sternum?

- Manubrium
- Body and
- Xiphoid process.

Q.30 What is jugular notch?

Also called suprasternal notch, present in middle of superior border of manubrium.

Q.31 What is the level of jugular notch?

It lies at level of intervertebral disc between T_2 and T_3 vertebra.

Q.32 What is sternal angle (angle of Louis)?

It is the angle formed at the junction of manubrium and body of sternum. It is convex forwards.

Q.33 What is the level of sternal angle?

Intervertebral disc between T_4 and T_5 vertebra.

Q.34 What is the clinical significance of sternal angle?

It is an important landmark for counting ribs as 2nd costal cartilages articulates with sternum of this level.

Q.35 Name the structures lying at level of sternal angle.

- Ascending aorta ends
- Arch of aorta begins and also ends
- Descending aorta begins
- Pulmonary trunk divides into two pulmonary arteries
- Marks the upper limit of base of heart
- Azygous vein opens into superior vena cava
- Trachea divides into two principal bronchi.

Q.36 Why the sternum is commonly used for getting a specimen of bone marrow?

Because cortical bone of sternum is very thin and subcutaneous. It is therefore easily accessible.

Q.37 What types of joints are present between different parts of sternum ?

Between manubrium and body (manubriosternal joint):

Secondary cartilaginous joint.

Between body and xiphisternum (xiphisternal joint):

Primary cartilaginous joint (Fig. 4.3).

Q.38 Which rib is attached to junction of body with xiphoid process?

Seventh costal cartilage

Q.39 What type of joint is sternocostal joint?

Synovial joint

Q.40 What is the level of xiphisternal joint?

9th thoracic vertebra.

Q.41 What is 'Funnel chest'?

Deformity of chest in which the body of sternum and xiphoid process is depressed. This predisposes to respiratory and cardiovascular disturbances.

Q.42 What is 'pigeon chest'?

It is condition in which deformity of chest occurs due to forward projection of sternum and flattening of chest on either side.

Q.43 What is 'ectopia cordis'?

In this, the sternum and adjoining parts of costal cartilages and ribs are missing. So that the heart can be seen from outside.

Thoracic Vertebrae

Q.44 What are the typical' and 'atypical' thoracic vertebrae?

Typical thoracic vertebrae: 2nd to 8th, they have common features.

Atypical vertebrae: 1st, 9th to 12th, they have special features.

Q.45 What are the structures attached to transverse process?

- *Lateral costotransverse ligament:* At tip.
- *Inferior costotransverse ligament:* On anterior surface.
- *Superior costotransverse ligament:* On lower border.
- *Intertransverse muscles:* Upper and lower borders.
- *Levator costae:* To posterior surface.

Q.46 What is the characteristic feature of a typical thoracic vertebrae?

It has costal facets on sides of vertebral bodies and transverse processes for articulation with ribs.

Q.47 What movement is possible in the thoracic spine?

Rotation, greater in lower thoracic region as compared to upper thoracic region.

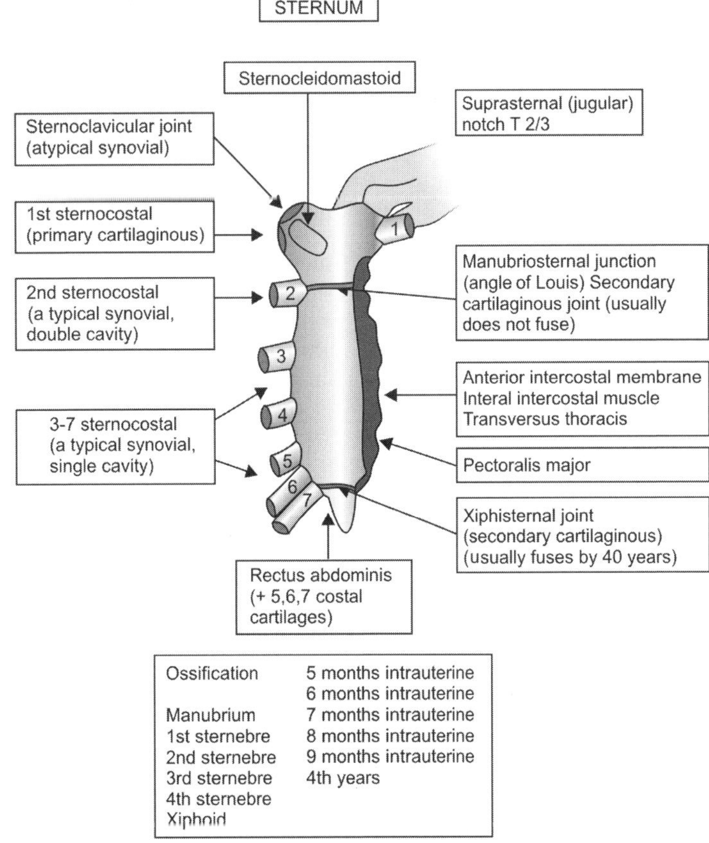

Ossification	5 months intrauterine
	6 months intrauterine
Manubrium	7 months intrauterine
1st sternebre	8 months intrauterine
2nd sternebre	9 months intrauterine
3rd sternebre	4th years
4th sternebre	
Xiphoid	

Fig. 4.3: Sternum

Intercostal Spaces

Q.48 What are intercostal spaces?

Gaps between ribs and their costal cartilages are called intercostal spaces.

Q.49 What are typical intercostal spaces?

The 3rd to 8th spaces are typical intercostal spaces. The blood and nerve supply of 3rd to 6th intercostal space is limited only to the thoracic while those of lower spaces extend into the abdomen.

Q.50 What are the contents of a typical intercostal space.

- *Muscles:*
  - External intercostal
  - Internal intercostal and
  - Transversus thoracis (innermost intercostal).
- Intercostal nerve
- Intercostal vessels and lymphatics (Fig. 4.4).

Q.51 What is the attachment and extent of external intercostal muscle?

- *Attachment:*
- *Origin:* Lower border of the rib above.
- *Insertion:* Outer lip of the upper border of the rib below.
  Fibers run downward and medially in anterior part and downwards and laterally in posterior part.
- *Extent:* From the tubercle of rib behind to its costochondral junction in front where it continues as external intercostal membrane.

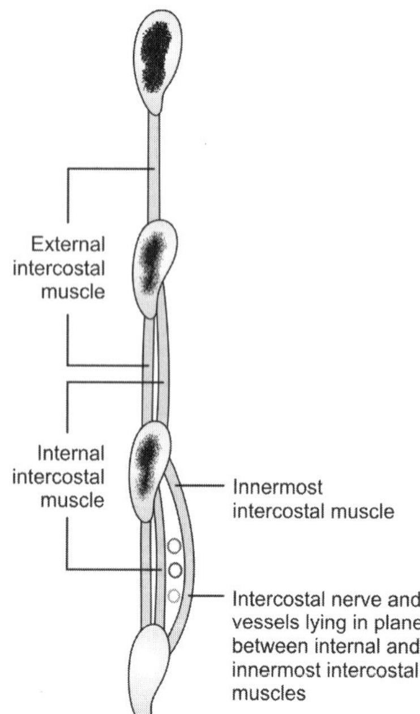

Fig. 4.4: Schematic section through intercostal spaces. The wall is gradually built up and all structures present are shown only in the lowest space

Q.52 What is the attachment and extent of internal intercostal muscle?

- *Attachment:*
  Origin: Floor of the costal groove of the rib above.
  Insertion: Inner lip of the upper border of the rib below. Fibers are at right angles to those of external intercostal.
- *Extent:* From the lateral margin of the sternum to the angle of the rib where it continues as the internal intercostal membrane.

Q.53 Name the muscles which comprise the transversus thoracic group of muscles.

- Subcostalis
- Intercostalis intimi and
- Sternocostalis.

Q.54 What is the attachment of muscles comprising transversus thoracic group of muscles?

They form the innermost layer of the muscles of the thoracic wall.

- *Subcostalis:* Present in posterior parts of the lower spaces. They are attached to the inner surface of rib near angle and to the inner surface of the second or third rib below.
- *Intercostalis intimi:* Present in the middle 2/4 of the upper spaces, except in the 1st space. They arise from inner surface of the upper rib and are inserted into the inner surface of the rib below.
- *Sternocostalis:* Present in the anterior part of the upper spaces, except in the 1st space. They arise from the lower part of the posterior surface of the body of the sternum and the xiphoid process and the adjacent costal cartilages (4th to 7th). They pass upwards and laterally and are inserted by slips to the costal cartilages of the 2nd to 6th ribs.
  The direction of fibers of these three parts is same as internal intercostal muscle.

Q.55 What is the position of neurovascular plane of thorax?

Between internal intercostal muscle and intercostalis intimi and posteriorly between pleura and internal intercostal membrane. The vein is highest, artery is in middle and nerve is lowest.

Q.56 What will happen if intercostal muscles are paralyzed?

There will be retraction of intercostal spaces during inspiration and bulging during expiration.

Q.57 How intercostal nerves are formed and how they are distributed?

These are the ventral primary rami of T_1 to T_{11} nerves. The ventral primary rami of T_{12} forms subcostal nerve. T_1 and T_2 supply the upper limb.

T_3 to T_6 supply thoracic wall (Typical intercostal nerves).

T_7 to T_{11} supply abdominal wall (Fig. 4.5).

Q.58 What are the branches of a typical intercostal nerve?

- *Communicating:*
  - White ramus communicans
    Connected to
  - Grey ramus communicans
    sympathetic ganglion.
- *Muscular:*
  - *Muscular branches:* Supplies intercostal muscles, serratus posterior superior.
  - *Collateral branch:* Supplies intercostal muscles, parietal pleura and periosteum of rib.
- *Cutaneous:*
  - *Lateral cutaneous branch:* Emerges at mid axillary line. Divides into anterior and posterior branches.
  - *Anterior cutaneous branch:* Emerges at lateral border of sternum.

Q.59 What is the characteristic feature of second thoracic nerve?

Its lateral cutaneous branch forms the *intercostobrachial nerve* which enters the upper limb and supplies the skin on medial side of upper arm.

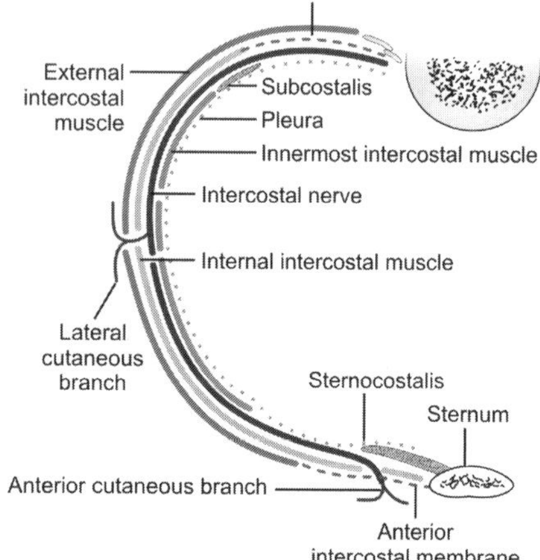

Fig. 4.5: Course and relations of a typical intercostal nerve

Q.60 Where the pain due to irritation of intercostal nerves is referred to?

To the front of chest or abdomen, i.e., at the peripheral termination of nerve.

Q.61 What is the course of pus from vertebral column around the thorax?

The pus may track along the course of neurovascular bundle and may point at the exit of cutaneous branches of intercostal nerve, i.e., lateral to erector spinae, in mid axillary line and just lateral to the sternum.

Q.62 Name the arteries of intercostal space.

- One posterior intercostal artery and
- Two anterior intercostal arteries.

Q.63 Name the branches of posterior intercostal arteries?

- Dorsal branch
- Muscular branches
- Collateral intercostal branch
- Lateral cutaneous branch and
- *Mammary branches:* Of 2nd, 3rd, and 4th arteries.
- Right bronchial artery: From right third posterior intercostal artery (Fig. 4.6).

Q.64 What is the origin of intercostal arteries?

- Posterior intercostal arteries:
  1st and 2nd: From superior intercostal artery which is a branch of costocervical trunk.
  3rd to 11th: From descending thoracic aorta.

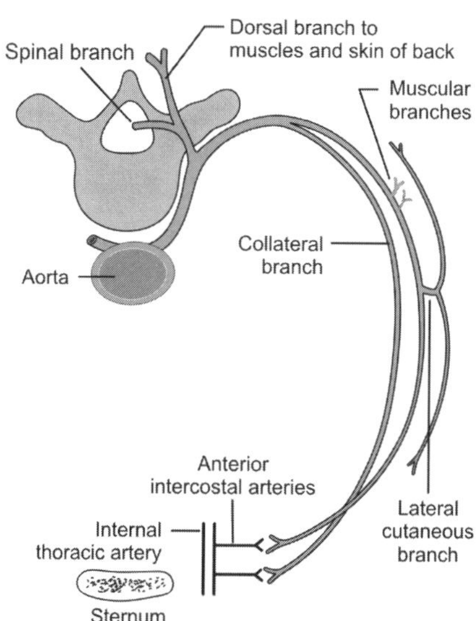

Fig. 4.6: Scheme to show course and branches of a typical posterior intercostal artery

- Anterior intercostal arteries:
  Of 1st to 6th space: From internal thoracic artery, which is a branch of first part of subclavian artery. *Of 7th to 9th space:* From musculophrenic artery, terminal branch of internal thoracic artery.
  10th and 11th spaces don't have anterior intercostal arteries.

Q.65 How the intercostal veins terminate?

- *Anterior intercostal veins:* Two in each of upper nine spaces.
  In upper six spaces: Drain into internal thoracic vein. *In lower three spaces:* Into musculophrenic vein.
- *Posterior intercostal vein:* One in each of eleven spaces. *Mode of termination is:*
  On left side:
  1st: Into left brachiocephalic vein.
  2nd and 3rd: Form left superior intercostal vein which drains into left brachiocephalic vein.
  4th to 8th: Into accessory hemiazygos vein.
  9th to 11th and subcostal veins: Into hemiazygos vein.
  On right side:
  1st: Into right brachiocephalic vein.
  2nd and 3rd: Form right superior intercostal vein which drains into azygos vein.
  4th to 11th and subcostal vein: Into azygos vein.

Q.66 Name the branches of internal thoracic artery.

- Pericardiophrenic
- Mediastinal
- Anterior intercostals of upper six spaces
- Perforating
- Superior epigastric and
- Musculophrenic (Fig. 4.7).

Q.67 What type of movement of ribs take place during the respiration?

The anterior end of ribs can move up during inspiration and down during expiration by rotation at costovertebral and costotransverse joints. This increases anteroposterior diameter of thorax. Angular movement at manubriosternal joint of upper six ribs leads to forward movement of ribs.

In inspiration, middle of rib is raised. This occurs at costotransverse and sternocostal joint at 7th to 10th ribs and increases transverse diameter of thorax. Elevation of ribs is produced by external intercostal and depression by internal intercostal during quiet breathing.

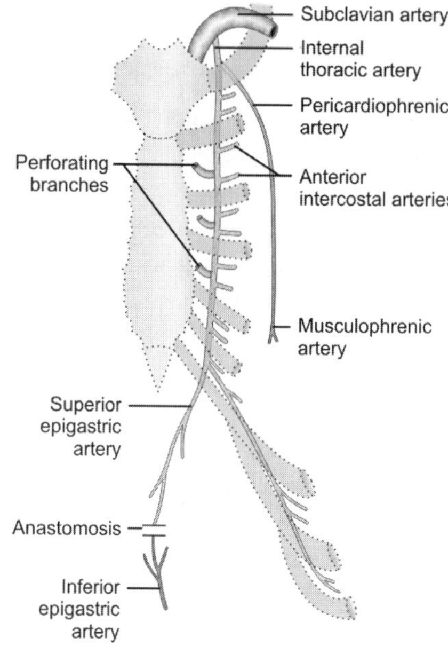

Fig. 4.7: Internal thoracic artery and its branches. The skeletal elements are drawn as if transparent

Pleurae

Q.68 What is pleura?

It is a serous membrane, lined by mesothelium. There are two pleural sacs, one on either side of mediastinum.

Q.69 What are the parts of pleura?

Outer layer: Parietal pleura.
Inner layer: Visceral pleura.

The two layers are continuous with each other at the hilum of lung. The two layers enclose between them a potential space known as *pleural cavity* (Fig. 4.8).

Q.70 What are the parts of parietal pleura?

- *Costal:* Lines the thoracic wall and is loosely attached to it by areolar tissue.
- *Diaphragmatic:* Lines upper surface of diaphragm.
- *Mediastinal:* Lines mediastinum.
- *Cervical:* Extends into neck and covers apex of lung. Covered by Sibson's fascia.

Q.71 What is the extent of cervical pleura in neck?

It extends two inches above the first costal cartilage and one inch above medial ½ of clavicle.

Q.72 What are the relations of cervical pleura?

Anteriorly: Subclavian artery and scalenus anterior muscle.

Posteriorly: Neck of first rib with its relations.

Medially: Large vessels of neck.

Laterally: Scalenus medius (Fig. 4.9).

Q.73 What is 'pulmonary ligament'? What are its functions?

It is parietal pleura surrounding the root of lung which hangs down as a fold called pulmonary ligament.

Functions:

- It provides dead space into which veins of lung can expand when venous return increases.
- Because of it, lung root can descend with the descent of diaphragm.

Q.74 What are the 'recesses of pleura'?

These are folds of parietal pleura, which act as reserve spaces, into which lungs can expand during deep inspiration.

Three in number:

- *Costomediastinal recess:* Present in cardiac notch of left lung anteriorly, between costal and mediastinal pleurae.
- *Costodiaphragmatic recess:* On both sides inferiorly, between costal and diaphragmatic pleurae (Fig. 4.10).

Q.75 What are the parts of lung not covered with visceral pleura?

At hilum and along the attachment of pulmonary ligament where it is continuous with parietal pleura.

Q.76 What is the nerve supply of pleura?

- *Parietal pleura:* Pain sensitive.
  Intercostal nerves: Costal and diaphragmatic pleurae at periphery.
  Phrenic nerves: Mediastinal and central part of diaphragmatic pleurae.
- *Visceral pleura:*
  Pain insensitive.
  Sympathetic nerves (T_2-T_5).

Q.77 What is the arterial supply of pleura?

Parietal pleura: Intercostal arteries, internal thoracic arteries and musculophrenic arteries.

Visceral pleura: Bronchial arteries.

Q.78 What is the lymphatic drainage of pleura?

Parietal pleura: Lymphatics drain into intercostal, internal mammary, mediastinal and diaphragmatic lymph nodes.

Visceral pleura: Drained by bronchopulmonary lymph nodes.

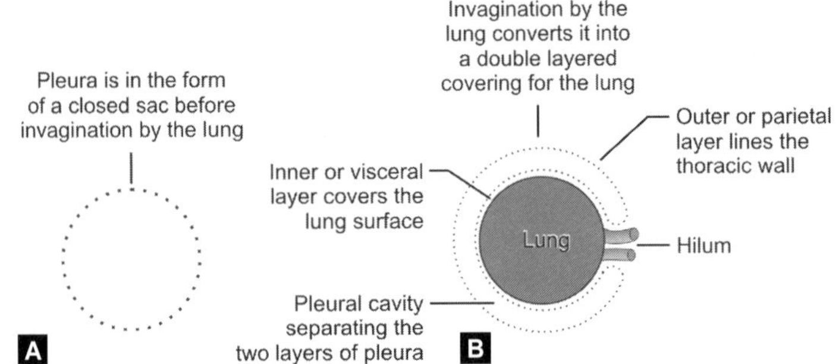

Fig. 4.8: Scheme to explain the basic relationship of the lung to pleura

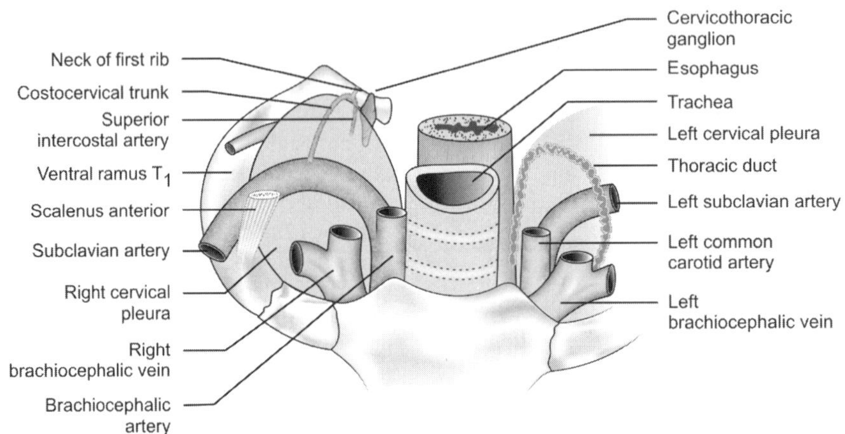

Fig. 4.9: Some structures in the root of the neck, related to the cervical pleura. Structures on the left side are shown only in part

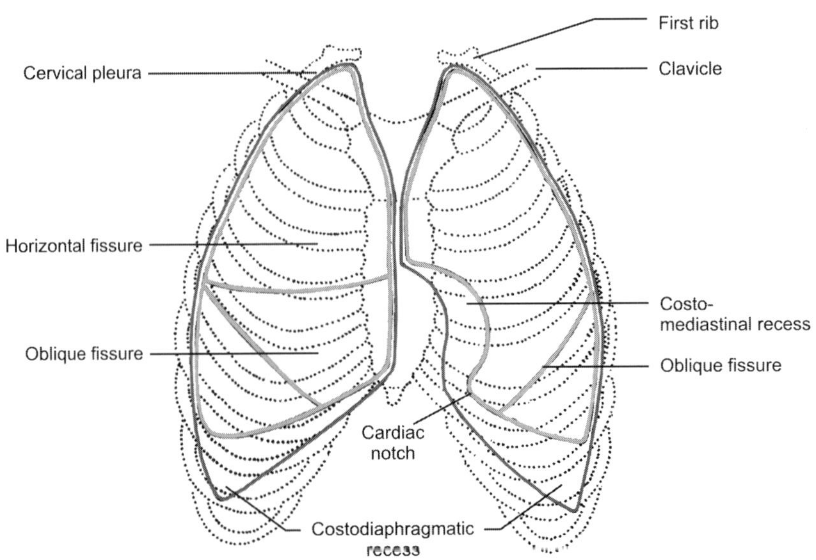

Fig. 4.10: Scheme to show the relationship of lines of pleural reflection and of the lungs, to the skeleton of the thorax

Q.79 What is the developmental origin of pleura?
Parietal pleura: Somatopleural layer of lateral plate mesoderm.

Visceral pleura: Splanchnopleural layer of lateral plate mesoderm.

Q.80 What is the surface marking of pleura?
Cervical pleura: Curved line forming a dome over the medial 1/3 of clavicle. The apex of curve lies 2.5 cm above the clavicle.

Anterior margin:
- *On right side:* From sternoclavicular joint downwards and medially to mid point of sternal angle, where it continues vertically downwards to mid point of xiphisternal joint.
- *On left side:* Same course up to fourth costal cartilage, where it arches and descends along sternal margin of 6th costal cartilage, about 3 cm from midline.

Inferior margin: Laterally from lower limit of anterior margin, so that it crosses the 8th rib in midclavicular line, 10th rib in mid axillary line and 12th rib at lateral border of sacrospinalis. Then horizontally to lower border of T_{12} vertebra, about 2 cm from midline.

Posterior margin: From a point 2 cm lateral to 12th thoracic spine to a point 2 cm lateral to 7th cervical spine.

Q.81 What are the places at which pleura descends below costal margin?
- Right costoxiphoid angle.
- Right costovertebral angle below 12th rib.
- Left costovertebral angle below 12th rib.

Q.82 What is pneumothorax?
A pneumothorax is produced by the presence of air in the pleural cavity.

Q.83 What is hemothorax?
It is the blood in the pleural cavity.

Q.84 What is empyema?
It is accumulation of pus in the pleural cavity.

Q.85 What is 'paracentesis thoracis' and from which site it is done?
It is the process of aspiration of any fluid from the pleural cavity.

Done in 6th intercostal space in mid axillary line.

Q.86 What are the structures pierced during paracentesis?
The structures pierced from outward to inwards in midaxillary line are: skin, fascia, serratus anterior muscle, intercostal muscles and parietal pleura to reach pleural cavity.

Q.87 What precaution should be taken during aspiration from pleural cavity?
Needle should be pricked in lower part of intercostal space to avoid injury to intercostal nerves and vessels in costal groove.

Lower Respiratory Tract

Trachea

Q.88 What is the extent of trachea?
Trachea extends from 6th cervical vertebra (lower border of cricoid cartilage) to lower border of 4th thoracic vertebra where it divides into right and left bronchi.

Q.89 What are the relations of cervical part of trachea?
Anteriorly:
- Isthmus of thyroid gland,
- Inferior thyroid veins below isthmus,
- Anastomosis between left and right superior thyroid arteries
- Jugular arch
- Pretracheal fascia
- Sternohyoid muscle
- Sternothyroid muscle
- Investing layer of deep cervical fascia
- Superficial fascia and
- Skin.

Posteriorly:
- Esophagus,
- Longus colli and
- Recurrent laryngeal nerve.

On each side:
- Thyroid gland
- Common carotid artery and
- Inferior thyroid artery.

Q.90 What are the relations of thoracic part of trachea.
Anterior:
- Manubrium sterni
- Sternothyroid and sternohyoid muscle
- Thymus
- Left brachiocephalic vein
- Inferior thyroid vein
- Aortic arch
- Brachiocephalic artery
- Left common carotid artery
- Deep cardiac plexus and
- Lymph nodes.

Posterior:
- Esophagus and
- Vertebral column.

Right side:
- Right pleura
- Right lung
- Right vagus
- Azygous vein
- Right brachiocephalic vein and
- Superior vena cava.

Left side:
- Arch of aorta
- Left common carotid artery
- Left subclavian artery and
- Left recurrent laryngeal nerve.

Q.91 What is the arterial supply of trachea?
Inferior thyroid arteries.

Q.92 What is the nerve supply of trachea?
Parasympathetic nerves: Vagus through recurrent laryngeal. It is
- Sensory
- Secretomotor
- Motor to tracheal muscle

Sympathetic nerves: Through middle cervical ganglion. It is vasomotor.

Q.93 What is the lymphatic drainage of trachea?
To
- Pretracheal lymph nodes and
- Paratracheal lymph nodes.

Q.94 What are the variations in the level of bifurcation of trachea with respiration?
Bifurcation of trachea, normally: Between T_4 and T_5 vertebra.

In deep inspiration: T_6 vertebra.

In expiration: T_4 vertebra.

Q.95 Where trachea can be palpated?
In suprasternal notch midway between sternal ends of two clavicles.

Q.96 What is 'tracheal tug'?
Arch of aorta lies in close relation to trachea and left bronchus. In aneurysm of aortic arch, a pull or drag is felt on the trachea which is known as 'tracheal tug'.

Q.97 How the trachea appear in an X-ray?
Since trachea is more radiolucent (because of air in it) than neighboring structures, it appears as a dark area passing downwards, backwards and slightly to the right.

Q.98 In what conditions the tracheostomy is done?
- In laryngeal obstruction.
- For removal of excessive secretions.
- For long continued artificial respiration.

Q.99 What is the commonest site for tracheostomy?
It is most commonly done in retrothyroid region after cutting the isthmus of thyroid

gland. Usually the second and third tracheal rings are cut.

Q.100 Why the tracheostomy is difficult and dangerous in children?

Because

- Neck is relatively short and left innominate vein may come up above suprasternal notch.
- Trachea is softer and more mobile, so it is not readily identified and isolated.

Q.101 What is the histological structure of trachea?

Trachea consists of following layers from within outwards:

- *Mucosa:* Lined by pseudostratified columnar ciliated epithelium. Lamina propria has mainly reticular fibers.
- *Submucosa:* Loose areolar tissue.
- *Cartilages and muscles:* C-shaped hyaline cartilages make the framework of trachea.

  Posterior gap has transverse fibers of smooth muscles and fibroelastic membrane.
- Adventitia.

Q.102 What is the advantage of posterior gap in tracheal cartilage?

Posteriorly, the oesophagus lies close to trachea so the esophagus can dilate into posterior membranous part, during passage of food bolus.

Bronchi

Q.103 What are the differences between right and left main bronchus?

Right main bronchus	*Left main bronchus*
• Wider, shorter (2.5 cm) more vertical	Narrower, longer (5 cm) and less vertical.
• Passes to root of lung at T5.	Passes to root of lung at T6
• Divides into 3 lobar bronchi	Divides into 2 lobar bronchi
• Relations:	
– Azygos vein arches over it from behind to reach superior vena cava	– It passes below the arch of aorta, in front of esophagus and descending aorta
– Right pulmonary artery lies first below and then anterior to it.	– Left pulmonary artery lies first anterior and then above it.
– Right upper pulmonary vein covers right principal bronchus	– Left upper pulmonary vein crosses in front of the bronchus.

Q.104 Why the foreign bodies and aspirated material tend to pass into right bronchus rather that into left?

Because of the greater width and more vertical course of the right bronchus.

Lungs (Figs 4.11 to 4.12A to E)

Q.105 What is a bronchopulmonary segment?

The main bronchus on each side gives off branches to each lobe of the lung, *lobar bronchus.* Each lobar bronchus then divides into *segmental bronchi,* each of which supplies a segment of the lung called a *bronchopulmonary segment.* Each segmental bronchus is accompanied by a branch of the pulmonary artery and a tributary of the pulmonary vein. The arteries lie posterolateral to the corresponding bronchi. Pulmonary veins tend to run between adjacent bronchopulmonary segments, therefore each vein may drain more than one segment.

Each bronchopulmonary segment is therefore a self contained, functionally independent respiratory unit of lung tissue.

These segments are wedge shaped with their apices at the hilum and bases at the lung surface. Each is surrounded by connective tissue continuous with that of the visceral pleura. There are also veins which run between the segments and are called intersegmental veins.

Q.106 Name the bronchopulmonary segments of the two sides.

Each bronchopulmonary segment receives its name from that of its supplying segmental bronchus.

Right side	*Left side*
Upper lobe • Apical segment • Posterior segment • Anterior segment *Middle lobe* • Lateral segment • Medial segment *Lower lobe* • Apical (superior) segment • Medial basal (cardiac) segment • Anterior basal segment • Lateral basal segment • Posterior basal segment	*Upper lobe* • Apico-posterior • Segment • Anterior segment *Lingular* • Superior segment • Inferior segment *Lower lobe* • Apical (superior) segment • Anterior basal segment • Lateral basal segment • Posterior basal segment

On the left side, the upper lobe bronchus gives off a combined apico-posterior segmental bronchus whereas on the right they arise separately as apical and posterior segmental bronchi.

The left upper lobe has a lingular segment which is equivalent to the right middle lobe.

On the right side there is a medial basal segmental bronchus which is absent on the left.

Q.107 Why the knowledge of the bronchial tree and bronchopulmonary segments is important?

It is important:

- During bronchoscopy
- For correct interpretation of bronchograms
- To determine the appropriate posture for promoting drainage of infected areas of lung

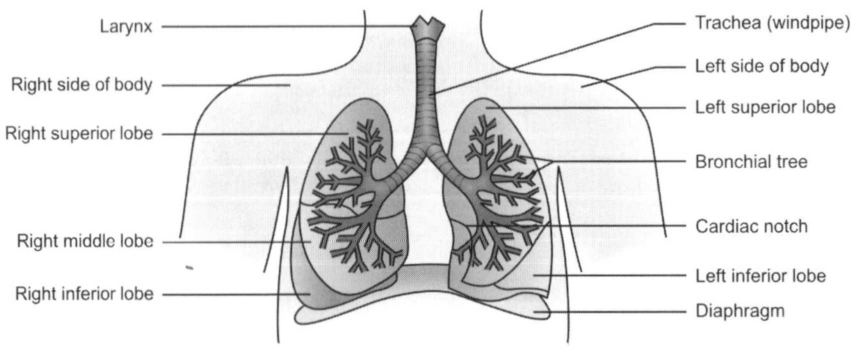

Fig. 4.11: The respiratory system

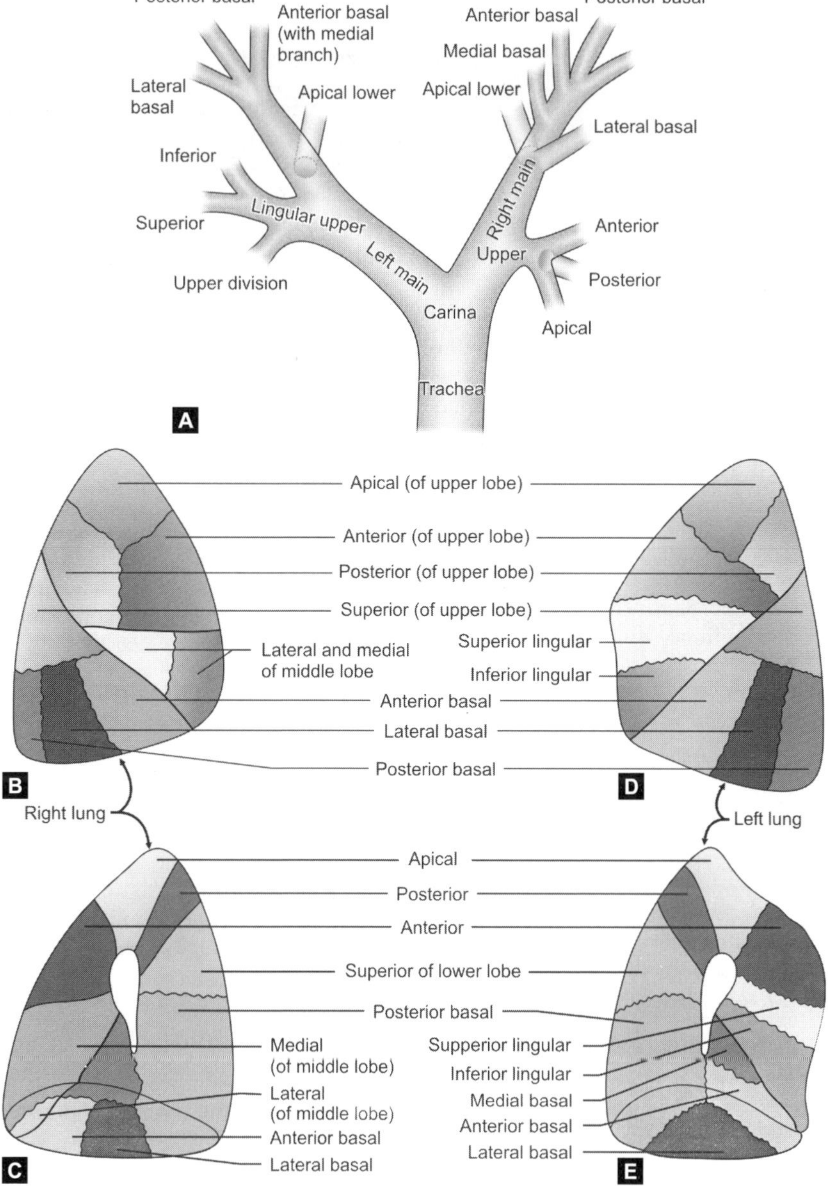

Figs 4.12A to E: (A) Scheme to show the bronchial tree as seen from the front, (B to E) Bronchopulmonary segments of the right and left lungs

- For surgical resection of a single or a number of diseased bronchopulmonary segments without affecting the function of the remaining segments.

Q.108 How do the bronchopulmonary segments drain?

Each segment is drained by a vein and an artery. The vein is located in the periphery, whereas the artery and its branches are located in the center of the segment.

Q.109 What are the differentiating features of the two lungs?

- The right lung has three lobes while the left lung has two lobes.
- The thin, sharp anterior border of the right lung is vertical while that of the left lung presents a cardiac notch.
- On the medial surface (mediastinal surface) of the lung, the cardiac impression is much deeper on the left than on the right.
- The right lung is wider than the left because of the smaller cardiac impression.
- The right lung is shorter than the left because of the higher position of the right dome of the diaphragm.

Q.110 Name the structures in root of lungs.

- Principal bronchus on left side and eparterial and hyparterial bronchi on right side.
- One pulmonary artery.
- Superior and inferior pulmonary vein.
- *Bronchial arteries:* One on right and two on left side.
- Bronchial veins.
- Anterior and posterior pulmonary plexuses of nerves.
- Lymphatics of lung.
- Bronchopulmonary lymph nodes.
- Areolar tissue.

Q.111 At what level the root of lungs lie?

Opposite body of T_{5-7} vertebra.

Q.112 What is the blood supply of the lungs?

The bronchial arteries and not the pulmonary arteries supply the lungs. This is so as the pulmonary arteries carry the deoxygenated blood.

Q.113 What are the relations of the structures at the root of the lung?

From above downwards (differs on two sides)

Right lung	*Left lung*
• Eparterial bronchus • Pulmonary artery • Hyparterial bronchus • Inferior pulmonary vein	• Pulmonary artery • Bronchus • Inferior pulmonary vein
From before backwards (similar on two sides): • Superior pulmonary vein, • Pulmonary artery and • Bronchus	

Q.114 What is the surface marking of the oblique fissure of the lung?

It corresponds approximately with the medial border of the scapula when the arm is raised above the shoulder.

The fissure may be represented by a line drawn obliquely from a point 2 cm, lateral to the 4th thoracic spine on the right side and at a slightly higher level on the left side to another point on the 5th rib in the midaxillary line and a third point on the 6th costal cartilage about 7.5 cm from midline.

Q.115 Where can the breath sound of the apical segment of the lower lobe be heard on auscultation?

Posteriorly below the upper end of the oblique fissure.

Q.116 Give the surface marking of the horizontal fissure of the right lung.
It corresponds approximately with a line drawn horizontally at the level of the 4th costal cartilage anteriorly. This line meets that of the oblique fissure in the mid axillary line.

Q.117 What is the 'lingula' of the left lung?
The upper lobe of the left lung corresponds with the upper and middle lobes of the right lung. The part of it which corresponds to the middle lobe is called the 'lingula' because it projects anteriorly to form the lingula (tongue-shaped structure) below the cardiac notch.

Q.118 What is the 'azygos lobe'?
The azygos vein is occasionally deeply embedded in the apex of the right lung, partly isolating its medial portion. This isolated medial portion of the right lung is referred to as the 'azygos lobe'.

Q.119 What is 'sequestration of lung'?
An area of lung not having any communication with the bronchial passages. Most frequently seen in lower lobe of left lung.

Q.120 What is the lymphatic drainage of the lung?
The lymphatics of the lung drain centripetally from the pleura towards the hilum into the bronchopulmonary lymph nodes. Efferents of these nodes drain into the tracheobronchial nodes which drain into the *paratracheal nodes and the mediastinal lymph trunks.* These lymph trunks drain directly into the brachiocephalic vein, or occasionally, indirectly via the right lymphatic duct or the thoracic duct.

Q.121 What are the structures related to the medial side of right lung?
The structures related to the medial side of right lung include (Fig. 4.13):
- Pulmonary veins
- Pulmonary artery
- Upper and main bronchus

Mediastinum

Q.122 Define mediastinum?
It is a median septum of thorax between two pleural cavities. Strictly speaking, it is septum between two lungs because mediastinal pleurae are also part of it.

Q.123 What are the boundaries of mediastinum?
Superiorly: Thoracic inlet
Inferiorly: Diaphragm

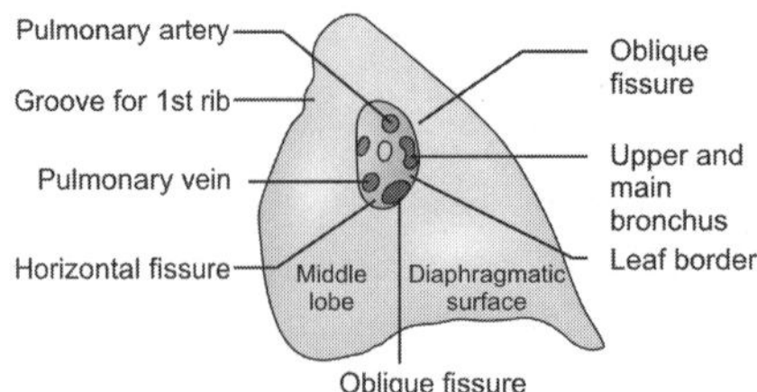

Fig. 4.13: Medial surface of right lung

Anteriorly: Sternum
On each side: Mediastinal pleura.

Q.124 What are the divisions of mediastinum?
Mediastinum is divided by an imaginary plane passing anteriorly through sternal angle and posteriorly through T_4 vertebra into:
- Superior mediastinum
- *Inferior mediastinum:* Subdivided by pericardium into:
  - *Anterior mediastinum:* In front of pericardium
  - *Middle mediastinum:* Pericardium and its contents
  - *Posterior mediastinum:* Behind pericardium.

Q.125 What are the boundaries of superior mediastinum?
Anteriorly: Manubrium sterni
Posteriorly: Upper 4 thoracic vertebrae.
Superiorly: Plane of thoracic inlet.
Inferiorly: Imaginary plane between superior and inferior mediastinum.
On each side: Mediastinal pleura.

Q.126 Name the contents of superior mediastinum.
Arteries:
- Arch of aorta,
- Brachiocephalic artery,
- Left common carotid artery and
- Left subclavian artery.

Veins:
- Right and left brachiocephalic veins,
- Upper ½ of superior vena cava and
- Left superior intercostal vein.

Muscles: Origin of
- Sternothyroid,
- Sternohyoid and
- Longus colli.

Nerves:
- Phrenic,
- Vagus,
- Cardiac and
- Left recurrent laryngeal.

Lymph nodes and lymphoid tissue:
- Thymus
- Thoracic duct and
- Lymph nodes.

Tubes:
- Trachea and
- Esophagus (Fig. 4.14).

Q.127 What are the contents of anterior mediastinum?
- Superior and inferior sternopericardial ligaments,
- Lymph nodes,
- Mediastinal branches of internal thoracic artery,
- Areolar tissue and
- Thymus.

Q.128 What is the arterial supply of thymus?
The thymus is supplied by inferior thyroid and internal thoracic arteries

Q.129 What is the function of thymus?
Thymus is primary lymphoid organ and it plays a vital role in making lymphocytes, immunologically competent T lymphocytes.

Q.130 Name the contents of middle mediastinum.
- Heart with pericardium
- Ascending aorta
- Pulmonary trunk
- Pulmonary arteries
- Bifurcation of trachea
- Principal bronchi
- Lower half of superior vena cava
- Terminal part of azygous vein
- Pulmonary veins
- Phrenic nerve
- Deep cardiac plexus and
- Tracheobronchial lymph nodes.

Q.131 What are the contents of posterior mediastinum?
- Descending thoracic aorta
- Azygous vein
- Hemizygous vein
- Accessory hemiazygous vein
- Vagus nerves
- Greater splanchnic nerve
- Lesser splanchnic nerve
- Least splanchnic nerve
- Thoracic duct
- Posterior mediastinal lymph nodes and
- Esophagus.

Q.132 What is mediastinal syndrome?
Compression of mediastinal structure by any growth gives rise to a group of symptoms known as mediastinal syndrome.

Q.133 How the pus in posterior mediastinum can enter the thighs?
The fascial sheath of psoas major muscle is open by its upper attachment to L_2 or L_1 vertebra. This upper edge forms medial

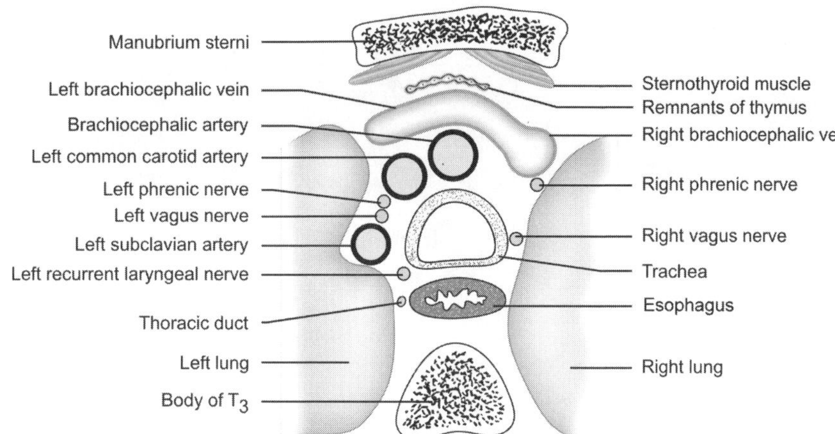

Fig. 4.14: Transverse section through the superior mediastinum just above the summit of the arch of the aorta to show some relations of the trachea

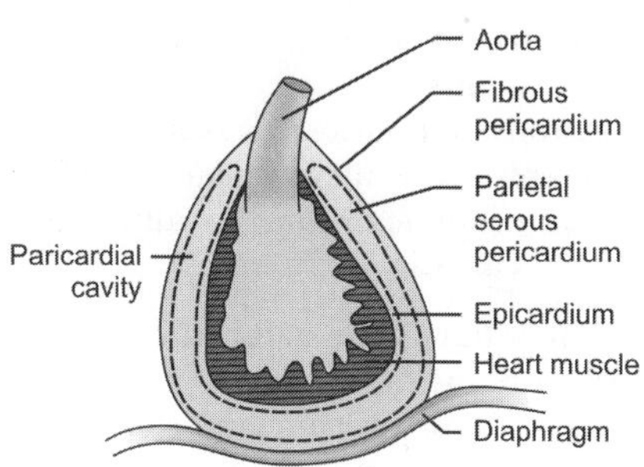

Fig. 4.15: Layers of pericardium

lumbocostal arch, from which vertebral part of diaphragm arises. So, psoas sheath opens into posterior mediastinum by a funnel-shaped orifice.

Pus in posterior mediastinum enters through funnel-shaped orifice and along the psoas sheath extends into thighs.

Q.134 Why the infection behind the prevertebral layer of deep cervical fascia cannot extend into posterior mediastinum?

The prevertebral layer of deep cervical fascia extends to the superior mediastinum and is attached to the 4th thoracic vertebra, so the neck infections behind this fascia cannot extend down beyond T_4.

Q.135 Infections between which layers of cervical fascia can extend into posterior mediastinum?

Posterior mediastinum is continuous through superior mediastinum with the neck between pretracheal and prevertebral layers of cervical fascia.

This region includes retropharyngeal space, spaces on each side of trachea and esophagus, space between trachea and esophagus.

Pericardium

Q.136 What is pericardium?

It is a fibroserous sac enclosing the heart and roots of great vessels.

Q.137 What are the parts of pericardium?

- *Fibrous pericardium:* Outer, single layered, tough and fibrous.
- *Serous pericardium:* Inner, double layered, thin (Fig. 4.15).

Q.138 What are the attachments of fibrous pericardium?

Fibrous pericardium is conical in shape.

Apex: Blunt and fused with roots of great vessels and pretracheal fascia.

Base: Blends with central tendon of diaphragm.

Anteriorly: By superior and inferior sternopericardial ligaments, attached to body of sternum.

Q.139 What are the different layers of serous pericardium?

- *Parietal pericardium:* Outer, fused with fibrous pericardium.
- *Visceral pericardium:* Inner, fused to heart except where it is separated from heart by blood vessels.
  Both layers are continuous at root of great vessels.

Q.140 What is pericardial cavity?

It is a potential space between parietal and visceral layers. It contains a thin layer of serous fluid.

Q.141 What is oblique sinus of pericardium?

It is a space behind heart between the left atrium, anteriorly and parietal pericardium, inferiorly.

Q.142 What are the boundaries of the oblique sinus of the pericardium?

The boundaries of the oblique sinus of the pericardium are as follows:

Anteriorly: Posterior surface of left atrium converted by serous pericardium.

Posteriorly: Posterior surface of left atrium covered by serous pericardium.

Left wall: Formed by pulmonary veins covered by serous pericardium.

Floor: It is open inferiorly.

Q.143 What is transverse sinus of pericardium?

It is a horizontal gap between ascending aorta and pulmonary trunk anteriorly and superior vena cava and atrium posteriorly.

On each side it opens into pericardial cavity.

Q.144 What are the boundaries of the transverse sinus of the pericardium?

The boundaries of the transverse sinus are as follows:

Anteriorly: Ascending aorta and pulmonary trunk.

Posteriorly: Anterior surface of the left labia

Roof: Reflections of serous pericardium from the posterior surface of the great arterial trunk to the left atria.

Floor: Floor is devoid of serous pericardium.

Q.145 How will you introduce your fingers into the transverse sinus of the heart?

To introduce the finger into the transverse sinus, the superior vena cava is used as a guide. The sinus is located anterior to it and so pass your finger in front of the S.V.C.

Q.146 What is surgical importance of transverse sinus?

Through this sinus a temporary ligature is given to occlude pulmonary trunk and aorta during cardiac operations.

Q.147 What is the developmental origin of sinuses of pericardium?

Transverse sinus: Develops due to degeneration of dorsal mesocardium.

Oblique sinus: Develops due to absorption of pulmonary veins into left atrium.

Q.148 What is the nerve supply of pericardium?

- *Fibrous and parietal pericardium:* By phrenic nerve. They are pain sensitive.
- *Visceral pericardium:* By autonomic nerves of heart.

Q.149 What is the arterial supply of pericardium?

- *Visceral layer:* By coronary arteries.
- *Fibrous and parietal layer:* By branches of internal thoracic, musculophrenic and descending thoracic aorta.

Q.150 What are the contents of the pericardium?

- Heart with cardiac vessels and nerves
- Ascending aorta
- Pulmonary trunk
- Lower half of superior vena cava
- Terminal part of inferior vena cava and
- Terminal part of pulmonary veins.

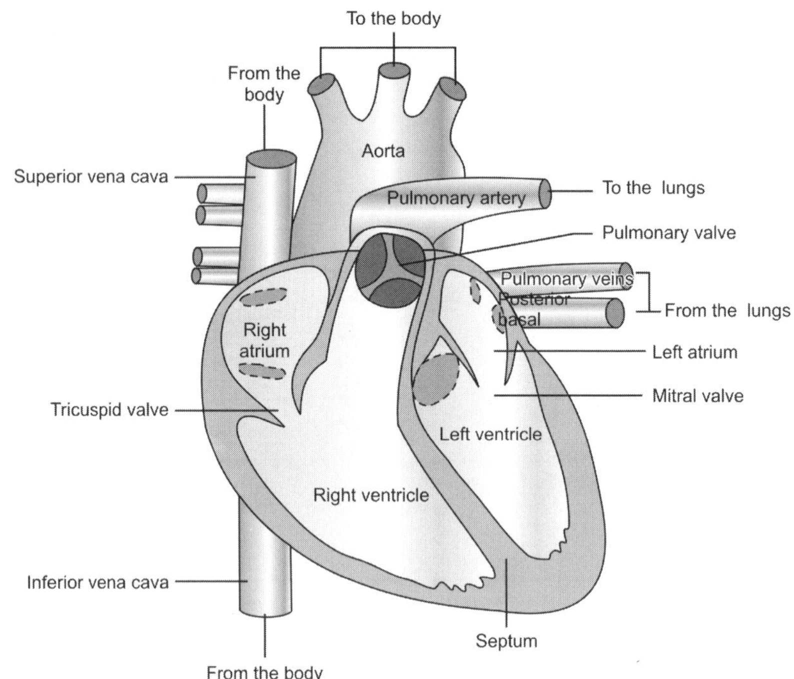

Fig. 4.16: Heart anatomy (interior view)

Heart

Q.151 What is the position of heart?

It is placed obliquely behind body of sternum and adjoining parts of costal cartilages of ribs.

1/3 of it lies to right and 2/3 of it lies to left of median plane.

Q.152 What are the divisions of heart?

Heart is composed of four chambers:

- Two atria: Right and left
- Two ventricles: Right and left.

  The atria are separated from ventricles by *coronary sulcus* (atrioventricular groove).

  The atria are separated by *interatrial groove* and the ventricles by *anterior and posterior interventricular grooves* (Fig. 4.16).

Q.153 Name the structures in anterior and posterior interventricular grooves.

In anterior interventricular groove:

- Interventricular branch of left coronary artery and
- Great cardiac vein.

In posterior interventricular groove:

- Interventricular branch of right coronary artery and
- Middle cardiac vein.

Q.154 Which chambers form the upper border of heart?

Two atria, chiefly left atrium.

Q.155 Which chambers form the left border of heart?

Mainly by left ventricle and

Partly by left auricle.

Q.156 Name the chambers forming the surfaces of heart?

Anterior surface:

- Mainly, by right ventricle and right auricle.
- Partly, by left ventricle and left auricle.

Inferior surface:

- Left 2/3 by left ventricle and
- Right 1/3 by right ventricle.

Left surface:

- Mostly by left ventricle and upper end by left auricle.

Posterior surface (Base):

- Mainly by left atrium,
- Small part by posterior part of right atrium (Fig. 4.17).

Q.157 What is right auricular appendage and what are its characteristic features?

It is the upper prolonged end of right atrium, which covers the root of ascending aorta. Externally, it is notched and interior is sponge like.

Q.158 What is the clinical importance of structure of right auricular appendage?

Its sponge like interior prevents free flow of blood and favor thrombosis which may dislodge to cause pulmonary embolism.

Q.159 What are the parts of right atrium and how they are developed?

Three parts:

- *Smooth posterior part (Sinus venarum):* Derived from right horn of sinus venosus.

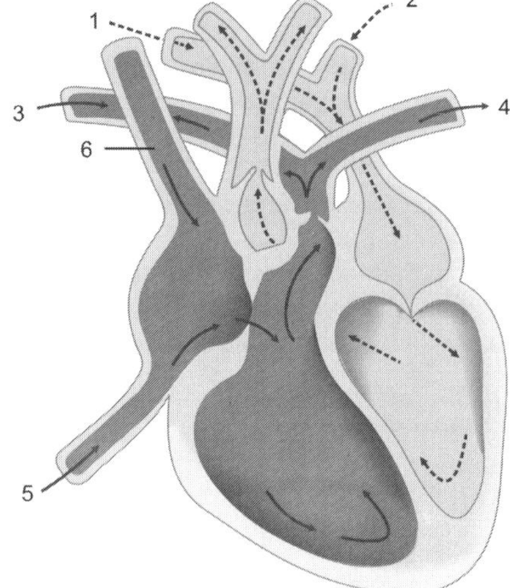

Fig. 4.17: Circulation inside heart: (1) From right lung; (2) From left lung; (3) To right lung; (4) To left lung; (5) Inferior vena cava; (6) Superior vena cava

  All large veins entering right atrium open in this part.
- *Rough anterior part including auricle (Atrium proper):* Derived from primitive atrial chamber.
- *Septal wall:* Derived from septum primum and septum secundum.

Q.160 Name the veins opening in the right atrium?

Ans.

- Superior vena cava
- Inferior vena cava
- Coronary sinus

- Anterior cardiac veins and
- Venae cordis minimi.

Q.161 What is eustachian valve?

It is a rudimentary valve guarding the opening of inferior vena cava. During embryonic life it guides the inferior caval blood to left atrium through foramen ovale.

Q.162 What are the features of septal wall of right atrium?

It has

- *Fossa ovalis:* Saucer shaped depression in lower part, formed by septum primum.
- *Limbus fossae ovalis:* It is prominent margin of fossa ovalis and represents free edge of septum secundum.

Q.163 What are the parts of right ventricle?

- *Inflowing part:* Rough and has muscular ridges called *trabeculae carneae.*
- *Outflowing part:* Smooth. Also called infundibulum. Opens into pulmonary trunk.
  The two parts are separated by a ridge, *supraventricular crest* and the inflow and outflow parts make an angle of about 90° with each other.

Q.164 What are the different types of trabeculae carneae?

- *Ridges:* Fixed elevations.
- *Bridges*: Fixed at ends but free in middle.
- *Papillary muscles:* Bases attached to ventricular wall and apex project into ventricular cavity and are connected to chordae tendineae.

Q.165 How the left atrium is developed?

- Greater part is smooth and is derived from absorption of pulmonary vein.
- Auricle develops from primitive atrial chamber.

Q.166 What are the parts of left ventricle?

Out flow part: Known as aortic vestibule. Opens into ascending aorta. Inflow part same as right ventricle. The inflow and outflow parts are at an acute angle.

Q.167 What is fossa lunata?

It is an impression in septal wall of left atrium, corresponding to fossa ovalis of right atrium.

Q.168 What are the parts of the interventricular septum?

Its right side is convex and buldges into right ventricle. Greater part of septum is thick and muscular and small area near posterior margin is membranous.

Q.169 What is the developmental origin of ventricles?

The ventricles develop from:

- Bulbus cordis and
- Primitive ventricle.

Q.170 How interventricular septum develops?

- *Muscular part:* Upgrowth from apex of heart.
- *Membranous part:* Downgrowth from interatrial septum.

Q.171 What type of valves is present in heart?

- *Atrioventricular valves:* One pair
  Right atrioventricular valve: Tricuspid valve, made up of three cusps.
  Left atrioventricular valve: Bicuspid or mitral valve, made up of two cusps.
- *Semilunar valves:* One pair
  Aortic and pulmonary valves. Each valve has 3 cusps.

Q.172 What is 'stenosis' of valve?

Narrowing of valve orifice due to fusion of valve cusps.

Q.173 What is 'incompetence' of valve?

The imperfect closure of the valve due to dilatation of valve orifice or stiffening of valve cusps.

Q.174 What are the septal defects?

These are the defects resulting from involvement of interatrial or interventricular septum.

- Atrial septal defects include ostium secundum and ostium primum defects. An ostium secundum defect lies high up in the atrial wall, while the ostium primum defect lies below. These result in communication between left and right atria.
- Interventricular septal defects which consist mainly of failure of development of membranous part. These are often associated with other septal defects.
- Complete failure of a septum to form, resulting in formation of common atrium or common ventricle or both.

Q.175 What is dextrocardia?

This is a congenital anomaly in which the heart position is reversed and it lies on the right side of the thorax. This may be associated with the reversal of all the intra-abdominal organs (*situs inversus*).

Q.176 What will be the effect of pulmonary stenosis?

There will be right ventricular hypertrophy because heart tries to force blood through the narrowed valve. This will be associated with congestion in the right atrium followed by secondary right atrial hypertrophy.

Q.177 What is Fallot's tetralogy?

This is the commonest cyanotic congenital anomaly of the heart and consists of (1) pulmonary stenosis, (2) right ventricular hypertrophy, (3) ventricular septal defect and (4) an overriding of the aorta over the septal defect. So, the aorta receives blood from both ventricles.

Q.178 What is complete transposition of great arteries?

It is a condition in which aorta arises from right ventricle and pulmonary trunk from left ventricle.

Blood Supply of Heart

Q.179 What is the origin of the right and left coronary arteries?

- The right coronary artery arises from the anterior aortic sinus. It is smaller than left.
- The left coronary artery arises from the left posterior aortic sinus.

Q.180 What are the branches of the right coronary artery?

- Marginal branch,
- Posterior (inferior) interventricular branch,
- Nodal branch
- Right atrial branch
- Infundibular and
- Terminal branches (Figs 4.18A to C).

Q.181 What are the branches of the left coronary artery ?

- Anterior interventricular branch.
- Branch to diaphragmatic surface of left ventricle and
- Left atrial branch.
  The continuation of the left coronary artery after anterior interventricular branch is called the *circumflex artery.*

Q.182 What is the distribution of the right coronary artery?

- Large part of the right ventricle except area adjoining anterior interventricular groove.
- Most of the right atrium.
- Part of the left ventricle, near interventricular septum.
- Posterior part of interventricular septum.
- SA node in 60% of the cases.
- AV node and bundle of His except part of left branch of AV bundle.

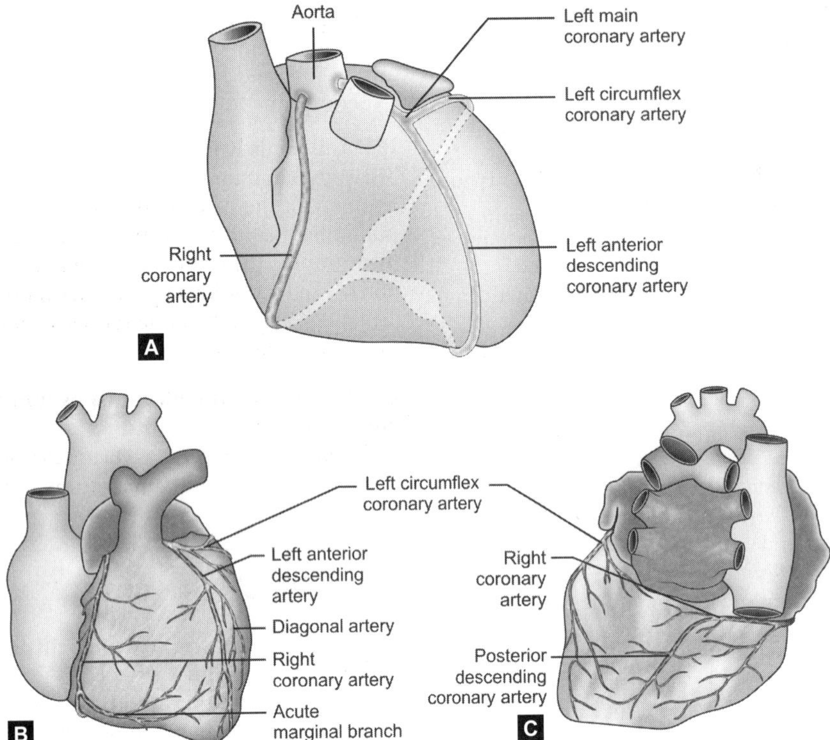

Figs 4.18A to C: (A) Schematic representation of the right and left coronary arteries, (B) Anterior view of the heart, (C) Posterior view of the heart

Q.183 What is the distribution of the left coronary artery?

- Large part of the left ventricle
- Right ventricle adjoining anterior interventricular groove
- Left atrium
- Anterior part of interventricular septum
- SA node in 40% of the cases
- Part of left branch of AV bundle.

  Both the interatrial and interventricular septa are supplied by branches of both coronary arteries.

Q.184 Do the coronary arteries anastomose?

They anastomose to a slight extent. The interventricular branches of the two coronary arteries anastomose near the apex of the heart and in the interventricular septum. Coronary arteries also anastomose with vasa vasora of aorta, internal thoracic artery and bronchial arteries.

Q.185 What is the clinical importance of the anastomosis between the coronary arteries?

The anastomosis between the branches of the coronary arteries is inadequate to compensate for the sudden occlusion.

A blockage therefore leads to death (infarction) of the affected cardiac tissue.

Q.186 What is 'angina pectoris'?

It is a clinical condition characterized by pain in front of the chest radiating to the ulnar side of the left arm and forearm. This is due to an incomplete obstruction of the coronary arteries.

Q.187 What are the tributaries of the coronary sinus?

- Great cardiac vein
- Middle cardiac vein
- Small cardiac vein
- Oblique vein of the left atrium
- Posterior vein of the left ventricle and
- Right marginal vein.

Q.188 What are Thebesian veins?

These are small veins present in all chambers of heart opening directly into cavity of chambers.

Q.189 How is the coronary sinus developed?

The coronary sinus is developed from the left horn of the sinus venosus.

Q.190 Where does the coronary sinus open?

Into posterior wall of right atrium.

Conducting System of the Heart

Q.191 What are the functions of conducting system of heart?

- It is responsible for initiating and maintaining normal cardiac rhythm.
- Ensures proper coordination of atrial and ventricular contractions (Fig. 4.19).

Q.192 What is the sinuatrial (SA) node and where is it located?

The SA node is the pacemaker of the heart. It is situated in the right atrium along the anterior margin of opening of the superior vena cava.

Q.193 What is the position of AV node?

The AV node lies in the wall of right atrium formed by interatrial septum near the opening of the coronary sinus. It receives impulse from SA node.

Q.194 What is atrioventricular bundle? What are its divisions?

AV bundle forms the connection between atrial and ventricular musculature. It begins at AV node and reaches posterior margin of membranous part of ventricular septum. Here it divides into left and right branches, which descend on left and right side of interventricular septum beneath endocardium. Each branch divides and subdivides to form Purkinje fibers, which terminate in ventricles.

Q.195 What is the 'moderator' band?

The 'moderator' band also called the *septo-marginal trabecula* (one of the trabeculae carneae) extends from the ventricular septum to the anterior papillary muscle. This is important as it carries the right branch of the atrioventricular bundle (bundle of His). It may assist in preventing over distension of ventricle.

Q.196 What is the nerve supply of the heart?

Nerve supply to heart is by:

- Parasympathetic fibers via vagus nerve. These are cardioinhibitory.
- Sympathetic fibers from T_{2-5} segments of spinal cord. There are cardioaccelerator and sensory.

  Both types of nerves form superficial and deep cardiac plexus and supply the heart.

Q.197 How is the superficial cardiac plexus formed? What are its branches?

The superficial cardiac plexus formed by:

- The inferior cervical cardiac branch of the left vagus and

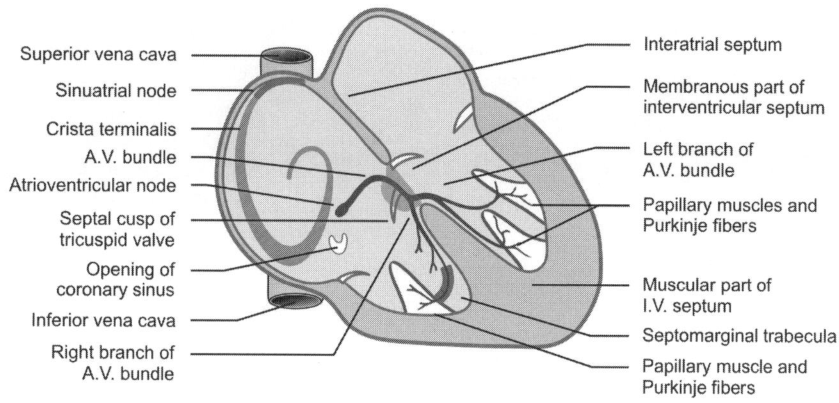

Fig. 4.19: Schematic view of the interior of the heart to show the parts of the conducting system

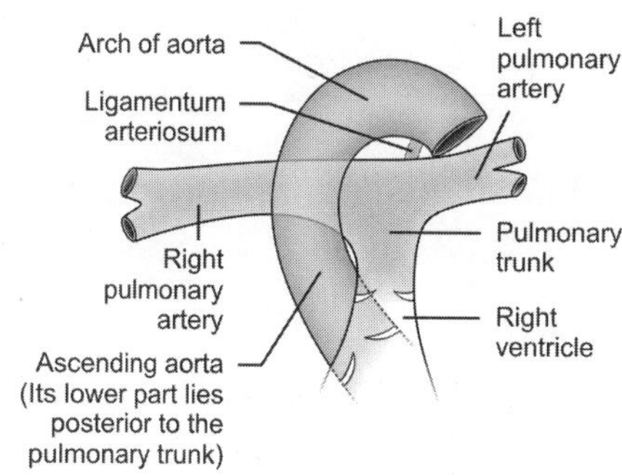

Fig. 4.20: Diagram to show the pulmonary trunk and pulmonary arteries, and their relationship to the aorta

- The superior cervical cardiac branch of left sympathetic trunk.

  It is located just below the arch of aorta close to ligamentum arteriosum. It gives branches to deep cardiac plexus, right coronary artery and left pulmonary plexus.

Q.198 How is the deep cardiac plexus formed and what is its distribution?

The deep cardiac plexus is formed by:

- Cardiac branches of the both vagus.
- Cardiac branches of both recurrent laryngeal nerves and
- The cardiac branches of cervical and thoracic branches of sympathetic trunk

  It gives branches to coronary and pulmonary plexuses and atria.

Major Blood Vessels of Thorax

Superior Vena Cava

Q.199 How superior vena cava is formed?

By the union of two brachiocephalic veins behind the lower border of first costal cartilage close to sternum.

Q.200 Name the tributaries of superior vena cava.

- Azygous vein,
- Mediastinal veins and
- Pericardial veins.

Q.201 What is the pathway for the collateral circulation in obstruction of superior vena cava?

- *If obstructed above opening of azygous vein:* Venous blood from upper half of body is returned through azygous vein and superficial veins of chest are dilated up to costal margin.
- *If obstructed below opening of azygous vein:* Venous blood is returned through inferior vena cava via femoral vein and superficial veins are dilated on chest and abdomen up to saphenous opening in thigh (Thoraco-epigastric vein).

Q.202 How superior vena cava is developed?

- *Upper half, up to opening of azygous vein:* Right anterior cardinal vein.
- *Lower half, below opening of azygous vein:* Right common cardinal vein.

Pulmonary Trunk and Arteries

Q.203 What is the course of pulmonary trunk?

It begins opposite the sternal end of left 3rd costal cartilage and upper end lies in front of fifth thoracic vertebra.

The bifurcation of pulmonary trunk lies below the arch of aorta (Fig. 4.20).

Q.204 What are the relations of right pulmonary artery?

Anterior: Ascending aorta, Superior vena cava and Upper right pulmonary vein.
Posterior: Esophagus and Right bronchus (Fig. 4.21).

Q.205 What are the relations of left pulmonary artery?

Posterior: Left bronchus and Descending aorta.
Superiorly: Connected to arch of aorta by ligamentum arteriosum.

Aorta

Q.206 What are the parts of aorta in thorax?

Ans.

- Ascending aorta,
- Arch of aorta and
- Descending thoracic aorta.

Q.207 What is the course of ascending aorta?

It begins at level of lower border of 3rd costal cartilage behind left half of sternum. It runs upwards, forwards and to right and continues as arch of aorta at sternal end of upper border of second right costal cartilage.

Q.208 What is aortic sinus?

It is dilatation of vessel wall at root of aorta above each cusp of aortic valve.

Q.209 Name the branches of ascending aorta?

- *Right coronary artery:* From anterior aortic sinus.
- *Left coronary artery:* From left posterior aortic sinus.

Q.210 What is the level of beginning and termination of arch of aorta?

It begins behind upper border of 2nd right sternochondral joint (lower border of T_4) and ends at lower border of body of 4th thoracic vertebra on left side.

Thus it begins and ends at same level but it begins anteriorly and ends posteriorly.

Q.211 What are the posterior and to the right relations of arch of aorta?

From behind forwards these are:

- Vertebral column
- Esophagus
- Trachea
- Superior vena cava
- Thoracic duct
- Left recurrent laryngeal nerve
- Deep cardiac plexus.

Q.212 Name the branches of arch of aorta.

- *Brachiocephalic artery:* Divides into right common carotid and right subclavian artery.
- Left common carotid,

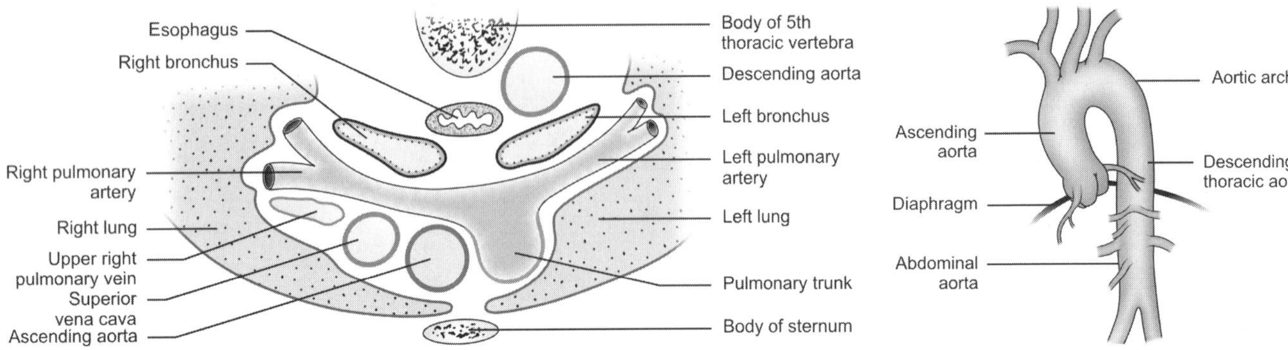

Fig. 4.21: Diagram to show the relations of the uppermost part of the pulmonary trunk, and of the pulmonary arteries (TS at level of vertebra T_5)

Fig. 4.22: Branches of the arch of the aorta

- Left subclavian.
- Thyroid ima
- Occasionally
- Vertebral artery (Fig. 4.22).

Q.213 Name branches of descending thoracic aorta.

- *Posterior intercostal arteries:* For 3rd-11th spaces, on both sides,
- *Subcostal arteries:* On both sides,
- Two left bronchial arteries,
- Esophageal branches,
- Pericardial branches,
- Mediastinal branches and
- Superior phrenic (Fig. 4.23).

Q.214 What is aortic aneurysm?

It is localized abnormal dilatation of aorta.

Q.215 What is coarctation of aorta?

It is the narrowing of aorta, occurring usually immediately beyond the origin of left subclavian artery.

It leads to hypertension above the narrowing e.g., arms, neck and head and hypotension below e.g., lower limb.

Q.216 What is the developmental origin of aorta?

- *Ascending aorta:* From truncus arteriosus.
- *Arch of aorta:*
  - From ventral part of aortic sac and its left horn and
  - Left fourth arch artery.
- *Descending aorta:*
  - From left dorsal aorta below attachment of fourth arch artery.
  - Fused median vessel.

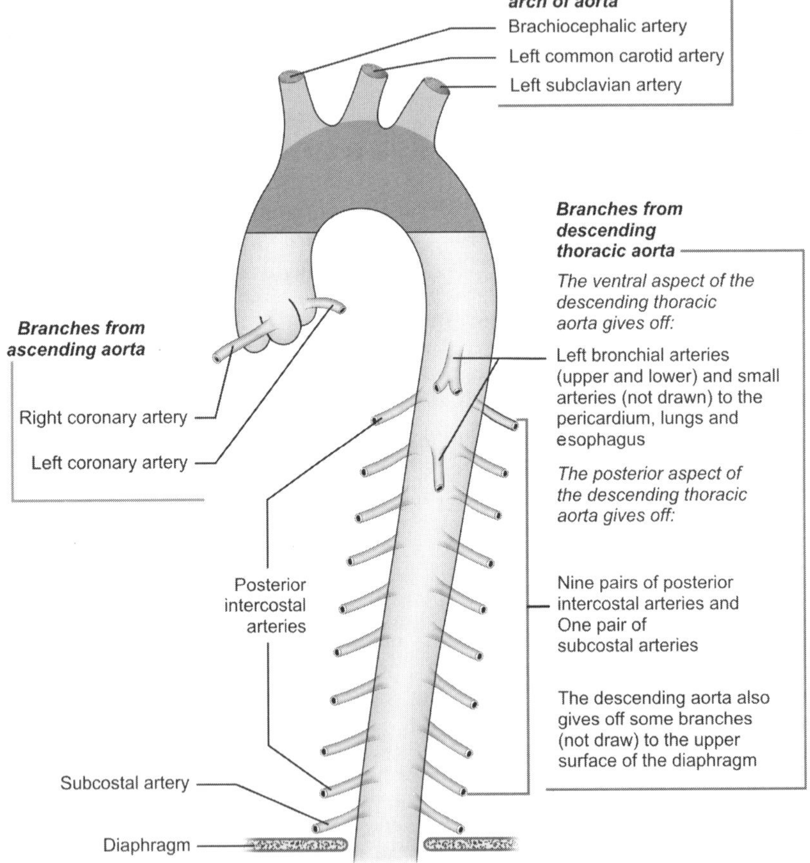

Fig. 4.23: Branches arising from the aorta in the thorax

Q.217 What is 'ductus arteriosus'?

It is communication present in fetal life connecting left pulmonary artery with aorta just distal to origin of left subclavian artery. After birth, it gets obliterated and forms ligamentum arteriosum.

Q.218 What will happen if ductus arteriosus remains patent?

It causes progressive enlargement of left ventricle and pulmonary hypertension.

Azygous and Hemizygous Veins

Q.219 How azygous vein is formed and terminates?

By the union of right ascending lumbar and right subcostal vein at level of T_{12} vertebra and terminates at level of T_4 vertebra into superior vena cava.

Q.220 Name the tributaries of azygous vein.
- Right posterior intercostal veins
- Right superior intercostal veins
- Hemiazygous veins: Present on left side and joins azygous vein at T_8 level
- Accessory hemiazygous vein: Present on left side and joins azygous vein at T_7 level
- Right bronchial veins
- Esophageal veins
- Mediastinal and pericardial veins
- Right ascending lumbar veins
- Right subcostal vein.

Q.221 Name the tributaries of hemiazygous vein.
- 9th-11th left posterior intercostal veins.
- Left ascending lumbar vein.
- Left subcostal vein.

Esophagus

Q.222 What is the length of esophagus?
25 cm.

Q.223 What is the extent of esophagus?
- It begins in neck at level of lower border of C_5 vertebra, i.e. at lower border of cricoid cartilage.
- It ends in abdomen at level of lower border of T_{11} vertebra, at cardiac orifice of stomach.

Q.224 What are the 'curvatures of esophagus'?
Esophagus shows
- Two side to side curvatures towards left.
  - At root of neck.
  - Esophageal opening in diaphragm.
- *Anteroposterior curvature:* Follows curvature of spine.

Q.225 What are the sites of esophageal constrictions?
- *At its commencement:* 6 inches from incisor teeth.
- *Where it is crossed by aortic arch:* 9 inches from incisor teeth.
- *Where it is crossed by left bronchus:* 11 inches from incisor teeth.
- *At its termination:* 15 inches from incisor teeth.

Q.226 Name the structures intervening between esophagus and vertebral column.
- Thoracic duct
- Vena azygos
- Hemiazygos vein
- Accessory hemiazygos vein
- Right posterior intercostal arteries.

Q.227 What are the divisions of esophagus?
The esophagus is divided into three parts:
- Cervical,
- Thoracic and
- Abdominal.

Q.228 What is the relation of recurrent laryngeal nerves to the esophagus in the neck?
The right and left recurrent laryngeal nerves lie anterolateral to the esophagus in corresponding grooves between it and the trachea.

Q.229 What is the blood supply of esophagus?
Arterial supply:
- *Cervical part:* Inferior thyroid artery.
- *Thoracic part:* Esophageal branches of aorta and bronchial arteries.
- *Abdominal part:* Esophageal branches of left gastric artery.

Venous drainage:
- *Cervical part:* Into brachiocephalic vein.
- *Thoracic part:* Into azygous vein.
- *Abdominal part:* Into portal vein through left gastric vein. Lower end is site of porto-systemic anastomosis.

Q.230 What is the nerve supply of esophagus?
Parasympathetic nerves: Sensorimotor and secretomotor.

Upper ½: Recurrent laryngeal nerve.

Lower ½: Esophageal plexus formed by two vagus nerves.

Sympathetic nerves: Vasomotor.

Upper ½: Fibers from middle cervical ganglion.

Lower ½: Fibers from upper 4 thoracic ganglia.

Q.231 What is 'achalasia cardia'?
It is condition of neuromuscular incoordination in which the lower end of esophagus fails to dilate when food is swallowed. As a result, food accumulates in the esophagus.

Q.232 What is the clinical importance of constrictions of esophagus?
- During endoscopy, these constrictions should be kept in mind.
- These are also the sites of development of strictures usually.

Q.233 What are esophageal varices and what is their clinical importance?
These are the dilatations of the esophageal veins in portal hypertension, which form anastomosis between azygos (systemic) and left gastric (portal) veins.

Clinical importance: These may rupture leading to severe hemorrhage.

Q.234 What is the effect of enlargement of left atrium on esophagus?
In mitral stenosis, enlargement of left atrium causes backward displacement of the esophagus, which can be seen in a barium swallow.

Q.235 How esophagus is developed?
The esophagus is developed from the part of foregut lying between pharynx and stomach. It is at first short but later on elongates with the descent of diaphragm and formation of neck.

Q.236 What are the characteristic histological features of esophagus?
From within outwards it is made up of:
- *Mucosa:* Lined by stratified squamous epithelium.
- *Submucosa:* Contains mucous glands.
- *Muscular layer:* Has external longitudinal and inner circular fibers. The muscle fibers are striated in upper two thirds and smooth in lower one third.
- Connective tissue sheath of areolar tissue.

Thoracic Duct

Q.237 What is the length of thoracic duct?
40 cm.

Q.238 What is the extent of thoracic duct?
Begins from cisterna chyli near lower border of T_{12} vertebra. Ends into angle of junction between left subclavian and left internal jugular vein at level of T_2 vertebra (Fig. 4.24).

Q.239 What are the relations of thoracic duct in aortic opening of diaphragm?
- *Anteriorly:* Diaphragm
- *Posteriorly:* Vertebral column
- *To the left:* Azygous vein
- *To the right:* Aorta.

Q.240 Name the tributaries of thoracic duct.
In thorax:
- Channels from posterior mediastinal and intercostal nodes.
- Left mediastinal trunk may drain.

At root of neck:
- Left jugular trunk,
- Left subclavian trunk.

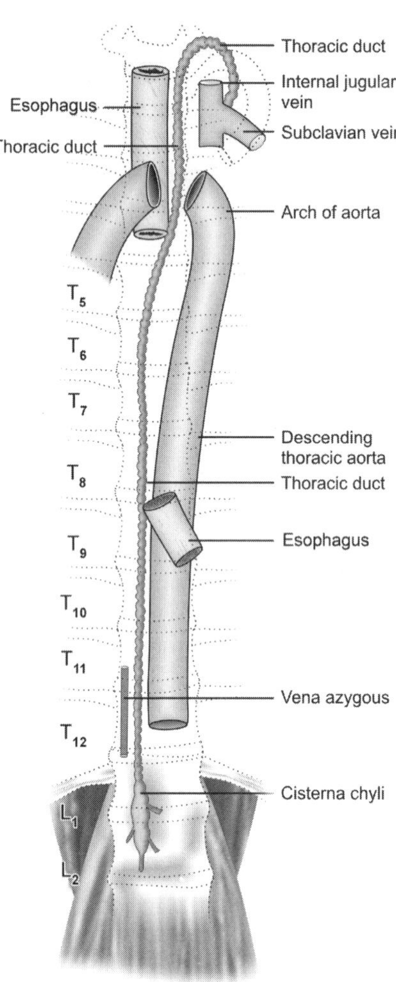

Fig. 4.24: Course and relations of the thoracic duct as seen from the front

Q.241 From which areas the thoracic duct drains lymph?

- Both halves of body below diaphragm and
- Left half above diaphragm.

Sympathetic Trunk

Q.242 What is the extent of sympathetic trunk?

Each trunk is placed on either side of vertebral column and extends from base of skull to coccyx below.

Q.243 What is the position of sympathetic trunk in relation to vertebral column?

In cervical region: Anterior to transverse processes of cervical vertebrae.

In thoracic region: Anterior to heads of ribs

In lumbar region: Anterolateral to lumbar vertebrae.

In sacral region: Anterior to sacrum.

The two join each other in front of the coccyx.

Q.244 What is the number of ganglia in cervical sympathetic trunk?

Three

Q.245 What is ganglion impar?

The lower fused ends of right and left sympathetic trunks are thickened in a midline ganglion called as ganglion impar.

Q.246 Where the cell body of sympathetic preganglionic neurons are present?

In the intermediolateral grey column of spinal cord in spinal segment T_1 to L_2.

Q.247 What is the location of sympathetic post ganglionic neurons?

Ganglia on sympathetic trunk.

Q.248 What is the number of ganglia in thoracic sympathetic trunk?

12, but may be reduced due to fusion of adjacent ganglia with one another.

Q.249 What is 'stellate ganglion'?

It is ganglion formed by fusion of first thoracic ganglion with inferior cervical ganglion.

Q.250 Name the branches of thoracic part of sympathetic trunk.

See Figure 4.25.

Lateral branches: Each ganglion is connected with corresponding spinal nerve, by white (preganglionic) and grey (postganglionic) rami communicantes.

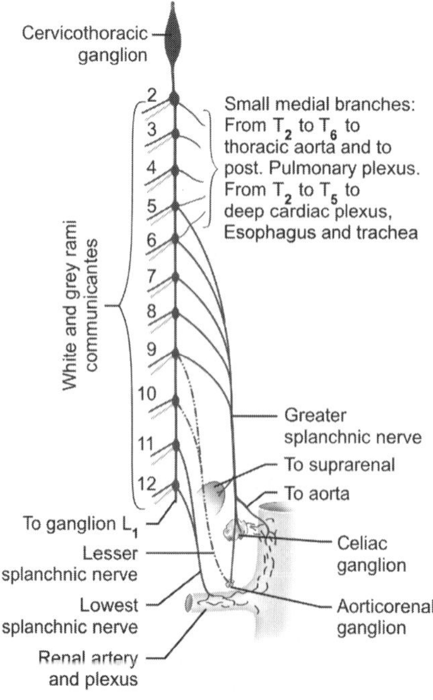

Fig. 4.25: Branches of the thoracic part of the sympathetic trunk

- *Medial branches:* (From five ganglia.)
  Pulmonary branches to pulmonary plexus
  Cardiac branches to upper cardiac plexus
  Aortic branches to arotic plexus
  Esophageal branches to esophageal plexus
- *Greater splanchnic nerve:* (By from lower seven ganglia)
  By roots from ganglia 5 to 9.
  Ends mainly in celiac ganglion
- *Lesser splanchnic nerve:*
  Preganglionic roots from ganglia 10 and 11.
  Ends mainly in aorticorenal ganglia.
- *Least splanchnic nerve:*
  By roots of ganglion 12.
  Ends in renal plexus

Phrenic Nerves

Q.251 How the phrenic nerve is formed?

Each nerve is formed by ventral primary rami of C_3, C_4 and C_5 spinal nerves. The contribution from C_4 is greatest.

Q.252 What is the distribution of the phrenic nerve?

Motor: To diaphragm.

Sensory:

- Proprioceptive fibers from diaphragm
- Sensory branches to pericardium and parietal pleura.
- Sensory branches to suprarenal glands, inferior vena cava and gallbladder.'

Diaphragm

Q.253 What is diaphragm? what are its attachments?

Diaphragm (Fig. 4.26) is a large muscle which forms a partition between the cavities of the thorax and abdomen origin. The origin of the diaphragm can be divided into sternal, costal and lumbar vertebral parts. The sternal part consists of two slips: right and left which arise from the back of xiphoid process. The costal part consists of broad slips one from the inner surface of each of the lower six ribs (7th to 12th) and their costal cartilages. The lumbar part consists of two crusa (right and left) that arise from the anterolateral aspects of the bodies of lumbar vertebrae and of fibers that arise (on either side) from tendinous arches called the lateral and medial arcuate ligaments. The right crus is larger than the left; it crusis from the bodies of vertebrae L_1, L_2 and L_3 and from the intervening intervertebral discs. The left crus similarly arises from vertebrae L_1 and L_2.

Insertion: From its extensive origin, the muscular fibres of the diaphragm run upwards and converge to the inserted on the margins of a large, flat, central tendon, which is located just below the pericardium and heart.

Q.254 Describe the apertures present in the diaphragm.

The apertures present in the diaphragm are as follows (Fig. 4.27):

- The aortic aperture: The aperture lies behind the medial arcuate ligament and in front of the disc between T_{12} and L_1. The aorta, therefore passes behind the diaphragm rather than through it.
- The aperture for esophagus: This is situated at the level of 10th thoracic vertebrae. The esophageal aperture also transmits the right and left gastric nerves.
- The aperture for inferior vena cava lies in the central tendon at the level of eighth thoracic vertebra. The vena caval opening also transmits the whole or part of right phrenic nerve.
- The left phrenic nerve passes through the muscular part of the diaphragm, to the left of the anterior follum of the central tendon. Numerous small veins also pass between the thorax and abdomen through small apertures in the central tendon.
- There are a number of small apertures present around the periphery of the diaphragm, in gaps between various slips of origin.

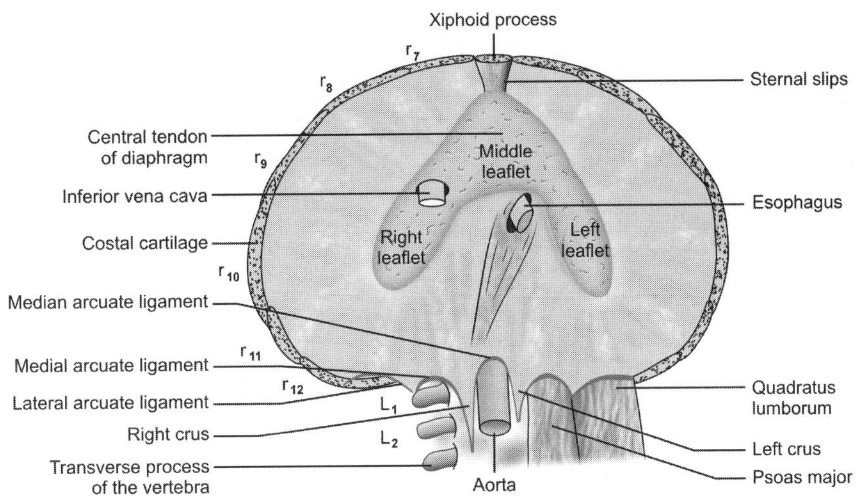

Fig. 4.26: Scheme to show attachments of the diaphragm

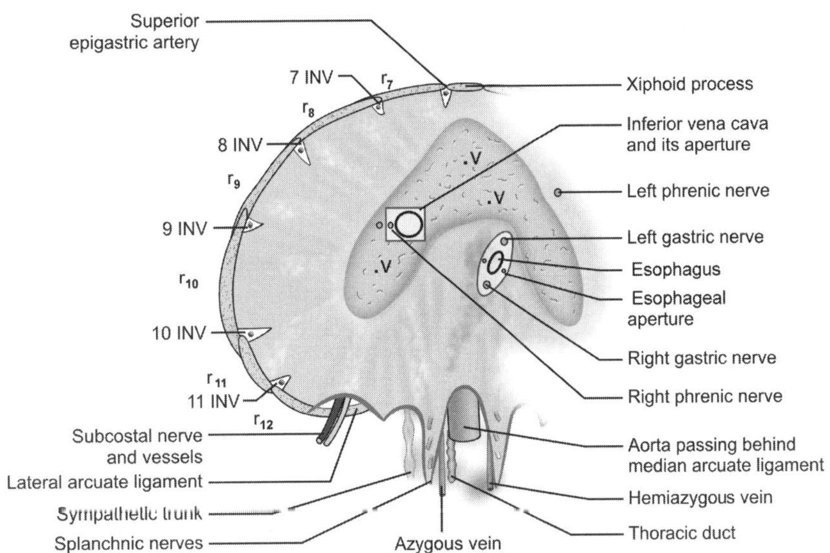

Fig. 4.27: Schematic diagram to show apertures in the diaphragm (INV= Intercostal nerve and vessels. v= small vein) r_7 to r_{12}= 7th to 12th ribs

Chapter 5

Abdomen

Anterior Abdominal Wall

Q.1 What is the position of umbilicus?

- In young adults at anterior median line at level of junction between L_3 and L_4 vertebra.
- It is lower in infants and in those with pendulous abdomen.

Q.2 What are different abdominal regions?

The abdomen can be divided into following nine regions (Fig. 5.1): In the midline from above downwards the regions are epigastrium (EPG); the umbilical region (UHB); and the hypogastrium (HVG), also known as pubic region. Lateral to the epigastrium, there is right hypochondrium (RH) and left hypochondrium (LH). Lateral to the umbilical region, there is right lumbar region (RL) and the left lumbar region (LL). Lateral to the hypogastrium, there is the right inguinal region (I), also called the right iliac fossa and the left inguinal region (LI), also called as the left iliac fossa.

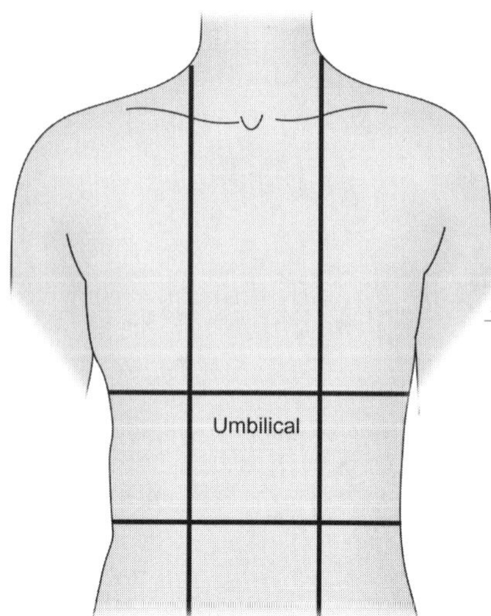

Fig. 5.1: Regions of the abdomen and the lines demarcating them

Q.3 What is developmental origin of umbilicus?

It is scar formed by the remnants of the root of umbilical cord.

Q.4 What happens if urachus remains patent?

Urinary fistula is formed so that urine may pass through umbilicus.

Q.5 What is exomphalos?

It is the persistence of physiological hernia of midgut loop outside the abdominal cavity.

Q.6 What is the importance of umbilicus?

Anatomical:

- It marks the watershed of body. The lymph and venous blood do not cross the umbilical plane.
- Supplied by T_{10} segment of spinal cord.
- Site of portacaval anastomosis.

Embryological:

Site of attachment of umbilical cord.

Clinical: Vitellointestinal duct may persist.

Q.7 What are the remnants of umbilical cord?

- Median umbilical ligament: Remnant of urachus.
- Lateral umbilical ligament: Formed by obliterated umbilical arteries.
- Ligamentum teres of liver: Remnant of left umbilical vein.

Q.8 What are the features of superficial fascia of abdominal wall?

Below the umbilicus, superficial fascia is divided into:

- Superficial fatty layer (fascia of Camper).
- Deep membranous layer (fascia of Scarpa).

Q.9 What are the attachments of fascia of Scarpa?

- It is continuous below with membranous layer of superficial fascia of perineum (Colles' fascia).
- The line of attachment passes over:
  - Along Holden's line (Lateral to pubic tubercle and extends for about 8 cm).
  - Pubic tubercle.
  - Body of pubis.
  - Deep fascia of adductor and gracilis.
  - Margins of pubic arch.
  - Posterior border of perineal membrane.
  - Above umbilicus, it merges with fatty layer.

Q.10 What are the contents of superficial fasica of abdominal wall?

- Fat.
  - Cutaneous nerves.
  - Cutaneous vessels.
  - Superficial lymphatics.

Q.11 Is there any deep fascia in anterior abdominal wall?

No. This absence of deep fascia allows expansion of abdominal wall.

Q.12 What is the cutaneous nerve supply of anterior abdominal wall?

- Anterior cutaneous nerves
  - 5 intercostal nerves (T_{7-11})
  - Subcostal nerve (T_{12})
  - Iliohypogastric (L_1)
- Lateral cutaneous nerves

  2 intercostal nerves (T_{10}-T_{11})

Q.13 What is the nerve supply of muscles of anterior abdominal wall?

- Lower six intercostal nerves (T_{6-11}) and subcostal nerve: Branches to external and internal oblique, transversus abdominis and rectus abdominis.
- Subcostal nerve (T_{12}): Also to pyramidalis
- Iliohypogastric nerve (L_1): Internal oblique and transversus abdominis muscle
- Ilioinguinal nerve (L_1): Internal oblique.

Q.14 What is the arterial supply of anterior abdominal wall?

- Branches of lower (10th and 11th) intercostal arteries: Branches of descending thoracic aorta.

- Branches of internal thoracic artery:
  - Superior epigastric
  - Musculophrenic.
- Branches of external iliac artery:
  - Inferior epigastric
  - Deep circumflex iliac
- Branches of subcostal artery: Branch of descending thoracic aorta.
- Branches of lumbar artery.
- Superficial branches of upper femoral artery: Superficial epigastric, superficial circumflex iliac and superficial external pudendal.

Q.15 What is the lymphatic drainage of anterior abdominal wall?

- Lymphatic drainage of skin
  - Above the umbilicus: Axillary nodes.
  - Below the umbilicus: Superficial inguinal lymph nodes.
- Lymphatic drainage of deeper tissues: External iliac nodes.

Q.16 Where the urine will collect in rupture of urethral bulb in perineum?

It will be collected in scrotum, perineum and penis and then lower abdomen deep to fibrous fascial plane. It does not extravasate into lower limb, because of attachment of membranous layer to the deep fascia of upper thigh along Holden's line.

Q.17 What is the drainage of cutaneous veins of anterior abdominal wall?

- Below umbilicus: Great saphenous vein into femoral vein which drains into inferior vena cava.
- Above umbilicus: Lateral thoracic vein to axillary vein which drains into superior vena cava.
- Few paraumbilical veins: Into left branch of portal vein along ligamentum teres in falciform ligament.

All these veins anastomose with each other.

Q.18 What is caput medusae and its clinical importance?

In portal vein obstruction, the superficial abdominal (cutaneous) veins are dilated for collateral circulation around the umbilicus in a radiating pattern.

In caput medusae the blood flows upwards above umbilicus and downwards below umbilicus.

Q.19 What is the clinical importance of thoracoepigastric vein?

It is a subcutaneous vein connecting the great saphenous vein with axillary vein. It becomes dilated and tortuous in vena caval obstructions to provide alternate channel for blood flow. It connects tributaries of lateral thoracic vein draining into axillary vein and superficial epigastric vein draining into great saphenous vein which in turn drains into femoral vein.

Clinical importance: In superior caval obstruction, blood in thoracoepigastric vein flows downwards and in inferior caval obstruction blood flows upwards.

Q.20 Name the muscles of anterior abdominal wall.

See Figures 5.2 to 5.5.

- External oblique
- Internal oblique
- Transversus abdominis
- Rectus abdominis
- Cremaster
- Pyramidalis.

Q.21 What are the functions of muscles of anterior abdominal wall?

- Support for abdominal viscera
- Expulsive acts: Helps in micturition, defecation, parturition, etc. by increasing the intra-abdominal pressure.
- Forceful expiratory acts: In coughing, sneezing, blowing.
- Movements of trunks:
  - Flexion of trunk: Recuts abdominis.
  - Lateral flexion: Oblique muscles.
  - Rotation of trunk: External oblique with internal oblique of opposite side.

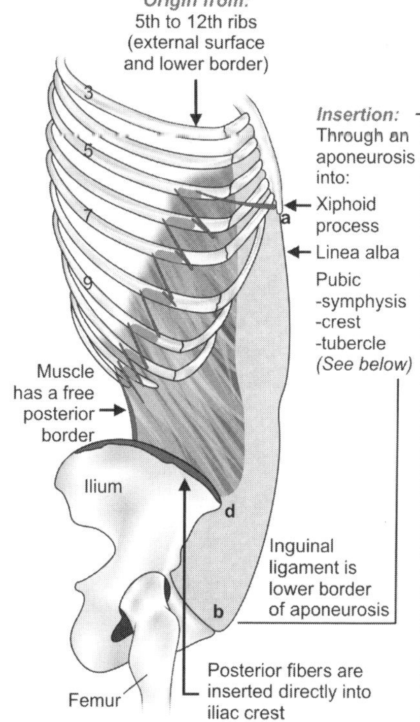

Fig. 5.2: Lateral view of the trunk to show the attachments of the external oblique muscle of the abdomen

- Pyramidalis tenses linea alba.
- Cremaster helps to suspend testis and can elevate it.

Q.22 What is the origin, insertion and nerve supply of pyramidalis muscle?

Pyramidalis is a small triangular muscle placed in front of rectus abdominis within its sheath.

Origin: Front of pubis and pubic symphysis.

Nerve supply: Subcostal nerve.

Q.23 What is cremasteric reflex? What is its clinical importance?

On stroking skin of upper part of medial side of thigh there is elevation of testis,

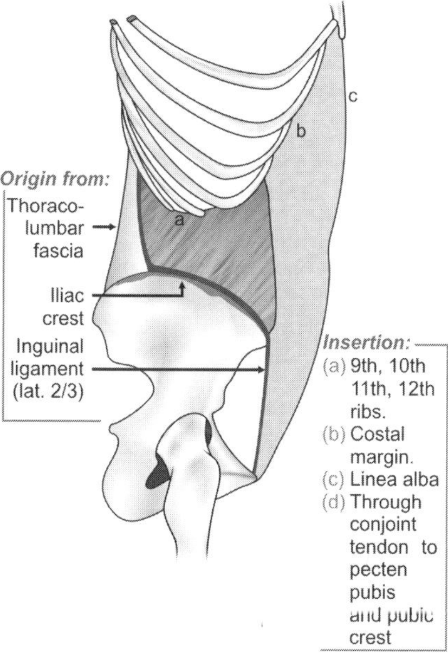

Fig. 5.3: Lateral view of the trunk to show attachments of the internal oblique muscle of the abdomen

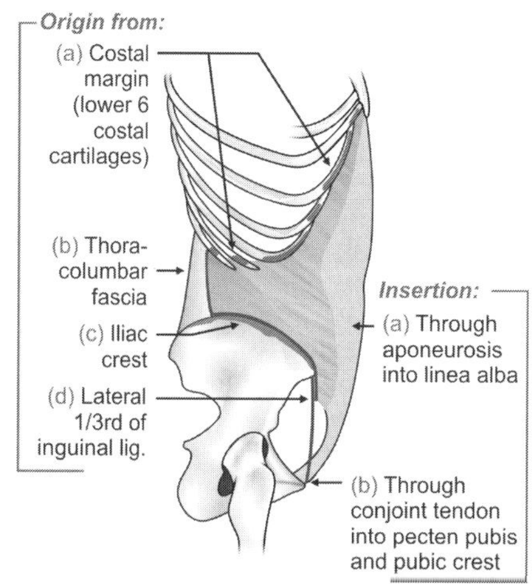

Fig. 5.4: Lateral view of the trunk to show the attachments of the transversus abdominis muscle

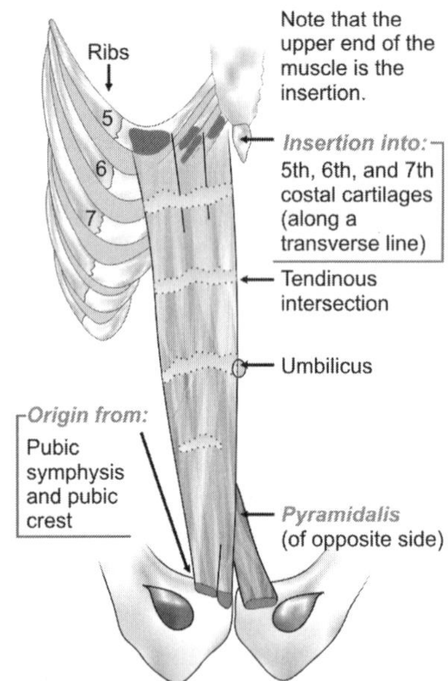

Fig. 5.5: Scheme to show the attachments of the rectus abdominis

due to reflex contraction of cremaster muscle. Reflex is more brisk in children.

Clinical importance: In upper motor neuron lesions above L_1 segment the reflex is lost.

Q.24 What is the ligament of Poupart?

Inguinal ligament.

Q.25 How inguinal ligament is formed?

Extension of lower border of external oblique aponeurosis, which is thickened and folded backwards (Fig. 5.6).

It extends from anterior superior iliac spine to pubic tubercle.

Q.26 What are the attachments to inguinal ligament?

Upper border:

- *Lateral* 2/3: Origin of internal oblique.
- *Medial* 1/3: Origin of transversus abdominis.

  Lower border: Fascia lata.

Q.27 Name the extensions of inguinal ligament.

- Pectineal part of inguinal ligament.
- Pectineal ligament (ligament of Cooper)
- Reflected part of inguinal ligament.

Q.28 How is conjoint tendon formed?

Formed by fusion of lower aponeurotic fibers of internal oblique and transversus abdominis.

It is attached to pubic crest and medial part of pecten pubis (Fig. 5.7).

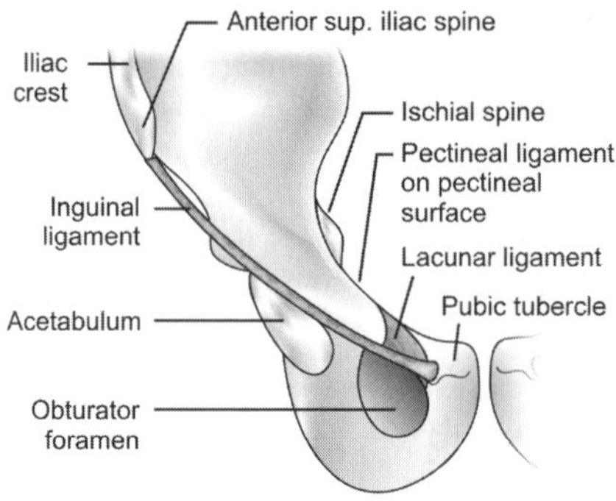

Fig. 5.6: Diagram to show the inguinal ligament and some related structures

Q.29 What is the function of conjoint tendon?

It guards the weak point of the superficial inguinal ring.

Q.30 What are the boundaries of lumbar triangle of Petit and what is its clinical importance?

Floor: Internal oblique muscle.

Below: Crest of ilium.

Laterally: External oblique.

Medially: Latissimus dorsi.

It is the site of the primary lumbar hernia.

Rectus Sheath

Q.31 What is rectus sheath?

It is an aponeurotic sheath covering rectus abdominis and pyramidalis muscle with their associated vessels and nerves.

Q.32 How is rectus sheath formed ?

- *Above costal margin:*

  Anterior wall: External oblique aponeurosis.

  Posterior wall: Deficient; Rectus lies directly on costal cartilages.
- Between costal margin and arcuate line:

  Anterior wall: External oblique aponeurosis, anterior lamina of internal oblique.

  Posterior wall: Posterior lamina of internal oblique,

  Aponeurosis of transversus abdominis (Fig. 5.8).
- *Below arcuate line:*

  Anterior wall: Aponeurosis of all three muscles of abdomen.

  Posterior wall: Deficient; Rectus muscle rests on fascia transversalis.

  Arcuate line (fold or Douglas) represents lower free margin of posterior wall of rectus sheath, at level midway between umbilicus and pubic symphysis.

Q.33 What are the 'tendinous intersections' of rectus abdominis?

- These are transverse fibrous bands which divide the muscle into smaller parts.
- Three in number: Present
  - Opposite umbilicus.
  - Opposite lower border of xiphoid.
  - In between 1 and 2.

  Sometimes intersections may be present below umbilicus.
- Traverse only the anterior half of muscle and are adherent to anterior wall of rectus sheath.

Q.34 What is the importance of tendinous intersections of rectus abdominis?

- They represent segmental origin of muscle.
- Functionally, they make the muscle more powerful by increasing the number of muscle fibers.

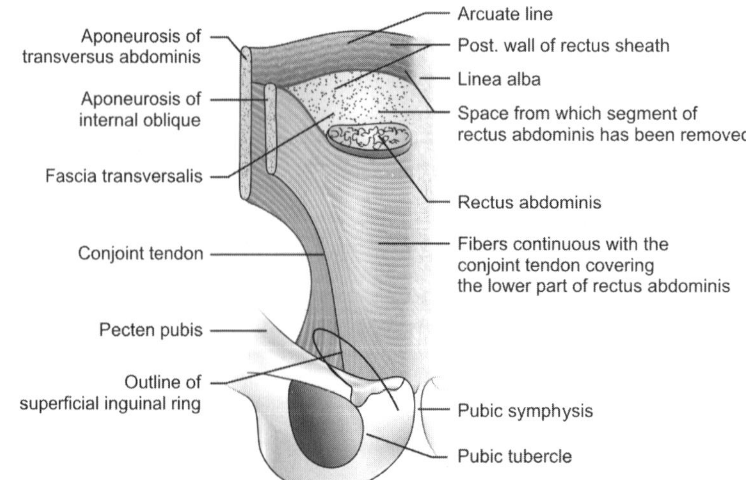

Fig. 5.7: Diagram to show the formation of the conjoint tendon. Some related structures are also shown

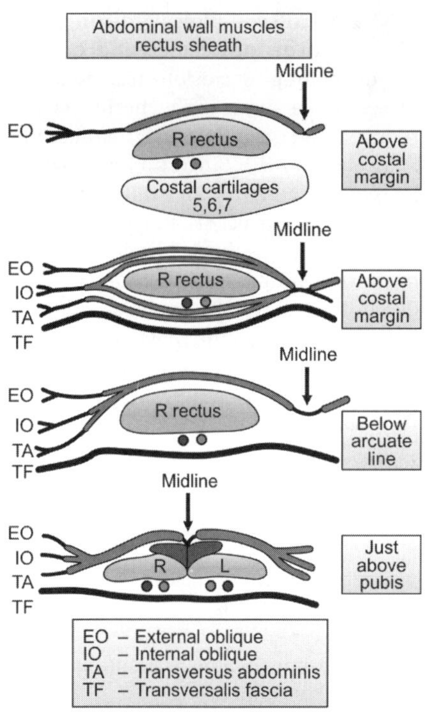

Fig. 5.8: Rectus sheath

Q.35 Where is the 'neurovascular plane' of abdomen is situated?

It lies between internal oblique and transversus muscle. Various abdominal nerves and vessels run in this plane.

Q.36 What are the function of rectus sheath?

- Support the abdominal viscera.
- Increases efficiency of rectus muscle by checking bowing during its contraction.

Q.37 What are the contents of rectus sheath?

- *Muscles:* Rectus abdominis
  Pyramidalis
- *Arteries:* Superior epigastric artery
  Inferior epigastric artery.
- *Veins:* Superior epigastric vein
  Inferior epigastric vein.
- *Nerves:* Lower 5-intercostal nerves
  Subcostal nerve.

Q.38 What is linea alba?

- It is a raphe formed by interlacing fibers of aponeuroses of three muscles forming rectus sheath.
  - It extends from xiphoid process to pubic symphysis.
  - Wider above and narrow below the umbilicus.
  - It separates the two rectus abdominis muscles from each other.

Q.39 What is divarication of recti?

- Seen in weak children and multipara women.
- There is weakness of linea alba, so the fingers can be insinuated between the two recti.

Q.40 Why supraumbilical median incision is given for surgery?

The incision through linea alba is given, because it is made of fibrous tissue only, so there is minimal blood loss.

It also does not cause damage to nerves.

Q.41 In the paramedian incision of rectus sheath, the rectus muscle is retracted laterally. Explain why?

- To avoid injury to nerves as they enter the rectus through its lateral border.
- On closing the incision, rectus slips back into its place.

Q.42 Why the trans-rectus incisions are not preferred during surgery?

Because the rectus receive its nerve supply laterally and muscle medial to incision is deprived of its innervation and hence undergoes atrophy.

Q.43 What is fascia transversalis?

Part of abdominopelvic fascia lining inner surface of transversus abdominis muscle and is separated from peritoneum by extraperitoneal tissue which is rich in fat.

Q.44 What are the prolongations of fascia transversalis?

- Over femoral vessels as anterior wall of femoral sheath.
- At deep inguinal ring, over spermatic cord as internal spermatic fascia.

Q.45 Why the cutting of one or two nerves supplying rectus produces clinical ill effects but not that of lateral abdominal muscles?

Because lateral abdominal muscles are supplied by a richly communicating network but the segmental nerve supply of rectus has little cross communications.

Inguinal Canal

Q.46 What is the position of inguinal canal?

In lower part of anterior abdominal wall, just above the medial half of inguinal ligament (Fig. 5.9).

It extends from deep to superficial inguinal ring, downwards and medially.

Q.47 What is the surface marking of deep inguinal ring?

- Situated half an inch above the mid-point between anterior superior iliac spine and pubic symphysis (mid-inguinal point).
- Oval opening in fascia transversalis.
- Larger in males.

Q.48 What is the surface marking of superficial inguinal ring?

- Just above and lateral to pubic crest.
- Triangular gap in external oblique aponeurosis.
- Medial to ring lie inferior epigastric vessels (Fig. 5.10).

Q.49 What are the boundaries of inguinal canal?

Anterior

Skin.
Superficial fascia.
External oblique aponeurosis, over entire length of canal.
Fibers of internal oblique in lateral 1/3 of canal.

Posterior

Fascia transversalis.
Extraperitoneal connective tissue.
Parietal peritoneum.
Conjoint tendon in medial 2/3.
Reflected part of inguinal ligament at medial end.

Roof

Arched fibers of internal oblique and transversus abdominis.

Floor

Union of inguinal ligament with fascia transversalis.
Lacunar ligament at medial end.

Q.50 Name the structures passing through inguinal canal.

- Spermatic cord in males.
- Round ligament of uterus in females.
- Ilioinguinal nerve in both sexes.

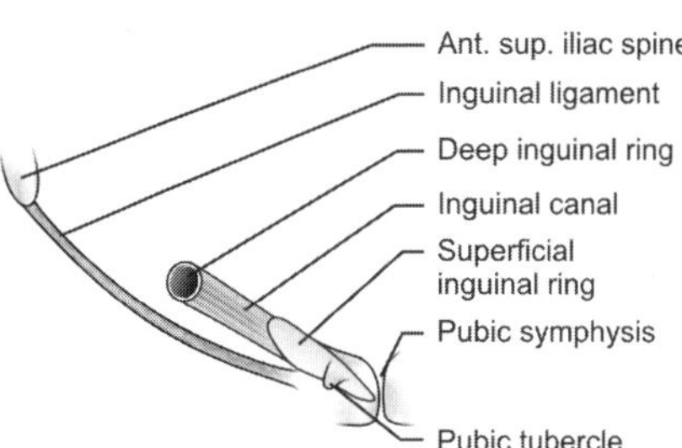

Fig. 5.9: Diagram to show the position of the inguinal canal

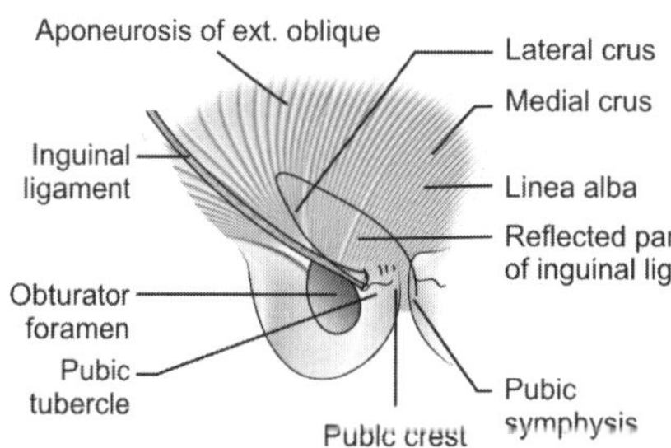

Fig. 5.10: Diagram to show the structure of the superficial inguinal ring

Q.51 Name the structures passing through deep inguinal ring.

Same as above except ilioinguinal nerve, which enters between external and internal oblique muscles and passes out through superficial inguinal ring.

Q.52 What are the boundaries of Hesselbach's triangle?

Laterally: Inferior epigastric artery.

Medially: Lateral border of rectus abdominis.

Inferiorly: Medial half of inguinal ligament.

It is divided into two unequal portions by obliterated umbilical artery.

Q.53 What is a hernia?

It is the protrusion of the contents of abdomen (usually gut) through an opening or weak area in wall of the body, e.g. femoral canal, inguinal canal, epiploic foramen.

Q.54 What are the factors which prevent the herniation through inguinal canal?

- The inguinal canal lies obliquely in abdominal wall, so deep and superficial ring do not lie opposite each other.
- Weakened posterior wall of canal due to deep ring is compensated by thickening of anterior wall by internal oblique muscle.
- Weakened anterior wall of canal due to superficial ring is compensated by presence of conjoint tendon and reflected part of inguinal ligament in posterior wall.
- With increased intra-abdominal pressure, anterior and posterior walls of canal get pressed together closing the canal.
- Contraction of internal oblique obliterates the canal, which it reinforces from above, front and behind.
- Contraction of external oblique closes the superficial ring.
- Contraction of cremaster pulls the spermatic cord upwards, making it thicker and closing the superficial ring.

Q.55 What are the different types of inguinal hernia?

- *Indirect (oblique) inguinal hernia:* Herniation occurs through the deep inguinal ring, lateral to inferior epigastric artery.
- *Direct inguinal hernia:* Occurs through the Hesselbach's triangle by pushing through the posterior wall of canal. Two types:
  - *Medial direct hernia:* Medial to obliterated umbilical artery.
  - *Lateral direct hernia:* Lateral to obliterated umbilical artery.

Q.56 What is incomplete and complete inguinal hernia?

Inguinal hernia is incomplete when it does not pass beyond the superficial inguinal ring. In complete hernia, the herniated gut descends in front of testis into tunica vaginalis.

Q.57 What are the main differences, between direct and indirect inguinal hernia?

Direct inguinal hernia	*Indirect inguinal hernia*
• Less frequent	More frequent
• Placed over the body of pubic bone.	Placed in the course of inguinal canal.
• Neck of the sac Medial to the inferior epigastric artery	Lateral to the inferior epigastric artery.
• *Spermatic cord:* Lies on its posterior	Lies behind it. and lateral sides.
• Usually acquired.	Usually congenital.

Q.58 How will you clinically distinguish an inguinal hernia from a femoral hernia?

An inguinal hernia lies above and medial to the medial end of inguinal ligament at its attachment to pubic tubercle. Femoral hernia lie below and lateral to pubic tubercle.

Q.59 In which sex:

- **Inguinal hernia common.**
- **Femoral hernia common and why?**
- In males because of greater diameter of deep inguinal ring.
- In females because of larger femoral ring due to broader pelvis and changes in tissues produced by the pregnancy.

Q.60 What can be the contents of a hernial sac?

- Omentum
- Intestine
- Portion of circumference of intestine
- Portion of bladder
- Fluid.

Q.61 What is Littre's hernia?

When the Meckel's diverticulum present in the hernial sac.

Q.62 What is a 'strangulated hernia'?

When the blood supply to hernial contents become impaired thus leading to the death of the tissue.

Q.63 How will you clinically distinguish a direct from an indirect inguinal hernia?

By deep ring occlusion test in cases of reducible hernia. The hernia is first reduced and deep (internal) inguinal ring is occluded with fingertip and patient is asked to cough while standing. If it is an indirect hernia, as the deep ring is occluded, it prevents hernial contents from descending into scrotum. But a direct hernia will protrude as contents herniate through the posterior wall of inguinal canal.

Q.64 What are the coverings of inguinal hernia?

Indirect hernia: From without inwards:

1. Skin.
2. Fascia of Camper
3. Fascia of Scarpa
4. External spermatic fascia
5. Cremasteric fascia
6. Internal spermatic fascia
7. Extraperitoneal areolar tissue
8. Parietal peritoneum

Direct hernia:

- Lateral: 1,2,3,4 same as above
- Fascia transversalis
- Extraperitoneal tissue
- Parietal peritoneum
- Medial: 1, 2, 3, 4 same as above
- Conjoint tendon
- Fascia transversalis
- Extraperitoneal tissue
- Parietal peritoneum

Q.65 What is the developmental origin of inguinal canal?

It represents the passage of gubernaculum through the abdominal wall.

It extends from caudal end of developing gonad to labioscrotal swelling.

Male External Genital Organs

Q.66 What are the parts of penis?

See Figures 5.11 and 5.12.

- *Root* (Attached part): Consists of two crura and one bulb.
  The two crura are attached to the inferior ramus of pubis and ramus of ischium and are covered by ischiocavernosus muscle. The bulb is covered by bulbospongiosus muscle.
- *Body* (Free part): Consists of two corpora cavernosa which are dorsal and one corpus spongiosum, ventral to corpora cavernosa and is traversed by penile urethra.

Q.67 What is Buck's fascia?

It is the membranous layer of superficial fascia of the penis.

Q.68 What is the arterial supply of penis?
- Deep artery of penis Branches of
- Dorsal artery of penis internal
- Artery of bulb of penis pudendal artery
- Superficial external pudendal artery: Branch of femoral artery.

Q.69 What is the lymphatic drainage of penis?
The glans penis drains into deep inguinal nodes and rest of penis into upper medial group of superficial inguinal lymph nodes.

Q.70 What is the developmental origin of penis?
The genital tubercle at cranial end of cloacal membrane, which lengthens to form phallus which enlarges to form penis.

Q.71 What is scrotum?
It is cutaneous bag containing testis, epididymis and lower part of spermatic cord.

Q.72 Name the structures forming layers of scrotum.
From without inwards:
- Skin.
- Dartos: Smooth muscle, closely adherent to skin.
- External spermatic fascia.
- Cremasteric fascia.
- Internal spermatic fascia.

Q.73 What is the blood supply of scrotum.
- Superficial pudendal Branch of
- Deep external pudenal femoral artery
- Scrotal branch of internal pudendal
- Cremasteric branch of inferior epigastric

Q.74 What is the nerve supply of scrotum?
- Anterior 1/3 of scrotum: By L_1 segment through
  - Ilioinguinal nerve
  - Genital branch of genitofemoral nerve.
- Posterior 2/3 of scrotum: By S_3 segment through
  - Posterior scrotal branch of pudendal nerve
  - Perineal branch of posterior cutaneous nerve of thigh.

Q.75 Why the extravasation of fluid into scrotal sac is bilateral?
Because the septum which divides scrotum into right and left compartments, is incomplete superiorly.

Q.76 What is the situation of testis?
- It is suspended in scrotum by spermatic cord.
- It lies obliquely, so that upper pole is tilted forwards and a little laterally and lower pole backwards and medially.
- Left testis is lower than the right.

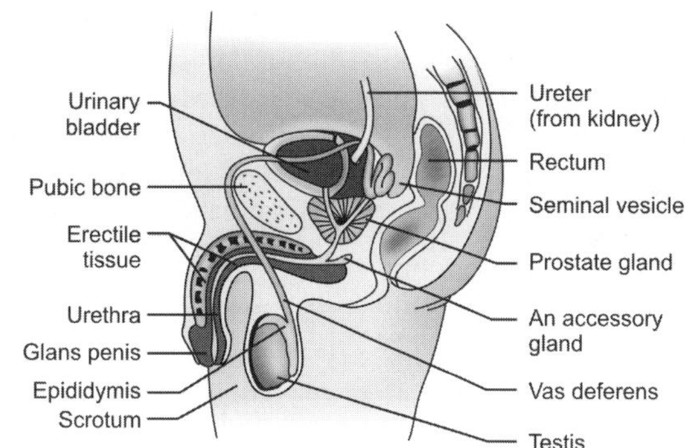

Fig. 5.11: Male urogenital system

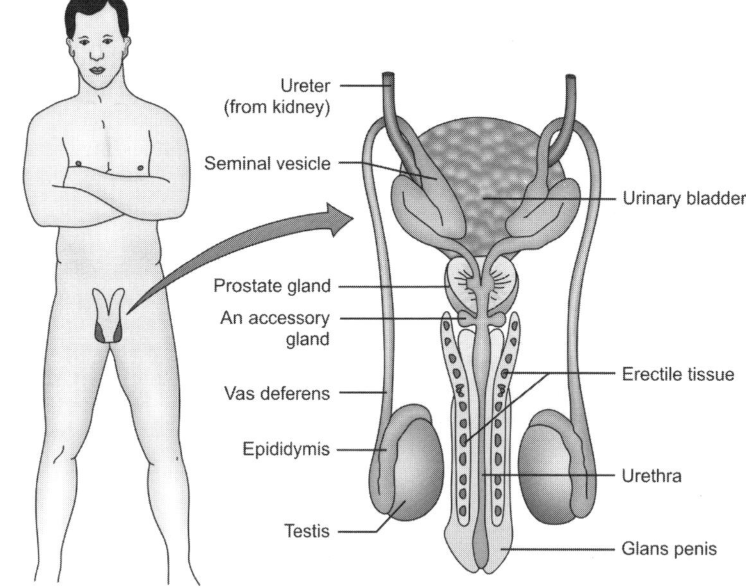

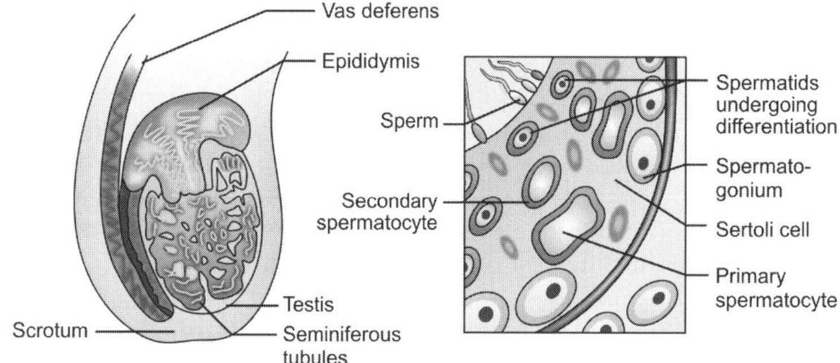

Fig. 5.12: Male genital system (magnified view)

Q.77 What is sinus of epididymis?
It is the extension of the cavity of tunica vaginalis between testis and epididymis from its lateral side, on posterior border.

Q.78 What is appendix of testis?
- Remnant of upper end of Mullerian duct.
- Minute, oval body at upper pole of testis just beneath the head of epididymis.
- Also called sessile hydatid of Morgagni.

Q.79 What are the coverings of testis?
From without inwards:
- Tunica vaginalis.
- Tunica albuginea.
- Tunica vasculosa.

Q.80 What is the arterial supply of testis?
Testicular artery: Branch of abdominal aorta. At posterior border of testis, it divides into branches:

- Small branches: Enter posterior border
- Larger branches: Pierce tunica albuginea and run on surface of testis to ramify on tunica vasculosa.

Q.81 What is pampiniform plexus?

- It is a plexus emerging from testis.
- The anterior part is arranged around testicular artery, middle part around ductus deferens and its artery and posterior part is isolated.
- At superficial inguinal ring, plexus condenses into 4 veins.
- At deep inguinal ring into 2 veins.
  Ultimately, one vein is formed which drains into inferior vena cava on right side and left renal vein on left side.

Q.82 What is the lymphatic drainage of testis?

Pre-aortic and para-aortic lymph nodes at level of L_2 vertebra (Fig. 5.13).

Q.83 What is the structure of testis?

Testis is divided into 200–300 lobules by septae passing from mediastinum testis to tunica albuginea each containing one to three seminiferous tubules. The tubules anastomose posteriorly into 20–30 straight tubules. These unite in mediastinum testis to form, Rete testis from which efferent ducts arise and pass into head of epididymis.

Q.84 What is the developmental origin of testis?

Testis arises from mesodermal genital ridge in posterior abdominal wall just medial to developing kidney and links up with epididymis and vas, which develop from mesonephric duct (Wolffian duct).

- Primordial germ cells: Are endodermal and derived from dorsal wall of yolk sac.
- Cells of Sertoli: Derived from celomic epithelium.
- Cells of Leydig: Derived from mesenchymal cells of mesonephros.

Q.85 What is gubernaculum testis?

It is a fibromuscular band attaching the testis to the bottom of scrotum. According to Hunter, gubernaculum forms the inguinal canal by its passage through abdominal wall.

It develops from a mesenchymal strand.

Q.86 What is processus vaginalis?

It is prolongation of peritoneal cavity projecting into scrotum. The testis in scrotum slides posterior to this and projects into it. Thus the testis is covered by peritoneum from front and sides. About the time of birth it obliterates, leaving the testis covered by tunica vaginalis.

Q.87 What are the positions of testis during its descent in fetal life?

3rd month: Reaches iliac fossa
7th month: Deep inguinal ring
During 7th month: Traverses inguinal canal
8th month: Reaches superficial inguinal ring

Beginning of 9th month; Descends into scrotum.

Q.88 Why the cervical lymph nodes become enlarged in tumors arising from testis?

Because of the plentiful communications of para-aortic lymph nodes in abdomen with those of thoracic region and which inturn communicate with cervical nodes.

Q.89 What is varicocele?

It is the dilatation of pamipiniform plexus of veins. It is commoner on left side because of left testicular veins compression by loaded sigmoid colon, left kidney tumor which invade renal veins and obstructs the drainage of left testicular veins, obstruction by angulation at site of entry of left testicular veins into left renal vein, in which pressure is higher than in inferior vena cava.

Q.90 What are ectopic testis?

The testis descends but is found in an unusual position, e.g. under the skin of front of abdomen, under skin of thigh in femoral canal, under skin of penis or in perineum behind scrotum.

In these cases, the cord is long (unlike the undescended testes).

Q.91 What is a hydrocele? What are the two different types of hydrocele?

Hydrocele occurs due to accumulation of fluid within tunica vaginalis of the scrotum or along the spermatic cord.

Hydroceles can be of two types: Communicating and non-communicating hydroceles (Figs 5.14 to 5.16).

Communicating hydrocele occurs due to incomplete closure of tunica vaginalis. As a result, there is a communication with the fluids of the abdominal cavity. As a result, there may be continuous variation in the size of hydrocele. This type of hydrocele is usually present at birth.

Non-communicating hydrocele: This type of hydrocele may be present at birth or develop years later for no obvious reason. It usually remains same in size or has a very slow rate of growth.

The pathophysiology of hydrocele is related to either increased fluid production or impaired thid absorption.

Q.92 What is monorchidism?

Developmental absence of a testicle.

Q.93 What is vas aberrans of Haller?

It is blind tube which lies between the tail of epididymis and commencement of vas.

Q.94 What is the length of epididymis?

When uncoiled 20 feet, but in coiled form the comma shaped body is only 1 inches long on posterolateral aspect of testis.

Q.95 What is developmental origin of appendix of epididymis?

Represents cranial end of mesonephric duct. Also known as pedunculated hydatid of Morgagni.

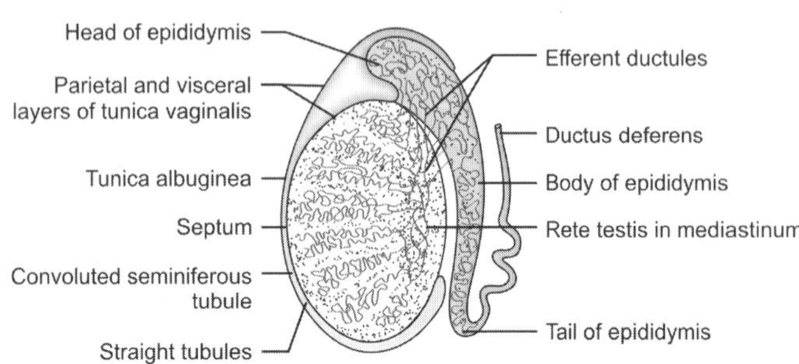

Fig. 5.13: Schematic coronal section through testis

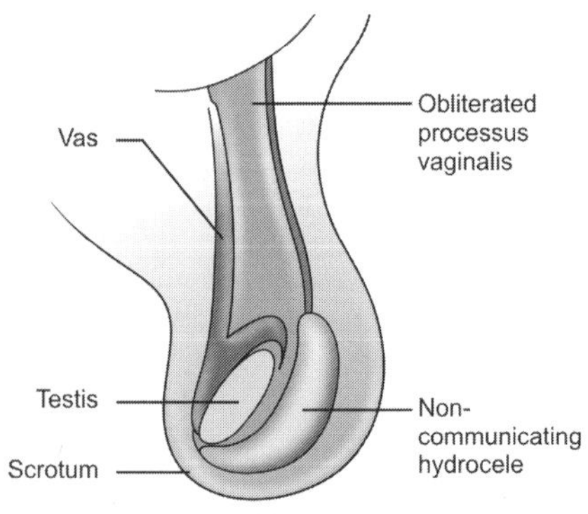

Fig. 5.14: Non-communicating hydrocele

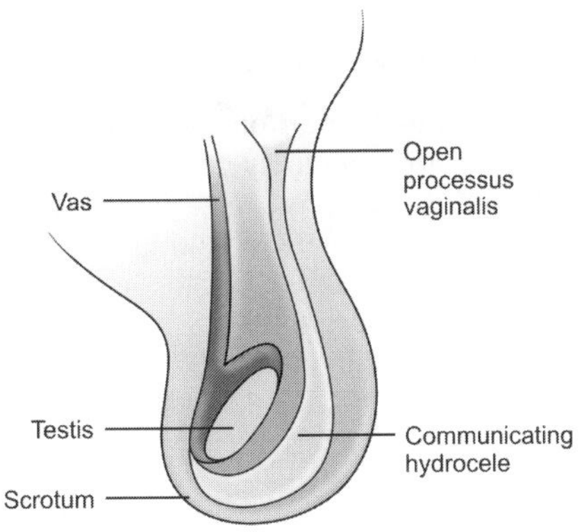

Fig. 5.15: Communicating hydrocele

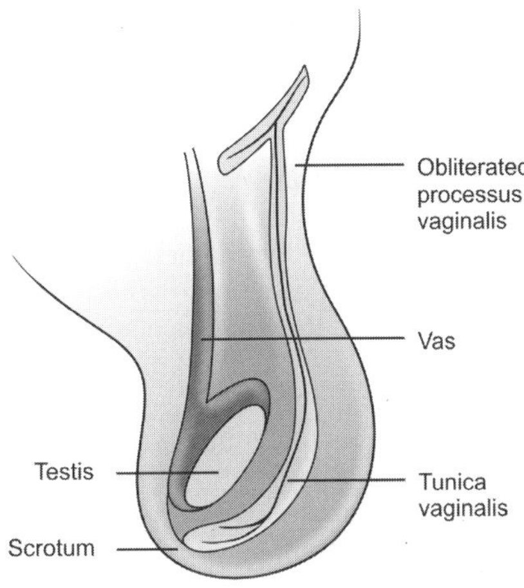

Fig. 5.16: Normal scrotum

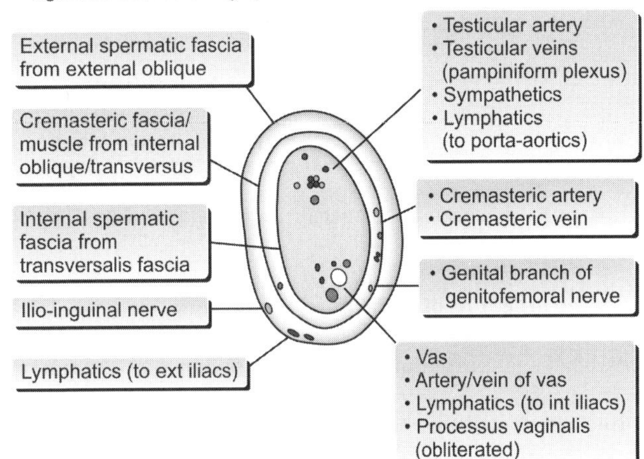

Fig. 5.17: Spermatic cord

Q.96 What is organ of Giraldes (Paradidymis)

Free tubules in spermatic cord above head of epididymis.

Represent caudal mesonephric tubules.

Q.97 What is the extent of spermatic cord?

It extends from the upper pole of testis, through the inguinal canal to the deep inguinal ring.

Q.98 What are the coverings of spermatic cord?

From within outwards (Fig. 5.17):

- Internal spermatic fascia: Derived from fascia transversalis.
- Cremasteric fascia: Derived from internal oblique and is fibromuscular.
- External spermatic fascia: Derived from external oblique aponeurosis. Present below the level of superficial inguinal ring.

Q.99 What are the constituents of spermatic cord?

- Vas deferens
- Veins: Pampiniform plexus
- Arteries: Testicular, cremasteric, artery of vas
- Nerves: Genital branch of genitofemoral nerve
  Testicular plexus of sympathetic nerves (T_{10})
  Sympathetic plexus around artery of vas.
- Lymphatics of testis
- Areolar tissue
- Remains of processus vaginalis.

Peritoneum (Figs 5.18A and B)

Q.100 What is peritoneum?

Peritoneum is a large serous membrane (sac) lining the abdominal cavity.

Q.101 What are the different parts of peritoneum?

The peritoneum is divided into:

- Outer layer, the parietal peritoneum.
- Inner layer, the visceral peritoneum.
- Folds of peritoneum, which suspend the viscera.
- Peritoneal cavity.

Q.102 What are the differences between parietal and visceral peritoneum?

Features	*Parietal peritoneum*	*Visceral peritoneum*
Position	Lines the inner surface of abdominal and pelvic walls (parieties) and lower surface of diaphragm.	Lines the outer surface of viscera.
Attachment	Loosely attached by extraperitoneal connective tissue.	Firmly adherent
Blood and nerve supply	Same as overlying parieties	Same as underlying viscera.
Pain sensitivity	Sensitive because of somatic innervation.	Insensitive because of autonomic innervation.
Development	Derived from somatopleural layer of lateral plate mesoderm	Derived from splanchnopleural layer of lateral plate mesoderm

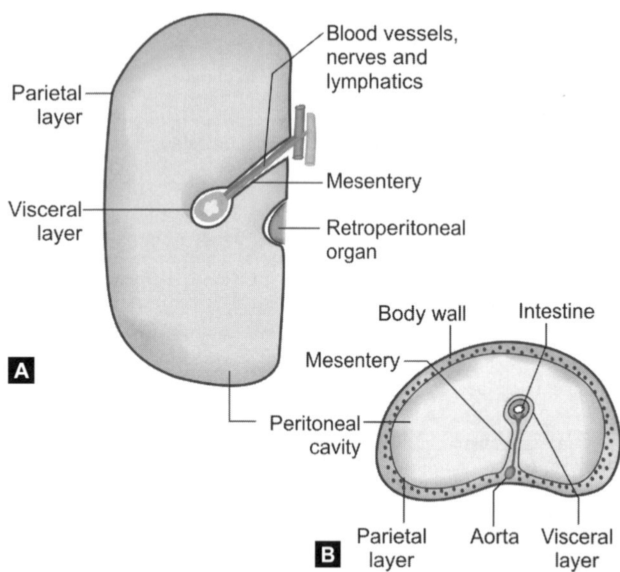

Figs 5.18A and B: Peritoneum—(A) sagittal section of abdomen, (B) TS of abdomen

Q.103 What are the functions of folds of peritoneum?

- These suspend the organs in abdominal cavity.
- Provide a degree of mobility to the organs.
- Provide media for the passage of vessels, nerves and lymphatics of the suspended organs.

Q.104 What are the different types of peritoneal folds?

The peritoneal folds are divided into 3 types:

- Omenta: Folds suspending the stomach.
- Mesentery: Folds suspending parts of small and large intestine.
- Ligaments.

Q.105 What is the peritoneal cavity?

It is a potential space lying between the parietal and visceral peritoneum.

Q.106 What are different parts of peritoneal cavity?

The peritoneal cavity is divided into two parts:

- *Greater sac:* Larger
- *Lesser sac:* Smaller, situated behind lesser omentum, stomach and liver. It also extends into interval between anterior and posterior parts of greater omentum. The two sacs communicate with each other through the epiploic foramen (foramen of Winslow).

Q.107 What are the retroperitoneal structures related to the lesser sac.

- Anterior surface of head, neck and body of pancreas
- Left kidney
- Left suprarenal gland
- Abdominal aorta, upper part
- Diaphragm
- Celiac trunk and its branches

Q.108 What are peritoneal fossae (recesses)?

These are small pockets of peritoneal cavity enclosed by small, inconstant of folds of peritoneum. More frequent in newborn babies and most of them become obliterated after birth. The largest of these is lesser sac. Smaller recesses are found in relation to duodenum, ileocecal region and sigmoid mesocolon.

Q.109 What is the 'policeman of abdomen'?

It is greater omentum hanging down from the greater curvature of stomach and covering the loops of intestine. It is called policeman of abdomen because it limits the spread of infection by moving to the site of infection and sealing it off from the surrounding areas.

Q.110 What are the contents of lesser omentum?

The right free margin of lesser omentum contains:

- Hepatic artery
- Portal vein
- Bile duct
- Hepatic plexus of nerves
- Lymph nodes and lymphatics.

  Along the lesser curvature of stomach and upper border of duodenum, it contains:
- Right and left gastric vessels
- Gastric lymph nodes and lymphatics
- Branches of gastric nerves.

Q.111 What are peritoneal ligaments?

These are the double layers of peritoneum connecting the viscera to each other or to the diaphragm or the abdominal wall or pelvic wall.

Example:

- Falciform ligament
- Right and left triangular ligaments
- Superior and inferior layers of coronary ligaments
- Gastrophrenic ligament
- Gastrosplenic ligament
- Lienorenal ligament
- Hepatogastric ligament
- Hepatoduodenal ligament
- Ligaments of the uterus and urinary bladder.

Q.112 What are the relations of root of mesentery?

It extends from duodenojejunal flexure on left side of L_2 vertebra to upper part of right sacroiliac joint and is about 15 cm long.

It crosses:

- Third part of duodenum
- Abdominal aorta
- Inferior vena cava
- Right ureter
- Right psoas major
- Right testicular (ovarian) vessels

Q.113 What are the contents of mesentery?

- Jejunal and ileal branches of superior mesenteric artery and veins
- Autonomic nerve plexus
- Lymphatics
- Lymph nodes
- Fat

Q.114 What is falciform ligament?

It is a sickle shaped fold of peritoneum which connects anterosuperior surface of liver to anterior abdominal wall and undersurface of diaphragm.

Q.115 What is the clinical importance of epiploic foramen?

Internal hernia can occur into lesser sac through the foramen. If the hernia becomes strangulated then it cannot be reduced by enlarging the foramen because of structures around it. So the gut is first aspirated and then reduced back through the epiploic foramen.

Q.116 At what level epiploic foramen lies?

It is situated at level of T_{12} vertebra, behind right free margin of lesser omentum.

Q.117 What are the boundaries of epiploic foramen?

Anteriorly: Right free margins of lesser omentum with structures in it.
Posteriorly: Inferior vena cava
Right suprarenal gland
T_{12} vertebra
Superiorly: Caudate process of liver
Interiorly: First part of duodenum and horizontal part of hepatic artery.

Q.118 What are subphrenic spaces?

These are the potential spaces below the diaphragm and are formed by reflections of peritoneum around liver, they are:

- *Intraperitoneal spaces:*
  - Left subphrenic space
  - Left subhepatic space (lesser sac)
  - Right subphrenic space
  - Right posterior (subhepatic) space
- *Extraperitoneal spaces:*
  - Right extraperitoneal space (bare area of liver)
  - Left extraperitoneal space around left suprarenal and upper pole of left kidney.

Q.119 Which part of the liver is related to superior recess of left subhepatic space?

Caudate lobe of liver.

Q.120 What is Morison's pouch? What is its clinical importance?

Right subhepatic space is known as Morison's pouch or hepatorenal pouch.
Clinical importance: This is the most dependent part of peritoneal cavity of abdomen proper in supine position. This is the commonest site of subphrenic abscess and also fluid effusions tend to accumulate here.

Q.121 What is rectouterine pouch (pouch of Douglas)?

This is the most dependent part of peritoneal cavity in sitting or standing position and of pelvic cavity in supine position. The floor of pouch is only 5.5 cm from the anus.

Q.122 What are the boundaries of rectouterine pouch?

Anteriorly: Uterus and posterior fornix of vagina.
Posteriorly: Rectum
Floor: Rectovaginal fold of peritoneum.

Q.123 What is clinical importance of rectouterine pouch?

This being the most dependent part of peritoneal cavity, so the pus tends to collect here and form the pelvic abscess.

Q.124 What is the clinical importance of peritoneal fossae?

Some of these may persist and may be the site of an internal hernia and strangulation.

Q.125 What is zygosis?

Some of the abdominal organs possess mesentery during the embryonic life, e.g. duodenum, ascending and descending colon, rectum. But due to fusion of their mesentery with peritoneum of posterior abdominal wall (zygosis) these become retroperitoneal.

Q.126 What are the different peritoneal fossae found?

- Lesser sac
- Duodenal fossae
- Superior duodenal fossa: Present in 50%
- Inferior duodenal fossa
- Para duodenal fossa: Present in 20%
- Duodenojejunal fossa: Present in 20%
- Retroduodenal fossa
- Mesentricoparietal fossa of Waldeyer
- Cecal fossae
- Superior ileocecal fossa
- Inferior ileocecal fossa
- Retrocecal fossa
- Intersigmoid fossa.

Q.127 What are the different retroperitoneal organs?

- Duodenum
- Ascending colon
- Descending colon
- Kidneys
- Rectum

Q.128 What is the developmental origin of peritoneum?

Parietal layer: From somatopleural layer of lateral plate mesoderm.
Visceral layer: From splanchnopleural layer of lateral plate mesoderm.

Q.129 What is the composition of peritoneal fluid?

- Water electrolytes and solutes derived from interstitial fluid of neighboring tissue and from plasma of adjacent blood vessels.
- Proteins.
- Desquamated mesothelial cells, macrophages, fibroblasts, lymphocytes.

Q.130 Why the irritation of peritoneum produces rigidity of abdominal muscles in that region?

The parietal peritoneum is supplied by somatic spinal nerves which also supply muscle and the skin of the parities, so, when parietal peritoneum is irritated the abdominal muscles are reflexly contracted, thus producing rigidity of abdominal wall in that region.

Q.131 What is ascites?

Collection of free fluid in peritoneal cavity.

Q.132 What is 'paracentesis abdominis' and from which site it is done?

It is the tapping of ascitic fluid.

It is done with a trocar and cannula by puncturing the abdomen either in median plane midway between umbilicus and pubic symphysis or at a point just above the anterior superior iliac spine.

Q.133 Why the herniation into paraduodenal fossa is associated with hemorrhoids (piles)?

The inferior mesenteric vein formed by superior rectal vein (which drains internal rectal venous plexus) runs in anterior wall of paraduodenal fossa, is compressed by the pressure of herniated gut.

Q.134 What are the functions of peritoneum?

- *Movement of viscera:* Peritoneum provides a slippery surface for free movement of abdominal viscera.
- *Protection of viscera:* Phagocytic cells of peritoneum guard against infections.
- *Absorption:* Fluid and solutes by mesothelium, which acts as a semipermeable membrane.
- *Healing and adhesions:* By transformation of mesothelium into fibroblasts.
- *Storage of fats:* Especially in peritoneal folds.

Q.135 What are the differences between male and female peritoneum?

In male:
Peritoneum is a closed sac lined by mesothelium.
In female:

- Peritoneal cavity communicates with exterior through uterine tubes.
- Peritoneum covering ovaries is lined by cubical epithelium.
- Peritoneum covering fimbria is lined by columnar ciliated epithelium.

Regions of Abdomen

Q.136 How is the abdomen divided into various regions?

Abdomen is divided into nine regions by:

- Two vertical planes: Right and left lateral planes. Passing from midinguinal point and crossing tip of ninth costal cartilage and passing up to midpoint between medial and lateral ends of clavicle (mid clavicular lines).

- Two horizontal planes:
  - *Transpyloric plane:* Passes through tip of 9th costal cartilage and lower border of L_1. It lies between upper border of manubrium sterni (suprasternal notch) and upper border of symphysis pubis.
  - *Transtubercular plane:* Passes through tubercle of iliac crest and body of L_5 vertebra near upper border.
  Two additional transverse planes have been described:
  - *Subcostal plane:* Can be used in place of transpyloric plane. Passes through the lower border of 10th costal cartilage, i.e. lowest part of costal margin and upper part of body of L_3.
  - *Supracristal plane:* Lies at the level of highest point of iliac crests and it passes posteriorly through spine of L_4 vertebra.

Q.137 What is 'linea semilunaris'?
It is curved line from the pubic tubercle to the tip of 9th costal cartilage, present on lateral edge of the rectus abdominis muscle.

Q.138 Name the structures lying at level of L_1 vertebra.
- Transpyloric plane
- Pylorus
- Duodenojejunal flexure
- Pancreas
- Hilum of kidneys.

Q.139 Name the structures lying at level of L_5 vertebra.
- Inter (trans) tubercular plane
- Common iliac veins end
- Inferior vena cava begins.

Q.140 Name the structures lying at level of L_2 vertebra.
- Spinal cord ends
- Thoracic duct begins
- Azygous vein begins.

Celiac Trunk

Q.141 Name the structures supplied by celiac trunk.
The celiac trunk supplies derivatives of foregut, i.e.
- Lower end of esophagus
- Stomach
- Upper 1½ parts of duodenum upto major duodenal papilla
- Liver
- Spleen
- Greater part of pancreas.

Q.142 How the celiac trunk develops?
The celiac trunk develops from one of the vitelline arteries (C_7 segment).

Q.143 What are the branches of celiac trunk?
- Left gastric
- Hepatic and
- Splenic arteries.

Q.144 What are the branches of hepatic artery?
- Gastroduodenal artery: It divides into:
  - Right gastroepiploic artery and
  - Superior pancreaticoduodenal artery
- Hepatic artery proper
- Right gastric artery
- Supraduodenal artery
- Cystic artery

Q.145 What are the branches of splenic artery?
- Pancreatic branches
- Short gastric arteries
- Left gastroepiploic artery
- Splenic branches

Superior Mesenteric Vessels

Q.146 Name the structures supplied by superior mesenteric artery.
Superior mesenteric artery supplies the derivatives of midgut, i.e.
- Lower 2½ parts of duodenum below the major duodenal papilla.
- Jejunum
- Ileum
- Appendix
- Cecum
- Ascending colon
- Right 2/3 of transverse colon
- Lower 1/2 of head of pancreas.

Q.147 At what level the superior mesenteric artery arises?
It arises from abdominal aorta at L_1 vertebra behind the body of pancreas.

Q.148 Name the branches of superior mesenteric artery?
- Inferior pancreaticoduodenal
- Jejunal
- Ileal
- *Ileocolic:* To terminal part of ileum, appendix, cecum and lower one-third of ascending colon.
- *Right colic:* To upper two-thirds of ascending colon.
- *Middle colic:* To right two-thirds of transverse colon.

Q.149 What are the relations of superior mesenteric artery?
- Above the root of mesentery.
  - *Anteriorly:* Body of pancreas and splenic vein
  - *Posteriorly:* Aorta, left renal vein, uncinate process of pancreas and third part of duodenum.
  - *To the right:* Superior mesenteric vein.
- Within the root of mesentery
  - *It crosses:* Inferior vena cava, Right ureter, Right psoas.
  - *To its right:* Superior mesenteric vein.

Q.150 What are the tributaries of superior mesenteric vein?
- Inferior pancreaticoduodenal
- Jejunal
- Ileal
- Ileocolic
- Right colic
- Middle colic and
- Right gastroepiploic vein.

Inferior Mesenteric Vessels

Q.151 What are the structures supplied by the inferior mesenteric artery?
It supplies the derivatives of hindgut, i.e.
- Left 1/3 of transverse colon
- Descending colon
- Sigmoid colon
- Rectum
- Anal canal above the pectinate line.

Q.152 What are the branches of inferior mesenteric artery?
- Left colic: Left one-third of transverse colon and descending colon.
- Sigmoid and
- Superior rectal artery.

Q.153 What is 'marginal artery'?
It is an arterial arcade situated along the concavity of colon formed by anastomosis between ileocolic, right colic, middle colic, left colic and sigmoid arteries. Vasa recta arise from the marginal artery and supply the colon.

Q.154 What is the clinical importance of marginal artery?
It forms extensive anastomosis, so it is capable of supplying the colon even in absence of one of the main feeding trunks. This fact is utilized in surgery of colon.

Portal Vein

Q.155 What is the characteristic feature of portal vein?

Portal vein is one vein which begins and also ends in capillaries, i.e. the vein formed from capillaries of an organ, enter another organ where they divide into another set of capillaries.

Q.156 Name the areas from which the blood is drained by the portal vein.

- Abdominal part of alimentary tract
- Spleen
- Gallbladder
- Pancreas

Q.157 What are the divisions of portal vein?

Portal vein is divided into 3 parts: Infraduodenal, retroduodenal and supraduodenal.

Q.158 What are the relations of different parts of portal vein?

- *Infraduodenal part:*
  Anteriorly: Neck of pancreas
  Posteriorly: Inferior vena cava
- Retroduodenal part:
  Anteriorly: First part of duodenum
  Gastroduodenal artery
  Common bile duct
  Posteriorly: Inferior vena cava.
  C. Supraduodenal part: Lies in the free margin of lesser omentum.
  Anteriorly: Bile duct
  Hepatic artery.
  Posteriorly: Inferior vena cava.

Q.159 How the portal vein forms and terminates?

Formation: Portal vein is formed at the level of L_2 vertebra behind the neck of pancreas, by union of superior mesenteric and splenic veins.

Termination: It ends at the right end of porta hepatis by dividing into a right and a left branch.

Q.160 What are the tributaries of portal vein?

- Splenic vein
- Superior mesenteric vein
- Left gastric vein
- Right gastric vein
- Superior pancreaticoduodenal vein
- Cystic vein
- Paraumbilical veins.

Q.161 What are the sites of portal-systemic communications? Name also the portal and systemic veins forming these.

Sites	*Portal vein*	*Systemic vein*
Umbilicus	Left branch of portal vein through paraumbilical vein	Veins of anterior abdominal wall
Lower end of esophagus	Esophageal tributaries of left gastric vein	Esophageal tributaries of the accessory hemiazygous vein
Anal canal	Superior rectal vein	Middle and inferior rectal vein
Bare area of liver	Hepatic vein	Phrenic and intercostal veins
Posterior abdominal wall	Veins of duodenum, ascending and descending colon	Retroperitoneal veins of posterior abdominal wall (renal, lumbar and phrenic veins)
Liver	Rarely ductus venosus remains patent and then connects left branch of portal vein	Inferior vena cava

Q.162 What is the importance of portal-systemic communications?

These communications form the important pathways of collateral circulation in portal obstruction and portal hypertension.

Q.163 What is the developmental origin of portal vein?

- *Infraduodenal part:* Part of left vitelline vein
- *Retroduodenal part:* Dorsal anastomosis between two vitelline veins
- *Supraduodenal part:* Part of right vitelline vein.

Spleen

Q.164 What is position of spleen in abdomen?

The spleen lies obliquely along the long axis of 10th rib. It lies mainly in left hypochondrium but the posterior end extends into epigastrium. It is directed downwards, forwards and laterally.

Q.165 What is the average size and weight of spleen?

Spleen is 1 inch thick, 3 inches broad, 5 inches long, 7 ounces in weight (Fig. 5.19).

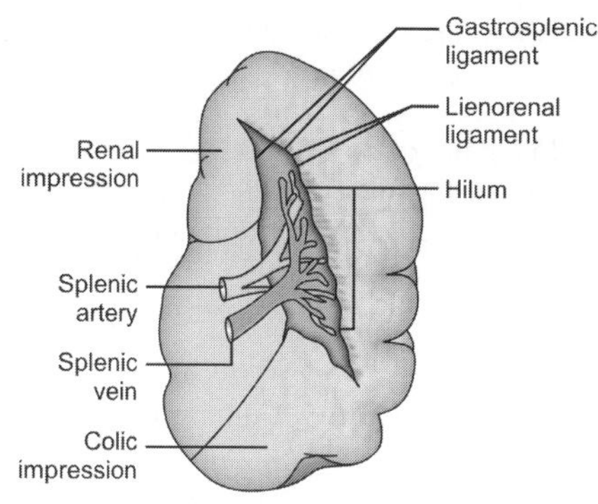

Fig. 5.19: Spleen

Q.166 What are the ends, borders and surfaces of spleen?

Ends:

- *Anterior:* Expanded, directed downwards, forwards and laterally.
- *Posterior:* Rounded, directed upwards, backwards and medially.

Borders:

- *Superior:* Sharp, notched near anterior ends, separates diaphragmatic surface from gastric impression.
- *Inferior:* More rounded and separates renal impression from diaphragmatic surface.
- *Intermediate border:* Rounded and separates gastric and renal impressions.

Surfaces:

- Diaphragmatic surface: convex.
- *Visceral surface:* Concave and has following impressions:
  Gastric impression: For fundus of stomach.
  Renal impression: For left kidney.
  Colic impression: For left flexure of colon.
  Pancreatic impression: For tail of pancreas.

Q.167 Name the structures lying at hilum of spleen?

The hilum transmits splenic vessels and nerves. It provides attachment to gastrosplenic and lienorenal ligaments.

Q.168 What are the peritoneal ligaments attached to spleen?

- Gastrosplenic ligament
- Lienorenal ligament
- Phrenico-colic ligament.

Q.169 What are the structures lying in the gastrosplenic ligament?

- Left gastroepiploic vessels
- Short gastric vessels
- Lymphatics
- Sympathetic nerves
- Fat.

Q.170 Name the structure lying in the lienorenal ligament?

It contains:
- Tail of pancreas
- Splenic vessels
- Pancreaticosplenic lymph nodes
- Lymphatics
- Sympathetic nerves
- Fat.

Q.171 What is the blood supply to the spleen?

Spleen is mainly supplied by splenic artery which is a branch of celiac trunk. It runs tortuously to the left along the superior border of the pancreas and finally enters into the splenic hilum (Fig. 5.20).

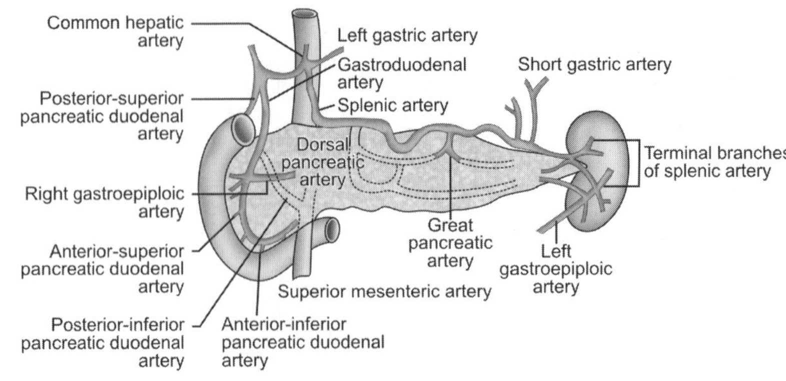

Fig. 5.20: Arterial supply of spleen

Q.172 Name the tributaries of splenic vein.
- Short gastric
- Left gastroepiploic
- Pancreatic and
- Inferior mesenteric vein.

Q.173 What are the functions of spleen?

The spleen has following functions:
- *Phagocytosis:* By the reticular cells, free macrophages and endothelial cells. They remove cell debris and old RBCs and other blood cells and micro-organisms.
- *Hemopoiesis:* Important during fetal life. lymphopoiesis continues throughout life.

Q.174 What is the developmental origin of spleen?

From left layer of cephalic part of dorsal mesogastrium, into a number of nodules which fuse to form a lobulated mass.

Q.175 What are the 'accessory spleen'?

These are the splenic nodules which have failed to fuse to form a lobulated mass.

Q.176 What is Kehr's sign?

Splenic infarction due to obstruction of branches of splenic artery, causes referred pain in left shoulder due to irritation of undersurface of diaphragm by effused blood.

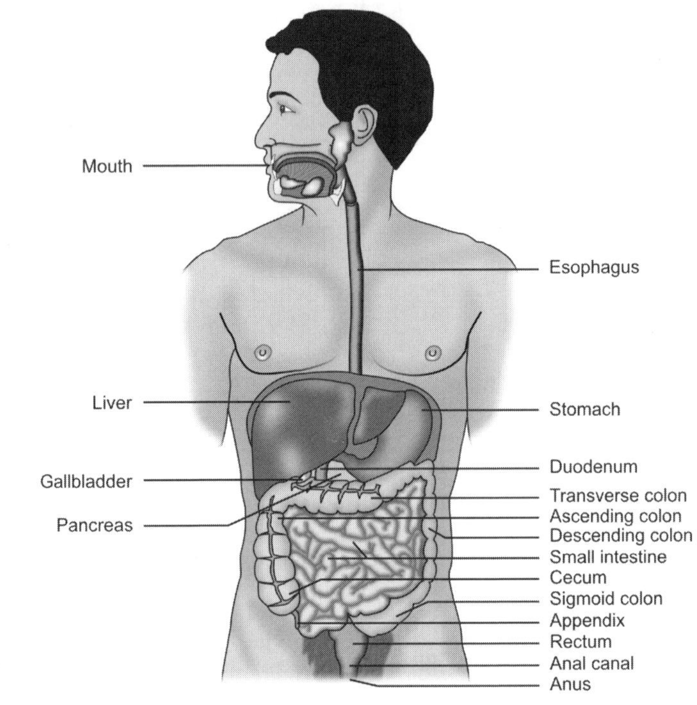

Fig. 5.21: Digestive system

Stomach

Q.177 What is the position of stomach?

It lies obliquely in upper and left part of abdomen, occupying epigastric, umbilical, left hypochondrium and left lumbar region (Fig. 5.21).

Q.178 What are the normal variations in capacity and shape of stomach?

Capacity: At birth: 30 mL
At puberty: 1000 mL
In adults: 1.5–2 litres

Shape: When empty: J-shaped
When distended: Pyriform

In obese: More horizontal (steerhorn stomach).

Q.179 What are the different parts of stomach?
- *Cardiac part:* Subdivided into:
  - *Fundus:* Part of stomach lying above the cardioesophageal junction.
  - Body.
  - *Pyloric part:* Subdivided into:
    Pyloric antrum
    Pyloric canal

Stomach has:
two orifices: Cardiac and pyloric
two curvatures: Lesser and greater
two surfaces: Anterior and posterior.

Q.180 How the pyloric orifice is recognised by a surgeon?

By:
- Circular groove (pyloric constriction) produced by pyloric sphincter which feels like a firm ring.
- Prepyloric vein: Lies anteriorly in pyloric constriction.

Q.181 What is the level of orifices of stomach?

Cardiac orifice (Physiological sphincter): T_{11} vertebra.

Pyloric orifice: L_1 vertebra.

Q.182 Name the 'bare areas' of stomach?
- Greater and lesser curvatures, along the peritoneal reflections.
- Triangular area on posterior surface close to cardiac orifice and related to the left crus of diaphragm.

Q.183 Name the structures forming the 'stomach bed.'

These structures are related to posterior surface of stomach.

- *Spleen:* Related to fundus and is separated by the cavity of greater sac.
- Other structures are separated by cavity of lesser sac:
  - Diaphragm
  - Left suprarenal
  - Left kidney
  - Splenic artery
  - Pancreas
  - Transverse mesocolon
  - Splenic flexure of colon.

Q.184 What is the blood supply of stomach?

Arterial supply (Fig. 5.22):

- *Left gastric artery:* Branch of celiac trunk
- *Right gastric artery:* Branch of gastroduodenal artery
- *Right gastroepiploic artery:* Branch of common hepatic
- *Left gastroepiploic artery:* Branch of splenic.
  Short gastric arteries: Branches of splenic.

Venous drainage: Into superior mesenteric and splenic veins, which pass into the portal vein.

Q.185 What is the lymphatic drainage of stomach?

For lymphatic drainage, the stomach is divided into four regions by imaginary lines as follows:

- Draw a vertical line just to the left of the cardioesophageal junction. It drains into pancreaticosplenic nodes, which drain into celiac nodes.
- Draw a vertical line separating pyloric part from the body. The area between two vertical lines is divided into upper 2/3 and lower 1/3 by a curved line parallel to greater curvature. Upper 2/3 is drained by left gastric nodes, which drain into celiac nodes and lower 1/3 drains into right gastroepiploic nodes which in turn drain into pyloric nodes, then hepatic nodes and finally celiac nodes.
  The pyloric part drains into pyloric, hepatic and left gastric nodes which inturn drain into celiac nodes.
  From celiac nodes, it passes to intestinal lymph trunk to reach cisterna chyli.

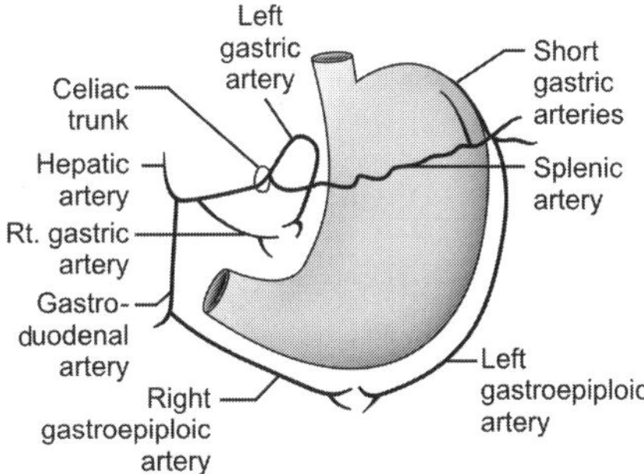

Fig. 5.22: Arteries supplying the stomach

Q.186 What is the nerve supply of stomach?

- *Sympathetic nerves:* T_6-T_{10} segments from celiac plexus. These are:
  - Vasomotor,
  - Motor to pyloric sphincter,
  - Chief pathway for pain sensation.
- *Parasympathetic nerves:* Vagus as:
  - Anterior gastric nerve (mainly left vagal fibers): Supplies anterior surface of fundus and body of stomach, pylorus and pyloric antrum.
  - Posterior gastric nerve (mainly right vagal fibers): Supplies posterior surface of fundus, body and pyloric antrum and gives a branch to celiac plexus.

  These are motor and secretomotor to stomach.

Q.187 What are 'nerve of Latarjet'?

Anterior and posterior vagi are also known as nerves of Latarjet.

Q.188 What are the functions of stomach?

- As a reservoir of food
- *Digestion:* Mainly breakdown of proteins to peptones
- *As antiseptic acid barrier:* By HCl
- *Self protection:* From HCl by mucus
- *Absorption:* Salt, water, alcohol and certain drugs
- Secretion of intrinsic factor of Castle.

Q.189 What is the advantage of highly selective vagotomy?

It does not cause stomach atony. Branches which supply the acid secreting body of stomach are only cut, thus preserving innervation and function of pyloric antrum.

Q.190 In which part of the stomach, X-ray shows gas?

Fundus of stomach which appears as a dark shadow below left dome of diaphragm.

Q.191 What is gastric canal? What is its clinical importance?

These are the mucosal folds (rugae) along the lesser curvature which are arranged longitudinally to form a canal.

Clinical importance: Gastric canal allow rapid passage of fluid along the lesser curvature to lower part before it spreads to other parts of stomach. Thus it is irritated most by the swallowed liquids and hence it is more vulnerable to peptic ulcer.

Q.192 What are the different types of glands in stomach?

- *Cardiac glands:* Tubular glands.
- *Glands of body and fundus:* Tubular glands.
- *Pyloric glands:* Convoluted tubular glands.

Q.193 What are the cell types present in glands of stomach?

- *Mucous cells:* Secrete mucous. Present in pyloric antrum and pyloric canal.
- *Zymogen cell (chief cells):* Present in glands of fundus and body. Secrete gastric enzymes.
- *Oxyntic cells (parietal cells):* Present in glands of fundus and body; secrete HCl
- *Mucous neck cells:* Present at neck of glands; secrete mucus.
- *Argentaffin cells:* Present at base of gastric glands of fundus. Secrete gastrin and serotonin.

Q.194 What is 'leather bottle stomach'?

Thickening of stomach wall due to proliferation of fibrous tissue especially in submucosa. The mucous membrane appears normal.

Q.195 What is the commonest site of gastric ulcer?

Gastric ulcers are usually found in distal part, near the lesser curvature (Fig. 5.23).

Q.196 What is the cause of fatal hemorrhage in perforated gastric ulcer?

Perforation of gastric ulcer on posterior wall of stomach can lead to erosion of splenic artery leading to fatal hemorrhage.

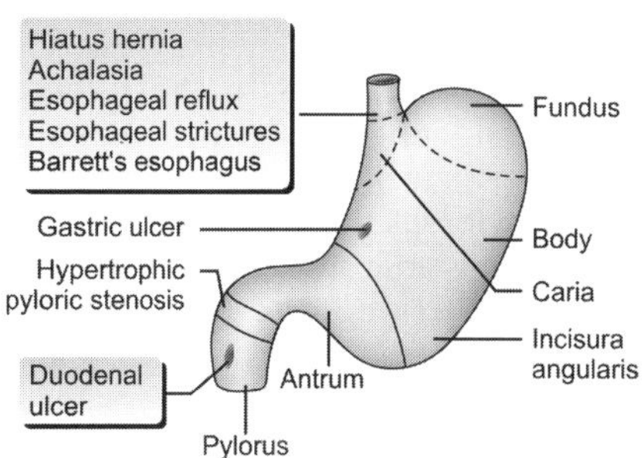

Fig. 5.23: Various pathologies affecting the stomach

Liver

Q.197 What are the anatomical lobes of the liver?
The liver is divided anatomically into two lobes, a right and a left by falciform ligament anteriorly and superiorly, by the fissure for ligamentum teres inferiorly and by the fissure for ligamentum venosum posteriorly.

Right lobe forms 5/6 part of liver and has two additional lobes:
- *Caudate lobe* on the posterior surface
- *Quadrate lobe* on the inferior surface and is rectangular in shape.

Left lobe forms 1/6 of liver.

Q.198 What is porta hepatis?
It is a deep transverse fissure, 5 cm long, on the inferior surface of right lobe of liver, between quadrate lobe below and front and caudate lobe above. Through it vessels, nerves and ducts pass to and from liver.

Lips of porta hepatis provide attachment to lesser omentum.

Q.199 What are the structures lying in the porta hepatis and relations of these within it?
Through porta hepatis portal vein, hepatic artery and hepatic plexus of nerve enter and right and left hepatic ducts and a few lymphatics pass out of the liver.

Within the porta hepatis, from behind forwards lie portal vein, hepatic artery and bile ducts.

Q.200 What are the boundaries of caudate lobe?
Caudate lobe is bounded on the right by the groove for inferior vena cava, on left by the fissure for ligamentum venosum and inferiorly by porta hepatis. Above it is continuous with the superior surface.

Q.201 What are the bare areas of the liver?
These are the parts of the liver not covered by the peritoneum. These include (Fig. 5.24):
- Main bare area: Situated on the posterior surface of the right lobe of liver, limited by coronary and right triangular ligament.
- Groove for inferior vena cava: Situated on the posterior surface of right lobe of liver, between caudate lobe and main bare area.
- Gallbladder fossa: On the inferior surface of the right lobe of the liver, on to the right of the quadrate lobe.
- Porta hepatis and
- Along the lines of reflection of peritoneum.

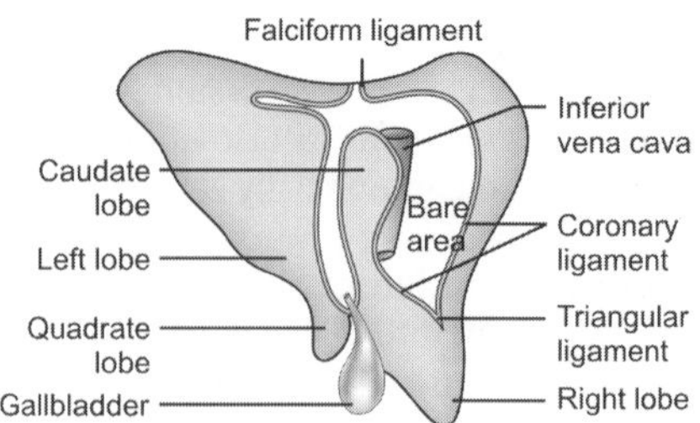

Fig. 5.24: Bare area of liver

Q.202 Which organ is related to main bare area of liver?
Right suprarenal gland.

Q.203 How is ligamentum teres formed?
Ligamentum teres is formed from left umbilical vein. It begins at umbilicus and ends by joining left branch of portal vein.

Q.204 What are the peritoneal ligaments of the liver?
Following are the peritoneal ligaments of the liver:
- *Falciform ligament:* Connecting antero-superior surface of liver to the anterior abdominal wall and under surface of diaphragm
- *Left triangular ligament:* Connecting superior surface of left lobe of liver to diaphragm.
- *Right triangular ligament:* Connects lateral part of posterior surface of right lobe of liver to the diaphragm.
- *Coronary ligament:* Encloses bare area of the liver, with superior and inferior layers.
- *Lesser omentum:* Attached to lips of porta hepatis.

Q.205 What are the boundaries of quadrate lobe of liver?
It is the part of visceral surface between fissure for ligamentum teres and fossa for gallbladder. It is bounded posteriorly and above by porta hepatis and anteriorly and below by inferior margin of liver.

Q.206 What are the relations of quadrate lobe of liver?
Quadrate lobe is related to lesser omentum, pylorus and first part of duodenum.

Q.207 What is ligamentum venosum?
It is the remnant of ductus venosus of fetal life. It is connected above to left hepatic vein near its entry into inferior vena cava and below to the left branch of portal vein, thus forming a bypass for blood during fetal life.

Q.208 What is the blood supply of the liver?
Liver receives 20% of its blood from hepatic artery and 80% from portal vein. Before entry, these divide into right and left branches. Within liver they divide to form segmental vessels and redivide into interlobular vessels, which run in portal canals. Further divisions open into the hepatic sinusoids. Thus in the hepatic sinusoids both arterial and venous blood mix.

The hepatic sinusoids drain into interlobular veins, which form sublobular veins and inturn form hepatic veins, which drain into inferior vena cava.

Q.209 What is 'portal triad'?
The interlobular branches of the hepatic artery and portal vein and an interlobular bile ductule together for a portal triad and lie within the portal canal.

Q.210 What are the functional lobes of the liver?
The liver is divided into two functional (physiological), right and left lobe, on the basis of intrahepatic distribution of hepatic artery, portal vein and biliary ducts. These lobes do not correspond to the anatomical lobes of the liver.

The physiological lobes are separated by a plane passing on the anterosuperior surface along a line joining the cystic notch to the groove for inferior vena cava, on the inferior surface the plane passes through gallbladder fossa and on the posterior surface through the middle of caudate lobe.

Each lobe is further divided and subdivided into segments.

Q.211 What is clinical importance of functional segments of liver?
The portal canals do not cross from one segment to the other, so the hepatic segments are of surgical importance.

Q.212 What is Riedel's lobe?
Sometimes, the lower border of the right lobe of liver, a little to right of gallbladder projects down as a tongue like process, this is known as Riedel's lobe.

Q.213 Why the gallbladder is rarely involved in the malignancy of liver?
Because of the absence of lymphatic pathways from liver to gallbladder.

Q.214 What is the developmental origin of liver?
From ventral surface of foregut as an outgrowth known as hepatic diverticulum close to point where it is continuous with yolk stalk. The diverticulum proliferates to form the liver. The connective tissue of liver is formed by the mesoderm of septum transversum.

Extra-Hepatic Biliary Apparatus

Q.215 What are the structures forming the extra-hepatic biliary apparatus?
It is formed by (Figs 5.25 and 5.26):
- Right and left hepatic duct
- Common hepatic duct
- Cystic duct and
- Bile duct.

Q.216 What are the parts of the gallbladder?
Gallbladder is divided into three parts:
- Fundus
- Body and
- Neck, it becomes continuous with cystic duct.

Q.217 What is a Hartmann's pouch?
It is the dilated posteromedial wall of the neck of gallbladder. It is directed downwards and backwards. Some regard it as pathological feature.

Q.218 What is the clinical importance of Hartmann's pouch?
The gallstones may become impacted in the pouch and cause obstruction.

Q.219 What is the capacity of gallbladder?
30 to 50 c.c., but is capable of 50 fold distention.

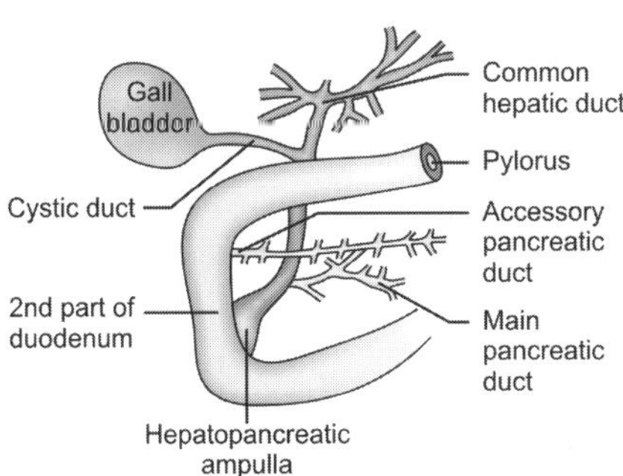

Fig. 5.25: Biliary tree

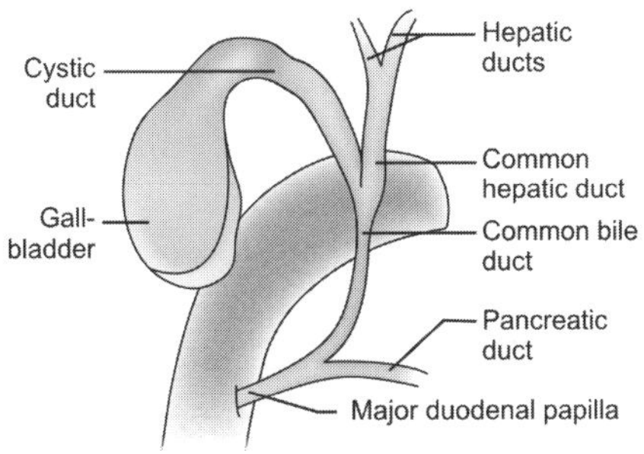

Fig. 5.26: Biliary system

Q.220 What is a common hepatic duct?
It is duct formed by the left and right hepatic duct. It is joined by cystic duct at an acute angle and then forms the common bile duct.

Q.221 What are accessory hepatic ducts and their clinical importance?
These are present in 15% subjects and arise usually from right lobe of liver. These terminate into gallbladder or common hepatic duct or bile duct.

If undetected, they are responsible for oozing of bile from wound after cholecystectomy i.e., removal of gallbladder.

Q.222 What is 'spiral valve of Heister'?
Spiral valve has 5–10 crescentic folds of mucous membranes in the cystic duct which are arranged spirally to form a valve-like structure.

Q.223 What are the relations of bile duct?
Bile duct is about 7 cm long. It lies from above downward in:
- Right margin of lesser omentum. It lies to right of hepatic artery and in front of portal vein.
- Behind first part of duodenum. The gastroduodenal artery lies to the left of bile duct.
- Behind head of pancreas and it lies in front of the inferior vena cava.

Q.224 What are the structures supplied by the cystic artery?
The cystic artery (usually a branch of right hepatic artery) supplies blood to gallbladder, cystic duct, hepatic ducts and upper part of the bile duct.

Q.225 What is the venous drainage of the gallbladder?
The superior surface of the gallbladder drains into hepatic veins through gallbladder fossa. Rest of gallbladder is drained by cystic veins.

Q.226 What are 'crypts of Luschka'?
The mucous membrane contains indentations of the mucosa that sink into the muscle coat, these are known as crypts of Luschka.

Q.227 What is caterpillar turn or 'Moynihan's hump'?
It is dangerous anomaly when the hepatic artery takes a tortuous course and the cystic artery is short. This tortuosity is known as caterpillar turn.

Q.228 What are the functions of gallbladder?
- Storage of bile
- Concentration of bile
- Regulates pressure in biliary system, to maintain normal choledochoduodenal mechanism.
- Secretion of mucin.
- Changing the reaction of bile: Bile excreted by liver has pH 8.2 and gallbladder changes the pH to 7.5–7.2.

Q.229 Why is referred pain felt over the right shoulder in acute cholecystitis?
The referred pain is felt at some other region having the same segmental innervation as the site of lesion (on right side). In acute cholecystitis, the under surface of the diaphragm is also inflamed. The pain sensation from under surface of diaphragm is carried by the phrenic nerve via C_4 spinal segment and skin over the shoulder is also supplied by the C_4 spinal segment.

Q.230 What is Courvoisier's law?
According to the Courvoisier's law the dilatation of the gallbladder occurs only in extrinsic obstruction of bile duct e.g., by carcinoma of head of pancreas. Intrinsic obstruction (e.g., by stones) does not cause any dilatation because of associated fibrosis.

Q.231 What is the role of gallbladder in typhoid fever?
The typhoid bacilli persist in the gallbladder and a carrier state develops and typhoid bacilli are disseminated to the population via the feces.

Q.232 What is Charcot's triad of cholecystitis?
Stone in bile duct causes:
- Intermittent biliary colic
- Intermittent jaundice following each colic and
- Intermittent fever.

Q.233 What are the characteristic histological features of gallbladder?
Gallbladder is composed of:
- Mucosa: Lined by tall columnar cells.
- Fibromuscular coat: Formed of interlacing bundles of smooth muscle fibers. Connective tissue, with abundant elastic fibers, lie between muscle bundles.
- No submucosa
- Serous coat with subserous areolar connective tissue.

Q.234 What is the developmental origin of extra hepatic biliary apparatus?
- Bile duct is formed by the narrowing of connection between hepatic diverticulum and foregut.
- Another ventral outgrowth from common bile duct forms the cystic duct and

gallbladder. Hepatic ducts are formed by lower end of hepatic diverticulum. The bile duct first opens into ventral wall of duodenum, later it migrates to dorsal (right) surface of duodenum to mesenteric border. This migration occurs due to differing rates of growth of duodenal walls.

Q.235 How the hemorrhage during cholecystectomy is controlled?
By compressing the hepatic artery, which gives off cystic branch, between finger and thumb where it lies in anterior wall of foramen of Winslow.

Q.236 Why the gangrene of gallbladder is uncommon in occlusion of cystic artery?
Because of rich secondary blood supply coming from liver bed.

Q.237 What are the developmental anomalies of gallbladder?
- Absence of gallbladder.
- Gallbladder may be septate.
- Double gallbladder with a single or separate cystic ducts.
- Floating gallbladder.

Q.238 What are the normal variations in bile ducts?
Normally: Cystic duct joins the common hepatic duct on right side to form common bile duct near upper border of duodenum.
Variations:
- Common hepatic and cystic ducts lie parallel before forming one duct.
- Cystic and common hepatic ducts unite behind pancreas.
- Cystic duct may join common hepatic duct in front or back of duodenum.
- Cystic duct may be absent, the common hepatic duct entering gallbladder and common bile duct leaving it.
- Accessary hepatic ducts present.

Duodenum

Q.239 What is the position of duodenum?
Duodenum lies above the level of umbilicus against L_{1-3} vertebrae, extending ½ inch to right and 1 inch to left of median plane. On either side of vertebral column, duodenum lies in front of psoas major muscle.

Q.240 What is the length of duodenum and what are its different parts?
Duodenum is a 10 inch long, curved around the head of the pancreas in form of 'C.' It is divided into 4 parts (Fig. 5.27):
- First (superior) part, 2 inches long.
- Second (descending) part, 3 inches long

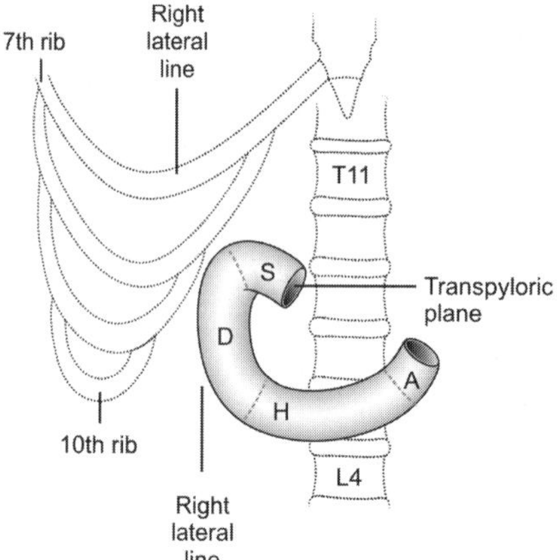

Fig. 5.27: Parts of the duodenum and their surface projection. S = superior part; D = descending part; H = horizontal part; A = ascending part

- Third (horizontal or inferior) part, 4 inches long
- Fourth (ascending) part, 1 inch long.

Q.241 What are the peritoneal relations of duodenum?
First part: The proximal 1 inch is suspended by lesser omentum above and greater omentum below, therefore it is moveable. Distal 1 inch is fixed because it is retroperi-toneal and is covered with peritoneum only anteriorly.

Second part: Retroperitoneal and fixed. Anteriorly crossed by peritoneum except in middle where it is related to transverse colon.

Third part: Also retroperitoneal and fixed. Covered by peritoneum anteriorly except where crossed by superior mesenteric vessels and root of mesentery.

Fourth part: Mostly retroperitoneal. Terminal part is moveable due to mesentery.

Q.242 What are the relations of first part of duodenum?
See Figures 5.28 to 5.30.
Anteriorly:
Quadrate lobe of liver and
Gallbladder.
Posteriorly:
Inferior vena cava,
Bile duct,
Portal vein and
Gastroduodenal artery
Superiorly:
Epiploic foramen
Inferiorly:
Head and neck of pancreas.

Q.243 What are the relations of second part of duodenum?
Medially:
Head of pancreas,
Bile duct and
Pancreatic ducts.
Laterally:
Right colic flexure
Anteriorly:
Right lobe of liver,
Transverse colon and transverse mesocolon and
Jejunum.
Posteriorly:
Anterior surface of right kidney near medial border.
Right renal vessels,
Right psoas major and
Inferior vena cava.

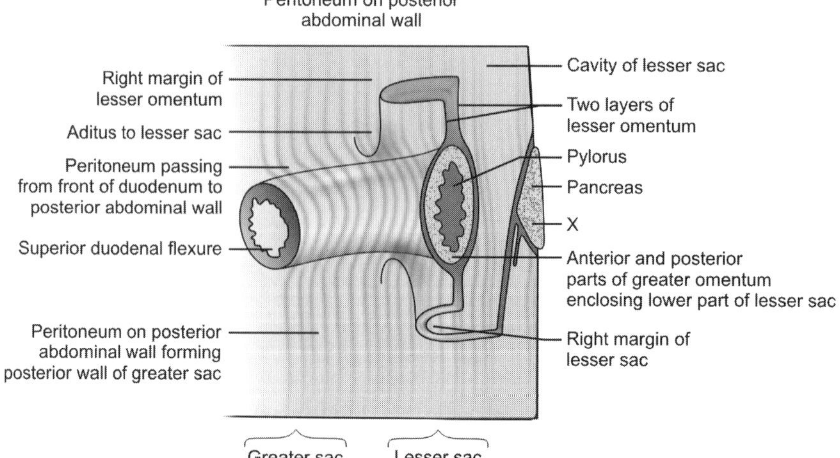

Fig. 5.28: Scheme to show the peritoneal relations of the superior part of the duodenum. Sections along axes YY and XX are shown in Figs 5.30 and 5.31 respectively. These diagrams are fundamental to the understanding of the boundaries of the lesser sac of peritoneum

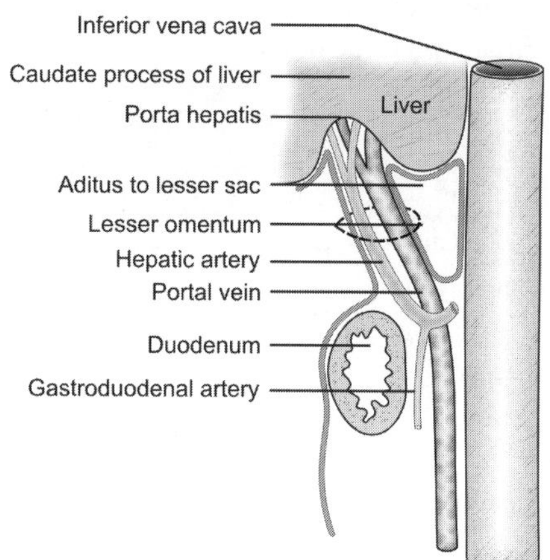

Fig. 5.29: Parasagittal section along axis YY in Fig. 5.28. Note the reflections of peritoneum. Also note how the portal vein and hepatic artery (which are at first retroperitoneal) come to lie between the two layers of the lesser omentum. The position of the bile duct (not shown) is similar to that of the portal vein. Finally, note the boundaries of the aditus to the lesser sac

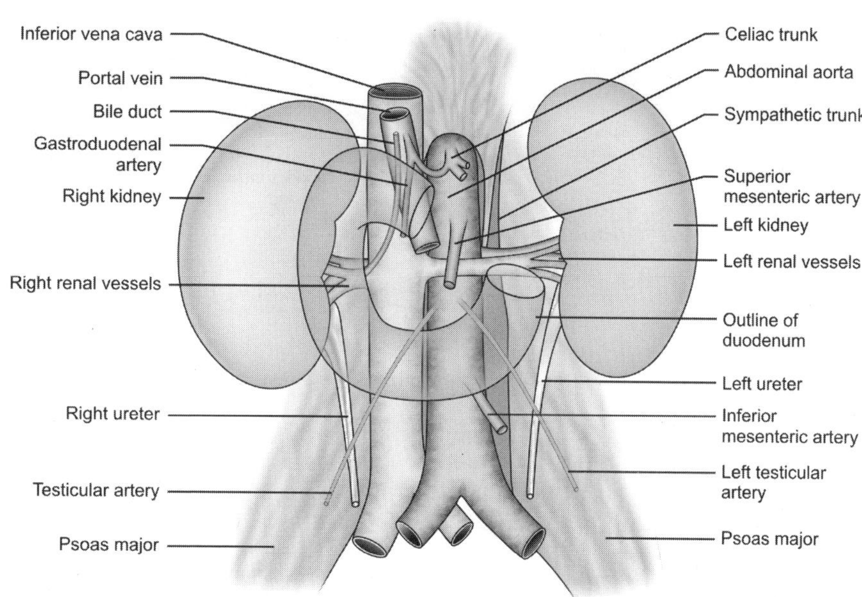

Fig. 5.31: Posterior relations of the duodenum. The duodenum is drawn as if it was transparent

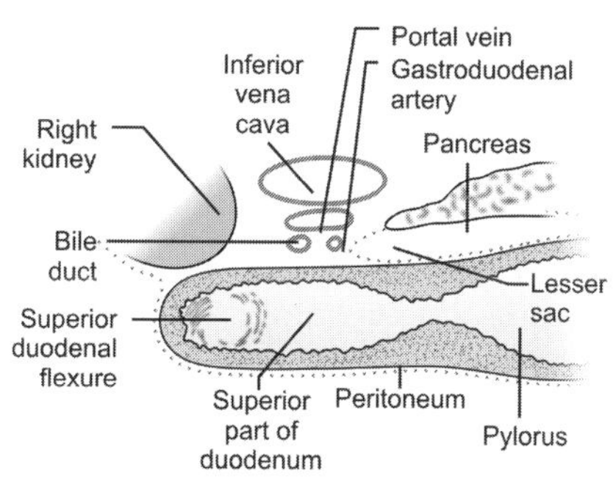

Fig. 5.30: Schematic transverse section through axis XX in Fig. 5.28 showing the posterior relations of the superior part of the duodenum

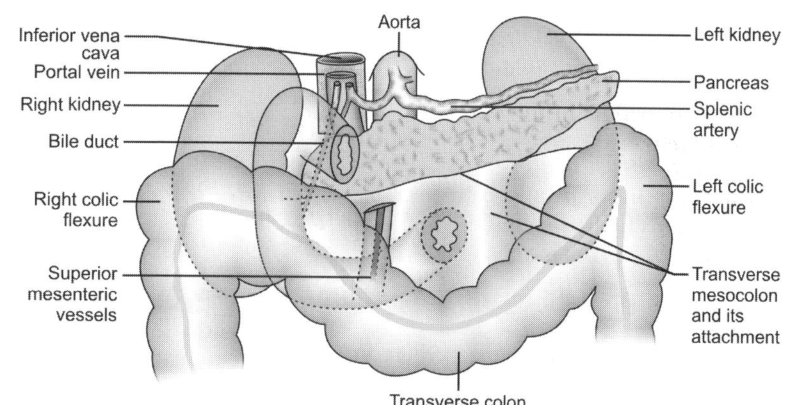

Fig. 5.32: Relationship of duodenum and pancreas to the transverse colon and its mesocolon

Q.244 Give relations of third part of duodenum.

Anteriorly:
- Superior mesenteric vessels and
- Root of mesentery.

Posteriorly:
- Right ureter
- Right psoas major
- Right testicular or ovarian vessels
- Inferior vena cava
- Abdominal aorta.

Superiorly:
- Head of pancreas.

Inferiorly:
- Coils of jejunum.

Q.245 Give structures related to fourth part of duodenum.

Figures 5.31 and 5.32.

Superiorly:
- Body of pancreas

To the Right:
- Upper part of root of mesentery and
- Aorta

To the Left:
- Left kidney and
- Left ureter.

Anteriorly:
- Transverse colon and mesocolon,
- Lesser sac and
- Stomach

Posteriorly:
- Left sympathetic trunk,
- Left psoas major,
- Left renal and testicular vessels and
- Inferior mesenteric artery.

Q.246 What is peculiar about development of duodenum?

The duodenum develops partly from foregut and partly from midgut. The junction of the two is in the second part of duodenum where the common bile duct opens, i.e. major duodenal papilla.

Q.247 What is the blood supply of duodenum?

Arterial supply: The part above the level of major duodenal papilla is supplied by superior pancreatico-duodenal artery and below it by the inferior pancreatico duodenal artery, branch of superior mesenteric artery (Fig. 5.33).

The first part is also supplied by right gastric, right gastroepiploic artery and branches of renal and hepatic artery.

Venous drainage: The veins drain into splenic, superior mesenteric and portal veins.

Q.248 What is the lymphatic drainage of the duodenum?

Most of the lymph drains into pancreatico-duodenal nodes. Some vessels drain into pyloric nodes and directly into hepatic

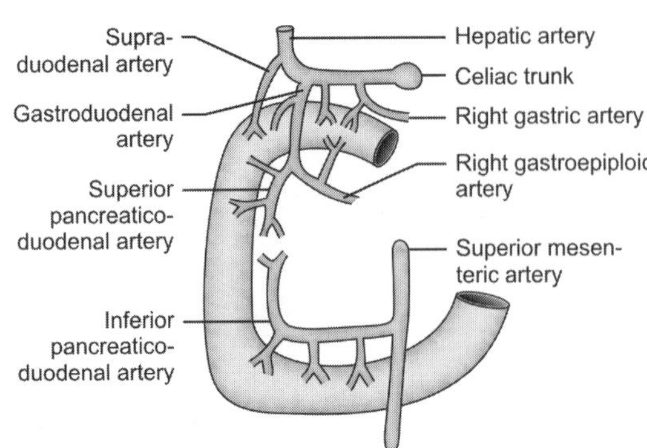

Fig. 5.33: Arterial supply of duodenum

nodes. All the lymph nodes drain into hepatic nodes which in turn drain into celiac nodes.

Q.249 What is ligament of Treitz?

It is fibromuscular band which supports the duodenojejunal flexure. It arises from right crus of diaphragm and is attached below to posterior surface of flexure and third and four parts of duodenum.

Q.250 What is typical of histology of ligament of Treitz?

It is made up of:

- Striated muscle fibers in upper part,
- Elastic fibers in middle part and
- Smooth muscle fibers in lower part.

Q.251 What is the importance of ligament of treitz?

- It marks the duodenojejunal junction.
- When it is attached only to flexure its contraction narrows duodenojejunal angle thus causing partial obstruction.

Q.252 What is 'duodenal cap' and its clinical importance?

In barium meal X-ray, the first part of duodenum is seen as a triangular homogenous shadow, known as Duodenal cap.

The duodenal cap is formed due to protrusion of pylorus into proximal half of first part of duodenum which is thus kept patent and filled with barium. Rest of duodenum shows floccular shadow.

Clinical importance: Persistent deformity of duodenal cap indicates chronic duodenal ulcer.

Q.253 What is the clinical importance of relations of duodenum?

- In barium meal X-ray, widening of duodenal loop, suggests carcinoma of the pancreas.
- In a duodenal ulcer (most common in first part), liver and gallbladder may be affected if the perforation of ulcer occurs or hemorrhage occurs, if gastroduodenal artery is affected in ulcers on posterior wall.
- Third part of duodenum may be obstructed by pressure from superior mesenteric artery.

Intestines

Q.254 What are the different parts of the small intestine?

Small intestine about 6 m long, is divided into:

- Upper fixed part: Duodenum 25 cm in length.
- Lower mobile part: Upper 2/5 forms jejunum and lower 3/5 forms ileum.

Q.255 What are 'valves of Kerckring'?

These are circular folds of mucous membrane which begin in second part of duodenum and extend upto proximal half of ileum.

These increase the absorptive surface area and also retard the passage of food.

Q.256 What are the different parts of large intestine?

The large intestine 1.5 m. long, is divided into:

- Appendix: 9 cm long
- Caecum: 6 cm long
- Transverse colon: 50 cm long
- Ascending colon: 15 cm long
- Descending colon: 25 cm long
- Sigmoid colon: 40 cm long
- Rectum: 12 cm long
- Anal canal: 3.8 cm long.

Q.257 What are the differences between small and large intestine?

	Small intestine	*Large intestine*
Calibre	Smaller	Wider
Sacculations	Absent	Present
Taenia coli	Absent	Present
Appendices epiploicae	Absent	Present
Fixity	Greater part is freely mobile	Greater part is fixed
Transverse mucosal folds	Permanent	Obliterated when longitudinal muscle coat relaxes
Villi	Present	Absent
Peyer's patches	Present in ileum	Absent

Q.258 What are the differences between jejunum and ileum?

See Figures 5.34 to 5.37.

Features	*Jejunum*	*Ileum*
Location	Occupies upper and left part of intestinal area	Occupies lower and right part of intestinal area
Lumen	Larger	Narrow
Mesentery	• Windows present	Windows absent
	• Fat less	Fat abundant
	• Arterial arcades 1 to 3	Arterial arcades: 5 to 6
	• Vasa recta longer and	Vasa recta shorter and more fewer
Circular	Larger and more mucosal folds	Smaller and sparse. closely set
Villi	Large, thick, more	Shorter, thinner and fewer
Peyer's patches	Absent	Present
Solitary lymphatic follicles	Fewer	More numerous

Q.259 What is 'taenia coli'?

These are ribbon-like bands formed by longitudinal muscle coat, present only in large intestine till terminal part of sigmoid colon.

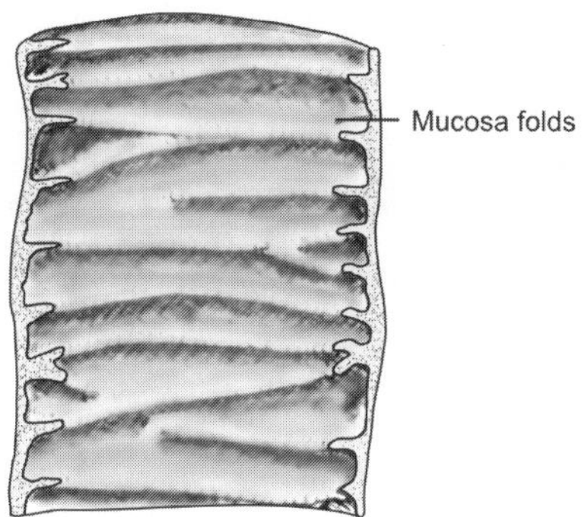

Fig. 5.34: Internal surface of part of jejunum

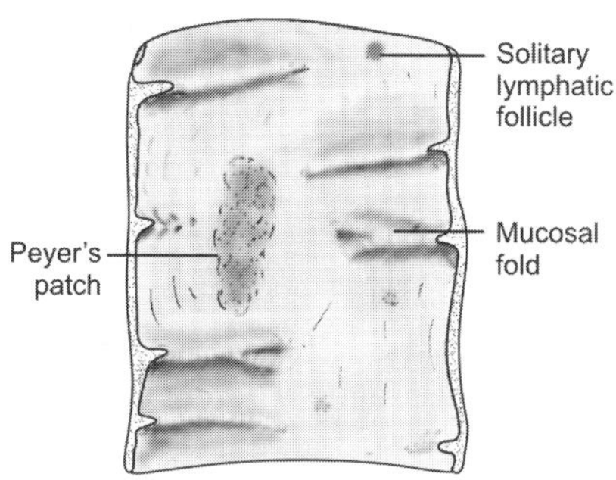

Fig. 5.35: Internal surface of part of ileum

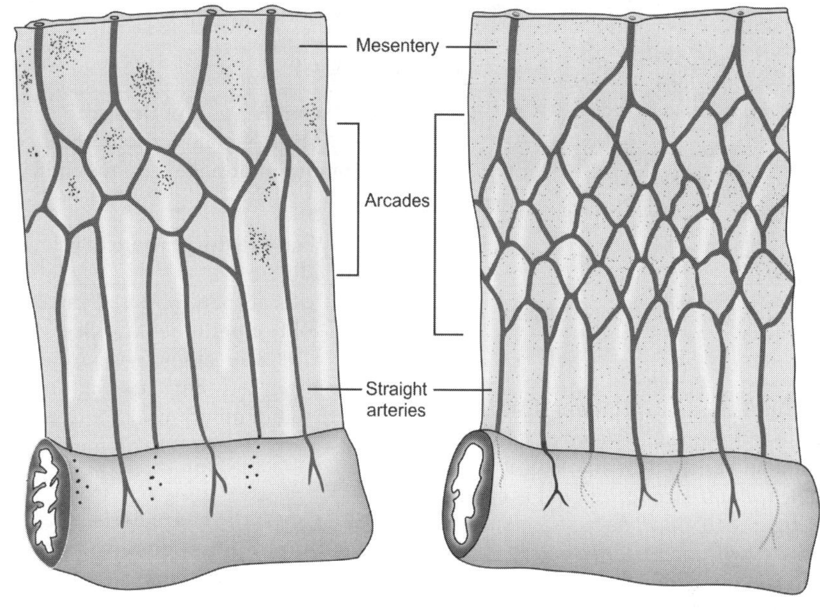

Figs 5.36 and 5.37: Comparison of the pattern of the arteries supplying the jejunum (Fig. 5.36) and the ileum (Fig. 5.37). Note that the arcades are fewer, and the straight arteries longer, in the jejunum. Fat is much more abundant in the mesentery of the ileum

Colon

Q.260 What are the functions of colon?

The functions of colon are:

- Lubrication of feces, by mucus.
- Absorption of salt, water and other solutes.
- Bacterial flora of colon synthesizes vitamin B.
- Mucoid secretion of colon has IgA antibodies which protect it from invasion by micro-organisms.
- The microvilli of some columnar cells serve a sensory function.

Q.261 What is phrenico-colic ligament?

It is a horizontal fold of peritoneum, attaching left colic flexure to the 11th rib. It supports the spleen and forms the partial upper limit of left paracolic gutter.

Q.262 What are the posterior relations of ascending colon?

- Right iliacus
- Iliac crest
- Right quadratus lumborum
- Right transversus abdominis
- Lateral cutaneous nerve of thigh
- Iliac branch of iliolumbar artery
- Right kidney
- Iliohypogastric nerve
- Ilioinguinal nerve

Q.263 What is attachment of transverse mesocolon?

It is attached to the anterior aspect of the head of pancreas and anterior border of body of pancreas.

Q.264 What are the posterior relations of the descending colon?

- Left kidney
- Left transversus abdominis
- Left quadratus lumborum
- Left iliacus
- Left psoas major
- Iliohypogastric nerve
- Ilioinguinal nerve
- Lateral cutaneous nerve of thigh
- Iliac branch of iliolumbar artery
  Just above, inguinal ligament it lies over
- External iliac artery
- Femoral nerve
- Genitofemoral nerve
- Testicular vessels

Q.265 What are the structures related to apex of sigmoid mesocolon?

- Bifurcation of left common iliac artery
- Left ureter.

Q.266 What is the features of sigmoid mesocolon.

The sigmoid mesocolon is shaped like an inverted 'V' and is attached to the posterior abdominal and pelvic walls.

Q.267 What is the parasympathetic innervation of the gut?

Parasympathetic nerve supply from pharynx to right two thirds of the transverse colon is through the vagus. The left one third of transverse colon, descending sigmoid colon, rectum and upper part anal canal are supplied through sacral part (S_{2-4}) of parasympathetic system by pelvic splanchnic branch.

Postganglionic parasympathetic neurons are located in myenteric and submucosal plexuses.

Q.268 What is the function of parasympathetic nerves to the gut?

- Stimulate the intestinal movement
- Inhibit the intestinal sphincters
- Secretomotor to the glands in mucosa.

Q.269 What are the fibers which carry the pain sensation from the gut?

Pain from most of the gut is carried by sympathetic nerves. Pain from pharynx and esophagus is carried by the vagus and from rectum and lower part of pelvic colon by parasympathetic pelvic splanchnic nerve.

Q.270 What is the blood supply to the colon?

The colon is mainly supplied by superior mesenteric artery, inferior mesenteric

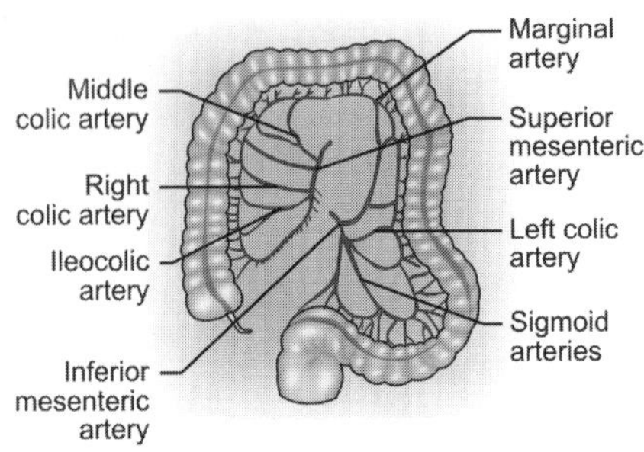

Fig. 5.38: Blood supply of colon

artery and branches of internal iliac artery. The superior mesenteric artery gives rise to middle colic, right colic and ileocolic arteries which supply the right colon and the right half of transverse colon. The branches of inferior mesenteric artery are: left colic, sigmoid and superior rectal (hemorrhoidal) arteries. These vessels supply the left half of the transverse colon to proximal rectum. The distal rectum is supplied by inferior and middle rectal (hemorrhoidal) arteries, which are the branches of the internal iliac artery (Fig. 5.38).

Q.271 What is the characteristic feature of arterial supply of transverse colon?
The right 2/3 of transverse colon develops from midgut, so it is supplied by superior mesenteric artery.

The left 1/3 is formed from hindgut, so it is supplied by inferior mesenteric artery.

Q.272 What is the lymphatic drainage of colon?
- Ascending and transverse colon drain into superior mesenteric group of preaortic nodes.
- Descending and sigmoid colon drain into inferior mesenteric group of preaortic nodes.

Appendix

Q.273 What are the dimensions of appendix?
The length of appendix varies from 2 to 20 cm, average about 9 cm. It is longer in children.

Q.274 What are the different positions of the appendix?
The base of the appendix is fixed but its tip can point in any direction. Depending on it following positions of the appendix are described.
- Retrocecal, commonest (60%)
- Pelvic (30%)
- Subcecal
- Preileal
- Postileal

Q.275 What is 'valve of Gerlach'?
It is indistinct semilunar fold of mucous membrane guarding the appendicular orifice.

Q.276 What are the peritoneal relations of appendix?
Appendix is suspended by a small, triangular fold of peritoneum called mesoappendix. Unlike other mesenteries the mesoappendix is not attached to the posterior abdominal walls but to the mesentery of the terminal part of the ileum.

Q.277 What is the characteristic feature of blood supply of appendix?
The appendix is supplied only by appendicular artery, a branch of ileocolic artery. It runs first in the free edge of appendicular mesentery and then distally along the wall of appendix.

Q.278 What is McBurney's point?
It is the point of maximum tenderness in acute appendicitis. It lies at the junction of medial 2/3 and lateral 1/3 of a line joining umbilicus to anterior superior iliac spine.

Q.279 What is Murphy's triad?
Appendicitis first causes pain around umbilicus. Then followed by vomiting and fever. The sequence of symptoms is known as Murphy's triad.

Q.280 Why the gangrene of appendix is common in acute infections?
Because appendicular artery supplying the appendix gets thrombosed and it has no collateral circulation.

Cecum

Q.281 What is the position of cecum?
It is situated in the right iliac fossa above the lateral half of inguinal ligament.

Q.282 What are communications to the cecum?
Cecum communicantes (Fig. 5.39):
- Superiorly with ascending colon
- Medially with ileum
- Posteromedially with appendix.

Q.283 What are the relations of cecum?
Anteriorly: Anterior abdominal wall
Posteriorly: Iliacus
Posas major
Lateral cutaneous nerve of thigh

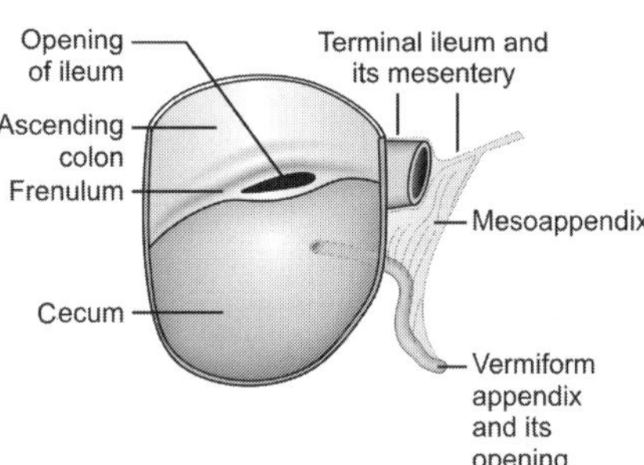

Fig. 5.39: Some features in the interior of the cecum seen after opening it

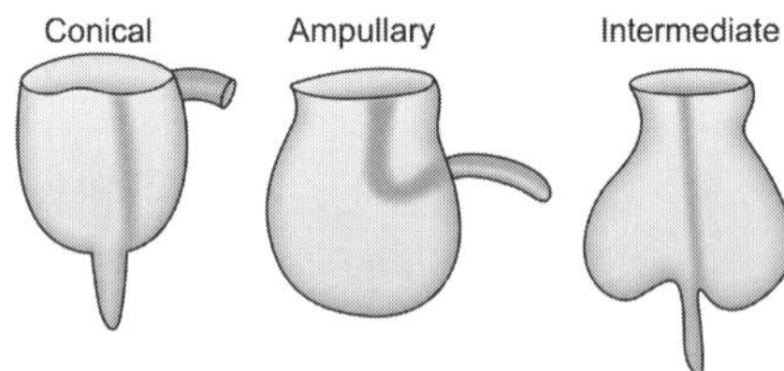

Fig. 5.40: Different forms of cecum

Q.284 What are the different shapes of cecum?
There are three types of cecum (Fig. 5.40):
- Conical type
- Ampullary type, commonest
- Intermediate type.

Meckel's Diverticulum

Q.285 What is Meckel's diverticulum?
It is persistent proximal part of the vitellointestinal duct, which normally disappears, during 6th week of intrauterine life.

Q.286 What is position of persistent Meckel's diverticulum?
It is situated 2 feet proximal to the ileocecal valve, attached to antimesenteric border of ileum.

Q.287 What is the clinical importance of Meckel's diverticulum?
- It may cause intestinal obstruction.
- Acute inflammation of diverticulum may resemble appendicitis.
- It is often the site of heterotrophic gastric mucosa with oxyntic cells.

Q.288 What is the effect of patent Meckel's diverticulum?
Small intestine contents being discharged at the umbilicus.

Q.289 Why the infections of Meckel's diverticulum are dangerous?
Because
- Its wall are thinner so, it perforates more easily.
- It lies in middle of peritoneal cavity, so more chances of widespread peritonitis.

Q.290 What will happen if vitello-intestinal duct persists as a fibrous band?
- This fibrous band passes from umbilicus to some part of mesentery or small gut.
- This band may cause compression of loop of gut under it.
- If attached to branch of mesenteric artery, which may be torn during abdominal operations.

Pancreas

Q.291 Why pancreas is called a 'double gland'?
Pancreas is called a double gland because it is partly exocrine and partly endocrine (Fig. 5.41).

Q.292 What are the secretions of the pancreas?
Exocrine part secretes pancreatic juice which has digestive functions.

Endocrine part secretes hormones, e.g. insulin, glucagon, etc.

Q.293 At what level the pancreas lie?
The pancreas lies across the posterior abdominal wall at the level of L_1 and L_2 vertebra.

Q.294 What is the shape and different parts of pancreas?
Pancreas is a J-shaped organ. It is divided into 4 parts:
- Head with the uncinate process,
- Neck,
- Body and
- Tail.

Q.295 What are the relations of head of pancreas?
Anterior surface:
Gastroduodenal artery
Transverse colon
Jejunum over area covered by peritoneum
Posterior surface:
Inferior vena cava
Renal veins
Right crus of diaphragm
Bile duct.
Superior border:
Superior pancreaticoduodenal artery.
Inferior border:
Third part of duodenum and
Inferior pancreaticoduodenal artery.
Right lateral border:
Second part of duodenum
Terminal part of bile duct.

Uncinate process is related anteriorly to superior mesenteric vessels and posteriorly to aorta.

Q.296 What are relations of neck of pancreas?
Anterior surface:
Peritoneum
Lesser sac
Pylorus
Posterior surface:
Beginning of portal vein.

Q.297 What are the relations of body of pancreas (Figs 5.42 and 5.43)?
Anterior surface:
Lesser sac
Stomach.
Posterior surface:
Aorta with origin of superior mesenteric artery
Left crus of diaphragm
Left suprarenal
Left kidney
Left renal vessels
Splenic vein
Inferior surface:
Duodenojejunal flexure
Coils of jejunum
Left colic flexure
Anterior and inferior surface are covered by the peritoneum.

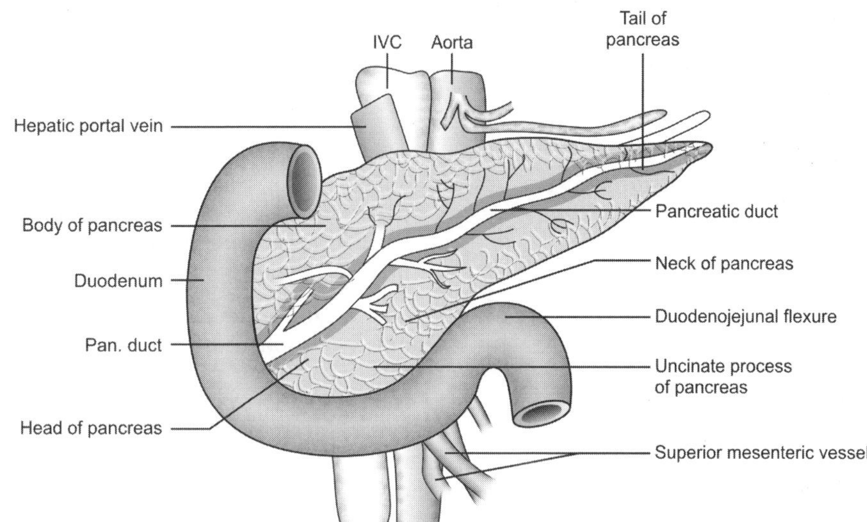

Fig. 5.41: Pancreas

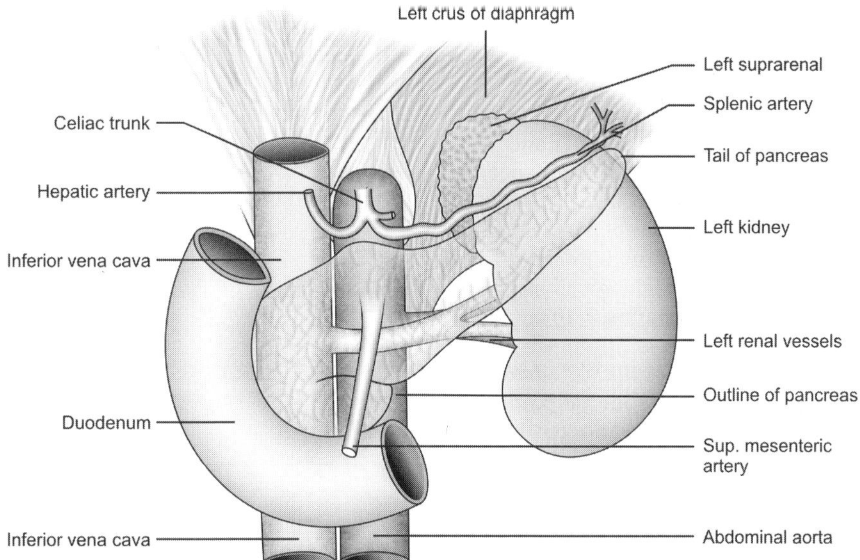

Fig. 5.42: Some posterior relations of the pancreas. The pancreas is shown only in outline. Additional posterior relations are shown in Figure 5.43

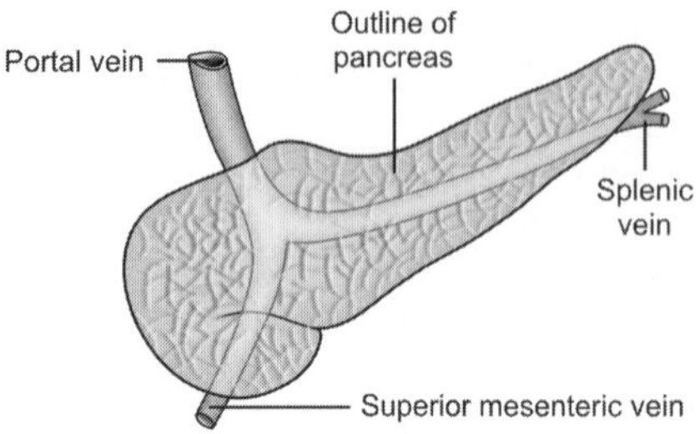

Fig. 5.43: Relationship of portal vein, superior mesenteric vein and splenic vein to the pancreas. Compare with Figure 5.42

Inferior border:
Superior mesenteric vessels.
Superior border:
Celiac artery
Hepatic artery
Splenic artery
Anterior border:
Provides attachment to root of transverse mesocolon.

Q.298 What are relations to tail of pancreas?

Tail of pancreas lies in lienorenal ligament and is related to gastric surface of spleen.

Q.299 What is the arterial supply of pancreas?

- Pancreatic branches of splenic artery.
- Superior pancreaticoduodenal artery, a branch of celiac trunk.
- Inferior pancreaticoduodenal artery, a branch of superior mesenteric artery.

Q.300 What is the venous drainage of the pancreas?

The pancreas drains into splenic, superior mesenteric and portal veins.

Q.301 What are the ducts draining the secretions of exocrine part of pancreas?

The two ducts carrying the exocrine secretion of pancreas are:

- *Main pancreatic duct (of Wirsung):* Joins with bile duct to form ampulla of Vater and open at major duodenal papilla in 2nd part of duodenum, 8–10 cm distal to pylorus.
- *Accessory pancreatic duct (of Santorini):* Opens at minor duodenal papilla in 2nd part of duodenum, 6–8 cm distal to pylorus.

Q.302 What is 'pseudopancreatic cyst'?

Anterior to pancreas lies the stomach, separated from it by the lesser sac. The sac may be closed off and distended with fluid either from perforation of posterior gastric ulcer or as a result of acute pancreatitis, thus forming pseudopancreatic cyst.

Q.303 Why carcinoma of head of pancreas is associated with obstructive jaundice?

The head of pancreas lies in the C-curve of the duodenum in relation to the opening of the common bile duct. Therefore, carcinoma of the head of the pancreas will cause the compression of the common bile duct and causes obstructive jaundice.

Q.304 What is the developmental origin of pancreas?

- From:
  - *Dorsal diverticulum from duodenum:* Larger.
  - *Ventral out pouching from side of common bile duct:* Smaller

  The ventral pouch rotates posteriorly to fuse with lower aspect of dorsal diverticulum, trapping the superior mesenteric vessels between two parts.
- Ducts of two segments communicate and that of smaller takes over the main pancreatic flow to form the main duct and the duct of larger portion persists as accessory duct.

Q.305 What are the common developmental anomalies of pancreas (Fig. 5.44)?

- *Annular pancreas:* Two segments of pancreas completely surround second part of duodenum.
- *Accessory pancreatic tissue:* In duodenum (usually), jejunum, wall of stomach and gallbladder.

Q.306 How are secretions of pancreas passed into the duodenum?

The pancreatic secretions are poured into the duodenum with the help of two ducts:

- *Main pancreatic duct*—This begins in the tail of pancreas and passes to the right through the body. At the neck of pancreas, it turns downwards and backwards and joins the bile duct just outside the duodenal wall. The walls of the bile and main pancreatic ducts join each other here, but their lumens remain separate as the ducts descend through the muscle wall and submucosa of the duodenum.
- *Accessory pancreatic duct*—This begins in the lower part of the head of pancreas. It runs upwards crossing in front of the main duct and opens into the duodenum at the minor duodenal papilla (which has a short distance above and in front of the major papilla).

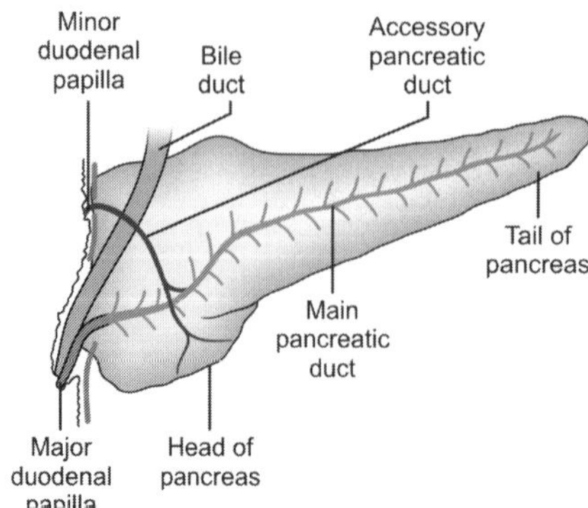

Fig. 5.44: Schematic diagram of the ducts of the pancreas

Suprarenal (Adrenal) Glands

Q.307 What is the position of adrenal glands?

- Posterior abdominal wall over the upper pole of kidneys behind the peritoneum.
- In front of crus of diaphragm opposite vertebral ends of 11th intercostal space and 12th rib.

Q.308 What are the parts of adrenal glands seen in cross section?

- *Cortex:* Outer. Mesodermal origin.
- *Medulla:* Inner. Neural crest origin. Volume of medulla is about one-tenth of cortex.

Q.309 What is chromaffin system?

- It is made up cells which have an affinity for salts of chromic acid.
- Develop from the neural crest.
- Cells secretes adrenaline and noradrenaline.

Q.310 What are the components of chromaffin system?

- Suprarenal medulla.
- Para-aortic bodies.
- Paraganglia.
- Small masses of cells among ganglia of sympathetic chain, splanchnic nerves and prevertebral autonomic plexus.

Q.311 What are the different layers of adrenal cortex?

- *Zona glomerulosa:* Outermost. Produce aldosterone
- *Zona fasciculata:* Middle. Produce hydrocortisone and other glucocorticoids.
- *Zona reticularis:* Inner most. Produce probably sex hormones.

Q.312 What is the blood supply of suprarenal glands?

Arterial supply:

- *Superior suprarenal artery:* Branch of inferior phrenic
- *Middle suprarenal artery:* Branch of abdominal aorta

- *Inferior suprarenal artery:* Branch of renal artery

Venous drainage:

- *Right suprarenal vein:* Drains into inferior vena cava
- *Left suprarenal vein:* Drains into left renal vein.

Q.313 Name of structures lying between two suprarenal glands.

- Crura of diaphragm.
- Aorta (abdominal).
- Celiac artery plexus.
- Inferior vena cava (Fig. 5.45).

Q.314 Compare the two suprarenal glands.

	Left	*Right*
Shape	Semilunar	Triangular
Size	Larger	Smaller
Position	Upper part of medial border of kidney	Upper part of anterior surface of kidney
Level	Lower	Higher
Hilum	Near lower end	Near upper end
Peritoneal relations	Separated from stomach by peritoneum	Only lower part related to peritoneum
Visceral relations:		
Anterior surface	*Superior:* Stomach *Inferior:* Pancreas splenic artery	*Medial:* Inferior vena cava *Lateral:* Part of bare area of liver
Posterior	*Medial:* Crus of *surface* *Lateral:* Kidney	*Inferior:* Kidney diaphragm *Superior:* Crus of diaphragm
Medial border	Left celiac ganglion, left inferior phrenic artery, left gastric artery	Right celiac ganglion, right inferior phrenic artery

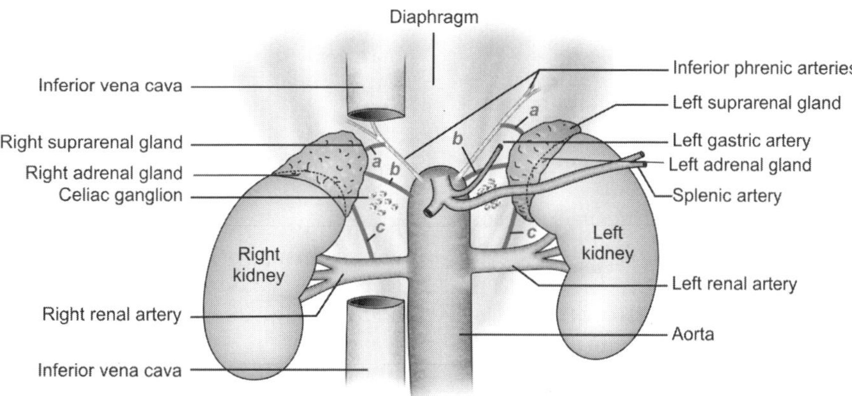

Fig. 5.45: Suprarenal glands and some related structures as seen from the front. a, b and c = superior, middle and inferior arteries to the suprarenal glands

Kidneys

Q.315 Where are the kidneys situated ?

The kidneys are situated retroperitoneally on the posterior abdominal wall on each side of the vertebral column. The right kidney is slightly lower than left and the left kidney is a little nearer to the median plane (Fig. 5.46).

Q.316 What is the extent of kidney in relation to vertebral column?

The kidneys vertically extend from upper border of T_{12} vertebra to centre of body of L_3 vertebra.

The right kidney is lightly lower than the left.

Q.317 What is the relation of transpyloric plane to kidneys?

Transpyloric plane passes through the upper part of hilus of right kidney and through lower part of hilus of the left.

Q.318 What are the measurements of normal kidney?

Each kidney is 11 cm long
6 cm broad
3 cm thick

Left kidney is a little longer and narrower.

Q.319 What are the anterior relations of the kidneys?

Right kidney:

Right suprarenal
Liver
Second part of duodenum
Hepatic flexure of colon
Small intestine
Hepatic and intestine surfaces are covered by peritoneum

Left kidney:

Left suprarenal
Stomach
Spleen
Pancreas
Jejunum

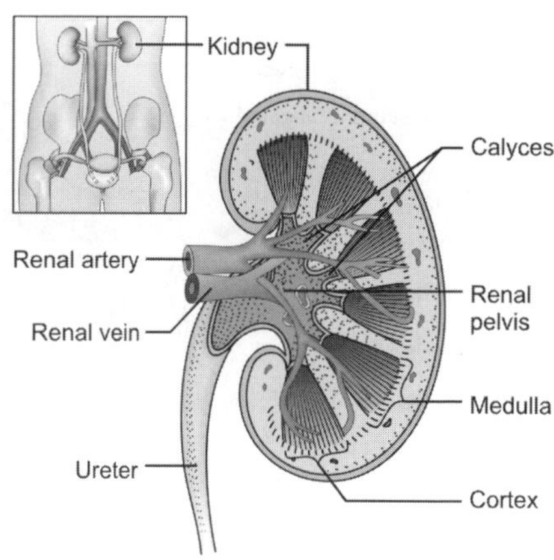

Fig.5.46: Excretory system

Splenic flexure
Descending colon and splenic vessels.
The gastric, splenic and jejunal surfaces are covered by peritoneum (Fig. 5.47).

Q.320 What are the posterior relations of kidney?

Both kidneys are related to:

- Diaphragm
- Medial and lateral arcuate ligaments
- Psoas major
- Quadratus lumborum
- Transversus abdominis
- Subcostal vessels and
- Iliohypogastric, subcostal and ilioinguinal nerves.

  The right kidney is also related to 12th rib and the left kidney to 11th and 12th ribs.

Q.321 What are the coverings of the kidneys?

From within outwards the coverings are:

- *Fibrous capsule:* Thin membrane, made up of white and yellow fibers and smooth muscle fibers (Fig. 5.48).
- *Perirenal (perinephric) fat:* Outer to the fibrous capsule. It is thickest at the borders of the kidney.
- *Renal fascia (fascia of Gerota):* Fibroareolar sheath around the perirenal fat.

  Superiorly, two layers of renal fascia first enclose the suprarenal gland in a separate compartment, then they fuse with each other and become continuous with fascia on undersurface of diaphragm.

  Inferiorly, the two layers remain separate and enclose ureter. Laterally, the two layers fuse and become continuous with fascia transversalis.

  Medially, anterior layer passes in front of renal vessels and fuses with connective tissue around aorta and inferior vena cava. The posterior layer, fuses with fascia covering quadratus lumborum and

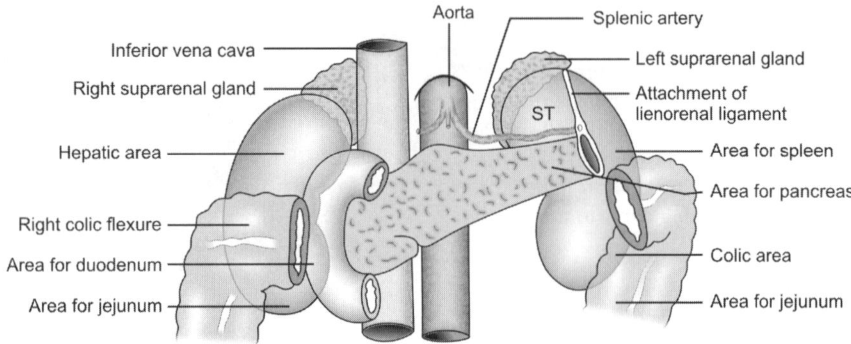

Fig. 5.47: Scheme to show the anterior relations of the right and left kidneys

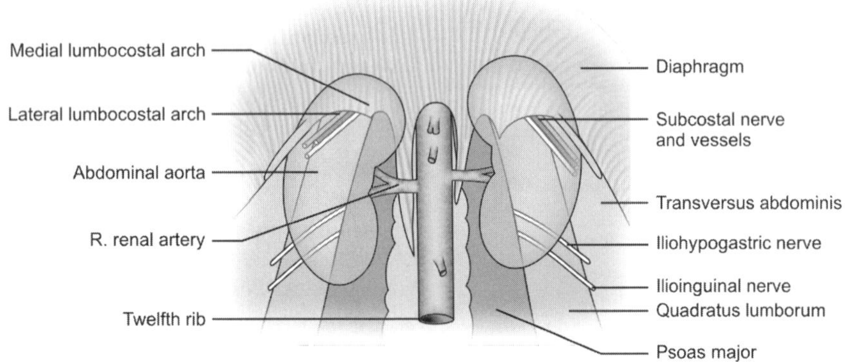

Fig. 5.48: Posterior relations of kidneys

psoas major. At medial border of kidney fascia forms a septum.
- *Pararenal (paranephric) fat:* Fat outer to renal fascia is more abundant posteriorly and towards the lower pole of the kidney.

Q.322 What is the fascia of Toldt and fascia of Zuckerkandl?
The anterior layer of renal fascia is known as fascia of Toldt and posterior layer as fascia of Zuckerkandl.

Q.323 What are the structures found at the hilus of kidney?
From before backwards:
- Renal vein
- Renal artery
- Pelvis of the ureter
- In 30% accessory renal artery.

Q.324 What are the vascular segments of the kidney?
Each renal artery at the hilus of the kidney divides into an anterior and posterior branch, which in turn divides into segmental arteries which supply a definite part (segment) of the kidney. In each kidney there are five segments, i.e. apical, upper, lower, middle and posterior (Fig. 5.49).

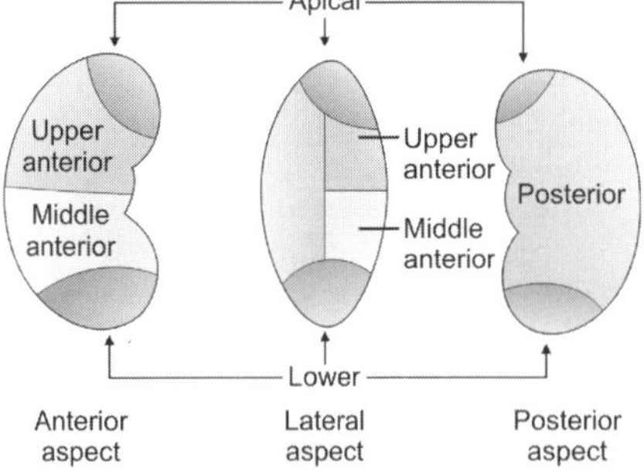

Fig. 5.49: Interior of the descending part of the duodenum showing the major and minor papillae. Note the transverse folds of the mucous membrane

Q.325 What is the clinical importance of vascular segments of kidney?
Each segmental artery is an end artery, so the vascular segments are independent units. So the intersegmental incisions are given for the removal of a part of the kidney.

Q.326 What is the direction of blood flow in ruptured kidney or pus in perinephric abscess?
First it causes distension of renal fascia and then downwards into pelvis within fascial compartment.

The mid-line attachment of renal fascia and fascial septum prevents extravasation to opposite side.

Q.327 What care should be taken in exposure of kidneys from behind when 12th rib is to be excised?
Push up the pleura which crosses the medial half of the 12th rib.

Q.328 How the kidneys are developed?
Kidneys are formed in the sacral region and then ascend upwards. The kidney develop from:
- *Mesonephric duct:* Gives rise to pelvis, calyces and collecting tubules.
- *Metanephros:* Develops into glomeruli and proximal part of renal duct system.

Q.329 Name the common congenital abnormalities of kidneys.
- Congenital polycystic kidney
- Horse-shoe kidney: Fusion of lower poles of two kidneys
- Congenital absence of one kidney
- Unilateral fused kidney
- Accessory kidneys
- *Pelvic kidneys:* Failure of ascent of kidney from lower lumbar or sacral region.

Ureter

Q.330 What are ureters?
These are pair of narrow, thick walled muscular tubes which convey urine from the kidneys to urinary bladder.

Q.331 What is the length of the ureter?
Each ureter is about 25 cm long, of which upper half lies in abdomen and lower half in pelvis.

Q.332 What is the course in the ureter?
The course of ureter is divided into two parts (Fig. 5.50):
- *In the abdomen:* Ureter begins at renal pelvis (Funnel-shaped dilatation) from the hilus of the kidney and descends along its medial border. It gradually narrows and becomes ureter proper at the lower pole of the kidney. It descends on psoas muscle and enters pelvis by crossing in front of the termination of common iliac artery.
- *In the pelvis:* It first runs downwards, backwards and laterally, following the anterior margin of greater sciatic notch. Opposite ischial spine it turns forwards and medially to reach base of urinary bladder.

  Ureter enters bladder wall obliquely and opens at the lateal angle of trigone. Its point of termination corresponds to the pubic tubercle.

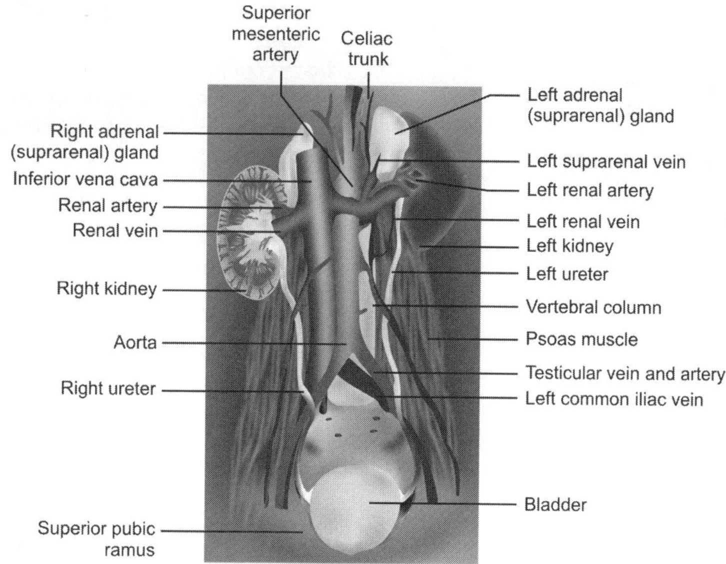

Fig. 5.50: Relations of abdominal parts of right and left ureters

Q.333 Name the sites at which constriction are present in ureter.

Three sites:

- Pelvic-ureteral junction (related to transverse process of L_2 verterbra).
- Brim of lesser pelvis (Related to sacroiliac joint) and
- At its passage through bladder wall (slightly medial to ischial spine).

Q.334 What is the clinical importance of constrictions of ureter?

A ureteric calculus is likely to lodge at one of these three levels as described in Q.332.

Q.335 What are structures crossing the abdominal part of right ureter from medial to lateral side?

- Genitofemoral nerve
- Testicular (ovarian) vessels
- Right colic vessels
- Ileocolic vessels
- Terminal part of superior mesenteric artery.

  Genitofemoral nerve crosses behind the ureter while others cross in front.

Q.336 What are the structures crossing the abdominal part of left ureter from medial to lateral side?

- Genitofemoral nerve
- Testicular (ovarian) vessels
- Left colic artery

Genitofemoral nerve crosses behind the ureter while others cross in front.

Q.337 What are the relations of ureter in its forward course in pelvis?

In males:

- *Ductus deferens:* Crosses ureter superiorly from lateral to medial side.
- *Seminal vesicle:* Below and behind ureter.
- *Vesical veins:* Surround terminal part ureter.

In females:

- Ureter lies in lower and medial part of broad ligament of uterus.
- *Uterine artery:* First above and in front of ureter and then crosses superiorly from lateral to medial side.
- Ureter lies above lateral fornix of vagina.
- Ureter lies 2 cm lateral to supravaginal part of cervix.
- Terminal part of ureter lies anterior to vagina.

Q.338 What is the arterial supply of ureter?

- *For upper part:* Renal artery, branches of gonadal and colic arteries.
- *For middle part:* Branches from aorta, gonadal and iliac arteries.
- *For lower part:* From vesical, middle rectal or uterine arteries.

Q.339 What is the nerve supply of ureter?

- *Sympathetic nerves:* T_{10}-L_1.
- *Parasympathetic nerves:* S_{2-4}.

  They reach through renal, aortic and both hypogastric plexus. Autonomic nerves to ureter are predominantly sensory in function.

Q.340 What is renal colic?

It is spasm of ureter by a stone. There is a sudden, agonizing pain in the loin.

Q.341 Where the pain of renal colic is referred?

Pain is referred to cutaneous area innervated by T_{11}-L_2.

Q.342 What is developmental origin of ureter?

From part of ureteric bud that lies between the pelvis of kidney and vesicourethral canal.

Q.343 What are the congenital abnormalities of ureter?

- Ureter may be duplicated.
- Ureter may end into prostatic urethra, vas deferens, seminal vesicles, vagina and rectum.
- Upper end of ureter may not be connected to kidney.
- Ureter may have diverticula.

Q.344 What is 'post caval' ureter? What is its clinical importance?

The right ureter instead of lying to right of inferior vena cava may pass behind it.

Clinical importance: May lead to compression of ureter and obstruction to flow of urine.

Q.345 How the reflux of urine from bladder into ureter is prevented?

The intravesical oblique course of ureter has valvular action which prevents the reflux of urine from bladder to ureter.

Diaphragm (Fig. 5.51)

Q.346 What is the origin of diaphragm?

Arise from periphery, in three parts:

- *Sternal:* Back of xiphoid process.
- *Costal:* Inner surfaces of cartilages and adjacent parts of lower six ribs.
- *Lumbar:* Medial and lateral lumbocostal arches and from lumbar vertebrae by right and left crura.

Q.347 What are lumbocostal arches?

These are tendinous arches in the fascia covering the muscles in posterior abdominal wall, e.g. medial lumbocostal arch (medial arcuate ligament) in fascia over upper part of psoas major and lateral lumbocostal arch (lateral arcuate ligament) in fascia over upper part of quadratus lumborum.

Q.348 What is the origin of crus of diaphragm?

Right crus: From anterolateal surface of body of $L_{1,2,3}$.

Left crus: From anterolateral surface of body of $L_{1,2}$.

The medial margins of two crura join to form the median arcuate ligament.

Q.349 What is the insertion of muscle fibers of diaphragm?

Trilobed central tendon, which lies below and is fused to the pericardium.

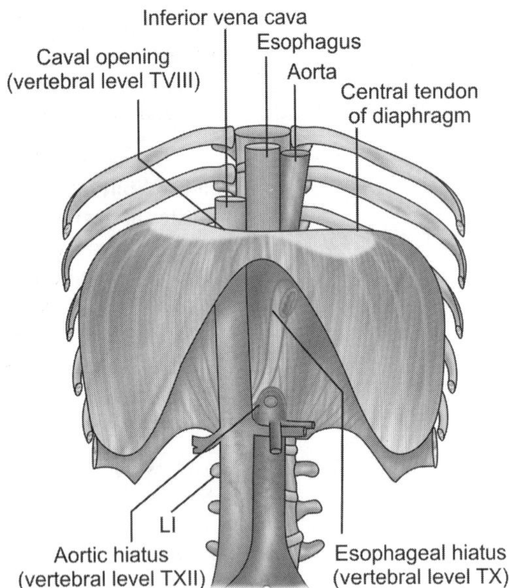

Fig. 5.51: The diaphragm

Q.350 What is the nerve supply of diaphragm?

Motor: Phrenic nerve ($C_{3,4}$).

Sensory: - Phrenic nerves: Central part.
- Lower six thoracic nerves: Peripheral part.

Q.351 What are the other structures supplied by phrenic nerve?

Sensory fibers to:

- *Pleura:* Mediastinal and diaphragmatic.
- *Pericardium:* Fibrous and parietal layer of serous pericardium.
- *Peritoneum:* Below central part of diaphragm.
- Through celiac plexus to falciform and coronary ligaments of liver, gallbladder, suprarenals and inferior vena cava.

Q.352 What are functions of diaphragm?

- Separates the thoracic and abdominal cavity.
- Principal muscle of inspiration.
- In all expulsive acts, e.g. sneezing, coughing, vomiting, defecation, etc. It provides additional power to each effort.

Q.353 What are the variations in position of diaphragm with posture?

Level of diaphragm is:

- Highest in supine position.
- Lowest in sitting position.
- Midway in standing.

Q.354 Name the structures passing through the opening of diaphragm.

- *Caval opening (T_8):* Slightly to right of median plane in the central tendon. Transmits inferior vena cava and half of the right phrenic nerve.
- *Esophageal opening (T_{10}):* Slightly to left of median plane. Transmits esophagus, right and left vagi, esophageal branches of left gastric artery with accompanying veins.
- *Aortic opening (T_{12}):* Central. Transmits (from right to left) vena azygous, thoracic duct and aorta. Aortic opening is deep to median arcuate ligament.
- *Smaller orifices in diaphragm:*
  - *Between xiphoid slip and that from 7th cartilage:* Superior epigastric vessels.
  - *Between slips from 7th and 8th costal cartilages:* Musculophrenic vessels. Also transmits 7th intercostal nerve and vessels.
  - *Between each papir of remaining slips:* One of lower five intercostal nerves and vessels.
  - *Behind lateral lumbocostal arch:* Subcostal nerve and vessels.
  - *Behind medial lumbocostal arch:* Sympathetic trunk.
  - *Each crus:* Greater, lesser and least splanchnic nerve. Left crus in addition is pierced by vena hemiazygous.
  - *Muscular part of diaphragm to the left of anterior folium of central tendon:* Left phrenic nerve.

Q.355 Why irritation of diaphragm causes pain in shoulder tip?

Because phrenic nerve and supraclavicular nerves have same root value, i.e. $C_{3,4}$.

Q.356 What is 'eventration' of diaphragm?

This is congenital defect, in which the high position of diaphragm occurs due to replacement of left half of diaphragm by fibrous membrane.

Q.357 What is foramen of Morgagni?

Also called space of Larry. It is space between the xiphoid and costal origins of diaphragm. Site of congenital hernia. More common on right side.

Q.358 What is foramen of Bochdalek?

This is the commonest site of congenital diaphragmatic defect in periphery of diaphragm in region of 10th and 11th ribs attachment. Defect is in posterolateral part of dome on left side of diaphragm resulting from failure of closure of pericardiopleural canal.

Q.359 What is the development origin of diaphragm?

Diaphragm is developed from:

- Septum transversum
- Pleuroperitoneal membrane
- Ventral and dorsal mesenteries of esophagus
- Mesoderm of body wall.

Posterior Abdominal Wall

Q.360 What are the different layers of thoracolumbar fascia?

- *Posterior layer:* Medially attached to lumbar and sacral spines and laterally blends with anterior layer. Covers erector spinae muscle.
- *Middle layer:* Medially attached to tips of transverse process of lumbar vertebrae and laterally blends with posterior layers. It separates erector spinae from quadratus lumborum muscle.
- *Anterior layer:* Medially attached to anterior surface of transverse processes of lumbar vertebrae and laterally blends with posterior layer. Covers anterior surface of quadratus lumborum muscle.

Q.361 What is the extent of the abdominal aorta?

It extends from lower border of T_{12} vertebra to front of L_4 where it terminates into left and right common iliac arteries.

Q.362 What are the branches of abdominal aorta?

- *Ventral branches:*
  - Celiac trunk
  - Superior mesenteric artery
  - Inferior mesenteric artery.

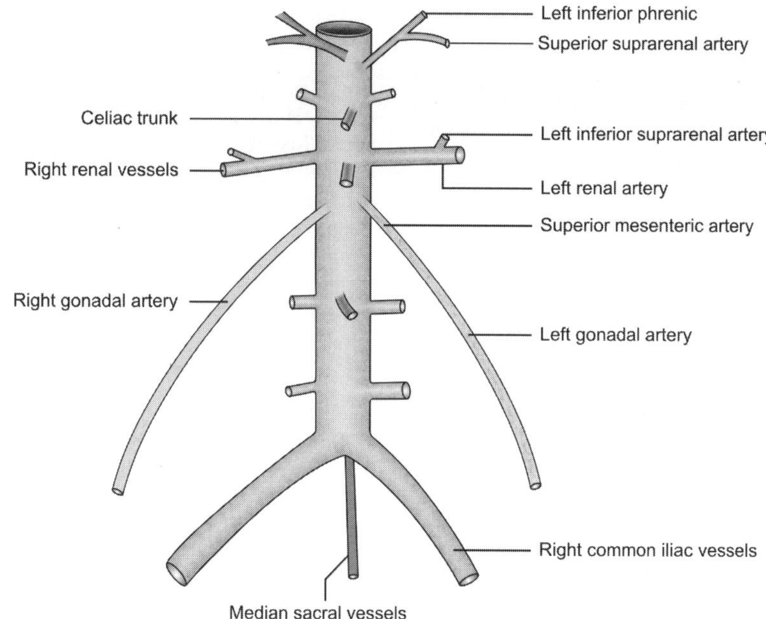

Fig. 5.52: Abdominal aorta

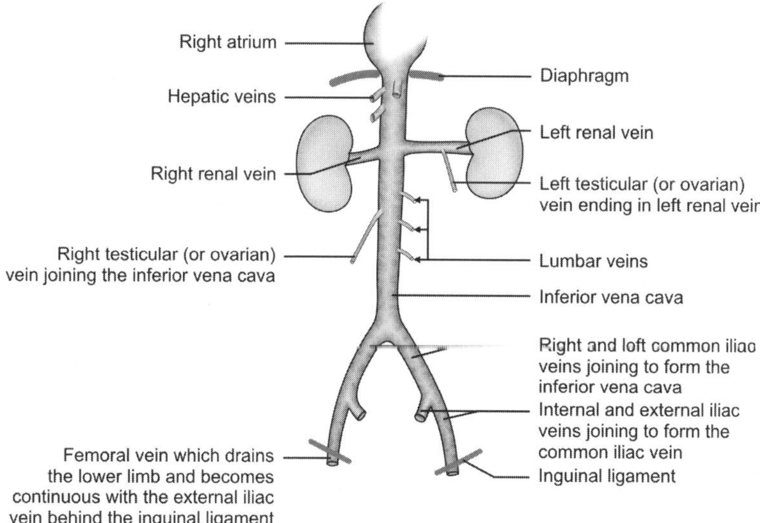

Fig. 5.53: Scheme to show the inferior vena cava and its tributaries

- *Dorsal branches:*
  - Lumbar
  - Median sacral.
- *Lateral branches:*
  - Inferior phrenic
  - Middle suprarenal
  - Renal
  - Testicular or ovarian.
- *Terminal branches:* Common iliac arteries (Fig. 5.52).

Q.363 What are the tributaries of inferior vena cava?

- Common iliac veins (Fig. 5.53)
- Third and fourth lumbar veins
- Right testicular or ovarian vein
- Renal veins
- Right suprarenal vein
- Hepatic veins
- Right inferior phrenic vein.

Q.364 What are the posterior relations of the gonadal arteries?

On both sides:

- Psoas major
- Genitofemoral nerve
- External iliac vessels.

On right side, in addition it is related to inferior vena cava.

Q.365 What is the extent of the common iliac arteries?

Each artery begins in front of body of fourth lumbar vertebra and it terminates in front of sacroiliac joint, at level of disc between fifth lumbar vertebra and sacrum.

Q.366 What are the branches of external iliac artery?

- Inferior epigastric artery
- Deep circumflex iliac artery
  Each artery continues into thigh as the femoral artery deep to the inguinal ligament.

Q.367 Which veins on left side open into left renal vein but corresponding veins on right side open into inferior vena cava?

- Inferior phrenic vein
- Suprarenal vein
- Testicular vein (ovarian vein)

Q.368 What are the tributaries of common iliac veins?

- Iliolumbar vein
- Median sacral vein

Q.369 What is the pathway by which the blood reaches heart in obstruction of inferior vena cava?

In obstruction of inferior vena cava communications between tributaries of inferior and superior vena cava undergo considerable enlargement. Veins involved from below are inferior epigastric, circumflex iliac and external pudendal. Blood from them passes into lateral thoracic, internal thoracic and posterior intercostal passing over abdominal wall. Communication is also established through azygous and hemiazygous veins and vertebral venous plexus.

Q.370 What is cisterna chyli?

It is a 5–7 cm long lymphatic sac, situated in front of $L_{1,2}$ vertebrae, to the right of abdominal aorta.

It continues upwards as thoracic duct.

Q.371 How the lumbar aplexus is formed?

- It is formed by ventral rami of upper four lumbar nerves.
- First lumbar nerve also receives a contribution from subcostal nerve.
- L_4 nerve gives a contribution to lumbosacral trunk ($L_{4,5}$) which forms part of sacral plexus.

Q.372 What are the branches of lumbar plexus?

- Iliohypogastric nerve (L_1),
- Ilioinguinal nerve (L_1),

- Genitofemoral nerve ($L_{1,2}$).
- Lateral cutaneous nerve of thigh (L_{2}, $_{3}$).
- Femoral nerve ($L_{2,3,4}$).
- Obturator nerve ($L_{2,3,4}$).
- Lumbosacral trunk (L_{4}, $_{5}$).

Q.373 Name the muscles of posterior abdominal wall?

- Psoas major
- Psoas minor
- Iliacus and
- Quadratus lumborum.

Q.374 What are the actions of psoas major?

- Helps in maintaining posture at hip. Balances trunk while sitting.
- With iliacus, flexor of hip joint.
- One psoas, causes lateral flexion of trunk on that side.
- Lateral rotation of hip.

Q.375 What are the boundaries of lower lumbar triangle?

Also called Petit's triangle, formed by latissimus dorsi and posterior border of external oblique muscle of abdomen. The base is formed by the iliac crest.

Perineum

Q.376 What are the boundaries of perineum?

Superficial (Fig. 5.54):

Anterior: Scrotum in male
Mons pubis in female

Posterior: Buttocks.

Lateral: Upper part of medial side of thigh.

Deep:

Anterior: Upper part of pubic arch
Arcuate pubic ligament.

Posterior: Tip of coccyx.

Lateral: Conjoined ischiopubic rami, Ischial tuberosity and Sacro-tuberous ligament.

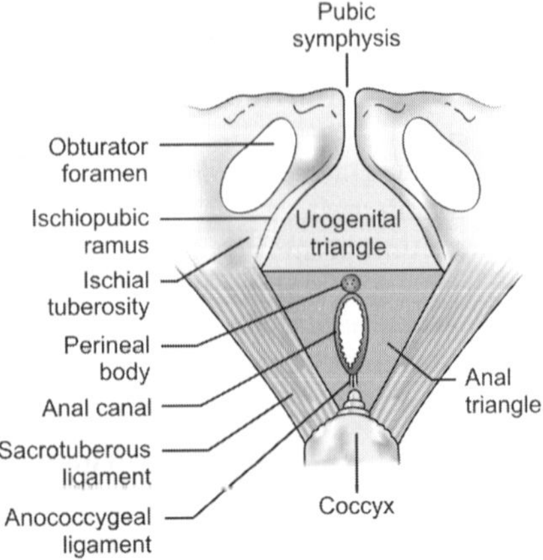

Fig. 5.54: Boundaries of the perineum

Q.377 What are the divisions of perineum?

An imaginary transverse line joining the anterior parts of ischial tuberosities divide rhomboid shaped perineum into two triangular regions:

- *Urogenital region:* Anterior
- *Anal region:* Posterior.

Q.378 What are the boundaries of urogenital triangle?

- *Apex:* By pubic symphysis.
- *On either side:* By ischiopubic ramus
- *Base:* Posteriorly, by imaginary line joining two ischial tuberosities.

Q.379 What are the boundaries of anal triangle?

- *Apex:* By coccyx
- *On either side :* Sacrotuberous ligament, Inferior margin of gluteus maximus, superficially
- *Base:* Imaginary line joining two ischial tuberosities.

Q.380 What is perineal body?

Fibromuscular structure in median plane about 1.25 cm in front of anal margin. Supports pelvic organs in female.

Q.381 Name the muscles forming perineal body.

Nine muscles:

- Unpaired:
  - External anal sphincter
  - Bulbospongiosus
  - Fibers of longitudinal muscles coat of rectal ampulla and anal canal.
- *Paired:*
  - Superficial transversus perinei
  - Deep transversus perinei
  - Levator ani.

Q.382 What is the clinical importance of perineal body?

In females, it may rupture during child birth, which if unrepaired may lead to prolapse of urinary bladder, uterus and rectum.

Q.383 What are the boundaries of ischiorectal fossa?

It is a wedge shaped space, on each side of anal canal below pelvic diaphragm between obturator internus and levator ani.

Base: Skin

Apex: Meeting of obturator fascia with inferior layer of pelvic fascia.

Anterior: Posterior border of perineal membrane.

Posterior: Lower border or gluteus maximus and sacrotuberous ligament.

Lateral wall: Obturator internus with fascia, medial surface of ischial tuberosity.

Medial wall: External anal sphincter, in lower part, levator ani fascia, in upper part.

Q.384 What are the contents of ischio-rectal fossa?

- Ischiorectal pad of fat.
- Inferior rectal nerve and vessels.
- Posterior scrotal (or labial in females) nerves and vessels.
- Perineal branch of S_4 nerve.
- Perforating cutaneous branches of $S_{2,3}$ nerves.
- Pudendal canal with internal pudendal vessels and pudendal nerve.

Q.385 Why the infections of perianal space are very painful but those of ischiorectal space are much less painful?

Fat in perianal space is tightly arranged in small loculi formed by complete septa therefore little swelling due to infections cause increased tension and pain, but in ischiorectal space fat is loosely arranged therefore swelling can occur without tension.

Q.386 Why the infections are more common in ischiorectal fossa?

Because of the presence of poorly vascularized fat in fossa this region is very vulnerable to infection. Infections usually reach the fossa from anal canal.

Q.387 Why the unilateral ischiorectal abscess if not drained becomes bilateral?

Because of extension of infection through the horse-shoe recess behind anal canal which connects the fossa of both sides.

Q.388 Why in debilitating disorders prolapse of rectum occurs?

Because of:

- Loss of fat from ischiorectal fossa which normally acts as support to rectum and anal canal.
- Weakness of perineal muscles forming perineal body.

Q.389 Why the abscesses of ischiorectal fossa can be drained by incision easily?

- Because the fossa has poor vascularity, so there is less blood loss.
- Fossa contains no important structures.

Q.390 Name the structures passing through gap between arcuate pubic and transverse perineal ligament.

- Deep dorsal vein of penis.
- Dorsal nerve of penis.

Q.391 How the deep perineal space is formed?

The deep perineal space is formed between superior and inferior fascia of the urogenital diaphragm.

Q.392 What are the contents of deep perineal space in males?

- Sphincter urethrae (Fig. 5.55)
- Deep transverse perinei
- Bulbourethral glands of Cowper
- Internal pudendal artery and its branches
- Dorsal nerve and perineal nerve of penis: Branches of pudendal nerve.

Q.393 What are the contents of superficial perineal space in male?

- Root of penis made up of bulb and right and left crura
- Bulbospongiosus muscle
- Ischiocavernosus muscle
- Superficial transverse perinei
- Branches of internal pudendal artery and pudendal nerve.

Q.394 Name the structures piercing the perineal membrane (inferior fascia of the urogenital diaphragm).

In males:

- Membranous urethra
- Branches of perineal nerve to superficial perineal muscles
- Ducts of bulbourethral glands
- Artery and nerve to the bulb (bilateral)
- Urethral artery (bilateral)
- Deep artery of penis (bilateral)
- Dorsal artery of penis (bilateral)
- Posterior scrotal nerves and vessels (bilateral).

In females:

- 1, 2 same as above
- Vagina
- Artery and neve to bulb of vestibule
- Deep artery of clitoris
- Dorsal artery of clitoris
- Posterior labial arteries and nerves.

Q.395 Name the structures forming urogenital diaphragm.

- Deep transverse perinei
- Superior fascia of urogenital diaphragm
- Inferior fascia of urogenital diaphragm
- Sphincter urethrae.

Q.396 Name the female external genital organs?

- Mons pubis
- Labia majora
- Labia minora
- Clitoris
- Vestibule of vagina having various openings
- Bulb of vestibule
- Greater vestibular glands (of Bartholin) (Fig. 5.56).

Q.397 What are the boundaries of gynecological perineum?

The area between posterior commissure (skin connecting prominent posterior ends of labia majora) and anus, constitutes gynecological perineum.

Q.398 What is the position of glands of Bartholin?

These are homologous with bulbourethral glands (of Cowper) in males. Lie in the superficial perineal space at vaginal orifice.

Duct of each gland opens at side of hymen, between hymen and labium minora.

Q.399 How pudendal canal is formed?

- By splitting of fascia lunata.
- Fascial wall of canal is fused with:
  Laterally: Obturator fascia
  Medially: Perineal fascia.
  Inferiorly: Falciform process of sacrotuberous ligament.
  Superiorly: Arches over ischiorectal fat and fused with inferior fascia of pelvic diaphragm.

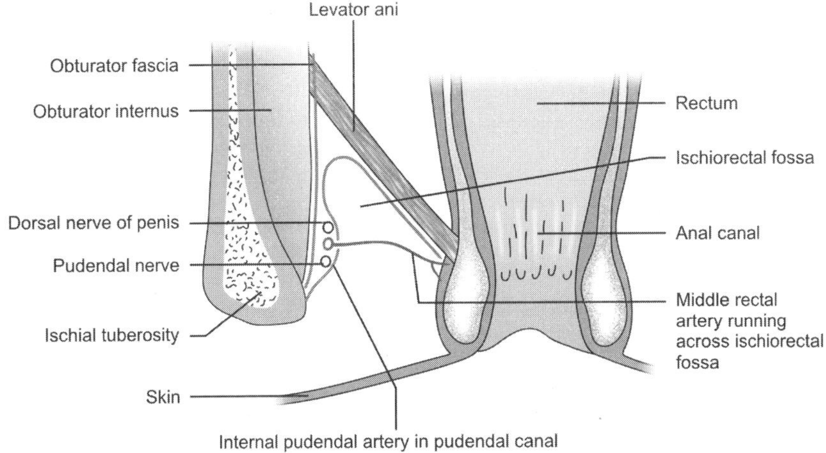

Fig. 5.55: Section through the ischiorectal fossa and the pudendal canal in plane *xy* shown in Figure 5.58

Q.400 What are the contents of pudendal canal (Alcock's canal)?

- Pudendal nerve ($S_{2,3,4}$)
- Internal pudendal vessels (Fig. 5.57).

Q.401 What are the structures supplied by pudendal nerve?

- *Inferior rectal nerve:* Supplies external anal sphincter, skin around anus and anal canal below pectinate line.
- *Perineal nerve:*
  - Posterior scrotal nerves: Posterior 2/3 of scrotum in males and posterior labial nerves (sensory) in females to lower one inch of vagina and labium majora.
  - *Muscular branches:* To urogenital muscles, anterior parts of external anal sphincter and levator ani. Nerve to bulbospongiosus supplies corpus spongiosum of penis and urethra.
- *Dorsal nerve of penis:* Supplies skin of body of penis and glans.

Q.402 Where the 'pudendal nerve block' given in vaginal operations?

Near the ischial spine by a needle passed through vaginal wall and then guided by a finger.

Q.403 What are the branches of internal pudendal artery?

- Inferior rectal artery
- Perineal artery (Fig. 5.58)
- Artery of bulb of penis
- Deep and dorsal arteries of penis

Urinary Bladder (Fig. 5.59)

Q.404 What are the variations in shape of urinary bladder?

- When empty, tetrahedral and lies within pelvis
- When fills, ovoid and extends into abdominal cavity.
  The maximum capacity of urinary bladder is about 500 mL.

Q.405 What are the variations in position of bladder with age?

In infants, at higher level, the internal urethral orifice lies at level of superior border of symphysis pubis.

Then orifice descends rapidly for first three years, then slowly from 4 to 9 years, then it again descends to adult position after puberty.

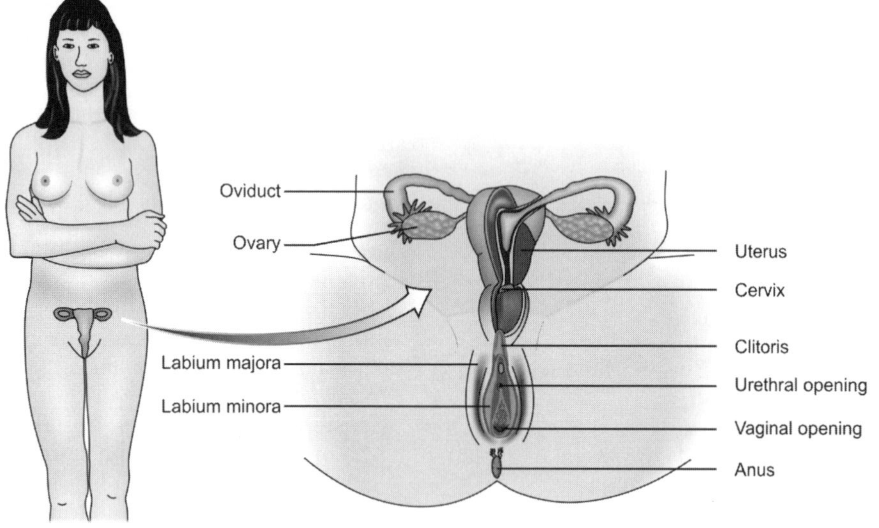

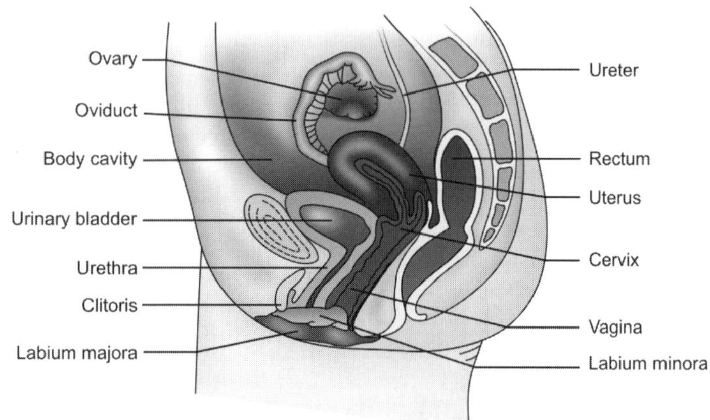

Fig. 5.56: Female urogenital system

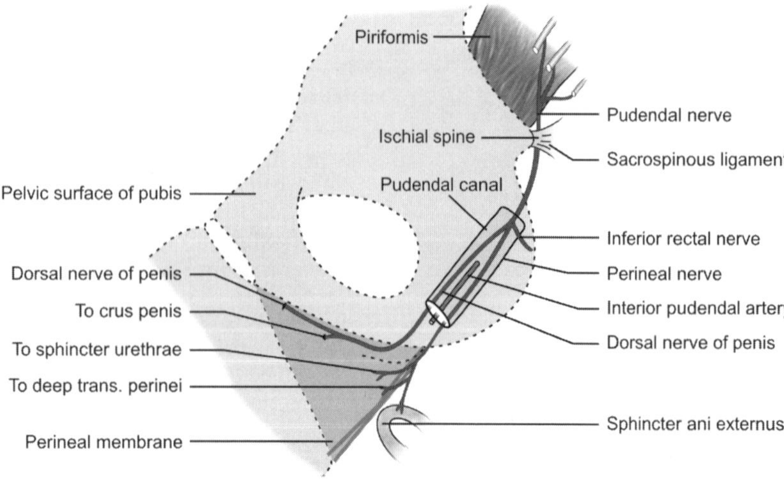

Fig. 5.57: Scheme to show the course and distribution of the pudendal nerve

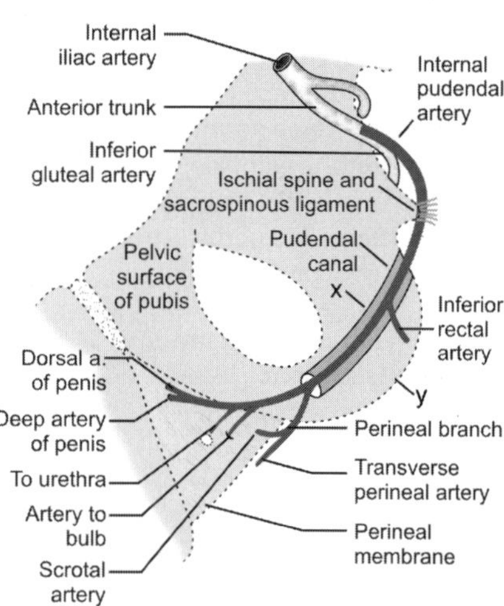

Fig. 5.58: Scheme to show course and branches of the internal pudendal artery

Q.406 What are the peritoneal folds of urinary bladder?

- Median umbilical fold
- *Median umbilical ligament (fold):* Connect superior vesical arteries to umbilicus
- *Lateral false ligament:* Connect superior surface of bladder to lateral wall of pelvis
- *Posterior false ligament:* Connect lateral margin of base of bladder to rectum.

Q.407 Name the ligaments formed by the pelvic fascia around urinary bladder?

- *Lateral true ligament:* Formed from fascia covering obturator internus
- Medial puboprostatic ligament
- Lateral puboprostatic ligament.
  In females, bands similar to puboprostatic ligaments are known as pubovesical ligaments. Formed from fascia over upper surface of levator ani.
- *Median umbilical ligament:* Remnant of urachus. Connects apex of urinary bladder to umbilicus.
- *Posterior ligament:* Connects base of bladder to lateral pelvic wall.

Q.408 What are the relations of base of urinary bladder?

In males:

- *Upper part:* Rectovesical pouch containing intestine.
- *Lower part:* Seminal vesicles, terminations of vasa deferentia.

In females:

- Cervix
- Vagina.

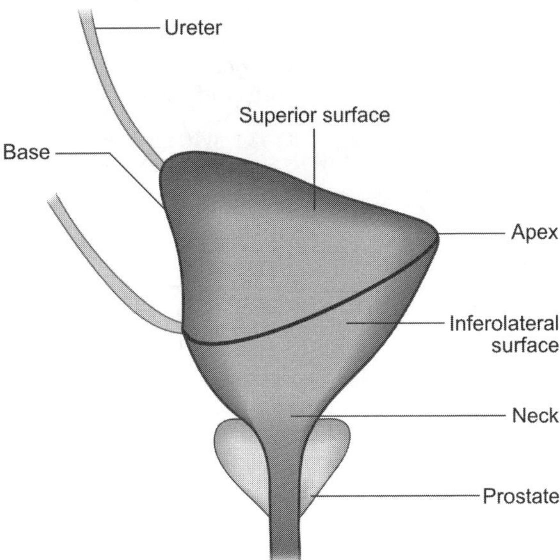

Fig. 5.59: Urinary bladder (male)

Q.409 What are the characteristic features of trigone of bladder?

- Mucosa is firmly attached to muscular coat.
- Has an internal urethral orifice, at anteroinferior angle and two ureteric openings, at postero-lateral angles.
- Uvula vesicae, is slight elevation just posterior to urethral orifice.
- Interureteric ridge forms base of trigone, which are the continuations of longitudinal muscle coat of two ureters.

Q.410 What are the boundaries of paravesical fossa?

Laterally, it is bound by ductus deferens in male and round ligament of uterus in female.

Q.411 What is the arterial supply of urinary bladder?

- Superior and inferior vesical arteries, branches of anterior trunk of internal iliac artery.
- Obturator and inferior gluteal arteries.
- In females, also the uterine artery and vaginal artery in place of inferior vesical artery

Q.412 What is the nerve supply of urinary bladder?

- *Parasympathetic efferent fibers (nerve erigentes $S_{2,3,4}$):* Motor to detrusor muscle and inhibitory to sphincter vesicae.
- *Sympathetic efferent fibers (T_{10} to L_2):* Inhibitory to detrusor and motor to sphincter vesicae.
- *Somatic pudendal nerve ($S_{2,3,4}$):* Supplies sphincter urethrae.
- *Sensory nerve:* Both parasympathetic and sympathetic nerve. They carry the sensation of pain and distension.

Q.413 What is fascia of Denonvilliers?

It is rectovesical fascia in males, separating rectum and triangular area between two ductus deferens at base of bladder.

Q.414 What is histological structure of urinary bladder?

Made up of three coats:

- *Serous coat.* Outer.
- *Muscular coat:* Forms detrusor muscle, consists of three layers of smooth muscle fibers, an external and internal longitudinal and a middle circular.
- *Mucosa:* Epithelium is of transitional variety. Submucosa and glands are absent.

Q.415 Why it is possible to drain a distended bladder through anterior abdominal wall without injuring the peritoneum?

In adults, the bladder is a pelvic organ. When it distends, its upper part cornea in contact with anterior abdominal wall above the pubic symphysis. As bladder ascends, the fold of peritoneum passing from anterior abdominal wall to superior surface of bladder also rises so no peritoneum intervenes between distended bladder and anterior abdominal wall. So it can be relieved by a needle just above the pubic symphysis.

Q.416 How urinary bladder is developed?

- *Epithelium of urinary bladder:* Endodermal, cranial part of vesicourethral canal.
- *Epithelium of trigone:* Mesodermal, absorbed mesonephric ducts.
- *Muscular and serous coat:* Splanchnopleuric mesoderm.

Q.417 What are common congenital anomalies of urinary bladder?

- Bladder may be duplicated.
- Sphincter vesicae may be absent.
- *Hourglass bladder:* Divided into two compartments by a constriction in middle of organ.
- Communication with rectum or vagina may exist.
- Congenital diverticula may be present.

Q.418 What is ectopia vesicae?

Congenital defect in which lower part of anterior abdominal wall and anterior wall of bladder does not develop. The cavity of bladder may be exposed on surface of body. Usually associated with epispadias (urethra opens on dorsal aspect of penis).

Q.419 What does the urachus presents?

The fibrous allantois, which extends from apex of bladder to umbilicus.

Q.420 What are lacunae of Luschka?

These are small cavities which may remain in urachus. One of these may enlarge to form a cyst.

Urethra

Q.421 What is the length of urethra?

In males: 18–20 cm

In females: 4 cm long.

Q.422 What are the part of urethra in male?

- *Prostatic part:* 3 cm.
- *Membranous part:* 1.5–2 cm. Passes through urogenital diaphragm.
- *Spongy (penile) part:* 15 cm.

Q.423 What are the features of floor of prostatic part?

- *Urethral crest (veru montanum):* Median longitudinal ridge on posterior wall.
- *Colliculus seminalis:* Elevation in middle part of crest. In midline has opening of blind sac, prostatic utricle. On either side of crest has opening of left and right ejaculatory ducts.
- *Prostatic sinuses:* On each side of crest, has opening of prostatic ducts (Fig. 5.60).

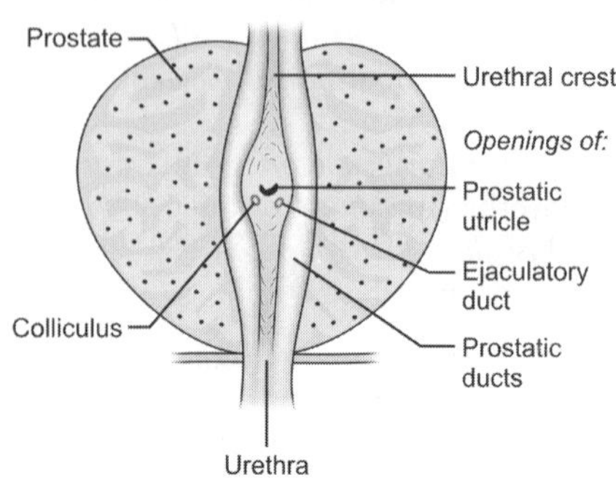

Fig. 5.60: Posterior wall of the prostatic urethra

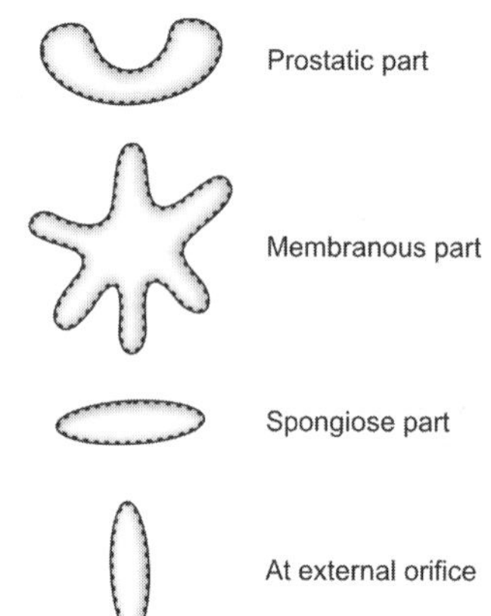

Fig. 5.61: Transverse sections through various parts of the male urethra to show the shape of its lumen

Q.424 Which is the narrowest part of urethra?
Narrowest part of male urethra is external orifice, otherwise membranous urethra is narrowest part.

Q.425 What is position of bulbourethral gland?
These are placed one on each side of membranous urethra. Their ducts open into penile urethra.

Q.426 What are the variations in shape of lumen of male urethra?
- *Prostatic part:* Semilunar (Fig. 5.61).
- *Membranous part:* Star shaped.
- *Spongy part:* Transverse, except external urethral orifice which is vertical slit.

Q.427 What are the characteristic features of sphincters of urethra?
- *Internal urethral sphincter* (Sphincter vesicae) (Fig. 5.62):
  - Involuntary.
  - Supplied by sympathetic nerve.
  - Made up of smooth muscle fibers with elastic and collagenous fibers.
- *External urethral sphincter* (Sphincter urethrae)
  - Voluntary.
  - Supplied by pudendal nerve.
  - Made up of striated muscle fibers.

Q.428 What is the lymphatic drainage of urethra?
Membranous and prostatic part drains into internal iliac lymph nodes. Penile part drains into superficial inguinal nodes.

Q.429 Which part of male urethra is ruptured during instrumentation?
Membranous part because it is narrowest and least dilatable.

Q.430 What is commonest cause of urethral stricture?
Gonococcal infection.

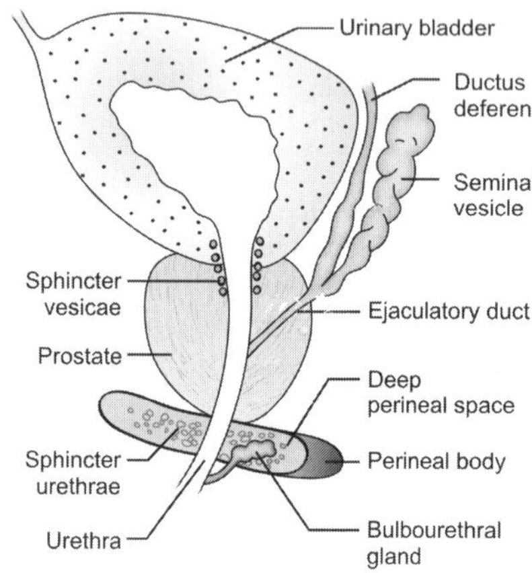

Fig. 5.62: Diagram showing the sphincters of the urethra, and the bulbourethral glands

Q.431 Why the instruments in urethra should be introduced with beak downwards?
Because immediately within external meatus, urethra dilates into a terminal fossa, whose roof bears a mucosal fold (lacuna magna) which may catch the tip of catheter.

Q.432 How the urethra is developed?
Female urethra: Caudal part of vesicourethral canal.
Male urethra:
- *From urinary bladder upto opening of ejaculatory ducts:* Endodermal (caudal part of vesicourethral canal).
- *Rest of prostatic urethra and membranous urethra:* Pelvic part of definitive urogenital sinus.
- Penile part except terminal part: Epithelium of phallic part of definitive urogenital sinus.
- *Terminal part of penile urethra:* From ectoderm.

Q.433 What is hypospadias?
Due to inability of urethral folds to unite anteriorly, the urethra opens on undersurface of penis.

Q.434 What is epispadias?
The urethral orifice opens on the dorsal aspect of penis.

Prostate

Q.435 What is the position of prostate?
It lies in lesser pelvis, below neck of urinary bladder, behind lower part of pubic symphysis and upper part of pubic arch and in front of ampulla of rectum.

Q.436 What are the lobes of prostate?
Five lobes (Figs 5.63 to 5.65):
- Anterior lobe
- Posterior lobe
- Median (middle or prespermatic) lobe
- Right and left lateral lobes.

Q.437 Name the structures lying within the prostate?
- Prostatic urethra.
- Prostatic utricle.
- Ejaculatory ducts.

Q.438 How the capsules of prostate are formed?
- False capsule: Outer and is derived from pelvic fascia.

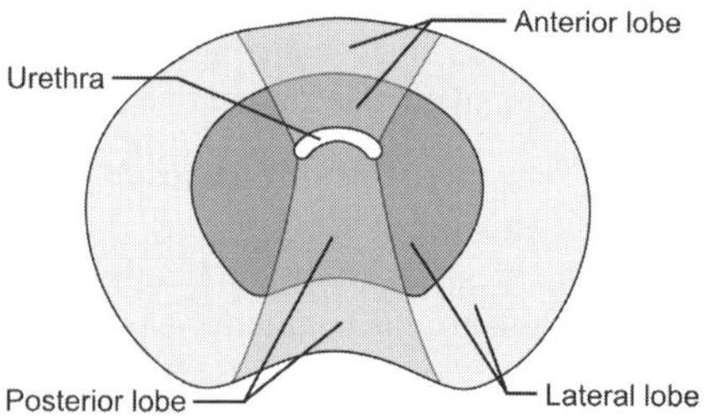

Fig. 5.63: Transverse section through the prostate to show its lobes

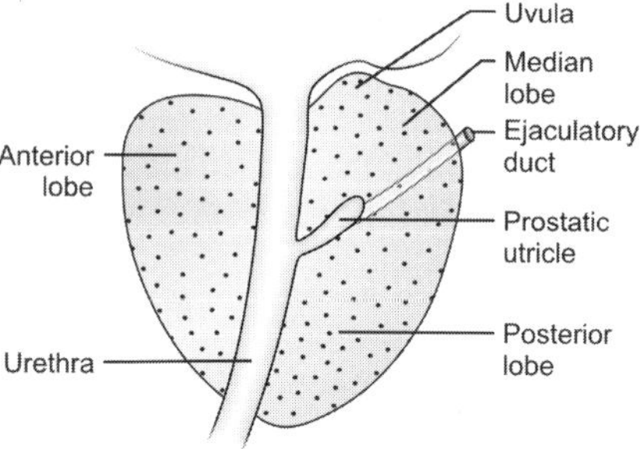

Fig. 5.64: Sagittal section through the prostate to show its lobes

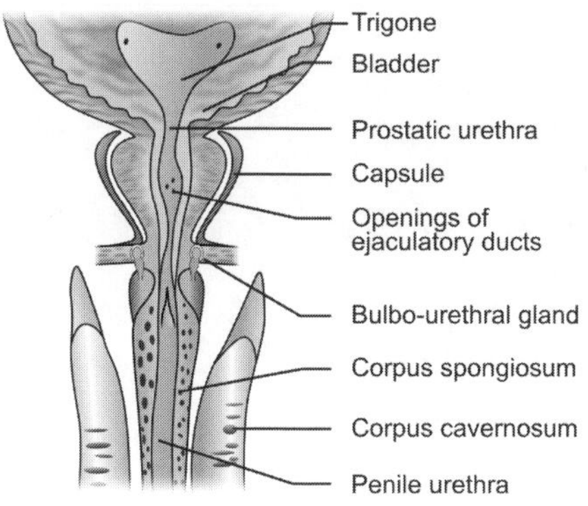

Fig. 5.65: Prostate gland

- True capsule: Inner to false capsule and is formed by condensation of fibromuscular peripheral part of gland.

Q.439 What is venous drainage of prostate?

- Veins form a plexus deep to false capsule, around sides and base of gland. The plexus communicates with vesical plexus, vertebral plexus and internal pudendal vein.
- Drains into vesical and internal iliac veins.

Q.440 What is the lymphatic drainage of prostate?

Prostatic lymphatics drain into internal and external iliac nodes.

Q.441 What is 'cave of Retzius'?

This is potential retropubic space separating pubic symphysis and anterior surface of prostate. This is filled with fat.

Q.442 What is the 'pathological capsule of prostate'?

In benign tumors of prostate, the normal peripheral part of gland becomes compressed into a capsule around the tumor mass.

Q.443 What are 'valveless vertebral veins of Bateson'? What is their clinical importance?

Some veins of prostatic plexus communicate with plexus of veins lying in front of vertebral bodies and neural canal. These veins are valveless.

Clinical importance: Because of these valveless veins, there is a retrograde spread of carcinoma of prostate to pelvis and vertebrae.

Q.444 Which part of the prostate is enlarged in benign hypertrophy?

Benign hypertrophy most commonly affects median lobe of prostate. This lobe enlarges upwards and forwards to produce projection on interior of urinary bladder just behind internal urethral orifice, thus obstructing it.

Q.445 Which part of the prostate is affected by the carcinoma?

Outer glandular zone.

Q.446 Why rectal involvement is uncommon in carcinoma of prostate?

Because fascia of Denonvilliers is rarely penetrated by carcinoma of prostate.

Q.447 What is the histology of prostate?

Prostate is composed of glands present in smooth muscle stroma.

Part of gland in front of urethra has dense muscular tissue and very little glandular tissue. The glands are made up of follicles, lined by columnar cells.

Q.448 How prostate is developed?

It develops from buds arising from prostatic urethra.

From epithelium: Secretory part.
From mesoderm: Inner glandular zone.
From endoderm: Outer glandular zone.
From mesenchyme: Muscle and connective tissue.

Q.449 What are the homologous of prostate in female?

Urethral glands and paraurethral glands of Skene.

Q.450 What do you understand by enucleation of adenoma of prostate?

In enucleation, plane between adenomatous mass and pathological capsule is cleaved, the tumour is removed and peripheral condensed prostatic tissue is left behind. The prostatic venous plexus, lying between true and false capsule, is not disturbed.

Ductus Deferens (Vas Deferens)

Q.451 What is the ampulla of vas?

- The dilated and tortuous part of vas behind the base of bladder.
- It has no lumen.

Q.452 How ejaculatory duct is formed?

By the union of lower end of seminal vesicle and ductus deferens, at the base of prostate.

Q.453 What are the constituents of secretion of seminal vesicle?

- Fructose
- Vesiculase enzyme
- Albumin.

  Secretion is slightly alkaline.

Q.454 Where do you palpate the vas deferens?

It is felt as a firm structure between thumb and finger as it lies within the spermatic cord at scrotal neck. As it is traced upwards it passes medial to pubic tubercle and then through external inguinal ring, which can be felt by invaginating scrotal skin with finger tip.

Q.455 What is the position of seminal vesicle?

The left and right seminal vesicles lie posterior to base of urinary bladder, between it and rectum.

Ovary

Q.456 What is the position of ovary ?

It lies in the ovarian fossa on lateral pelvic wall, just below and behind lateral part of uterine tube (Fig. 5.66).

Q.457 What are the boundaries of ovarian fossa?

Inferior: Obliterated umbilical artery.
Posterior: Ureter and internal iliac vessels
Anterior: External iliac vessels (Fig. 5.67).

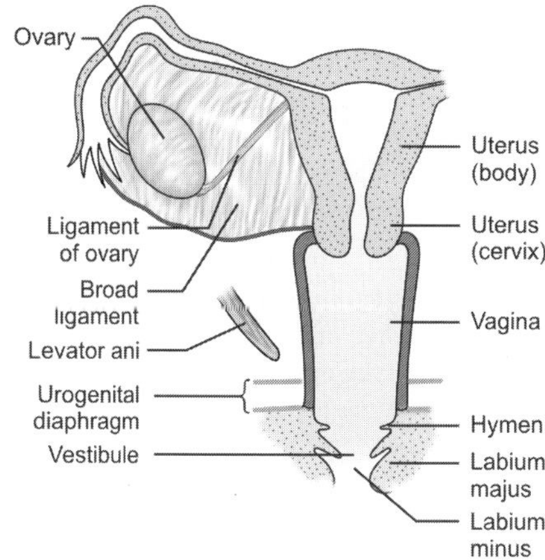

Fig. 5.66: Scheme to show the female reproductive organs

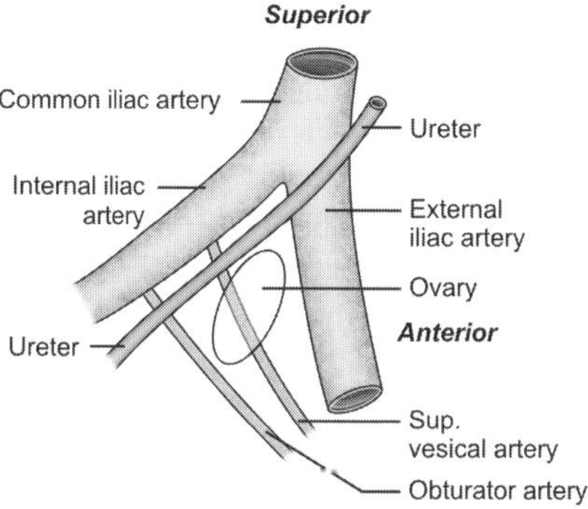

Fig. 5.67: Boundaries of ovarian fossa

Q.458 What are the peritoneal relations of ovary?

- Ovary is entirely covered with peritoneum except along the anterior border where the two layers of peritoneum are continuous with the posterior layers of broad ligament of uterus, as mesovarium.
- Lateral part of broad ligament, from infundibulum of tube and upper pole of ovary to external iliac vessels forms suspensory ligament of ovary.

Q.459 What is the blood supply of ovary?

- *Arterial supply:*
  - Ovarian artery.
  - Branches of uterine artery.
- *Venous drainage:* Pampiniform plexus which condense into a single vein, near the pelvic inlet.
  Right ovarian vein, drains into inferior vena cava.
  Left ovarian vein, drains into left renal vein.

Q.460 What is the histological structure of ovary?

Ovary is made up of (from within outwards):

- *Medulla:* Vascular connective tissue having vessels, nerves and lymphatics.
- *Cortex:* Has various stages of development of ovarian follicles.
- *Tunica albuginea:* Thin layers of connective tissue.
- *Germinal epithelium:* Made up of cuboidal cells, derived from peritoneum.

Q.461 What is ovulation (Fig. 5.68)? What are its indicates?

Release of one or more ova from one of the ovaries during each menstrual cycle is termed ovulation. Development of an ovum followed by the process of ovulation is shown in the Figure 5.68.

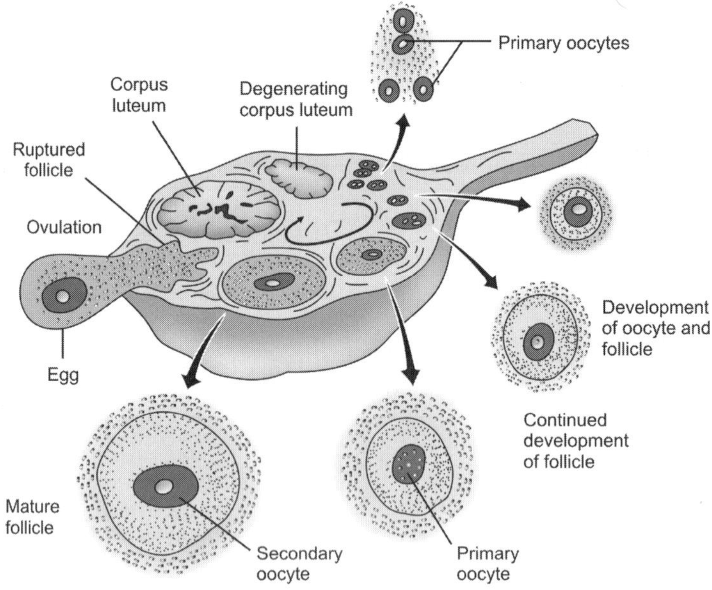

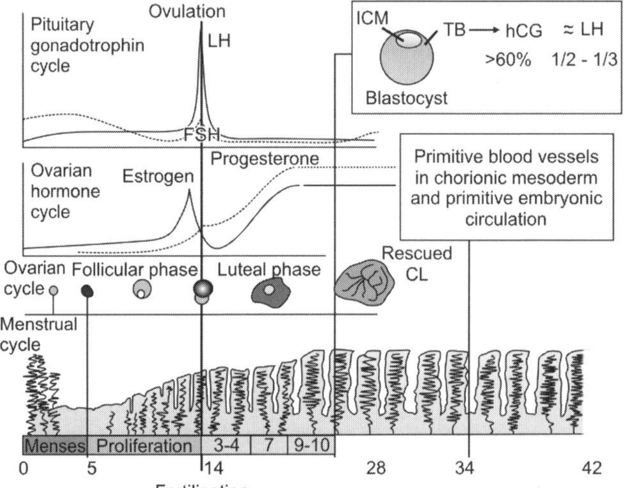

Fig. 5.68: Ovulation

Uterine Tubes (Fallopian Tubes)

Q.462 What are the parts of uterine tube?

- *Infundibulum (fimbriated):* Opens into peritoneal cavity by abdominal ostium.
- *Ampulla:* Forms lateral 2/3 of tube. Thin walled and wider lumen.
- *Isthmus:* Forms medial 1/3 tube. Thick walled and narrow lumen.
- *Uterine (interstitial):* Lies within uterine wall and opens into uterine cavity by uterine ostium.

Q.463 What are the relations of uterine tube with ovary?

- Near lateral pelvic wall, ampulla is related to anterior and posterior borders, upper pole and medial surface of ovary.
- One of the fimbria is long and is attached to tubal (upper) pole of ovary, it is known as ovarian fimbria.

Q.464 What is mesosalphinx?

It is part of broad ligament between mesovarium and uterine tube.

Q.465 What is blood supply of uterine tube?

Arterial supply:

- *Medial 2/3:* Uterine artery.
- *Lateral 1/3:* Ovarian artery.
  Venous drainage: Into pampiniform plexus and uterine veins.

Q.466 What is histological structure of uterine tubes?

They are made up of following coats:

- *Outer:* Serous coat
- *Middle:* Muscular coat has circular muscles.
- *Inner:* Mucous membrane is lined by ciliated columnar cells and nonciliated secretory cells. The mucous membrane forms folds which fill up the lumen of tube.

Q.467 How uterine tubes are developed?

From unfused parts of paramesonephric ducts. The point of invagination of duct remain as the abdominal openings.

Q.468 At what site the fertilization of ovum takes place?

In the ampulla of the fallopian tube.

Q.469 What is tubectomy?

Female sterilization, in which 2 to 3 cm long segment of tube is excised and cut ends are ligated.

Uterus

Q.470 What are the parts of uterus?
- *Body:* upper 2/3.
- *Cervix:* lower 1/3.
  The upper 1/3 of cervix forms 'isthmus'.

Q.471 What are the parts of cervix?
- *Vaginal:* Projects into vagina.
- *Supravaginal.*
  The cervical canal (cavity of cervix) extends from internal os above to external os below, where it opens into vagina.

Q.472 What are 'arbor vitae'?
The mucous membrane of cervical canal is thrown into fold and oblique furrows which pass away from anterior and posterior vertical ridges.

Q.473 What are the angulations of uterus?
- *Angle of anteversion:* Forward angulation between cervix and vagina. About 90 degrees.
- *Angle of anteflexion:* Forward angulation between body and cervix. About 120 to 125 degrees.

Q.474 Name the structures attached to lateral border of body of uterus.
- Broad ligament.
- Uterine tube at upper end.
- *Round ligament of uterus:* Anteroinferior to tube.
- *Ligament of ovary:* Posteroinferior to tube.

Q.475 What are the supports of uterus?
The uterus is prevented from sagging down by a number of factors. They are classified into (Figs 5.69 and 5.70):
- *Primary Supports:*
  - *Muscular:*
    Pelvic diaphragm including levator ani muscles and pelvic fascia lining them
    Perineal body
    Urogenital diaphragm.
  - *Fibromuscular:*
    Uterine axis
    Pubocervical ligaments

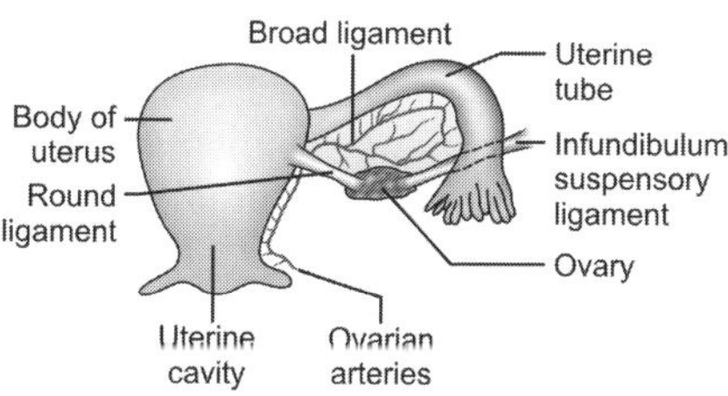

Fig. 5.69: Supporting ligaments of the uterus

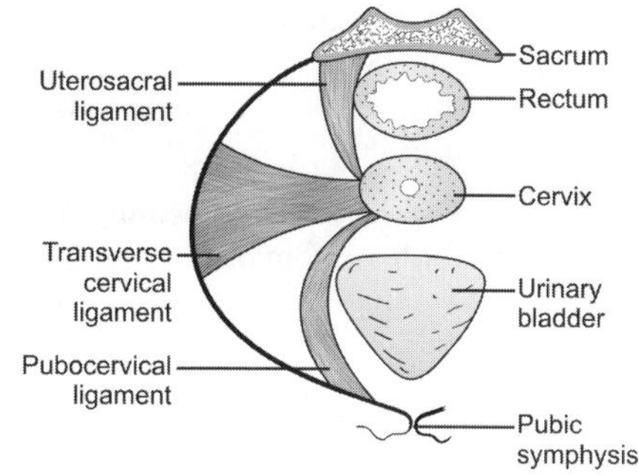

Fig. 5.70: Scheme to show some ligaments of the uterus

    Transverse cervical ligament (Mackenrodt or cardinal ligament)
    Uterosacral ligament
    Round ligament of uterus.
- *Secondary supports (of doubtful value):*
  Broad ligaments
  Uterovesical fold
  Rectovaginal fold.

Q.476 How uterine axis supports the uterus in maintaining its position?
The anteversion prevents uterus from sagging through vagina. Any rise in intra-abdominal pressure tends to push uterus against bladder, which further accentuates anteversion. Angle of anteversion is maintained by uterosacral and round ligaments.

Q.477 What is canal of Nuck?
Round ligament of uterus in inguinal canal, in fetal life is accompanied by a process of peritoneum, which if persists, after birth is known as canal of Nuck.

Q.478 What are the parts of broad ligament of uterus?
- *Mesosalphinx:* Between tube and ovarian ligament.
- *Mesometrium:* Below ovarian ligament.
- Suspensory ligament of ovary
- Mesovarium.

Q.479 What are contents of broad ligament of uterus?
- *Tube:* Uterine tube.
- *Ligaments:* Round ligament of uterus
  Ligament of ovary
- *Vessels:* Uterine vessels
  Ovarian vessels.
- *Nerves:* Uterovaginal plexus
  Ovarian plexus.
- *Embryological remnants:*
  Epoophoron
  Duct of epoophoron (Gartner's duct).
  Paroophoron.
- Lymphatics and lymph nodes.
- Fibroareolar tissue (parametrium).

Q.480 What is the arterial supply of uterus?
- Uterine arteries, mainly
- Ovarian arteries.

Q.481 Name the structures supplied by uterine artery?
- Uterus
- Vagina
- Medial 2/3 of uterine tube
- Ovary
- Ureter and
- Contents of broad ligament.

Q.482 What is the histological structure of uterus?
Wall of uterus is made up of 3 layers:
- *Outer:* Perimetrium (derived from peritoneum)
- *Middle:* Myometrium (muscular) consists of 3 layers:
  External: Longitudinal fibers
  Middle: Muscle fibers interlace
  Inner: Circular fibers.
- *Inner:* Endometrium, consist of surface epithelium, glands and stroma. In cervix submucosa is absent, so epithelium and glands come in direct contact with myometrium.

Q.483 What is histological difference between two parts of cervix.
Vaginal portion is covered by squamous epithelium which becomes continuous with columnar cells of cervical canal at external os.

Q.484 Describe the course of uterine artery and its distribution to uterus.
Uterine artery is branch of anterior trunk of internal iliac artery. It runs downwards and forward and when reaches parametrium, it turns medially towards uterus.

It reaches uterus at level of internal os, where it turns upwards at right angles and runs a spiral course along lateral border of uterus to uterine cornu. During vertical part it gives branches, which run transversely into myometrium (arcuate arteries).

From these arise radial arteries at right angles and they reach basal layers of endometrium (basal arteries).

Basal arteries give rise to terminal spiral and straight arterioles of endometrium.

Q.485 What is the lymphatic drainage of uterus?
Lymphatics of uterus form three intercommunicating plexuses, which drain into:
- *Upper lymphatics from fundus and upper part of body:* Aortic and superficial inguinal nodes.

- *Middle lymphatics from lower part of the body:* External iliac nodes.
- *Lower lymphatics from cervix:* External and internal iliac and sacral nodes.

Q.486 What are the changes in uterus with age?

- *In fetal life:* Cervix is larger than body of uterus.
- *At puberty:* Uterus enlarges. The body grows more than cervix so it acquires its pyriform shape.
- *During menstruation:* Uterus slightly enlarged and more vascular.
- *During pregnancy:* Uterus enormously enlarged.
- *After pregnancy:* Regresses to normal size but thickness of wall and size of cavity remains larger.
- *In older age:* Uterus smaller and denser in texture.

Q.487 How is uterus divided into upper and lower segment during pregnancy?

The uterus is divided into upper uterine segment consisting of fundus and greater part of the body and lower segment consisting of lower part of body and cervix. The upper one-third of the cervix is known as isthmus.

Q.488 What is the developmental origin of uterus?

Epithelium of uterus develops from fused paramesonephric ducts. Myometrium from surrounding mesoderm.

The unfused part of paramesonephric duct embedded in myometrium forms fundus.

Q.489 What are the common anomalies of uterus?

- Uterus may be duplicated or absent.
- Lumen of uterus may be divided by a septum.
- One half of uterus may be absent (unicornuate uterus).
- Uterus may remain rudimentary.

Q.490 What are the advantages and disadvantages of midline incision made in the uterus?

The midline part of uterus is least vascular part, so there is less bleeding during surgery but the wound also heals poorly due to poor vascularity.

Q.491 What precaution should be taken in relation to ureter while removing uterus (hysterectomy)?

At supravaginal cervix, ureter lies just above the level of lateral fornix and below uterine vessels as these pass within broad ligament. In hysterectomy, ureter may be accidentally divided when clamping the uterine vessels, especially when pelvic anatomy is distored.

Q.492 What is the fate of mesonephric ducts and tubules in female?

They form a number of vestigeal structures.

- *Epoophoron:* Represent cranial mesonephric tubules.
- *Paroophoron:* Represent caudal mesonephric tubules.
- *Duct of epoophoron:* Represents mesonephric duct.

Q.493 How vesicular appendix is developed?

From cranial part of paramesonephric duct.

Vagina

Q.494 What is the position and extent of vagina?

It is situated behind bladder and urethra and in front of rectum and anal canal.

It extends from vulva to uterus.

Q.495 What are the variations in shape of lumen of vagina?

- *At upper end:* Circular.
- *In middle part:* Transverse.
- *At lower end:* H-shaped.

Q.496 What are relations of vagina?

Anterior wall: 8 cm long
Upper half: Base of bladder.
Lower half: Urethra
Posterior wall: 10 cm long
Upper 1/4 Separated from rectum by pouch of Douglas.
Middle 1/2: Separated from rectum by loose connective tissue.
Lower 1/4: Separated from anal canal by perineal body.
Lateral wall: On each side
Upper 1/3: Transverse cervical ligament in which are embedded vaginal veins and ureter crossed by uterine artery.
Middle 1/3: Levator ani
Lower 1/3: Urogenital diaphragm
Bulb of vestibule
Bulbospongiosus
Greater vestibular glands.

Q.497 What are the fornices of vagina?

The upper part of vagina is converted into a circular groove by protruding cervix, which is divided into four parts known as vaginal fornices: anterior, posterior and two lateral fornices. Anterior fornix is shallowest and posterior fornix deepest.

Q.498 What is the arterial supply of vagina?

- Vaginal branch of internal iliac, mainly.
- *Upper part:* Also by cervicovaginal branch of uterine artery.
- *Lower part:* Also by middle rectal and internal pudendal arteries.
  These vessels form anterior and posterior midline vessels called vaginal azygous arteries.

Q.499 What is the lymphatic drainage of vagina?

- *Upper 1/3:* External iliac nodes.
- *Middle 1/3:* Internal iliac nodes.
- *Lower 1/3:* Medial group of superficial inguinal nodes.

Q.500 What is the nerve supply of vagina?

- *Upper 2/3:* Pain insensitive
  Supplied by sympathetic ($L_{1,2}$) and parasympathetic ($S_{2,3}$) nerves.
- *Lower 1/3:* Pain sensitive.
  Supplied by inferior rectal and posterior labial branches of pudendal nerve.

Q.501 What is the characteristic feature of lining epithelium of vagina?

Vagina is lined by stratified squamous epithelium and has no glands. It is lubricated partly by cervical mucus and partly by desquamated vaginal epithelial cells.

Q.502 What important information can be obtained by per vaginal (PV) examination?

The condition of:

- *Vagina:* Abnormalities of entrance or walls.
- *Urethra:* Can be rolled against symphysis in anterior wall.
- *Rectum:* If it contains tumor, foreign body or feces.
- Cervix and external os
- Vaginal fornices
- *Rectovaginal pouch:* By finger in posterior fornix. If it contains uterus, prolapsed ovaries, tumors or abdominal collection of fluid.
- *Ureter:* If thickened or has a stone, can be rolled against pelvic bone just before it enters bladder.
- To determine the diagonal conjugate of pelvis, to assess whether pelvis is large enough to transmit fetal head.

Q.503 What is hymen?

A fold of mucous membrane at lower end of vagina which partially closes it.

Q.504 What is the developmental origin of vagina?
The vagina is formed by development of lumen within vaginal plate, which is formed by:
- Endodermal cells of urogenital sinus which proliferate to form sinovaginal bulbs and
- Proliferation of mesodermal cells at lower end of uterovaginal canal.

Q.505 What are the abnormalities of vagina?
- Vagina may be duplicated.
- Lumen of vagina may be subdivided by septum.
- Vagina may be absent.
- Vagina may have abnormal communications with rectum and urinary bladder.

Rectum

Q.506 What is the length of rectum?
12 cm.

Q.507 What is the situation and extent of rectum?
Situation: In posterior part of lesser pelvis, in front of lower three pieces of sacrum and coccyx.

Extent: From S_3 vertebra (rectosigmoid junction) to 2–3 cm, in front and a little below the tip of coccyx (anorectal junction).

Q.508 What are the curves of rectum and what is their position?
Rectum lies in median but shows two types of curvatures:
- *Anteroposterior curves:*
  - Sacral flexure of rectum follows concavity of sacrum and coccyx.
  - Perineal flexure of rectum is backward bent at anorectal junction.
- *Lateral curves*
  - Upper lateral curve is convex to right.
  - Middle lateral curve is convex to left.
  - Lower lateral curve is convex to right.

Q.509 What is rectal ampulla?
The lower dilated part of rectum is called rectal ampulla.

Q.510 What are the peritoneal relations of rectum?
- *Upper 1/3:* Covered with peritoneum in front and at sides.
- *Middle 1/3:* Covered only in front.
- *Lower 1/3:* Devoid of peritoneum.

Q.511 What are different types of folds of mucous membrane of rectum?
- *Longitudinal folds:* Present in lower part of rectum and they are obliterated by distension.
- *Transverse folds (Houston's valves):* Are permanent and most marked when rectum is distended.
  These are formed by infoldings of mucous membrane containing circular and longitudinal muscle coat.
- *Upper fold:* From right or left wall, near upper end of rectum.
- *Middle fold:* From anterior and right walls, at rectal ampulla.
- *Lowest fold:* 2.5 cm below middle fold and projects from left wall.
- Sometimes fourth fold is present 2.5 cm above middle fold.

Q.512 What are the relations of the rectum?
Anteriorly:
- Lower part of sacrum
- Coccyx
- *Piriformis muscle:* Left and right
- *Coccygeus muscle:* Left and right
- *Levator ani muscle:* Left and right
- Sympathetic truama
- Median sacral artery
- Lateral sacral artery
- Medial sacral artery
- Pudendal nerve
- Ganglion impar.

Posteriorly:
In male: To urinary bladder, seminal vesicles, ductus deferens lower ends of ureters and prostate.
In female: To vagina and lower part of uterus.
In both sexes: Sigmoid colon, ileum
Laterally: Coccygeus muscle: Left and right,
Levator ani muscle: Left and right ileum

Q.513 What are the boundaries of the pararectal fossa?
It is bound laterally in males by sacrogenital fold and in females by rectouterine fold.

Q.514 What is the blood supply of rectum?
- *Anterior supply:*
  - *Superior rectal artery:* Continuation of inferior mesenteric artery.
  - *Middle rectal arteries:* Arise from anterior division of internal iliac artery.
  - *Median sacral artery:* Arise from aorta.
- *Venous drainage:*
  - *Superior rectal veins:* Continues upward as inferior mesenteric vein.
  - *Middle rectal veins:* Drain into internal iliac veins.

Q.515 What is the lymphatic drainage of rectum?
- *From upper half:* To inferior mesenteric nodes through pararectal and sigmoid nodes.
- *From lower half:* To internal iliac nodes.

Q.516 What are the supports of rectum?
- *Pelvic floor.*
- *Fascia of Waldeyer:* Suspends lower part of rectal ampulla to sacrum.
- *Lateral ligaments of rectum:* Condensation of pelvic fascia.
- *Rectovesical fascia:* Extends from rectum to seminal vesicles and prostate in front.
- Pelvic peritoneum and related vascular pedicles.

Q.517 What structures are palpated on per rectum (PR) examination?
In a normal patient:
- *In males:* Anteriorly (from below upwards): Bulb of penis and membranous urethra, prostate, seminal vesicles and base of urinary bladder.
- *Female:* Anteriorly, vagina and uterus.
- *In both sexes:* Posteriorly: coccyx and lower part of sacrum.
  Laterally: Ischial spine and ischial tuberosity, ischiorectal fossa.
  Abnormalities which can be detected include:
- *Within lumen:* Fecal impaction, foreign body.
- *In the wall:* Rectal growths, strictures but not hemorrhoids unless thrombosed.
- *Outside rectal wall:* Pelvic bony tumors, collections of fluid or tumors in pouch of Douglas and rectovesical pouch, abnormalities of prostate and seminal vesicles; distended bladder, pelvic appendix, ureters, uterine tubes and ovary.

Q.518 How rectum is developed?
From primitive rectum, i.e. dorsal subdivision of cloaca.

Anal Canal (Fig. 5.71)

Q.519 What is the position of anal canal?
Anal canal is situated in perineum between two ischiorectal fossae.

Q.520 What is the extent of anal canal?
Extends from anorectal junction which lies 2–3 cm in front and slightly below the tip of coccyx to anus, about 4 cm below and in front of tip of coccyx in cleft between two buttocks.

Q.521 What are the relations of anal canal?
- *Anteriorly.* Perineal body
  In males: Membranous urethra, bulb of penis

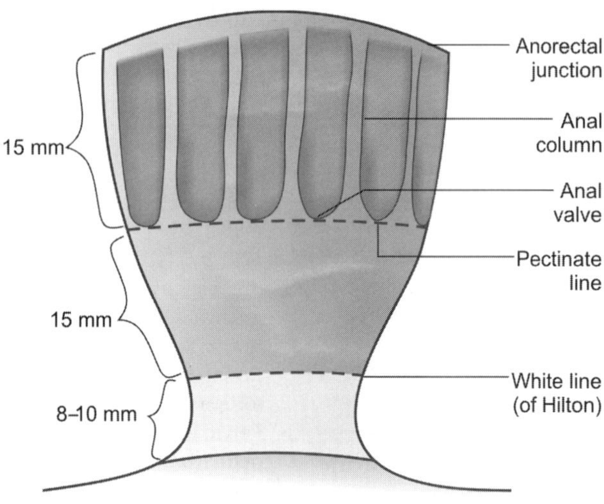

Fig. 5.71: Interior of anal canal

In females: Lower end of vagina
- *Posteriorly:* Anococcygeal ligament
  Tip of coccyx.
- *Laterally:* Ischiorectal fossa.
- *All around:* Sphincter muscles.

Q.522 What are the divisions of anal canal? What are the characteristic features of each part?

- *Upper part:*
  15 mm long, upto pectinate line.
  Lined by columnar epithelium.
  Mucous membrane shows: 6–10 longitudinal folds.
  - *Anal columns:* Vertical mucosal folds
  - *Anal valves:* Small crescentic folds connecting lower ends of adjoining anal columns.
  - *Anal sinuses:* Small pockets above anal valves.
  - *Pectinate line:* Circular line of attachment of anal valves.
- *Middle part:*
  15 mm long
  Between pectinate line and white line of Hilton's.
  Stratified squamous epithelium lining
  No sweat or sebaceous glands or hair.
  Anal columns are not present
  Submucosa has dense connective tissue.
- *Lower part:*
  8 mm long
  Lined by true skin
  Has sweat and sebaceous glands.

Q.523 What is the distribution of anal glands?

Anal glands are present in submucosa and they open above each anal valve into anal sinus. Opening of glands on anal mucosa is referred to as anal crypts.

Q.524 What is the position of Hilton's line?

At level of interval between subcutaneous part of external anal sphincter and lower border of internal anal sphincter. Felt as groove on digital examination.

Q.525 What are the parts of external anal sphincter?

Made of striated muscle. Three parts:
- *Deep part:* Surrounds upper part of internal anal sphincter i.e., above pectinate line and is fused with puborectalis.
  Arise from anococcygeal ligament.
  Inserted into perineal body where fibers decussate.
- *Superficial part:* Elliptical in shape.
  Arise from terminal part of coccyx.
  Fibers surround internal sphincter in lower part between pectinate part and white line of Hilton and are inserted into perineal body.
- *Subcutaneous part:* Below internal sphincter.
  No bony attachment.
  Surrounds lower part of anal canal.

Q.526 What are the features of internal sphincter?

Involuntary.

Formed by thickened circular muscle coat.

Surrounds upper ¾ of anal canal, above the subcutaneous part upto white line of Hilton. Deep to external sphincter.

Q.527 How the conjoint longitudinal coat is formed?

It is formed by fusion of puborectalis with longitudinal muscle coat of rectum at anorectal junction.

It lies between external and internal sphincters.

Divides into fibroelastic septa, which are attached to skin around anus and submucosa below anal valves (corrugator cutis ani).

Q.528 What is 'anorectal ring'?

It is a muscular ring at anorectal junction formed by fusion of puborectalis, deep external and internal sphincter.

Q.529 Name surgical spaces in and around anal canal.

- *Ischiorectal fossa:* On each side.
- *Perianal space:* Around anal canal below white line.
- *Submucous space:* Above white line, between mucous membrane and internal sphincter.

Q.530 What is the blood supply of anal canal?

- *Arterial supply:*
  - *Superior rectal artery (continuity of inferior mesenteric artery):* Above pectinate line
  - *Inferior rectal artery, branch of internal pudendal artery:* Below pectinate line
  - *Median sacral artery:* To posterior part of anorectal junction and anal canal.
- *Venous drainage:*
  - *Internal rectal venous plexus:* In submucosa. It drains into superior rectal vein.
    It communicates with external rectal plexus.
  - *External rectal venous plexus:* Outside muscular coat. Drained by inferior and middle rectal vein.
  - *Anal veins:* Arranged radially around anal margin.

Q.531 What is the characteristic of venous drainage of anal canal?

Superior rectal vein is a tributary of portal system, middle and inferior rectal veins drain into systemic veins. These veins anastomose with each other. Blood from the portal system can pass into systemic circulation through these anastomoses if the portal venous pressure rises leading to their dilatation.

Q.532 What is the lymphatic drainage of anal canal?

The upper part of anal canal drains into internal iliac nodes and lower part into superficial inguinal nodes.

Q.533 What is the developmental origin of anal canal?

It develops partly from ectoderm (proctoderm) and partly from endoderm (dorsal division of cloaca). The junction is indicated by the pectinate line (anal valves).

Q.534 What is the nerve supply of anal canal?

- *Above pectinate line:*
  Inferior hypogastric plexus (Sympathetic).
  Pelvic splanchnic (Parasympathetic).
- *Below pectinate line:* Inferior rectal (Somatic).
- Sphincters:
  Internal sphincter: Sympathetic and parasympathetic.
  External sphincter: Inferior rectal and perineal branch of fourth sacral nerve.

Q.535 What is the cause of imperforate anus?

Failure of anal membrane to break at pectinate line at the end of eight weeks of intrauterine life.

Q.536 What are the other congenital anomalies of anal canal?

- Anal stenosis.
- Anal agenesis.
- Anorectal agenesis.

Q.537 What is the cause of rectal incontinence?

- Damage to anorectal ring.
- Damage of the nerves supplying the muscles of anorectal ring.

Q.538 What is anal fissure?

Rupture of one of anal valves, usually by passage of dry hard stool.

May be painful, if skin is also involved.

Q.539 What is fistula in ano?

It is an abnormal epithelialized track connecting anal canal with the exterior.

Caused by an abscess around anus.

Q.540 What are hemorrhoids (piles)?

These are the dilatations of rectal venous plexus. They are of two types:

- External hemorrhoids are the dilated anal veins around the anal margin.
  These occur below the pectinate line and are very painful.
- Internal hemorrhoids are the dilated internal rectal venous plexus. These occur above pectinate line.

Two types:

- *Primary piles:* Occur at 3, 7 and 11 o'clock position, representing three main radicles of superior rectal vein in the anal columns.
- *Secondary piles:* These are dilatations in other positions of the lumen.

Q.541 How the piles are caused?

Following factors are thought to play a role in the causation of piles.

- *Hereditary:* Frequently associated with varicose veins.
- *Anatomical causes:* Collecting radicles of the superior rectal vein lies unsupported in the very loose submucous connective tissue of the rectum.
  These veins pass through muscular tissues and are liable to be constricted by its contraction during defecation. This increases pressure within them.
- *Morphological causes:* Valves are absent in the portal system. Hence, the whole burnt of the pressure of the portal vein is borne by the columns of mucous membrane in anal canal and produces a high pressure in lower rectum and anal canal.
- *Exciting causes:* Straining during constipation or over purgation.

Nerves, Muscles, Fascia and Artery of Pelvis

Q.542 Name the branches of internal iliac artery.

- Branches of anterior division:

In males:

- Superior vesical
- Obturator
- Middle rectal
- Inferior vesical
- Inferior gluteal and
- Internal pudendal.

In females: Same as above except inferior vesical is replaced by vaginal artery. Also uterine artery.

- *Branches of posterior division:*
  - Iliolumbar
  - Lateral sacral
  - Superior gluteal.

Q.543 Name the tributaries of internal iliac vein.

Veins arising outside pelvic wall:

- Superior gluteal
- Inferior gluteal
- Internal pudendal
- Obturator
- Lateral sacral.

Veins arising from venous plexuses of pelvic viscera:

- Rectal venous plexus
- Prostatic venous plexus
- Vesical venous plexus
- Uterine venous plexus
- Vaginal venous plexus.

Q.544 How sacral plexus is formed?

- *Lumbosacral trunk:* Formed by descending branch of L_4 and whole of L_5.
- Ventral rami of $S_{1,2,3}$ and part of S_4 nerves.

Q.545 Describe the levator ani muscle.

Levator ani is a brood thin muscle which is situated on the side of pelvis and supports the viscera in pelvic cavity (Fig. 5.72).

Origin: The L ani muscle originates from the following structures (front to back)

- Pelvic surface of the body of the pubis
- Obturator fascia
- Spine of ischium

Insertion

- The anterior most fibers pass across the sides of prostate (in males) and sides of the vagina (in females) to end into the perineal body.
- The intermediate fibers pass across the sides of the rectum and become continuous with these of the opposite side side behind the anorectal junction.
- The posterior most fibers of lateral ani are attached to the coccyx and to a fibrous band called the anococcygeal ligament. The posterior margin of the muscle is continuous with the coccyx.

Q.546 What are the parts of levator ani?

- *Pubococcygeus:*
  - Anterior fibers from levator prostate in male and pubovaginalis in female.
  - Middle fibers form puborectalis.
  - Posterior fibers form pubococcygeus proper.
- Iliococcygeus.

Q.547 What is the insertion of levator ani?

Perineal body.

Q.548 Where the pain of pelvic organs is referred to?

The autonomic supply which is sensory to pelvic organs is by $S_{2,3,4}$ spinal segments, which also gives cutaneous nerves to perineum. Therefore disease of pelvic organs causes referred pain in perineum.

Q.549 What is hiatus of Schwalbe and what is its clinical importance?

Levator ani arises from pubic bone in front, ischial spine behind and obturator fascia between these points. Sometimes, it arises from a tendinous sling, which is attached to bone in front and behind and not to fascia at all. Thus, a potential gap exists between sling and obturator fascia, called hiatus of Schwalbe.

Clinical importance: Pelvic peritoneum may herniate through it into ischiorectal fossa.

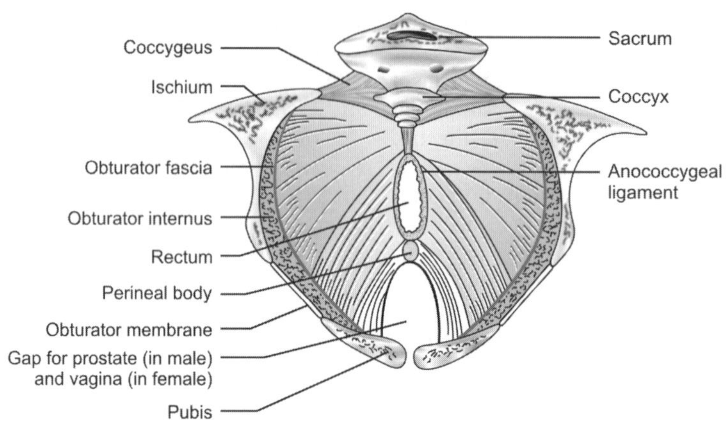

Fig. 5.72: Scheme to show the arrangement of the levator ani and coccygeus muscle

Joints of Pelvis

Q.550 Name the joints of pelvis.
- Lumbosacral joint
- Sacrococcygeal joint
- Intercoccygeal joint
- Sacro-iliac joint
- Pubic symphysis.

Q.551 Name the ligaments of sacrococcygeal joint.
- Ventral
- Deep dorsal
- Superficial dorsal
- Sacrococcygeal ligament
- Lateral
- Intercornual ligament.

Q.552 What variety of joint pubic symphysis is?
Secondary cartilaginous joint.

Q.553 Name the ligaments of sacroiliac joint.
- Ventral
- Dorsal sacroiliac ligament
- Interosseous
- Sacrotuberous ligament
- Sacrospinous ligament.

Osteology of Abdomen and Pelvis

Q.554 How the sacrum is formed?
By the fusion of 5 sacral vertebrae (Fig. 5.73).

Q.555 What is the anatomical position of sacrum in the body?
- Pelvic surfaces downwards and forwards.
- Upper surface of body of first sacral vertebra slopes forwards at an angle of about 30 degrees.

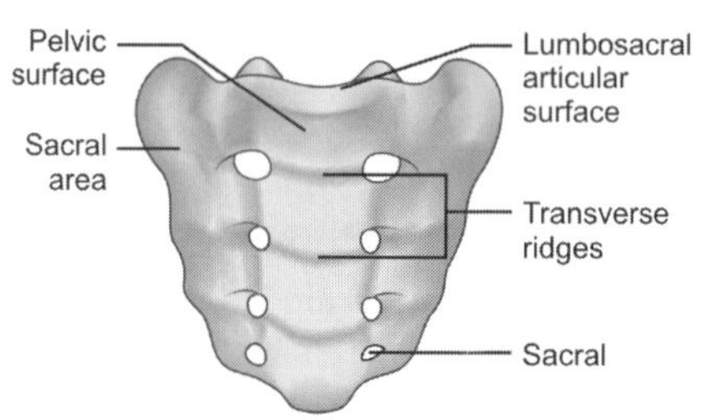

Fig. 5.73: Sacrum

- Upper end of sacral canal is directed upwards and slightly backwards.

Q.556 What are the relations and attachments of ala of sacrum?
- Smooth medial part (Fig. 5.74): Related to
  - Sympathetic chain,
  - Median sacral vessels
  - Right and left sacral vessels
  - Superior rectal vessels and cumbosacral trunk

  All are covered by psoas major muscle.
- Rough lateral part:
  - Origin to iliacus
  - Attachment to iliolumbar ligament
- Margins: Ventral sacroiliac ligament.

Q.557 What are the relations of pelvic surface of sacrum?
- *Median sacral vessels:* In median plane.
- *Sympathetic trunk:* Medial margin of pelvic foramina.
- *Peritoneum:* In front of upper 3 pieces, interrupted obliquely by medial limb of sigmoid mesocolon.
- *Rectum:* In front of lower 3rd pieces, separated at S_3 by bifurcation of superior rectal artery.

Q.558 How 'sacral hiatus' is formed?
By failure of fusion of laminae of S_5 vertebrae posteriorly.

Q.559 Name the structures emerging at sacral hiatus.
- 5th sacral nerve.
- Coccygeal nerves.
- Filum terminale.

Q.560 What is the origin of erector spinae?
It has a linear U shaped origin from dorsal aspect of sacrum. The medial limb of U is attached to spinous tubercles and lateral limb to the transverse tubercles.

Q.561 How will you identify lumbar vertebra?
- Large size of body
- Absence of costal facets on body.

Q.562 Name the structures attached to spine of lumbar vertebra.
- Posterior layer of lumbar fascia Interspinous ligament
- Supraspinous ligament
- Erector spinae muscle
- Multifidus muscle
- Interspinalis muscle.

Q.563 What is 'sacralisation of lumbar vertebra'?
Fusion of L_5 vertebra or its transverse process on one or both sides with sacrum.

Q.564 What is 'spina bifida'?
Failure of posterior fusion of two halves of neural arch with each other resulting in a bony gap.

Through the gap, meninges and spinal cord may herniate out in the mid-line.

Q.565 How will you differentiate between male and female sacrum?

	Male sacrum	*Female sacrum*
Length and breadth	Longer and narrower	Shorter and wider
Body and ala	Transverse diameter of body of S_1 larger than that of ala	Transverse diameter of body of S_1 is equal to the width of the ala
Auricular surface	Dorsal concavity is less marked	Dorsal concavity is more marked
Pelvic curve	Sacrum more uniformly curved, concavity is shallower	Curvature is irregular, concavity is deeper

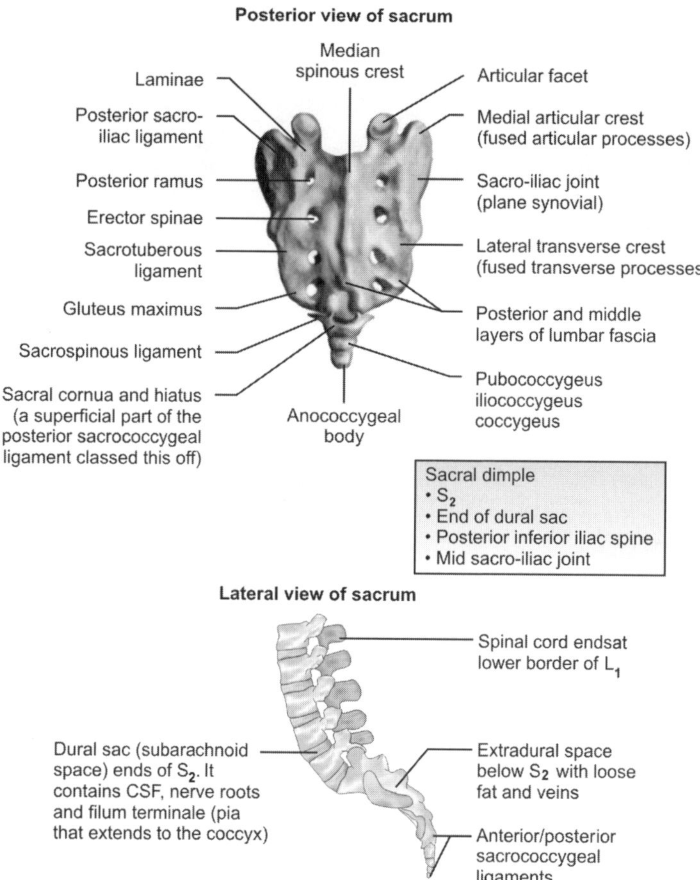

Fig. 5.74: Sacrum posterior attachments and dural sac

Bony Pelvis

Q.566 How the bony pelvis is placed in anatomical position?

- Pelvic surface of pubic symphysis faces backwards and upwards.
- Plane of pelvic inlet faces forwards and upwards at an angle of 50°–60° with horizontal.
- Plane of pelvic outlet makes an angle of 15° with horizontal.
- Upper end of sacral canal is directed upwards.

Q.567 What are the parts of pelvis?

The pelvis is divided by pelvic brim (pelvic inlet) into two parts:

- *Greater (false) pelvis:* upper part and
- *Lesser (true) pelvis:* lower part.

Q.568 What are boundaries of pelvic inlet (superior pelvic aperture)?

Anteriorly: Upper margin of pubic symphysis.
Posteriorly: Sacral promontory.
On each side: Anterior margin of ala of sacrum, and linea terminales which includes arcuate line of ilium, pectineal line of pubis and pubic crest.

Q.569 What are the boundaries of inferior pelvic aperture?

Anteriorly: Pubic arch
Posteriorly: Lateral margin of sacrum and coccyx
On each side: Ischial tuberosity, lesser sciatic notch, ischial spine and greater sciatic notch.
Lateral margin is formed by sacrotuberous ligament.

Q.570 What is the axis of pelvis?

This is an imaginary line joining the central points of anteroposterior diameters of pelvic outlet and inlet.

It is J-shaped, directed first downwards and backwards and then downwards and forwards.

Q.571 How will you differentiate between male and female pelvis (Fig. 5.75)?

	Male pelvis	*Female pelvis*
General features	Heavier and stronger *Bony markings:* More prominent.	Lighter and thinner Less prominent.
Ilia	Less vertical Greater in height *Iliac fossae:* Deeper *Intercristal diameter:* Greater	More vertical Lesser in height Shallower Smaller.
Pelvic inlet	Smaller and heart shaped. Pubic tubercles nearer because pubic crest is narrower. *Body of S_1 vertebra forms:* More than width of lateral part	Larger and more circular. Pubic tubercles wider apart because pubic crest is longer. Equal to width of lateral part
Pelvic outlet	Smaller *Sub-pubic angle:* 50-60° *Ischial tuberosities:* Less everted *Coccyx:* Less vertical	Larger 80-85° More everted More vertical
Sciatic notches	*Greater sciatic notch:* Narrower *Ischial spines:* Closer and inturned	 Wider Wider apart.
Pelvic walls	*Concavity of sacrum:* Shallower *Sacrum:* Long and narrow *Obturator foramen:* Larger and ovoid. *Acetabulum:* Larger and faces less forwards *Puboischial index:* <90	 Deeper Short and wide Smaller and triangular Smaller and faces more forwards. >90
Pelvic cavity	Longer and more conical	Shorter and cylindrical.

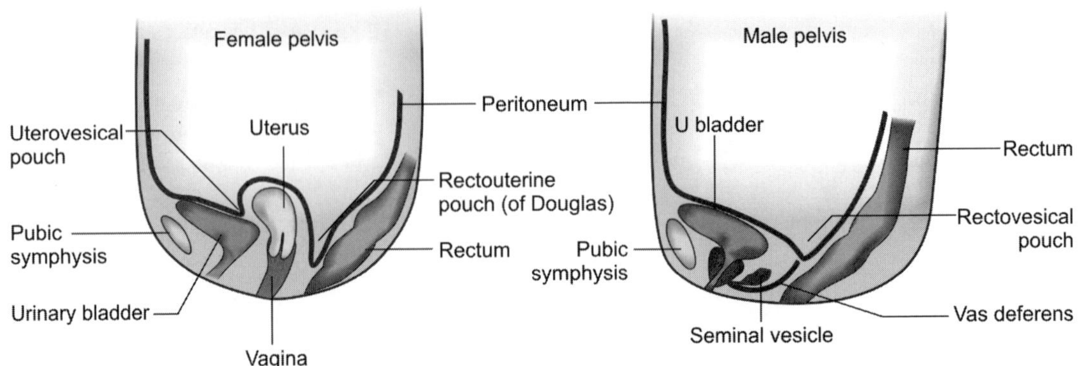

Fig. 5.75: Relations of urinary bladder (Sagittal section of female pelvis and male pelvis)

Q.572 What is diagonal conjugate of pelvis?

Diagonal conjugate is anteroposterior diameter between sacral promontory and pubic symphysis. This is normally at least 11.5 cm.

Q.573 What is the normal shape of female pelvis and what are the different variations in it.

- *Gynaecoid:* Normally pelvic inlet is oval and transverse diameter is slightly larger than anteroposterior (AP) diameter. Seen in 41 percent.
- *Android:* Resembles triangular male pelvis and transverse diameter is more posteriorly placed than in the female. Seen in 33 percent.
- *Anthropoid:* Long and narrow pelvis. AP diameter is more than transverse diameter. Seen in 24 percent.
- *Platypelloid:* AP diameter is small but transverse diameter of inlet is normal, i.e. pelvis appears flattened.

Chapter 6

Head and Neck

Scalp

Q.1 What is the extent of scalp?

Anterior: Supraorbital margins.

Posterior: External occipital protuberance and superior nuchal lines

On each side: Superior temporal lines.

Q.2 Name the layers of scalp.

- Skin.
- Superficial fascia.
- Epicranial aponeurosis with occipitofrontalis muscle.
- Loose areolar tissue and
- Pericranium (periosteum) (Fig. 6.1).

Q.3 Why the wounds of scalp bleed profusely?

Because of:

- Rich blood supply of scalp and
- The torn vessels fail to retract because of attachment to fibrous fascia.

Q.4 Why the wound of scalp heal rapidly?

Because more vascular the area, the more rapid is healing.

Q.5 Why the infections of superficial fascia of scalp cause much pain?

Because it is dense and fibrous so, little swelling causes much increase in tension.

Q.6 What is the 'dangerous area of scalp' and why it is so called?

Subaponeurotic space (loose areolar tissue). Because:

- Emissary veins which open here, may transmit the infection from scalp to intracranial venous sinuses.
- Bleeding in this space causes generalized swelling of the scalp and may extend anteriorly into root of nose and eyelids, causing black eye.

Q.7 Why the bleeding or pus collection beneath the periosteum is not extensive?

Because the periosteum adheres to the suture lines of skull bones, so the collection of blood or pus outlines the affected bone *(cephalhematoma).*

Q.8 What is 'safety valve' hematoma?

In children, dura and pericranium are more intimately attached to skull. So, in fractures of vault of skull tearing of both dura and pericranium occurs and the intracranial hemorrhage may make its way through line of fracture and collect in subaponeurotic space of scalp. No signs of compression of brain develop until subaponeurotic space is full of blood, such a collection of blood is termed safety valve hematoma.

Q.9 How the hemorrhage from blood vessels of scalp is arrested?

By pressing with the fingers firmly down on to the skull on either side of the wound, thus compressing the vessels.

Q.10 Which wounds of the scalp gape?

In scalp, skin and epicranial aponeurosis are firmly adherent and fibers of aponeurosis run anteroposteriorly. A wound of scalp does not gape unless epicranial aponeurosis is divided. Anteroposterior cuts also do not gape because of the direction of fibers.

Face

Q.11 Why the wounds of face bleed profusely?

Because of its rich vascularity.

Q.12 Why the edema in nephrotic syndrome appears first on face and eyelids?

Because, here the skin is very lax, which facilitates rapid spread of edema fluid.

Q.13 Why do the wounds of face tend to gape?

Because the facial muscles are inserted into skin making it thick and elastic.

Q.14 Why the facial muscles are called 'muscles of expression'?

They are subcutaneous muscles and they work under a fine control to bring about

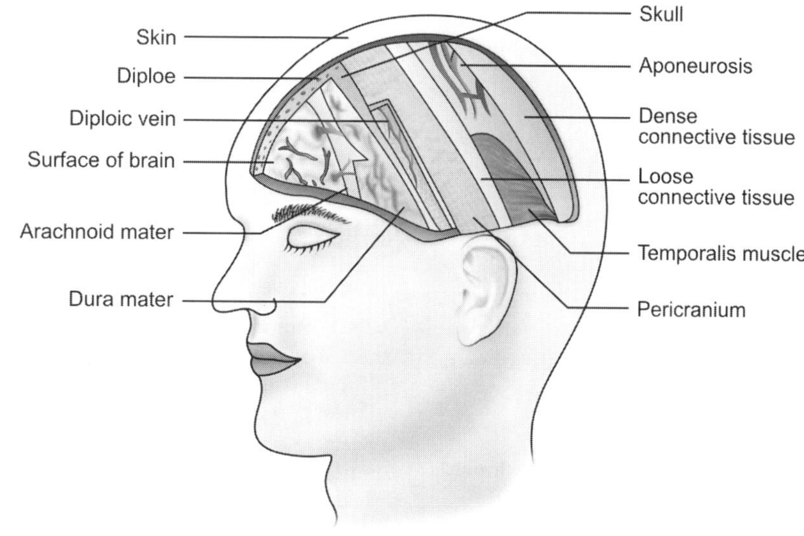

Fig. 6.1: Scalp

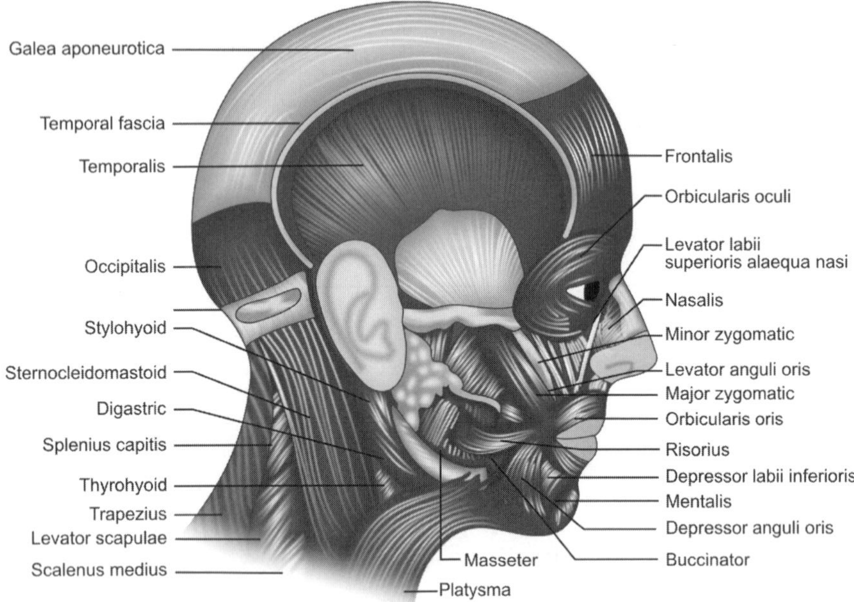

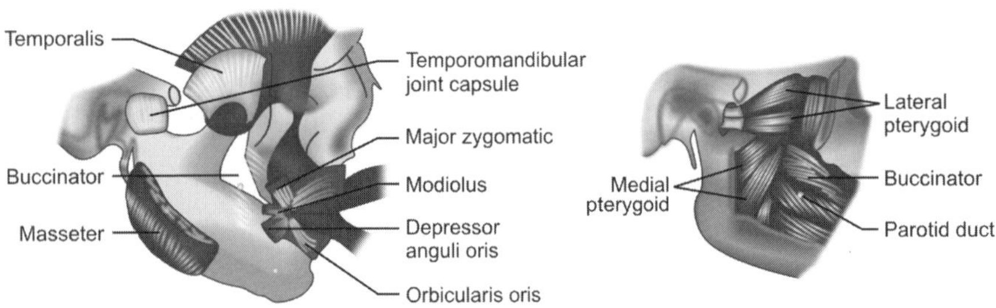

Fig. 6.2: Muscles of the head and neck

different shades of facial expressions (Fig. 6.2).

Q.15 Name the muscle producing transverse wrinkles on bridge of nose.
Procerus.

Q.16 Name the facial muscle forming the cheek.
Buccinator.

Q.17 Name the muscle producing vertical wrinkles on forehead.
Corrugator supercilii.

Q.18 What is the developmental origin of facial muscles?
Second branchial arch.

Q.19 What is the nerve supply of facial muscles?
Facial (VII cranial) nerve.

Q.20 Why headache occurs in sinusitis and cold?
Because of the same sensory nerve supply of face, nasal cavity and paranasal air sinuses, i.e., trigeminal nerve.

Q.21 What are the branches of facial artery?
In the neck:
- Ascending palatine
- Tonsillar
- Glandular
- Submental

In the face:
- Inferior labial
- Superior labial
- Lateral nasal
- Angular artery: Terminal part.

Q.22 How the facial vein is formed?
By supratrochlear and supraorbital vein near medial angle of eye.

Q.23 Which is the 'dangerous area of face'? Why it is so called?
- Upper lip and
- Lower part of nose.

Because infections of these sites are very common, which may spread in retrograde direction in facial vein and cause infection and thrombosis of the cavernous sinus through deep connections of the facial vein.

Orbit

Q.24 Name the different layers of eyelid.
- Skin
- Superficial fascia (has no fat).
- Palpebral part of orbicularis oculi muscle.
- Palpebral fascia.
- Tarsal glands.
- Palpebral conjunctiva (Fig. 6.3A and B).

Q.25 What are the glands found in eyelid?
- *Zeis's glands:* Large sebaceous glands of cilia. Found at lid margin.
- *Ciliary glands of Moll:* Sweat glands. Present at lid margin.
- *Meibomian glands (tarsal glands):* Sebaceous glands. Present in posterior surface of tarsi.

Q.26 What are the modifications of palpebral fascia?
- *Tarsal plates, in the lids:* Tarsal plates are attached to orbital margin by orbital septum.
- *Palpebral ligament, at the angles:* Attached to walls of orbit, just inside orbital margin.

Q.27 What is the nerve supply of eyelids?
- *Upper eyelid and whole of bulbar conjunctiva:* Supratrochlear, infratrochlear, lacrimal and supraorbital nerves (branches of ophthalmic nerve).
- *Lower eyelid:* Infraorbital and infratrochlear nerve (branch of maxillary nerve).

Q.28 What is chalazion?
It is chronic inflammation of tarsal gland, causing a localized swelling.

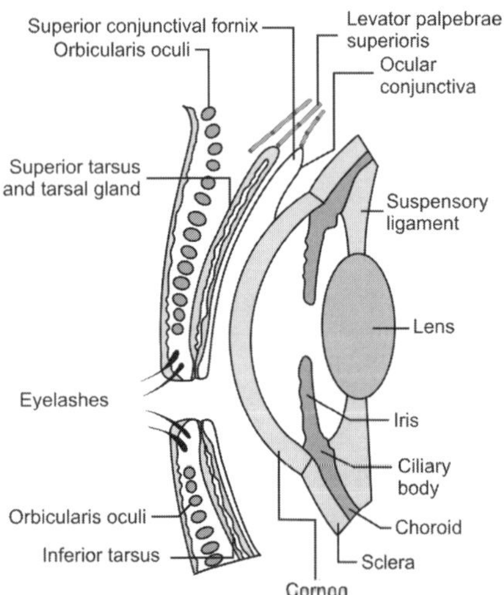

Fig. 6.3A: Schematic sagittal section through the eyelids and anterior part of the eyeball

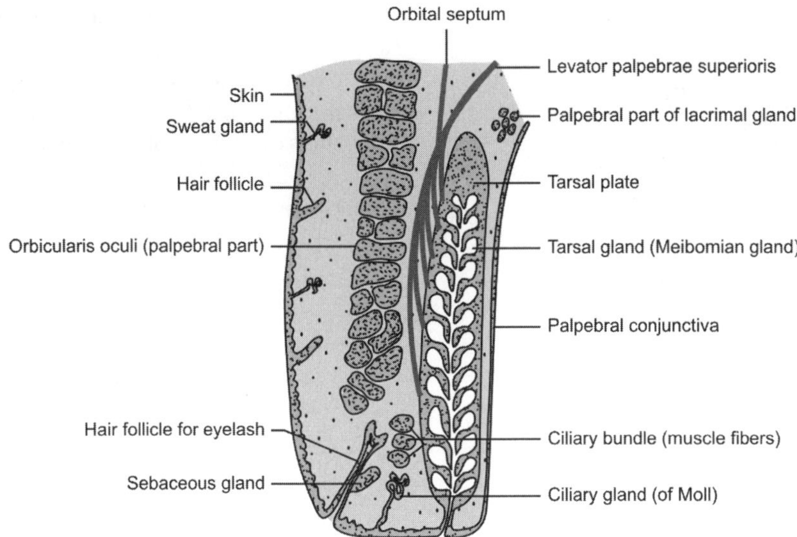

Fig. 6.3B: Schematic sagittal section through the upper eyelid

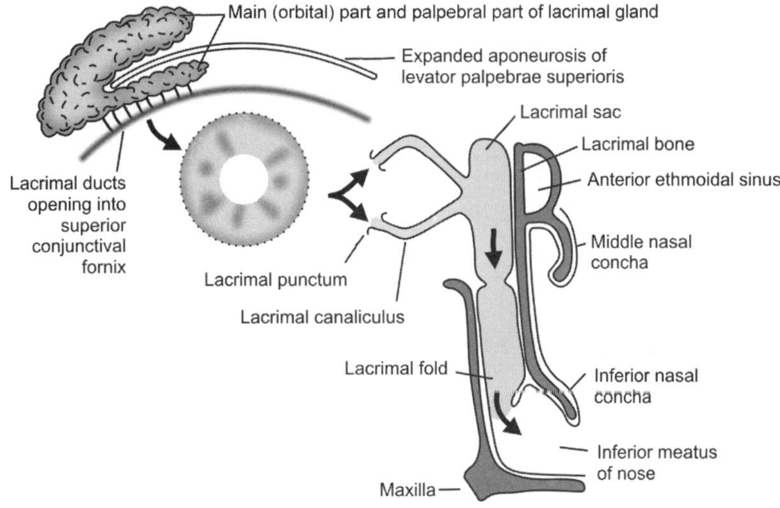

Fig. 6.4: Scheme to show the parts of the lacrimal apparatus. The arrows indicate the direction of flow of lacrimal fluid

Q.29 What is stye?

It is infection of Zeis's gland. The lid margin is edematous and gland is swollen and painful.

Q.30 What are the constituents of lacrimal apparatus?

- Lacrimal gland and its ducts.
- Conjunctival sac
- Lacrimal puncta and lacrimal canaliculi
- Lacrimal sac and
- Nasolacrimal duct (Fig. 6.4).

Q.31 What is the nature of lacrimal gland?

Exocrine and serous.

Q.32 What are the parts of lacrimal gland?

- *Orbital part:* Larger, in lacrimal fossa in upper lateral part of orbit.
- *Palpebral part:* Smaller, in upper eyelid. It drains into superior conjunctival fornix through twelve ducts.

Q.33 Why the removal of palpebral part of gland is equivalent to functional removal of whole gland?

Because the ducts of orbital part also pass through the palpebral part. So when palpebral part is removed the secretions of orbital part cannot be drained.

Q.34 What is the advantage of blinking of lids?

It helps to spread the lacrimal fluid in front of eye and deep surface of lids, thus keep the conjunctiva and cornea moist.

Q.35 What is valve of Hasner?

It is a fold of mucous membrane at the lower end of nasolacrimal duct.

Q.36 What is epiphora?

Leakage of tears down the face, due to the blockade of nasolacrimal duct.

Q.37 What is Tenon's sheath?

It is a thin membranous sheath around the eyeball. Extends from optic nerve to sclerocorneal junction. Eyeball can move freely within it.

Q.38 Name the structures piercing fascial sheath of eyeball.

- Tendons of extraocular muscles
- Ciliary vessels
- Ciliary nerves.

Q.39 What is suspensory ligament of Lockwood?

It is thickened Tenon's capsule in lower part.

Formed by union of margins of sheath of inferior rectus and inferior oblique with medial and lateral check ligaments.

Q.40 Name the extraocular muscles?

Voluntary muscles:

Rectus:
- Superior rectus
- Inferior rectus
- Medial rectus
- Lateral rectus

Oblique:
- Superior oblique
- Inferior oblique

Levator palpebrae superioris (Fig. 6.5).

Involuntary muscles:
- Superior tarsal
- Inferior tarsal
- Orbitalis (Fig. 6.5).

Q.41 What is the origin of rectus muscles?

They arise from corresponding part of common tendinous ring which surrounds optic canal and encloses a part of superior orbital fissure.

Q.42 What is the origin of oblique muscles?

Superior oblique: Body of sphenoid above and medial to optic canal.

Inferior oblique: Anterior and medial part of floor of orbit from maxilla just lateral to nasolacrimal groove.

Q.43 Which muscles produce the different movements of eyeball?

Upwards: Superior rectus
Inferior oblique

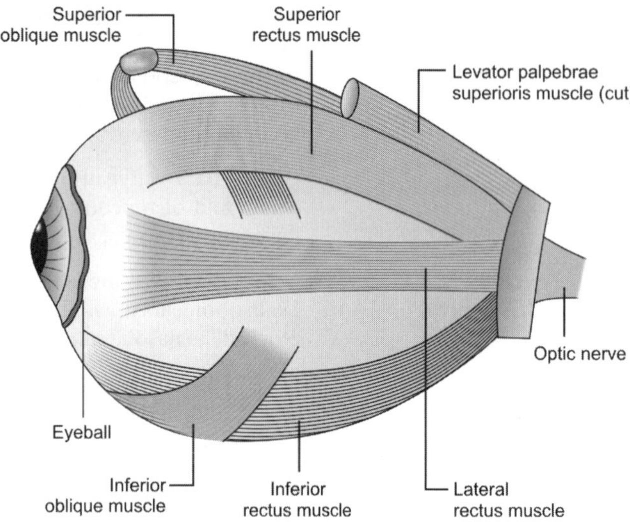

Fig. 6.5: Muscles of the eye

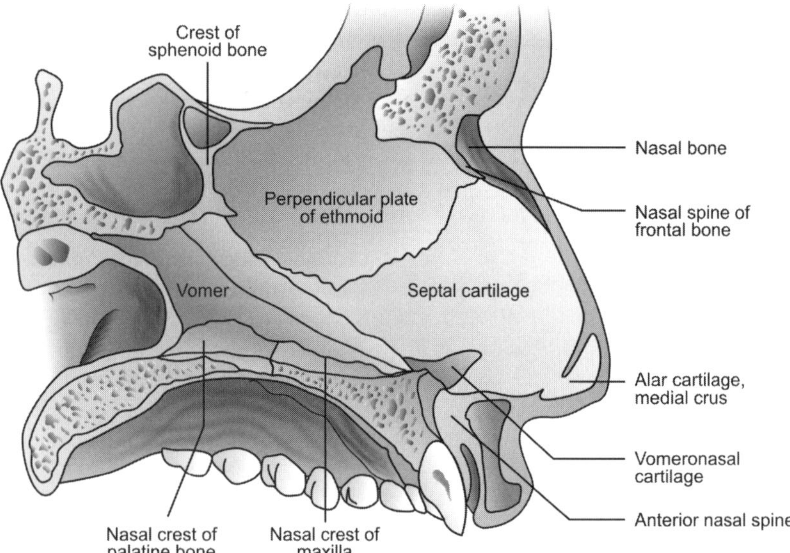

Fig. 6.6: Skeletal basis of nasal septum

Downwards:
- Inferior rectus
- Superior oblique

Inwards:
- Medial rectus
- Superior rectus
- Inferior rectus

Outwards:
- Lateral rectus
- Superior oblique
- Inferior oblique

Extorsion:
- Inferior oblique
- Inferior rectus

Intorsion:
- Superior oblique
- Superior rectus.

Q.44 What is the nerve supply of extraocular muscles?

- *Superior oblique:* Trochlear nerve
- *Lateral rectus:* Abducent nerve
- *Superior, inferior and medial rectus, inferior oblique and levator:* Oculomotor nerve.
- *Involuntary muscles:* By sympathetic fibers.

Q.45 What are conjugate movements of eye?

The normal co-ordinated movements of both eyes are called conjugate movements. These are usually horizontal and vertical.

Q.46 What is squint?

It is the abnormal deviation of eye due to weakness or paralysis of a muscle.

Nose and Paranasal Air Sinuses

Q.47 Name the structures forming nasal cavity.

Roof: From anterior to posterior
- Nasal part of frontal bone
- Nasal bone
- Nasal cartilages
- Cribriform plate of ethmoid
- Anterior surface of body of sphenoid bone.

Floor: Palatine process of maxilla, Horizontal plate of palatine.

Q.48 Name the structures forming nasal septum.

Bones:
- Vomer
- Perpendicular plate of ethmoid
- Margins by nasal spine of frontal, rostrum of sphenoid and nasal, palatine and maxilla.

Cartilage: Septal cartilage
Inferior nasal cartilage.

Cuticular part: Lower end, formed by skin (Fig. 6.6).

Q.49 Which the main artery supplying mucous membrane of nose?

Sphenopalatine branch of maxillary artery.

Q.50 What are the arteries forming the Little's area?

It is the anteroinferior part of nasal septum containing anastomosis between:

Superior labial branch of facial artery and sphenopalatine artery

Large capillary network.

Q.51 What is the clinical importance of Little's area?

It is the commonest site of bleeding from the nose.

Q.52 Name the bones forming the lateral wall of nose.

- Nasal
- Frontal process of maxilla
- Lacrimal
- Labyrinth of ethmoid with superior and middle conchae
- Inferior nasal conchae
- Perpendicular plate of palatine bone
- Medial pterygoid plate (Fig. 6.7).

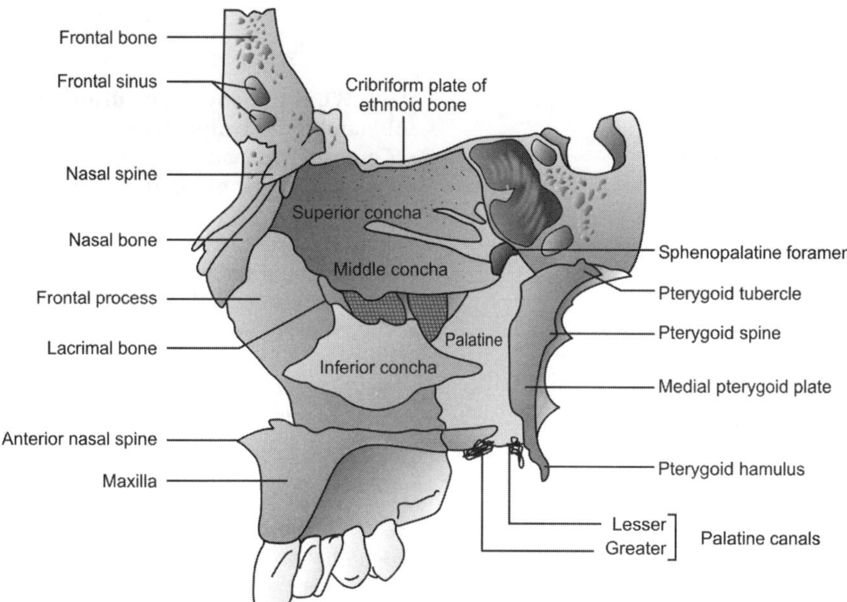

Fig. 6.7: Lateral wall of the nasal cavity with mucous membrane

Q.53 What are 'nasal conchae'?
These are shelf like bony projections from lateral wall of nose directed downwards and medially.

There in number: Superior, middle and inferior.

Q.54 Name the bones forming the 'nasal conchae'.
Superior and middle conchae is formed by ethmoid bone while inferior concha is an independent bone.

Q.55 What are the meatuses of nose?
These are the passages beneath the overhanging conchae.

Q.56 Name the paranasal air sinuses.
Frontal,
Maxillary,
Sphenoidal and
Ethmoidal.

Q.57 Name the openings in middle meatus.

- Opening of frontal air sinus, through frontonasal duct above ethmoidal infundibulum.
- Opening of maxillary air sinus, in lower part of hiatus semilunaris.
- Opening of middle ethmoidal air sinus, on bulla ethmoidalis.
- Opening of anterior ethmoidal air sinus, into ethmoidal infundibulum.

Q.58 Name the openings in superior meatus.
Opening of posterior ethmoidal air sinus.

Q.59 Name the openings in inferior meatus.
Nasolacrimal duct.

Q.60 Where does sphenoidal air sinus open?
Sphenoethmoidal recess above the superior concha.

Q.61 What are the characteristic features of frontal sinus?
Frontal sinuses are contained in frontal bone.

These are separated from each other by a median bony septum and each is further broken up by incomplete septa.

Each sinus drains into anterior part of middle nasal meatus.

Q.62 What is C.S.F. rhinorrhea?
A fracture of frontal bone, tearing the dura and pia mater, causes communication between nasal cavity and subarachnoid space and C.S.F., may trickle through nostril on the affected side.

Q.63 What is antrum of Highmore?
Maxillary sinus is also called as the antrum of Highmore.

Q.64 Why frontal sinusitis often leads to maxillary sinusitis?
Because secretions drainage out of frontal sinus flow towards the opening into maxillary sinus.

Q.65 Why maxillary sinusitis is more likely to be chronic?

- Because the level of opening of maxillary sinus into nose is placed at a higher level than the floor of the sinus so natural drainage is difficult.
- Sinuses open into nasal cavity by narrow openings, so slight swelling of mucosa or thick secretions can block the outflow of secretions, that accumulate within sinus.

Triangles of Neck

Q.66 Name the structures palpable anteriorly in the midline of the neck.

- Body of hyoid bone
  - Adam's apple (thyroid cartilage)
  - Arch of cricoid cartilage
  - Tracheal ring
  - Isthmus of thyroid gland.
  - Suprasternal notch (Figs 6.8 and 6.9).

Q.67 What is platysma?
It is a subcutaneous muscle forming a thin fleshy sheath running upwards and medially on the neck from deltoid and pectoral fasciae to the base of mandible. It is supplied by cervical branch of facial nerve.

Functions:

- Helps in releasing pressure of the skin on superficial veins.
- Pulls the angle of mouth downwards.

Q.68 What is 'jugular arch'?
A transverse channel in the suprasternal space connecting the two anterior jugular veins.

Q.69 What is the position of 'subhyoid bursa'? What is its function?
Position: Between posterior surface of body hyoid bone and thyrohyoid membrane.

Function: Lessens friction between above two structures during swallowing.

Q.70 What are the boundaries of 'anterior triangular' of neck?
Anterior: Anterior median line of neck.

Posterior: Anterior border of sternomastoid.

Base: Base of mandible and a line joining angle of mandible to mastoid process.

Apex: Manubrium sterni.

Q.71 What are the subdivisions of anterior triangle?

- Submental,
- Digastric,
- Carotid and
- Muscular triangle.

Q.72 What are the areas drained by submental lymph nodes?

- Superficial tissues below the chin
- Central part of lower lip and the adjoining gum.

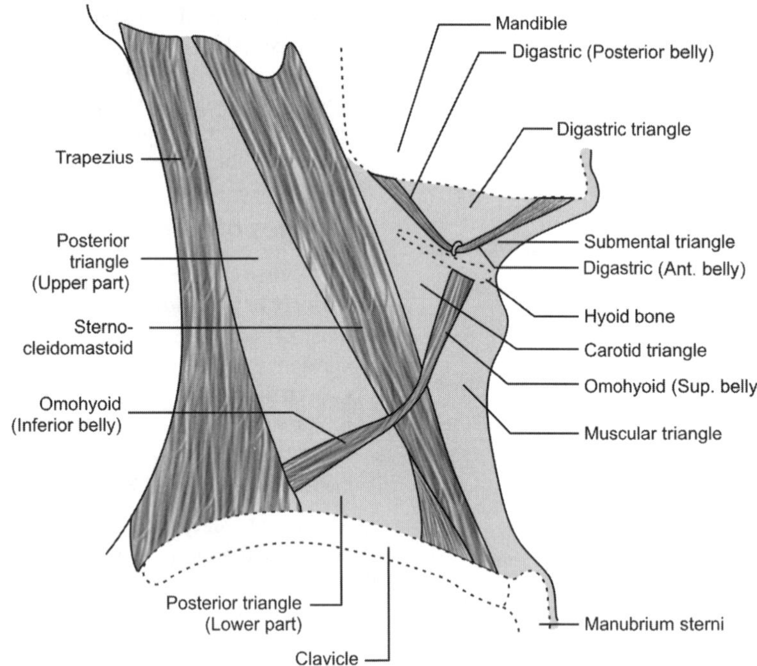

Fig. 6.8: Triangles of the neck

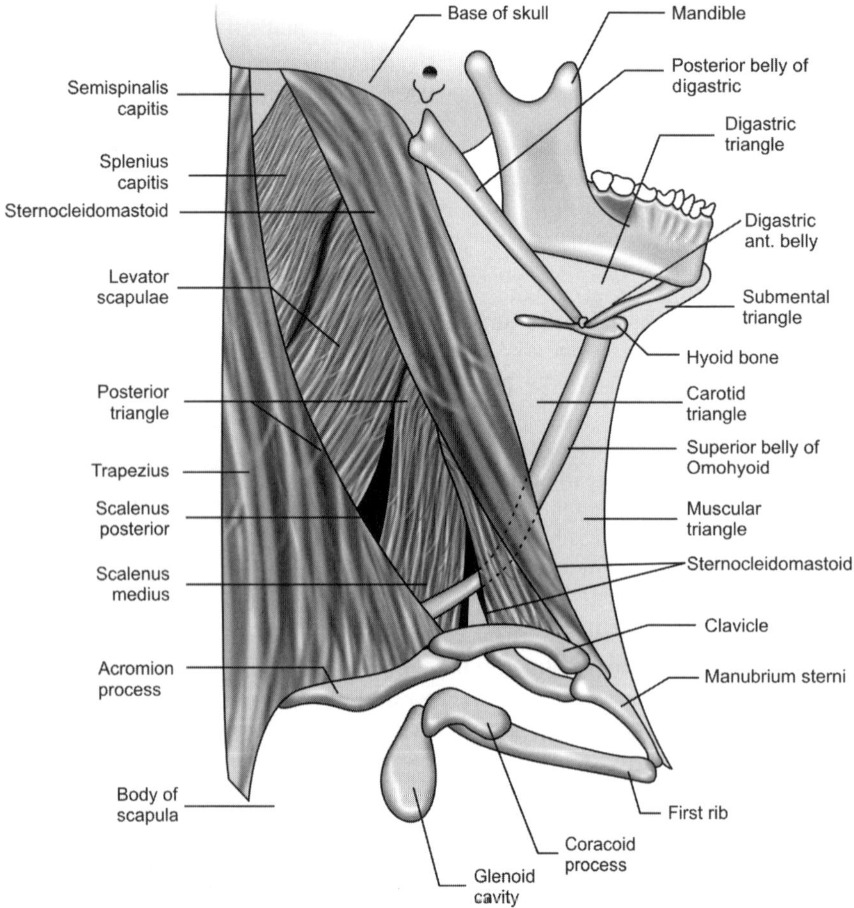

Fig. 6.9: Triangles of the neck. Also see Fig. 6.8

- Anterior part of floor of mouth
- Tip of tongue.

Q.73 What are the areas drained by submandibular lymph nodes?

- Centre of forehead
- Nose with frontal, maxillary and ethmoidal air sinuses
- Inner canthus of eye
- Upper lip and anterior part of cheek with underlying gum and teeth
- Outer part of lower lip with lower gum and teeth
- Anterior 2/3 of tongue (excluding tip)
- Floor of mouth.
- Efferents from submental nodes.

Q.74 What are the boundaries of digastric triangle?

Anteroinferiorly:	Anterior belly of digastric
Posteroinferiorly:	Posterior belly of digastric stylohyoid
Base:	Base of mandible and a line joining angle of mandible to mastoid process
Roof:	Skin
	Superficial fascia: Has platysma and cervical branch of facial nerve
Deep fascia:	Splits to enclose submandibular gland
Floor:	Mylohyoid, Hyoglossus and middle constrictor muscle.

Q.75 Name the various structures lying over hyoglossus muscle.

- Lingual nerve
- Submandibular ganglion
- Submandibular duct
- Mylohyoid nerve
- Hypoglossal nerve
- Stylohyoid muscle

Q.76 Name the various structures passing deep to hyoglossus muscle.

- Glossopharyngeal nerve
- Stylohyoid ligament
- Lingual artery

Q.77 Name the structures passing between external and internal carotid arteries.

- Styloglossus.
- Stylopharyngeus.
- Glossopharyngeal nerve.
- Pharyngeal branch of vagus nerve.
- Styloid process
- Part of parotid gland.

Q.78 What are the boundaries of carotid triangle?

Superiorly:	• Posterior belly of digastric. • Stylohyoid
Anteroinferiorly:	• Superior belly of omohyoid
Posterioly:	• Anterior border of sternomastoid
Roof:	• Skin Superficial fascia having platysma, cervical branch of facial nerve and transverse cutaneous nerve of neck. • Investing layer of deep fascia
Floor:	• Thyrohyoid, Hyoglossus, Middle constrictor and Inferior constrictor.

Q.79 Name the contents of carotid triangle.

- *Arteries:* Common carotid, Internal carotid and External carotid and its branches.
- *Veins:* Internal jugular, Common facial, Pharyngeal, Lingual and Superior thyroid.
- *Nerves:* Vagus, Superior laryngeal, Hypoglossal and Sympathetic chain.
- Carotid sheath with its contents
- Deep cervical lymph nodes.

Q.80 What is the position, structure and function of carotid sinus?

Position: At termination of common carotid artery as slight dilatation.

Characteristics: Media thin, Adventitia thick,

Rich innervation by plexus formed by glossopharyngeal (mainly), sympathetics and vagus nerve.

Function: As a baroreceptor helps to regulate blood pressure.

Q.81 What is the position, structure and function of carotid bodies?

Position: Present at bifurcation of common carotid artery near carotid sinus.

Structure: Has characteristic glomus cells, similar to neurons.

Also has sympathetic and parasympathetic postganglionic neurons and afferent nerve terminals from glossopharyngeal nerve.

Function: Act as chemoreceptors to monitor oxygen and carbon dioxide levels in blood by reflexly controlling the rate and depth of respiration.

Q.82 What are the branches of external carotid artery?

Anteriorly:	Superior thyroid, Lingual and Facial.
Posteriorly:	Occipital and Posterior auricular.
Medial:	Ascending pharyngeal.
Terminal:	Maxillary and Superficial temporal.

Q.83 Name the infrahyoid muscles.

- Sternohyoid,
- Sternothyroid,
- Thyrohyoid and
- Omohyoid (Fig. 6.10).

Q.84 Name the suprahyoid muscles.

- Mylohyoid,
- Digastric,
- Stylohyoid and
- Geniohyoid.

Q.85 What is the nerve supply of digastric muscle?

Anterior belly by mylohyoid branch of inferior alveolar nerve (V nerve) and posterior belly by facial nerve.

Q.86 What are the attachments of omohyoid muscle?

It has two bellies:

- *Inferior belly:* Origin from upper border of scapula near scapular notch and ends deep to sternocleidomastoid by joining intermediate tendon.
- *Superior belly:* From intermediate tendon to lower border of body of hyoid bone.

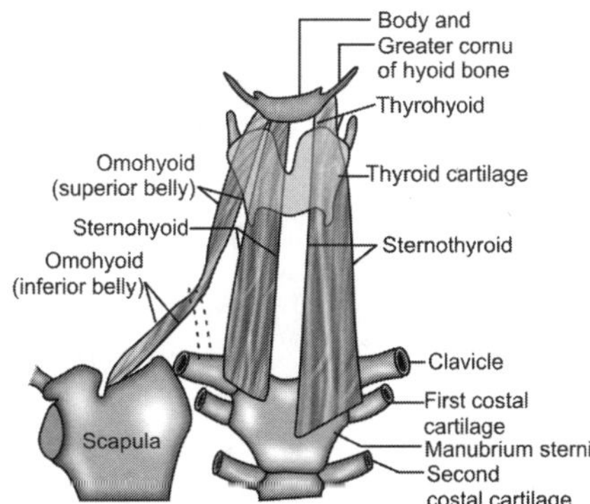

Fig. 6.10: Infrahyoid muscles

Q.87 What are the boundaries submental triangle?

Base is formed by hyoid bone.

Above and laterally on each side by anterior belly of digastric muscle.

Floor by mylohyoid muscle.

So half of it lies on each side of midline of neck.

Q.88 What are the boundaries of muscular triangle?

Posteroinferiorly: Sternocleidomastoid muscle

Posterosuperiorly: Superior belly of omohyoid

Anteriorly: Anterior midline of neck.

Q.89 What are the contents of muscular triangle?

- Infrahyoid muscle
- Thyroid gland
- Larynx and trachea
- Carotid sheath and its contents
- Brachiocephalic artery

Q.90 What are the boundaries of suboccipital triangle?

Superomedially:	Rectus capitis posterior major, rectus capitis posterior minor
Superolaterally:	Obliquus capitis superior
Inferiorly:	Obliquus capitis inferior.
Roof:	*Medially:* Fibrous tissue
Laterally:	Longissimus capitis
Floor:	Posterior arch of atlas Posterior atlanto-occipital membrane.

Q.91 What are the contents of sub-occipital traingle?

- Third part of verterbal artery
- Dorsal ramus of C_1
- Suboccipital plexus of veins
- Greater occipital nerve.

Q.92 What are the contents of 'suprasternal space of Burns'?

- Sternal head of sternomastoid,
- Jugular venous arch,
- Interclavicular ligament and
- Lymph node.

Q.93 What are the structures traversing supraclavicular space?

- External jugular vein,
- Supraclavicular nerves,
- Cutaneous vessels and,
- Lymphatics.

Q.94 What are the contents of carotid sheath?
- Common cartoid artery,
- Internal carotid artery,
- Internal jugular vein and
- Vagus nerve.

Q.95 Why the infections behind the prevertebral fascia do not extend to the posterior mediastinum?
Because the prevertebral fascia is attached to the fourth thoracic vertebra, which limits the downward spread of infection.

Q.96 What are the boundaries of posterior triangle?
Anterior: Posterior border of sterno-mastoid.
Posterior: Anterior border of trapezius.
Base: Middle 1/3 of clavicle.
Apex: Point where trapezius and sternomastoid meet.
Roof: Investing layer of deep cervical fascia.
Floor: Prevertebral layer of deep cervical fascia covering the muscles.

Q.97 What are the structures present in floor of posterior triangle below deep cervical fascia?
Semispinalis capitis,
Splenius capitis
Levator scapulae,
Scalenus posterior,
Scalenus medius and
Scalenus anterior.

By inferior belly of omohyoid in lower part it is divided into upper part (occipital triangle) and lower part (supraclavicular triangle)

Q.98 What are the contents of posterior triangle of neck?
- Cutaneous branches of cervical plexus:
  - Supraclavicular
  - Lesser occipital
  - Greater auricular
  - Transverse cutaneous
- Muscular branches from cervical plexus:
  - Levator scapulae
  - Trapezius
- Spinal accessory nerve
- Trunks of brachial plexus
- Branches of brachial plexus:
  - Nerve to rhomboids
  - Nerve to serratus anterior
  - Nerve to subclavius
  - Suprascapular nerve
- Subclavian artery
- Transverse cervical artery
- Occipital artery.

Q.99 What are 'signal nodes'?
These are lymph nodes which are enlarged in the malignant growths of distant places e.g., left supraclavicular nodes in malignancy of stomach, testes and other abdominal organs.

Q.100 What is the origin of sterno-mastoid?
- *Sternal head:* From superolateral part of front of manubrium sterni (a).
- *Clavicular head:* Medial 1/3 of superior surface of clavicle (b and c) (Fig. 6.11).

Q.101 What is the nerve supply of sterno-mastoid?
Motor: Spinal accessory nerve.
Sensory: Ventral rami of $C_{2,3}$.

Q.102 What is the action of sternocleidomastoid muscle?
When muscle of one side contacts, the head is tilted to same side and face is rotated to opposite side. When muscles of both sides act together, the head and neck are flexed.

Q.103 What is 'torticollis'?
Also known as wryneck.
The head is bent to one side and chin points to the otherside.

Q.104 How torticollis occurs?
It occurs due to spasm of muscles supplied by spinal accessory nerve i.e., sternomastoid and trapezius.

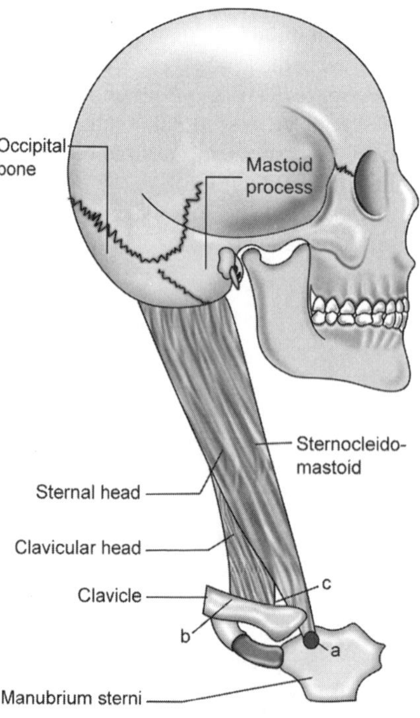

Fig. 6.11: Scheme to show the attachments of the sternocleidomastoid muscle

Mouth

Q.105 What are the divisions of oral cavity?
- Vestibule
- Oral cavity proper.

Q.106 What the boundaries of vestibule?
External: Lips and cheeks
Internal: Teeth and gums

Q.107 How frenulum of lip is formed?
It is formed by a median fold of mucous membrane between lips and gums.

Q.108 What is the lymphatic drainage of lips?
Central part of lower lip drains into submental nodes and rest of lip to sub-mandibular nodes.

Q.109 What are the boundaries of oropharyngeal isthmus (isthmus of fauces)?
Superior: Soft palate
Inferior: Tongue
On each side: Palatoglossal arches.

Q.110 What do you understand by diphyo-dont teeth?
Two sets of teeth are present.
First dentition: Milk or deciduous teeth.
Second set: Permanent teeth.

Q.111 What is the dental formula for deciduous teeth?
Incisor 2/2, Canine 1/1, Molar 3/3.
Total No. 20.

Q.112 What is the dental formula for permanent teeth?
Incisor 2/2, Canine 1/1, Premolar 2/2, Molar 3/3. Total No. 32.

Q.113 Which is first permanent tooth to appear?
First molar at 6 years of age.

Q.114 Which are wisdom teeth and what age they appear?
Third molar teeth and they appear at age of 17 years or above.

Q.115 What are the parts of a tooth?
3 parts:
- *Crown:* Projecting above gum
- *Root:* Embedded in jaw beneath the gum
- *Neck:* Between crown and root.

Salivary Glands

Q.116 Name the salivary glands.
- Parotid,
- Submandibular,
- Small glands in tongue, palate, cheeks and lips.

Q.117 What is the type of salivary glands?
Exocrine.

Parotid Gland

Q.118 What is the position of parotid gland?
Between ramus of mandible and sternomastoid, below the external acoustic meatus.

Q.119 What are the attachments of 'parotid capsule'?
It is formed by splitting of investing layer of deep cervical fascia between angle of mandible and mastoid process.

Superficial lamina: Attached above to zygomatic arch.

Deep lamina: To styloid process, mandible and tympanic plate. Also forms stylomandibular ligament.

Q.120 Name the structures within parotid gland.
- *Facial nerve:* Enters through posteromedial surface and divides into branches which emerge from anteromedial surface.
- *External carotid artery:* Enters through posteromedial surface and divide into branches.
- *Retromandibular vein:* Formed within parotid gland by superficial temporal and maxillary veins, superficial to the artery (Fig. 6.12).

Q.121 Name the branches of external carotid artery (ECA) within the parotid gland.
- Posterior auricular artery,
- Superficial temporal and
- Maxillary.

Q.122 Name the structures pierced by parotid duct.
- Buccal pad of fat,
- Buccopharyngeal fascia and
- Buccinator muscle.

Q.123 Where the parotid duct opens?
In vestibule of mouth opposite the crown of upper second molar tooth.

Q.124 What are the nerves supplying the parotid gland?
- *Parasympathetic nerve through auriculotemporal nerve:* Secretomotor.
- *Symphathetic nerve from plexus around ECA:* Vasomotor
- *Auriculotemporal nerve:* Sensory
- *Greater auricular nerve (C_2 fibers):* Sensory for parotid fascia.

Q.125 Why the parotid swellings are painful?
Because the parotid fascia is very dense and unyielding. Therefore, it cannot stretch on parotid swelling and causes increased tension beneath fascia.

Q.126 What is the nature of parotid gland?
Purely serous.

Q.127 How parotid gland is removed surgically?
In two parts, superficial and deep, in order to preserve the facial nerve.

Submandibular and Sublingual Gland

Q.128 What is the position of submandibular gland?
Anterior parts of digastric triangle.

Q.129 What are the parts of submandibular gland?
Superficial part: Large and superficial to mylohyoid.

Deep part: Small and deep to mylohyoid.

The two parts are continuous round the posterior border of mylohyoid (Fig. 6.13).

Q.130 Name the structure separating the posterior end of submandibular gland and parotid gland.
Stylomandibular ligament.

Q.131 What is the lymphatic drainage of submandibular gland?
Submandibular lymph nodes which in turn drain into jugulo-omohyoid nodes.

Q.132 Where is the opening of submandibular duct?
In the floor of mouth, on the summit of sublingual papilla, at the side of frenulum of tongue.

Q.133 What is the nerve supply of submandibular gland?
Secretomotor: Chorda tympani
Sensory: Lingual nerve
Vasomotor: Sympathetic fibers from plexus on facial artery.

Q.134 What is the nature of submandibular gland?
Mixed, but predominantly serous.

Q.135 Why the incision for removal of submandibular gland is placed more than 1 inch below the angle of mandible?
Because marginal mandibular nerve, a branch of VII nerve passes 1 inch behind angle of jaw before arching upwards over body of mandible.

Q.136 Which artery is likely to be injured in surgery on submandibular gland?
Facial artery.

Q.137 What is the nature of sublingual gland?
Mixed, but predominantly mucous (Fig. 6.14).

Q.138 Where the ducts of sublingual gland upon?
About 15 ducts which open on summit of sublingual fold in floor of mouth.

Q.139 What is the blood supply of sublingual glands?
From sublingual branch of lingual artery and submental branch of facial artery.

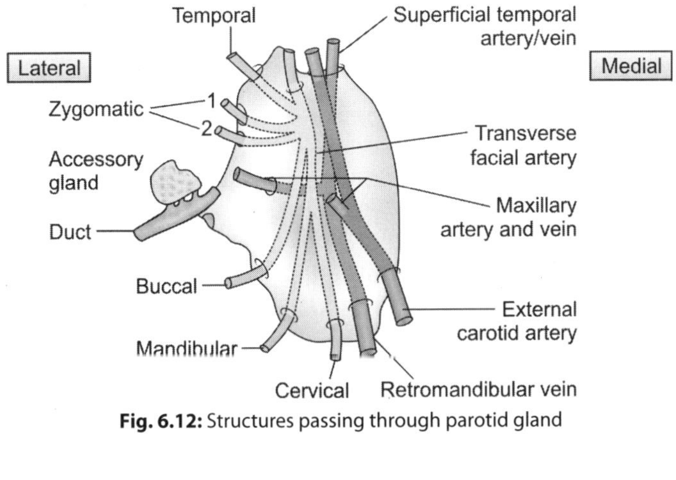

Fig. 6.12: Structures passing through parotid gland

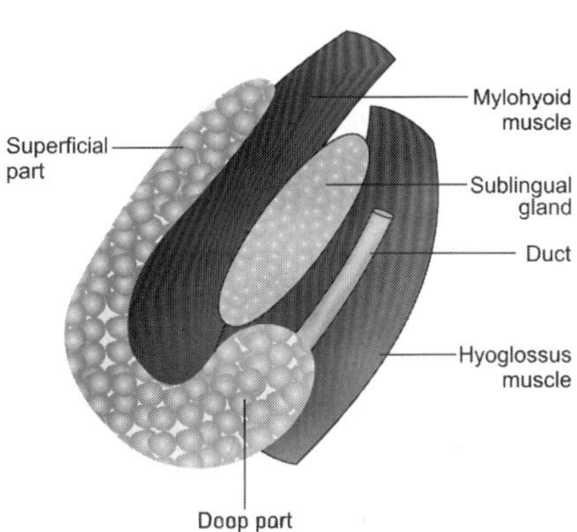

Fig. 6.13: Submandibular gland and its relations

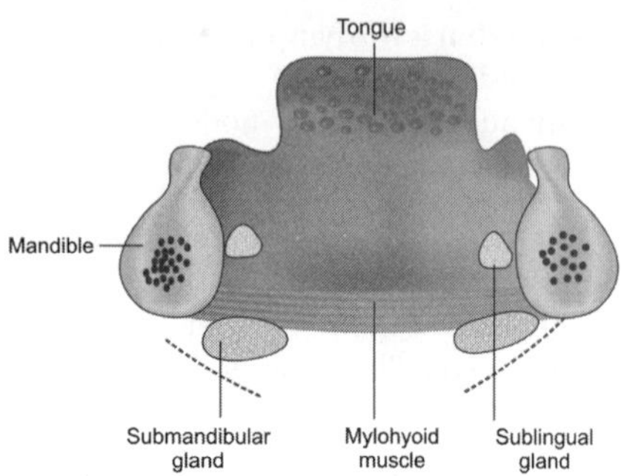

Fig. 6.14: Sublingual gland

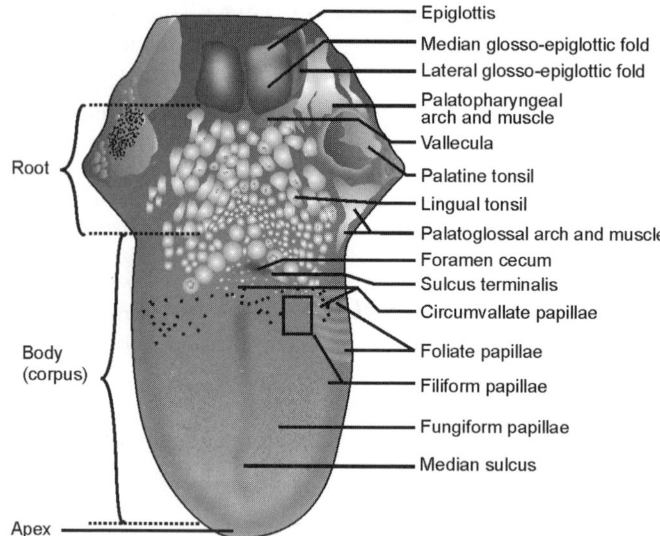

Fig. 6.15: Tongue

Q.140 What is the developmental origin of salivary glands?

Parotid arises as an ectodermal outgrowth from buccal epithelium in relation to line along which maxillary and mandibular processes fuse i.e., just lateral to angle of mouth to form cheek.

Sublingual and submandibular glands are endodermal in origin, arising in relation to linguo-gingival sulcus.

Tongue (Fig. 6.15)

Q.141 What are the parts of tongue?

- *Anterior 2/3:* Oral part
- *Posterior 2/3:* Pharyngeal part
  At junction is a V-shaped groove with apex of V pointing backwards *sulcus terminalis,* with a median pit, *foramen cecum.*

Q.142 What are papillae and what are their type?

Papillae are projections of mucous membrane situated in anterior 2/3 of tongue on dorsal surface.

Types:

- *Vallate papillae*: 8-12 in number.
  Situated immediately in front of sulcus terminalis. Each is a cylindrical projection surrounded by a circular sulcus.
- *Fungiform papillae:* Numerous.
  Near tip and margins of tongue.
  Each has a narrow pedicle and large rounded head.
- *Filiform papillae:* Most numerous.
  Covers presulcal area of dorsum of tongue, pointed and covered with keratin.
- *Foliate papillae:* On lateral margin of posterior past. Appear as vertical folds but are not true papillae.
- *Papillae simplex:* Are microscopic. They are not surface projections.

Q.143 Name the muscles of tongue.

Tongue is divided into two halves by a midline fibrous septum.

Each half has:

- Four intrinsic muscles.
  - Superior longitudinal,
  - Inferior longitudinal,
  - Transverse and
  - Vertical.
- Four extrinsic muscles
  - Genioglossus,
  - Hypoglossus,
  - Styloglossus and
  - Palatoglossus.

Q.144 What is the lymphatic drainage of tongue?

- *Submental nodes:* Drain tip of tongue.
- *Jugulodigastric, jugulo-omohyoid and other deep cervical nodes:* Posterior 1/3, posterior marginal part of anterior 2/3 and central part of anterior 2/3.
- *Submandibular nodes:* Tip of tongue, anterior marginal part of anterior 2/3.
  Submental nodes and submandibular nodes inturn drain into jugulodigastric, jugulo-omohyoid and other deep cervical nodes.
  Lymphatics from near midline can pass either to left or right side lymph nodes.

Q.145 Why jugulo-omohyoid node is called 'lymph node of tongue?

Because it drains most of the lymph from the tongue.

Q.146 What is the nerve supply of tongue?

Motor nerves:

- *All muscles except palatoglossus:* Hypoglossal nerve.
- *Palatoglossus:* Cranial part of accessory nerve.

Sensory:

- *Anterior 2/3:*
  - *Lingual nerve:* General sensory.
  - *Chorda tympani:* Special sensory.
- *Posterior 1/3:* Glossopharyngeal nerve, general and special sensory.
- *Posterior most part:* Vagus (superior laryngeal).

Q.147 What are the functions of tongue?

- Taste,
- Speech,
- Mastication and
- Deglutition.

Q.148 Where the taste buds are situated?

- *Vallate papillae:* Most numerous on sides of papillae.
- Foliate papillae.
- Posterior 1/3 of tongue.

Q.149 How the tongue is developed?

- *Anterior 2/3:*
  First branchial arch by two lingual swellings and one tuberculum impar.
- *Posterior 1/3:* Third arch by cranial half of hypobranchial eminence.
- *Posterior part:* Fourth arch.
- *Muscles:* Occipital myotomes.
- *Connective tissue:* Local mesenchyme

Q.150 How the bleeding from lacerated tongue is stopped?

By applying pressure posterior to the area of laceration, because lingual artery supplying it runs forwards.

Q.151 What does foramen caecum represents?

The site of downgrowth of thyroglossal duct.

Q.152 What is the clinical importance of attachment of genioglossus to the genial tubercles of mandible?

In unconscious patient or during general anaesthesia, the tongue may fall back and obstruct the respiratory passage. So, the advantage of this attachment is taken by pulling the mandible forwards which prevents the falling back of tongue.

Q.153 What are the developmental anomalies of tongue?

- *Macroglossia:* Large tongue.
- *Microglossia:* Small tongue.
- *Bifid tongue:* Non fusion of two linguinal swellings.
- Surface of the tongue may be fissured.

Palate (Fig. 6.16)

Q.154 Name the muscles of soft palate.

- Tensor palati,
- Levator palati,
- Musculus uvulae,
- Palatoglossus and
- Palatopharyngeus.

Q.155 What is the arterial supply of soft palate?

- Greater palatine branch of maxillary,
- Ascending palatine branch of facial and
- Palatine branch of ascending pharyngeal.

Q.156 What is the nerve supply of soft palate?

Motor nerves:

- *Tensor palati*: Mandibular nerve
- *Other muscles:* Pharyngeal plexus (cranial part of accessory through vagus).

General sensory nerves:

- Greater and lesser palatine nerves
- Glossopharyngeal nerve

Special sensory nerve: Lesser palatine nerve

Secretomotor: Lesser palatine nerve.

Q.157 What are the functions of soft palate?

It controls the opening of the pharyngeal and oropharyngeal isthmus, during chewing, coughing, sneezing, speech and swallowing.

Q.158 How is palatine aponeurosis formed? What is its importance?

It is an expanded fibrous band of tensor palati. It is attached anteriorly to posterior edge of hard palate and in midline the aponeurosis of two sides fuse with each other.

It forms fibrous basis of soft palate and gives attachment to other muscles of palate.

Q.159 What is Passavant's muscle? What is its importance?

It consists of horizontal fibers of palatopharyngeus at the level of hard palate, which meet with those of opposite side. These contract and form a Passavant's ridge at junction of nasopharynx with oropharynx. Acting along with levator palati, it closes the pharyngeal isthmus preventing food from entering nasopharynx.

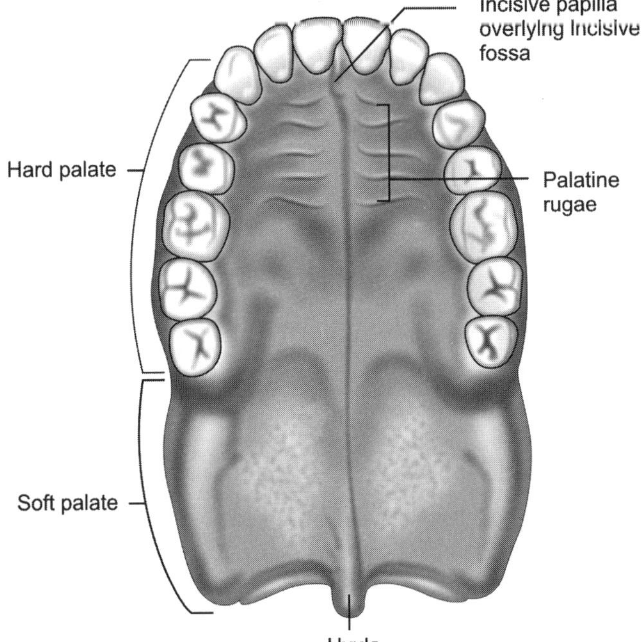

Fig. 6.16: Palate

Q.160 How is palate developed?

By

- Two palatal processes of maxillary process and
- Frontonasal process, which is a median structure. These processes fuse and form palate.

  The mesoderm of palate undergoes intramembranous ossification to form the hard palate. But, the ossification does not extend into posterior most portion, which remains as soft palate.

Q.161 Which part of the palate is formed by the frontonasal process?

It forms the triangular anterior part of hard palate which bears the incisor teeth.

Q.162 What is cleft palate?

This results from defective fusion of various components of palate. It results in communication between mouth and nose.

Pharynx

Q.163 What is length of pharynx (Fig. 6.17)?

12 cm.

Q.164 What is the extent of pharynx?

Superiorly: Base of skull including posterior part of body of sphenoid and basilar part of occipital bone.

Inferiorly: C_6 vertebra or lower border of cricoid cartilage.

Q.165 What are the attachments of pharynx on each side?

- Medial pterygoid plate,
- Pterygomandibular raphe,
- Mandible,
- Tongue,
- Hyoid bone and
- Thyroid and cricoid cartilages.

Q.166 What are the parts of pharynx?

- Nasopharynx,
- Oropharynx and
- Laryngopharynx.

Q.167 What are the characteristic features of nasopharynx?

- Respiratory in function.
- Wall are rigid and non-collapsible.
- Lined by columnar ciliated epithelium.

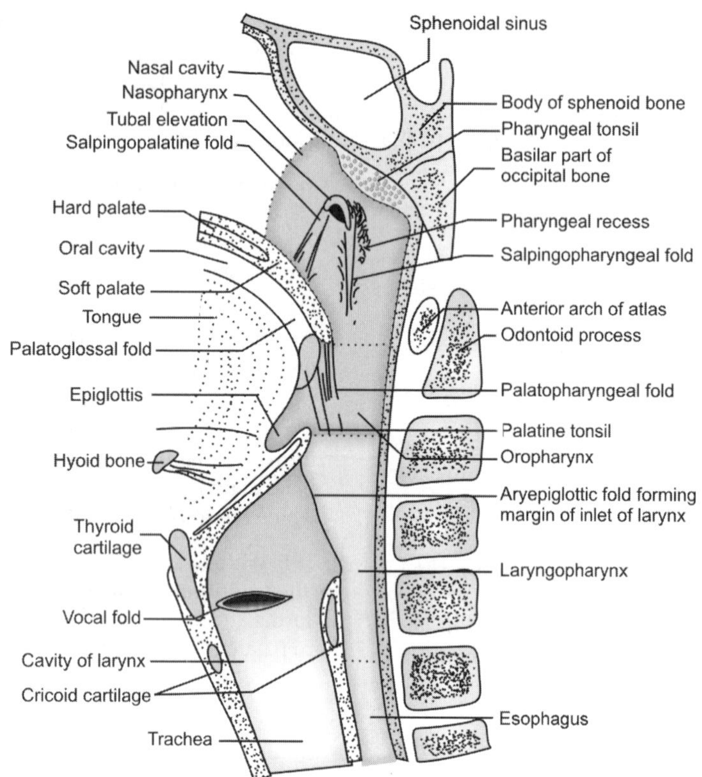

Fig. 6.17: Schematic median section through the pharynx and neighboring structures to show its lateral wall. The limits of the subdivisions of the pharynx are indicated in dotted lines

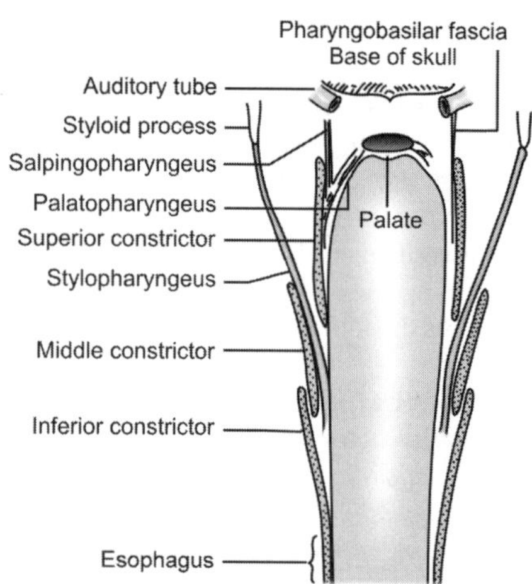

Fig. 6.18: Schematic coronal section to show the arrangement of muscles of the pharynx

Q.168 What are the features of lateral wall of nasopharynx?
- Pharyngeal opening of auditory tube
- Tubal elevation around the opening
- Salpingopharyngeal fold
- Salpingopalatine fold
- Pharyngeal recess.

Q.169 What is the clinical importance of pharyngeal recess (fossa of Rosenmuller)?
It forms a flat pocket. A catheter missing the tubal opening may enter recess and perforate the pharyngobasilar fascia and enter the ICA (internal carotid artery).

Q.170 What structures form the junction of nasopharynx and oropharynx?
- Lower border of soft palate and
- Passavant's muscle.

Q.171 What is isthmus of fauces?
Also known as oropharyngeal isthmus and it represents the junction of oropharynx and oral cavity.

Q.172 Which muscle helps in closure of oropharyngeal isthmus?
Palatoglossus.

Q.173 What is the extent of laryngeal part of pharynx?
From third to sixth cervical vertebra.

Q.174 Where does the junction of oropharynx and laryngopharynx lie?
Upper border of epiglottis.

Q.175 What is the position of 'piriform fossa'?
It is present on each side of inlet of larynx. The fossa is bounded medially by aryepiglottic fold and laterally by thyroid cartilage and thyrohyoid membrane.

Beneath mucosa of fossa lie internal laryngeal nerve.

Q.176 What is the clinical importance of piriform fossa?
A foreign body may lodge here. If removed, damage to internal laryngeal nerve may occur leading to anesthesia in supraglottic part of larynx. Thus, leading to aspiration pneumonia.

Q.177 What are the different layers forming wall of pharynx?
- *Mucosa:* Lined by squamous epithelium except nasopharynx.
- *Submucosa.*
- *Pharyngobasilar fascia:* Fibrous sheath. Present between mucous membrane and layer of muscle.
- *Muscular coat:* Has outer circular layer and inner longitudinal layer.
- *Buccopharyngeal fascia.*

Q.178 Name the muscles of pharynx.
Longitudinal layer:
- Stylopharyngeus,
- Salpingopharyngeus and
- Palatopharyngeus.

Circular layer:
- Superior constrictor,
- Middle constrictor and
- Inferior constrictor (Fig. 6.18).

Q.179 What is the nerve supply of the pharynx?
Motor fibers:
- *Stylopharyngeus:* Glossopharyngeal nerve.
- *Palatopharyngeus:* Cranial part of accessory nerve.
- *Salpingopharyngeus and constrictors of pharynx:* Pharyngeal branch of vagus through pharyngeal plexus.

Sensory: Glossopharyngeal and vagus.

Nasopharynx: Maxillary nerve.

Taste: Internal laryngeal branch of vagus.

Secretomotor: Greater petrosal nerve.

Q.180 What is pharyngeal plexus?
It is a plexus of nerve, present beneath the buccopharyngeal fascia.

Formed by:
- Pharyngeal branch of vagus (cranial accessory fibers)
- Pharyngeal branches of glossopharyngeal
- Pharyngeal branches of superior cervical sympathetic ganglion.

Q.181 What is the arrangement of constrictors of pharynx?
These are arranged like three flower pots one above the other, the lower constrictor overlapping the upper one.

Q.182 What is 'sinus of Morgagni'?
It is a semilunar gap between base of skull and upper border of superior constrictor.

Q.183 What are the structures passing through sinus of Morgagni?
- Auditory tube,
- Levator palati muscle and
- Ascending palatine artery.

Q.184 What are the structures passing through gap of superior and middle constrictor?
- Glossopharyngeal nerve and
- Stylopharyngeus.

Q.185 Name the structures passing between middle and inferior constrictor.
- Internal laryngeal nerve and
- Superior thyroid artery

Q.186 Name the structures passing through gap between inferior constrictor and esophagus.
- Recurrent laryngeal nerve and
- Inferior laryngeal vessels.

Q.187 What is Killian's dehiscence?
This is a weak part in posterior wall of pharynx, between thyropharyngeal and cricopharyngeal part of inferior constrictor.

Q.188 What is the clinical importance of Killian's dehiscence?
Pharyngeal diverticula may be formed due to out-pouching at dehiscence.

Tonsil

Q.189 What is the position of tonsil?
Tonsil occupies tonsillar fossa between diverging palatoglossal fold in front and palatopharyngeal fold behind (Fig. 6.19).

Q.190 What are the structures forming tonsillar bed?
- Pharyngobasilar fascia.
- Superior constrictor and palatopharyngeus muscle.
- Buccopharyngeal fascia.
- In lower part, styloglossus and 9th cranial nerve.

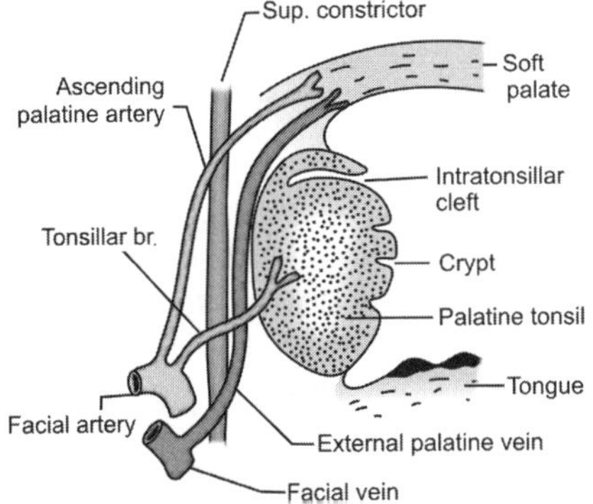

Fig. 6.19: Coronal section through the palatine tonsil

Q.191 What are the boundaries of pharyngomaxillary space? What is its clinical importance?
It is a triangular space bound by
- *Superiorly:* Base of skull
- *Medially:* Superior constrictor of pharynx
- *Laterally:* Fascia covering medial pterygoid and submandibular gland.
  Importance: Infection can spread into it from palatine tonsil or peritonsillar abscess.

Q.192 What is 'plica triangularis'?
It is triangular vestigial fold of mucous membrane covering anteroinferior part of tonsil.

Q.193 What is Waldeyer's ring?
It is a lymphatic ring which guards entry to digestive and respiratory passages.

It is formed by six masses of lymphoid tissue:
- 2 palatine tonsils,
- 2 tubal tonsils,
- 1 pharyngeal tonsil and
- 1 lingual tonsil.

These are connected together by scattered lymphoid tissue (Fig. 6.20).

Q.194 What is adenoid?
Pathological enlargement of pharyngeal tonsil.

Q.195 How does a tonsil differ from a lymph node?
- Lined by squamous epithelium.
- No subcapsular lymph space.
- No afferent lymphatics.
- Do not have a complete capsule.

Q.196 What is the arterial supply of tonsil?
- Tonsillar branch of facial, mainly
- Ascending palatine branch of facial
- Dorsal lingual branch of lingual
- Ascending pharyngeal branch of external carotid and
- Greater palatine branch of maxillary.

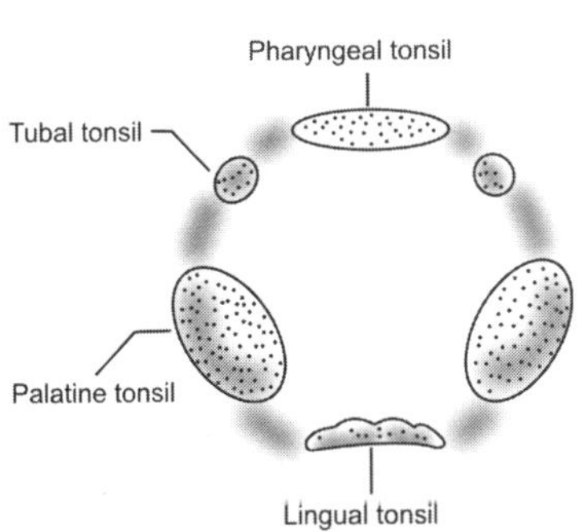

Fig. 6.20: Waldeyer's ring of lymphoid tissue

Q.197 Hemorrhage during tonsillectomy occurs from injury to which vessels?
It can result from injury to ascending palatine branch of facial artery, which is separated from tonsil only by superior constrictor muscle or external palatine vein descending on lateral side of tonsil from soft palate between capsule and superior constrictor.

Q.198 What is the lymphatic drainage of tonsils?
Jugulo-digastric node.

Q.199 How hemorrhage after tonsillectomy is checked?
By removal of clot from the raw bed.

Q.200 Why tonsillitis causes referred pain in the ear?
Because of the common nerve supply, i.e., by glossopharyngeal (IX cranial) nerve.

Q.201 How palatine tonsil develops?
These develop in relation to lateral part of second pharyngeal pouch by endodermal proliferation and lymphocyte collection.

Larynx

Q.202 What is the extent of larynx?
From root of tongue to trachea.

In front of C_{3-5} vertebra.

Q.203 Name the cartilages forming the skeletal framework of larynx.
Upaired:
- Thyroid,
- Cricoid and
- Epiglottic.

Paired:
- Arytenoid,
- Corniculate and
- Cuneiform (Figs 6.21 and 6.22).

Q.204 What is Adam's apple?
Also called laryngeal prominence. It is formed by fusion of anterior borders of lamina of thyroid cartilage.

It is more prominent in males.

Q.205 Name the structures attached to oblique line of thyroid cartilage.
Sternothyroid, thyrohyoid and inferior constrictor.

Q.206 What is the histological type of laryngeal cartilages?
Thyroid, cricoid and bases of arytenoid: Hyaline. Ossify after 25 years of age.

Rest: Fibrocartilage. Never ossify.

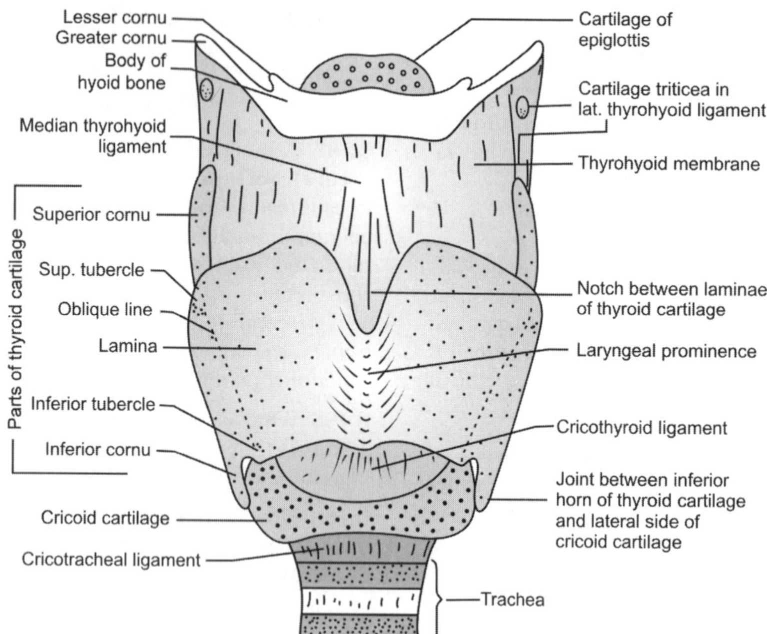

Fig. 6.21: Cartilages of the larynx as seen from the front

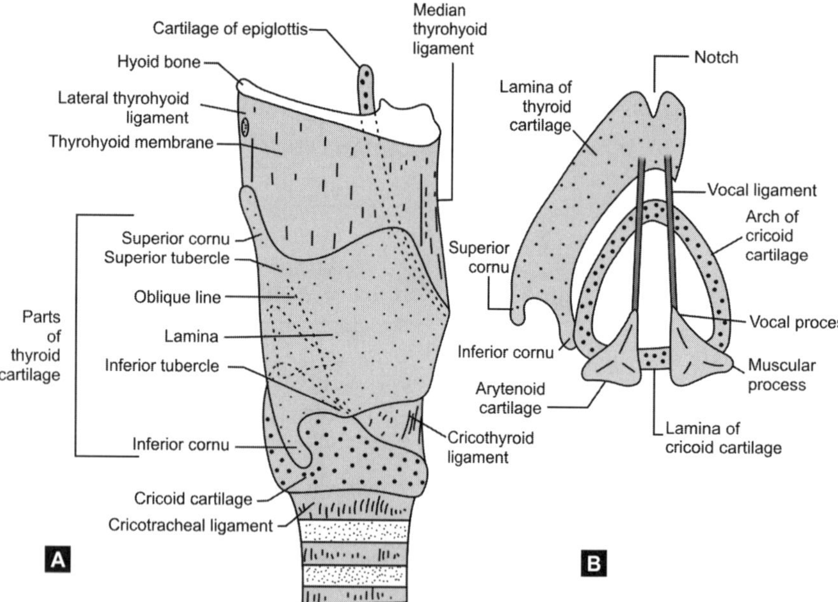

Figs 6.22A and B: Cartilages of the larynx: (A) Seen from the lateral side. (B) Seen from above

Q.207 What are different laryngeal joints and what are their movements?

- *Cricothyroid joint:* Synovial joint. Between inferior cornu of thyroid cartilage and lateral side of arch of cricoid.
  Rotatory movements around transverse axis and gliding movements.
- *Cricoarytenoid joint:* Synovial joint. Between base of arytenoid and upper border of cricoid.
  Rotatory movements around a vertical axis and gliding movements.

Q.208 Name laryngeal ligaments and membranes.

- Thyrohyoid membrane, thickens to form median and lateral thyrohyoid ligament
- Cricotracheal ligament
- Thyroepiglottic ligament
- Anterior cricothyroid ligament
- Hypoepiglottic ligament
- Cricovocal membrane
- Vocal ligament.

Q.209 What are the boundaries of inlet of larynx?

Anterior: Epiglottis
Posterior: Inter-arytenoid fold of mucous membrane
On each side: Aryepiglottic fold.

Q.210 Name the cartilages lying within aryepiglottic fold.

Corniculate and cuneiform cartilages.

Q.211 What are the parts of larynx?

- *Vestibule of larynx:* Lying above vestibular folds
- *Sinus of larynx:* Between vestibular and vocal folds
- *Infraglottic part:* Below vocal folds.

Q.212 What is the characteristic feature of laryngeal mucous membrane?

- Anterior surface and upper 1/2 of the posterior surface of epiglottis, upper parts of aryepiglottic folds and vocal folds are lined by stratified squamous epithelium. Rest of laryngeal mucous membrane is covered with columnar ciliated epithelium.
- Mucous membrane is loosely attached except to vocal ligament and posterior surface of epiglottis.
- Mucous glands are absent over vocal cord.

Q.213 Name the intrinsic muscles of larynx?

a. Muscles that open or close the inlet of larynx:
  - *Oblique arytenoids:* Closes the inlet of larynx.
  - *Thyroepiglottic:* Opens the inlet of larynx.
  - *Aryepiglottic:* Closes the inlet of larynx.
- Muscles that open or close the glottis:
  - *Posterior cricoarytenoid:* Opens glottis. Both anterior and posterior triangular part, wide open in forced respiration.
  - *Lateral cricoarytenoid:* In whispring, anterior part of glottis is closed but posterior part in open.
  - *Transverse cricoarytenoid:* During speech both vocal fold and arytenoid cartilage are close together.
- Muscles that increase or decrease the tension of vocal fold:
  - *Cricothyroid:* Tense the vocal fold
  - *Thyroarytenoid:* Relax the vocal fold
  - *Vocalis:* Tenses the vocal fold. Also called as tuning fork of larynx.

Q.214 Which intrinsic muscle of larynx is unpaired?
Transverse arytenoid.

Q.215 What is the nerve supply of larynx?
- *Motor:*
  Cricothyroid: External laryngeal nerve
  Other intrinsic muscles: Recurrent laryngeal neve.
- *Sensory:*
  Mucous membrane up to vocal cord: Internal laryngeal nerve.
  Mucous membrane below vocal cord: Recurrent laryngeal nerve.

Q.216 What is the effect of lesion of 'external laryngeal nerve'?
Weakness of phonation due to loss of tightening effect of cricothyroid on vocal cord.

Q.217 What is the effect of lesion of 'internal laryngeal nerve'?
Anesthesia of mucous membrane in supraglottic part of larynx, so the foreign bodies can readily enter larynx.

Q.218 What is the effect of lesion of 'recurrent laryngeal nerve'?
- *When bilateral:* Complete loss of phonation. Difficulty in breathing.
  Vocal cords lie in between adduction and abduction.
- *When unilateral:* Phonation possible because opposite vocal cord compensates.

Q.219 What is 'Semon's law?
In progressive lesions of recurrent laryngeal nerve, abductors of vocal cord are first to be paralyzed and last to recover, as compared to adductors.

But in functional paralysis of larynx, adductors are first paralyzed.

Q.220 Why edema of larynx causes suffocation?
Because tissue fluid cannot move downwards due to firm attachment of mucous membrane to vocal ligament and thus, causing obstruction.

Thyroid Gland (Fig. 6.23)

Q.221 What is the situation of thyroid?
In front and sides of lower part of neck.

Q.222 What is the extent of thyroid?
- $C_{5,6,7}$ T_1 vertebrae.
- Middle of thyroid cartilage to fifth tracheal ring.

Q.223 Name the capsules of thyroid?
True capsule: Condensation of connective tissue of gland.

False capsule: From pretracheal fascia.

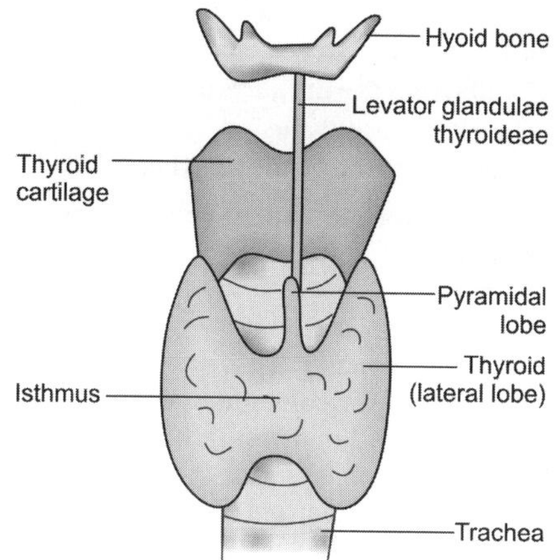

Fig. 6.23: The thyroid gland

Q.224 Why the thyroid is removed along with true capsule?
To avoid hemorrhage because the capillary plexus is present deep to true capsule.

Q.225 What is 'isthmus'? What are its relations?
It is part of thyroid gland connecting two thyroid lobes in lower part.

Extent: Lies against II, III and IV tracheal ring.

Relations:

Anterior surface:
- Strenothyroid
- Sternohyoid
- Anterior jugular
- veins
- Fascia
- Skin.

Posterior surface: II, III and IV tracheal ring.

Upper border: Anastomosis between superior thryoid arteries.

Lower border: Inferior thyroid veins leave gland.

Q.226 What is the arterial supply to thyroid?
- *Superior thyroid artery:* Supplies upper 1/3 of lobes and upper 1/2 of isthmus. Branch of external carotid.
- *Inferior thyroid artery:* Supplies lower 2/3 of lobes and lower 1/2 of isthmus. Branch of thyrocervical trunk.
- *Thyroidea ima artery*: From brachiocephalic trunk.
- *Accessory thyroid arteries:* From vessels to esophagus and trachea.

Q.227 What is the venous drainage of thyroid?
- Superior thyroid vein,
- Middle thyroid vein,
- Inferior thyroid vein,
- Sometimes, fourth thyroid vein (of Kocher).

The veins form a plexus deep to true capsule of gland.

Q.228 What is goiter?
Any enlargement of the thyroid gland.

Q.229 In partial thyroidectomy, why the posterior part of lobes are left behind?
- To avoid risk of removal of parathyroids and
- To avoid postoperative myxedema.

Q.230 Why thyroid moves with deglutition?
Because thyroid is attached to the larynx (cricoid cartilage) by the suspensory ligament of Berry.

Q.231 What are the precautions to be taken during thyroidectomy?
- Ligate superior thyroid artery near gland to avoid injury to external laryngeal nerve.
- Ligate inferior thyroid artery away from gland to save recurrent laryngeal nerve.
- Remove thyroid along with its true capsule to avoid injury to venous plexus.

Q.232 How the thyroid is developed?
Immediately behind *tuberculum impar* (a midline swelling in mandibular arches) in floor of pharynx a diverticulum called *thyroglossal duct* develops, which grows down into neck and its tip bifurcates and proliferates to form thyroid gland.

The developing thyroid also fuses with caudal pharyngeal complex.

Q.233 Name the common anomalies of thyroid?
- Pyramidal lobe present.
- Isthmus may be absent.
- One of the lobes may be absent.
- Thyroid gland may be found in abnormal position, i.e. any where in its path of descent, e.g. in tongue, above or below hyoid.
- Thyroglossal duct may persist and lead to the formation of thyroglossal cyst and fistula.

Q.234 Why enlarging thyroid tends to grow downward?
Because the sternothyroid muscles, which cover the thyroid gland infront, are attached above to the thyroid cartilage, limit the upward expansion of thyroid.

Q.235 Why the enlargements of thyroid produces compression symptoms earlier?
The thyroid is enclosed in pretracheal fascia which is much denser in front than behind. The enlarging gland therefore tends to push backwards, burying itself round the sides of trachea and esophagus

and compress or displace them, with resulting difficulty in breathing and swallowing.

Parathyroid Glands (Fig. 6.24)

Q.236 What is the number of parathyroid glands?
Four.

Two superior and two inferior.

Q.237 What is the position of parathyroid glands?
Superior parathyroids: Usually lies at middle of posterior border of lobe of thyroid above the level at which inferior thyroid artery crosses recurrent laryngeal nerve.

Inferior parathyroids: Usually below inferior thyroid artery near lower end of posterior border of thyroid gland.

Q.238 What type of cells are present in parathyroid glands?
- Chief cells or principal cells: Majority of cells.
- Oxyphil or eosinophil cells.

Q.239 How parathyroids are developed?
Superior parathyroids: From endoderm of fourth pharyngeal pouch.

Inferior parathyroids: From endoderm of third pharyngeal pouch.

The inferior parathyroids are carried down by the descending thymus, while superior parathyroids are prevented from going down because of its relationship to thyroid.

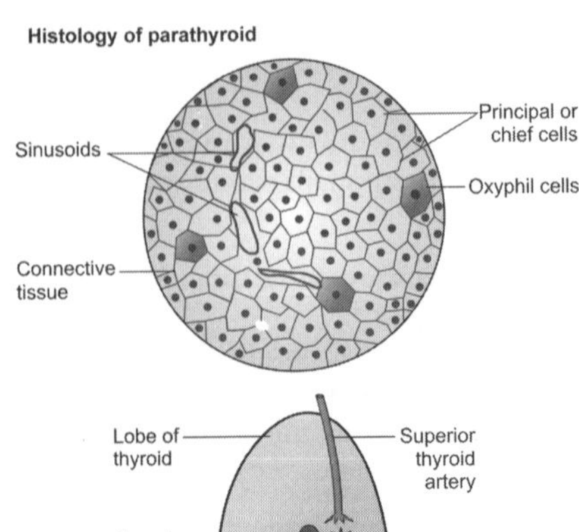

Fig. 6.24: Position of parathyroid glands on posterior surface of thyroid lobe

Ear

Q.240 What are the parts of external ear?
- Auricle (pinna) and
- External acoustic meatus (Fig. 6.25).

Q.241 What is the nerve supply of auricle?
- *Sensory:*
  Lateral surface:
  - *Anterosuperior part including tragus:* Auriculotemporal (branch of mandibular division of trigeminal nerve)
  - *Posteroinferior part including lobule:* Greater auricular ($C_{2,3}$)

  Cranial surface
  - *Upper 1/3:* Lesser occipital (C_2)
  - Lower 2/3: Greater auricular ($C_{2,3}$)
  - *Eminentia conchae:* Auriculotemporal nerve

  Concavity of conchae on external surface: Auricular branch of vagus.
- *Motor*
  To auricular muscles: Facial nerve.

Q.242 What is the shape of external acoustic meatus?
It follows a S-shaped course. Cartilaginous part, first passes medially, forwards and upwards. It then passes medially, backwards and upwards. Bony part runs medially, forwards and downwards.

Q.243 What is the nerve supply of external acoustic meatus?
Anterior wall and roof: Auriculotemporal nerve.

Posterior wall and floor: Auricular branch of vagus nerve.

Q.244 What are the parts of external acoustic meatus?
- Pars externa,
- Pars media and
- Pars interna.

Q.245 What is the length of external acoustic meatus?
24 mm. Outer 8 mm is cartilaginous and inner 16 mm is bony.

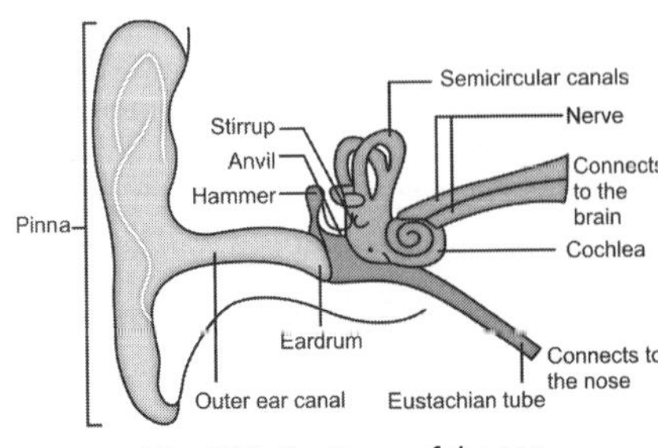

Fig. 6.25: Anatomy of the ear

Q.246 What are ceruminous glands?
These are modified sweat glands in skin of external acoustic meatus. Secrete yellow-brown ear wax.

Q.247 Why sometimes syringing of ear produces sudden death?
Due to irritation of auricular branch of vagus, reflex cardiac inhibition occurs leading to death.

Q.248 Why does the pain of external ear radiate to temporomandibular joint and teeth of lower jaw?
Because all these structures are supplied by the branches of mandibular nerve (branch of trigeminal nerve).

Q.249 What are the parts of tympanic membrane?
- *Pars flaccida:* Small triangular area above malleolar folds.
- *Pars tensa:* Greater part of membrane below malleolar folds.

Q.250 Why the infections of external ear are very painful?
Because the skin is firmly adherent to the underlying bone and cartilage, so the little swelling due to infection causes pain.

Q.251 What is 'umbo'?
It is point of maximum convexity on inner surface of tympanic membrane, at the tip of handle of malleus.

Q.252 What are the different layers of tympanic membrane?
From lateral to medial:
- Skin
- Fibrous layer
- Mucous membrane

Q.253 What is the nerve surface of tympanic membrane?
External surface: Auriculotemporal nerve and auricular branch of vagus.
Internal surface: Tympanic branch of glossopharyngeal nerve.

Q.254 What is the position of middle ear?
It is narrow air space situated in the petrous part of temporal bone between the external and internal ears.

Q.255 What are the communications of middle ear?
Anterior wall: Nasopharynx through auditory tube.
Posterior wall: Mastoid antrum through aditus antrum.

Q.256 Name the contents of middle ear?
- *Ear ossicles:* Malleus, incus and stapes.
- *Ligaments of ear ossicles.*
- *Muscles:* Tensor tympani and stapedius.

- *Vessels:* Supplying and draining the middle ear.
- *Nerves:* Chorda tympani and tympanic plexus.
- *Air.*

Q.257 What is the arterial supply of middle ear?

Mainly by:
- Anterior tympanic branch of maxillary.
- Posterior tympanic branch of posterior auricular.

Also by:
- Superior tympanic branch of middle meningeal,
- Inferior tympanic from ascending pharyngeal and
- Tympanic branch from artery of pterygoid canal.

Q.258 What is the length of auditory tube?

36 mm

Outer bony part: 12 mm

Inner cartilaginous part: 24 mm.

Q.259 What is the direction of auditory tube?

Downward, forward and medially.

Q.260 Which is the narrowest part of auditory tube?

Isthmus, the junction of bony and cartilaginous part.

Q.261 At what time the auditory tube opens?

During deglutition and swallowing of saliva.

Q.262 Name the muscle responsible for opening the auditory tube during deglutition?

Tensor palati.

Q.263 Name the structures in infratemporal fossa which are separated from eustachian tube by tensor palati.

- Mandibular nerve
- Chorda tympani nerve
- Middle meningeal artery
- Otic ganglion.

Q.264 How the throat infections spread to the middle ear?

Through the auditory tube. More commonly seen in children because auditory tube is shorter and wider in children.

Q.265 What is the function of auditory tube?

It maintains atmospheric pressure in the middle ear cavity, thus the air pressure on the two sides of tympanic membrane are equalized.

Q.266 Why meningitis is common in children suffering from middle ear infection?

In children, roof of middle ear presents a gap at unossified petrosquamous suture where middle ear is in direct contact with the meninges.

Q.267 What are the functions of middle part?

- Transmission of sound waves from external to the internal ear by ear ossicles.
- The intensity of sound waves is increased by ossicles, without any change in frequency.

Q.268 What is the nerve supply of muscles of middle ear?

- *Tensor tympani:* Mandibular nerve (branch of trigeminal nerve).
- *Stapedius:* Facial nerve.

Q.269 What is the function of muscles of middle ear?

They help to damp down the intensity of high pitched sound waves and thus protect the internal ear.

Q.270 What are the types of joints between ear ossicles?

- Between incus and malleus:
  Incudomalleolar joint: Saddle joint.
- Between incus and stapes:
  Incudostapedial joint: Ball and socket joint.

Both are synovial joints.

Q.271 Name the structures in angle between anterior and lateral walls of middle ear.

- Petrotympanic fissure: Anterior ligament of malleus passes through it.
- Anterior canaliculus: Chorda tympani passes through it.

Q.272 Name the structure producing the promontory on medial wall of middle ear cavity.

It is produced by the basal turn of the cochlea.

Q.273 Name the structure attached to fenestra vestibuli (oval window) on medial wall of middle ear.

Base of stapes and scala vestibuli of cochlea of internal ear.

Q.274 Name the structure attached to fenestra cocheal (round window) on medial wall of middle ear.

Lower part of cavity of cochlea (scala tympani) opens into it and it is closed by secondary tympanic membrane.

Q.275 What are the parts of internal ear?

- *Bony labyrinth:* Consist of
  - Cochlea,
  - Vestibule and
  - Semicircular canals.
- *Membranous labyrinth:* Consist of
  - Duct of cochlea,
  - Utricle and saccule and
  - Semicircular ducts.

Membranous labyrinth is filled with endolymph and is separated from bony labyrinth by perilymph (Figs 6.26 and 6.27).

Q.276 Name the ducts connecting the saccule and duct of cochlea.

Ductus reuniens.

Q.277 What is the importance of aqueduct of cochlea?

It is a opening in medial wall of scala tympani just near fenestra cochleae, which leads into a canal.

It represents the communication between perilymph and subarachnoid fluid.

Q.278 Where do the semicircular ducts open?

Utricle.

Q.279 What is the blood supply of internal ear?

Arterial supply: By labyrinthine artery, which is a branch of anterior inferior cerebellar artery. Also some branches from stylomastoid artery which also supplies middle ear.

Venous drainage: Into superior petrosal or transverse sinus.

Q.280 What are the functions of internal ear?

Cochlear portion: Hearing

Vestibular part: Equilibrium.

Semicircular canals act as *kinetic labyrinth* while utricle and saccule as *static labyrinth.*

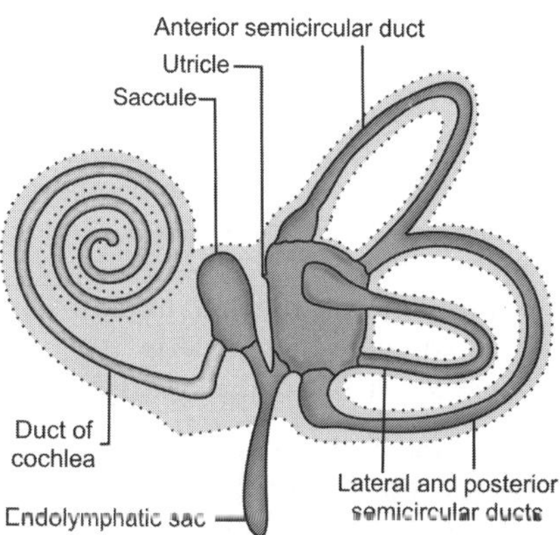

Fig. 6.26: Interior of the bony labyrinth as seen from the lateral side

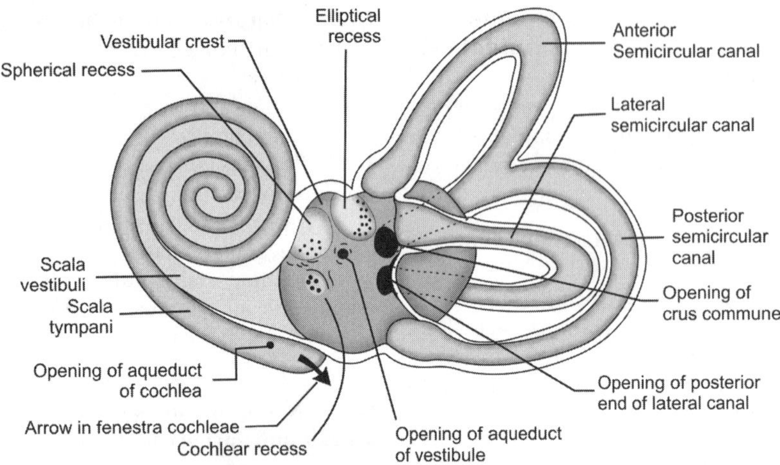

Fig. 6.27: Scheme to show the parts of the membranous labyrinth. Note the ampullated ends of the semicircular ducts

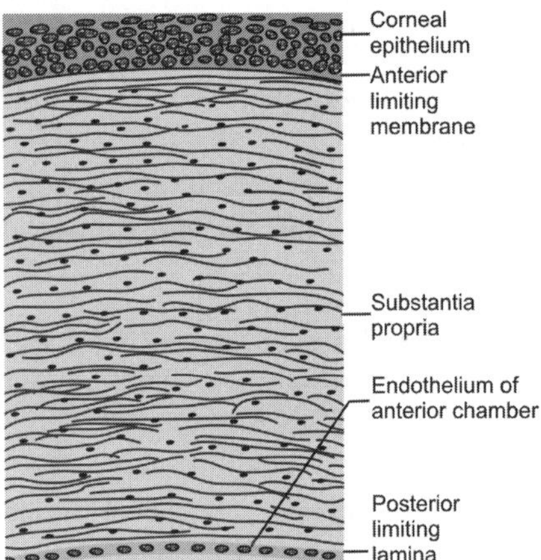

Fig. 6.29: Diagram of a section through the cornea to show its layers

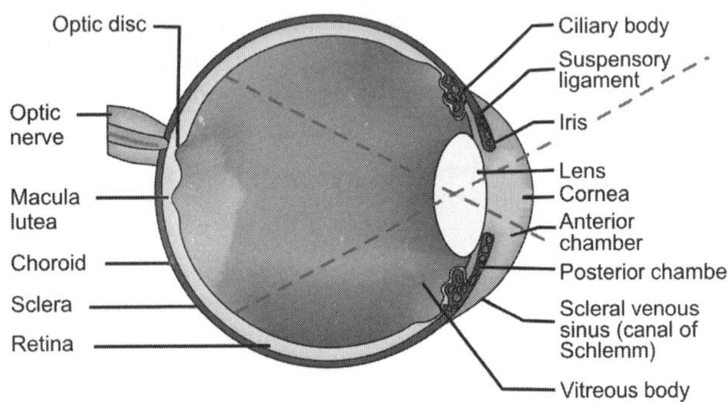

Fig. 6.28: Structure of eye-focusing

Q.281 What are the receptor cells for hearing and where they are located?
The receptors are neuroepithelial hair cells situated on the organ of Corti in duct of cochlea, just above basilar membrane.

Q.282 What are the receptors for equilibrium and where they are located.
Receptor cells are hair cells located on macula of utricle and saccule (for static balance) and on crista of ampulla of semicircular ducts (for kinetic balance).

Eye

Q.283 Name the different layers of eye.
- Outer or fibrous coat consists of sclera and cornea.
- Middle or vascular coat comprises choroid, ciliary body and iris.
- Inner or nervous coat, retina (Fig. 6.28).

Q.284 What is the diameter of eyeball?
The posterior five sixths has a diameter of about 24 mm. The anterior one sixth is much more convex and represents part of sphere having a diameter of 15 mm.

Q.285 Name the refractive media of eye.
From before backwards:
- Cornea,
- Aqueous humour,
- Lens and
- Vitreous body.

Q.286 What is 'lamina fusca of sclera'?
It is thin layer of delicate tissue between choroid and sclera.

Q.287 Name the structures piercing sclera.
- Optic nerve,
- Long ciliary nerves and arteries,
- Short ciliary nerves and arteries
- Venae verticosae.

Q.288 What are the layers of cornea seen histologically?
From before backwards:
- Corneal epithelium (Stratified squamous),
- Bowman's membrane,
- Substantia propria,
- Descemet's membrane and
- Endothelium of anterior chamber (Fig. 6.29).

Q.289 Name the membrane separating the choroid from retina.
Membrane of Bruch.

Q.290 What are the types of muscle fibers in ciliary body and what is their function?
- *Radial fibers.*
  Function: Relax the suspensory ligament of lens, so the lens bulges and becomes more convex for near vision.
- *Circular fibers.*
  Function: Also relax suspensory ligament of lens.

Q.291 What is ora serrata?
The retina proper ends anteriorly, just behind the sclerocorneal junction in a wavy line called as ora serrata. It also represents junction of choroid with ciliary body. Anterior to ora serrata retina continues as double layered epithelium lining the inner surface of ciliary body and posterior surface of iris.

Q.292 What is the nerve supply of ciliary muscle?
Parasympathetic nerves through third cranial nerve.

Q.293 What are the muscles of iris?
Has smooth muscle consisting of
- Sphincter pupillae: Has circular muscle fibers and its contraction narrows the pupil
- Dilator pupillae: Has radial muscle fibers.

Q.294 What is the nerve supply of muscles of iris?
- *Sphincter pupillae:* Parasympathetic nerve. Preganglionic neurons in Edinger Westphal nucleus give axons to oculomotor nerve and its branches reach the ciliary ganglion. Postganglionic fibers reach muscle through short ciliary nerves.

b. *Dilator pupillae:* Sympathetic nerve. Preganglionic neurons in T_1 segment. Postganglionic neurons in superior cervical sympathetic ganglion.

Q.295 What is the dioptric power of lens?
15 dioptres.

Q.296 What is the total dioptric power of eye?
58 dioptres.

Q.297 At which layer of eye maximum refraction takes place.
Corneal surface.

Q.298 What is 'fovea centralis'?
This is the centre of macula. This is thinnest part of retina containing only cones and is the site of maximum acuity of vision.

Q.299 Name the layers of retina.
From without inwards:
- Outer pigmented layer
- Layer of rods and cones
- External limiting membrane

Q.300 What is the arterial supply of retina?
Supplied by an artery, central artery of retina. In optic disc, it divides into branches which supplies deeper layer of retina upto bipolar cells. Branches of central artery of retina are end arteries. Rods and cones with nuclei are supplied by diffusion from capillaries.

Q.301 Describe the circulation of aqueous humour.

Secreted into posterior chamber from capillaries of ciliary process
↓
Through pupil enters anterior chamber
↓
Filters through spaces of iridocorneal angle (trabecular spaces)
↓
Enters sinus venosus sclerae
↓
Anterior ciliary veins

Q.302 What is hyaloid fossa?
It is depression in vitreous body on which the posterior surface of the lens lies.

Pituitary Gland (Hypophysis)

Q.303 What is the position of pituitary gland?
It lies in floor of middle cranial fossa in a depression on superior surface of body sphenoid called sella turcica.

It is suspended from floor of third ventricle of brain by a narrow stalk called infundibulum (Fig. 6.30).

Q.304 What are the relations of pituitary gland?
Anterosuperior: Optic chiasma
Inferior: Sphenoid air sinus
Lateral side: Cavernous sinus.

Q.305 What are the cell types in pars anterior and what are their secretions?
- Acidophil cells (Alpha cells)
  - Growth hormone
  - Prolactin
- Basophil cells (Beta cells)
  - Adrenocorticotropic hormone
  - Thyrotropic hormone
  - Gonadotrophic hormones.
- Chromophobe cells: Granules are absent and some are stem cells which give rise to chromophil cells.

Q.306 Which hormone is produced by pars intermedia?
Melanocyte stimulating hormone.

Q.307 Which hormone is produced by pars posterior?
- Antidiuretic hormone (vasopressin)
- Oxytocin.

Q.308 Where the hormones of pars posterior are synthesized?
In the nuclei of the hypothalamus, vasopressin in supraoptic nucleus and oxytocin in paraventricular nucleus of hypothalamus. These secretions pass down the axons through infundibulum into pars posterior.

Q.309 What is the function of gonadotrophic hormones?
- *Follicle stimulating hormone:* In females stimulate the growth of ovarian follicles and secretion of estrogens by the ovaries. In males, it stimulates spermatogenesis.
- *Luteinizing hormone:* In females, stimulates maturation of corpus luteum and secretion by it of progesterone.

In males, it is called *interstitial cell stimulating hormone* and stimulates the production of androgens by interstitial cells of testes.

Q.310 What is hypothalamohypophyseal portal system? What is its significance?
In this two sets of capillaries are present between arteries and veins. One of these is in median eminence and upper infundibulum and second set is in sinusoids of pars anterior.

Clinical significance: Neurons in hypothalamus produce releasing factors for hormones of adenohypophysis in capillaries of median eminence and infundibulum. These are carried by portal system to pars anterior which in turn stimulates to release appropriate hormones.

Q.311 What are 'pituicytes'?
These are cells in pars posterior.

Q.312 What is the developments origin of hypophysis?
Adenohypophysis: Develops from Rathke's pouch, which arises from ectoderm lining roof of primitive mouth (stomadeum).
Neurohypophysis: Develops as downgrowth from floor of third ventricle.

Blood Vessels of Head and Neck

Q.313 What are the branches of subclavian artery?
- Vertebral artery,
- Internal thoracic,
- Thyrocervical trunk,
- Costocervical trunk and
- Dorsal scapular. In 1/3 cases it arises with superficial cervical from thyrocervical trunk.

Q.314 What are the tributaries of subclavian vein?
- External jugular,
- Dorsal scapular,
- Thoracic duct on left and
- Right lymphatic duct on right.
- Sometimes, anterior jugular vein (Fig. 6.31).

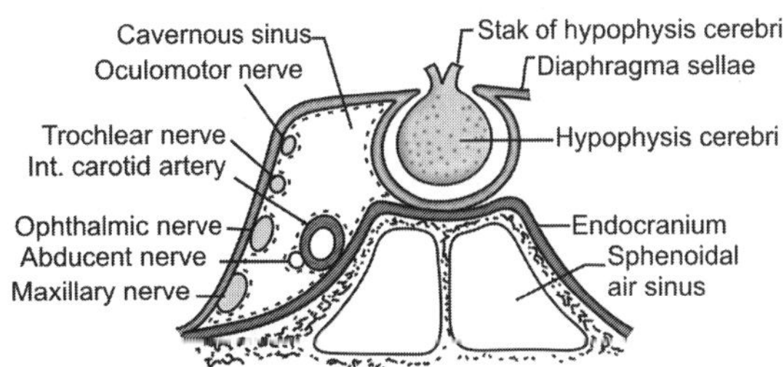

Fig. 6.30: Corneal section through hypophyseal fossa, cavernous sinus and diaphragma sellae

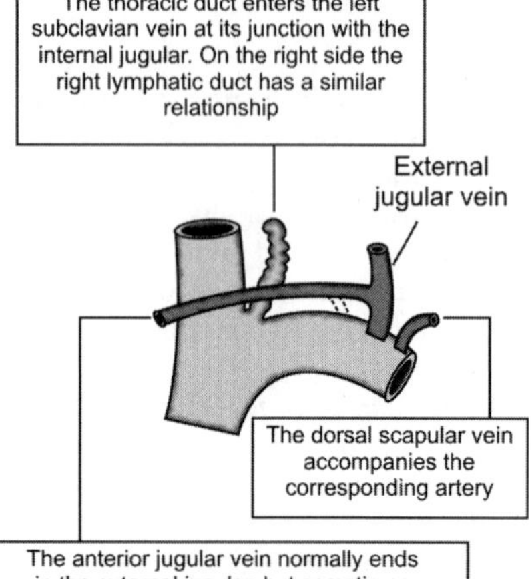

Fig. 6.31: Tributaries of the subclavian vein

Q.315 Name the tributaries of internal jugular vein?
- Inferior petrosal sinus,
- Sigmoid sinus,
- Common facial vein,
- Lingual vein,
- Pharyngeal vein,
- Superior thyroid vein and
- Middle thyroid vein.
- Sometimes, occipital vein (Fig. 6.32).

Q.316 Name the tributaries of brachiocephalic vein.

Brachiocephalic vein is formed by internal jugular vein and subclavian vein.

Right brachiocephalic:
- Vertebral.
- Internal thoracic.
- Inferior thyroid.
- First posterior intercostal

Left brachiocephalic:

1-4: Same as above.
- Left superior intercostal.
- Thymic veins.
- Pericardial veins.

Q.317 Name the branches of internal carotid artery .
- *Cervical part:* No branches.
- *Petrous part:*
  - Caroticotympanic artery.
  - Pterygoid branch.
- *Cavernous part:*
  - Cavernous branches to trigeminal ganglion.
  - Superior hypophyseal.
  - Inferior hypophyseal.
- *Cerebral part:*
  - Ophthalmic.
  - Anterior cerebral.
  - Middle cerebral.
  - Posterior communicating.
  - Anterior choroidal.
  - Meningeal.

Questions on
- *External carotid:* In chapter Triangles of neck.
- *Venous sinuses:* In chapter Meninges of brain and CSF (Fig. 6.33).

Q.318 Name the branches of vertebral artery.
- *Cervical branches:*
  - Spinal.
  - Muscular.
- *Cranial branches:*
  - Meningeal.
  - Posterior spinal.
  - Anterior spinal.
  - Posterior inferior cerebellar.
  - Medullary (Fig. 6.34).

Q.319 Name the branches of superficial temporal artery.
- Frontal
- Parietal
- Anterior auricular
- Middle temporal
- Zygomatico-orbital
- Transverse facial (Fig. 6.35)

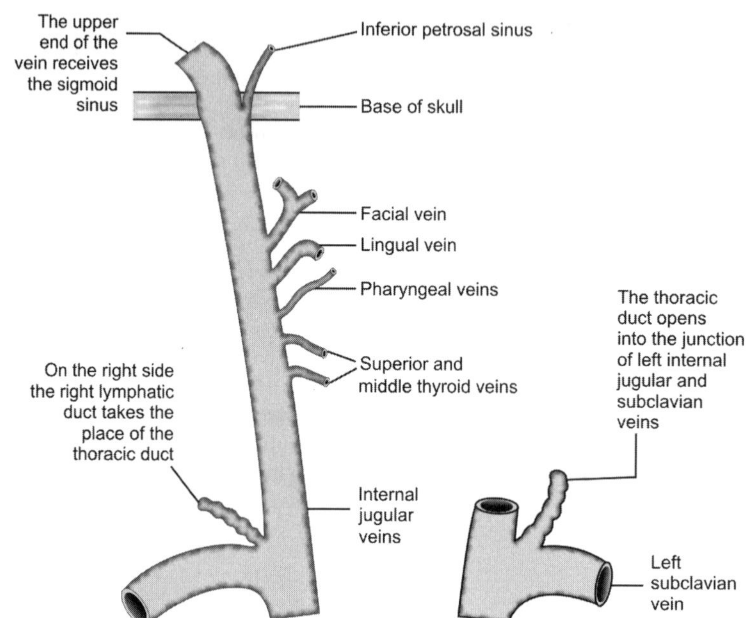

Fig. 6.32: Scheme to show the tributaries of the internal jugular vein

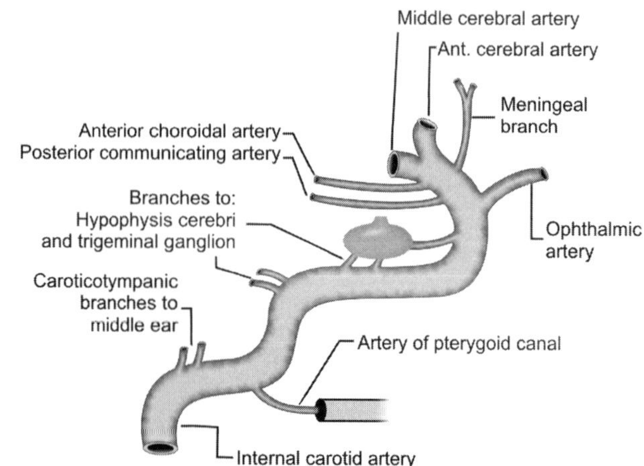

Fig. 6.33: Scheme to show the branches given off by the internal carotid artery

Q.320 How the retromandibular vein is formed?

It is formed by union of superficial temporal and maxillary vein behind ramus of mandible.

Q.321 How the external jugular vein is formed and what are its tributaries?

It is formed by union of posterior division of retromandibular vein and posterior auricular vein. Other tributaries are:
- Posterior external jugular
- Transverse cervical
- Suprascapular
- Anterior jugular.

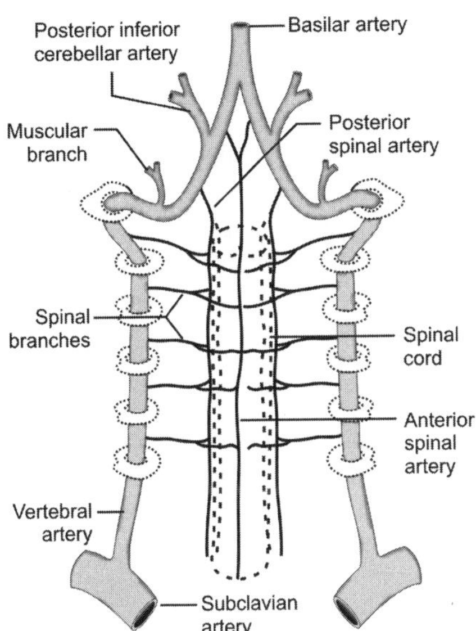

Fig. 6.34: Scheme to show the branches of the vertebral artery

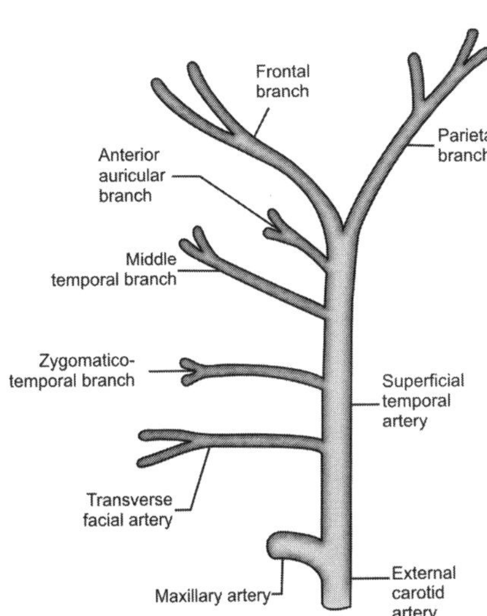

Fig. 6.35: Scheme to show the branches of the superficial temporal artery

Q.322 Name the branches of maxillary artery.

First part:
- Deep auricular
- Anterior tympanic
- Middle meningeal
- Accessory meningeal
- Inferior alveolar

Second part:
- Deep temporal
- Pterygoid
- Masseter
- Buccal.

Third part:
- Posterior superior alveolar
- Infraorbital
- Greater palatine
- Pharyngeal
- Artery of pterygoid canal
- Sphenopalatine (Fig. 6.36).

Nerves of Head and Neck

Q.323 What is the characteristic feature of cervical spinal nerves?

In spine, each spinal nerve lies below the numerically corresponding vertebra. But the upper seven cervical nerves lie above the numerically corresponding vertebrae. The eighth cervical nerve lies below C_7 vertebra.

Q.324 How the greater occipital nerve is formed?

It is formed by the medial branch of the dorsal ramus of second cervical nerve.

Q.325 How the cervical plexus is formed?

By ventral rami of C_{1-4} spinal nerves.

Q.326 What are the branches of cervical plexus?

- *Cutaneous branches:*
  - Lesser occipital nerve (C_2)
  - Greater auricular nerve ($C_{2,3}$)
  - Transverse cutaneous nerve of neck ($C_{2,3}$)
  - Supraclavicular nerves ($C_{3,4}$)
- *Muscular branches:*
  - To prevertebral muscles: Rectus capitis lateralis and anterior from C_1, longus capitis from C_{1-3}, longus colli from C_{2-4}.
  - To sternocleidomastoid (C_2).
  - To levator scapulae, scalenus medius and trapezius ($C_{3,4}$).
  - Phrenic nerve (C_{3-5}) to diaphragm.
  - To infrahyoid muscles through hypoglossal nerve and ansa cervicalis.

Q.327 (a) How is 'ansa cervicalis' formed?
(b) What is its distribution?

- By union of two roots formed by C_1 through hypoglossal nerve and $C_{2,3}$ superficial to common carotid artery.

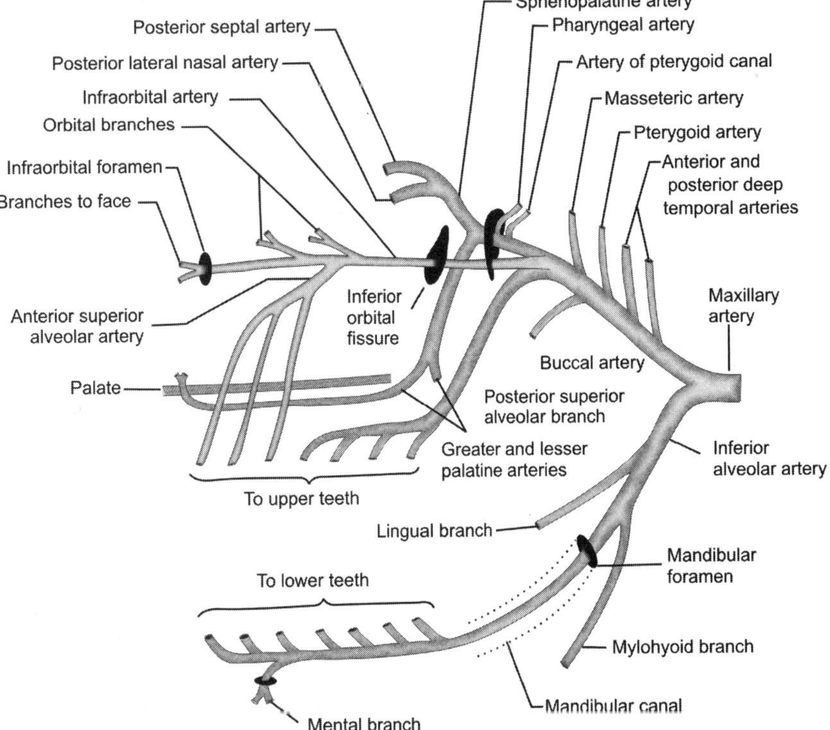

Fig. 6.36: Branches of the maxillary artery

b. All infrahyoid muscles except thyrohyoid.

(Questions on cranial nerves: See Cranial Nerves in chapter of CNS)

Joints of Head and Neck Temporomandibular Joint (Jaw Joint)

Q.328 What type of joint is temporomandibular joint (T-M joint)?
Condylar variety of synovial joint.

Q.329 Name the ligaments of T-M joint.
- Fibrous capsule,
- Articular disc,
- Lateral ligament,
- Sphenomandibular ligament and
- Stylomandibular ligament.

Q.330 What are the characteristic features of articular disc?
- It divides joint into an upper and a lower compartment.
- It has a concavoconvex superior surface and a concave inferior surface.
- *It has five parts:*
  - Anterior extension,
  - Anterior band,
  - Intermediate thin zone,
  - Thick posterior band and
  - Posterior bilaminar region.

Q.331 What is the developmental origin of sphenomandibular ligament?
It is remnant of dorsal (cephalic) end of Meckel's cartilage.

Q.332 Name the structures piercing sphenomandibular ligament.
Mylohyoid nerve and vessels.

Q.333 What is the nerve supply of T-M joint?
- Auriculotemporal nerve and
- Masseteric nerve.

Q.334 Name the muscles of T-M joint (muscles of mastication).
- Masseter,
- Temporalis,
- Lateral pterygoid and
- Medial pterygoid.

Q.335 What is the developmental origin of muscles of mastication?
Mesoderm of first branchial arch.

Q.336 What are the structures passing between two heads of lateral pterygoid?
- Maxillary artery and
- Buccal branch of mandibular nerve.

Q.337 What are the movements of T-M joints?
- Depression of jaw (opening of mouth)
- Elevation of jaw
- Protraction
- Retraction
- Rotatory movements (chewing)

Q.338 Describe movements which occur in chewing.
Head of one side of mandible with articular disc
- Glides forwards.
- Rotates around a vertical axis passing immediately behind the head of opposite side.
- Glides backwards.
- Rotates in opposite direction as head of opposite side comes forward.

Q.339 What are the different muscles producing movements of jaw joint?
- *Depression:* Lateral pterygoid of both sides with digastric, geniohyoid and mylohyoid
- *Elevation:* Masseter, temporalis and medial pterygoid of both sides
- *Protraction:* Lateral and medial pterygoids with masseter
- *Retraction:* Posterior fibers of temporalis with digastric and geniohyoid
- *Chewing:* Medial and lateral pterygoids of each side acting alternately.

Atlanto-occipital Joint

Q.340 What is the variety of atlanto-occipital joint?
Ellipsoid variety of synovial joint.

Q.341 What are the articular surfaces of Atlanto-occipital joint?
From above: Occipital condyles.
From below: Superior articular facets of atlas vertebra.

Q.342 What are the movements, possible at this joint?
- Flexion,
- Extension and
- Lateral flexion.

Atlanto-axial Joint

Q.343 What type of this joint is?
- *Median atlanto-axial joint:* Pivot synovial joint
- *Lateral atlanto-axial joint:* Plane synovial joint.

Q.344 Name the ligaments connecting atlas, axis and occipital bones.
- Anterior longitudinal ligament
- Anterior atlanto-occipital membrane
- Posterior atlanto-occipital membrane
- *Ligamentum flavum:* Between atlas and axis
- *Membrana tectoria:* Upward extension of posterior longitudinal ligament.
- Ligaments connecting dens of axis with occipital bone.

Q.345 Name the ligaments connecting the dens of axis with the occipital bone.
- Cruciate ligament
- Apical ligament of dens and
- Alar ligaments (left and right)

Q.346 Which movement takes place at atlanto-axial joint?
Side to side movement.

Q.347 How the death in hanging occurs?
Due to dislocation of dens of axis following rupture of transverse ligament of atlas, which then crushes the spinal cord and medulla, i.e. vital centers.

Bones of Head and Neck Skull

Q.348 What is total number of bones in skull (Fig. 6.37)?
Twenty two.

Q.349 How the skull is held in normal anatomical position?
By considering any one of the following two planes:
- *Reid's base line:* Horizontal line between infraorbital margin and center of external acoustic meatus.
- *Frankfurt horizontal plane:* Joint infraorbital margin and upper margin of external acoustic meatus.

Q.350 What are sutures?
The joints between the various skull bones. These are immovable and fibrous in type.

Q.351 What are different sutures seen in superior view (norma verticalis) of skull.
- *Coronal:* Between frontal and two parietal bones.
- *Sagittal:* Between two parietal bones.
- *Lambdoid:* Between occipital and two parietal bones.
- *Metopic:* In 3–8% between two halves of frontal bones.

Q.352 What is vertex?
It is the highest point on sagittal suture.

Q.353 What is bregma?
Meeting point between coronal and sagittal sutures. In fetal skull, represents anterior fontanelle, which closes at the age of 1½ years.

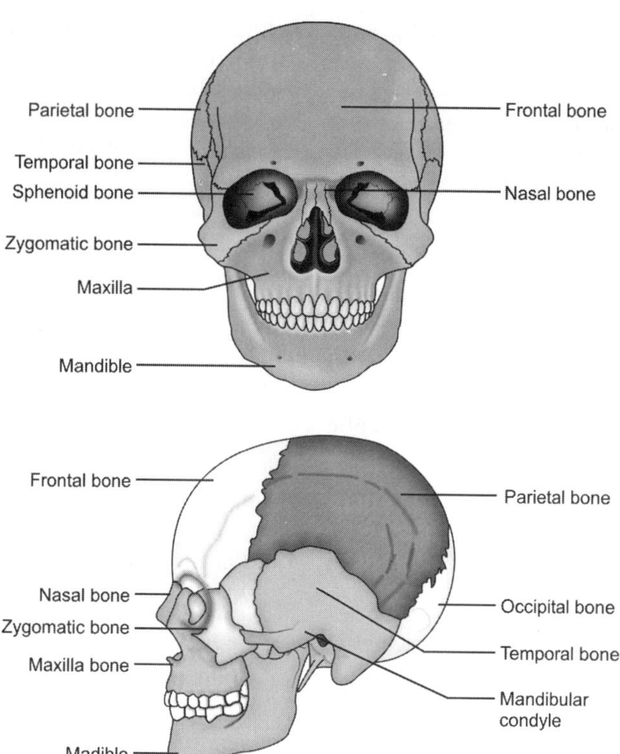

Fig. 6.37: Bones of the human skull

Q.354 What is lambda?
Meeting point between sagittal and lambdoid sutures. In fetal skull, represent posterior fontanelle, which closes by the age of 2–3 months.

Q.355 Name the structures attached to external occipital protuberances?
Trapezius, in upper part.
Ligamentum nuchae, in lower part.

Q.356 What is 'occipital point'?
Median point just above inion (most prominent part of external occipital protuberance) which is farthest from glabella.

Q.357 What is 'glabella'?
The meeting point of two superciliary arch in midline.

Q.358 What is 'nasion'?
Median point at the root of nose, where internasal suture meets with the frontonasal suture.

Q.359 Name the bones forming medial orbital margin.
Frontal process of maxilla, mainly.
Nasal part of frontal bone, in upper part.

Q.360 Name the sutures present in norma frontalis?
- Internasal
- Frontonasal
- Nasomaxillary
- Lacrimo-maxillary
- Fronto-maxillary
- Inter-maxillary
- Zygomatico-maxillary
- Zygomatico-frontal.

Q.361 What is 'jugal point'?
Anterior end of upper border of zygomatic arch.

Q.362 What are the boundaries of suprameatal triangle?
Above: Supramastoid crest.
Front: Posterosuperior margin of external auditory meatus.
Behind: Vertical tangent to posterior margin of meatus.

Q.363 What is the significance of tegmen tympani?
It is thin plate of bone formed by petrous temporal bone. It separates the middle cranial fossa from the cavities of middle ear, auditory tube and mastoid antrum.

Q.364 Name the structures attached to mastoid process.
From before backwards:
- Sternomastoid,
- Splenius capitis,
- Longissimus capitis and
- Posterior belly of digastric attached to mastoid notch.

Q.365 What are the relations of styloid process?
Laterally: Parotid gland.
Medially: Internal jugular vein.
Base: Ensheathed by tympanic plate. Related to facial nerve.
Apex: Posterior border of ramus of mandible, laterally.

Q.366 Name the structures attached to styloid process.
Muscle:
- Styloglossus,
- Stylohyoid and
- Stylopharyngeus.

Ligaments:
- Stylohyoid and
- Stylomandibular.

Q.367 What is 'pterion'?
H-shaped suture seen in norma lateralis formed by the frontal, parietal, sphenoid and temporal bone.

Q.368 What is 'asterion'?
The point at which parietomastoid suture (formed by mastoid part of temporal bone with parietal bone) and occipitomastoid suture (formed by mastoid part of temporal bone with occipital bone) meet.

Q.369 Name the structures related to pterion.
- Middle meningeal vein,
- Anterior division of middle meningeal artery and
- Stem of lateral sulcus of brain.

Q.370 What are the boundaries of temporal fossa?
Superior: Temporal line
Inferior: Zygomatic arch
Floor: By parts of frontal, parietal, squamous temporal and greater wing of sphenoid bone.
Anterior wall: Mainly by temporal surface of zygomatic bone. Also by greater wing of sphenoid and frontal bone.

Q.371 What is the origin, insertion and nerve supply of temporalis muscle?
Origin: Floor of temporal fossa
Insertion: Coronoid process of mandible
Nerve supply: Deep temporal nerve, branch of anterior division of mandibular nerve.

Q.372 What are the boundaries of infratemporal fossa?
Roof: Mainly by greater wing of sphenoid. Also squamous temporal bone.
Medial: Pterygoid process of sphenoid
Anterior: Posterior surface of maxilla.

Q.373 What are the communications of infratemporal fossa?

- *Temporal fossa:* Superiorly through gap between zygomatic arch and rest of skull.
- *Pterygopalatine fossa:* Through pterygopalatine fissure.
- *Orbit:* Anteriorly through inferior orbital fissure.

Q.374 What are the contents of 'infratemporal fossa'?

- *Muscles:* Lower part of temporalis, Lateral pterygoid, Posterior part of buccinator and medial pterygoid.
- *Ligaments:* Sphenomandibular.
- *Arteries:* First and second part of maxillary artery and their branches.
- *Veins:* Pterygoid plexus of veins and Maxillary vein.
- *Nerves:* Mandibular, Chorda tympani and Maxillary.

Q.375 Name the bones forming hard palate.

Anterior 2/3: Palatine processes of maxilla.

Posterior 1/3: Horizontal plates of palatine bones.

Q.376 Name the structures attached to lateral pterygoid plate.

Lateral surface: Lower head of lateral pterygoid.

Medial surface: Deep head of medial pterygoid.

Posterior border: Pterygospinous ligament.

Q.377 Why the upper surface of body of sphenoid is called 'sella turcica'?

Because it is hollowed out in the form of a 'Turkish saddle'.

Q.378 What lodges the hypophysis (pituitary gland)?

Hypophyseal fossa of sphenoid bone.

Q.379 What are the boundaries of 'superior orbital fissure'?

Above and medially: Lesser wing of sphenoid.

Below and laterally: Greater wing of sphenoid.

Q.380 What are the boundaries of inferior orbital fissure?

Above and laterally: Greater wing of sphenoid.

Below and medially: Orbital surface of the maxilla.

Q.381 What are the relations of 'spine of sphenoid'?

Lateral: Auriculotemporal nerve.

Medial: Chorda tympani nerve. Auditory tube.

Tip: Attachment to sphenomandibular ligament.

Anterior: Origin of fibers of tensor palati.

Q.382 Name the structures passing through internal acoustic meatus.

- 7th and 8th cranial nerves.
- Labyrinthine vessels.

Q.383 What are the boundaries of optic canal?

Lateral: Lesser wing of sphenoid.

Medial: Body of sphenoid.

Q.384 What structures are transmitted through optic canal?

- Optic nerve,
- Meningeal sheath of optic nerve and
- Ophthalmic artery.

Q.385 Name the structures passing through superior orbital fissure?

It is divided into three parts by a common tendinous ring of Zinn. It transmits:

- *Lateral part:*
  - Lacrimal nerve, branch of ophthalmic division
  - Frontal nerve, of trigeminal nerve
  - Trochlear nerve
  - Recurrent branch of ophthalmic artery and
  - Superior ophthalmic vein.
- *Middle part:*
  - Upper and lower division of oculomotor nerve,
  - Nasociliary nerve, branch of ophthalmic division of V nerve and
  - Abducent nerve.
- *Medial part:*
  - Inferior ophthalmic vein.

Q.386 What structure passes through foramen rotundum?

Maxillary nerve.

Q.387 What structures passes through foramen ovale?

- Mandibular division of V nerve,
- Lesser petrosal nerve,
- Accessory meningeal artery and
- Emissary vein connecting cavernous sinus to pterygoid venous plexus.

Q.388 Name the structures passing through foramen spinosum.

- Middle meningeal artery,
- Meningeal branch of mandibular nerve and
- Emissary vein.

Q.389 What are the boundaries of foramen magnum?

Anterior: Basiocciput.

Posterior: Squamous occipital.

On each side: Condylar part of occipital bone.

Q.390 Name the structures passing through foramen magnum.

- *Through wider posterior part:*
  - Lower part of medulla.
  - Tonsils of cerebellum.
  - Meninges.
- *Through narrow anterior part:*
  - Apical ligament of dens.
  - Membrana tectoria.
  - Superior band of cruciform ligament.
- *Through subarachnoid space:*
  - Spinal accessory nerve.
  - Vertebral arteries.
  - Sympathetic plexus.
  - Posterior spinal arteries.
  - Anterior spinal arteries.

Q.391 Name the structure transmitted by mastoid foramen.

Emissary vein connecting sigmoid sinus to occipital vein.

Q.392 Name the structure transmitted by sphenopalatine foramen.

Nasopalatine nerve and vessels.

Q.393 What are the contents of Hypoglossal canal?

- Hypoglossal nerve.
- Meningeal branch of ascending pharyngeal artery.
- Emissary vein.

Q.394 Name the structures passing through Jugular foramen?

- *Through the anterior part:*
  - Inferior petrosal sinus
- *Through the middle part:*
  - IX, X and XI cranial nerves.
  - Meningeal branch of ascending pharyngeal artery.
- *Through the posterior part:*
  - Lower end of sigmoid sinus
  - Meningeal branch of occipital artery.
  - Emissary vein connecting sigmoid sinus to occipital vein.

Q.395 Name the structures transmitted by carotid canal.

- Internal carotid artery (ICA)
- Venous and sympathetic plexuses around ICA.

Q.396 Name the structures transmitted by Inferior orbital fissure.

- Maxillary nerve
- Zygomatic nerve

- Orbital branches of pterygopalatine ganglion
- Infraorbital vessels
- Emissary vein connecting inferior ophthalmic vein to pterygoid plexus.

Q.397 Deep petrosal and greater petrosal nerve passes through which foramen of skull?
Foramen lacerum.

Q.398 Name the structures passing through stylomastoid foramen?
- Facial nerve,
- Stylomastoid branch of posterior auricular artery.

Q.399 What are Wormian bones?
These are small irregular bones found in region of fontanelles and are formed by the additional ossification centers.

Q.400 Which skull bone is formed by intramembranous ossification?
Frontal, parietal, zygomatic, palatine, nasal, lacrimal, maxilla and vomer.

Q.401 Which skull bone is formed partly in cartilage and partly in membrane?
Occipital, sphenoid, temporal and mandible.

Q.402 Which skull bone is formed entirely in cartilage?
Ethmoid and inferior nasal concha.

Q.403 What is cephalic index?
It is the ratio of the breadth (widest diameter) and length (longest diameter) of the skull.

Q.404 What is the clinical importance of cephalic index?
- In subdivision of the human population into different races.
- *In medicological practice:* To know the race to which the person belonged when skull and other bones are found.

Q.405 What is 'scaphocephaly'?
Boat shaped skull resulting from early union of sagittal suture.

Q.406 What is 'acrocephaly'?
Pointed skull as a result of early union of coronal suture.

Mandible (Figs 6.38A and B)

Q.407 Name the structures transmitted by mental foramen.
Mental nerve and vessels.

Q.408 Name the structures present in the submandibular fossa.
- Submandibular salivary gland,
- Submandibular lymph nodes and
- Facial artery.

Q.409 What structures are attached to genial tubercles?
- *Upper genial tubercle:* Origin of genioglossus.
- *Lower genial tubercle:* Origin of geniohyoid.

Q.410 Name the structures attached to angle of mandible.
- Stylomandibular ligament: To angle and posterior border of ramus.
- Masseter: To lateral surface of ramus and angle.
- Medial pterygoid: To medial surface of ramus and angle.

Q.411 Name the structures related to neck of mandible.
On lateral aspect: Parotid gland below attachment of lateral ligament of jaw joint.

On medial aspect:
- Auriculotemporal nerve, above.
- Maxillary artery, below.

Q.412 What are the changes in position of mental foramen with age?
The mental foramen at birth, opens below sockets for deciduous molar teeth near lower border.

The foramen gradually shifts upwards and in adults, it opens midway between upper and lower borders. In old age, due to absorption of alveolar border, mental foramen lies close to alveolar border.

Q.413 What is characteristic feature of ossification of mandible?
It ossifies partly in membrane and partly in cartilage.

Parts ossifying in membrane: Body of mandible except incisive part and lower half of ramus up to mandibular foramen.

Parts ossifying in cartilage: Incisive part below incisor teeth, coronoid and condyloid processes and upper half of ramus above mandibular foramen.

Q.414 How will you determine the sex to which the mandible belongs?

Features	*Male*	*Female*
• General size	Larger and thicker	Smaller and thinner
• Height of body	Greater	Lesser
• Angle of mandible	– Lesser – Everted	Greater Inverted
• Chin	Quadrilateral	Rounded
• Inferior border of body of mandible	Irregular	Smooth curve
• Condyles	Larger	Smaller

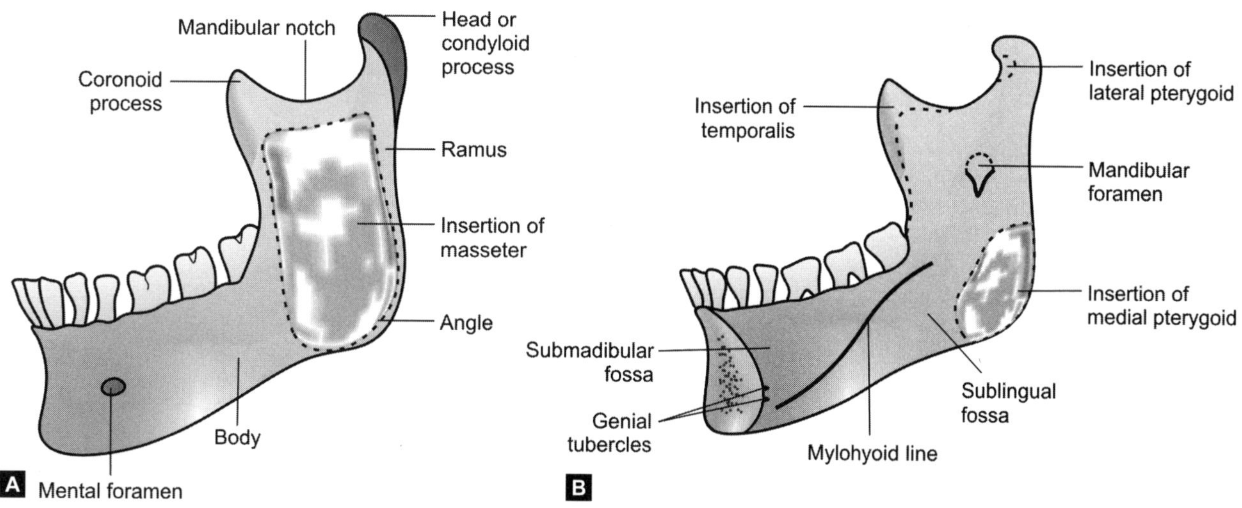

Figs 6.38A and B: Mandible

Q.415 What is the commonest site of fracture of body of mandible?
At the level of canine socket.

Hyoid Bone

Q.416 What is the level of hyoid bone?
C_3 vertebra behind and base of mandible in front.

Q.417 Name the structures attached to anterior surface of body of hyoid.
Insertion to: Geniohyoid and mylohyoid.
Origin to: Hyoglossus.
Below mylohyoid: Investing fascia.

Q.418 Name the structures attached to lower border of hyoid body?
Sternohyoid: Medial.
Omohyoid, superior belly: Lateral.
Thyrohyoid: Below omohyoid.
Pretracheal fascia.

Q.419 What structures are attached to greater cornua of hyoid?
Upper surface:
- Middle constrictor: Medial.
- Hyoglossus: Lateral.
- Stylohyoid: Lateral to hyoglossus.
- Fibrous pully for digastric tendon.

Medial border: Thyrohyoid membrane.
Lateral border: Thyrohyoid muscle and Investing fascia.

Q.420 What structures are attached to lesser cornua of hyoid?
- Stylohyoid ligament and
- Middle constrictor muscle.

Q.421 What is the developmental origin of hyoid bone?
Upper half of body and lesser cornua: Cartilage of second pharyngeal arch
Lower half of body and greater cornua: Cartilage of third pharyngeal arch.

Cervical Vertebrae (Fig. 6.39)

Q.422 What are the differences between cervical, thoracic and lumbar vertebrae?

	Cervical	*Thoracic*	*Lumbar*
• Foramen trans-versarium	Present	Absent	Absent
• Costal facet	Absent	Present	Absent
• Vertebral body	Oval	Triangular	Oval
Increase in size from above downwards			
• Upper and lower surfaces of vertebral body	Concave	Flat	Flat
• Vertebral foramen	Triangular and large	Small and circular	Triangular
• Pedicles	Long, directed backwards and laterally	Directly and laterally back-wards	Thick and short, directed backwards
• Spinous process	Short and bifid	Long and project down-ward	Large quad-rangular and almost horizontal
• Lamina	Long trans-sversely, and narrow vertically	Short tran-versely Broad verti-cally and overlap	Short and broad but do not overlap
• Transverse process	Short	Large with blut ends	Small with tapering ends
• Facets – Superior facet	Flat Back-ward and upwards	Flat Back-ward, slightly up-wards and laterally	Vertical Backward medially, and has mamillary process
– Inferior facet	Forward and down-ward	Forward slightly down-ward and medially	Forward and laterally
– Articular pillar	Present	Absent	Absent

Contd...

Q.423 What is the structure of intervertebral disc?
Each disc is made up of:
- *Annulus fibrosus:* Outer part. Superficial part is made up of collagen fibers and deeper part by fibrocartilage.
- *Nucleus pulposus:* Inner part. In young, it is soft and gelatinous but is gradually replaced by fibrocartilage. It is remnant of notochord.

Q.424 What are variations in thickness and shape of intervertebral discs in different parts of vertebral column.
- The discs are thinnest in upper thoracic region and thickest in lower lumbar region.
- In cervical and lumbar regions, discs are thicker in front than behind in thoracic region discs are flat.

Q.425 What are functions of the intervertebral disc?
- They transmit weight.
- Act as shock absorbers.
- Provide resilience to spine.
- Constitute one fifth of length of vertebral column.
- Contribute to formation of curves of the spine.

Q.426 Which ligament of spine is made up of elastic tissue?
Ligamentum flavum.

Q.427 Which movements is possible in the thoracic spine?
Rotation. Greater in lower thoracic region as compared to upper thoracic region.

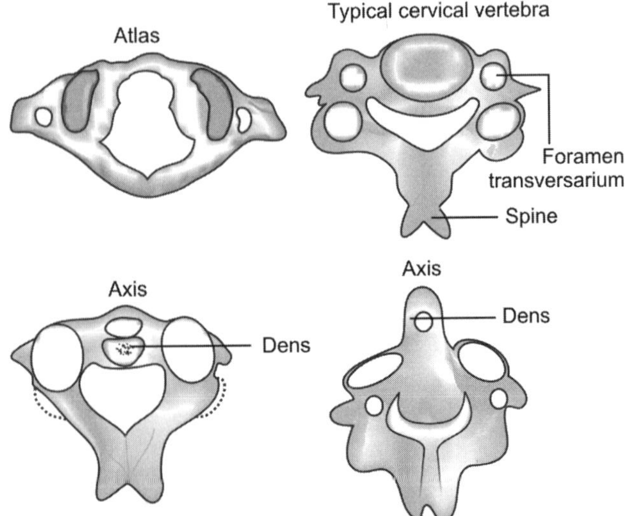

Fig. 6.39: Cervical vertebrae

Q.428 What type of joint is formed between vertebral articular processes?
Synovial joints.

Q.429 Name the structures transmitted by foramen transversarium?
- Vertebral artery,
- Vertebral vein and
- Branch from inferior cervical ganglion.
- In C_7 vertebra, transmits only accessory vertebral vein.

Q.430 Name the structures attached to anterior tubercle?
Origin of:
- Scalenus anterior,
- Longus capitis and
- Oblique part of longus colli.

Q.431 What muscles are arising from posterior tubercle.
- Scalenus medius
- Scalenus posterior
- Levator scapulae
- Splenius cervicis
- Longissimus cervicis
- Iliocostalis cervicis.

Q.432 Name the muscles arising from spine.
- Interspinalis
- Semispinalis thoracis
- Semispinalis cervicis
- Spinalis cervicis
- Multifidus.

Q.433 How will you identify atlas [C_1 vertebra]?
- Ring shaped
- No body
- No spine.

Q.434 Name the ligaments attached to atlas.
- *Anterior longitudinal ligament:* Anterior tubercle
- *Ligamentum nuchae:* Posterior tubercle tip
- *Ligamentum flavum:* Lower border of posterior arch
- *Transverse ligament:* Medial surface of lateral mass.

Q.435 Name the structures related to groove on superior surface over the posterior arch of atlas.
- Vertebral artery
- Vertebral vein plexus
- Plexus of sympathetic nerve fibers
- First cervical nerve.

Q.436 Name the structures passing through spinal canal of atlas.
- Spinal cord,
- Meninges,
- Spinal part of accessory nerve
- Anterior and posterior spinal arteries.

Q.437 Name the ligaments attached to posterior surface of body of axis.
- Posterior longitudinal ligament
- Membrana tectoria
- Vertical limb of cruciate ligament.

Q.438 What is the clinical importance of C_7 vertebra?
Spine of C_7 can be felt through the skin because it is long, thick and horizontal (C_7 is also known as *vertebra prominens). So, it serves as an important anatomical landmark.*

Chapter 7

Central Nervous System

Q.1 What are the divisions of the nervous system?
Anatomically the nervous system is made up of:

- *Central nervous system (CNS):* Consisting of the
  - Brain and
  - Spinal cord
- *Peripheral nervous system (PNS):* Consisting of the nerves and ganglia outside of the brain and spinal cord. It is divided into:
  - Somatic (cerebrospinal) nervous system.
  - Autonomic (splanchnic) nervous system (Figs 7.1 and 7.2).

Q.2 What are the constituents of the somatic nervous system?
It consists of 12 pairs of cranial nerves and 31 pairs of spinal nerves.

Q.3 What are the functions of somatic nervous system?
It is concerned with the response of body to external environment.

Q.4 Name the constituents of 'autonomic nervous system'?
It consists of sympathetic and parasympathetic nervous system.

Q.5 What are the functions of autonomic nervous system?
It is mainly concerned with control of internal environment of body, e.g. regulation of heart, bronchial tree, gut and glands of alimentary tract.

Q.6 What is the main difference between somatic and autonomic nervous system?
The efferent fibers of somatic nervous system reach the effectors without interruption while the efferent fibers of ANS first relays in a ganglion and then postganglionic fibers pass to the effectors.

Spinal Cord

Q.7 What is the extent of spinal cord ?
It extends from the upper border of atlas vertebra to the lower border of L_1 (Fig. 7.3).

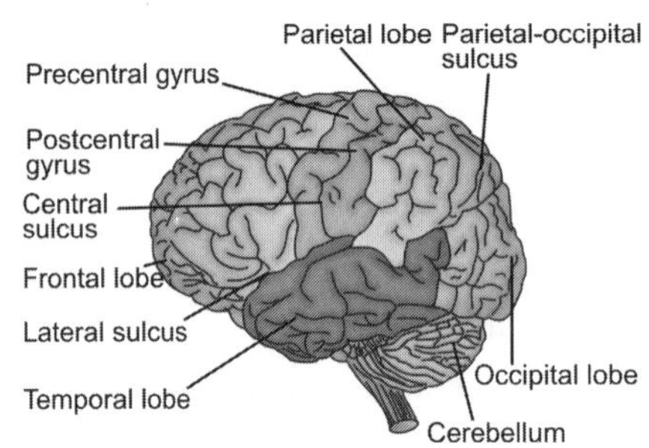

Fig. 7.1: External anatomy of brain

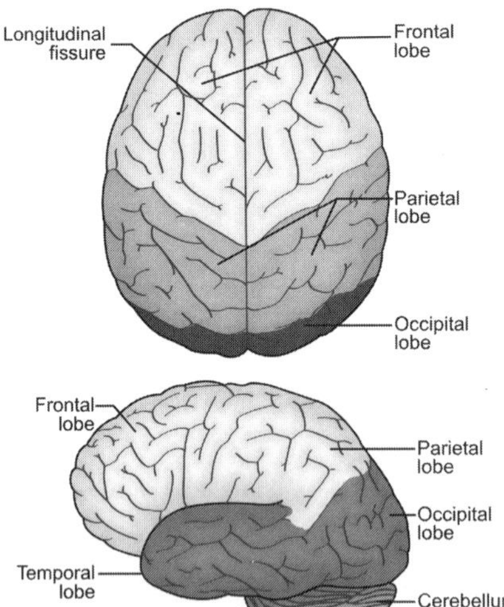

Fig. 7.2: The lobes of human brain

Q.8 How the spinal cord develops?
It develops from caudal tubular part of neural tube, which gradually increases in length.

Q.9 What are the age changes in the length of the spinal cord?
Up to 3rd month of fetal life: Spinal cord occupies full extent of vertebral canal.

At birth: At level of L_3 vertebra.

At adolescence: At level of intervertebral disc between L_1 and L_2 vertebra.

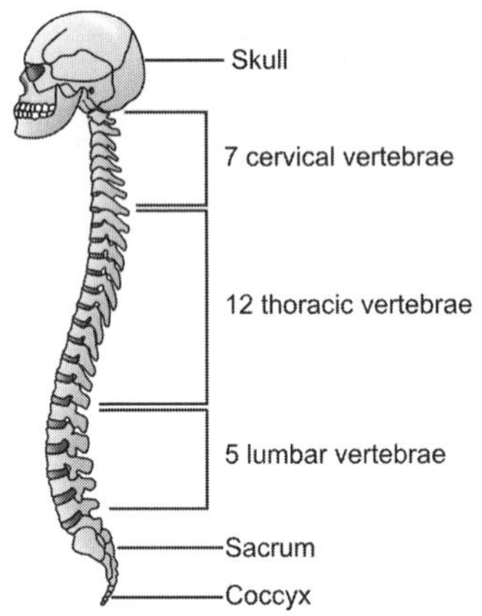

Fig. 7.3: Skull and spine

Q.10 Name the arteries supplying spinal cord?
See Figure 7.4.

- *Anterior spinal artery:* One, in anterior median fissure
- *Posterior spinal artery:* Two, along postero-lateral sulcus, i.e. along the line of attachment of dorsal nerve roots.
- *Arterial vasocorona:* Arterial plexus in pia mater covering the spinal cord.
- *Radicular arteries:* Reach cord along roots of spinal nerves.

Q.11 Which artery supplies the greater part of cross-section of spinal cord?
Anterior spinal artery.

Q.12 What is the venous drainage of spinal cord?
The veins draining spinal cord are arranged in six longitudinal channels. Anteromedian and posteromedian lying in midline and anterolateral and posterolateral that are paired. These are interconnected by a plexus, venous vasocorona, these drain into radicular veins which inturn drain into epidural venous plexus and which drains into external vertebral venous plexus through intervertebral and basivertebral veins.

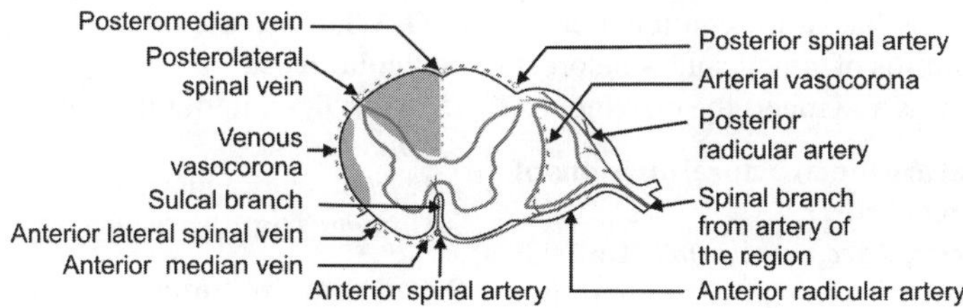

Fig. 7.4: Blood vessels supplying the spinal cord. In the left half of the figure, the area shaded green is supplied by the posterior spinal artery; the areas shaded pink is supplied by the arterial vasocorona; and the area shaded yellow is supplied by the anterior spinal artery

Q.13 What are 'arteries of Adamkiewicz'?
These are the anastomotic arteries between anterior and posterior spinal arteries at the level of T_1 and T_{11}.

Q.14 Name the 'descending' and 'ascending' tracts of spinal cord?
Descending tracts: Motor in function:
- Lateral corticospinal
- Anterior corticospinal
- Rubrospinal
- Vestibulospinal
- Olivospinal
- Tectospinal
- Medial reticulospinal
- Lateral reticulospinal.

Ascending tracts: Sensory in functions:
- Fasciculus gracilis
- Fasciculus cuneatus
- Posterior spinocerebellar
- Anterior spinocerebellar
- Lateral spinothalamic
- Anterior spinothalamic
- Spinotectal
- Spino-olivary.

Q.15 What is the function of interneurons?
- Axon of an interneuron may form number of branches, which synapse with a number of efferent neurons. So, an impulse in single afferent neuron may result in effector response by a number of efferent neurons.
- Afferent impulse from different afferent neurons may converge on single afferent neuron through interneurons. These impulses may be facilitatory or inhibitory.
- Afferent neuron may form contact with efferent neuron in opposite half of spinal cord or in higher or lower segment of cord through interneuron.

Q.16 Trace the pathway of the posterior column ascending tract.
Receptors: Sensory end organs in various tissues. Peripheral process of dorsal root neurons form the afferent fibers of peripheral nerves.

First order neuron: Central processes of neurons in dorsal nerve root ganglia. The fibers ascend in spinal cord as posterior column tracts, up to lower part of medulla and end in nucleus gracilis and nucleus cuneatus.

Second order neuron: Neurons in nucleus gracilis and nucleus cuneatus. The axons cross the midline (sensory decussation) and run upwards as medial lemniscus to end in thalamus, passing through medulla, pons and midbrain.

Third order neuron: Neurons in thalamus. Gives axons to somatosensory area of cerebral cortex passing through internal capsule and corona radiata.

Q.17 What are the tracts of posterior column?
- Fasciculus gracilis and
- Fasciculus cuneatus.

Q.18 What are the sensations carried by the posterior column?
- Deep touch and pressure
- Tactile localization
- Tactile discrimination
- Stereognosis
- Sense of vibration
- Sense of position and movements of different parts of body (proprioceptive impulses).

Q.19 What are the sensations carried by spinothalamic tracts?
- *Anterior spinothalamic tract:* Sensation of crude touch and pressure.
- *Lateral spinothalamic tract:* Sensation of pain and temperature.

Q.20 What are the functions of spinocerebellar tracts?
These carry proprioceptive impulses arising in muscle spindles, Golgi tendon organs and other proprioceptive receptors of lower limbs.

Dorsal tract: Impulses concerned with fine coordination of muscles controlling posture and movements of individual muscles.

Ventral tract: Concerned with movements of limb as a whole.

Q.21 What is the function of various descending spinal tracts?
These influence the activity of ventral column neurons both alpha and gamma, through internuncial neurons, affecting both contraction and tone of skeletal muscle.

They also influence the transmission of afferent impulses through ascending tracts.

Q.22 What are 'ligamenta denticulata'? What is their function?
These are toothed processes extending from pia to dura, pushing the arachnoid before them. They leave the pia midway between anterior and posterior nerve roots and serve to suspend the spinal cord in midline.

Q.23 What is 'conus medullaris'?
It is lower end of spinal cord which is conical. The apex of conus continues downwards as *filum terminale,* up to first coccygeal space.

Q.24 What is 'cauda equina'?
The spinal cord gives rise to spinal nerves which pass out through intervertebral foramina. Below L_1 vertebra, nerve roots become more and more oblique to reach respective intervertebral foramina. The bundle of lumbar and sacral nerve roots below termination of spinal cord is termed cauda equina.

Q.25 What is cauda equina syndrome?
Compression of cauda equina gives rise to flaccid paraplegia, saddle anesthesia, which is known as cauda equina syndrome.

Q.26 What is tabes dorsalis?
It is degenerative disease of posterior columns and posterior nerve roots, which is characterized by loss of proprioception (position sense).

Q.27 What is Brown-Séquard's syndrome?
It occurs in hemisection of spinal cord. It is characterized by:
- Paralysis of affected side below the lesion (corticospinal tract).
- Loss of proprioception and fine discrimination on affected side below lesion (fasciculus cuneatus and gracilis).
- Loss of pain and temperature sense on opposite side below lesion (spinothalamic tract).

Q.28 At what site lumbar puncture is done?
Lumbar puncture is done to withdraw CSF from subarachnoid space at level between L_3 and L_4 vertebra.

Brain

Q.29 What are the different parts of brain?
The brain is divided into three parts:
- Forebrain (prosencephalon)
- Midbrain (mesencephalon)
- Hindbrain (rhombencephalon).

Q.30 How the brain develops?
The brain develops from cranial part of neural tube. The cavity of developing brain shows three dilatations. Craniocaudally, these are prosencephalon, mesencephalon and rhombencephalon (Fig. 7.5).

Forebrain

Q.31 What are the subdivisions of forebrain?
- *Telencephalon:* Made of 2 cerebral hemi-spheres and their cavity, i.e. lateral ventricles.
- *Diencephalon (thalamencephalon):* Made of thalamus, metathalamus, epithalamus and its cavity third ventricle.

Telencephalon

Q.32 What are the different lobes of cerebrum?
- Frontal lobe
- Parietal lobe
- Occipital lobe
- Temporal lobe.

This division is done by:

Three sulci: Central, lateral, occipito-parietal.

Two imaginary lines: One from parieto-occipital sulcus to preoccipital notch and second is backward continuation of posterior ramus of lateral sulcus before it turns upwards and meets the first line.

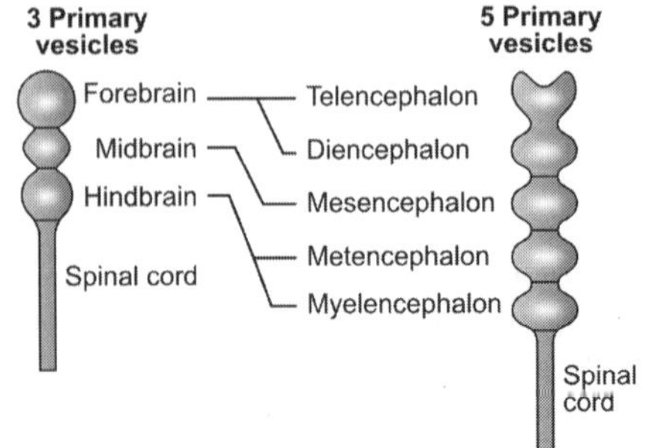

Fig. 7.5: Development of brain

Q.33 What are the structural divisions of the cerebral cortex?
- *Allocortex (archipallium):* Consist of piriform area and hippocampal formation. Made up of 3 layers.
- *Isocortex (neopallium):* Consist of granular and agranular cortex. Made up of 6 layers.

Q.34 What are the functional divisions of the cerebral cortex?
The cortex is divided into motor and sensory areas.

Q.35 What is the motor area of cerebral cortex?
In precentral gyrus on superolateral surface and in anterior part of paracentral lobule on medial surface of cerebral hemisphere. It corresponds to areas 4 of Brodmann.

Q.36 How the lateral and anterior corticospinal tracts are formed?
The corticospinal fibers from the cerebral cortex descend and at lower end of medulla 80% cross to opposite side forming the lateral tract. Fibers which do not cross form the anterior corticospinal tract and at appropriate levels of spinal cord cross to the opposite side. So both tracts ultimately connect cerebral cortex of one side with opposite half of spinal cord ending in ventral grey column neurons.

Q.37 How the body parts are represented in the cerebral cortex?
- The body is represented upside downwards with the legs at top and head at bottom.
- In motor area, angle of mouth, thumb, finger movements are represented by the larger areas.

Q.38 What is Broca's area?
It is motor speech area No. 44, 45 which controls the speech. It lies in inferior frontal gyrus. Lesion of this area produces 'motor aphasia'.

Q.39 What is effect of lesion at area 6, 8 of frontal lobe?
Loss of horizontal conjugate movements of the eyes.

Q.40 What is the 'sensory area'?
It is located in postcentral gyrus and corresponds to areas 1, 2 and 3 of Brodmann.

Q.41 What are the boundaries of interpedicular fossa?
This area lies anterior to midbrain.

Boundaries:
In front: Optic chiasma
On sides: Optic tracts.

Q.42 What are boundaries of anterior perforated substance?
It is a triangular area lying on each side of optic chiasma.

Boundaries:
Anterolateral: Lateral olfactory stria
Posterolateral: Uncus

Q.43 Name the constituents of the limbic system.
- Olfactory nerve, bulb, tract, striae and trigone
- Anterior perforated substance
- Piriform lobe
- Anterior part of parahippocampal and cingulate gyri
- Hippocampal formation
- Amygdaloid nuclei
- Septal region
- Fornix, stria terminalis, stria habenularis.

Q.44 Where the hippocampus is situated?
The hippocampus forms a longitudinal projection occupying greater part of floor of inferior horn of lateral ventricle. It is the superior limb of the 'S' of the cerebral cortex that lies between the choroid fissure above and hippocampal fissure below.

Q.45 What are the communications of lateral ventricle?
Each lateral ventricle communicates with the third ventricle through an interventricular foramen or foramen of Monro.

Q.46 What are the different parts of the lateral ventricle?
Each lateral ventricle is made up of:
- Central part
- Three horns: Anterior, posterior and inferior.

Q.47 What are the constituents of white matter of cerebrum?
It consists of myelinated fibers which connect various parts of cortex and other parts of the CNS.

Q.48 What are the different types of fibers of white matter?
Three types:
- *Association fibers:* Connect different cortical areas of same side.
- *Projection fibers:* Connect cerebral cortex to other part of CNS, e.g. brainstem, spinal cord by various tracts.

- *Commissural fibers:* Connect corresponding parts of two sides.

Q.49 What are the different commissures of cerebrum?
- *Corpus callosum:* Largest, connecting who cerebral hemispheres.
- Anterior commissure
- Posterior commissure
- Commissure of fornix
- Habenular commissure
- Hypothalamic commissure
- Commissures of cerebellum.

Q.50 What are the different parts of corpus callosum?
- *Genu:* Anterior end, connects two frontal lobes by forceps minor fibers.
- *Rostrum:* Connects orbital surfaces of two frontal lobes.
- *Trunk.*
- *Splenium:* Posterior end, thickest. Connects two occipital lobes by forceps major.

Diencephalon

Q.51 Name the parts of diencephalon.
- *Thalamus* (dorsal thalamus)
- *Metathalamus:* Medial and lateral geniculate bodies.
- *Epithalamus:* Pineal body and habenular nuclei.
- Hypothalamus
- *Subthalamus* (ventral thalamus).

Q.52 Name the cavity of diencephalon.
Third ventricle

Q.53 What are the different parts and nuclei of thalamus?
White matter:
- *Stratum zonale:* Covers superior surface
- *External medullary lamina:* Covers lateral surface
- *Internal medullary lamina:* Y-shaped dividing grey matter into medial, lateral and anterior part.

Grey matter:
- Anterior nucleus
- Medial dorsal nucleus
- *Lateral nuclei:* Divided into:
  - Ventral group: Has anterior, intermediate posterolateral and posteromedial nucleus
  - Lateral group: Has dorsal, posterior nucleus and pulvinar
- *Intralaminar nuclei:* In internal medullary lamina. Most important is centromedian nucleus.
- Midline nuclei
- Reticular nucleus.

Q.54 What are the afferents and efferents to the thalamus?
Afferents:
- Cerebral cortex
- Corpus striatum
- Cerebellum
- Reticular formation: Carry visceral impulses.
  - Amygdaloid complex: Carry olfactory impulses
  - Medial lemniscus
  - Spinothalamic tracts
  - Trigeminothalamic tracts. 6, 7, and 8 carry exteroceptive and proprioceptive impulses.
  - Solitariothalamic tract: Taste sensation
  - Hypothalamus

Efferents:
- Cerebral cortex: To sensory area 3, 2, 1
- Corpus striatum
- Reticular formation
- Hypothalamus.

Q.55 What are the functions of thalamus?
- Capable of appreciating painful and thermal stimuli.
- Through RAS participates in maintenance of state of wakefulness and alertness.
- Great sensory relay station on pathway of sensory impulses to cerebral cortex, except for sense of smell, visual and auditory impulses.
- Integration of impulses from sensory system, cerebral cortex, striatum, cerebellum, hypothalamus, reticular formation.
- With hypothalamus and frontal lobe, control emotions and behavior.

Q.56 What is the characteristic feature of structure of lateral geniculate body?
It is six layered. Layers 1,4 and 6 receive contralateral optic fibers and layers 2,3 and 5 ipsilateral fibers.

Q.57 What is pineal body?
It is a small conical body projecting downwards in posterior wall of third ventricle, just above the superior colliculi of midbrain.

Q.58 What is 'brain sand'?
Calcareous concretions appear in pineal body after 17 years of life and form aggregations. These are called brain sand or corpora arenacea. They appear as radio-opaque structures in X-ray.

Q.59 What is the clinical importance of Pineal concretions?
Normally the pineal concretions appear as midline structure in X-ray. They are shifted in cases of head injury.

Q.60 What is the function of pineal body?
- It produces hormone melatonin, synthesized from serotonin.
- It acts as biological clock which produce circadian rhythms in various parameters.
- It's secretion has regulatory influence on pituitary, thyroid, parathyroids, adrenals and gonads.
  Release of pineal secretions need sympathetic stimulation.

Q.61 Which cranial nerve is likely to be paralysed in tumors of pineal body?
It presses on tectum of midbrain damaging the oculomotor nucleus and leading to paralysis of oculomotor nerve.

Q.62 Why hypothalamus is called the 'head ganglion of the autonomic nervous system'?
Because it controls the various autonomic activities of the body. Sympathetic by caudal part and parasympathetic by cranial part.

Q.63 What is the position of hypothalamus?
It is present at base of brain and forms the floor and lateral wall of third ventricle. It is related to:

Anteriorly: Lamina terminalis
Posteriorly: Subthalamus and Tegmentum of midbrain.
Laterally: Internal capsule and Subthalamus.

Q.64 Which structures are related to hypothalamus in floor of third ventricle?
- Tuber cinereum,
- Infundibulum and
- Mammillary bodies.

Q.65 How does preoptic region differs from rest of hypothalamus?
Preoptic region is a derivative of telencephalon.

Q.66 What is function of habenular nuclei?
These are regarded as cell stations in olfactory and visceral pathway.

Q.67 What are the functions of hypothalamus?
- *Endocrine control:* By releasing or release inhibiting hormones, it regulates the functions of various endocrine glands of body.
- *Neurosecretion:* Oxytocin and ADH are secreted by hypothalamo-hypophyseal tract to posterior pituitary.
- Control of sexual behavior and reproduction through anterior pituitary.

- Regulation of food and water intake: Lateral zone is responsible for hunger, thirst and drinking and medial zone for satiety.
- Temperature regulation
- Control of emotional behavior.
- Maintains circadian rhythm of body: By suprachiasmatic nucleus.
- Control of autonomic functions.

Q.68 What are the recesses of third ventricle?

These are the extensions of the cavity of third ventricle. These are:

- Supraspinal
- Pineal
- Infundibular
- Optic

Q.69 Name the structures forming lateral wall of third ventricle.

From above downwards

- Medial surface of thalamus
- Hypothalamic sulcus
- Medial surface of hypothalamus.

Q.70 Name the structures present in floor of third ventricle.

- Optic chiasma
- Tuber cinerium
- Infundibulum
- Mammillary bodies
- Posterior perforated substance and
- Tegmenta of midbrain.

Internal Capsule

Q.71 What are the different parts of internal capsule?

- *Anterior limb:* Between caudate nucleus and lentiform nucleus.
- *Posterior limb:* Between thalamus and lentiform nucleus.
- *Genu:* Bend between two limbs.
- *Retrolentiform part:* Behind lentiform nucleus
- *Sublentiform part:* Below lentiform nucleus.

Q.72 What is the arrangement of corticospinal fibers in posterior limb of internal capsule?

The arrangement of fibers from anterior to posterior is upper limb, trunk and then lower limb.

Q.73 Where is the clinical importance of blood supply of internal capsule?

Lateral striate artery (*Charcot's artery*) supplying internal capsule is the commonest site of hemorrhage in cases of hypertension and it leads to the paralysis of opposite half of body (hemiplegia), depending on which side is involved in hemorrhage.

Basal Ganglia

Q.74 What are morphological divisions of corpus striatum?

The putamen and caudate nucleus form neostriatum, globus pallidus forms paleostriatum and amygdaloid body forms archistriatum.

Q.75 What is the function of corpus striatum?

It is an important integrating center in motor activity.

Midbrain

Q.76 What are the subdivisions of midbrain?

- Crus cerebri,
- Substantia nigra,
- Tegmentum and
- Tectum and its cavity, cerebral aqueduct.

Q.77 What is tectum?

It is the posterior part of midbrain. It is made up of 4 colliculi, a pair of superior and a pair of inferior.

Q.78 What are the characteristic features of substantia nigra?

It is a lamina of grey matter, made of deeply pigmented nerve cells.

Afferents are from motor cortex and collaterals of sensory tracts.

Efferents pass to corpus striatum and tegmentum.

Q.79 What are the contents of crus cerebri?

- *Middle 2/3:* Pyramidal tract
- *Medial 1/6:* Frontopontine fibers
- *Lateral 1/6:* Temporopontine, parietopontine and occipitopontine fibers.

Q.80 What are the connections and functions of superior colliculus?

Connections:

Afferents: From retina (visual),
Spinal cord (tactile),
Inferior colliculus (auditory),
Occipital cortex (modulating).

Efferents: To retina,
Spinal cord (Tectospinal),
Brain stem nuclei, Tegmentum.

Function: Control reflex movements of eyes, head and neck in response to visual stimuli.

Hindbrain

Q.81 What are the subdivisions of hindbrain?

- *Metencephalon,* made up of pons and cerebellum.
- *Myelencephalon,* made up of medulla oblongata. Fourth ventricle is the cavity of hindbrain.

Q.82 What are the constituents of brainstem?

- Midbrain,
- Pons and
- Medulla.

Q.83 Which cranial nerves are attached to brainstem?

- Third and fourth nerves emerge from surface of midbrain.
- Fifth nerves emerges from pons.
- Sixth, seventh and eighth nerves emerge at junction of pons and medulla.
- Ninth, tenth, eleventh and twelfth nerves emerge from surface of medulla.

Q.84 Name the cranial nerve nuclei in pons.

- Skin nerve nucleus
- Seventh nerve nucleus
- Vestibular and cochlear nuclei
- Main sensory nucleus, motor nucleus and spinal nucleus of trigeminal nerve.

Q.85 What are the connections and functions of reticular formation of brainstem?

It is connected to all parts of nervous system directly or indirectly. It receives impulses from motor and other areas of cerebral cortex and relays them to spinal cord by lateral and medial reticulospinal tract. It is also connected to cerebellum and thalamus. Fibers to thalamus constitute ascending reticular activating system.

Functions:

- Involved in fine control of movements.
- Influences conduction through somatosensory, visual and auditory pathway.
- Regulation of respiratory and cardiovascular control.
- Controls activity of adenohypophysis and neurohypophysis through hypothalamus.
- Control of pineal body.
- Through ascending reticular activating system maintains a state of alertness.

Q.86 What is medial longitudinal bundle?

Association tract, which coordinates movements of eyes, head and neck in response to stimulation of 8th cranial nerve.

The nuclei of 3rd, 4th, 5th, 6th and 11th cranial nerve are interconnected by the bundle.

Q.87 What is the effect of unilateral lesion in lower part of pons?
Crossed or alternate hemiplegia, i.e. paralysis of face on one side and limbs on the other side.

Q.88 Name the lobes of cerebellum.
- Anterior lobe,
- Middle lobe and
- Flocculonodular lobe.

Q.89 What is vermis and what are its parts?
It joins the two cerebellar hemispheres.
Parts:
- Lingula
- Central lobule — In anterior lobe
- Culmen
- Declive
- Folium
- Tuber
- Pyramid — In middle lobe
- Uvula
- Nodule: In flocculonodular lobe.

Q.90 What are phylogenetic divisions of cerebellum?
Archicerebellum: Oldest.
Flocculonodular lobe and lingula.
Paleocerebellum: Anterior lobe minus lingula.
Pyramid and uvula of middle lobe
Neocerebellum: Middle lobe minus pyramid and uvula.

Q.91 Name the contents of superior cerebellar peduncle.
Afferent tracts (fibers entering the cerebellum):
Anterior spinocerebellar
Superior spinocerebellar
Efferent tracts (fibers leaving the cerebellum):
Cerebellorubral
Cerebellothalamic
Cerebelloreticular (partly from densate nucleus and partly from fastigial nucleus).

Q.92 Name the contents of inferior cerebellar peduncle.
Afferent:
- Posterior pontocerebellar
- Cuneocerebellar
- Olivocerebellar
- Parolivocerebellar
- Reticulocerebellar
- Vestibulocerebellar
- Anterior external arcuate

Efferent:
- Cerebellovestibular
- Cerebelloolivary
- Cerebelloreticular.
- Some cerebellospinal and cerebello-nuclear.

Q.93 What are contents of middle cerebellar peduncle?
Afferent: Pontocerebellar.

Q.94 Name the cranial nerve nuclei in floor of fourth ventricle.
- Hypoglossal nucleus
- Dorsal nucleus of vagus
- Nucleus of tractus solitarius
- Inferior and medial vestibular nuclei.

Q.95 How the lateral wall of fourth ventricle is formed?
- *Upper part:* Superior cerebellar peduncle.
- *Lower part:* Inferior cerebellar peduncle and Gracile and cuneate tubercles.

Q.96 How the floor of fourth ventricle is formed?
By
- Posterior surface of pons
- Posterior surface of upper part of medulla.

Q.97 What is the effect of lesion in medulla oblongata?
The medulla contains the vital centers, i.e. respiratory, cardiac and vasomotor center. The lesion in medulla will lead to the failure of vital functions especially, respiratory failure.

Blood Supply of Brain

Q.98 Name the arteries supplying the brain.
- Internal carotid arteries and its branches.
- *Vertebral arteries:* At lower border of pons join to form *basilar artery.*

Q.99 What is artery of Heubner?
Recurrent branch of anterior cerebral artery which supplies anterior perforated substance.

Q.100 What are the branches of basilar artery?
- Superior cerebellar artery
- Anterior inferior cerebellar artery
- Pontine branches
- Labyrinthine artery.

Q.101 Why the macular vision is often spared in thrombosis of posterior cerebral artery?
Because part of visual area responsible for macular vision lies in the region where area of distribution of middle and posterior cerebral artery meet and this area may receive supply directly from middle cerebral artery or through anastomoses with branches of posterior cerebral artery.

Q.102 Which arterior of the brain are end arteries?
Long and short perpendicular branches of cortical arteries are end arteries.

Q.103 What is the arterial supply of cerebellum?
- Superior cerebellar artery
- Anterior inferior cerebellar : Arises from basilar artery
- Posterior inferior cerebellar: Branch of vertebral artery.

Q.104 What is 'circle of Willis' (Fig. 7.6)?
It is an arterial circle formed at the base of brain by interconnections between the main arteries supplying brain (Fig. 7.6).

Q.105 What is clinical importance of 'circle of Willis'.
- It helps in equalizing pressure in arteries of two sides.
- It also helps in maintaining blood supply to different parts of brain, if the main artery of one side is obstructed.

Q.106 Name the arteries forming 'circle of Willis'?
Anteriorly: Anterior communicating artery, joining two anterior cerebral arteries.
Posteriorly: Basilar artery as it divides into two posterior cerebral arteries.
On each side: Anterior cerebral, internal carotid, posterior communicating and posterior cerebral arteries.

Q.107 Name the structures forming the blood-brain barrier.
- Capillary endothelium and its basement membrane
- Arachnoid layer of perivascular sheath
- Perivascular space
- Pial layer of perivascular sheath
- Neuroglia and ground substance of brain.

Q.108 Thrombosis of central branches of cerebral arteries result in infarction. Why?
Because they are end arteries.

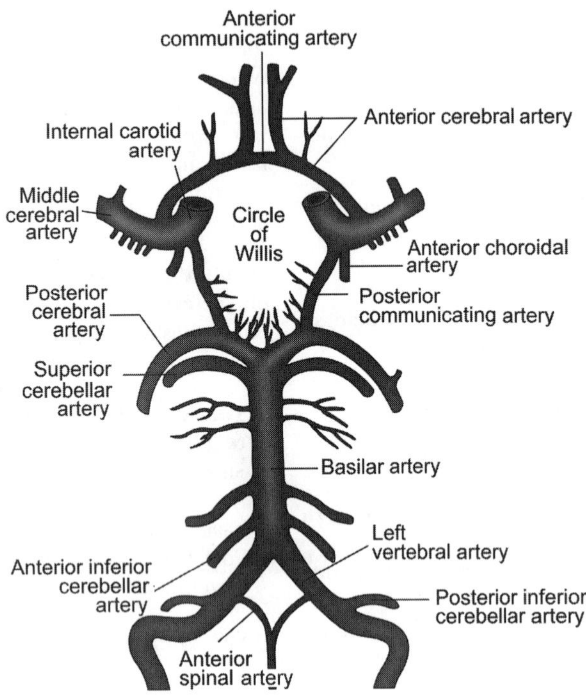

Fig. 7.6: Circulus arteriosus (circle of Willis)

Q.109 What is the arterial supply of cerebral cortex?
- Cortical branches of anterior, middle and posterior cerebral arteries.
- Motor area by anterior and middle cerebral artery.
- Auditory area and speech area by middle cerebral artery.
- Visual area by posterior cerebral artery.

Q.110 What are the characteristics of veins supplying the cerebrum?
- Vessel walls are devoid of muscle
- No valves are present
- Some open into cranial venous sinuses against direction of blood flow in sinus.

Q.111 How the basal vein is formed?
By union of anterior cerebral vein, deep middle cerebral veins and some inferior striate vein.

Q.112 How great cerebral vein is formed and terminates?
Formed by union of two internal cerebral veins and ends in straight sinus.

Meninges and Cerebro-Spinal Fluid (CSF)

Q.113 What are meninges?
These are the layers of connective tissue covering the brain and spinal cord.

Q.114 What are the layers of the meninges?
The meninges consist of three membranous layers.
- *Dura mater:* Outer most
- *Arachnoid:* Middle
- *Pia mater:* Inner most.

The dura mater is also known as 'pachymeninges'. The arachnoid and pia mater are together known as 'leptomeninges'. The subarachnoid space between arachnoid and pia mater contains cerebrospinal fluid.

Q.115 What is the developmental origin of meninges?
Leptomeninges: From neural crest.
Pachymeninges: From mesoderm surrounding neural tube.

Q.116 What are the layers of the dura mater?
Dura mater is the thickest and toughest membrane covering the brain and consists of two layers:
- *Endosteal layer:* Outer. Serves as internal periosteum (endocranium).
- *Meningeal layer:* Inner. Provides the protective membrane to brain.

These two layers are fused to each other except where venous sinuses are enclosed between them.

Q.117 To what structures the endosteal layer is attached.
It is attached to:
- Inner surface of cranial bones by fibrous and vascular processes.
- To pericranium through sutures and foramina.
- To periosteal lining of orbit through the superior orbital fissure.

Q.118 What are the structures covered by the endosteal layer other than brain?
It provides tubular sheaths for cranial nerves and fuses with epineurium except optic nerve, where sheaths are derived from meninges.

Q.119 What are meningocytes and what is their function?
Meningocytes are mesothelial cells found in:
- Fibrous tissue of dura
- Arachnoid sheath enveloping posterior root ganglia of spinal cord.
- Arachnoid sheath covering the stalk of choroid plexus of lateral ventricle.
- Lying free in cerebrospinal spaces.

Functions:
- Excretion of CSF into cerebral sinuses.
- Phagocytosis of foreign particles.
- Repair of dural defects.
- Production of bile pigments.

Q.120 Name the coverings of spinal cord.
- *Spinal dura mater:* Represents meningeal layer of cerebral dura mater
- Arachnoid mater
- Spinal pia mater.

Q.121 What are the folds of dura mater?
These are formed by the meningeal layer of the dura mater around brain. These projects inwards and divide the cranial cavity into different compartments.
These are:
- Falx cerebri
- Tentorium cerebelli
- Falx cerebelli
- Diaphragma sellae.

Q.122 What is the 'falx cerebri'?
It is a fold of dura mater, which is sickle shaped and occupies the median longitudinal fissure between two cerebral hemispheres.

Q.123 What are the venous sinuses enclosed by the falx cerebri?
- The upper convex margin encloses the, superior sagittal sinus.
- The lower concave free margin encloses, inferior sagittal sinus (Fig. 7.7).

Q.124 What is 'tentorium cerebelli'?
It is a tent shaped fold of dura mater, forming roof of the posterior cranial fossa. It separates cerebellum from the occipital lobes of the cerebrum. It lies at right angles to falx cerebri.

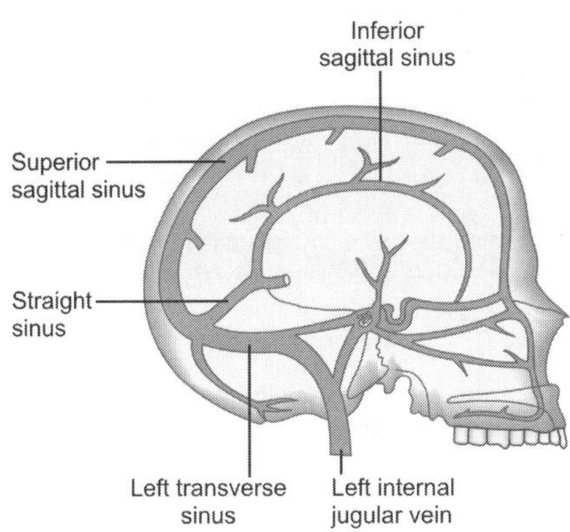

Fig. 7.7: Venous sinuses of brain

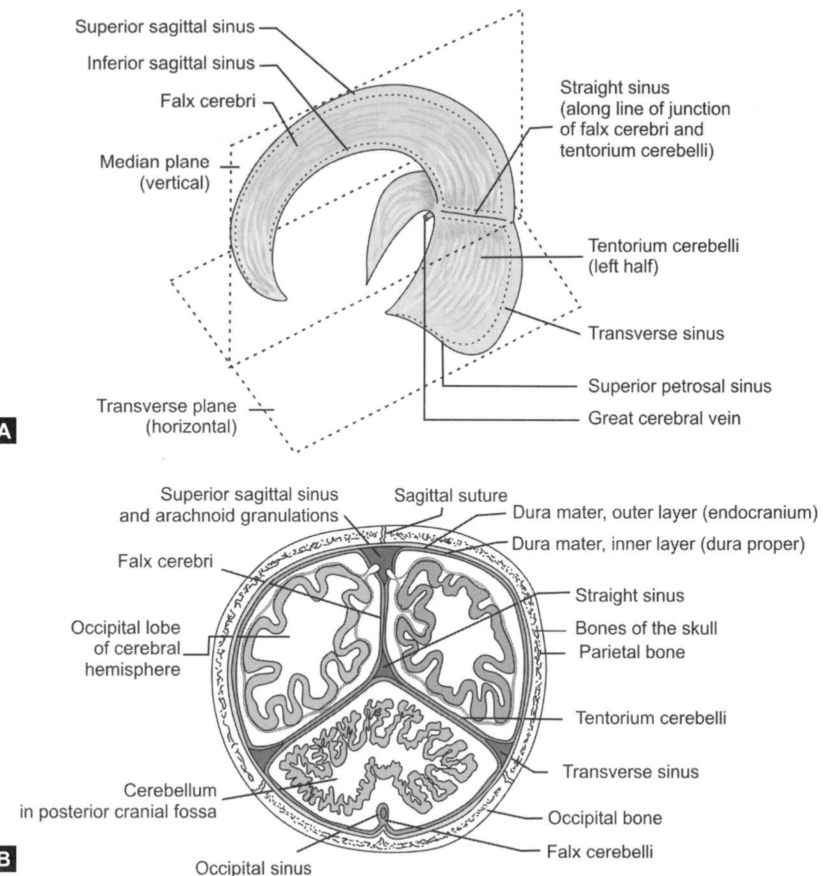

Figs 7.8A and B: (A) Scheme to show the orientation of the falx cerebri and tentorium cerebelli. Note the related venous sinuses, (B) Coronal section through posterior part of skull to show the relationship of the falx cerebri and tentorium cerebelli to each other and to the venous sinuses of the region

Q.125 What are sinuses enclosed by the tentorium cerebelli?

The attached margin encloses the transverse sinus in posterior part and superior petrosal sinus in anterolateral part (Figs 7.8A and B).

Q.126 Where the straight sinus is situated?

At the junction of lower edge of falx cerebri with upper surface of tentorium cerebelli. It lies between left and right layers of falx cerebri which become continuous with corresponding half of upper layer of tentorium cerebelli. The lower layer of tentorium cerebelli passes transversely across midline without interruption.

Q.127 How the straight sinus is formed and terminates?

Anteriorly it receives inferior sagittal sinus and great cerebral vein and posteriorly it terminates by becoming continuous with transverse sinus of side opposite to that with which the superior sagittal sinus is continuous, usually left side.

Q.128 What is 'trigeminal cave'?

Trigeminal or Meckel's cave is recess of dura mater, formed by the inferior layer of tentorium cerebelli, over the trigeminal ganglion, on anterior surface of petrous temporal bone.

Q.129 What is 'flax cerebelli'?

It is small sickle shaped fold in sagittal plane projecting forwards into posterior cerebellar notch (Fig. 7.9).

Q.130 Which sinus is enclosed by falx cerebelli?

Occipital sinus, lies along posteriorly attached part.

Q.131 What is 'diaphragma sellae'?

It is circular horizontal fold of dura mater forming the roof of hypophyseal fossa, in middle cranial fossa.

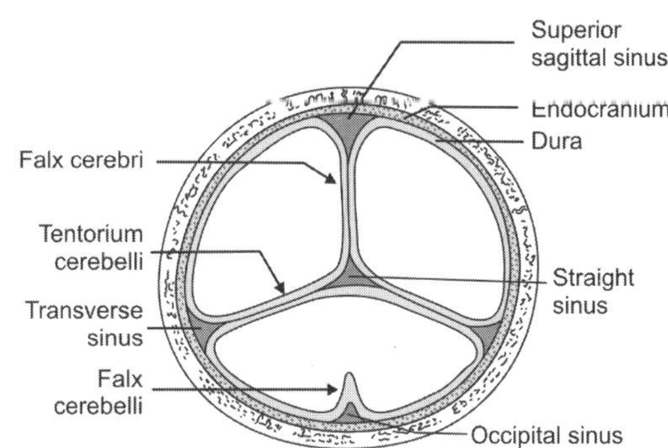

Fig. 7.9: Coronal section through posterior part of falx cerebri, and tentorium cerebelli. The falx cerebelli is also shown

Q.132 What is the structure transmitted by central aperture of diaphragma sellae?

Pituitary stalk (infundibulum).

Q.133 What are the characteristics of venous sinuses of the skull?

- They lie between 2 layers of the dura mater.
- They are lined by endothelium only. Muscular coat is absent.
- The receive:
  - Venous blood and
  - CSF
- No valves are present
- Blood flow is regulated by the emissary veins.

Q.134 What are different venous sinuses of the skull?

The venous sinuses of skull can be divided into two broad groups:

- *Paired:*
  - Cavernous
  - Superior petrosal
  - Inferior petrosal
  - Transverse
  - Sigmoid
  - Sphenoparietal
  - Petrosquamous and
  - Middle meningeal sinus
- *Unpaired:*
  - Superior sagittal
  - Inferior sagittal
  - Straight
  - Occipital
  - Anterior intercavernous
  - Posterior intercavernous
  - Basilar plexus veins.

Q.135 Where is the cavernous sinus is situated?

It is situated in middle cranial fossa on either side of body of sphenoid bone.

Q.136 What are the relations of cavernous sinus?

The relations can be divided into 3 broad subdivisions (Fig. 7.10):

- Structures in lateral wall of sinus:
  - Oculomotor nerve
  - Trochlear nerve
  - Ophthalmic nerve
  - Maxillary nerve
  - Trigeminal ganglion.
- Structures lying outside the sinus:

Superiorly: Optic tract
Internal carotid artery
Anterior perforated substance.

Inferiorly: Foramen lacerum
Junction of body and greater wing of sphenoid.

Medially: Hypophysis (pituitary gland)
Sphenoidal air sinus.

Laterally: Temporal lobe with uncus.

Anteriorly: Superior orbital fissure
Apex of orbit.

Posteriorly: Apex of petrous temporal
Crus cerebri of midbrain.

- Structures passing through the centre of sinus:
- Internal carotid artery
- Abducent nerve.

Q.137 Name the communications of cavernous sinus.

These are:

- Into transverse sinus through superior petrosal sinus.
- Into internal jugular vein through inferior petrosal sinus and venous plexus around internal carotid artery.
- Into pterygoid plexus of veins through emissary veins.
- Into facial vein through superior ophthalmic vein.
- Communication between two sinuses by anterior and posterior intercavernous sinuses and basilar plexus of veins.

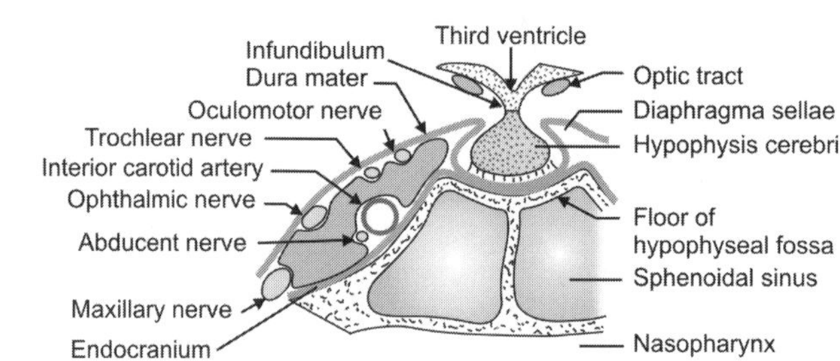

Fig. 7.10: Coronal section through the cavernous sinus showing the internal carotid artery and related structures

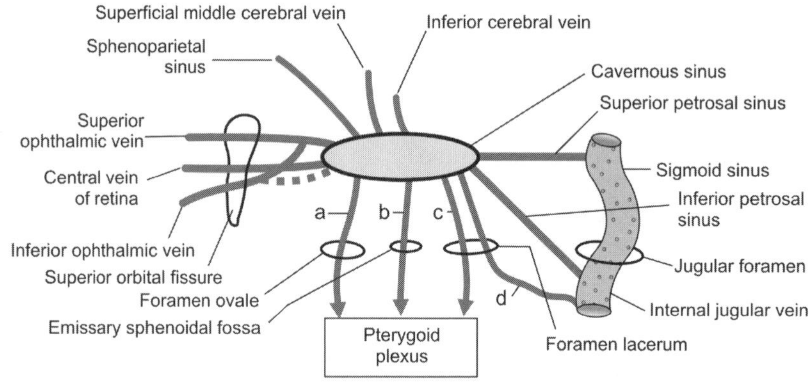

Fig. 7.11: Scheme to show the tributaries of the cavernous sinus. ***a, b, c,*** and ***d*** are emissary veins

Q.138 What are the tributaries of cavernous sinus?

- *From meninges:*
  - Sphenoparietal sinus
  - Frontal trunk of middle meningeal vein.
- *From brain:*
  - Superior middle cerebral vein
  - Inferior cerebral veins.
- *From orbit:*
  - Superior ophthalmic vein.
  - Inferior ophthalmic vein.
  - Central vein of retina (Fig. 7.11).

Q.139 What is the commonest cause of thrombosis of cavernous sinus?

Infection of the danger area of face, nasal cavities and paranasal air sinuses.

Q.140 What do you understand by term 'confluence of sinuses'?

This is the posterior dilated end of superior sagittal sinus lying on right side of internal occipital protuberance. It continues as corresponding transverse sinus and it is connected to opposite transverse sinus and straight sinus and drains the occipital sinus.

Q.141 What is the characteristic feature of pia mater?

It is a highly vascular layer and is closely adherent to brain extending into the sulci, but the arachnoid mater does not do so and jumps across the sulci. So the subarachnoid space extends into the sulci.

Q.142 What are arachnoid villi?

These are the finger like processes of the arachnoid tissue which project into cranial venous sinuses. Their function is to absorb CSF into bloodstream.

Q.143 What are Pacchionian bodies?

Also called arachnoid granulations. These are aggregations of arachnoid villi clumped together. Found in adults. They are most numerous in relation to superior sagittal sinus.

Q.144 What is 'tela chorioidea' and 'choroid plexuses'? What is their importance?

The folds of highly vascular pia mater projecting into ventricles are tela chorioidea.

Cavity of ventricle is lined by ependyma. The masses of pia mater covered by ependyma are referred to as choroid plexuses.

Importance: At these sites the CSF is secreted into ventricles of brain.

Q.145 What are 'Cisterns'? What is their functions?

These are communicating pools formed by the subarachnoid space at base of brain and around the brainstem.

Function: These reinforce the protective effect on the vital centers in the medulla.

Q.146 What are the communications of subarachnoid space?

It communicates with ventricular system of brain by:

- *Foramen of Magendie:* Median, single
- *Foramen of Luschka:* Lateral, two.

Q.147 What is CSF?

It is a clear fluid found in subarachnoid space of brain and spinal cord, ventricular system of brain and central canal of spinal cord.

Q.148 Where is CSF formed?

It is formed by choroid plexuses of ventricles of brain by an active secretory process.

Q.149 How a sample of CSF obtained?

- Lumbar puncture.
- Cisternal puncture.
- Ventricular puncture.

Q.150 What is hydrocephalous?

It is the dilatation of the ventricular system and enlargement of head due to obstruction of flow of CSF within ventricular system in children.

Q.151 What is Queckenstedt's test?

Done to detect whether there is a blockade to the circulation of CSF in subarachnoid space of spinal cord.

Anatomical basis of test: Any increase in intracranial pressure, raises the pressure of CSF. This increase is transmitted to CSF in spinal subarachnoid space.

Compression of both internal jugular veins above the sternal ends of clavicles dams back blood in skull and so raises the intracranial pressure. Should a part of spinal subarachnoid space be completely cut off from above by a tumor, this increase of pressure will not be transmitted to the part of subarachnoid space below tumor.

Q.152 What is the commonest cause of extradural hemorrhage?

Rupture of anterior division of middle meningeal artery.

Q.153 What is the commonest cause of subdural hemorrhage?

Rupture of superior cerebral vein at its entry into superior sagittal sinus.

Cranial Nerves (Fig. 7.12)

Q.154 Name the cranial nerves.

There are 12 pairs of cranial nerves:

- Olfactory
- Optic
- Oculomotor
- Trochlear
- Trigeminal
- Abducent
- Facial
- Vestibulocochlear (auditory)
- Glossopharyngeal
- Vagus
- Accessory and
- Hypoglossal.

Q.155 How the cranial nerves are classified?

- *Purely sensory:* I, II and VIII
- *Purely motor:* III, IV, VI and XII
- *Mixed:* V, VII, IX, X and XI.

Olfactory Nerve

Q.156 Trace the pathway of olfactory nerve.

Consists of two neurons:

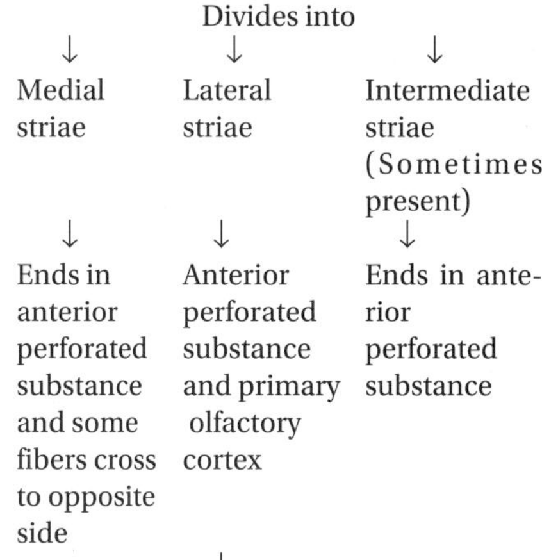

Q.157 What is the characteristic feature of olfactory nerve?

The fibers of olfactory nerve are central process of olfactory cells and not peripheral process of central ganglion cells.

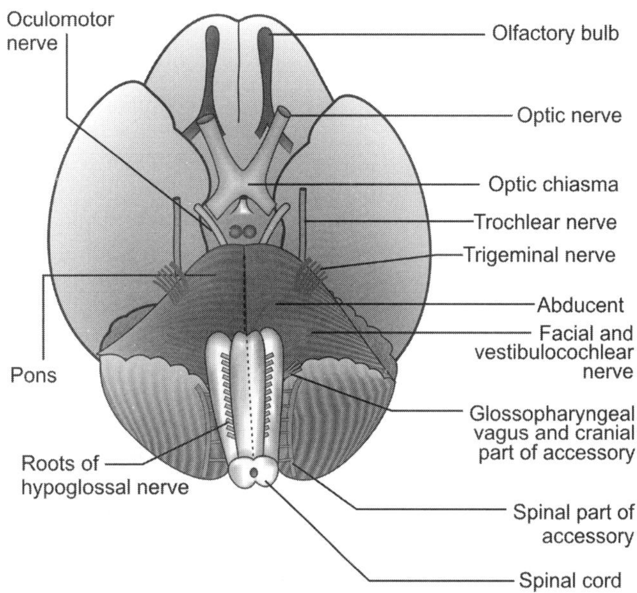

Fig. 7.12: Cranial nerves

Q.158 What is cacosmia?
It is a condition in which a person imagines of non-existent odors.

Q.159 What is the cause of unilateral anosmia (loss of sensation of smell)?
Frontal lobe tumor.

Q.160 What is the cause of bilateral anosmia?
Head injury leading to damage to both olfactory nerves.

Optic Nerve

Q.161 What is the length of optic nerve?
40 mm horizontally and 25 mm vertically is in orbit, 5 mm in optic canal and 10 mm in cranial cavity.

Q.162 What are the relations of intraorbital part of optic nerve?
Intraorbital part:
- Surrounded by four recti
- Ciliary ganglion: Lateral
- Ophthalmic artery: Inferolateral in posterior part and then crosses above the nerve from lateral to medial side
- Nasociliary nerve: Crosses from medial to lateral side above the nerve.
- Branch of oculomotor nerve to medial rectus: Crosses from medial to lateral side below the nerve.
- Central artery of retina: Below.

Q.163 Trace the optic pathway.
Axons of ganglion cells of retina
↓
Optic nerve
Enters through optic canal
↓
Optic chiasma
(Decussation of fibers occur)
↓
Optic tract
(Has fibers from nasal half of macula and retina of opposite side and temporal half of same side)
↓

Lateral Root	Medial Root
↓	↓
Terminates in lateral geniculate body	Terminates in superior colliculus, pretectal nucleus and
↓	
Optic radiation	Hypothalamus

↓
Pass through retrolentiform part of internal capsule
↓
Visual area of cerebral cortex No. 17, 18,19

Q.164 Trace the pathway of light reflex.
Retina
↓
Optic nerve
↓
Optic chiasma
↓
Optic tract
↓
Lateral geniculate body and pretectal nucleus
↓
Edinger-Westphal nucleus of III cranial nerve
↓
III cranial nerve
↓
Ciliary ganglion
↓
Short ciliary nerve
↓
Constrictor pupillae muscle

Q.165 What is consensual light reflex?
Constriction of pupil of other eye when the light is flashed on one eye.

Q.166 Why does the consensual light reflex occurs?
- Fibers of each optic nerve enter both optic tracts as a result of partial crossing in chiasma.
- Fibers from each optic tract end in both pretectal nuclei.
- Fibers from each pretectal nucleus end in both Edinger-Westphal nucleus.

Q.167 What is the pathway for accommodation reflex?

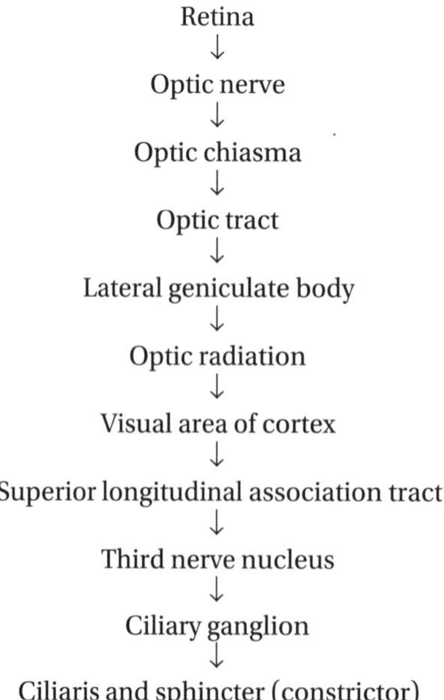

Q.168 What are the characteristic features of optic nerve?
- It is not a true cranial nerve but is brain tract which has developed as a lateral diverticulum of forebrain.
- It is incapable of regeneration after section because it lacks neurilemmal sheath.
- Nerve is enclosed in all the three meningeal sheaths.
- Myelin sheaths is formed by glial cells as in brain and not by Schwann cells.

Q.169 How the fibers from optic tract terminate in lateral geniculate body?
Fibers from same eye end in laminae 2,3 and 5 and from opposite eye end in laminae 1, 4 and 6 of lateral geniculate body. Macular fibers end in central and posterior part.

Q.170 What is Argyll-Robertson pupil?
It is a condition in which pupillary light reflex is absent but the accommodation reflex is present. It is caused by cerebral syphilis. Lesion is in pretectal nuclei.

Q.171 What are the effects of lesions of different parts of visual pathway?

Site	*Effect*
Retina	Scotoma (loss of corresponding field)
Optic nerve	Blindness of same side Consensual light reflex retained
Optic chiasma	
Peripheral lesion:	Binasal hemianopia (Bilateral)
Central lesion	Bitemporal hemianopia.
Optic tract, lateral geniculate body, optic radiation	Homonymous hemianopia (Loss of temporal field of one side and nasal field of other side). No macular sparing.
Visual cortex	Homonymous hemianopia. Macular sparing.

Q.172 Trace the pathway for corneal reflex.

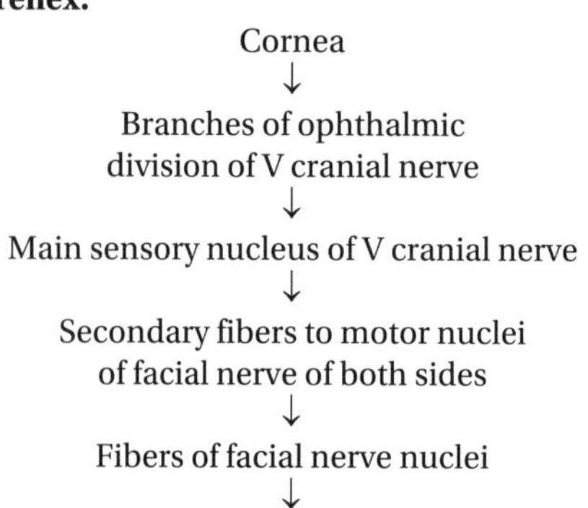

Oculomotor Nerve

Q.173 What are the functional components of oculomotor nerve?

- *General visceral efferents (parasympathetic):* For constriction of pupil and accommodation.
- *Somatic efferent:* For movements of eyeball.
- *General somatic afferent:* For proprioceptive impulses from muscles of eyeball.

Q.174 What is the position, subdivisions and structures supplied by nerves of oculomotor nucleus?

Position: At level of superior colliculus in ventromedial part of central grey matter of midbrain, ventral to aqueduct. The right and left nuclei fuse to form a midline complex.

Subdivisions:

- *Edinger-Westphal nucleus:* For ciliaris and sphincter pupillae muscle in a ciliary ganglion.
- *Ventromedial nucleus:* For superior rectus of both sides.
- *Dorsolateral nucleus:* For inferior rectus of same side.
- *Intermediate nucleus:* For inferior oblique of same side.
- *Ventral nucleus:* For medial rectus of same side.
- *Caudal central nucleus:*For levator palpebrae superioris of both sides.

Q.175 Name the connections of oculomotor nucleus.

To:

- Pretectal nuclei of both sides.
- Pyramidal tracts of both sides.
- IV, VI and VIII nerve nuclei.
- Tectobulbar tract.

Q.176 What are the relations of oculomotor nerve in superior orbital fissure?

Nasociliary nerve lies in between and abducent nerve inferolateral to, the two rami of oculomotor nerve.

Q.177 What is ciliary ganglion and what is its position, connections and branches?

It is a peripheral ganglion in course of oculomotor nerve. Has preganglionic fibers from Edinger-Westphal nucleus.

Position: Near apex of orbit between optic nerve and tendon of lateral rectus muscle.

Connections:

- *Motor root:* From nerve to inferior oblique.
- *Sensory root:* From nasociliary nerve.
- *Sympathetic root:* Branch from internal carotid plexus.

Branches: Short ciliary nerves 8–10 pierce sclera.

Q.178 What is Weber's syndrome?

It is a midbrain lesion causing:

- Paralysis of 3rd cranial nerve of same side.
- Hemiplegia of opposite side.

Q.179 What are the effects of infranuclear lesion of 3rd cranial nerve?

- Ptosis (drooping of upper eyelid).
- Lateral squint (Outward deviation of eye ball by lateral rectus and downwards by superior oblique).
- Mydriasis (dilatation of pupil).
- Cycloplegia (loss of accommodation).
- Proptosis (abnormal protrusion of the eyeball).
- Diplopia (double vision).
- Loss of light reflex and accommodation reflex.

Trochlear Nerve

Q.180 Name the functional components of IV cranial nerve.

- *Somatic efferent:* For movement of eyeball.
- *General somatic afferent:* For proprioceptive impulses from superior oblique muscle.

Q.181 What is the position of trochlear nucleus?

In ventromedial part of central grey matter of midbrain at the level of inferior colliculus, ventral to aqueduct. Fibers from nucleus cross and emerge on posterior surface of brainstem just below inferior colliculus.

Q.182 What is the effect of lesion of IV cranial nerve?

Diplopia occurs on looking downwards.

Trigeminal Nerve

Q.183 What are the functional components of trigeminal nerve?

- *General somatic afferent:* From skin and mucosa and proprioceptive from muscle.
- *Special visceral efferent:* Supplies muscles derived from mesoderm of first branchial arch.

Q.184 What is the position of trigeminal nerve nucleus?

It is made up of:

- *Main sensory nucleus:* In upper part of pons.
- *Spinal nucleus:* Extends from pons down into the upper two segments of spinal cord.
- *Mesencephalic nucleus:* Extends from upper end of main nucleus into midbrain.
- *Motor nucleus:* In dorsal part of upper pons.

Q.185 What is the position of trigeminal ganglion?

The ganglion is placed in depression called trigeminal impression in anterior aspect of petrous temporal bone and is enclosed in pouch like recess of dura mater.

Q.186 What are the divisions of trigeminal nerve?

- *Ophthalmic nerve:* Sensory
- *Maxillary nerve:* Sensory
- *Mandibular nerve:* Mixed

Q.187 What is the distribution of trigeminal nerve?

- *Motor:* Muscles of mastication.
- *Sensory:*
  - Skin of head and face
  - Mucous membrane of mouth, nose and paranasal air sinuses.

Q.188 Name the divisions of ophthalmic nerve and structures supplied by it.

- *Frontal nerve:* By supratrochlear and supraorbital divisions supply upper eyelid, scalp up to lambdoid suture and skin of forehead in lower and medial part.
- *Lacrimal nerve:* To lacrimal gland and lateral part of conjunctiva and skin of upper eyelid.
- *Nasociliary nerve:* Eyeball, to ciliary ganglion, medial half of lower eyelid, mucosa and skin of nose and dura of anterior cranial fossa.

Q.189 Name the divisions of maxillary nerve and its distribution.

- *Zygomatic nerve:* Zygomatico-temporal and zygomatico-facial branches supply skin of temple and cheek.
- *Superior alveolar nerves:* Teeth of upper jaw.
- *Greater and lesser palatine nerves:* Mucous membrane of hard and soft palates and tonsil.
- *Nasal branch:* Mucous membrane of nose.
- *Sphenopalatine branch:* Nasal septum.
- *Pharyngeal branch:* Mucosa of nasopharynx.
- *Meningeal branch:* Dura mater of middle cranial fossa.

- *Palpebral branch:* Lower eyelid
- *Nasal branch:* Skin on lateral side of nose
- *Superior labial:* Skin of upper lip and part of the cheek.

Q.190 What is the distribution of mandibular nerve?

- Before division to anterior and posterior trunk:
  - Nerve to medial pterygoid: Supplies medial pterygoid muscle and gives a branch to optic ganglion.
  - Nerve to tensor palati and tensor tympani.
  - *Meningeal branch:* To dura mater of middle cranial fossa.
- Anterior trunk:
  - *Buccal nerve:* Skin of cheek and mucous membrane on its inner aspect.
  - Nerve to masseter, temporalis and lateral pterygoid
- Posterior trunk:
  - *Auriculotemporal nerve:* Sensory to skin of temple, auricle, external auditory meatus and tympanic membrane and secretomotor fibers to parotid gland.
  - *Lingual nerve:* Mucous membrane of floor of mouth and anterior 2/3 of tongue and secretomotor fibers to sublingual and submandibular salivary gland.
  - *Inferior alveolar nerve:* Teeth and lower jaw, skin over chin and lower lip and nerve to mylohyoid and anterior belly of digastric.

Q.191 What is the effect of complete unilateral lesion of trigeminal nerve?

Unilateral anesthesia of face and anterior part of scalp, auricle and mucous membrane of nose, mouth and anterior two-thirds of tongue, with paralysis and wasting of muscles of mastication on affected side.

Q.192 What is 'trigeminal neuralgia'?

It is the disease of unknown etiology in which there is sudden severe pain in the area of distribution of trigeminal nerve.

Abducent Nerve

Q.193 Name the functional components of adbucent nerve.

- *Somatic efferent:* For lateral movement of eyeball.
- *General somatic afferent:* For proprioceptive impulses from lateral rectus muscle.

Q.194 What is position of VI cranial nerve nucleus?

Upper part of floor of fourth ventricle beneath facial colliculus.

Q.195 What is effect of paralysis of abducent nerve?

- Medial squint
- Diplopia.

Q.196 What is Raymond syndrome?

It is a pons lesion causing contralateral hemiplegia and paralysis of abducent nerve on same side.

Facial Nerve

Q.197 What is the origin of facial nerve?

By two roots in lateral part of groove between lower border of pons and upper border of medulla.

Q.198 What are the functional components of facial nerve?

- *Special visceral efferent:* Motor to muscles of facial expression and elevation of hyoid bone, which arise from mesoderm of second branchial arch.
- *General viscerent efferent:* Secretomotor to submandibular and sublingual glands, lacrimal gland and glands of nose, palate and pharynx.
- *Special visceral afferent:* Carries taste sensation from anterior 2/3 of tongue and palate.
- *General somatic afferent:* For proprioceptive impulses from muscles supplied.

Q.199 What is nervus intermedius?

Sensory root of facial nerve because it is attached between motor root (medially) and vestibulocochlear nerve (laterally).

Q.200 What is the position of geniculate ganglion?

It is present in the course of facial nerve through the substance of petrous temporal bone.

Q.201 What is the position of facial nerve nucleus?

In reticular formation of pons, medial to spinal nucleus of the trigeminal nerve.

Q.202 What is the position, connections and branches of pterygopalatine ganglion?

It is peripheral autonomic ganglion of the cranial parasympathetic outflow.

Position: Present in pterygopalatine fossa and is suspended from maxillary nerve by two ganglionic branches.

Connections:

- Motor (Parasympathetic) root: Nerve of pterygoid canal.
- Sympathetic root: From internal carotid plexus pass through ganglion without relay.
- Sensory root: From maxillary nerve several branches pass through it without relay.

Branches:

- Secretomotor fibers to lacrimal gland and glands of nasal and palatine mucosa from postganglionic fibers of nerve of pterygoid canal.
- Orbitalis muscle by orbital branch by sympathetic nerves.
- Sensory root: From palate, nose and pharynx. Taste fibers from soft palate.

Q.203 What is the position, connections and branches of submandibular ganglion?

It is a peripheral autonomic ganglion of the cranial parasympathetic outflow.

Position: It lies over hyoglossus muscle suspended from lingual nerve by two or more roots.

Connections:

- Parasympathetic root: Lingual nerve
- Sympathetic root: From facial artery plexus, pass through ganglion without relay.

Branches:

- Secretomotor fibers to submandibular and sublingual salivary glands by parasympathetic fibers.
- Blood vessels of submandibular and sublingual glands by sympathetic plexus.

Q.204 What are the branches of facial nerve and structures supplied?

- Within the facial canal.
  - *Greater petrosal nerve:* Arises from geniculate ganglion. Joins deep petrosal nerve at foramen lacerum, to form nerve of pterygoid canal. Supply glands of nose, palate and pharynx and lacrimal gland. Also carries taste sensation from palate.
  - Nerve to stapedius muscle.
  - *Chorda tympani:* Joins lingual nerve. Supplies:
  - Secretomotor fibers to submandibular and sublingual glands.
  - Carries taste sensation from anterior 2/3 of tongue.
- At exit from stylomastoid foramen.
  - *Posterior auricular:* Supplies auricularis posterior, occipitalis and intrinsic muscles on back of auricle.

- *Digastric branch:* To posterior belly of digastric.
- *Stylohyoid branch:* To stylohyoid muscle.

- Terminal branches within parotid gland.
  - *Temporal branches:* Supply auricularis anterior and superior, intrinsic muscles on lateral side of ear, frontalis, orbicularis and corrugator supercilli.
  - *Zygomatic branches:* To orbicularis oculi.
  - *Buccal branches:* To buccal muscles.
  - *Mandibular branch:* To muscles of lower lip and chin.
  - *Cervical branch:* Supplies platysma.
  - *Communicating branches:* To trigeminal and vagus nerve to supply part of skin of auricle (Figs 7.13A and B).

Q.205 What is Bell's palsy?

It is the infranuclear lesion of facial nerve, in which the whole of face is paralyzed on same side. Face becomes asymmetrical and is drawn to the normal side.

Q.206 Why in the supranuclear lesion of facial nerve, only lower part of face is paralyzed?

Because the lower facial muscles have a unilateral cortical representation through opposite pyramidal tract but the upper facial muscles have a bilateral representation through pyramidal tracts of both sides.

Vestibulocochlear Nerve

Q.207 Name the nuclei of origin of vestibulocochlear nerve.

- Dorsal and ventral cochlear nuclei, situated in relation to inferior cerebellar peduncle.
- Superior, inferior, medial and lateral vestibular nuclei, situated laterally in pons and medulla.

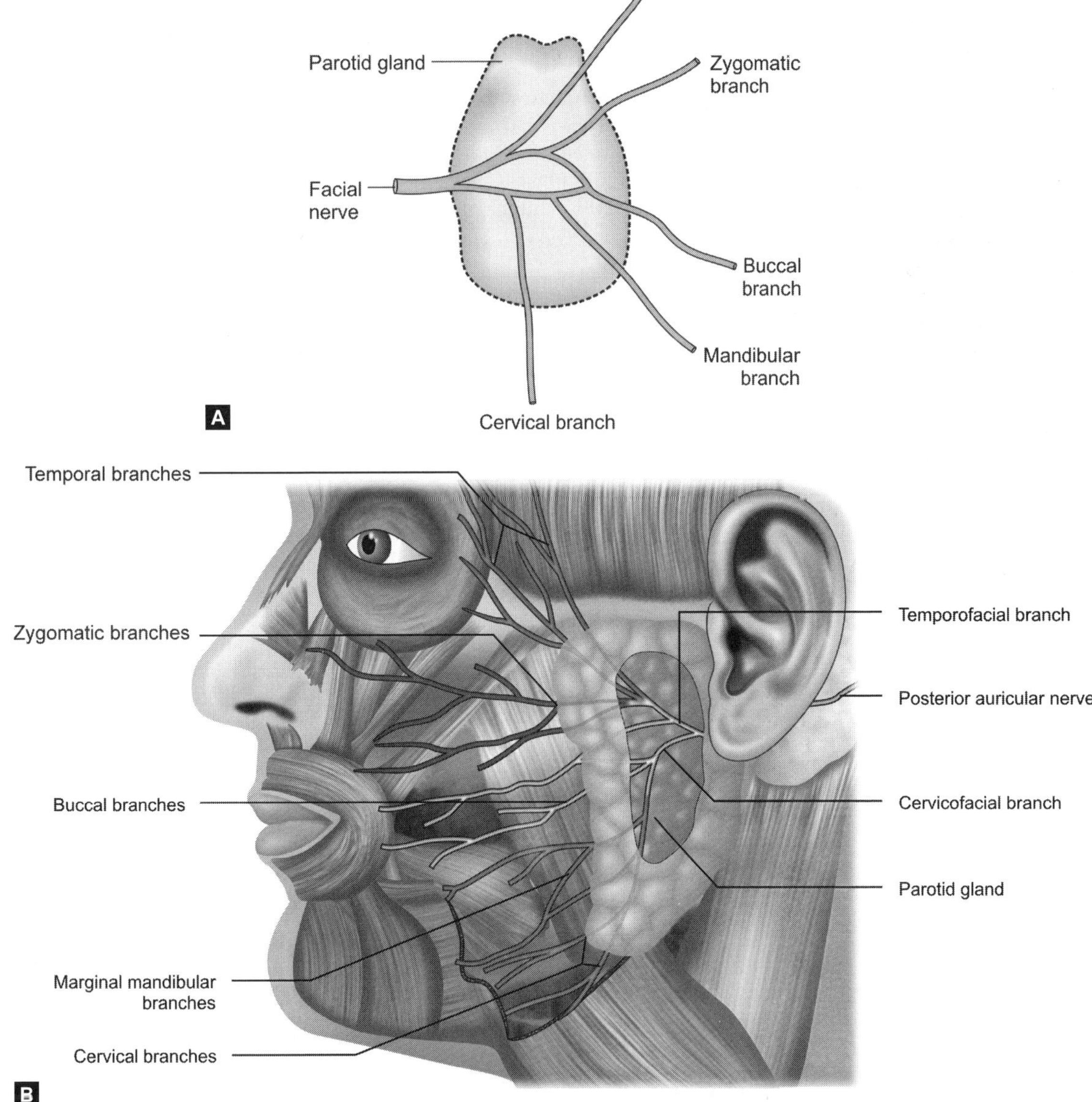

Figs 7.13A and B: Branches of facial nerve in the face

Q.208 What are the part of VIII cranial nerve?

- *Cochlear nerve:* Nerve of hearing.
- *Vestibular nerve:* Nerve of equilibrium (balance).

Q.209 Trace the auditory and balance pathways (Fig. 7.14).

- *Auditory pathway:*
  Receptors: Hair cells of organ of Corti.
  First order sensory neuron: Spiral ganglion of bipolar cells (in a canal around modiolus). Central processes of ganglion form cochlear nerve, which terminates in dorsal and ventral cochlear nuclei.
  Second order neurone: Neurons in cochlear nuclei. Axons of these pass to dorsal part of pons and most of them cross to the opposite side. The crossing fibers of two sides form trapezoid body. These end in superior olivary complex.
  Third order neuron: Arise from superior olivary complex and form an ascending bundle called lateral lemniscus and end in inferior colliculus of midbrain from which fibers reach the medial geniculate body. From it, acoustic radiations pass to acoustic area of cerebral cortex (Area 41,42) via sublentiform part of internal capsule.
  The fibers from superior olivary complex also reach dorsal and ventral cochlear nuclei of inferior cerebellar peduncle.
- *Equilibrium pathway:*
  Receptor: Hair cells in macula of saccule, utricle and crista of ampullae of semicircular canals.
  First order neuron: Vestibular ganglion of bipolar neurons. Central processes of ganglion form vestibular nerve.
  Second order neuron: Vestibular nuclei. These send fibers to:
  - Archicerebellum
  - Motor nuclei of brainstem (of III, IV and VI nerve).
  - Anterior horn cells of spinal cord.

Q.210 Why the unilateral injury to cochlear nerve do not greatly affect auditory acuity?

Because auditory radiations to cortex are bilaterally distributed.

Q.211 What is the effect of lesion of vestibular nerve?

Vertigo, ataxia and nystagmus.

Glossopharyngeal Nerve

Q. 212 Name the functional components of IX cranial nerve.

- *Special visceral efferent:* Motor to stylopharyngeus. This muscle develops from mesoderm of third branchial arch
- *General visceral efferent:* Secretomotor to parotid.
- *General visceral afferent:* Sensory to mucous membrane of pharynx, tonsil, soft palate and posterior 1/3 of tongue.
- *Special visceral afferent:* Taste sensation from posterior 1/3 of tongue.
- *General somatic afferent:* Proprioceptive impulses from stylopharyngeus and skin of the auricle.

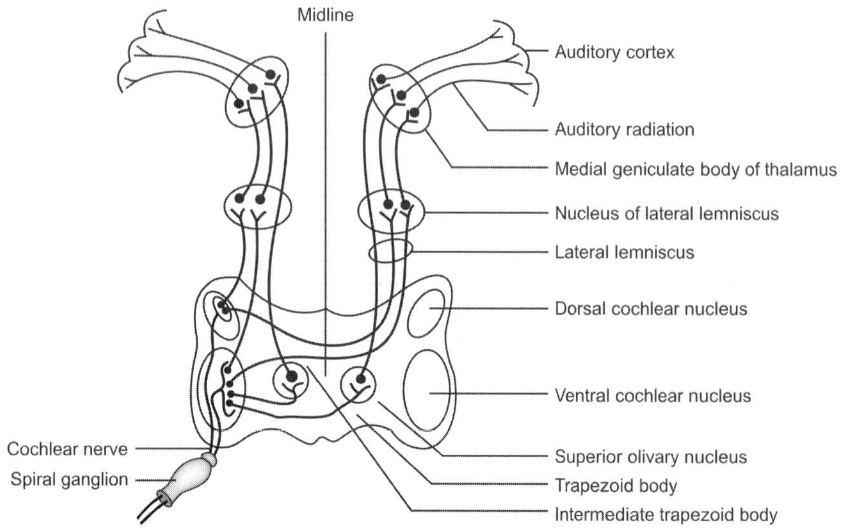

Fig. 7.14: Auditory pathway. Blue = First order neuron. Red = Second order neuron. Green = Third order neuron. Black = Auditory radiation

Q.213 Name the nuclei of origin of ninth nerve?

- Nucleus ambiguus.
- Nucleus of tractus solitarius
- Inferior salivatory nucleus.

Q.214 What are branches of IX cranial nerve?

- *Tympanic:* To middle ear, auditory tube, mastoid air cells and lesser petrosal nerve to parotid gland via otic ganglion.
- *Carotid:* To carotid body and carotid sinus
- *Pharyngeal:* Forms pharyngeal plexus
- *Muscular:* To stylopharyngeus
- *Tonsillar:* Supply palatine tonsils and soft palate
- *Lingual:* Taste and general sensations from posterior 1/3 of tongue.

Q.215 What is the position, connections and branches of otic ganglion?

It is a peripheral autonomic ganglion of the cranial parasympathetic outflow.

Position: Present just below the foramen ovale medial to trunk of mandibular nerve. It is connected to nerve to medial pterygoid muscle.

Connections:
- Parasympathetic root: Lesser petrosal nerve, part of tympanic branch.
- Sympathetic root: From plexus on middle meningeal artery and pass through ganglion without relay.
- Motor root: Through nerve to medial pterygoid, branch of mandibular nerve and pass through ganglion without relay.

Branches:
- Secretomotor fibers to parotid gland via a branch connecting otic ganglion to auriculotemporal nerve, which itself gives a parotid branch.
- Sympathetic fibers to parotid gland through auriculotemporal nerve.
- Motor fibers to tensory tympani and tensor palati muscles.

Vagus Nerve

Q.216 Name the functional components of X cranial nerve.

- *General visceral efferent:* Parasympathetic fibers to thoracic viscera and greater part of gastrointestinal tract.
- *Special visceral efferent:* To musculature of pharynx, larynx and soft palate, derived from branchial arches.
  Superior laryngeal branch is nerve of fourth arch and recurrent laryngeal branch is nerve of sixth arch.

- *General visceral afferent:* Branches to pharynx, larynx, trachea and esophagus and thoracic and abdominal viscera
- *Special visceral afferent:* Carries taste sensation from posterior most part of tongue and epiglottis.
- *General somatic afferent:* To skin of auricle.

Q.217 Name the nuclei of vagus nerve.

- Nucleus ambiguus
- Nucleus of tractus solitarius
- Dorsal nucleus of vagus.

Q.218 Name the ganglia on vagus and what are their connections.

Superior ganglion: In jugular foramen.
Connected to IX and XI nerves and
Superior cervical ganglion of sympathetic chain.
Inferior ganglion: Near base of skull.
Connected to XII nerve, superior cervical ganglion and
Loop between C_1 and C_2 nerves.

Q.219 Name the branches of vagus. What are the structures supplied by these?

From superior ganglion:

- *Meningeal:* Dura of posterior cranial fossa.
- *Auricular:* Conchae and root of auricle. Posterior ½ of external auditory meatus and
  Outer surface of tympanic membrane.

In Neck (From inferior ganglion):

- *Pharyngeal:* Has mainly fibers of cranial accessory nerve. Forms pharyngeal plexus. Supplies muscles of pharynx and soft palate except tensor palati.
- *Carotid:* To carotid body.
- *Superior laryngeal nerve:* It divides into:
  - *External laryngeal:* Inferior constrictor and circothyroid muscle.
  - *Internal laryngeal:* Sensory to larynx up to vocal fold.
- *Right recurrent laryngeal nerve:*
  - To intrinsic muscles of larynx except cricothyroid.
  - Sensory to larynx below vocal fold.
  - Sensory branches to trachea, esophagus and inferior constrictor.
  - To deep cardiac plexus.
- *Cardiac:* To superficial and deep cardiac plexus.

In abdomen: The two vagus nerves are distributed to stomach and celiac, hepatic and renal plexuses.

Q.220 What is the effect of lesion of vagus nerve?

- Nasal regurgitation of swallowed liquids
- Nasal twang in voice
- Hoarseness of voice
- Flattening of palatal arch
- Cadaveric position of vocal cord and
- Dysphagia.

Accessory Nerve

Q.221 Name the roots of the accessory nerve.

Two roots: Cranial and spinal roots.

Q.222 Where do the roots of accessory nerve arise?

The *cranial root* arises from the lower part of nucleus ambiguus. The *spinal root* arises from the lateral part of anterior grey column of the cervical part, C_{1-5} of the spinal cord.

Q.223 What is the functional component of IX cranial nerve?

Special visceral efferent: Supplies the muscles derived from branchial arches.

Q.224 How does the spinal root enter the cranial cavity?

The spinal rootlets of the accessory nerve unite to form a trunk which ascends in the vertebral canal and enters the cranial cavity through the foramen magnum.

Q.225 Why is accessory nerve called 'accessory'?

It is accessory to the vagus nerve, hence the name. The cranial root is in fact a part of the vagus nerve.

Q.226 What is the distribution of the cranial accessory nerve?

It is distributed via branches of the vagus to the muscles of the soft palate (except the tensor palati), pharynx (except the stylopharyngeus) and intrinsic muscles of the larynx.

Q.227 What is the distribution of the spinal accessory nerve?

It supplies the sternomastoid and trapezius muscles.

Q.228 What will be the effects of a complete lesion of the spinal accessory nerve?

There will be paralysis of the sternomastoid and trapezius muscles (lower motor neurone type of paralysis). The patient will not be able to rotate his head to the healthy side (due to paralysis of sternomastoid) and he will not be able to shrug the affected shoulder nor will he be able to raise the arm above the head (due to paralysis of trapezius).

Hypoglossal Nerve

Q.229 What is the position of hypoglossal nucleus?

It is present in medulla extending into both open and closed parts of the medulla.

Q.230 What is the distribution of the hypoglossal nerve?

- Hypoglossal is motor nerve to all muscles of the tongue except the palatoglossus.
- Branches of hypoglossal nerve containing fibers of C_1 nerve.
  - *Meningeal branch:* To meninges of posterior cranial fossa.
  - *Descending branch:* Upper root of ansa cervicalis.
  - To thyrohyoid and geniohyoid.

Q.231 What will be the effects of cutting this nerve on one side?

There will be ipsilateral lower motor neurone type of paralysis of muscles of the tongue. On asking the patient to protrude his tongue, it will deviate to the paralyzed side.

Q.232 How will you differentiate nuclear lesion from an infranuclear lesion of the hypoglossal nerve?

In addition to features of the infranuclear lesion (flaccid paralysis and wasting of muscles) there will also be fasciculations in the muscles of the tongue on the affected side. There will be wrinkling of the mucous membrane of the tongue due to wasting of muscles and their fasciculations.

DO YOU KNOW?

- The only movable skull joint is the temporomandibular joint and allow chewing. All other bones are fixed to each other by joints known as sutures which are also journal only in skull
- Most of the basal ganglion are telencephalic in origin.

Section 2

PHYSIOLOGY

CHAPTERS

Chapter 8

General Physiology

Q.1 Define cell.
Cell is defined as the structural and functional unit of living body. Figure 8.1 shows the detail structure of a cell.

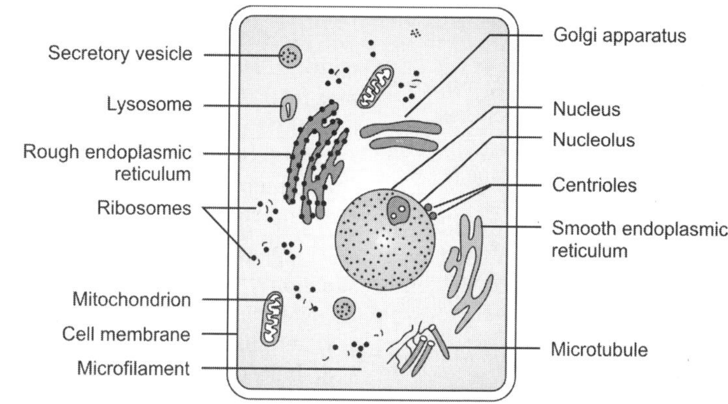

Fig. 8.1: Structure of the cell

Q.2 What is the composition of the cell membrane?
The cell membrane contains proteins (55%), lipids (40%) and carbohydrates (5%).

Q.3 Name the structural models of cell membrane. Mention the accepted one.
- Danielli-Davson model
- Unit membrane model
- Fluid mosaic model.

The fluid mosaic model is the accepted one.

Q.4 What are the layers of the cell membrane?
One central lipid layer and two outer protein layers. Figure 8.2 shows the lipid layer of cell membrane.

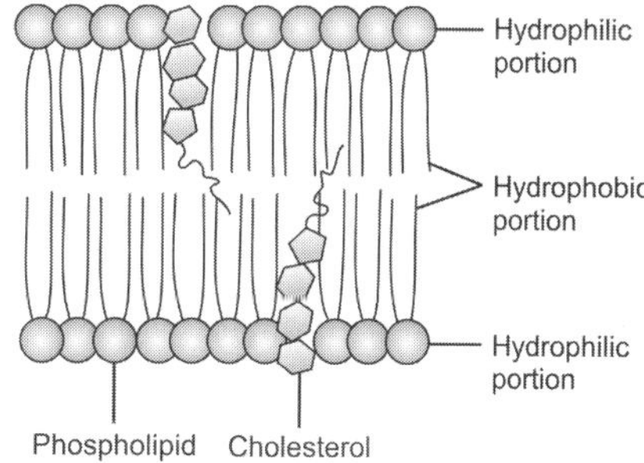

Fig. 8.2: Lipid layer of the cell membrane

Q.5 What is the characteristic feature of the lipid layer of cell membrane? What is its advantage?
Lipid layer of the cell membrane is fluid in nature. Because of this, the portions of the cell membrane move from one point to another point along the surface of the cell.

The advantage of this is that the materials dissolved in lipid layer can move to all the areas of the cell membrane.

Q.6 Name the types of proteins present in the cell membrane.
- Integral proteins
- Peripheral proteins.

Q.7 What are the functions of proteins in the cell membrane?
Proteins:
- Provide structural integrity to the cell membrane
- Form the channels through which the water soluble substances can diffuse
- Function as carrier proteins, which help in transport of substances across the cell membrane
- Form the enzymes
- Function as the receptor proteins for the hormones.

Q.8 Name the carbohydrates present in the cell membrane.
- Glycoproteins—attached to proteins
- Glycolipids—attached to lipids.

Q.9 What is the functional importance of carbohydrates in the cell membrane?
Carbohydrate molecules are negatively charged. So, these molecules do not allow the negatively charged particles to move out of the cells. This helps in the maintenance of resting membrane potential.

Q.10 Name the cytoplasmic organelles which are bound with limiting membrane.
- Endoplasmic reticulum
- Golgi apparatus
- Lysosome
- Peroxisome
- Centrosome and centrioles
- Secretory vesicles
- Mitochondria
- Nucleus.

Q.11 Name the cytoplasmic organelles which are not bound with limiting membrane.
Ribosomes and cytoskeleton.

Q.12 What is endoplasmic reticulum?
Endoplasmic reticulum is the interconnected network of tubular and microsomal vesicular structures in the cytoplasm.

Q.13 Name the types of endoplasmic reticulum. Mention the function of each.
- Rough or granular endoplasmic reticulum
  - to which the ribosomes are attached. It is concerned with:
    i. Synthesis of proteins in the cell
    ii. Degradation of toxic substances.
- Smooth or a granular endoplasmic reticulum—to which the ribosomes are not attached. It is concerned with:
  i. Synthesis of lipids and steroids
  ii. Storage and metabolism of calcium
  iii. Degradation of toxic substances.

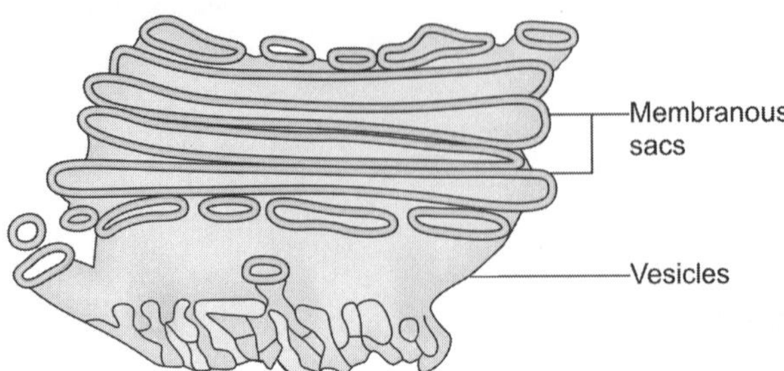

Fig. 8.3: Golgi apparatus

Q.14 What are the functions of Golgi apparatus?
Processing, packing, labeling and delivery of proteins and lipids (Fig. 8.3).

Q.15 What are the functions of lysosomes?
- Degradation of macromolecules like bacteria
- Degradation of worn out organelles
- Secretory function.

Q.16 What are the lysozymes?
Lysozymes are the hydrolytic enzymes present in lysosomes.

Q.17 What are peroxisomes and what are their functions?
Peroxisomes are the membrane limited vesicles derived from the endoplasmic reticulum.
Peroxisomes are concerned with:
- Degradation of toxic substances like hydrogen peroxide
- Oxygen utilization
- Breakdown of excess fatty acids
- Acceleration of gluconeogenesis from fats
- Degradation of purine to uric acid
- Formation of myelin and bile acids.

Q.18 What is the other name of mitochondrion? What are the functions of mitochondrion?
The other name of mitochondrion is 'power house' of the cell.
Functions of mitochondrion:
- Production of energy
- Synthesis of ATP
- Initiation of apoptosis.

Q.19 What are the functions of ribosomes?
Ribosomes are concerned with protein synthesis. The ribosomes attached to rough endoplasmic reticulum are involved in the synthesis of hormonal proteins, lysosomal proteins and proteins of the cell membrane. The free ribosomes are concerned with synthesis of protein in hemoglobin, and proteins present in peroxisomes and mitochondria.

Q.20 What is cytoskeleton of the cell? What are the protein components of cytoskeleton?
The cytoskeleton of the cell is a complex network of structures in various sizes present throughout the cytoplasm.
Protein components of cytoskeleton:
- Microtubules
- Intermediate filaments
- Microfilaments.

Q.21 What are the functions of cytoskeleton?
Cytoskeleton is concerned with:
- Determination of shape of the cell
- Stability of cell shape
- Cellular movements.

Q.22 What are microtubules? What are their functions?
Microtubules are tubular organelles without limiting membrane present in the cytoplasm. These tubules are formed by the tubulin molecules.
Microtubules:
- Determine the shape of the cell
- Give structural strength to the cell
- Act like conveyer belts which allow the movement of granules, vesicles, protein molecules and some organelles like mitochondria to different parts of the cell
- Form the spindle fibers which separate the chromosomes during mitosis
- Are responsible for the movements of centrioles and the complex cellular structures like cilia.

Q.23 What are microfilaments? What are their functions?
Microfilaments are nontubular thread like organelles present in the cytoplasm of the cell. The microfilaments in ectoplasm are made up of actin molecules and the filaments in endoplasm are made up of actin and myosin molecules.
Microfilaments:
- Give structural strength to the cell
- Provide resistance to the cell against the pulling forces
- Are responsible for cellular movements like contraction, gliding and cytokinesis (partition of cytoplasm during cell division).

Q.24 List the functions of nucleus.
- Control of all activities of the cell
- Synthesis of RNA
- Formation of ribosomal subunits
- Sending genetic instruction to the cytoplasm through mRNA for protein synthesis
- Control the cell division through genes
- Storage of hereditary information (in genes) and transformation of this information from one generation of the species to the next.

Q.25 What is DNA?
DNA (deoxynucleic acid) is a nucleic acid present in nucleus and mitochondria of cell.

Q.26 What is RNA? What are the types of RNA?
RNA (ribonucleic acid) is a nucleic acid derived from DNA.
RNA is of three types:
- Messenger RNA
- Transfer RNA
- Ribosomal RNA.

Q.27 Define gene.
A gene is a portion of DNA molecule that contains the message or code for the synthesis of a specific protein from amino acids.

Q.28 Define transcription and translation.
Transcription is the copying of genetic code from DNA to RNA. Translation is the process by which protein synthesis occurs in the ribosome of the cell under the direction of genetic instruction given by mRNA.

Q.29 What are growth factors? Name some growth factors.
Growth factors are proteins which act as cell signaling molecules like cytokines and hormones.
Growth factors:
- Platelet derived growth factor
- Colony stimulating factors
- Nerve growth factors
- Neurotropins
- Erythropoietin
- Thrombopoietin
- Insulin like growth factors
- Epidermal growth factor
- Basic fibroblast growth factor
- Myostatin.

Q.30 What is apoptosis?
Apoptosis is the programmed cell death under genetic control.

Q.31 What is necrosis?
Necrosis is the uncontrolled and unprogrammed death of cells due to unexpected and accidental damage.

Q.32 Define cell junction.
The cell junction is the connection between the neighboring cells or the contact between the cell and extracellular matrix.

Q.33 Classify cell junctions.
- Occluding junctions—tight junctions
- Communicating junctions—gap junctions and chemical synapse
- Anchoring junctions—adherence junctions, focal adhesions, desmosomes and hemidesmosomes.

Q.34 What are the proteins present in tight junctions?
- Tight junction membrane proteins or integral membrane proteins—occludin, claudin and junctional adhesion molecules (JAMs)
- Scaffold (platform) proteins or peripheral membrane proteins or cytoplasmic plaque proteins—cingulin, symplekin and ZO-1, 2, 3.

Q.35 What are the functions of tight junction?
- Strength and stability to the tissues
- Selective permeability
- Fencing function
- Maintenance of cell polarity
- Formation of blood-brain barrier.

Q.36 What are connexons or connexins?
Connexons or connexins are the protein subunits present in gap junctions.

Q.37 What are the functions of gap junction?
- Allows the passage of small molecules, ions and chemical messengers
- Helps in propagation of action potential from one cell to another cell.

Q.38 What is the basic difference between passive transport and active transport?
The basic difference between the passive transport and active transport is that the passive transport does not require expenditure of energy and the active transport requires expenditure of energy.

Q.39 What are the types of passive transport or diffusion?
- Simple diffusion
- Facilitated diffusion.

Q.40 Explain simple and facilitated diffusion briefly.
Simple diffusion occurs through lipid layer and protein layer of the cell membrane. The lipid soluble substances like oxygen, carbon dioxide and alcohol are transported through lipid layer. The water soluble substances like electrolytes are transported through protein layer.

The facilitated diffusion is also known as carrier mediated diffusion because it involves the help of a carrier protein present in the cell membrane. The substances with larger molecules like glucose and amino acids are attached to the carrier protein and are transported into the cell.

Q.41 Name the factors affecting the diffusion of substances across the cell membrane.
- Permeability of cell membrane
- Temperature
- Concentration gradient or electrical gradient
- Solubility of the substances
- Thickness of cell membrane
- Size of the molecules and ions
- Charge of the ions.

Diffusion is directly proportional to permeability of cell membrane, temperature, concentration gradient or electrical gradient and the solubility of the substances. It is inversely proportional to thickness of the cell membrane, size of the molecules and ions and charge of the ions.

Q.42 Name types of active transport. Explain them briefly.
- *Primary active transport:* In this, the energy is liberated from break down of ATP. The electrolytes like sodium, potassium, calcium, hydrogen and chloride are transported by this method.
- *Secondary active transport:* In this type of active transport, a carrier protein is involved in transport of a substance like sodium ion and this carrier protein is capable of transporting another substance along with the primary substance. The energy is derived from process involved in the transport of the primary substance.

Q.43 Name the types of secondary active transport. Explain them briefly.
- *Co-transport:* In this, along with the primary substance like sodium, the carrier protein carries another substance. Substances like glucose and amino acids are transported by this method
- *Counter transport:* In this mechanism, the substances are carried in exchange of the primary substance like sodium. The different counter transport mechanisms are sodium-calcium counter transport, sodium-hydrogen counter transport, sodium-magnesium counter transport, sodium-potassium counter transport, calcium-magnesium counter transport, calcium-potassium counter transport, chloride-bicarbonate counter transport and chloride-sulfate counter transport.

Q.44 What is bulk flow? Give example.
The movement of large number of molecules of a substance in bulk along the concentration gradient is known as bulk flow.

The example is the diffusion of respiratory gases across the respiratory membrane.

Q.45 Define and classify endocytosis.
Endocytosis is the process by which the larger molecules (which cannot enter the cell by means of active or passive transport) are transported into the cell. Endocytosis is of two types: pinocytosis and phagocytosis.

Q.46 What is pinocytosis? Give example.
The movement of larger particles by means of evagination of the cell membrane is called pinocytosis. It is otherwise known as 'cell drinking'.
Example: Transport of macromolecules like bacteria and antigens.

Q.47 What is phagocytosis? Give example.
The process by which the particles larger than the macromolecules are engulfed into the cells is called phagocytosis. It is also known as 'cell eating'.
Example: Transport of larger bacteria, larger antigens and other larger foreign bodies inside the cell.

Q.48 Name the cells which show phagocytosis.
Neutrophils, monocytes and tissue macrophages.

Q.49 Define homeostasis.
The maintenance of constant internal environment is known as homeostasis.

Q.50 What are the mechanisms involved in homeostatic control system? Explain them briefly.
The homeostatic control system is mainly by the feedback mechanisms:
- *Negative feedback:* If the activity of a particular system increases, it will be immediately regulated by reduction (example—thyroxin secretion). The negative feedback controls most of the homeostatic mechanisms.

- *Positive feedback:* When the activity of a particular system increases, it will be further increased (examples—formation of prothrombin activator during coagulation, secretion of oxytocin during milk ejection reflex and the pain produced during labor). Positive feedback is less common than the negative feedback. However, it has its own significance particularly during emergency conditions.

Q.51 What is the normal pH of ECF?

The normal pH of ECF is 7.4. It varies between 7.38 and 7.42 in physiological conditions.

Q.52 How is the pH of ECF and plasma determined?

To determine the pH of ECF, the concentrations of bicarbonate ions and carbon dioxide dissolved in the fluid are measured. The pH is calculated by using Hendersön-Hasselbalch equation.

The pH of plasma is determined by pH meter.

Q.53 What are the mechanisms which regulate acid base balance?

- Blood buffer system
- Respiratory mechanism
- Renal mechanism.

Q.54 What are the buffer systems in the body?

- Bicarbonate buffer system
- Phosphate buffer system
- Protein buffer system.

Q.55 What is basic mechanism involved in the regulation of acid base balance by respiratory system?

The respiratory system regulates acid base balance by regulating carbon dioxide content in the blood.

Q.56 What is basic mechanism involved in the regulation of acid base balance by kidney?

Kidney regulates acid base balance by secretion of hydrogen ions and retention of bicarbonate ions.

Q.57 What are the disturbances of acid base status?

- *Acidosis:* When hydrogen ion concentration increases, it leads to reduction in pH. It is called acidosis.
- *Alkalosis:* When hydrogen ion concentration decreases, it leads to increase in pH. It is known as alkalosis.

Q.58 Classify acidosis. Explain briefly.

- Respiratory acidosis that occurs during respiratory disturbances. This is due to the increase in the partial pressure of carbon dioxide above 60 mm Hg in the arterial blood.
- Metabolic acidosis that occurs during metabolic disturbances. It is due to the excessive accumulation of organic acids like lactic acid, acetoacetic acid and beta hydroxy butyric acid.

Q.59 Name some conditions when respiratory acidosis occurs.

Respiratory acidosis occurs in conditions leading to hypoventilation like:

- Airway obstruction (as in bronchitis)
- Lung diseases (like fibrosis)
- Respiratory center depression (by anesthetics, sedatives, etc.)
- Extrapulmonary thoracic diseases (like kyphosis and scoliosis)
- Neural diseases (poliomyelitis)
- Paralysis of respiratory muscles.

Q.60 Name some conditions when metabolic acidosis occurs.

- Lactic acidosis (as in circulatory shock)
- Ketoacidosis (as in diabetes mellitus)
- Uric acidosis (as in renal failure)
- Acid poisoning
- Renal tubular acidosis
- Loss of excess of bicarbonate ions (as in diarrhea).

Q.61 What are the types of alkalosis? Explain briefly.

- Respiratory alkalosis that occurs during respiratory disturbances. It is due to the reduction in the partial pressure of carbon dioxide (< 20 mm Hg) in arterial blood.
- Metabolic alkalosis that occurs during metabolic disturbances. It is due to the excessive loss of hydrogen ions from the body.

Q.62 Name some conditions when respiratory alkalosis occurs.

Hyperventilation is the primary cause for loss of excess carbon dioxide from the body leading to respiratory alkalosis. Hyperventilation occurs in hypoxic conditions, cerebral disturbances, and psychological and emotional trauma.

Q.63 Name some conditions when metabolic alkalosis occurs.

Metabolic alkalosis is due to loss of excess hydrogen ions that occurs in:

- Vomiting and diarrhea

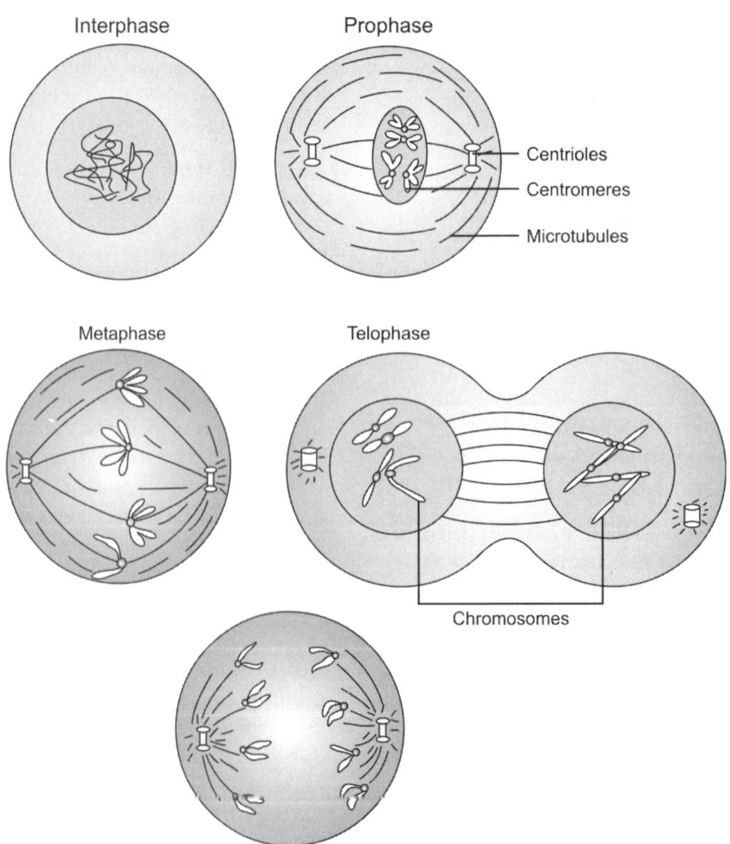

Fig. 8.4: Mitosis of an animal cell

- Endocrine disorders (Cushing's syndrome, Conn's syndrome)
- Diuretic therapy.

Q.64 The intracellular fluid has more sodium content than potassium. Is this statement correct?

No, it is the other way around, i.e. intracellular fluid has more potassium content than sodium.

Q.65 Give a brief description of chromosome.

Chromosomes are fine threadlike structures forming the chromatin in the nucleus, and are made up of specific tiny structures called genes. The various hereditary peculiarities of the cell are passed on from one generation to another by the chromosomes through the genes of which the chromosomes are composed of.

Q.66 What is the number of chromosomes in somatic cells of human being?

There are 46 number of chromosomes, i.e. 22 pairs of somatic chromosomes and one pair of sex chromosomes.

Q.67 What is the difference between mitotic and meiotic cell division?

In mitotic cell division the number of chromosomes remains the same (Fig. 8.4), whereas in meiotic cell division the number of chromosomes is halved.

Q.68 Name the various phases in mitotic division of a cell.

These are interphase, prophase, metaphase, cleavage stage, anaphase and telophase.

Q.69 What is DNA composed of?

DNA has phosphoric acid (deoxyribose) and four nitrogenous bases, i.e. two purine (adenine and guanine) and two pyramidines (thymine and cystosine).

Chapter 9

Blood and Body Fluids

Q.1 How much is the volume of total body water (TBW) in a normal young adult?
Males: 60 - 65% of body weight. *Females:* 50 - 55%. Normally, TBW is about 40 liters in a person weighing 70 kg.

Q.2 Name the compartments of body fluid.
- Intracellular fluid (ICF) present inside the cells forming about 55% of the TBW, i.e. 22 liters
- Extracellular fluid (ECF) present outside the cells forming about 45% of TBW, i.e. 18 liters.

Q.3 How is ECF distributed?
ECF is distributed in five subunits:
- Interstitial fluid and lymph—20%
- Plasma—7.5%
- Fluid in bones—7.5%
- Fluid in connective tissues—7.5%
- Transcellular fluid—2.5%. Transcellular fluid includes cerebrospinal fluid, intraocular fluid, digestive juices, serous fluid (like intrapleural fluid, pericardial fluid and peritoneal fluid), synovial fluid and fluid in urinary tract.

Q.4 What are the main differences between ECF and ICF?
- *Composition:* ECF contains more of sodium, chlorides and bicarbonates whereas ICF contains more of potassium, magnesium, phosphates, sulfates and proteins
- *Volume:* The quantity of ECF is less (18 liters) and that of ICF is more (22 liters)
- *pH:* The pH of ECF is 7.4 and that of ICF is 7.0.

Q.5 Name the method by which volume of body fluids is measured.
Indicator (dye) dilution method.

Q.6 What are the qualities (characteristics) of the marker substance?
Marker substance:
- Must be nontoxic
- Must mix well with fluid compartment within reasonable time
- Should not be excreted rapidly
- Should not change the color of body fluid
- Should not alter the volume of body fluid.

Q.7 Which type of marker substances is used to measure TBW? Give examples.
The marker substances which can move freely into all the compartments of the body fluid are used to measure TBW.
Examples: Deuterium oxide, tritium oxide and antipyrine.

Q.8 Which type of marker substances is used to measure ECF volume? Give examples.
The substances which remain within the compartments of ECF and do not enter inside the cells are used to measure ECF volume.
Examples: Radioactive ions of sodium, chloride, bromide, sulfate and thiosulfate, and nonmetabolizable saccharides like inulin, mannitol and sucrose.

Q.9 What are sodium space, chloride space, inulin space and sucrose space?
Some of the marker substances like sodium, chloride, inulin and sucrose, which are used to measure ECF volume move widely throughout all the sub-compartments of ECF. The measured volume of ECF by using these substances is called sodium space, chloride space, inulin space or sucrose space.

Q.10 How is the ICF volume measured?
The volume of ICF cannot be measured directly because there is no substance, which can enter the cells without mixing with ECF. So, the ICF volume can be measured only by indirect method, i.e. by measuring the volume of TBW and ECF.

Thus, ICF volume = TBW - ECF volume.

Q.11 Which type of substance is used to measure plasma volume? Give examples.
Plasma volume can be measured by using marker substances, which bind strongly with plasma proteins and do not diffuse into interstitium.
Examples: Radioactive iodine (^{131}I) and Evans blue (T-1824).

Q.12 How is interstitial fluid volume measured?
It cannot be measured directly. It is calculated from the values of ECF volume and plasma volume. Interstitial fluid volume = ECF volume - Plasma volume.

Q.13 What are the features of severe and very severe dehydration?
Severe dehydration: Decrease in blood volume, decrease in cardiac output and hypovolemic shock.
Very severe dehydration: Damage of organs like brain, liver and kidneys, mental depression, confusion, renal failure and coma.

Q.14 What is overhydration (hyperhydration, water excess or water intoxication)?
It is the condition in which the water content in body increases enormously.

Q.15 List the causes of overhydration.
- Heart failure
- Renal disorders
- Hypersecretion of ADH
- Administration of large quantities of medications and fluids
- Underdeveloped kidney in first month of infancy
- Swimming practice during infancy
- Consumption of excess water (> 8 liters/day).

Q.16 What is the difference between plasma and serum?
Plasma is fluid portion of the blood obtained without clotting while serum is the fluid obtained after clotting. Serum is thus plasma without fibrin.

Q.17 What is the normal concentration of plasma protein?
It is 6.4–8.3 gm/100 mL of blood.

Q.18 How hypoproteinemia produces edema?
Hypoproteinemia → decrease in capillary oncotic pressure → decrease in filtration

at arterial end → decrease in absorption of fluid at venous end → abnormal collection of fluid in interstitial spaces → edema.

Q.19 What is A/G ratio?
It is the ratio of albumin to globulin. Normally it is 1.7:1.

Q.20 What is the average daily production of plasma proteins?
It is about 15 gm/day.

Q.21 Can any of the plasma proteins pass through capillary endothelium?
Capillary endothelium normally is impermeable to plasma proteins though in some diseases like glomerulonephritis, nephrotic syndrome, etc. albumin can pass through the capillary membranes.

Q.22 What are the features of severe conditions of overhydration?
Delirium, seizures and coma.

Q.23 How much is the volume of the blood in a normal young healthy adult?
5 liters.

Q.24 What is the normal pH of the blood?
7.4.

Q.25 What is normal viscosity of the blood?
Normally, the blood is five times more viscous than water.

Q.26 What is the cause for the viscosity of the blood?
Presence of red blood cells and plasma proteins.

Q.27 What is the composition of blood?
Blood consists of many components these include (Flow chart 9.1):
55% Plasma.
45% Blood cells of these 99% are erythrocytes (RBC) and
1% Leukocytes (WBC and platelets).

Flow chart 9.1: Constituents of blood

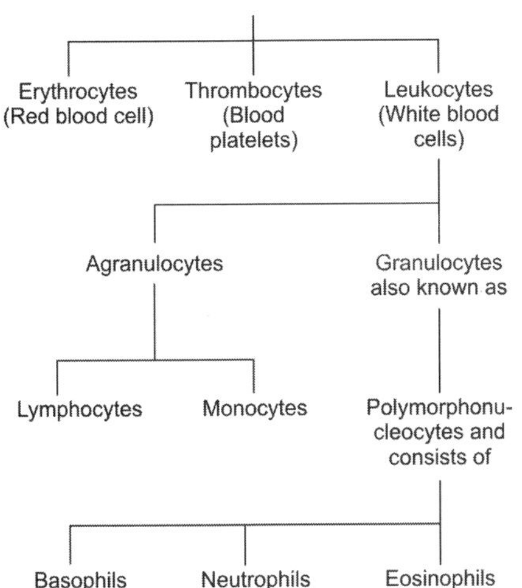

Q.28 Name the organic substances of plasma.
- Plasma proteins—albumin, globulin and fibrinogen
- Amino acids
- Carbohydrates—glucose
- Fats—triglycerides, cholesterol and phospholipids
- Internal secretions—hormones
- Enzymes
- Nonprotein nitrogenous substances—ammonia, creatin, creatinine, xanthine, hypoxanthine, urea and uric acid
- Antibodies.

Q.29 Name the inorganic substances and gases of plasma.
The inorganic substances are sodium, calcium, potassium, magnesium, bicarbonate, chloride, phosphate, iodide, iron and copper. The gases present in blood are oxygen and carbon dioxide.

Q.30 What are the formed elements of the blood?
The formed elements of the blood are the blood cells:
- Erythrocyte or red blood cell (RBC)
- Leukocyte or white blood cell (WBC)
- Platelet (thrombocyte).

Q.31 What is hematocrit? What is the other name for it? What is its normal value?
The volume of RBCs in the blood expressed in percentage is called hematocrit. It is otherwise called packed cell volume (PCV).
Normal value: 45%.

Q.32 How is hematocrit determined?
Hematocrit is determined by using Wintrobe's tube or hematocrit tube. Blood is mixed with anticoagulant (EDTA), filled in this tube and centrifuged for 30 minutes at a speed of 3000 revolutions per minute (RPM). Then the tube is taken out and reading is taken.

Q.33 What are the different layers noticed in the hematocrit tube after centrifuging?
- The upper clear supernatant fluid is plasma and it is normally 55%
- The lower red colored column is packed red blood cells, which is about 45%
- In between the plasma and red blood cells, there is a thin white buffy coat, which is formed by the collection of WBCs and platelets.

Q.34 What is the composition of serum?
It contains all the substances, which are present in the plasma except fibrinogen. The fibrinogen is converted into fibrin during the process of clotting. That is why serum is usually expressed as plasma minus fibrinogen.

Q.35 Give the normal values of plasma proteins.

Total plasma proteins	:	7.3 gm%
Albumin	:	4.7 gm%
Globulin	:	2.3 gm%
Fibrinogen	:	0.3 gm%

Q.36 Name the methods by which the plasma proteins are separated.
- Precipitation method
- Salting out method
- Electrophoretic method
- Cohn's fractional precipitation method
- Ultracentrifugation method
- Immunoelectrophoretic method.

Q.37 What are the functions of plasma proteins?
The plasma proteins:
- Help in coagulation of blood (fibrinogen)
- Play important role in defense mechanism against invading organism (gamma globulin)
- Help in transport of hormones (albumin and globulin)
- Maintain the osmotic pressure of the blood (albumin plays important role)
- Regulate the acid-base balance in blood (buffering action)
- Provide viscosity to the blood
- Help in the erythrocyte sedimentation rate
- Help in maintaining the suspension stability of the red blood cells
- Along with leukocytes, the plasma proteins produce trephone bodies in tissue culture
- Act as reserve proteins during conditions like starvation.

Q.38 Name the conditions when hypoproteinemia occurs.
- Diarrhea
- Hemorrhage
- Burns
- Pregnancy
- Malnutrition
- Prolonged starvation
- Cirrhosis of liver
- Chronic infections like chronic hepatitis or chronic nephritis.

Q.39 Name the conditions when hyperproteinemia occurs.
- Dehydration
- Hemolysis

- Acute infections like acute hepatitis or acute nephritis
- Respiratory distress syndrome
- Excess of glucocorticoids
- Leukemia
- Rheumatoid arthritis
- Alcoholism.

Q.40 What is plasmapheresis?
Plasmapheresis is the experimental procedure done in animals to demonstrate the importance of plasma proteins.

Q.41 What is therapeutic plasma exchange?
Therapeutic plasma exchange is the process of plasmapheresis. It is used as a blood purification procedure for an effective temporary treatment of many autoimmune diseases like myasthenia gravis, thrombocytopenic purpura, etc.

Q.42 Which is the most common site for bone marrow biopsy?
Body of sternum between 2nd and 3rd ribs.

Q.43 What is the normal average ratio of WBC to RBC in human blood?
It is 1:700 (WBC:RBC).

Q.44 Why erythrocytes are red in color?
Because of the presence of hemoglobin.

Q.45 What is normal RBC count?

Adult males : 5 millions/cu mm of blood
Adult females : 4.5 millions/cu mm of blood.

Q.46 What is the normal size of RBC?

Diameter : 7.2 microns (μ)
Thickness : 2.2 μ in periphery and 1 μ in the center
Surface area : 120 square μ
Volume : 90 cubic μ.

Q.47 What is the normal shape of RBC?
When seen from front : Spherical
When seen from side : Biconcave or dumb bell shaped
The reason for the dumb bell shape of RBC is the thicker periphery (2.2 μ) and thinner center (1 μ).

Q.48 What are the advantages of the biconcave shape of RBC?

- It helps in equal and rapid diffusion of oxygen and other substances into the interior of the cell
- It provides large surface area for absorption or removal of different substances
- It offers minimal tension on the membrane when the volume of cell alters
- While passing through minute capillaries, these cells can squeeze through the capillaries very easily without being damaged.

Q.49 What is the fate of hemoglobin?
Old and inactive red cells are ingested by the RES and are broken into globin and iron. Globin and iron are reused whereas the porphyrin moiety of iron is converted into biliverdin and thence bilirubin which are excreted into bile and ultimately excreted mostly through the feces and partly through the urine.

Q.50 What is rouleaux formation?
When the blood is taken out of the blood vessel and allowed to stand without movement, the RBCs pile up one above the other like the pile of coins. This is known as rouleaux formation.

Q.51 What is polycythemia?
Increase in RBC count is called polycythemia.

Q.52 What is physiological polycythemia? Name some conditions when it occurs.
Increase in the number of RBC in physiological conditions is known as physiological polycythemia.
It occurs in:

- Age—infancy
- Sex—in males
- High altitude
- Muscular exercise
- Emotional conditions
- Increased environmental temperature
- After meals.

Q.53 Describe pathological polycythemia in brief.
The abnormal increase in RBC count is called pathological polycythemia.
It is two types:

- Primary polycythemia or polycythemia vera—persistent increase in RBC count occurs because of malignancy of bone marrow
- Secondary polycythemia—increase in number of RBC because of diseases other than the bone marrow diseases, i.e. it is secondary to some pathological conditions.

Q.54 Name the diseases when secondary polycythemia occurs.

- Respiratory diseases
- Congenital heart disease
- Ayerza's disease
- Chronic carbon monoxide poisoning
- Poisoning by chemicals like phosphorus and arsenic
- Repeated mild hemorrhages.

Q.55 What are the physiological conditions when RBC count decreases?

- After sleep
- During pregnancy
- At high barometric pressure.

Q.56 What are the possible variations in the size of RBC?

- Microcytes—decrease in the size of RBC as in the case of iron deficiency anemia
- Macrocytes—increase in the size of RBC as in the case of megaloblastic anemia
- Anisocytosis—unequal sizes of RBC as in the case of pernicious anemia.

Q.57 What are the abnormal shapes of RBC?

- Crenation—shrunken cell
- Spherocytosis—globular form
- Elliptocytosis—elliptical shape
- Sickle shape—crescent shape
- Poikilocytosis—unusual shapes.

Q.58 What is the normal life span of RBC?
120 days.

Q.59 How is the life span of RBC determined?
By radioisotope method.

Q.60 What is the fate of RBC after its life span?
After the life span of 120 days, the RBC is destroyed and heme and globin are released. From heme, iron and bilirubin are released. The iron is stored as ferritin. The globin part is stored as protein.

Q.61 Name the main conditions in which blood viscosity rise.
It is during acidosis, hypercalcemia and hyperglycemia.

Q.62 How is the biconcavity of RBCs maintained?
It is maintained by the presence of a contractile layer of a lipoprotein molecule "spectrin" in a fibrillar manner below its cell membrane.

Q.63 What are the advantages of biconcave shape of RBC?
These are:

- It can withstand considerable changes of osmotic pressure by altering its cell volume and thereby prevent hemolysis.
- Allows easy passage of RBC through narrow capillaries by folding itself.
- Facilitates quick and optimal exchange of gases in and out of hemoglobin.

Q.64 What are the advantages of having no nucleus, no mitochondria and no ribosome in RBC.
It is to accommodate more amount of hemoglobin and also to decrease the use of O_2 by its own structure and thereby increases the availability of O_2 to the other cells.

Q.65 What are the disadvantages of nonnucleated RBC?
- It cannot multiply.
- It cannot synthesize necessary enzymes so has less life span.

Q.66 How does RBC survive for 120 days though it has no nucleus, mitochondria and ribosomes?
For energy supply RBCs depend on glucose metabolism only, which comes through facilitated diffusion. These glucoses are oxidized by cytoplasmic enzymes already present inside the cells to get the energy for their activity. When these cytoplasmic enzymes are exhausted, i.e. after 120 days, it dies.

Q.67 Which is the principal cation in RBC?
It is potassium ion.

Q.68 Why RBC is stained pink by Leishman's stain though it has no ribosomes in their cytoplasm?
It is because of presence of hemoglobin.

Q.69 Mention the site of RBC formation.
- In fetus—bone marrow, spleen, liver and thymus gland.
- After birth—red bone marrow of long bones like sternum, vertebrae, etc.

Q.70 What is the site of production of heme of Hb?
It is in mitochondria.

Q.71 Name the common methods of Hb estimation.
These are:
Sahli's hemoglobinometer method, Haldane hemoglobinometer method, Oxy-Hb colorimetric and also cyano methemoglobin colorimetric method.

Q.72 Why is Sahli's method so accurate?
- In this method reduced Hb in the blood is not converted into acid hematin, thereby the value obtained is less than the total Hb content in the blood.
- This method depends on person's color vision. As the color vision varies from person to person result may also vary.

Q.73 In which form iron is stored in reticuloendothelial cells?
Ferritin and hemosiderin form.

Q.74 Mention the varieties of hemoglobin with special reference to difference of each.
These are as follows:
- Adult Hb (HbA) - Contains 2α and 2β globin chain.
- Adult Hb (HbA_2) - Contains 2α and 2δ chain.
- Fetal Hb (HbF) - Contains 2α and 2γ chain
- HbS - Contains 2α and 2β chain but in b chain glutamate of 6th position is replaced by a valine residue.

Q.75 Define erythropoiesis.
Erythropoiesis is the process which involves origin, development and maturation of red blood cells. Figure 9.1 shows the various stages of erythropoiesis.

Q.76 What are the sites of erythropoiesis?
- Fetal life:
  - Mesoblastic stage—from mesenchyme of yolk sac during the first 2 months
  - Hepatic stage—from liver, spleen and lymphoid organs from 3rd month
  - Myeloid stage—from bone marrow and liver during last trimester
- Children:
  - From the red bone marrow in all the bones.
- Adults:
  - From the membranous bones and the head of long bones.

Q.77 What are the changes taking place in the cell during the process of erythropoiesis?
- Reduction in size of the cell (from the diameter of 25 μ to 7.2 μ).
- Disappearance of nucleoli and nucleus
- Appearance of hemoglobin
- Change in the staining properties of the cytoplasm.

Q.78 What are stem cells? What are the different types of stem cells?
The primitive cells in the bone marrow which give rise to blood cells are called stem cells. The different stem cells: Uncommitted pluripotent hemopoietic stem cells develop into committed pluripotent hemopoietic stem cells which give rise to lymphoid stem cells and colony forming blastocystes.

Lymphoid stem cells develop into lymphocytes.

The colony forming blastocytes are of three types:
- Colony forming unit—Erythrocytes (GFU-E) which develop into the red blood cells

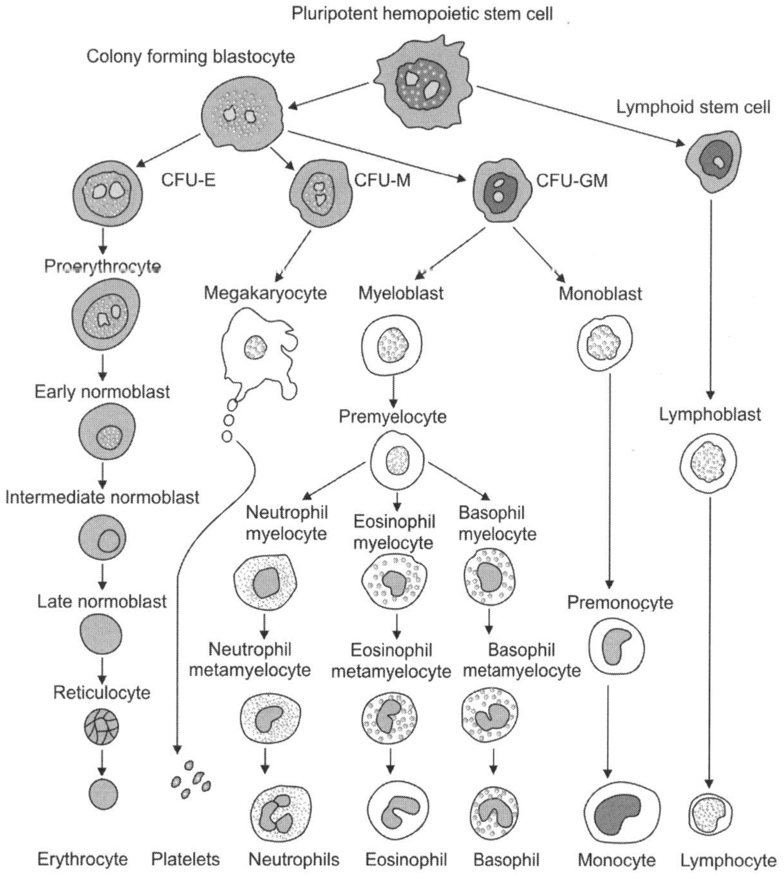

Fig. 9.1: Stages of erythropoiesis.
(CFU-E = Colony forming unit—Erythrocyte, CFU-M = Colony forming unit—Megakaryocyte, CFU-GM = Colony forming unit—Granulocyte/Monocyte)

- Colony forming unit—Granulocytes/ Monocytes (GFU-GM) from which the granulocytes and monocytes develop
- Colony forming unit—Megakaryocytes (CFU-M), which give rise to the platelets.

Q.79 Name the stages of erythropoiesis.

- Proerythroblast
- Early normoblast
- Intermediate normoblast
- Late normoblast
- Reticulocyte
- Matured red blood cell.

Q.80 In which stage, nucleoli disappear?

Early normoblast stage.

Q.81 In which stage, hemoglobin appears?

Intermediate normoblast stage.

Q.82 In which stage, nucleus disappears? How does the nucleus disappear?

Nucleus disappears in late normoblast stage and it disappears by the process called pyknosis.

Q.83 What is the normal reticulocyte count?

In newborn baby	: 2 to 6% of red blood cells
In adults	: 1% or less than 1% of red blood cell.

Q.84 Why the reticulocyte is called the immature red blood cell?

The reticulocyte has large quantity of hemoglobin and nucleus is absent. It is larger than the red blood cell, round in shape with remnants of disintegrated organelles. So, it is called the immature red blood cell.

Q.85 How long does it take for the complete development of red blood cells?

It takes 7 days for the formation and maturation of red blood cells. It takes 5 days up to the stage of reticulocyte and 2 more days for the development of matured red blood cells.

Q.86 What are the factors necessary for erythropoiesis?

- General factors—erythropoietin, thyroxine, growth inducers (interleukin-3) and vitamins B, C and D
- Maturation factors—vitamin B_{12}, intrinsic factor of Castle, and folic acid
- Factors necessary for hemoglobin formation—proteins, iron, copper, cobalt, nickel and vitamins.

Q.87 At what serum bilirubin level jaundice occurs in adults and infants?

- *In adult*: If serum bilirubin increases beyond 2 mg% it results in jaundice.
- *In infants*: If serum bilirubin increases beyond 5 mg% it results in jaundice.

Q.88 Why jaundice is first detected in the eyes?

It is because of whiteness of sclera. Sclera has a protein known as elastin which has high affinity to bind bilirubin. So even in low grade of jaundice bilirubin can get bind with sclera.

Q.89 Why does stool darken on standing in air?

It is due to the conversion of residual urobilinogens (colorless) to colored urobillins.

Q.90 What is prehepatic jaundice? Why it does result in unconjugated hyperbilirubinemia?

When jaundice occurs due to increased formation of bilirubin it is called as prehepatic or hemolytic jaundice. In this case the liver is unable to conjugate the large amounts of bilirubin produced resulting in unconjugated hyperbilirubinemia.

Q.91 What is posthepatic jaundice? Why it does result in conjugated hyperbilirubinemia?

If the jaundice occurs due to biliary obstruction it is known as posthepatic jaundice. In this case the conjugated bilirubin produced in the liver regurgitates back into blood instead of flowing out into the duodenum. This is why it results in conjugated hyperbilirubinemia.

Q.92 What is hepatic jaundice? Why it does usually result in conjugated hyperbilirubinemia?

The jaundice due to the impairment of all steps of bilirubin metabolism in liver is known as hepatic jaundice. The commonest cause is infective hepatitis. In this case the excretion of bilirubin is worstly affected that results in conjugated hyperbilirubinemia.

Q.93 Define hemoglobin.

Hemoglobin is the coloring matter or the chromoprotein of the red blood cells.

Q.94 What is normal hemoglobin content in the blood?

Males: 15 g%

Females: 14.5 g%.

Q.95 What are the functions of hemoglobin?

- Transport of respiratory gases
- Buffer action.

Q.96 What are the components of hemoglobin?

Hemoglobin consists of a protein (globin) and iron containing pigment (heme). The pigment part of heme is porphyrin.

Q.97 What are the types of hemoglobin?

- Adult hemoglobin (HbA) $\rightarrow H_2\beta_2$
- Fetal hemoglobin (HbF) $\rightarrow H_2\gamma_2$

Q.98 What are the differences between adult hemoglobin and fetal hemoglobin?

Structural difference: Adult hemoglobin has got two alpha chains and two beta chains whereas the fetal hemoglobin has two alpha chains and two gamma chains

Functional difference: Fetal hemoglobin has got more affinity for oxygen than the adult hemoglobin.

Q.99 Name the abnormal hemoglobin.

- Hemoglobinopathies—hemoglobin S, C, E and M
- Hemoglobin in thalassemia and related disorders—hemoglobin G, H, I, Bart's, Lepore and constant spiring.

Q.100 Name the abnormal hemoglobin derivatives.

- Carboxyhemoglobin
- Methemoglobin or ferrihemoglobin
- Sulfhemoglobin.

Q.101 What is the quantity of iron in the body?

About 4 g.

Q.102 How is iron transported in the blood?

Iron is transported in blood as transferrin (iron combines with beta globulin called apotransferrin and forms transferrin).

Q.103 How is iron stored in the body?

Large quantity of iron is stored in reticuloendothelial cells and liver hepatocytes and small quantity is stored in other cells. In the cell cytoplasm, iron combines with a protein to form apoferritin. This is converted into ferritin and stored in the cytoplasm. A small quantity of iron is also stored in the form of hemosiderin.

Q.104 What is the morphological classification of anemia?

The classification of anemia according to the size and color (hemoglobin content) of RBC is known as morphological classification.

By this, the anemia is classified into four types:

- Normocytic normochromic anemia—the size and the hemoglobin content of

the cells are normal but the number is reduced
- Macrocytic normochromic anemia—the cells are larger with normal hemoglobin content but the number is reduced
- Macrocytic hypochromic anemia—the cells are larger in size with less hemoglobin content
- Microcytic hypochromic anemia—cells are smaller with less hemoglobin content.

Q.105 What is the etiological classification of anemia?

The classification of anemia depending upon the cause is known as etiological classification.

By this, anemia is classified into five types:
- Hemorrhagic anemia—due to blood loss
- Hemolytic anemia—due to destruction of large number of red blood cells
- Nutrition deficiency anemia - due to lack of nutritive substances like iron, protein, and vitamins C and B_{12}
- Aplastic anemia—due to destruction of bone marrow
- Anemia of chronic diseases - due to sustained diseases.

Q.106 What is the commonest form of anemia in the world?

It is iron deficiency anemia.

Q.107 What is the etiology of pernicious anemia?

It is due to deficiency of hematinic principle, i.e. lack of castle's intrinsic factor, resulting in failure of absorption of vitamin B_{12} from diet through ileum.

Q.108 Name the physiological and pathological condition of anemia?

Physiological—Pregnancy.

Pathological—Thalassemia, spherocytosis, malaria, iron deficiency, etc.

Q.109 What will happen if folic acid is given to pernicious anemic patient?

The administration of folic acid to pernicious anemic patient will improve the blood picture but it cannot protect against neuropathy which is due to deficiency of vit-B_{12}.

Q.110 What is the role of iron in the body?

These are: synthesis of hemoglobin, myoglobin and cytochromes.

Q.111 What is the blood and bone marrow picture in iron deficiency anemia?

The RBCs are microcytic, hypochromic and MCH, MCHC and total RBC count is reduced, whereas the bone marrow shows proliferation of the precursor cells with a larger proportion of the mature forms. Some of the precursor cells may show scanty, polychromatic cytoplasm with a pyknotic nucleus, i.e. the cytoplasmic maturity is less than nuclear maturity.

Q.112 What is the blood and bone marrow picture in megaloblastic anemia?

The blood picture is characterized by macrocytosis, anisocytosis, poikilocytosis, neutropenia with over matured neutrophils and also thrombocytopenia. Whereas in bone marrow all the RBC precursors show megaloblastic changes that includes:
- Larger cell with larger nucleus
- More reticular chromatin
- Normal hemoglobinization of cytoplasm.

Q.113 What are the effects of foliate deficiency?

This results in defective erythropoiesis resulting in megaloblastic anemia.

Q.114 Compare and contrast the folic acid, vit-B_{12} and iron deficiency anemia.

The comparison is made in Table 9.1.

Q.115 Give characteristics features of (i) iron deficiency anemia (ii) pernicious anemia.

- *Characteristics feature of iron deficiency anemia* is as follows:
  - Microcytic hypochromic RBC
  - MCV, MCH, MCHC and CI decreases
  - RBC count decreases or remains normal
  - Normoblastic hyperplasia of bone marrow
  - Normal WBC and platelet count
  - Soft, brittle and spoon shaped nail
  - Angry red tongue and dysphagia
  - Irritability, loss of concentration, headache, impotence
  - Early breathlessness, palpitation
- *Characteristics feature of pernicious anemia* is as follows:
  - Microcytic normochromic RBC
  - Marked decrease in RBC count and Hb concentration
  - Lemon yellow colored skin due to anemic paleness and mild jaundice
  - MCV, MCH increases and MCHC remains normal
  - Increase in reticulocyte count
  - Low grade hemolytic jaundice
  - Increase in serum iron concentration
  - Paresthesia, i.e. numbness, tingling, burning sensation ataxia, etc.

Q.116 What is sickle cell anemia?

Anemia due to presence of sickle shaped RBCs that contain abnormal hemoglobin (hemoglobin S) is called sickle cell anemia.

Q.117 What is the role of intrinsic factor of Castle?

Intrinsic factor of Castle is essential for the absorption of vitamin B_{12} (extrinsic factor) from the intestine. Vitamin B_{12} and the intrinsic factor of Castle together form the hematinic principle, which is necessary for the maturation of red blood cells.

Q.118 What is pernicious or Addison's anemia?

Anemia due to deficiency of vitamin B_{12} or intrinsic factor of Castle is known as pernicious or Addison's anemia.

Q.119 What is the morphology of RBC in pernicious anemia? Why?

In pernicious anemia, the RBCs are macrocytic and normochromic. This is because of lack of vitamin B_{12} and/or intrinsic factor which are necessary for maturation of red blood cells. So, the cells are not matured and remain larger in size.

Q.120 What is megaloblastic anemia?

Anemia due to deficiency of folic acid is known as megaloblastic anemia.

Table 9.1: Comparative study between deficiency of iron, folic acid and vitamin B_{12}

Iron deficiency anemia	*Folic acid deficiency anemia*	*Vit-B_{12} deficiency anemia*
RBCs are microcytic and hypochromic	RBCs are macrocytic	RBCs are macrocytic
Bone marrow shows hyperplasia of red cell precursors of erythroblasts	Bone marrow shows megaloblastic changes and presence erythroblasts	Bone marrow shows megaloblastic changes and presence of megaloblasts
Nuclear maturation of RBC is normal	Nuclear maturation of RBC is impaired	Nuclear maturation of RBC is impaired
No associated neurological hazards	No associated neurological hazards	Associated neurological hazards

Q.121 What is the morphology of RBC in megaloblastic anemia? Why?
Megaloblastic anemia is due to the lack of folic acid, which is essential for the synthesis of DNA in red blood cells. Because of deficiency of folic acid, DNA synthesis becomes defective. So, the cells are not matured and are macrocytic and hypochromic in nature.

Q.122 What is ESR? What is its normal value?
ESR or erythrocyte sedimentation rate is the rate at which the RBCs settle down when the blood is allowed to stand.
Normal value:
Males : 3 to 7 mm in one hour
Females : 5 to 9 mm in one hour
Infants : 0 to 2 mm in one hour.

Q.123 How is ESR determined?
By Westergren's method or Wintrob's method.

Q.124 What is the clinical importance of determining ESR?
Determination of ESR helps for diagnosis and prognosis. It has got more of prognostic value than the diagnostic importance.

Q.125 Name the physiological conditions when ESR increases.
ESR increases in infants, children and females. In females, it further increases during menstruation and pregnancy.

Q.126 Name the pathological conditions when ESR increases.
Tuberculosis, some types of anemia, rheumatoid arthritis, rheumatic fever and liver diseases.

Q.127 Name the pathological conditions when ESR decreases.
Allergic conditions, sickle cell anemia, peptone shock and polycythemia.

Q.128 What is packed cell volume (PCV)?
Packed cell volume is the volume of RBCs packed at the bottom of a hematocrit tube when the blood is centrifuged. The normal value is 40 to 45% in males and 38 to 42% in females. (Refer Questions 31 - 33 of this section for further details)

Q.129 Name the condition when PCV increases.
Polycythemia.

Q.130 Name the conditions when PCV decreases.
Anemia and pregnancy.

Q.131 What are blood indices?
Blood indices are the values, which indicate the size, volume and the hemoglobin content of RBCs.

Q.132 Define and give the normal values of blood indices.
- *Mean corpuscular volume (MCV):* The average volume of a single red blood cell—78 to 90 cu μ.
- *Mean corpuscular hemoglobin (MCH):* The quantity or amount of hemoglobin present in one red blood cell—27 to 32 pg
- *Mean corpuscular hemoglobin concentration (MCHC):* The concentration of hemoglobin in one red blood cell—30 to 38%
- *Color index (CI):* The ratio between percentage of hemoglobin and the percentage of RBCs in the blood—0.8 to 1.2.

Q.133 What is the advantage of blood indices?
Blood indices help to determine the type of anemia.

Q.134 What is hemolysis?
The breakdown of blood cells especially the RBCs is known as hemolysis.

Q.135 What is fragility? Which RBCs are more fragile?
The susceptibility of the RBCs to breakdown (hemolysis) is known as fragility. Older RBCs are more fragile and are easily broken down.

Q.136 Enumerate the variations of osmotic fragility of RBC.
Osmotic fragility is decreased during acholuric jaundice and some anemias. Whereas it is increased during hereditary spherocytosis, deficiency of glucose-6-phosphate dehydrogenase, cobra bite, etc.

Q.137 What happens if the RBC is kept in hypotonic and hypertonic saline?
- In hypotonic solution, water moves inside the RBC cell due to concentration gradient → swelling up of RBC → increase of RBC volume → rupture (hemolysis) of RBC takes place.
- In hypertonic solution, water moves out of the cell → shrinkage of the cell (crenated).

Q.138 What are hemolysins? Name the hemolysins.
Hemolysins or hemolytic agents are the substances, which cause the breakdown of red blood cells.
Hemolysins are:
- Chemical substances - alcohol, benzene, chloroform, ether, acid, alkalis like ammonia, bile salts, saponin and poisons like arsenical preparations, carbolic acid nitrobenzene and resin
- Toxins from bacteria
- Venom of poisonous snakes like cobra
- Hemolysins from normal tissues.

Q.139 What are the differences between the RBC and WBC?
Structural differences: RBC is nonnucleated, biconcave in shape (round in front view) and red in color whereas WBC is nucleated, irregular in shape and colorless.
Functional differences: RBC transports respiratory gases and plays an important role in buffer function whereas WBC plays important role in defense of the body.

Q.140 Classify WBCs.
WBCs are classified into granulocytes and agranulocytes depending upon the presence or absence of granules in the cytoplasm.

Q.141 What are trephones?
These are the substances prepared by leukocytes from plasma proteins that help in tissue nutrition.

Q.142 What is the average period for normal development of neutrophils?
12 days.

Q.143 Why the neutrophils are called polymorphs?
Because they have multilobed nucleus.

Q.144 Why neutrophils are so named?
This is a misnomer because they are not stained by the neutral stain rather by the mixed (both acidic and basic) stain like Leishman stain.

Q.145 Which stain is generally used to stain the peripheral blood smear?
It is Leishman stain.

Q.146 Which is the largest cell in the peripheral blood?
It is monocyte (diameter 15-20 μm).

Q.147 What is respiratory burst?
Within seconds of stimulation neutrophils sharply increase their oxygen uptake which is known as respiratory burst.

Q.148 What is Cook-Arneth count? What is its significance?
Counting of neutrophils on the basis of the number of lobes of their nuclei is called Cook-Arneth count.

Clinical significance: It represents the maturity of neutrophils. If shift to the left occurs that indicates the hyperactive bone marrow whereas hypoactive bone marrow is indicated by shift to the right.

Q.149 How do you differentiate between neutrophils, eosinophils and basophils? (Table 9.2)

Q.150 How do you differentiate small and large lymphocyte? Which one is more mature? (Table 9.3)

Small lymphocyte is more mature.

Q.151 How do you differentiate large lymphocyte with monocyte? (Table 9.4)

Q.152 What is Schilling index?

Arranging and counting of all leukocytes according to their age is known as Schilling index.

Q.153 What are the body's Ist line of defense and where they are located?

It is monocyte macrophage system or RES. They are located in almost all the tissues but in different form, e.g.

- In skin and subcutaneous tissues - Histocytes
- Lungs - Alveolar macrophages
- Intestine - Lymphoid tissue
- Liver and spleen pulp - Kupffer cells.

Q.154 Which lobed neutrophils are most active?

It is three lobed neutrophil (N_3).

Q.155 What do you mean by shift to the left and what is its significance?

In Arneth count, if $N_1 + N_2 + N_3$ becomes greater than 80 percent then it is known as shift to left or regenerative shift. It indicates the hyperactive bone marrow.

Q.156 What is shift to the right? What is its significance?

In Arneth count, if $N_4 + N_5 + N_6$ is greater than 20 percent it is called as shift to right or degenerative shift which indicates the hypoactive bone marrow.

Q.157 Is trilobed eosinophil possible?

Yes, 15 percent of eosinophils are trilobed.

Q.158 What are the stages of phagocytosis of WBC?

These are as follows:

Diapedesis → chemotaxis → opsonization and then phagocytosis → which causes degranulation → then inflammatory response → finally stops or limits inflammation.

Q.159 Classify the lymphocytes.

Histologically—two types: Small and large lymphocytes.

Functionally—two types: T-lymphocytes (responsible for cellular immunity) and B-lymphocytes (responsible for humoral immunity).

Q.160 Name the agranulocytes and describe them briefly.

The types of agranulocytes are as follows (Fig. 9.2):

- Monocytes, which are the largest of all the WBCs with clear cytoplasm. The nucleus is kidney shaped and it is either in the center of the cell or pushed to one side and large amount of cytoplasm is seen.
- Lymphocytes, which also have clear cytoplasm. The nucleus is oval or kidney shaped and occupies the whole of cytoplasm. Depending upon the size, the lymphocytes are classified into small and large lymphocytes.

Q.161 Give the total count and differential count of WBCs.

Total WBC count ranges between 4,000 and 11,000/cu mm of blood.

Differential WBC count:

Neutrophils : 50 to 70% (3000 to 6000/cu mm)
Eosinophils : 2 to 4% (150 to 450/cu mm)
Basophils : 0 to 1% (0 to 100/cu mm)
Monocytes : 2 to 6% (200 to 600/cu mm)
Lymphocytes : 20 to 30% (1500 to 2700/cu mm)

Q.162 Name the granulocytes (Fig. 9.2) and describe them briefly.

The types of granulocytes are as follows (Fig. 9.3):

- Neutrophil with fine granules, which take both acidic and basic stain (violet). It has multilobed nucleus
- Eosinophil with coarse granules, which stain bright red or orange with eosin. It has bilobed nucleus

Table 9.2: Neutrophils, eosinophils and basophils

S. No.	Parameter	Neutrophils	Eosinophils	Basophils
1.	***Size of the cell***	10–14 μm	10–16 μm	10–14 μm
2.	***Number of lobes in nucleus***	Multilobed	Usually bilobed	bilobed
3.	***Shape of nucleus***	Irregular	Spectacle shaped	Usually 'S' shaped
4.	***Color of granules in cytoplasm***	Pinkish	Brick red	Purple
5.	***Nature of granules***	Very fine	Coarse	Very coarse, making the nucleus obscure
6.	***Number of granules***	Few	Very dense	Small in number

Table 9.3: Small and large lymphocytes

S. No.	Parameter	Small lymphocyte	Large lymphocyte
1.	***Size of the cell***	Almost equal to the RBC size	Almost twice of RBC
2.	***Amount of cytoplasm***	Very thin layer of cytoplasm present only in periphery	Plenty compared to small lymphocyte
3.	***Shape of nucleus***	Round	Round or oval

Table 9.4: Large lymphocyte and monocyte

S. No.	Parameter	Large lymphocyte	Monocyte
1.	***Size***	Twice of RBC	Almost thrice of RBC
2.	***Shape of nucleus***	Round or oval	Indented or kidney-shaped
3.	***Position of nucleus***	Central	Eccentric
4.	***Amount of cytoplasm***	More than half of the cell	Less than half of the cell

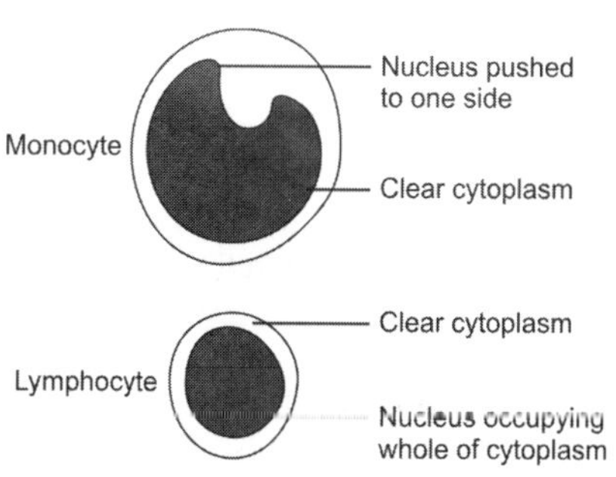

Fig. 9.2: Different types of agranulocytes

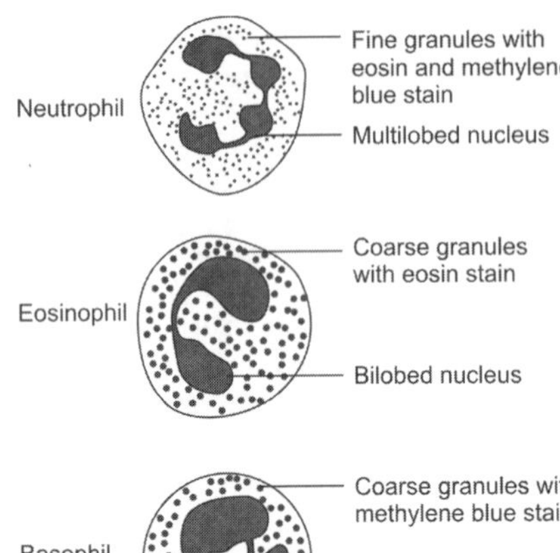

Fig. 9.3: Different types of granulocytes

- Basophil with coarse granules, which stain purple with methylene blue. It also has bilobed nucleus.

Q.163 What is leukocytosis? Name some physiological conditions when leukocytosis occurs.

Increase in WBC count is known as leukocytosis.

Physiologically, it is found in infants, children and males. It also occurs in high altitudes, during muscular exercise, during emotional conditions and in pregnancy.

Q.164 Name some pathological conditions when leukocytosis occurs.

- Infections
- Allergic conditions
- Common cold
- Tuberculosis
- Glandular fever.

Q.165 What is leukopenia? Name some pathological conditions when leukopenia occurs.

Decrease in WBC count is called leukopenia. It occurs in:

- Anaphylactic shock
- Cirrhosis of liver
- Disorders of spleen
- Pernicious anemia
- Typhoid and paratyphoid
- Viral infections.

Q.166 What is neutrophilia? Name some pathological conditions when it occurs.

Increase in neutrophil count is called neutrophilia or neutrophilic leukocytosis. It occurs in:

- Acute infections
- Metabolic disorders
- Injections of foreign proteins
- Injections of vaccines
- Poisoning by chemicals and drugs like lead, mercury, camphor, benzene derivatives, etc.
- Poisoning by insect venom
- After acute hemorrhage.

Q.167 Define neutropenia and mention its causes.

Neutropenia can be defined as a clinical condition characterized by the reduction of both differential and absolute neutrophil count.

Causes

Viral infection like typhoid.

Paratyphoid, AIDS, kala-azar, bone marrow depression, etc.

Q.168 What is eosinophilia? Name some pathological conditions when it occurs.

Increase in eosinophil count is called eosinophilia.

It occurs in:

- Allergic conditions
- Asthma
- Blood parasitism
- Intestinal parasitism
- Scarlet fever.

Q.169 What is eosinopenia and when does it occur?

Eosinopenia can be defined as the reduction in absolute eosinophil count below 50/cu.mm of blood.

It can be caused by ACTH or glucocorticoid therapy, acute stressful illness and acute pyogenic infection.

Q.170 What is basophilia? Name some pathological conditions when it occurs

Increase in basophil count is called basophilia.

It occurs in:

- Smallpox
- Chickenpox
- Polycythemia vera.

Q.171 What is monocytosis? Name some pathological conditions when it occurs.

Increase in monocyte count is called monocytosis.

It occurs in:

- Tuberculosis
- Syphilis
- Malaria
- Kala azar
- Glandular fever.

Q.172 Mention the causes of monocytopenia.

Bone marrow depression is the major cause of monocytopenia.

Q.173 Name some pathological condition in which lymphocytopenia occurs.

Low blood lymphocyte count, also known as lymphocytopenia usually occurs in patients with steroid therapy.

Q.174 What is lymphocytosis? Name some pathological conditions when it occurs.

Increase in lymphocyte count is called lymphocytosis.

It occurs in:

- Diphtheria
- Infections
- Hepatitis
- Mumps
- Rickets
- Syphilis
- Thyrotoxicosis
- Tuberculosis.

Q.175 What are the properties of WBC?

- Diapedesis—process of squeezing through the narrow blood vessels
- Amoeboid movement—movement by protruding the cytoplasm
- Chemotaxis—movement due to the attraction by chemical substances called chemoattractants released from the affected tissues
- Phagocytosis—process by which the foreign bodies are engulfed

Q.176 What are the functions of neutrophils?

Neutrophils provide first line defense along with monocytes. The neutrophils move to the site of infection by diapedesis and engulf the foreign bodies by phagocytosis. The enzymes like proteases, myeloperoxidases, elastases and metalloproteinases present in the neutrophils destroy the foreign invaders. Neutrophils secrete platelet activating factor.

Q.177 What are the chemical substances present in the granules and cell membrane of neutrophils?

Granules: Enzymes like proteases, myeloperoxidases, elastases and metalloproteinases and the antibody like substances called defensins.

Cell membrane: Dihydronicotinamide adenine dinucleotide phosphate oxidase (NADPH oxidase).

All these substances help the neutrophils to destroy the foreign bodies.

Q.178 What are the functions of eosinophils?

The eosinophils play an important role in defense mechanism by detoxification, disintegration and removal of foreign proteins. Eosinophils also act against the parasites.

Q.179 Name the chemical substances present in the granules of eosinophils.
Eosinophil peroxidase, major basic protein, eosinophil cationic protein, eosinophil derived neurotoxin and cytokines.

Q.180 What are the functions of basophils?
Basophils prevent intravascular clotting by secreting heparin and play an important role in healing processes after inflammation and allergy.

Q.181 Name the chemical substances present in the granules of basophils.
Histamine, heparin, hyaluronic acid, proteases, myeloperoxidase and cytokine.

Q.182 What are mast cells? What is their function?
Mast cells are large tissue cells resembling the basophils. These cells are present in the bone marrow and around the cutaneous blood vessels but do not enter the circulation.

Mast cells play an important role during allergy and anaphylaxis by secreting substances like heparin, histamine, serotonin and hydrolytic enzymes.

Q.183 What are the functions of monocytes?
Monocytes provide first line defense along with neutrophils. These cells wander freely through all the tissues. The matured monocytes move into the tissues and become tissue macrophages. The macrophages engulf the foreign particles by phagocytosis and destroy them.

Q.184 What are the functions of lymphocytes?
Lymphocytes protect the body by providing immunity.

Q.185 What is pus? What are the pus cells?
Many WBCs are destroyed while attacking the invading organisms. These dead WBCs along with plasma, liquefied tissue cells and RBCs combine to form a liquid product called pus. The dead WBCs are called the pus cells.

Q.186 What is leukopoiesis?
Leukopoiesis is the origin, development and maturation of WBCs.

Q.187 What is leukemoid reaction?
Extreme increase of TLC (>50,000/cumm of blood) characterized with elevated level of leukocyte alkaline phosphatase due to severe infection is known as leukemoid reaction. Its difference with leukemia is that in case of leukemia alkaline phosphatase level is reduced whereas here it is increased significantly.

Q.188 What is leukemia? What is its difference with leukocytosis?
Leukemia is a group of malignant neoplasms resulting from uncontrolled proliferation of hemopoietic leukocytic stem cells of bone marrow and lymphoid tissue. In this case, TLC becomes much higher than leukocytosis, i.e. 1–3 lac/cumm and number of immature cells are dominant.

Q.189 What are the types of immunity?
- Innate immunity or inborn immunity—present from the birth itself like the resistance given by the stomach against the pathogens entering through the food.
- Acquired immunity—developed in the body when exposed to a new invading organism.

Q.190 What are the types of acquired immunity?
Cellular immunity and humoral immunity.

Q.191 Which are the cells responsible for acquired immunity?
Lymphocytes are responsible for acquired immunity. T lymphocytes provide cellular immunity and B lymphocytes provide humoral immunity.

Q.192 What are T lymphocytes?
Lymphocytes which are processed in thymus and taking part in cellular immunity are called T lymphocytes.

Q.193 What are B lymphocytes? Why these cells are called so?
Lymphocytes which are processed in bone marrow and liver and taking part in humoral immunity are called B lymphocytes. These cells were first discovered in the bursa of Fabricius in birds and hence the name B lymphocytes.

Q.194 Where are the T cells and B cells stored?
After being processed, the T cells and B cells migrate and get stored in the lymphoid tissues present in the lymph nodes, spleen, bone marrow and gastrointestinal tract.

Q.195 What are the different types of T cells?
- Helper T cells
- Cytotoxic or killer T cells
- Suppressor T cells
- Memory T cells.

Q.196 What are the different types of B cells?
Plasma cells and memory B cells.

Q.197 What are antigens? What are the types of antigens?
The antigens are the protein substances, which induce specific immune reactions in the body.
Types of antigens:
- Self antigens or autoantigens
- Nonself antigens.

Q.198 What are the self antigens?
The antigens present in cells of our own body are known as self antigens.

Q.199 What are the nonself antigens?
Nonself antigens are the antigens, which enter the body from outside through some bacteria, virus, fungus, transplanted organs, transfused incompatible blood cells, allergens, etc.

Q.200 What are the antigen presenting cells?
The cells, which expose or present the antigen of invading organisms to the lymphocytes are called antigen presenting cells.

The macrophages and dendritic cells are the antigen presenting cells.

Q.201 Name the two types of helper T cells. Mention their functions.
Helper-1 cells are concerned with cellular immunity and secrete
- Interleukin-2 which activates other T cells
- Gamma interferon that stimulates the cytotoxic cells, macrophages and natural killer cells.

Helper-2 cells are concerned with humoral immunity and secrete interleukins 4 and 5 which are concerned with
- Activation of B cells
- Proliferation of plasma cells
- Antibody production by plasma cells.

Q.202 What are the functions of cytotoxic T cells?
The cytotoxic T cells:
- Attack the invading organisms and destroy them by releasing cytotoxic substances like lysosomal enzymes
- Destroy cancer cells, transplanted cells and other foreign bodies
- Destroy even body's own tissues which are affected by foreign bodies.

Q.203 What are the disadvantages of the actions of cytotoxic T cells?
The cytotoxic T cells are otherwise called killer T cells because these cells destroy the invading organisms. But, at the same time, the cytotoxic T cells may attack the cells in transplanted heart or kidney

leading to rejection of the transplanted tissues. These cells may destroy even the tissues affected by the invading organisms.

Q.204 What is the role of suppressor T cells?

Suppressor T cells or regulatory cells suppress the action of killer cells so that, the destruction of body's own tissues is prevented. The suppressor T cells also suppress the activities of helper T cells.

Q.205 What is the importance of memory T cells?

Some of the T cells activated by the antigens of invading organism move to the lymphoid tissues and remain there. These cells are called memory T cells. When the body is attacked by the same organism for the second time, these memory cells recognize the organism and immediately activate the other T cells so that, the invading organism is destroyed quickly and effectively.

Q.206 What is humoral immunity? Which are the cells responsible for it?

The immunity provided by the antibodies is known as humoral immunity. B lymphocytes are responsible for it.

Q.207 What is the role of plasma cells in humoral immunity?

Plasma cells produce antibodies against the antigens of invading organisms. The antibodies which are also called immunoglobulins destroy the invading organisms.

Q.208 Name the immunoglobulins secreted by the plasma cells.

Immunoglobulins secreted by the plasma cells are IgA, IgD, IgE, IgG and IgM.

Q.209 What are the mechanisms of action of immunoglobulins?

Immunoglobulins destroy the invading organisms by two mechanisms:

- Direct action
- Through complement system.

Q.210 What are the direct actions of immunoglobulins?

The direct actions by which the immunoglobulins destroy the foreign bodies are agglutination, precipitation, neutralization and lysis.

Q.211 What is complement system?

The system of enzymes that enhances or accelerates various activities during the fight against the invading organisms is called complement system. Apart from the direct actions the immunoglobulins can destroy the invading organism through this system also.

Q.212 What is natural killer cell (NK cell)? What are its functions?

NK cell is a large granular cell with a nucleus. It is considered as the third type of lymphocyte.

NK cell:

- Destroys virus
- Destroys viral infected or damaged cells, which might form tumors
- Destroy the malignant cells
- Secretes cytokines—interleukin-2, interferons, colony stimulating factor and tumor necrosis factor-α.

Q.213 What is autoimmune disease? Name some autoimmune diseases.

When the immune system fails in the body, antibodies are produced against body's own tissues and destroy them. This is known as autoimmune disease.

Some of the autoimmune diseases are insulin dependent diabetes mellitus (IDDM), myasthenia gravis, Hashimoto's thyroiditis, Graves' disease and rheumatoid arthritis.

Q.214 Define and classify immunization.

Immunization is the method of preparing the body to fight against a specific disease. It is of two types:

- Passive immunization—produced by administration of serum or gamma globulins from a person who is already immunized to a non-immune person
- Active immunization—acquired by activating immune system of the body.

Q.215 Define and classify cytokines.

Cytokines are the hormone like small proteins acting as intercellular messengers by binding to specific receptors of target cells.

Cytokines are of six types:

- Interleukins
- Interferons
- Tumor necrosis factors
- Chemokines
- Defensins
- Cathelicidins.

Q.216 What are platelets? What is the normal platelet count?

Platelets or thrombocytes are small, colorless and nonnucleated formed elements of the blood.

The normal platelet count is 2.5 lakhs (250,000)/cu mm of blood (Fig. 9.4).

Q.217 Name the organic substances present in the platelets.

- Contractile proteins—actin, myosin, thrombosthenin.
- von Willebrand factor
- Fibrin stabilizing factor
- Platelet derived growth factor
- Platelet activating factor
- Vitronectin
- Thrombospondin.

Q.218 What are the events involved in hemostasis?

These are:

- Vasoconstriction
- Formation of temporary hemostatic plug.
- Conversion of temporary hemostatic plug into secondary or definitive hemostatic plug by fibrin.

Q.219 How the primary hemostatic plug is formed?

It is represented by the following sequences:

Platelets adhesion → platelets activation → platelets aggregation → activation of phospholipase A_2 → release of arachidonic acid from membrane phospholipids → release of thromboxane A_2 and prostacyclin → this ultimately causes adhesion of more and more platelets and then platelets are aggregated with each other to seal the rupture of blood vessels temporarily.

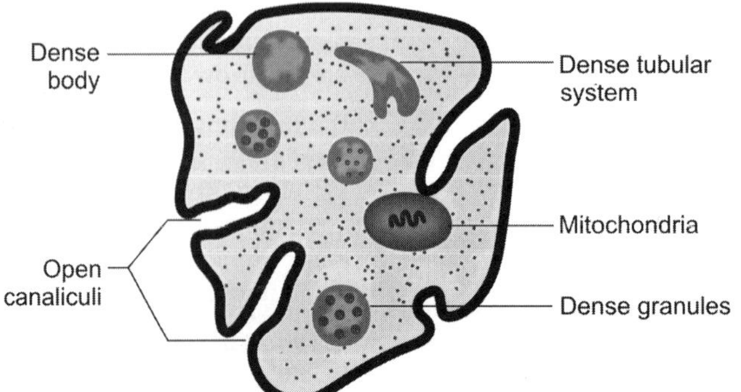

Fig. 9.4: Platelet

Q.220 What are the principal causes of hemorrhagic state in the body?

These are:

- Defect in the blood vessels due to infection, allergy, etc.
- Defect in platelets (purpura)
- Defect in clotting mechanism.

Q.221 What is thrombocytosis? Name some conditions when thrombocytosis occurs.

Increase in platelet count is known as thrombocytosis.

It occurs in:

- Allergic conditions
- Asphyxia
- Hemorrhage
- Bone fractures
- Surgical operations
- Splenectomy
- Rheumatic fever
- Trauma.

Q.222 What is thrombocytopenia? Name some conditions when thrombocytopenia occurs.

Decrease in platelet count is known as thrombocytopenia.

It occurs in:

- Acute infections
- Acute leukemia
- Aplastic anemia and pernicious anemia
- Chicken pox
- Small pox
- Splenomegaly
- Scarlet fever
- Typhoid
- Tuberculosis.

Q.223 What is thrombocythemia? Name some conditions when thrombocythemia occurs.

The condition with persistent and abnormal increase in platelet count is called thrombocythemia.

It occurs in:

- Carcinoma
- Chronic leukemia
- Hodgkin's disease.

Q.224 Name the properties of platelets.

Adhesiveness, aggregation and agglutination.

Q.225 What are the functions of platelets?

Platelets:

- Are responsible for blood clotting
- Are responsible for clot retraction
- Prevent blood loss during hemorrhage, by causing vasoconstriction and sealing the wound by plug formation
- Help in the repair of endothelium of damaged blood vessels
- Play a role in defense mechanism by agglutination and phagocytosis.

Q.226 What is hemostasis?

The arrest of bleeding is called hemostasis.

Q.227 Name the stages of hemostasis.

Stage I: Vasoconstriction caused by serotonin secreted by platelets.

Stage II: Formation of platelet plug caused by ADP and thromboxane A_2 secreted from platelets.

Stage III: Coagulation of blood.

Q.228 Define coagulation of blood.

When blood is shed out or collected in a container, it loses its fluidity and becomes a jelly-like mass after few minutes. This is known as coagulation or clotting of blood.

Q.229 Name the clotting factors.

- Fibrinogen
- Prothrombin
- Thromboplastin
- Calcium
- Labile factor (proaccelerin or accelerator globulin)
- Presence has not been proved
- Stable factor
- Antihemophilic factor (antihemophilic globulin)
- Christmas factor
- Stuart-Prower factor
- Plasma thromboplastin antecedent
- Hegman factor (contact factor)
- Fibrin stabilizing factor (fibrinase).

Q.230 Which is the inorganic ion necessary for blood clotting?

Calcium ion (factor IV).

Q.231 Mention the role of Ca^{++} in clotting mechanism.

Except for the first 2 steps in the intrinsic pathway calcium ions are required for the promotion of all reactions involved in both intrinsic and extrinsic pathway.

Q.232 Why Christmas factor is called so?

Christmas factor was named after the patient in whom it was discovered.

Q.233 Why blood does not clot during circulation?

Blood does not clot during circulation because:

- Clotting factors are in inactive form
- Smooth endothelial lining of the blood vessels does not allow the blood clotting
- Continuous flow does not allow the blood clotting
- The natural anticoagulant called heparin in the blood prevents clotting during circulation.

Q.234 What are the three stages of blood clotting?

Stage I: Formation of prothrombin activator

Stage II: Conversion of prothrombin into thrombin

Stage III: Conversion of fibrinogen into fibrin.

Q.235 What are the components of blood clot?

The blood clot consists of the RBCs, WBCs and the platelets entrapped in the fibrin meshwork.

Q.236 What is clot retraction?

30 to 45 minutes after formation, the blood clot contracts and a straw colored fluid called serum oozes out of it. This process is called clot retraction. (Refer Question 34 of this section for details of serum).

Q.237 What are the substances necessary for clot retraction?

The contractile proteins actin, myosin and thrombosthenin present in cytoplasm of platelets are necessary for clot retraction.

Q.238 What is lysis of clot? How is it brought?

The destruction or dissolution of blood clot is known as lysis of clot. It is brought out by a substance called plasmin.

Q.239 What is anticoagulant?

A substance that prevents or prolongs blood clotting is called anticoagulant.

Q.240 Name some anticoagulants, which can be used *in vivo* (inside the body).

Heparin, dicoumarol, warfarin and EDTA.

Q.241 What is the mechanism of action of heparin?

Heparin prevents blood clotting by:

- Suppressing activity of thrombin (antithrombin activity)
- Activating antihrombin
- Removing thrombin from circulation
- Inactivating other clotting factors.

Q.242 Name the anticoagulants, which are used *in vitro*.

Heparin, EDTA, oxalates and citrates.

Q.243 Define bleeding time.

The time interval from oozing of blood after injury till the arrest of bleeding is called bleeding time.

Q.244 What is the normal bleeding time? In which disease it is prolonged?

The normal bleeding time is 1 to 3 minutes. It is prolonged in purpura.

Q.245 Define clotting time.

The time interval between oozing out of blood after injury and clot formation is called clotting time.

Q.246 What is the normal clotting time? In which disease it is prolonged?

The normal clotting time is 3 to 8 minutes. It is prolonged in hemophilia.

Q.247 What are indications of BT and CT?

These are:

- Frequent and persistent bleeding from minor injuries.
- Before the minor/ major surgeries.
- In case of family history of bleeding.

Q.248 Which aspects of hemostasis are tested by BT and CT?

BT is to test for platelet function whereas CT is to test the abnormalities (if any) in clot formation. That is why in hemophilia BT is normal but CT is prolonged as in hemophilia, temporary hemostatic plug is formed because of normal functioning of platelets but they are washed off by the flowing blood as definitive hemostatic plug, i.e. clot is not formed.

Q.249 Mention the conditions when BT and CT is prolonged.

BT is increased during thrombocytopenic purpura, allergic and also senile purpura, infection like typhus, bacterial endocarditis, deficiency of vitamin C, etc.

CT is prolonged in hemophilia, afibrinogenemia, vitamin-K deficiency, liver disease, etc.

Q.250 What are the physiological and pathological variations of platelet count?

Physiological

Increase of count—in severe exercise and high altitude.

Decrease of count—in newborn babies and after menstruation.

Pathological

Increase of count—Severe hemorrhage and removal of spleen.

Decrease of count—Bone marrow depression, acute septic fever, aplastic anemia, toxemia, autoimmuno-destruction of platelets, AIDS, etc.

Q.251 What is the basic difference between intrinsic and extrinsic system of blood clotting?

In the intrinsic system, injury to blood cells like platelets, releases phospholipid that activate different clotting factors to induce clotting. Whereas in the extrinsic system injury to blood vessels or nearby tissues releases tissue thromboplastin which induces clotting mechanism by activating different clotting factors by cascade mechanism (Fig. 9.5).

Q.252 What is prothrombin time? What is its significance?

Prothrombin time is the test for prothrombin activity and thereby it is a test for testing the extrinsic system of blood coagulation. Normal value of prothrombin time is 11–16 sec and it is increased in liver failure and deficiency of vitamin-K. It is generally used to monitor patients with anticoagulants therapy to adjust its dose.

Q.253 Name the bleeding disorders.

Hemophilia, purpura and von Willebrand disease.

Q.254 What is hemophilia?

Hemophilia is sex linked inherited bleeding disorder with prolonged clotting time and normal bleeding time.

Q.255 What are the types of hemophilia? And what is the cause for each?

Hemophilia is of two types:

- Hemophilia A or classical hemophilia. It is due to the deficiency of clotting factor VIII (antihemophilic factor).
- Hemophilia B or Christmas disease. It is due to the deficiency of clotting factor IX (Christmas factor).

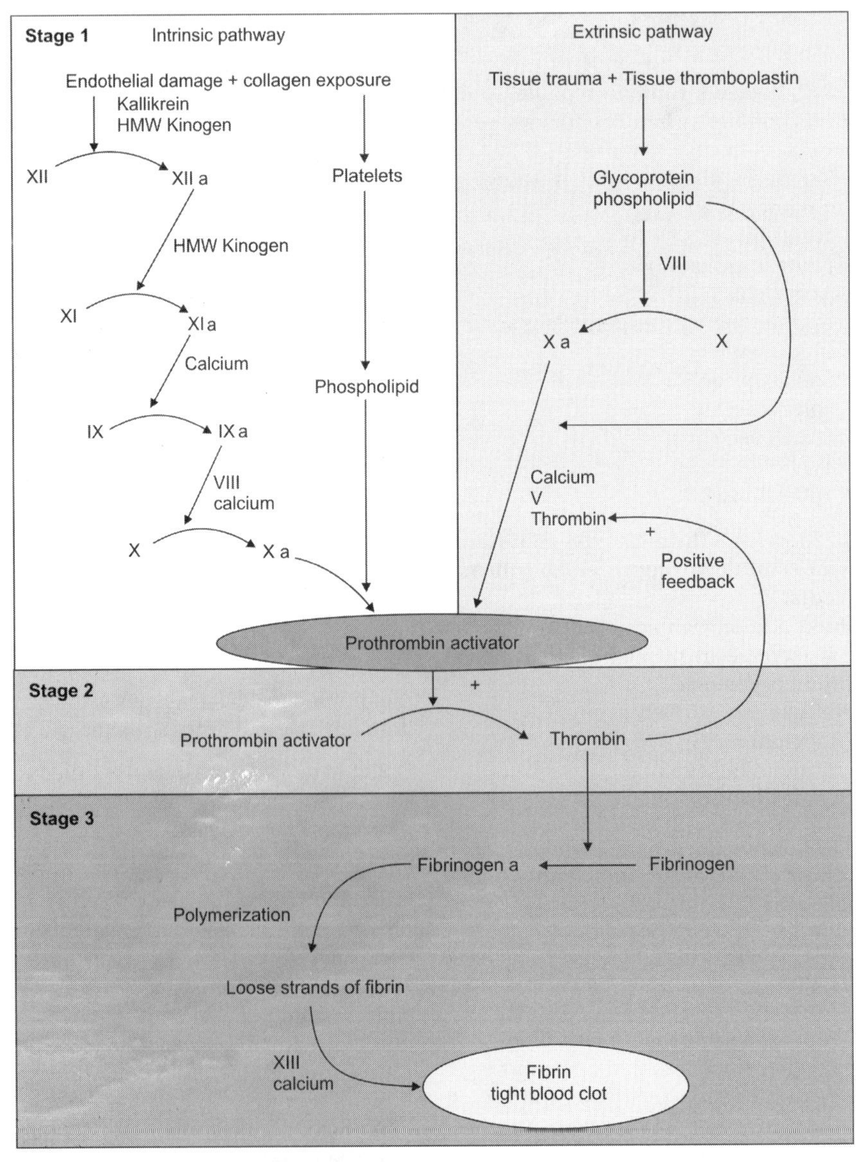

Fig. 9.5: Stages of blood coagulation.
a = activated + = thrombin induces formation of more thrombin

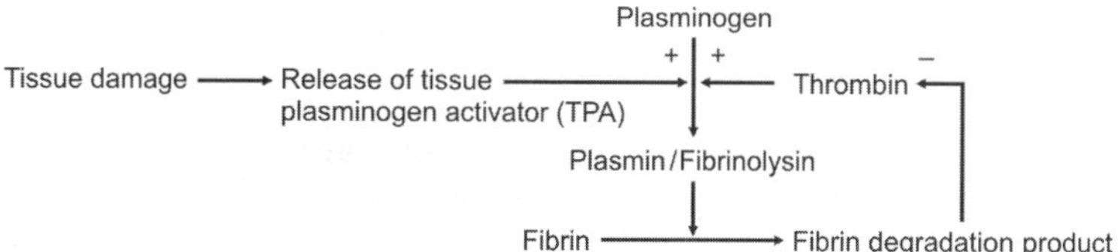

Fig. 9.6: Clotting mechanism and fibrinolytic system in the body

Q.256 How a balance is maintained between the clotting mechanism and fibrinolytic system in the body?
Factors that initiate clotting mechanism also stimulate the dissolution of clot (fibrinolysis) by the following mechanism (Fig. 9.6).

Q.257 What is purpura? What are its causes?
The purpura is purple colored petechial hemorrhagic condition with bruises in the skin due to the degradation of Hb over a period of time. The causes are thrombocytopenia, allergy, old age, functional platelet defects, etc.

Q.258 What is the difference between thrombocytopenia and thrombasthenia?
Reduction of platelet count below 1.5 lakh/cumm of blood is known as thrombocytopenia whereas impairment of platelet functions due to presence of abnormal platelets are known as thromboasthenia.

Q.259 Name two well known vascular causes of bleeding?
Scurvy and Cushing syndrome.

Q.260 Why does vit-K deficiency cause bleeding tendency?
This is due to the facts that Vitamin K deficiency results in low plasma levels of both procoagulants as well as some anticoagulants. These proteins are called vitamin K dependent proteins.

Q.261 Why does blood become incoagulable following violent death?
In case of violent death, the blood remains in fluidic and incoagulable in nature due to fibrinolysis resulted due to adrenaline induced rapid release of plasminogen activators from endothelial cells.

Q.262 What is the difference between rouleaux formation and agglutination?
Rouleaux formation is simply stacking of RBCs without any hemolysis whereas in agglutination there is antigen-antibody reaction on the red cells resulting in hemolysis of RBC.

Q.263 Name the cold antibodies present in our body.
ABO antibodies are the cold antibodies because they act best at low temperature, i.e. between 5°C and 20°C.

Q.264 Name the warm antibodies present in our body?
Rh-antibody is the warm antibody because they act best at normal body temperature, i.e. 37°C.

Q.265 How is hemophilia differentiated from purpura by simple laboratory test?
In hemophilia, the clotting time is prolonged whereas in purpura the bleeding time is prolonged.

Q.266 What is von Willebrand disease? What is its cause?
von Willebrand disease is the condition associated with excessive bleeding even with a mild injury.

It is due to the deficiency of von Willebrand factor. This factor is a protein necessary for the adherence of platelets to endothelium of blood vessel during hemostasis. If there is deficiency of this factor, the platelets do not adhere and this leads to excessive bleeding even with mild injury.

Q.267 What is thrombosis?
Intravascular clotting is known as thrombosis.

Q.268 What is thrombus?
The solid mass of intravascular clot is called thrombus.

Q.269 What is embolism?
Embolism is the process in which the thrombus or a part of it gets detached, travels in the blood stream, and obstructs the blood flow to any part of the body.

Q.270 Name the causes for thrombosis.
- Injury to blood vessel
- Rough endothelial lining
- Sluggish flow of blood
- Agglutination of red blood cells
- Presence of toxic substances like mercury and snake venom
- Congenital absence of protein C.

Q.271 What is Landsteiner's law?
Landsteiner's law states that "if an agglutinogen is present in red blood cell of a person, the corresponding agglutinin must be absent in the plasma" and "if an agglutinogen is absent in the red blood cell, the corresponding agglutinin will be present in the plasma". According to Landsteiner's law, blood group is classified as 'A', 'B', 'AB' and 'O' depending upon the presence or absence of agglutinogen (antigen) in the red blood cell. This grouping is also known as ABO system.

Q.272 Name the agglutinogen (antigen) and agglutinin (antibody) present in ABO system.
The agglutinogen and agglutinin present in ABO system are as follows (Fig. 9.7):

In 'A' group : Agglutinogen is 'A' and agglutinin is beta (anti 'B')
In 'B' group : Agglutinogen is 'B' and agglutinin is alpha (anti 'A')
In 'AB' group : Both 'A' and 'B' agglutinogens are present but no agglutinin
In 'O' group : No agglutinogen is present but both alpha and beta agglutinins are present.

Q.273 Who is universal donor? Why?
Person with 'O' group blood is called universal donor because his blood does not contain any agglutinogens in his blood.

Usually, during transfusion of blood, the RBCs of the donor (which contains agglutinogen) agglutinate with the agglutinin present in recipient's plasma. Since 'O' group blood does not contain any agglutinogen it can be given to any blood group person without the risk of agglutination. So he is known as universal donor.

Q.274 Who is universal recipient? Why?
Person with 'AB' blood group is called universal recipient, because, his blood does not contain any agglutinin in his plasma.

Usually, during blood transfusion, the donor's agglutinogen will agglutinate with recipient's agglutinin. But, 'AB' group blood does not contain any agglutinin in plasma and, so the person with 'AB' group can receive blood from persons with any other blood group. So, this person is called universal recipient.

Q.275 What is cross matching? What is its importance?
Matching (or blood typing or determination of blood group) is done by mixing the

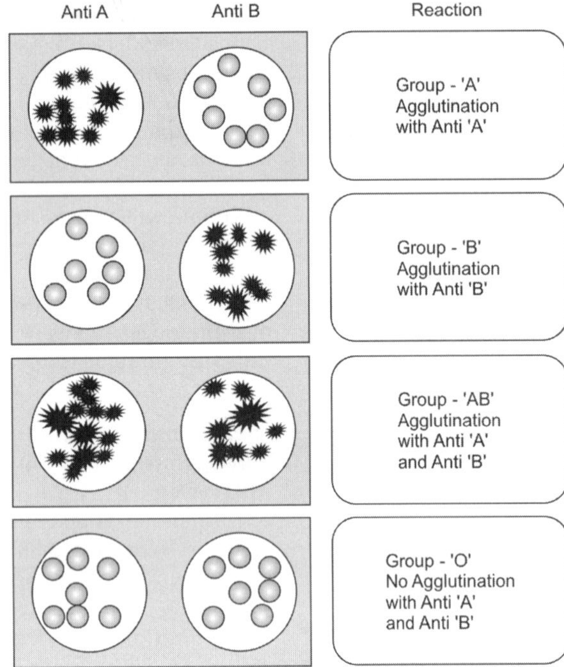

Fig. 9.7: Determination of blood groups

recipient's RBCs with test sera. In cross matching, the serum of the recipient and the RBCs of the donor are mixed.

Cross matching is always done before blood transfusion. If agglutination of the RBCs from a donor occurs during cross matching, the blood from that person is not used for transfusion.

Q.276 What is Rh factor? Why is it called so?

Rh factor is an antigen present in the red blood cell. It was first found in rhesus monkey and hence it is called Rh factor.

Q.277 How is Rh blood type classified?

Rh blood type is classified depending upon the presence or absence of Rh factor (antigen) in the RBCs. If Rh factor is present, the person is called Rh positive and if Rh factor is absent, the person is called Rh negative.

Q.278 In what way Rh type is different from ABO system?

In ABO system of blood grouping, there is natural corresponding antibody (agglutinin) whereas, in Rh typing, there is no natural corresponding antibody.

Q.279 What are the complications (transfusion reactions) of mismatched blood transfusion?

- Agglutination
- Hemolysis
- Jaundice
- Cardiac shock
- Renal shut down.

Q.280 Why the transfusion reactions do not occur when Rh negative person is given Rh positive blood for the first time? And what happens if the same person is given Rh positive blood for the second time?

There is no antigen in Rh negative blood and there is no antibody in the Rh positive person. So, when Rh positive blood is given to Rh negative person for the first time, there is no reaction. But, the Rh antibody develops and remains in his blood. So, when the same person receives Rh positive blood for the second time, the transfusion reactions occur.

Q.281 Name the hemolytic diseases of newborn.

- Erythroblastosis fetalis
- Hydrops fetalis
- Kernicterus.

Q.282 Explain erythroblastosis fetalis briefly.

It is the complication developed in the fetus of Rh negative mother. When the mother is Rh negative and father is Rh positive, the fetus may be Rh positive. The placental barrier does not allow Rh antigen (D antigen) to move from fetal blood into mother's blood. So, there is no complication and the child escapes. But, during delivery of the child, due to the severance of umbilical cord, the Rh antigen from the fetal blood enters the mother's blood. This causes development of antibody in mother's blood.

During second pregnancy, the Rh antibody from mother's blood enters fetus since, the placental barrier permits the Rh antibody. If this fetus also is Rh positive, agglutination occurs in fetal blood leading to complications like severe hemolysis, jaundice and anemia. This condition is called erythroblastosis fetalis.

Q.283 What are the complications of hemolysis in Rh positive fetus apart from presence of erythroblastic cells?

The other complications are hydrops fetalis and kernicterus.

Q.284 Name the blood groups other than ABO group.

Lewis blood group, MNS blood group, Auberger group, Diego group, Bombay group, Duffy group, Lutheran group, P group, Kell group, I group, Kidd group and Sulter Xg group.

Q.285 What is the importance of determining blood group?

Determination of blood groups helps in
- Safe blood transfusion
- Medicolegal cases
- Paternity test
- Prevention of complications like erythroblastosis fetalis.

Q.286 Name the conditions when blood transfusion is essential.

Hemorrhage, trauma, burns and anemia.

Q.287 Why the stored blood is not suitable for transfusing WBCs and platelets to a recipient?

It is because the blood stored for more than 24 hours does not contain active WBCs and platelets.

Q.288 The term universal donor and universal recipient are no longer valid. Justify

In both the cases complications can also be produced due to mismatching of Rh factors and other blood groups.

Q.289 What changes RBCs undergo during cold storage?

Cold storage results following changes:
- Appearance of spherocytic RBC due to net increase in volume of cell.
- Increase in tendency of hemolysis.

Q.290 What are the precautions to be taken before the transfusion of blood?

- Donor must be healthy without any infectious diseases like syphilis, hepatitis and AIDS
- Only compatible blood must be transfused
- Both matching and cross matching must be done.

Q.291 What are the precautions to be taken while transfusing blood?
- Apparatus must be sterile
- Temperature of the blood must be same as body temperature
- Transfusion must be done slowly to avoid the load on the heart.

Q.292 What is blood substitute? Name some commonly used blood substitutes.
The substance infused in the body instead of whole blood is known as blood substitute. The commonly used blood substitutes are human plasma, 0.9% sodium chloride solution, 5% glucose solution and some colloids like gum acacia, isinglass, albumin and animal gelatin.

Q.293 What is exchange transfusion or replacement transfusion? What is its significance?
Exchange transfusion is the procedure which involves the removal of patient's blood and replacing it with fresh donor blood or plasma. It is an important life saving procedure usually done to decrease or remove the effects of severe jaundice or changes in the blood like sickle cell anemia.

Q.294 What is the normal blood volume?
5 liters in a young healthy adult weighing about 70 kg.

Q.295 Name some pathological conditions when blood volume decreases.
- Hemorrhage
- Fluid loss
- Hemolysis
- Anemia
- Obesity
- Hypothyroidism.

Q.296 Name some pathological conditions when blood volume increases.
- Hyperthyroidism
- Hyperaldosteronism
- Cirrhosis of liver
- Congestive heart failure.

Q.297 How is blood volume regulated?
Blood volume is regulated by renal mechanism and hormonal mechanism which are controlled by hypothalamus.

Q.298 What is reticuloendothelial system?
Reticuloendothelial system is a system of primitive cells, which play an important role in formation of blood cells, destruction of blood cells and defense mechanism of the body.

Q.299 What is macrophage?
Macrophage is a large cell which has the property of phagocytosis.

Q.300 What are the two types of cells (macrophages) found in reticuloendothelial system?
- Fixed cells—tissue macrophages (fixed histiocytes) present in pleura, omentum, mesentery, endothelium of blood sinusoids, reticulum of spleen and liver, meningocytes, microglia in brain, lungs and subcutaneous tissue
- Wandering cells—free histiocytes present in blood (neutrophils and monocytes) and solid tissues like connective tissue.

Q.301 What are the tissue macrophages?
The fixed reticuloendothelial cells present in the tissues are called tissue macrophages or fixed histiocytes.

Q.302 What are the functions of reticuloendothelial system?
Most of the functions of reticuloendothelial system are carried out by tissue macrophages. The functions are:
- Phagocytosis
- Secretion of bactericidal agents
- Secretion of interleukins
- Secretion of tumor necrosis factors
- Secretion of transforming growth factor
- Secretion of colony stimulation factor
- Secretion of platelet derived growth factor
- Removal of carbon particles and silicon
- Destruction of senile RBC
- Destruction of hemoglobin
- Hemopoietic function.

Q.303 What is splenomegaly and hypersplenism?
Enlargement of spleen is called splenomegaly and increased activities of spleen is called hypersplenism.

Q.304 Name some causes of splenomegaly.
- Infectious diseases
- Inflammatory diseases
- Pernicious anemia
- Liver diseases
- Hematological disorders
- Cysts in spleen
- Hodgkin's disease
- Glandular fever.

Q.305 What is asplenia?
Absence of normal functions of spleen is called asplenia.

Q.306 What is lymphatic system? And what is lymph?
Lymphatic system is a closed system of lymph channels or lymph vessels. And lymph is a tissue fluid.

Q.307 What is the composition of lymph?
Lymph contains 96% of water and 4% of solids. Solids are organic and inorganic substances.
Organic substances:
- Proteins—albumin, globulin, fibrinogen, prothrombin, clotting factors, antibodies and enzymes
- Lipids—chylomicrons and lipoproteins
- Carbohydrate—glucose
- Amino acids
- Nonprotein nitrogenous substances—urea and creatinine.

Inorganic substances:
Sodium, calcium, potassium, chlorides and bicarbonates.

Q.308 What do you mean by secretors and nonsecretors?
The A and B antigens are also present in other tissues like liver, pancreas, kidney, etc. and also in body fluids like saliva, semen, etc. The individuals who have high concentration of these antigens in their body fluids are called secretors and those having low concentration of these antigens in their body fluid are known as nonsecretors.

Q.309 What are human leukocyte antigens (HLA) and what is their importance?
HLA are the antigens present on the surface of WBCs. HLA typing is done before tissue or bone marrow transplant since they are responsible for early rejection of transplant.

Q.310 Name the factors promoting erythropoiesis.
These are hypoxia, iron, porphyrin, traces of copper, cobalt, protein, vitamin-C, thyroxin, hematinic principle, maturation factor, etc.

Q.311 Why we do not make an immunological response of our own body proteins?
Body has got the capacity to identify it's own 'self' protein against of which no immunological response is evoked.

Q.312 What are the functions of lymph?
- Return of proteins from tissue spaces to blood
- Redistribution of fluid in the body

- Removal of substances like toxins and bacteria
- Maintenance of structural and functional integrity of tissues
- Serves as the route for absorption of fat
- Transport of lymphocytes.

Q.313 What are the functions of lymph nodes?

- Filtration of lymph
- Destruction of bacteria and toxic substances by acting like defense barriers.

Q.314 Name the main extracellular fluid cation.

It is sodium.

Q.315 Name the main anions in extracellular fluid.

These are chloride and bicarbonate.

Q.316 Which is the main intracellular fluid cation?

It is potassium.

Q.317 What is average pH of extracellular fluid?

It is 7.4.

Q.318 What is the normal average capillary pressure at the arterial end?

It is 32 mm Hg.

Q.319 What is the normal average capillary pressure at the venous end?

It is 12 mm Hg.

Q.320 Name the various of RES in the body.

These are Kupffer cells in liver: endothelium and reticulum cells in spleen; reticulum cells in bone marrow and lymph glands; histiocytes in serous membrane, and cells in capsules of suprarenal glands and anterior pituitary body.

Q.321 Of the lymph and the tissue fluid which has more protein content?

It is the lymph. Protein content of tissue fluid is almost negligible.

Q.322 What is the rate of lymph flow?

It is 1-1.5 mL/min. The direction of lymph flow is shown in Figure 9.8.

Q.323 What is the most important function of lymphatic circulation?

It is the removal of proteins from interstitial spaces, without which it is very difficult for the person to survive.

Q.324 What is tissue fluid?

Tissue fluid is the interstitial fluid that forms about 20% of the total body water.

Q.325 What are the functions of tissue fluid?

- It acts as a medium for exchange of various substances between the cells and the blood in capillaries.
- It functions as a medium for exchange of respiratory gases.

Q.326 How is tissue fluid formed?

Tissue fluid is formed by means of a process called filtration.

Q.327 How is volume of tissue fluid regulated?

The volume of tissue fluid is regulated by the process of reabsorption.

Q.328 What is edema?

The swelling due to excessive accumulation of fluids in the tissues is called edema.

Q.329 Name the types of edema.

- Intracellular edema—collection of fluid inside the cell
- Extracellular edema—collection of fluid outside the cell.

Q.330 What are the causes for intracellular edema?

- Malnutrition
- Poor metabolism
- Inflammation of tissue.

Q.331 What are the causes for extracellular edema?

- Increased capillary pressure
- Decreased amount of plasma proteins
- Obstruction of lymph flow.

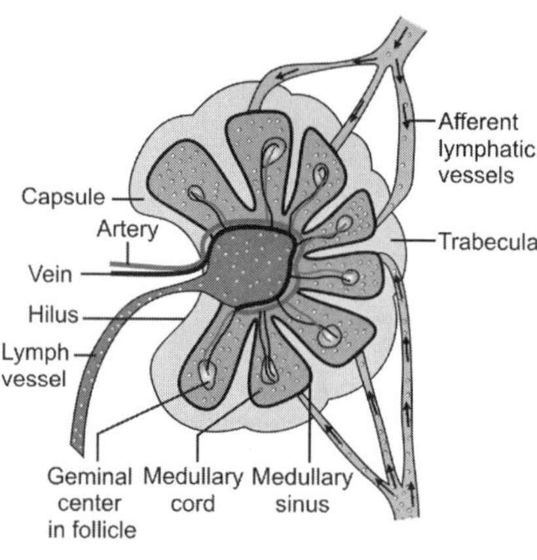

Fig. 9.8: Direction of lymph flow

Q.332 Name some common clinical conditions when extracellular edema occurs.

- Heart failure
- Renal disease
- Hypoproteinemia.

Q.333 What is pitting edema?

When the area of edema is pressed by a finger, displacement of fluid occurs producing a depression or pit. The pit remains for few seconds to one minute till the fluid flows back into that area. This type of edema is called pitting edema.

Q.334 What is nonpitting edema? What is its cause?

When the area of edema is pressed by a finger, there is no displacement of fluid or development of a depression or pit and the area remains hard. This type of edema is called nonpitting edema. This occurs because the accumulated fluid is bound in a proteoglycan meshwork, which is hard. So, the fluid is not displaced when the area is pressed. The nonpitting edema also occurs due to swelling of the cells or clotting of interstitial fluid in the presence of fibrinogen.

Chapter 10

Muscle Physiology

Q.1 How are the muscles classified?
By three methods:
- Depending upon the structure – striated and nonstriated muscles
- Depending upon the control – voluntary and involuntary muscles
- Depending upon the function – skeletal muscle, cardiac muscle and smooth muscle.

Q.2 Which are the striated muscles?
Skeletal muscles and cardiac muscles are striated muscles.

Q.3 What is the difference between the skeletal, cardiac and smooth muscles?
The difference between skeletal, cardiac and smooth muscles is shown in Table 10.1.

Table 10.1: Differentiating features of skeletal, cardiac and smooth muscles

Features	*Skeletal muscle*	*Cardiac muscle*	*Smooth muscle*
Location	In association with bones	In the heart	In the visceral organs
Shape	Cylindrical and unbranched	Branched	Spindle shaped unbranched
Length	1–4 cm	80–100 μ	50–200 μ
Diameter	10–100 μ	15–20 μ	2–5 μ
No. of nucleus	More than one	One	One
Cross striations	Present	Present	Absent
Myofibrils	Present	Present	Absent
Sarcomere	Present	Present	Absent
Troponin	Present	Present	Absent
Sarcotubular system	Well developed	Well developed	Poorly developed
T tubules	Long and thin	Short and broad	Absent
Depolarization	Upon stimulation	Spontaneous	Spontaneous
Fatigue	Possible	Not possible	Not possible
Summation	Possible	Not possible	Possible
Tetanus	Possible	Not possible	Possible
Resting membrane potential	Stable	Stable	Unstable
For trigger of contraction, calcium binds with	Troponin	Troponin	Calmodulin
Source of calcium	Sarcoplasmic reticulum	Sarcoplasmic reticulum	Extracellular
Speed of contraction	Fast	Intermediate	Slow
Neuromuscular junction	Well defined	Not well defined	Not well defined
Action	Voluntary action	Involuntary action	Involuntary action
Control	Only neurogenic	Myogenic	Neurogenic and myogenic
Nerve supply	Somatic nerves	Autonomic nerves	Autonomic nerves

Q.4 What is the nerve supply of different types of muscles?
Skeletal muscle is supplied by somatic nerves. Cardiac and smooth muscles are supplied by autonomic nerve fibers.

Q.5 What are myofibrils?
Myofibrils are the thin parallel filaments present in sarcoplasm of the muscle fiber.

Q.6 What is sarcomere?
The structural and functional unit of skeletal muscle is known as sarcomere. It extends between two 'Z' lines.

Q.7 Discuss in short microscopic structure of voluntary muscle cell.
A muscle cell (Fig. 10.1) consists of alternate transverse dark (anisotropic) A-band, and light (isotropic) I-band. A-band has in its center a region of low refractive index (H-band or Hensen line), and I-band a line of high refractive index (Z-line or Dobie line).

Q.8 What is 'A' band in the muscle? Why is it called so?
'A' band is the dark band present in the myofibrils of the muscle. It is anisotropic to polarized light; i.e. if polarized light is passed through this area of the muscle, the light rays are refracted at different directions. So this band is called 'A' band.

Q.9 What is 'I' band in the muscle? Why is it called so?
'I' band is the light band present in the myofibrils of the muscle. It is isotropic to polarized light, i.e. when polarized light is passed through this area of the muscle, all the light rays are refracted at the same angle. So this band is called 'I' band.

Q.10 What are the myofilaments?
Myofilaments are the thread-like protein filaments present in the sarcomere.

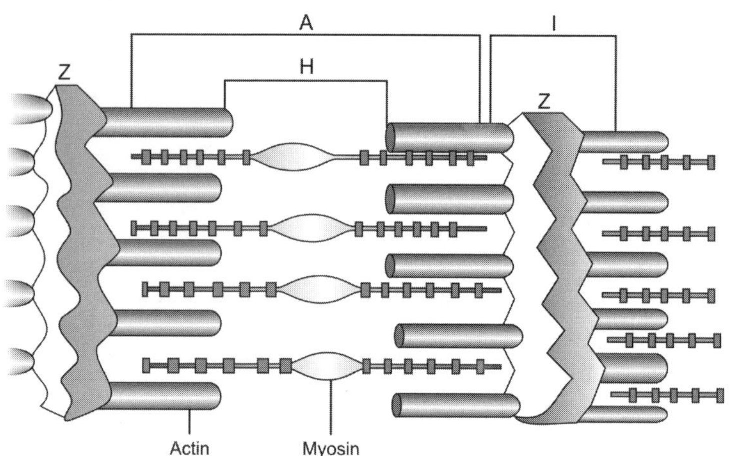

Fig. 10.1: Microscopic structure of voluntary muscle

Myofilaments are of two types, actin filaments and myosin filaments.

Q.11 What are the myofilaments present in 'A' band?

Myosin filaments and part of actin filaments.

Q.12 What are the myofilaments present in 'I' band?

Actin filaments.

Q.13 Explain the features and situation of myofilaments briefly.

Actin filaments are thin filaments with diameter of 20 Å and extend from either side of the 'Z' lines, run across 'I' band and enter into 'A' band up to 'H' zone. Myosin filaments are thick filaments with diameter of 115 Å and are situated in the center of 'A' band.

Q.14 What are the components of actin and myosin filaments?

The actin filament consists of three types of proteins called actin, tropomyosin and troponin.

The myosin filament consists of myosin molecules.

Q.15 What are the contractile elements of the skeletal muscle?

The contractile elements of the skeletal muscle are the muscle proteins namely myosin, actin, tropomyosin and troponin.

Q.16 What is 'H' zone? And what is 'M' line?

'H' zone is a light area in the middle of 'A' band. 'M' band is the middle part of myosin filaments situated in the middle of 'H' zone.

Q.17 What is sarcotubular system? What are its components?

Sarcotubular system is a system of membranous tubular structures present in the skeletal muscle fiber (Fig. 10.2).

The components of this system are 'T' tubules (transverse tubules) and 'L' tubules (longitudinal tubules). 'L' tubule is otherwise called sarcoplasmic reticulum.

Q.18 What is the functional importance of sarcotubular system?

The 'T' tubules are responsible for rapid transmission of action potential through the muscle fiber. The "L" tubules store a large quantity of calcium ions.

Q.19 What are the organic substances present in skeletal muscle?

- Proteins—actin, myosin, tropomyosin, troponin, actinin, desmin, nebulin, titin and myoglobulin
- Carbohydrates—glycogen and hexophosphate
- Lipids—neutral fat, cholesterol, lecithin and steroids
- Nitrogenous substances - ATP, adenylic acid, carnosine, carnitine, creatine, phosphocreatine, urea, uric acid, xanthine and hypoxanthine.

Q.20 Name the properties of skeletal muscle.

The properties of skeletal muscle are excitability, contractility and muscle tone.

Q.21 Define excitability.

The response of the living tissue to a stimulus in the form of physicochemical change is known as excitability.

Q.22 Define stimulus. What are the types of stimulus?

Stimulus is an agent or influence that brings about the response in an excitable tissue.

Stimulus is of four types—mechanical, electrical, thermal and chemical stimulus.

Q.23 Name the qualities of a stimulus.

- Intensity or strength
- Duration.

Q.24 What is strength duration curve? What is its other name?

Strength duration curve (Fig. 10.3) is the curve that demonstrates the relationship between the strength and the duration of stimulus.

It is also known as excitability curve.

Q.25 What is rheobase?

Rheobase is the minimum strength of the stimulus that is required to excite the tissue.

Q.26 What is utilization time?

Utilization time is the minimum time required to excite the tissue when a stimulus with rheobasic strength (threshold strength of stimulus) is applied.

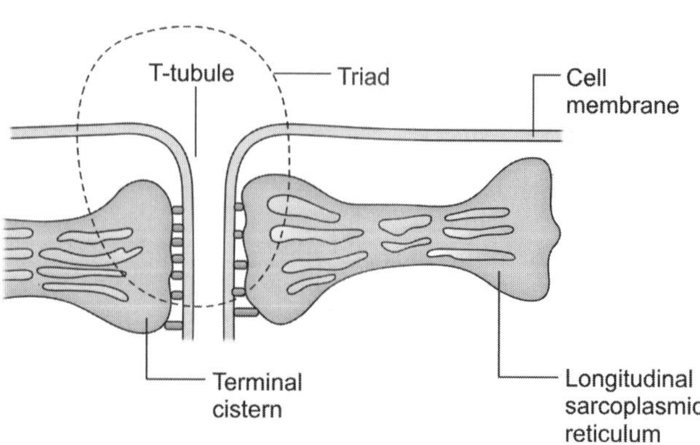

Fig. 10.2: Sarcotubular system

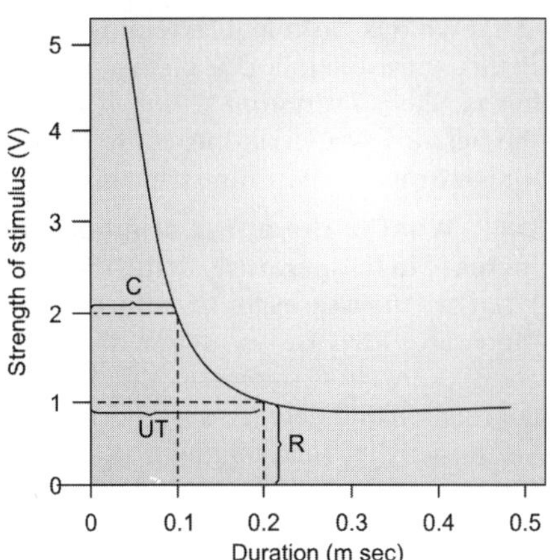

Fig. 10.3: Strength–duration curve. R = Rheobase. UT = Utilization time. C = Chronaxie

Q.27 What is chronaxie?
Chronaxie is the minimum time required to excite the tissue when a stimulus with double the rheobasic strength is applied.

Q.28 What is the importance of chronaxie?
Chronaxie helps to determine the excitability of the tissue. Longer the chronaxie, lesser is the excitability.

Q.29 Name some conditions when chronaxie increases.
- Paralysis of muscles
- Neural diseases.

Q.30 What are the types of muscular contractions?
- Isotonic contraction
- Isometric contraction.

Q.31 Define isotonic contraction and give example.
Isotonic contraction is the type of contraction in which the tension remains the same and change occurs only in the length of the muscle fibers.

Example is the contraction of the biceps muscle during simple flexion of arm.

Q.32 Define isometric contraction and give example.
Isometric contraction is the type of contraction in which the length of the muscle fibers remains the same and change occurs only in the tension.

Example is contraction of arm muscles while pulling any heavy object.

Q.33 What is preload?
It is the load on a muscle in a relaxed state.

Q.34 What is afterload?
It is the load that the muscle must generate to overcome the higher pressure.

Q.35 What are the different periods in a simple muscle twitch?
- Latent period—between the point of stimulus and point of contraction
- Contraction period—between the point of contraction and point of maximum contraction
- Relaxation period—between the point of maximum contraction and point of maximum relaxation.

Q.36 Give the normal duration of different periods of a simple muscle twitch.

Latent period	=	0.01 sec
Contraction period	=	0.04 sec
Relaxation period	=	0.05 sec
Total twitch period	=	0.10 sec

Q.37 Why is the contraction period shorter than relaxation period?
Contraction period is shorter than relaxation period because the contraction is an active process and relaxation is a passive process.

Q.38 Define latent period.
Latent period is defined as the time interval between the point of stimulus and point of contraction.

Q.39 What are the causes for latent period?
- It is the time taken for the impulse to travel along the nerve from the place of stimulation to the muscle
- It is the time taken for the initiation of chemical changes
- It is the delay in the conduction of impulse at the neuromuscular junction
- It is the time taken for the release of neurotransmitter at the neuromuscular junction
- It is the time taken to overcome the viscosity of the muscle
- It is the time taken to overcome the inertia of the instruments in experimental conditions.

Q.40 Name some conditions when the latent period is prolonged.
- Cold conditions
- During onset of fatigue
- When the load on the muscle is increased.

Q.41 When does the latent period decrease?
Latent period decreases when temperature is increased.

Q.42 Classify the skeletal muscles depending upon the contraction time. Give examples.
- Slow or red muscles, which have longer contraction time. Examples: back muscles
- Fast or pale muscles which have shorter contraction time. Examples: hand muscles and ocular muscles.

Q.43 What are the differences between red and white muscle fibers?
The differences between red and pale muscles are described in Table 10.2.

Q.44 What are the factors affecting the force of contraction of the muscle within physiological limits?
- Increase in the strength of stimulus
- Increase in the number of stimulus
- Temperature
- Load.

Q.45 Classify the stimulus depending upon the strength.
- Subminimal stimulus
- Minimal stimulus
- Submaximal stimulus
- Maximal stimulus
- Supramaximal stimulus.

Q.46 What is threshold stimulus?
Threshold or minimal stimulus is the stimulus with minimum strength required to cause minimum response in the tissues.

Q.47 What are the effects of two successive stimuli on muscle?
- Beneficial effect
- Superposition
- Summation.

Q.48 What is beneficial effect?
When two stimuli are applied to a muscle one after another in such a way that the second stimulus falls after the relaxation period of the first twitch, two separate contractions are recorded and the force of second contraction is greater than that of the first contraction. This is known as beneficial effect.

Q.49 What is the cause for beneficial effect?
Increase in the temperature during first contraction decreases the viscosity of muscle. So, the force of second contraction is more.

Q.50 What is superposition?
While applying two successive stimuli, if the second stimulus falls during relaxation of the first twitch, the first curve is superimposed by the second curve. This is called superposition or incomplete summation.

Q.51 What is summation?
When two stimuli are applied one after another and if the second stimulus falls during the contraction period or second half of the latent period, two contractions

Table 10.2: Differentiating features of red and pale muscles

S. No.	Red (Slow muscle)	Pale (Fast muscle)
1.	Type I fibers are more	Type II fibers are more
2.	Myoglobin content is high. So, it is red	Myoglobin content is less. So, it is pale
3.	Sarcoplasmic reticulum is less extensive	Sarcoplasmic reticulum is more extensive
4.	Blood vessels are more extensive	Blood vessels are less extensive
5.	Mitochondria are more in number	Mitochondria are less in number
6.	Response is slow with long latent period	Response is rapid with short latent period Response is rapid with short latent
7.	Contraction is less powerful	Contraction is more powerful
8.	This muscle is involved in prolonged and continued activity as it undergoes sustained contraction	This muscle is not involved in prolonged and continued activity as it relaxes immediately
9.	Fatigue occurs slowly	Fatigue occurs quickly
10.	Depends upon cellular respiration for ATP production	Depends upon glycolysis for ATP production

are summed up, giving single contraction which is bigger and broader than simple muscle curve. This is known as summation or complete summation.

Q.52 Define fatigue.
The decrease in the response of the muscle due to repeated stimuli is known as fatigue.

Q.53 What are the causes of fatigue?
- Exhaustion of acetylcholine
- Accumulation of metabolites like lactic acid and carbon dioxide
- Lack of nutrients like glycogen
- Lack of oxygen.

Q.54 Mention the order of site (seat) of fatigue in the intact body.

First site of fatigue	: Cerebral cortex (Betz cells)
Second site of fatigue	: Motor neuron in spinal cord
Third site of fatigue	: Neuromuscular junction
Fourth site of fatigue	: Muscle.

Q.55 How to prove that the neuromuscular junction is the first site of fatigue in frog's muscle nerve preparation?
In the isolated muscle nerve preparation, nerve is stimulated continuously and the curves are recorded till the fatigue occurs, i.e. till the muscle fails to respond to the stimulus. Then, immediately the muscle is stimulated directly. A response is noticed in the form of curve. This shows that the muscle is not yet fatigued. The nerve cannot be fatigued. So, the site where fatigue must have occurred is the neuromuscular junction.

Q.56 Is fatigue a reversible or irreversible phenomenon?
Fatigue is a reversible phenomenon.

Q.57 What are the causes for recovery from fatigue?
- Removal of metabolites
- Formation of acetylcholine at the neuromuscular junction
- Availability of nutrients
- Availability of oxygen.

Q.58 What is tetanus?
Summation or complete fusion of muscular contractions due to repeated stimuli is known as tetanus. Tetanus is defined as the sustained contraction of muscle due to repeated stimuli. Figure 10.4 demonstrates genesis of tetanus and its curves.

Q.59 What is clonus?
When the frequency of stimuli is not sufficient to cause tetanus, the fusion of contraction is not complete. This is known as clonus or incomplete tetanus.

Q.60 What is the frequency of stimuli to cause tetanus and clonus?
Frog muscle:
Frequency of stimuli to cause tetanus = 40/sec.
Frequency of stimuli to cause clonus = 35/sec.
Human muscle:
Frequency of stimuli to cause tetanus = 60/sec.
Frequency of stimuli to cause clonus = 55/sec.

Q.61 What is pathological tetanus?
Pathological tetanus is a disease caused by bacillus *Clostridium tetani.* It affects the nervous system and its common features are muscle spasm and paralysis.

Q.62 What is the effect of moderate increase in temperature on the muscle? What are the causes for the effect?
Moderate increase in temperature to about 30 to 40°C, increases the force of contraction and decreases all the periods, i.e. the activity is accelerated.
Causes:
- Increase in excitability of the muscle
- Acceleration of chemical processes
- Decrease in the viscosity of the muscle.

Q.63 What is the effect of decrease in temperature on the muscle? What are the causes for the effect?
Decrease in temperature to about 10°C, reduces the force of contraction and increases all the periods, i.e. the activity is slowed down.
Causes:
- Decrease in excitability of the muscle
- Slowness of the chemical processes
- Increase in the viscosity of the muscle.

Q.64 What is the effect of very high temperature on the muscle?
When the temperature increases above 60°C, heat rigor occurs.

Q.65 What is heat rigor? What is its cause?
Stiffening and shortening of the muscle fibers because of high temperature is called heat rigor.
It is due to the coagulation of muscle proteins.

Q.66 Is heat rigor reversible?
Heat rigor is not reversible.

Q.67 What is cold rigor? Is it reversible?
Stiffening and shortening of the muscle fibers due to extreme cold is called cold rigor and it is reversible.

Q.68 What is rigor mortis? What is the cause for it?
The rigidity that develops after death is called rigor mortis.
Cause: After death there is loss of ATP. Relaxation cannot occur because of lack of ATP and that is the cause of rigor mortis.

Q.69 What is free load? Give an example.
Free load or fore load is the load which acts on the muscle freely even before the onset of contraction of the muscle.
Example: Filling water from a tap by holding the bucket in hand.

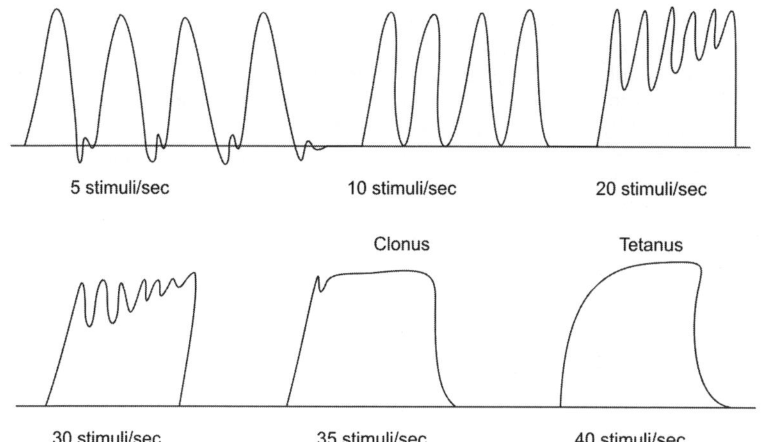

Fig. 10.4: Genesis of tetanus and tetanus curves

Q.70 State whether the muscle works better in after loaded condition or in free loaded condition. Why?

Muscle works better in free loaded condition than in the after loaded condition. Because, in free loaded condition the initial length of the muscle fibers increases even before the onset of muscular contraction. And according to Frank Starling's law, the force of contraction of muscle is directly proportional to initial length of the muscle fiber within physiological limits.

Q.71 What is optimum load?

Optimum load is the load at which the work done by the muscle is maximum.

Q.72 What is refractory period?

Refractory period is the period at which the muscle does not show any response to a stimulus.

Q.73 What are the types of refractory period?

- Absolute refractory period—the period during which the muscle does not show any response at all, whatever may be the strength of stimulus
- Relative refractory period—the period during which the muscle shows some response if the strength of stimulus is increased to maximum.

Q.74 What is the duration of absolute and relative refractory periods in skeletal muscle?

Absolute refractory period extends for 0.005 sec, i.e. during the first half of latent period. Relative refractory period extends for 0.005 sec, i.e. during the second half of latent period. Thus, the duration of refractory period in skeletal muscle is 0.01 sec.

Q.75 What is the duration of absolute and relative refractory periods in cardiac muscle?

Absolute refractory period is 0.27 sec, i.e. it extends throughout contraction period. Relative refractory period is 0.25 sec, i.e. it extends during the first half of relaxation period. Thus, totally the refractory period in cardiac muscle extends for about 0.52 sec. It is very long compared to that of skeletal muscle.

Q.76 What is muscle tone?

The muscle fibers always maintain a state of slight contraction with certain degree of vigor and tension. This is known as muscle tone or tonus.

Q.77 How is the tone maintained in skeletal and cardiac muscle?

Skeletal muscle: Maintenance of tone is neurogenic and it is under the influence of gamma motor neuron system. *Cardiac muscle*: Maintenance of tone is purely myogenic and it is by the muscle itself.

Q.78 Name the changes taking place during muscular contraction.

- Electrical changes
- Physical changes
- Histological changes
- Chemical changes
- Thermal changes.

Q.79 What is resting membrane potential (RMP)?

The potential difference between inside and outside of the cell across the cell membrane under resting conditions is known as RMP. It is negative inside and positive outside.

Q.80 What are the mechanisms involved in the ionic basis of RMP?

Two transport mechanisms are involved in the ionic basis of RMP.

- Sodium-Potassium pump
- Selective permeability of the cell membrane.

Q.81 How much is the RMP in skeletal muscle?

RMP in skeletal muscle is - 90 mV.

Q.82 What is action potential?

Series of electrical changes taking place in cell when stimulated is known as action potential (Fig. 10.5).

Q.83 What are the properties of action potential?

Action potential:

- Is propogative
- Is biphasic
- Obeys all or non law
- Summation is not possible
- Shows refractory period.

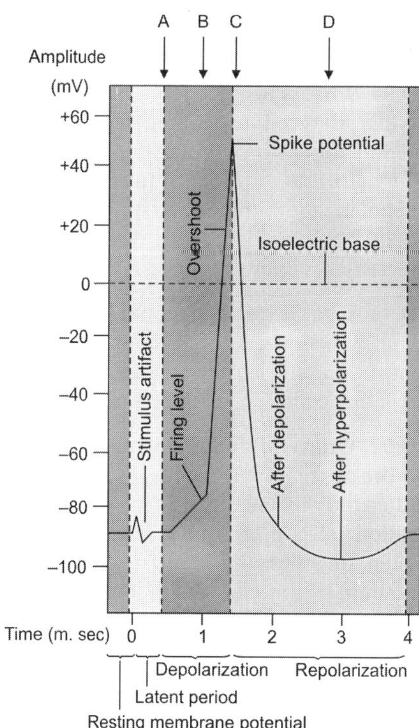

Fig. 10.5: Action potential in a skeletal muscle (A = Opening of few Na^+ channels, B = Opening of many Na^+ channels, C = Closure of Na^+ channels and opening of K^+ channels, D = Closure of K^+ channels)

Q.84 What are the phases of action potential?
- Depolarization
- Repolarization.

Q.85 What is depolarization?
When stimulated, the resting membrane potential is lost in the cell. Interior of the cell becomes positive (up to +55 mV) and exterior becomes negative. This is known as depolarization.

Q.86 What is the cause for depolarization?
Depolarization is due to opening of sodium channels and rush of sodium ions into the cell.

Q.87 Why the depolarization is short lived?
Because of the rapid inactivation and closure of sodium channels.

Q.88 What is repolarization?
The restoration of negativity inside the cell and positivity outside is known as repolarization.

Q.89 What is the cause for repolarization?
Repolarization is due to opening of potassium channels and efflux of potassium ions from inside to outside the cell.

Q.90 What is firing level?
When the cell is stimulated, depolarization starts slowly. After the initial slow depolarization up to—15 mV, the rate of depolarization increases suddenly. The point at which the rate of depolarization increases is known as firing level.

Q.91 What is spike potential?
During action potential, the rapid depolarization and rapid repolarization are together called spike potential.

Q.92 What is after depolarization? What is the cause for it?
After rapid repolarization, slow repolarization takes place and this is known as after depolarization or negative after potential.

It is due to decrease in the rate of potassium efflux.

Q.93 What is after hyperpolarization? What is the cause for it?
When repolarization occurs, it does not stop at the level of resting membrane potential but goes beyond that level causing more negativity inside the cell. This is known as after hyperpolarization or positive after potential.

Unlike sodium channels, the potassium channels remain opened for a longer duration allowing large number of potassium ions to move out of the cell. So, the interior of the cell becomes more negative than the resting level.

Q.94 What is graded potential (graded membrane potential or graded depolarization)?
Stimulation of the receptors, synapse or neuromuscular junction produces some mild change (mild depolarization) in the membrane potential. It loses its intensity as it starts spreading. This potential change is called graded potential.

Q.95 What are the properties of graded potential?
Graded potential:
- Is non propagative
- Is monophasic
- Does not obey all or non law
- Summation is possible
- Has no refractory period.

Q.96 What is patch clamp technique?
Patch clamp technique is the method to measure the ionic currents across the biological membranes.

Q.97 What is the molecular basis of muscular contraction?
When muscle is stimulated, action potential develops leading to the development of excitation contraction coupling and formation of actomyosin complex. This makes the actin filaments to slide over the myosin filaments leading to the contraction of the muscle.

Q.98 What is excitation contraction coupling? What is responsible for it?
The process involved in between the excitation and the contraction of the muscle is known as excitation contraction coupling. Calcium is responsible for it.

Q.99 What is Ratchet theory? What are the other names for it?
Ratchet theory explains the mechanism involved in the sliding of actin filaments over the myosin filaments during the muscular contraction.

The other names for it are sliding theory and walk along theory.

Q.100 What is power stroke?
Tilting of the head of myosin towards the arm and dragging the active filament along with it is called power stroke.

Q.101 What are the changes taking place in the sarcomere during contraction of muscle?
- Length of sarcomere decreases and 'Z' lines come close
- Length of 'I' band reduces because of overlapping of actin filaments from opposite ends
- 'H' zone disappears
- Length of 'A' band, actin filaments and myosin filaments remains same.

Q.102 How does the relaxation of muscle take place?
After contraction, the calcium ions are actively pumped back into the sarcotubular reticulum from the sarcoplasm. Decreased calcium content in sarcoplasm leads to detachment of calcium ions from troponin. This causes release of myosin from actin and the relaxation of muscle occurs.

Q.103 What are the chemical changes taking place during muscular contraction?
- Glycolysis and liberation of energy
- Changes in pH.

Q.104 What are the sources of energy for muscular contraction?
The energy for muscular contraction is obtained by the break down of adenosine triphosphate (ATP) and resynthesis of ATP from creatine phosphate and glycolytic pathway.

Q.105 What is glycolytic pathway or Embden–Meyerhof pathway? How many molecules of ATP are formed in this pathway?
Breakdown of glycogen into pyruvic acid is called glycolytic pathway or Embden–Meyerhof pathway. Two molecules of ATP are formed in this pathway.

Q.106 Amongst the aerobic glycolysis and anaerobic glycolysis, which one is better and why?
Aerobic glycolysis is better because greater amount of energy is liberated during this process.

Q.107 How many molecules of ATP are formed during carbohydrate metabolism?
38 molecules of ATP are formed during carbohydrate metabolism, i.e. during break down of each glycogen molecule. 2 molecules are formed during glycolysis and 2 molecules are formed during Krebs cycle. The remaining 34 molecules of ATP

are formed by utilization of hydrogen atoms which are released during Krebs cycle.

Q.108 Explain the changes in pH of the muscle during contraction.

In resting condition the reaction is alkaline with a pH of 7.3. During onset of contraction, muscle becomes acidic due to break down of ATP. During later part of contraction, the muscle becomes alkaline due to resynthesis of ATP from creatine phosphate. And at the end of contraction, once again it becomes acidic due to the formation of pyruvic acid and lactic acid.

Q.109 What are the different stages of heat production during muscular contraction?

Heat is produced in three stages during muscular contraction

- Resting heat
- Initial heat
- Recovery heat.

Q.110 What is neuromuscular junction?

The junction between the motor nerve ending and muscle fiber is known as neuromuscular junction.

Q.111 What are the parts of neuromuscular junction?

- Axon terminal with motor end plate
- Presynaptic membrane
- Synaptic cleft
- Postsynaptic membrane
- Subneural clefts.

Q.112 What are the important structures present in axon terminal?

Synaptic vesicles containing neurotransmitter and the mitochondria.

Q.113 What is the neurotransmitter secreted in neuromuscular junction?

Acetylcholine.

Q.114 What is the effect of Ca-ions and Mg-ions on the release of acetyl choline from motor nerve terminals?

Ca-ions serve to stimulate the release of acetylcholine while Mg-ions inhibit this release.

Q.115 Where is acetylcholinesterase present in neuromuscular junction? What is its action?

Acetylcholinesterase is present in the basal lamina of synaptic cleft in the neuromuscular junction. It destroys acetylcholine.

Q.116 Name the important events taking place during neuromuscular transmission.

- Release of acetylcholine
- Action of acetylcholine
- Development of end plate potential
- Destruction of acetylcholine.

Q.117 What is end plate potential?

The change in electrical potential in neuromuscular junction is called end plate potential. It is a slight depolarization up to—60 mV.

Q.118 What are the differences between end plate potential and action potential?

End plate potential differs from action potential by its properties viz.

- It is nonpropagative
- It is monophasic
- It does not obey all or none law.

Q.119 What is the significance of end plate potential?

The significance of end plate potential is that it causes the development of action potential in the muscle fiber.

Q.120 What is miniature end plate potential?

When a small quantum of acetylcholine is released from synaptic vesicle, it produces a weak end plate potential up to—0.5 mV. This is called miniature end plate potential.

Q.121 Name some neuromuscular blockers.

Bungarotoxin, succinylcholine, carbamyl choline and botulinum toxin.

Q.122 Name some drugs, which can stimulate the neuromuscular junction.

Neostigmine, physostigmine and diisopropyl fluorophosphate.

Q.123 What do you understand by 'oxygen debt'?

During muscular exercise oxygen demand increases, but muscle can keep on contracting anaerobically. The amount of oxygen required for muscle recovery after this is called the 'oxygen debt'.

Q.124 What are the smooth muscles?

Smooth muscles are the nonstriated in-voluntary muscles, which form the contractile elements of various organs in the body.

Q.125 What are the types of smooth muscle fibers?

- Multiunit smooth muscle fibers
- Visceral smooth muscle fibers.

Q.126 Name the muscle proteins present in the smooth muscles.

Actin, myosin, and tropomyosin. Troponin or troponin like substance is absent in smooth muscles.

Q.127 Name the substance that initiates the contraction of smooth muscles.

Calmodulin initiates the contraction of smooth muscle along with calcium.

Q.128 What are the differences between the electrical activity of smooth muscle and skeletal muscle?

- In smooth muscle, the resting membrane potential is low ranging between - 50 and - 70 mV whereas in skeletal muscle it is - 90 mV.
- Three types of action potential occur in smooth muscle (spike potential, spike potential with slow wave rhythm and action potential with plateau). But in skeletal muscle only one type of action potential occurs.

Q.129 What is tonus or tone in smooth muscles? What is it due to?

Tonus or tone is a state of partial contraction maintained by the smooth muscles of some visceral organs.

It is due to the tonic contraction of the smooth muscle without action potential.

Q.130 What is the difference between the nerve supply of smooth muscles and skeletal muscles?

Smooth muscles are supplied by autonomic nerve fibers (sympathetic and parasympathetic fibers) whereas the skeletal muscles are supplied by somatic nerve fibers.

Q.131 What is electromyogram (EMG)? What is its use?

Electromyogram (EMG) is the record of the electrical activity of the muscle.

It is useful in the diagnosis of neuromuscular diseases.

Q.132 What do you mean by muscle cramps?

Muscular cramps are involuntary, localized painful contractions of muscles often relieved by stretching the affected muscles.

Q.133 What do you understand by muscular fasciculation?

Muscular fasciculation is spontaneous contraction of motor units, which is visible through the skin as fine ripping movement in the relaxed muscles.

Q.134 What is myopathy?

Myopathy is a neuromuscular disease in which progressive dysfunction of muscle fiber occurs leading to muscular weakness.

Q.135 What is myasthenia gravis?
Myasthenia gravis is a muscular disease characterized by extreme weakness of muscles due to inability of neuromuscular junction to transmit the impulses from nerve to muscle.

Q.136 What is the cause for myasthenia gravis?
Myasthenia gravis is an autoimmune disease in which the body develops antibodies against its own acetylcholine receptors. The antibodies destroy the acetylcholine receptors. So even if the acetylcholine is released, it cannot act because of the destruction of the receptors. So the neuromuscular transmission is affected leading to weakness of the muscles.

Q.137 What is strength of the muscle?
The maximum force that can be developed during contraction is known as strength of the muscle.

Q.138 What is power of the muscle?
The amount of work done by the muscle in the given unit of time is called power of the muscle.

Q.139 What is endurance of the muscle?
The capacity of the muscle to withstand the power produced during activity is known as endurance.

Chapter 11

Digestive System

Q.1 What are the different layers of gastrointestinal (GI) tract?
Layers of GI tract from outside to inside:
- Serous coat
- Muscular coat
- Submucous coat
- Mucus coat.

Q.2 What are the nerves supplying GI tract?
GI tract is supplied by two types of nerve fibers:
- Intrinsic nerves:
  - Auerbach's or myenteric nerve plexus present in the muscular layer
  - Meissner's plexus or submucus nerve plexus situated in between the muscular and submucus layers.
- Extrinsic nerves:
  - Sympathetic nerve fibers
  - Parasympathetic nerve fibers.

Q.3 Name the major salivary glands in human beings.
- Parotid glands
- Submaxillary or submandibular glands
- Sublingual glands.

Q.4 What are the properties of saliva?

Volume	:	1000 to 1500 mL/day
Reaction and pH	:	Slightly acidic with a pH of 6.35 to 6.85
Specific gravity	:	1.002 to 1.012.

Q.5 Name the organic substances present in saliva.
- Salivary enzymes:
  - Amylase (ptyalin), maltase, lingual lipase, lysozyme, phosphatase, carbonic anhydrase and kallikrein.
- Other organic substances:
  - Proteins—mucin and albumin
  - Blood group antigens
  - Free amino acids
  - Nonprotein nitrogenous substances—urea, uric acid, creatinine, xanthine and hypoxanthine.

Q.6 Name the inorganic substances present in saliva.
Sodium, potassium, calcium, bicarbonates, bromide, chloride, fluoride and phosphate.

Q.7 What are the nerves supplying the salivary gland?
Salivary glands are supplied by parasympathetic and sympathetic nerves.

Parasympathetic nerves to parotid gland arise from the inferior salivatory nucleus and reach the parotid gland by passing through glossopharyngeal nerve. The parasympathetic nerves to submandibular and sublingual glands arise from the superior salivatory nucleus and reach the glands by passing through the facial nerve.

Sympathetic nerves to the salivary glands arise from lateral horns of first and second thoracic segments in spinal cord and reach the glands through the postganglionic fibers of superior cervical ganglion.

Q.8 What are the effects of stimulation of parasympathetic nerve fibers to salivary glands?
Stimulation of parasympathetic nerve fibers to salivary glands causes vasodilatation and increase in secretion of watery saliva.

Q.9 What are the effects of stimulation of sympathetic nerve fibers to salivary glands?
Stimulation of sympathetic nerve fibers to salivary glands causes vasoconstriction and decrease in secretion of saliva that is thick and rich in mucus.

Q.10 How salivary secretion is regulated?
Salivary secretion is regulated by reflex phenomenon in which both conditioned and unconditioned reflexes are involved.

Q.11 Name some conditions when hyposalivation occurs.
Temporary hyposalivation occurs in emotional conditions like fear, fever and dehydration. Permanent hyposalivation occurs in sialolithiasis, congenital absence of salivary glands and Bell's palsy.

Q.12 What is xerostomia?
Dryness of the mouth due to hyposalivation or absence of salivary secretion (aptyalism) is called xerostomia.

Q.13 Name some conditions when hypersalivation occurs.
- Decay of tooth or neoplasm of mouth or tongue
- Diseases of esophagus, stomach and intestine
- Neurological disorders like cerebral palsy and mental retardation
- Parkinsonism
- Psychological and psychiatric conditions
- Nausea and vomiting.

Q.14 What is chorda tympani syndrome?
It is the condition characterized by sweating while eating.

Q.15 What are the gastric glands? Mention the types of gastric glands.
Gastric glands are the exocrine glands of the stomach, which secrete gastric juice.
Types of gastric glands:
- Fundic glands situated in the body and fundus
- Pyloric glands situated in pyloric part
- Cardiac glands situated in the cardiac region.

Q.16 Name the substances secreted by different cells of gastric gland.
- *Chief or pepsinogen cells:* Enzymes—pepsinogen, rennin, lipase, gelatinase and urease
- *Parietal or oxyntic cells:* Hydrochloric acid and intrinsic factor of Castle

- *Mucus neck cells:* Mucin
- *G cells:* Gastrin
- *Enterochromaffin (EC) or Kulchitsky cells:* Serotonin
- *Enterochromaffin-like (ECL) cells:* Histamine.

Q.17 What are the properties of gastric juice?

Volume	:	1200 mL/day
Reaction and pH	:	Highly acidic with a pH of 0.9 to 1.2
Specific gravity	:	1.002 to 1.004.

Q.18 What is the cause for the high acidity of gastric juice?
Gastric juice is highly acidic because of hydrochloric acid.

Q.19 Name the organic substances present in gastric juice.
- Enzymes—pepsin, rennin, gastric lipase, gelatinase and urease
- Other organic substances—mucus and intrinsic factor of castle.

Q.20 What are the functions of gastric mucus?
Mucus:
- Protects the stomach wall from irritation or mechanical injury
- Prevents the digestive action of pepsin on the wall of the stomach
- Protects the gastric mucosa from hydrochloric acid of gastric juice.

Q.21 Briefly explain the secretion of hydrochloric acid in stomach.
Hydrochloric acid is formed in the canaliculus of the parietal cells of the gastric glands. In the parietal cell, carbon dioxide combines with water to form carbonic acid. Carbonic acid dissociates into hydrogen and bicarbonate ions immediately. The whole reaction is accelerated by the enzyme carbonic anhydrase. The bicarbonate ion diffuses from the cell to the extracellular fluid in exchange for chloride ions. The hydrogen and chloride ions move from the cell into the canaliculus and combine to form hydrochloric acid.

Q.22 What are the functions of gastric juice?
- Digestion of proteins and lipids
- Hemopoietic function—intrinsic factor helps in erythropoiesis
- Protective function—mucus protects the wall of stomach from proteolytic enzymes and hydrochloric acid
- Antibacterial action—hydrochloric acid destroys the microorganisms entering the gastrointestinal tract through diet
- Activator function—hydrochloric acid activates pepsinogen into pepsin.

Q.23 What are the actions of pepsin?
Pepsin acts on proteins and converts them into proteoses, peptones and polypeptides. It also causes curdling and digestion of milk (casein).

Q.24 How is pepsinogen converted into pepsin?
Pepsinogen is converted into pepsin by acid medium provided by hydrochloric acid.

Q.25 What is rennin?
Rennin is a milk curdling enzyme present in animals.

Q.26 Name the factors regulating the secretion of hydrochloric acid in stomach.
Gastrin, histamine and vagal stimulation increase the secretion of hydrochloric acid. Secretin, gastric inhibitory polypeptide and peptide YY inhibit the acid secretion.

Q.27 Briefly explain Pavlov's pouch.
It is a small part of stomach that is incompletely separated from the main portion and made into a bag like pouch. Russian scientist Pavlov devised it. This pouch is fully innervated with both sympathetic and parasympathetic nerve supply intact. It is useful to study the hormonal and nervous regulation of gastric juice.

Q.28 What is sham feeding?
Sham feeding means false feeding, i.e. the animal eats the food but the food does not reach the stomach. This is done by cutting the esophagus transversely and the cut ends are brought out by making a hole in the neck. So, when the animal swallows the food, it comes out. It is useful to demonstrate the unconditioned reflex during cephalic phase of gastric secretion.

Q.29 Name the phases of gastric secretion.
- Cephalic phase
- Gastric phase
- Intestinal phase.

Q.30 What is cephalic phase of gastric secretion?
The sight, smell or thought of food or the presence of food in the stomach stimulates the secretion of gastric juice. This is known as cephalic phase of gastric secretion because the impulses are sent from head. It is purely under nervous control and operates through conditioned and unconditioned reflexes.

Q.31 Briefly explain the gastric phase of gastric secretion.
The secretion of gastric juice when food enters the stomach is called gastric phase. It is under nervous and hormonal control. Nervous control is operated through local myenteric reflex and vagovagal reflex. The hormonal control is operated through secretion of gastrin.

Q.32 Briefly explain intestinal phase of gastric secretion.
When the chyme reaches the small intestine from the stomach, initially there is secretion of gastric secretion due to the action of gastrin. Later, the gastric secretion is inhibited due to enterogastric reflex and GI hormones like secretin, cholecystokinin, gastric inhibitory polypeptide (GIP), vasoactive intestinal polypeptide (VIP), polypeptide YY and somatostatin.

Q.33 What are the effects of alcohol and caffeine on gastric secretion?
Alcohol and caffeine stimulate gastric secretion.

Q.34 What is fractional test meal (FTM)?
It is one of the methods of gastric analysis. After overnight fasting, a sample of gastric juice is collected. Then a test meal is given and the samples of gastric juice are collected at the interval of 15 minutes for about 2½ hours. All the samples are analyzed for peptic activity and gastric acidity.

Q.35 How is gastric juice collected in human beings?
By using Ryle's tube.

Q.36 What is gastric atrophy?
Gastric atrophy is the condition in which the muscles of the stomach shrink and become weak.

Q.37 What is Zollinger-Ellison syndrome?
It is the condition characterized by secretion of excess hydrochloric acid in stomach.

Q.38 What are the basic structures of exocrine part of pancreas?
The alveoli or acini are the basic structures of exocrine part of pancreas.

Q.39 Name the pancreatic duct. How does it open into the intestine?
Pancreatic duct is called Wirsung's duct. It joins the common bile duct and forms ampulla of Vater that opens into the duodenum.

Q.40 What are the properties of pancreatic juice?

Volume	:	500 to 800 mL/day
Reaction and pH	:	Highly alkaline with a pH of 8 to 8.3
Specific gravity	:	1.010 to 1.018.

Q.41 What is the cause for high alkalinity of pancreatic juice?

Presence of large quantity of bicarbonate is responsible for the high alkalinity of pancreatic juice.

Q.42 Name the enzymes present in pancreatic juice.

- *Proteolytic enzymes:* Trypsin, chymotrypsin, carboxypeptidases, nuclease, elastase and collagenase
- *Lipolytic enzymes:* Pancreatic lipase, cholesterol ester hydrolase, phospholipases A and B, collapse and bile salt-activated lipase
- *Amylolytic enzyme:* Pancreatic amylase.

Q.43 What are the functions of pancreatic juice?

- *Digestive functions:* Digestion of proteins, lipids and carbohydrates
- *Neutralizing action:* Neutralization of acidity of chyme in intestine.

Composition of pancreatic juice is given in Figure 11.1.

Q.44 How is trypsinogen converted into trypsin?

Trypsinogen is converted into trypsin by the enzyme enterokinase. Once formed trypsin also converts trypsinogen into trypsin by means of autocatalytic action.

Q.45 What are the actions of trypsin?

- *Digestion of proteins:* It converts proteins into proteoses and polypeptides
- *Curdling of milk:* It converts caseinogens in the milk into casein
- *Acceleration of blood clotting*
- *Activation of other enzymes of pancreatic juice:* It converts chymotrypsinogen into chymotrypsin and procarboxypeptidases into carboxypeptidases.

Q.46 What are the actions of chymotrypsin?

Chymotrypsin:

- Hydrolyses the proteins into polypeptides
- Digests milk.

Q.47 What is the action of carboxypeptidase?

Carboxypeptidase converts polypeptides into amino acids.

Q.48 What is the importance of pancreatic lipase?

Pancreatic lipase is the strongest lipolytic enzyme in the gastrointestinal tract. 80% of the fat is digested by this enzyme. Absence of pancreatic lipase leads to steatorrhea.

Q.49 Name the hormones, which increase the secretion of pancreatic juice.

Gastrin, secretin and cholecystokinin.

Q.50 What is the effect of secretin on pancreatic juice?

Secretin causes secretion of large amount of watery juice with high concentration of bicarbonate ion.

Q.51 What is the effect of cholecystokinin on pancreatic juice?

Cholecystokinin causes secretion of pancreatic juice with more amount of enzymes.

Q.52 What is steatorrhea?

Steatorrhea is the condition in which large quantity of undigested fat is excreted in feces. It is due to the lack of pancreatic lipase.

Q.53 What is biliary system or extrahepatic biliary apparatus?

It is the system formed by structures present outside the liver. It includes gallbladder and the extrahepatic bile ducts namely, right and left hepatic ducts, common hepatic duct, cystic duct and common bile ducts.

Q.54 What are the sources of blood supply to liver?

Liver receives blood from two sources, the hepatic artery and portal vein.

Q.55 What is the importance of hepatic portal vein?

Hepatic portal vein brings deoxygenated blood from stomach, intestine, spleen and pancreas to liver. Deoxygenated blood contains large amount of monosaccharides and amino acids.

Q.56 What is bile?

Bile is a golden yellow or greenish fluid produced by liver.

Q.57 What are the properties of bile?

Volume	:	1200 mL/day
Reaction and pH	:	Alkaline with pH of 8 to 8.6
Specific gravity	:	1.010 to 1.011.

Q.58 Name the organic substances present in bile.

Bile salts, bile pigments, cholesterol, fatty acids, lecithin, and mucin.

Q.59 What are the bile salts?

Bile salts are the sodium and potassium salts of bile acids. Bile acids are cholic acid and chenodeoxycholic acid.

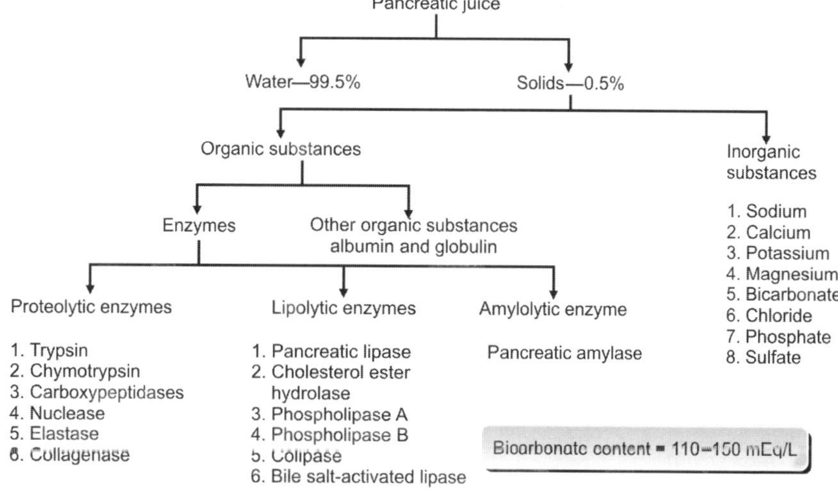

Fig. 11.1: Composition of pancreatic juice

Q.60 Explain briefly the formation of bile salts.

The primary bile acids namely, cholic acid and chenodeoxycholic acids are formed in liver and enter the intestine. Due to the bacterial action in intestine, the cholic acid is converted into deoxycholic acid and chenodeoxycholic acid is converted into lithocholic acid. Deoxycholic acid and lithocholic acid are called secondary bile acids. Now, these two acids from intestine enter the liver through enterohepatic circulation. In liver, the secondary bile acids are conjugated with glycine and taurine forming glycocholic acid and taurocholic acid. These two conjugated bile acids combine with sodium or potassium salt to form bile salts.

Q.61 Name the functions of bile salts.

- Emulsification of fat
- Absorption of fats
- Choleretic action
- Cholagogue action
- Laxative action
- Prevention of gallstone formation.

Q.62 What are the bile pigments?

Bile pigments are bilirubin and biliverdin and these pigments are the excretory products of bile.

Q.63 How are the bile pigments formed?

When the old red blood cells are destroyed in the reticuloendothelial system, hemoglobin is released. It is broken into globin and heme. Heme is split into iron and the pigment biliverdin. Biliverdin is reduced to bilirubin.

Q.64 Explain briefly the circulation of bile pigments.

Bilirubin formed in reticuloendothelial system is released into blood. It is called free bilirubin. Through blood it reaches the liver. There, the free bilirubin is conjugated by glucuronic acid to form conjugated bilirubin. Conjugated bilirubin is excreted through bile into the intestine. From intestine, 50% of conjugated bilirubin enters the liver via enterohepatic circulation and excreted through bile. Remaining 50% of conjugated bilirubin is converted into urobilinogen. Urobilinogen is excreted through urine as urobilin and through feces as stercobilinogen.

Q.65 What is enterohepatic circulation?

The flow of blood from intestine to liver through portal vein is known as enterohepatic circulation. Bile salts and bile pigments are transported through enterohepatic circulation.

Q.66 Name the functions of bile.

- Digestive function
- Absorptive function
- Excretory function
- Laxative action
- Antiseptic action
- Choleretic action
- Maintenance of pH in GI tract
- Prevention of gallstone formation
- Lubrication function
- Cholagogue action.

Q.67 Name the functions of liver.

- Storage function
- Synthetic function
- Secretion of bile
- Metabolic function
- Excretory function
- Heat production
- Hemopoietic function
- Hemolytic function
- Inactivation of hormones and drugs
- Defensive and detoxification functions.

Q.68 What are the changes taking place in the bile when it is stored in gallbladder?

- Reduction in volume due to reabsorption of water
- Concentration of bile due to reabsorption of water and electrolytes
- Reduction of pH of bile from 8–8.6 to 7–7.6
- Addition of mucin.

Q.69 What are the differences between liver bile and gallbladder bile?

- Liver bile is dilute and gallbladder is concentrated
- The pH of liver bile (8–8.6) is more than the pH in gallbladder bile (7–7.6)
- Concentration of bile salts, bile pigments, cholesterol, fatty acids and lecithin is less in liver bile and more in gallbladder bile
- Mucin is absent in liver bile and present in gallbladder bile
- Sodium, chloride, and bicarbonate are more in liver bile than in gallbladder bile
- Calcium and potassium are less in liver bile than in gallbladder bile.

Q.70 What is the normal bilirubin content in blood and at what level jaundice occurs?

Normal bilirubin content in blood is 0.5 to 1.5 mg%. When it exceeds 2 mg% jaundice occurs.

Q.71 What are the types of jaundice?

Jaundice is classified into 3 types depending upon the causes.

- Prehepatic or hemolytic jaundice—due to excessive destruction of red blood cells
- Hepatic or hepatocellular jaundice—due to damage of hepatic cells
- Posthepatic or obstructive jaundice—due to obstruction of bile duct.

Q.72 What are the causes for prehepatic jaundice?

- Liver failure
- Renal disorder
- Hypersplenism
- Burns
- Infections like malaria
- Hemoglobin abnormalities like sickle cell anemia or thalassemia.

Q.73 What are the causes for hepatic jaundice?

- Infection (infective jaundice) by virus resulting in hepatitis (viral hepatitis)
- Alcoholic hepatitis
- Cirrhosis of liver
- Exposure to toxic materials.

Q.74 What are the causes for posthepatic jaundice?

- Gallstones
- Cancer of biliary system or pancreas.

Q.75 What is cholelithiasis?

Formation of gallstone is called cholelithiasis. Gallstone is formed by the precipitation of cholesterol. Cholesterol in gallbladder bile combines with bile salts and lecithin. Now, cholesterol becomes soluble in water and it is precipitated forming crystals. To these crystals, bile pigments and calcium ions get attached forming gallstones.

Q.76 What are the causes for gallstone formation?

- Reduction in bile salts and/or lecithin
- Excess of cholesterol
- Disturbed cholesterol metabolism
- Excess of calcium ions due to increased concentration of bile
- Damage or infection of gallbladder epithelium
- Obstruction of bile flow from the gallbladder.

Q.77 What are crypts of Lieberkuhn?

Crypts of Lieberkuhn are the intestinal glands.

Q.78 What are the cells present in the intestinal glands?

- Columnar epithelial cells called enteroctyes, which secrete enzymes
- Argentaffin or enterochromaffin cells which secrete intrinsic factor of Castle
- Goblet cells which secrete mucus
- Paneth cells which secrete defensins (cytokines).

Q.79 What are Brunner's glands?

Brunner's glands are the mucus glands present in the first part of duodenum.

Q.80 What is succus entericus?
Digestive juice secreted by small intestine is called succus entericus or small intestinal juice.

Q.81 What are the properties of succus entericus?

Volume	:	1800 mL/day
Reaction and pH	:	Alkaline with a pH of 8.3

Q.82 Mention the composition of succus entericus.
- Succus entericus contains water, organic and inorganic substances
- Organic substances are enzymes, mucus, intrinsic factor and defensins
- Inorganic substances are sodium, calcium, potassium, bicarbonate, chloride, phosphate and sulfate.

Q.83 What are the enzymes present in succus entericus?
- *Proteolytic enzymes:* Peptidases—amino peptidase, dipeptidase and tripeptidase
- *Lipolytic enzyme:* Lipase
- *Amylolytic enzymes:* Sucrase, maltase, lactase, dextrinase, trehalase
- Enterokinase.

Q.84 What are the functions of succus entericus?
- Digestive function by enzymes
- Protective function by mucus
- Activator function by enterokinase
- Hemopoietic function by intrinsic factor
- Hydrolytic function by water.

Q.85 What are the functions of small intestine?
- Mechanical function
- Secretory function
- Hormonal function
- Digestive function
- Activator function
- Hemopoietic function
- Hydrolytic function
- Absorptive function.

Q.86 How is succus entericus collected?
Succus entericus is collected by using multilumen tube.

Q.87 What are the properties of large intestinal juice?
Large intestinal juice is a watery fluid and highly alkaline with the pH of 8.0.

Q.88 What is the composition of large intestinal juice?
Large intestinal juice contains water and solids. Solids are organic and inorganic substances. Organic substances are albumin, globulin, mucin, urea and debris of epithelial cells. Inorganic substances are sodium, calcium, potassium, bicarbonate, chloride, phosphate and sulfate.

Q.89 What are the functions of large intestinal juice?
- Neutralization of acids
- Lubrication activity.

Q.90 What are the functions of large intestine?
- *Absorptive function:* Absorption of water, electrolytes, glucose, alcohol and drugs like anesthetic agents, sedatives and steroids
- *Formation of feces*
- *Excretory function:* Excretion of mercury, lead, bismuth and arsenic
- *Secretory function:* Secretion of mucin, chloride and bicarbonate
- *Synthetic function:* Synthesis of folic acid, vitamin B_{12} and vitamin K.

Q.91 What are the causes of constipation?
- Dietary causes—lack of fiber or water
- Irregular bowel habits
- Spasm of sigmoid colon
- Many types of diseases
- Dysfunction of myenteric plexus in large intestine (megacolon).

Q.92 What is megacolon or Hirschsprung's disease?
Dysfunction of myenteric plexus in large intestine causes constipation and accumulation of large quantity of feces in colon. This leads to distension of colon to a diameter of 4 to 5 inches. This condition is known as megacolon or Hirschsprung's disease.

Q.93 What are the significances of mastication?
- Breakdown of foodstuffs into smaller particles
- Mixing of saliva with food substances
- Lubrication and moistening of dry food by saliva so that, the bolus can be easily swallowed
- Appreciation of taste of the food.

Q.94 What is deglutition?
Swallowing of food is known as deglutition. In this process, the masticated food from the mouth enters the stomach via pharynx and esophagus.

Q.95 What are the stages of deglutition?
- Oral stage—entrance of food into pharynx from mouth.
- Pharyngeal stage—entrance of food into esophagus from pharynx.
- Esophageal stage—entrance of food into stomach from esophagus.

Q.96 Explain in brief how the entrance of bolus through different passages other than esophagus is prevented.
- Return of bolus back into the mouth is prevented by the position of tongue against the roof of the mouth and the high intraoral pressure
- Movement of bolus into nasopharynx is prevented by elevation of soft palate
- Movement of bolus into the larynx is prevented by:
  - Approximation of vocal cords
  - Forward and upward movement of larynx
  - Backward movement of epiglottis to close the larynx causing deglutition apnea.

Q.97 What is deglutition apnea or swallowing apnea?
The temporary arrest of breathing during the pharyngeal stage of deglutition is called deglutition apnea or swallowing apnea.

Q.98 What is the significance of deglutition apnea?
Deglutition apnea prevents entrance of bolus into larynx during swallowing.

Q.99 What are the movements of esophagus during deglutition?
Movements of esophagus during deglutition are the primary and secondary peristaltic contractions. Some times tertiary contraction may also occur.

Q.100 What is dysphagia? What are its causes?
Difficulty in swallowing is called dysphagia. Its causes:
- Mechanical obstruction of esophagus
- Decreased movement of esophagus
- Muscular disorders.

Q.101 What is esophageal achalasia?
It is a neuromuscular disease characterized by accumulation of food in esophagus. It is because the lower esophageal (cardiac) sphincter fails to relax during swallowing.

Q.102 What is gastroesophageal reflux disease (GERD)?
GERD is a disorder characterized by regurgitation of acidic gastric content into esophagus.

It is due to the weakness or incompetence of cardiac sphincter.

Q.103 Define peristalsis.
Peristalsis is the wave of contraction followed by wave of relaxation that travels in aboral direction.

Q.104 What is the significance of peristalsis?
By peristalsis, the contents are propelled along the gastrointestinal tract.

Q.105 Trace the pathway for deglutition reflex.
The receptors are present in the pharynx. Afferent fibers pass through glossopharyngeal nerve. Center is in medulla oblongata. Efferent fibers pass through hypoglossal, glossopharyngeal and vagus nerves. Effectors are the muscles of pharynx and esophagus.

Q.106 What are the types of movements of stomach?
- Hunger contractions which occur when the stomach is empty
- Peristalsis when the stomach is filled with food.

Q.107 What is receptive relaxation?
Relaxation of the upper part of the stomach when bolus enters the stomach from esophagus is called receptive relaxation.

Q.108 What are the factors influencing emptying of stomach?
- Volume of gastric content
- Consistency of gastric content
- Chemical composition of gastric content
- pH of gastric content
- Osmolar concentration of gastric contents.

Q.109 What are the factors, which inhibit gastric emptying?
- Nervous factor—enterogastric reflex
- Hormonal factors—hormones VIP, GIP, secretin and cholecystokinin.

Q.110 What is enterogastric reflex?
When the chyme enters the intestine, the gastric muscle is inhibited and the gastric movements are reduced or stopped. It causes stoppage of gastric emptying. This is known as enterogastric reflex and it involves vagus nerve.

Q.111 What are the movements involved in vomiting?
Vomiting involves antiperistalsis in intestine, stomach and esophagus, relaxation of lower and upper esophageal sphincters, closure of glottis and contraction of abdominal muscles.

Q.112 Trace the pathway for vomiting.
Receptors are mostly in the gastrointestinal tract. Afferent fibers are vagus and sympathetic afferent fibers. Center is in medulla oblongata near tractus solitarius. Efferent fibers are the fibers of V, VII, IX, X and XII cranial nerves and spinal nerves. Effectors are the muscles of gastrointestinal tract and abdominal muscles.

Q.113 What are the movements of small intestine?
- Mixing movements—segmentation movements and pendular movements
- Propulsive movements—peristaltic movements and peristaltic rush
- Peristalsis in fasting (migrating motor complex)
- Movements of villi.

Q.114 What is peristaltic rush? What is its cause?
Peristaltic rush is a powerful peristaltic contraction that begins in duodenum, passes through entire length of small intestine and reaches ileocecal valve. It is by excessive irritation of intestinal mucus membrane or extreme distention of intestine.

Q.115 What is the significance of peristaltic rush?
Peristaltic rush sweeps the contents of small intestine into colon and thus it relieves the small intestine off irritant substances or excessive distention.

Q.116 What is peristalsis in fasting or migrating motor complex?
It is the most powerful peristaltic contraction involving a large portion of stomach or intestine during the period of fasting or several hours after the meals. It starts in stomach and runs through the entire length of small intestine.

Q.117 What is the significance of migrating motor complex?
It sweeps the excessive digestive secretions into the colon and prevents the accumulation of secretions in stomach and small intestine.

Q.118 What are the movements of large intestine?
- Mixing movements—segmentation contractions
- Propulsive movements—mass peristalsis.

Q.119 What is the significance of mass peristalsis or mass movement?
It propels the feces from colon towards anus.

Q.120 What is gastrocolic reflex?
The distention of stomach with entrance of food causes contraction of colon and entrance of feces into rectum. This is known as gastrocolic reflex.

Q.121 What is the nerve supply to internal and external anal sphincters?
Internal anal sphincter that is formed by smooth muscle fibers is innervated by parasympathetic fibers via pelvic nerve. The external anal sphincter that is formed by skeletal muscle fibers is innervated by somatic nerve fibers via pudendal nerve.

Q.122 Trace the pathway for defecation reflex.
Receptors are in rectum. Afferent fibers pass via pelvic nerve. Center is in sacral segment of spinal cord. Efferent fibers pass via pelvic nerve. Effectors are muscles of rectum and internal sphincter.

Q.123 What is the importance of pudendal nerve?
The pudendal nerve always keeps the external anal sphincter constricted. During defecation reflex, the pudendal nerve is inhibited by impulses arising from cerebral cortex and this causes relaxation of external anal sphincter and defecation.

Q.124 What are the gastrointestinal hormones?
Gastrointestinal hormones are the local hormones secreted in the stomach and intestine.

Q.125 Name the cells secreting the gastrointestinal hormones.
APUD (amine precursor uptake and decarboxylation) cells present in the gastrointestinal tract secrete the gastrointestinal hormones.

Q.126 Name the hormones secreted by stomach.
Gastrin, GIP, somatostatin and motilin.

Q.127 Name the hormones secreted by small intestine.
Secretin, cholecystokinin, GIP, VIP, glucagon, glicentine, GLP-2. somatostatin, pancreatic polypeptide, peptide YY, neuropeptide Y, motilin and substance P.

Q.128 What is the function of gastric inhibitory peptide (GIP)?
GIP is secreted by duodenum and inhibit the stomach motility and secretion.

Q.129 What is the function of gastrin releasing peptide (GRP)?
GRP stimulates the release of gastrin from G cell.

Q.130 What is migrating myoelectric complex (MMC)?
It is the propulsive movements initiated during fasting which beings in the stomach and moves undigested material

from stomach to small intestine and finally into colon.

Q.131 Which are the enzymes secreted in stomach in inactive form?

Inactive form		*Active form*
Trypsinogen	—enterokinase→	Trypsin
Chymotrypsinogen	—trypsin→	Chymo-trypsin
Procarboxy-peptidase	—trypsin→	Carboxy-peptidase

Q.132 What are micelle?
They are water soluble sphere with a lipid soluble interior.

Q.133 What are the functions of micelle?
They help in digestion, transport and absorption of lipid soluble substance from duodenum to distal ilium.

Q.134 What are stercobilin?
Produced from metabolism of bilirubin by interstinal bacteria. It gives brown color to stool.

Q.135 What is the action of parasympathetic and sympathetic nerve supply on salivary glands?
The parasympathetic fibers are secretomotor in action, while the sympathetic fibers are vasoconstrictor in action.

Q.136 What are the digestive enzymes present in gastric juice?
These are pepsinogen, renin, and lipase.

Q.137 What are the function of HCl in gastric juice?
HCl is bactericidal in action; it hydrolyses the food and acids in digestion. It activates pepsinogen. It also helps in iron and calcium absorption.

Q.138 What is the action of pepsinogen?
Pepsinogen is activated into pepsin which along with HCl converts protein into peptones and proteoses.

Q.139 Which are the digestive enzymes present in the pancreatic juice?
These are trypsinogen, chymotrypsinogen, amylase (amylopsin), and (stypsin).

Q.140 What is the function of trypsinogen and chymotrypsinogen?
Trypsinogen is activated by enterokinase into trypsin, which in turn is activated chymotrysinogen into chymotrypsin. These convert proteoses and peptones up to dipeptides stage.

Q.141 Name the various enzymes present in the succus entericus?
These are erepsin (peptidase), nuclease, nucleosidase, arginase, amylase, maltase, sucrase, lactase and enterokinase.

Q.142 What is the role of nuclease, nucleosidase and nucleotidase?
These are concerned with digestion of nucleoproteins.

Q.143 What is the action of arginase?
Arginase converts arginine into urea and ornithine.

Q.144 What is the action of sucrase?
Sucrase splits sucrose into fructose and glucose.

Q.145 What is the action of maltase?
Maltase converts maltose into two molecules of glucose.

Q.146 What is the action of lactase?
Lactase converts lactose into glucose and galactose.

Q.147 What are the actions of secretin?
Secretin:
- Causes secretion of watery juice with more water and bicarbonate ions
- Inhibits secretion of gastric juice
- Inhibits motility of stomach
- Causes constriction of pyloric sphincter
- Increases the potency of action of cholecystokinin on pancreatic secretion.

Q.148 What are the actions of cholecystokinin?
Cholecystokinin:
- Contracts gallbladder
- Causes secretion of pancreatic juice with large amount of enzymes
- Accelerates the activity of secretin
- Increases the secretion of enterokinase
- Inhibits the gastric motility
- Increases the motility of intestine and colon
- Augments contraction of pyloric sphincter
- Plays an important role in satiety by suppressing hunger
- Induces drug tolerance to opioids.

Q.149 How is carbohydrate digested?
Carbohydrate digestion starts in the mouth by ptyalin and continues in the stomach where gastric amylase also acts. Final digestion occurs in small intestine by pancreatic amylase, sucrase, maltase, lactase, dextrinase and trehalase.

Q.150 How is carbohydrate absorbed from small intestine?
Carbohydrate is absorbed from small intestine mainly as monosaccharides (glucose, galactose and fructose).

Q.151 How is protein digested?
Protein digestion starts only in the stomach. Pepsin breaks proteins into proteoses, peptones and large polypeptides. In the small intestine, final digestion of proteins occurs because of proteolytic enzymes in pancreatic juice and succus entericus.

Q.152 How is protein absorbed from small intestine?
Protein is absorbed from small intestine mainly as amino acids.

Q.153 How is lipid digested?
Lipid digestion starts in the stomach by gastric lipase. But it is a very weak lipolytic enzyme. In the small intestine, most of the lipid is digested by pancreatic lipase. Succus entericus also contains lipase but it is very weak and its action is negligible.

Q.154 Name the bile pigments present in the bile juice.
These are bilirubin and biliverdin.

Q.155 What is the nature of bile pigments? Is it excretory or secretory?
Bile pigments are excretory products of bile.

Q.156 What is the cholagogue?
Cholegogue is the agent, which tends to increase the bile flow and its expulsion from biliary passages into the intestines.

Q.157 What is ptyalism?
Ptyalism is excessive salivation produced reflexly by irritation of mouth or esophagus or by drugs.

Q.158 What is normal daily secretion of bile?
It is 0.5 to 1 liter.

Q.159 What is hepatocrinin?
It is hormone found in intestinal extract and acts as a stimulant for bile secretion.

Q.160 What is the role of bile salts in lipid digestion?
The lipid molecules are not soluble in water due to the surface tension. So, the lipids cannot be digested by any lipolytic enzymes. Due to the detergent action of bile salts in small intestine, the lipid molecules become water soluble. This action of bile salts is known as emulsification. During this, the bile salts convert the lipid substances into micelles. The emulsified fat molecules in micelles are easily digested by lipolytic enzymes.

Q.161 How is lipid absorbed from small intestine?

Lipid is absorbed from small intestine in two forms:

- In the form of fatty acids which are absorbed into blood by diffusion.
- In the form of chylomicrons, which contain triglycerides, and cholesterol esters. Because of the larger size, chylomicrons cannot pass through membrane of blood capillaries. And, these lipid materials are absorbed into lymph vessels and transferred into blood from lymph.

Q.162 What are the importance of HDL and LDL?

HDL (good cholesterol) carries cholesterol and phospholipids from tissues and organs back to the liver for degradation and elimination. It prevents the deposition of cholesterol on the walls of arteries by carrying cholesterol away from arteries to liver. High level of HDL indicates a healthy heart, because it reduces the blood cholesterol level.

LDL (bad cholesterol) carries cholesterol and phospholipids from the liver to muscles, other tissues and organs such as heart. It is responsible for deposition of cholesterol on walls of arteries causing atherosclerosis. High level of LDL increases the risk of heart disease.

Q.163 What is lipid profile?

The lipid profile is a group of blood tests which are carried out to determine the risk of coronary artery diseases (CAD).

Q.164 What are the tests involved in lipid profile? Give the normal values.

- Total cholesterol (200–240 mg%)
- Triglyceride (150–200 mg%)
- HDL (40–60 mg%)
- LDL (60–100 mg%)
- Total cholesterol—HDL ratio (2–6).

Chapter 12

Renal Physiology and Excretion

Q.1 What are the functions of kidney?

The primary function of kidney is homeostasis, i.e. the maintenance of internal environment. Various functions of kidney:

- Role in homeostasis—by the formation of urine and excretion of water, electrolytes and waste products through urine
- Hemopoietic function
- Endocrine function
- Regulation of blood pressure
- Regulation of blood calcium level.

Q.2 Name the layers of kidney (Fig. 12.1).

- Outer cortex containing renal corpuscles and convoluted tubules
- Inner medulla containing tubular and vascular structures arranged in the form of medullary pyramids
- Renal sinus containing renal pelvis, major calyces, minor calyces, branches of nerves and arteries, tributaries of veins, loose connective tissue and fat.

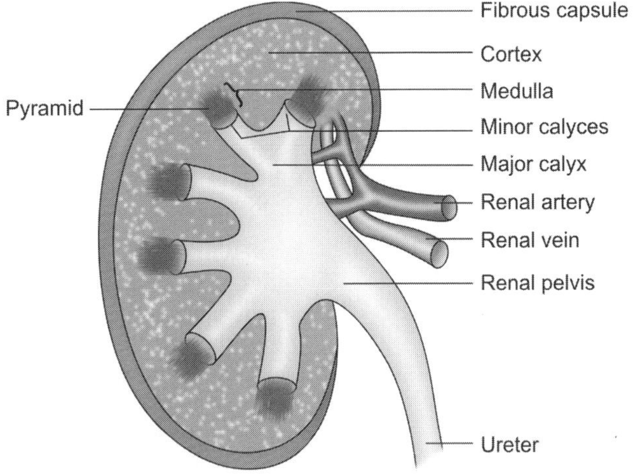

Fig. 12.1: Longitudinal section of human kidney

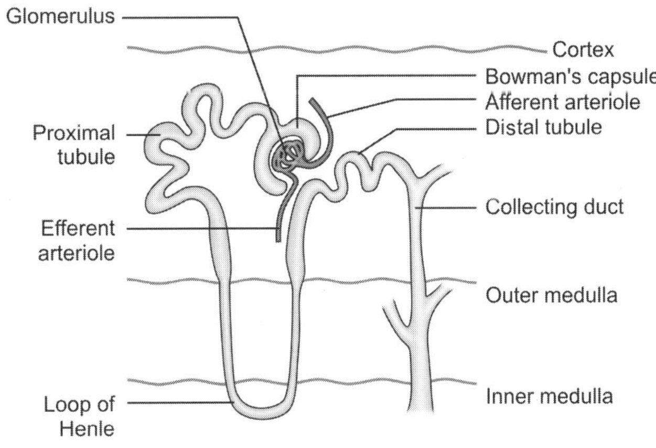

Fig. 12.2: Structure of nephron

Q.3 What are uriniferous tubules? Name their parts.

Uriniferous tubules are the tubular structures forming the parenchyma of kidney.

Each uriniferous tubule consists of nephrons and collecting ducts.

Q.4 Define nephron.

Nephron is defined as structural and functional unit of kidney (Fig. 12.2).

Q.5 How many nephrons are present in each kidney?

1 to 1.3 million nephrons.

Q.6 What are the two types of nephrons?

- Cortical or superficial nephrons whose renal corpuscles are situated in the outer part of cortex
- Juxtamedullary nephrons whose renal corpuscles are situated in the inner part of cortex near medulla.

Q.7 What are the parts of nephron?

- Renal corpuscle or Malpihigian corpuscle
- Tubular portion or renal tubule.

Q.8 What are the structures of renal corpuscle?

- Glomerulus
- Bowman's capsule that encloses the glomerulus.

Q.9 What is glomerulus?

Glomerulus is a tuft of capillaries formed from the afferent arteriole and drained by efferent arteriole.

Q.10 What are the layers of Bowman's capsule?

- Inner visceral layer
- Outer parietal layer.

Q.11 What are podocytes?

Podocytes are the epithelial cells of visceral layer of Bowman's capsule, which are connected to basement membrane by means of foot like projections called pedicles.

Q.12 What are the parts of renal tubule?

- Proximal convoluted tubule
- Loop of Henle that includes the thick descending limb, thin descending limb, hairpin bend, thin ascending limb and thick ascending limb
- The distal convoluted tubule.

Q.13 What is the unique feature of the wall of proximal convoluted tubule?

The wall of proximal convoluted tubule is formed by brush bordered cuboidal epithelial cells.

Q.14 What is the advantage of brush bordered cuboidal epithelial cells in proximal convoluted tubule?

The brush bordered cuboidal epithelial cells increase the surface area for reabsorption.

Q.15 What is juxtaglomerular apparatus?

It is a specialized organ situated near the glomerulus of each nephron.

Q.16 What are the parts of juxtaglomerular apparatus?

The juxtaglomerular apparatus is formed by three different parts (Fig. 12.3):

- Macula densa
- Extraglomerular mesangial cells
- Juxtaglomerular cells.

Q.17 What is macula densa?

Macula densa is the part of distal convoluted tubule near the afferent arteriole, which is formed by tightly packed cuboidal epithelial cells.

Q.18 What are extraglomerular mesangial cells?

Extraglomerular mesangial cells are the special type of agranular or lasis cells situated in the triangular region bound by afferent arteriole, efferent arteriole and macula densa.

Q.19 What are juxtaglomerular cells?

Juxtaglomerular cells are the specialized type of smooth muscle cells present in the afferent arteriole before it enters the Bowman's capsule. This part of afferent arteriole is thickened like a cuff called polar cushion or polkissen.

Q.20 What are the functions of juxtaglomerular apparatus?

- Secretion of renin
- Secretion of other substances—prostaglandin, cytokines, and thromboxane A_2
- Regulation of glomerular blood flow and glomerular filtration rate.

Q.21 What is the role of renin in the body?

Renin converts inactive angiotensinogen into angiotensin I. Angiotensin I is converted into angiotensin II by the converting enzyme. Angiotensin II is converted into angiotensin III by angiotensinases. Angiotensin III is converted into angiotensin IV (Fig. 12.4).

Q.22 Name the factors which stimulate renin secretion.

- Decreased arterial blood pressure
- Reduction in ECF volume
- Increased sympathetic activity
- Decreased load of sodium and chloride in macula densa.

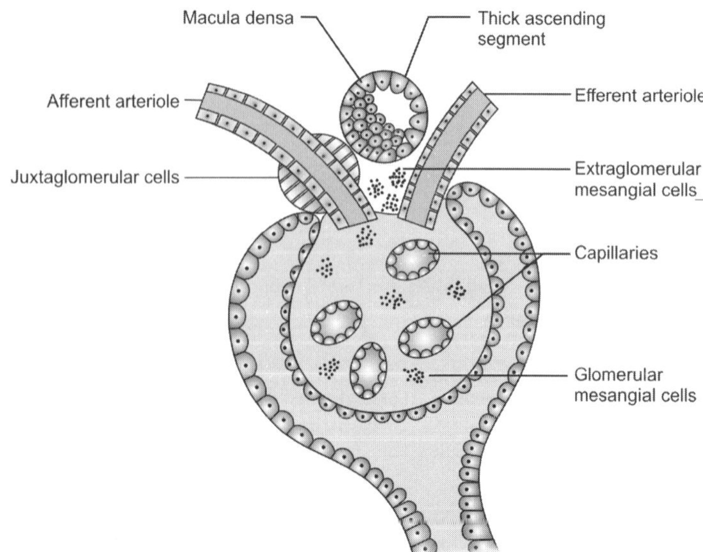

Fig. 12.3: Juxtaglomerular apparatus

Q.23 What are the functions of angiotensins?

Angiotensin I is physiologically inactive.

Angiotensin II:

- Increases blood pressure
- Increases aldosterone secretion
- Regulates glomerular filtration rate
- Inhibits response of baroreceptor reflex

Angiotensins III and IV:

- Increase the blood pressure
- Increase the aldosterone secretion.

Q.24 How much of blood is supplied to both the kidneys?

1300 mL/minute.

Q.25 How is renal blood flow measured?

By renal clearance test using para aminohippuric acid.

Q.26 What is autoregulation? What are the mechanisms involved in renal autoregulation?

The intrinsic ability of an organ to regulate its own blood flow is called autoregulation.

Renal autoregulation involves myogenic response and tubuloglomerular feedback.

Q.27 What are the special features (peculiarities) of renal circulation?

- Renal arteries arise directly from aorta
- Kidneys receive maximum amount of blood (1,300 mL/minute) next to liver (1,500 mL/minute)
- Whole blood passes through glomerulus
- Renal circulation has a portal system
- The capillary pressure in glomerulus is very high (60 mm Hg)
- Peritubular capillaries form low pressure bed
- The autoregulation is well established in kidney.

Q.28 What is the normal urinary output?

1 to 1.5 L/day.

Q.29 Name the processes involved in urine formation.

The processes involved in formation of urine are:

- Glomerular filtration
- Tubular reabsorption
- Tubular secretion or excretion.

Figures 12.5A and B should the mechanism of formation of urine.

Q.30 What is glomerular filtration?

When the blood passes through the glomerular capillaries, the plasma is filtered into the Bowman's capsule. This process is called glomerular filtration and the filtered fluid is called glomerular filtrate.

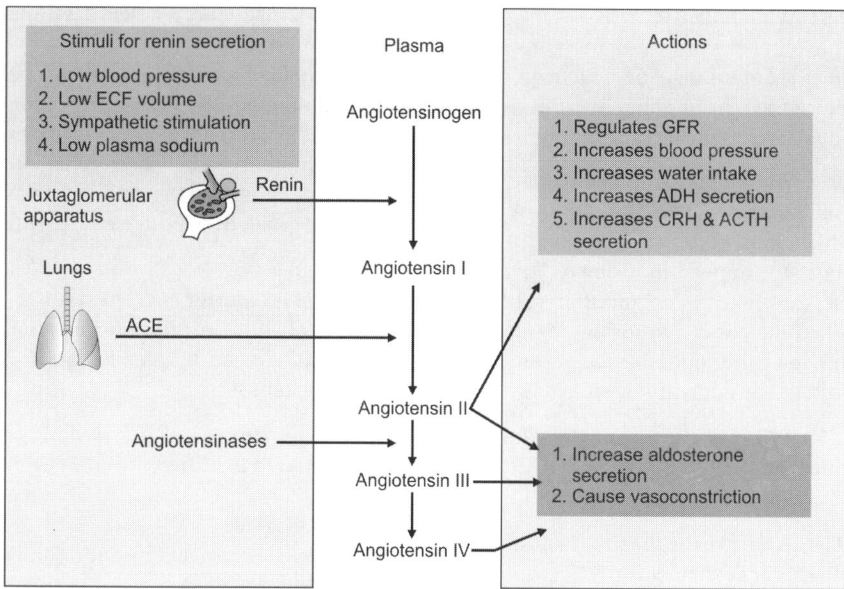

Fig. 12.4: Renin–angiotensin system

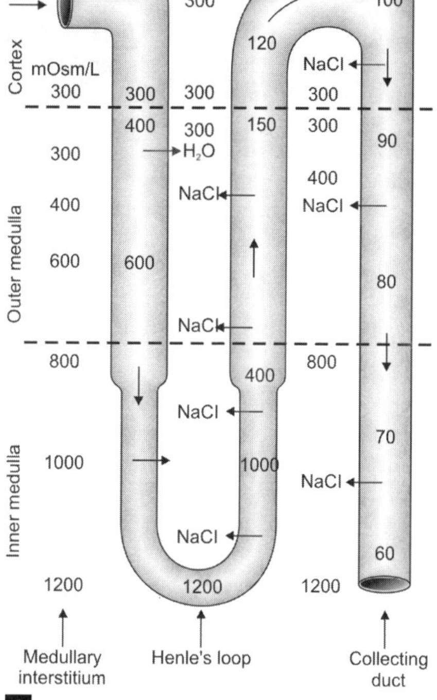

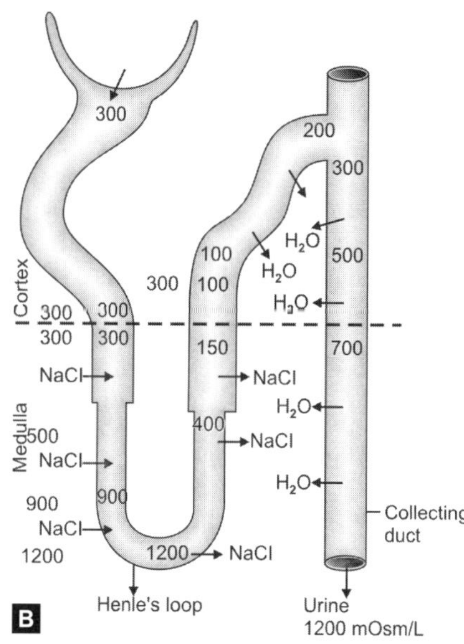

Figs 12.5A and B: (A) Mechanism for the formation of dilute urine. Numerical values indicate osmolarity (mOsm/L), (B) Role of ADH in the formation of concentrated urine. ADH increases the permeability for water in distal convoluted tubule and collecting duct. Numerical values indicate osmolarity (mOsm/L)

Q.31 What is the composition of glomerular filtrate?

The glomerular filtrate is the plasma without plasma proteins, i.e. all the substances present in the plasma are present in glomerular filtrate also except plasma proteins.

Q.32 Why glomerular filtration is called ultrafiltration?

Glomerular filtration is called ultrafiltration because even the minute particles are filtered from glomerular capillary into Bowman's capsule.

Q.33 Define glomerular filtration rate.

The total amount of filtrate formed in all the nephrons of both the kidneys per unit time is known as glomerular filtration rate (GFR).

Q.34 What is the normal value of GFR?

125 mL/minute or 180 L /day.

Q.35 What is filtration fraction?

The fraction or part of the renal plasma that becomes the filtrate is called filtration fraction. Or, it is the ratio of renal plasma flow and glomerular filtration rate that is expressed in percentage.

Q.36 What is the normal filtration fraction?

15 to 20%.

Q.37 Name the pressures, which determine the GFR.

- Glomerular capillary pressure (60 mm Hg)
- Colloidal osmotic pressure in the glomeruli (25 mm Hg)
- Hydrostatic pressure in the Bowman's capsule (15 mm Hg).

The glomerular capillary pressure favors filtration. Colloidal osmotic pressure and hydrostatic pressure oppose or prevent filtration.

Q.38 What is net or effective filtration pressure? How much is it?

The balance between the pressure favoring filtration and pressures opposing filtration is known as net or effective filtration pressure.

Effective filtration pressure = 60 – (25 + 15) mm Hg. Normally it is 15 to 20 mm Hg.

Q.39 What is Starling's hypothesis?

Starling's hypothesis states that the net filtration through capillary membrane is proportional to the hydrostatic pressure difference across the membrane minus the oncotic pressure difference.

Q.40 What is filtration coefficient?

Filtration coefficient is the GFR in terms of net filtration pressure. It is the glomerular filtration rate per mm Hg of effective filtration pressure.

Q.41 Name the factors affecting GFR.

- Tubuloglomerular feedback
- Glomerular capillary pressure
- Colloidal osmotic pressure
- Hydrostatic pressure in Bowman's capsule
- Renal blood flow
- Constriction of afferent arteriole
- Constriction of efferent arteriole
- Systemic arterial pressure
- Sympathetic stimulation
- Surface area of capillary membrane
- Permeability of capillary membrane
- Contraction of glomerular mesangial cells
- Hormonal and other factors.

Q.42 What is tubular reabsorption?
When the glomerular filtrate passes through the renal tubule, large quantity of water, electrolytes and other substances are reabsorbed back into the blood in peritubular capillaries. This process is known as tubular reabsorption.

Q.43 Why the tubular reabsorption is called selective reabsorption?
The cells of the renal tubule selectively reabsorb the substances present in the glomerular filtrate according to the need of the body. So, the tubular reabsorption is called selective reabsorption.

Q.44 Name the substances reabsorbed in proximal convoluted tubule.
Glucose, amino acids, sodium, potassium, calcium, bicarbonates, chlorides, phosphates, uric acid and water.

Q.45 Name the substances reabsorbed in loop of Henle.
Sodium and chloride.

Q.46 Name the substances reabsorbed in distal convoluted tubule.
Sodium, bicarbonate and water.

Q.47 What are the high threshold substances?
The substances which are completely reabsorbed from the renal tubules and do not appear in urine under normal conditions are known as high threshold substances. These substances appear in urine only if their concentration in plasma is very high or in renal diseases when reabsorption is inhibited.
Examples: Glucose, amino acids and vitamins.

Q.48 What are low threshold substances?
The substances which are reabsorbed only to a minimum extent and appear in urine even in normal condition, are known as low threshold substances.
Examples: Uric acid, phosphates, etc.

Q.49 What are non-threshold substances?
The metabolic end products which are not at all reabsorbed from renal tubules and appear in urine irrespective of their plasma level are known as non-threshold substances.
Example: Creatinine.

Q.50 What is tubular maximum (Tm)?
The maximum rate at which a substance is reabsorbed from the renal tubule is called tubular maximum (Tm).

Q.51 What is TmG?
The tubular maximum for glucose, i.e. the maximum rate at which glucose is reabsorbed from renal tubule is called TmG. It is about 380 mg/minute.

Q.52 What is threshold value?
The blood level of a substance below which it is completely reabsorbed and does not appear in urine is known as the threshold value for that substance. When the concentration increases above that level in blood, the excess amount is excreted through urine.

Q.53 What is the renal threshold for glucose?
180 mg%.

Q.54 What are the mechanisms involved in tubular reabsorption?
- Active reabsorption
- Passive reabsorption.

Q.55 Name the substances reabsorbed actively from renal tubules.
Sodium, calcium, potassium, phosphates, sulfates, bicarbonates, glucose, amino acids, ascorbic acid, uric acid and ketone bodies.

Q.56 Name the substances reabsorbed passively from renal tubules.
Chloride, urea and water.

Q.57 How is water reabsorbed from renal tubules?
By two ways:
- Obligatory water reabsorption in proximal convoluted tubule
- Facultative water reabsorption in distal convoluted tubule.

Q.58 What are the substances secreted into renal tubules?
Potassium is secreted in distal convoluted tubule and collecting duct. Ammonia is secreted in proximal convoluted tubule. Hydrogen ions are secreted in proximal and distal convoluted tubules.

Q.59 What are the factors, which determine the concentration of urine?
- Medullary gradient
- ADH mechanism.

Q.60 What is medullary gradient?
Medullary gradient is the gradual increase in the osmolarity of medullary interstitial fluid from 300 milliosmoles/L near the cortex up to 1,200 milliosmoles/L at the innermost part of medulla.

Q.61 How is medullary gradient developed and maintained?
The medullary gradient is developed and maintained by counter current mechanism. The development of medullary gradient is because of counter current multiplier and the maintenance of medullary gradient is because of counter current exchanger.

Q.62 What is counter current system?
The flow of fluid in opposite directions through 'U' shaped tubules is known as counter current system.

Q.63 Name the divisions of counter current system in kidney.
- Counter current multiplier that is formed by loop of Henle
- Counter current exchanger that is formed by vasa recta.

Q.64 Why the loop of Henle is called counter current multiplier?
Loop of Henle is called counter current multiplier because it is responsible for the increase or multiplication of osmolarity in medullary interstitium.

Q.65 Why vasa recta is called counter current exchanger?
Vasa recta is called counter current exchanger because it helps to exchange the sodium ions between the ascending limb and descending limb of loop of Henle by which the hyperosmolarity of medullary interstitium and medullary gradient are maintained.

Q.66 What are the special features of vasa recta, which help it to act as counter current exchanger?
- It has got an ascending limb and a descending limb
- Only 5% blood flowing to kidney passes through vasa recta
- The velocity of blood flow through vasa recta is very less.

Q.67 How does the final concentration of urine occur?
The final concentration of urine occurs under the influence of antidiuretic hormone (ADH). ADH increases the water reabsorption in distal convoluted tubule and collecting duct and causes concentration of urine.

Q.68 How is urine acidified?
Urine is acidified by the secretion of hydrogen ions in distal convoluted tubule.

Q.69 How are the hydrogen ions secreted in renal tubules?
Hydrogen ions are secreted in exchange of sodium ions and by the formation of ammonia in the renal tubules.

Q.70 Name the renal function tests.
- Routine examination of urine
- Examination of blood
- Examination of urine and blood.

Q.71 Define plasma clearance.
Plasma clearance is the amount of plasma that is cleared off a substance in a given unit of time.

Q.72 What are the advantages of determining plasma clearance?
Determination of plasma clearance helps to measure:
- Glomerular filtration rate
- Renal plasma flow
- Renal blood flow.

Q.73 Name the substances used to measure glomerular filtration rate and renal plasma flow by plasma clearance.
Inulin is used to measure glomerular filtration rate and para aminohippuric acid is used to measure renal plasma flow.

Q.74 Classify renal disorders.
- Acute renal failure
- Chronic renal failure.

Q.75 What is detrusor muscle?
The smooth muscle forming the body of urinary bladder is known as detrusor muscle.

Q.76. Mention the differences between the internal and external urethral sphincters.
- Internal urethral sphincter is formed by smooth muscle but the external urethral sphincter is formed by skeletal muscle
- Internal sphincter is innervated by sympathetic and parasympathetic fibers of autonomic nervous system, whereas, the external sphincter is innervated by somatic nerve fibers
- The internal sphincter functions under reflex control and the external sphincter is under voluntary control.

Q.77 Name the nerves supplying urinary bladder and sphincters.
Detrusor muscle and internal sphincter are supplied by parasympathetic fibers (pelvic nerve) and sympathetic fibers (hypogastric nerve). External sphincter is supplied by somatic nerve fibers (pudendal nerve).

Q.78 What is the action of parasympathetic nerve on urinary bladder and internal sphincter?
When stimulated, the parasympathetic (pelvic) nerve causes contraction of detrusor muscle and relaxation of internal sphincter leading to micturition. Hence it is called the nerve of micturition or nerve of emptying.

Q.79 What is the action of sympathetic nerve on urinary bladder and internal sphincter?
Stimulation of the sympathetic (hypogastric) nerve causes relaxation of detrusor muscle and constriction of internal sphincter. This helps in filling of urinary bladder and so it is called the nerve of filling.

Q.80 What is the action of pudendal (somatic) nerve on external sphincter?
The pudendal (somatic) nerve is always active and keeps the external sphincter constricted. When urine enters the urethra from bladder, the pudendal nerve is inhibited and the external sphincter relaxes leading to micturition. Thus, the pudendal nerve is responsible for voluntary control of micturition.

Q.81 What is cystometrogram?
Cystometrogram is the graphical recording of pressure changes in relation to volume changes in the urinary bladder while filling.

Q.82 What is intravesical pressure?
The pressure in the urinary bladder is known as intravesical pressure.

Q.83 When does the desire for micturition arise?
Desire for micturition arises when about 300 mL of urine is collected in urinary bladder and the intravesical pressure increases to about 10 to 15 cm H_2O.

Q.84 What is the maximum amount of urine collected in the bladder and intravesical pressure up to which the voluntary control of micturition is possible?
Voluntary control of micturition is possible up to 600 to 700 mL of urine collection in the urinary bladder at which the intravesical pressure is about 35 to 40 cm H_2O. When the volume of urine in the bladder increases beyond 700 mL, the pressure rises to 40 cm H_2O. Now, the voluntary control of micturition fails.

Q.85 Explain briefly the micturition reflex.
Micturition reflex occurs in two phases. Initially, when 300 to 400 mL of urine is collected in the urinary bladder, the stretch receptors in the wall of the bladder are stimulated. This leads to contraction of detrusor muscles and relaxation of internal sphincter and urine flows into the urethra from the urinary bladder.

In the second phase, when urine flows through urethra, the stretch receptors present in urethra are stimulated. This leads to inhibition of pudendal nerve, relaxation of external sphincter causing voiding of urine.

Q.86 Trace the pathway for micturition reflex.
First phase:

Receptors	– stretch receptors in the wall of urinary bladder.
Afferent fibers	– pass through pelvic nerve.
Center	– sacral segments of spinal cord.
Efferent fibers	– pass through pelvic nerve.
Response	– the contraction of detrusor muscles and relaxation of internal sphincter.

Second phase:
When urine flows through urethra, stretch receptors present in urethra are stimulated and send impulses through afferent fibers of pelvic nerve. These impulses inhibit pudendal nerve resulting in relaxation of external sphincter and voiding of urine.

Q.87 What is dialysis?
Dialysis means diffusion of solutes from an area of higher concentration to the area of lower concentration through a semipermeable membrane. And, this is the principle of artificial kidney.

Q.88 What is dialysate?
Dialysate is the dialyzing fluid that is used in artificial kidney. Through this fluid, the blood is purified during dialysis.

Q.89 What is the composition of dialysate?
Dialyzing fluid contains less quantity of sodium, potassium and chloride than in patient's blood. It contains more quantity of glucose, bicarbonate and calcium. It does not contain urea, uric acid, sulfate, phosphate and creatinine.

Q.90 What are diuretics?
Diuretics are the substances that increase the urine output.

Q.91 What are the glands present in skin?
- Sebaceous glands which secrete sebum
- Sweat glands which secrete sweat.

Q.92 What is the function of sebaceous glands?
Sebaceous glands secrete an oily substance called sebum that has antibacterial action, antifungal action and protective function. Sebum also prevents heat loss.

Q.93 What are sweat glands? Name them.
Sweat glands are the skin glands, which secrete sweat. Sweat glands are of two types, eccrine glands and apocrine glands.

Q.94 What are the functional differences between eccrine and apocrine glands?
Eccrine glands function throughout life since birth and secrete clear watery sweat. These glands play major role in temperature regulation. Apocrine glands start functioning only during puberty and secrete thick and milky sweat. These glands do not play any role in temperature regulation.

Q.95 Name the nerves supplying the sweat glands.
Eccrine glands are supplied by sympathetic cholinergic fibers whereas, apocrine glands are supplied by sympathetic adrenergic fibers.

Q.96 What is the normal body temperature?
37°C (98.6°F).

Q.97 What is core temperature?
The average temperature in deeper tissues of the body is known as core temperature and it is always more than the oral or rectal temperature. It is about 37.8°C (100°F).

Q.98 What are the pathological variations of body temperature?
- Hyperthermia—abnormal increase in body temperature
- Hypothermia—decrease in body temperature.

Q.99 What is heat balance?
The difference between heat produced in the body and the heat lost from the body is called heat balance.

Q.100 How is heat produced in the body?
By:
- Metabolic activities
- Muscular activity
- The actions of hormones
- Radiation of heat from environment
- Shivering.

Q.101 How is heat lost from the body?
By:
- Conduction
- Radiation
- Convection
- Evaporation
- Panting.

Q.102 Name the centers for temperature regulation in hypothalamus.
- Heat loss center in the anterior hypothalamus
- Heat gain center in the posterior hypothalamus.

Q.103 How is loss of heat from the body increased?
By secretion of sweat due to peripheral vasodilatation.

Q.104 How is heat increased in the body?
By the prevention of heat loss and by increase in heat production.

Chapter 13

Endocrinology

Q.1 What is a hormone?

Hormone is a chemical messenger that is secreted usually by a ductless (endocrine) gland (Fig. 13.1) and also by some other structures like kidney and heart.

Q.2 Classify the chemical messengers.

- Endocrine messengers—classical hormones secreted by endocrine glands
- Neurocrine messengers—neurotransmitters released from nerve endings
- Paracrine messengers which diffuse from control cells to target cells
- Autocrine messengers which control the source cells which secrete them.

Q.3 Classify the classical hormones.

Classical hormones are classified by their chemical nature:

- Steroid hormones
- Protein hormones
- Hormones derived from the amino acid tyrosine.

Q.4 Classify the hormones citing examples of each.

Hormones are classified into 3 major classes:

- *Steroids*—Like adrenocortical hormones, sex hormones and vit-D_3
- *Proteins and polypeptides*—Like anterior and posterior pituitary hormones, hypothalamic hormones, parathyroid hormones, calcitonin, insulin, glucagon, gastrin, secretin and angiotensin.
- *Amino acid derivatives*—Epinephrine, norepinephrine, thyroxine.

Q.5 Name the hormones secreted by following organs.

- *Hypothalamus*—Releasing hormones, like GnRH, TRH, CRH, etc.
- *Anterior pituitary*—TSH, ACTH, GH, FSH, LH, prolactin.
- *Posterior. pituitary*—ADH and oxytocin.
- *Thyroid*—thyroxin, Triiodothyronine and thyrocalcitonin
- *Parathyroid*—Parathormone (PTH).

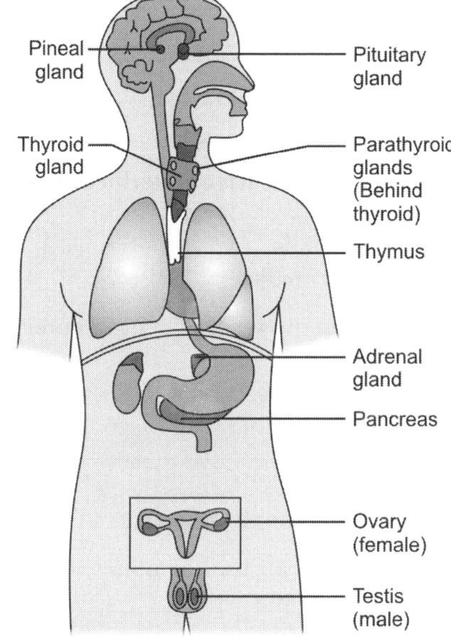

Fig. 13.1: Major endocrine glands

- *Adrenal cortex*—Cortisol, corticosterone, aldosterone, androgens, estrogens and progesterone.
- *Adrenal medulla*—Adrenaline and noradrenaline.
- *Testis*—Testosterone.
- *Ovary*—Estrogen and progesterone.
- *Placenta*—HCG, estrogen, progesterone, HPL.
- *GIT*—Gastrin, secretin, motilin, substance-P, cholecystokinin.
- *Kidney*—Erythropoietin, Vit-D_3, medullipin.
- *Heart*—ANF (Atrial natriuretic factor).

Q.6 What do you mean by long loop feedback, short loop feedback and ultrashort loop feedback?

- When peripheral gland hormones or substances from tissue metabolism exert negative feedback control on both the hypothalamus and anterior pituitary hormones, it is known as long loop feedback mechanism.
- When anterior pituitary hormones exert the negative feedback control over the synthesis and release of hypothalamic releasing hormones, it is known as short loop feedback mechanism.
- When hypothalamic releasing hormones after its secretion inhibit their own synthesis further and release it is known as ultrashort loop.

Q.7 What do you mean by hypothalamohypophyseal portal vessels and hypothalamo-hypophyseal fiber tract?

- The glandular part of pituitary gland has vascular connections with hypothalamus through a set of portal vessels through which hypothalamic releasing hormones enter into adenohypophysis to regulate their secretion. These portal blood vessels are known as hypothalamo-hypophyseal portal tract.
- Whereas the neurohypophysis is connected with hypothalamus by hypothalamo-hypophyseal fiber tracts from supraoptic and paraventricular nuclei of anterior hypothalamus. These tracts are known as hypothalamo-hypophyseal fiber tract through which those above mentioned nuclei pass the oxytocin and vasopressin into posterior pituitary gland for storage.

Q.8 How do you classify anterior pituitary gland cells histologically.

Anterior pituitary gland cells are classified histologically in Figure 13.2.

Q.9 Where are the hormonal receptors situated in the target cell?

The receptors of catecholamines and protein hormones are situated in the cell membrane. The receptors of steroid hormones are in the cytoplasm. And, the receptors of thyroid hormones are situated in the nucleus.

Q.10 Name the mechanism of action of different types of hormones.

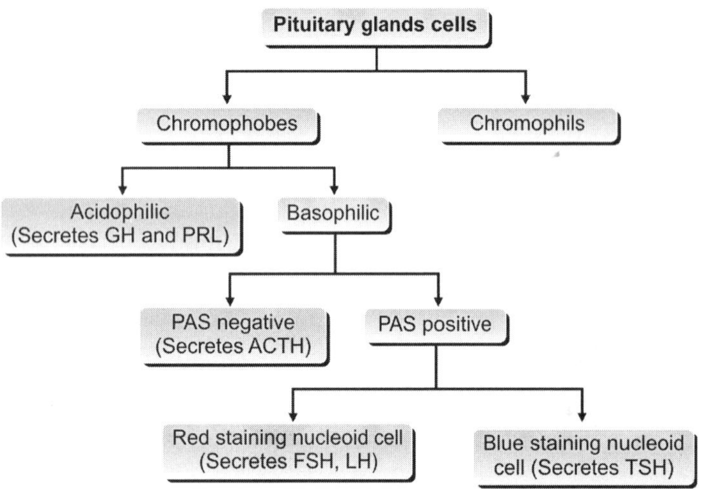

Fig. 13.2: Histological classification of anterior pituitary gland cells

Hormones act by any of the following mechanisms:
- By altering the permeability of cell membrane—neurotransmitters
- By activating the intracellular enzymes and formation of second messenger—protein hormones and catecholamines
- By activating the genes—thyroid and steroid hormones.

Q.11 What is second messenger?
The substance through which the hormonal actions are executed is known as second messenger.

Q.12 Name some second messengers.
Cyclic AMP, calcium, calmoduline, inositol triphosphate (IP_3), diacylglycerol (DAG) and cyclic GMP are second messengers.

Q.13 What are G proteins?
G proteins or guanosine nucleotide binding proteins are the membrane proteins to which the receptor proteins are attached in most of the target cells.

Q.14 Name the major endocrine glands in the body.
Pituitary gland, thyroid gland, parathyroid gland, adrenal glands, islets of Langerhans in pancreas and gonads (ovaries in females and testes in males).

Q.15 Where is pituitary gland (hypophysis) situated?
Pituitary gland (hypophysis) is situated at the base of the brain in sella turcica.

Q.16 What are the two parts of pituitary gland?
- Anterior pituitary or adenohypophysis
- Posterior pituitary or neurohypophysis.

Fig. 13.3 shows the parts of pituitary gland.

Q.17 Name the parts of anterior pituitary.
- Pars distalis
- Pars tuberalis
- Pars intermedia.

Q.18 Name the types of cells in anterior pituitary.
- Somatotropes
- Corticotropes
- Thyrotropes
- Gonadotropes
- Lactotropes.

Q.19 Enumerate the hormones secreted by anterior pituitary.
- Growth hormone
- Thyroid stimulating hormone
- Adrenocorticotropic hormone
- Follicle stimulating hormone
- Luteinizing hormone
- Prolactin.

Q.20 What are the gonadotropic hormones?

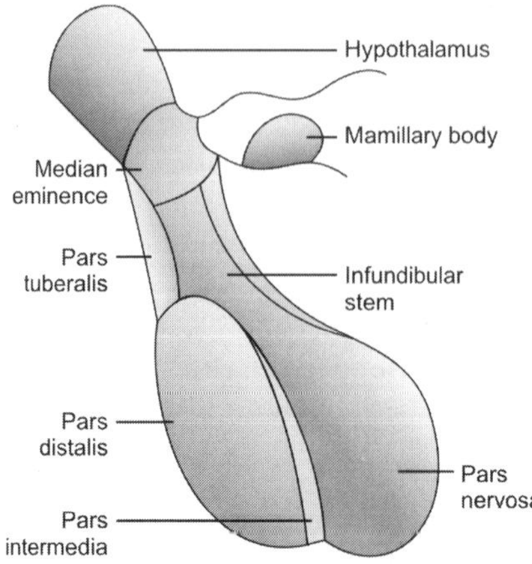

Fig. 13.3: Parts of pituitary gland
(1) Adenohypophysis (2) Neurohypophysis

Follicle stimulating hormone and luteinizing hormone are together called gonadotropic hormones or gonadotropins because of their action on gonads.

Q.21 How is anterior pituitary regulated?
Anterior pituitary is regulated by hypothalamus by the secretion of releasing and inhibitory hormones, which reach the anterior pituitary through hypothalamo hypophyseal portal vessels.

Q.22 Name the releasing hormones, which regulate anterior pituitary.
- Growth hormone releasing hormone
- Growth hormone releasing polypeptide
- Thyrotropic releasing hormone
- Corticotropin releasing hormone
- Gonadotropin releasing hormone.

Q.23 Name the inhibitory hormones, which control anterior pituitary.
- Growth hormone inhibitory hormone or somatostatin
- Prolactin inhibitory hormone.

Q.24 What are the metabolic effects of growth hormone?
Growth hormone acts on protein, carbohydrate and fat metabolism.
- On protein metabolism—it increases protein synthesis
- On carbohydrate metabolism—it increases conservation of sugar
- On fat metabolism—it increases mobilization of fat from fat depots and utilization of fat.

Q.25 How does growth hormone increase protein synthesis?
Growth hormone increases the protein synthesis by
- Increasing amino acid transport through cell membrane
- Increasing RNA translation
- Increasing transcription of DNA to RNA
- Decreasing the catabolism of proteins.

Q.26 How does growth hormone act as protein sparer?
Growth hormone acts as protein sparer by mobilizing fats from fat depots and making them available for energy production so that the proteins are not broken down.

Q.27 How does growth hormone increase the blood sugar level?
Growth hormone increases the blood sugar level by:
- Decreasing the peripheral utilization of glucose
- Increasing the deposition of glycogen in the cells and saturating the cells with glycogen

- Decreasing the uptake of glucose by the cells.

Q.28 What is the effect of growth hormone on bones?

In fetus, the growth hormone is responsible for the differentiation and development of bone cells. During childhood till puberty, growth hormone increases the length and thickness of bone. After puberty when the head of the bone fuses with shaft, the growth hormone increases the thickness of bones.

Q.29 How is secretion of growth hormone regulated?

Growth hormone secretion (Fig. 13.4) is regulated by hormones secreted by hypothalamus:

- Growth hormone releasing hormone
- Growth hormone releasing polypeptide
- Growth hormone inhibitory hormone (somatostatin).

Whenever the blood level of growth hormone decreases, hypothalamus secretes growth hormone releasing hormone, and growth hormone releasing polypeptide which in turn act on pituitary and increase the secretion of growth hormone.

When blood level of growth hormone increases, it is controlled by negative feedback mechanism. Hypothalamus secretes growth hormone inhibitory hormone which decreases or stops the secretion of growth hormone.

Q.30 Differentiate somatotropin, somatostatin and somatomedins.

Somatotropin is the growth hormone (GH) secreted by somatotroph cells of anterior pituitary. Somatostatin is the growth hormone inhibiting hormone released from hypothalamus and also found in nerve endings of brain, cells of antrum of stomach and in cells of pancreatic islets of Langerhans. Somatomedins are growth factors, synthesized and released from liver (mainly), kidneys, muscle, etc. in response to growth hormones and play role on skeletal growth mainly.

Q.31. Why the GH is known as protein sparer?

It decreases protein and amino acid catabolism by increasing fat catabolism. This is why it is known as "***protein sparer***".

Q.32. Why the growth stops after adolescence?

At the time of adolescence there is fusion between shaft and each end of epiphysis and thus GH cannot promote the increase

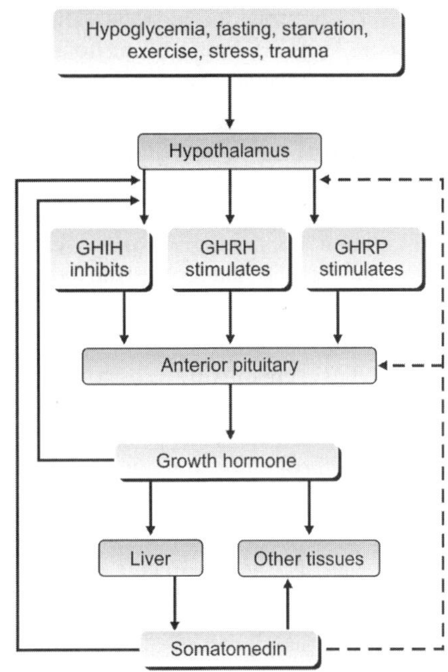

Fig. 13.4: Regulation of GH secretion. GHIH = Growth hormone inhibitory hormone. GHRH = Growth hormone releasing hormone. GHRP = Growth hormone releasing polypeptide. Growth hormone and somatomedin stimulate hypothalamus to release GHIH . Somatomedin inhibits anterior pituitary directly. Solid blue line = stimulation/secretion. Dashed line = inhibition

of growth of long bone at epiphyseal end plate. This results in no growth of long bones after adolescence.

Q.33 What are the actions of follicle stimulating hormone (FSH)?

In females: FSH causes development of Graafian follicle and activates the theca cells in the follicle to secrete estrogen.

In males: It acts along with testosterone to accelerate the process of spermeogenesis.

Q.34 What are the actions of luteinizing hormone (LH)?

In females: LH causes maturation of vesicular follicle into Graafian follicle along with FSH. It also causes ovulation and is responsible for the formation and secretory activity of corpus luteum.

In males: This hormone is known as interstitial cell stimulating hormone (ICSH) because, it stimulates the interstitial cells of Leydig in testes and causes secretion of testosterone.

Q.35 What are the actions of prolactin?

Prolactin acts on the mammary gland and prepares it for production and secretion of milk.

Q.36 What is β-lipotropin?

It is a polypeptide hormone found recently to be secreted from anterior pituitary. It mobilizes fat from adipose tissue and promotes lipolysis. It also forms the precursor of endorphins.

Q.37 Name the hormones of posterior pituitary.

- Antidiuretic hormone (ADH)
- Oxytocin.

Q.38 Which is the source of secretion of posterior pituitary hormones?

Posterior pituitary hormones are secreted from hypothalamus. ADH is secreted mainly from supraoptic nucleus and oxytocin is secreted mainly from paraventricular nucleus of hypothalamus.

Q.39 How do ADH and oxytocin reach the posterior pituitary from hypothalamus?

ADH and oxytocin, which are secreted from hypothalamic nuclei, reach the posterior pituitary through the nerve fibers of hypothalamohypophyseal tract.

Q.40 What are the actions of ADH?

- It increases the water reabsorption from the distal convoluted tubule and collecting duct and helps in final concentration of urine
- In higher doses ADH causes vasoconstriction and increases the blood pressure.

Q.41 Why ADH is called so?

Since this hormone prevents diuresis by reabsorption of water from distal convoluted tubule and collecting duct, it is called antidiuretic hormone (ADH).

Q.42 How is ADH secretion regulated?

ADH secretion is regulated by the volume and osmolar concentration of ECF. ADH secretion is stimulated by decrease in ECF volume and increase in the osmolar concentration of ECF.

Q.43 Which are the sites of action of oxytocin?

Mammary glands and uterus.

Q.44 What is the action of oxytocin on mammary glands?

Oxytocin causes ejection of milk by contracting the myoepithelial cells of mammary glands.

Q.45 What is milk ejection reflex? Why is it called neuroendocrine reflex?

When the infant suckles mother's nipple, the impulses produced from the touch receptors on and around the nipple pass through somatic afferent nerve fibers and reach the paraventricular and supraoptic nuclei of hypothalamus via cerebral cortex. Now, oxytocin is released into the blood. When the hormone reaches the mammary glands, it causes ejection of milk (Fig. 13.5).

As this reflex is initiated by nervous factors and completed through hormonal action, it is called neuroendocrine reflex. During this reflex, large quantity of oxytocin is secreted by positive feedback mechanism.

Q.46 What is the action of oxytocin on pregnant uterus?

Oxytocin causes contraction of uterus and helps in the expulsion of fetus during labor. Due to the movement of fetus through cervix during the onset of labor, the receptors on the cervix are stimulated and discharge the impulses. These impulses are carried to cerebral cortex by somatic nerve fibers. Cerebral cortex sends impulses to hypothalamus causing the release of oxytocin into blood. And oxytocin enhances labor by causing contraction of uterus.

This is a neuroendocrine reflex. During labor a large quantity of oxytocin is released by means of positive feedback mechanism.

Q.47 What is the action of oxytocin on nonpregnant uterus?

On nonpregnant uterus, oxytocin increases the uterine contractions during sexual intercourse and facilitates the transport of sperms through uterine cavity towards the fallopian tube.

Q.48 What is gigantism?

Abnormal increase in the height of the body in children due to hypersecretion of growth hormone (before the closure of epiphysis) is called gigantism.

Q.49 What is acromegaly?

Acromegaly is the enlargement, thickening and broadening of bones due to hypersecretion of growth hormone in adults (after the closure of epiphysis).

Q.50 What are the important features of acromegaly?

- Facial features: Acromegalic face or guerrilla face with protrusion of supraorbital ridges, broadening of nose, thickening of lips, wrinkles on forehead and protrusion of lower jaw (prognathism)
- Enlargement of hands and feet with kyphosis
- Bulldog scalp and overgrowth of body hair
- Enlargement of visceral organs
- Hyperactivity of other endocrine glands
- Hyperglycemia and glycosuria resulting in diabetes mellitus
- Hypertension.

Q.51 What is acromegalic gigantism?

If the hypersecretion of growth hormone starts in children resulting in gigantism and if it continues after puberty leading to acromegaly, the condition is known as acromegalic gigantism.

Q.52 What is Cushing's disease?

It is a disease characterized by obesity. It is due to hypersecretion of ACTH.

Q.53 What are the features of Cushing's disease?

Refer Q. 164.

Q.54 What is dwarfism?

The stunted growth in children due to lack of secretion of growth hormone is known as dwarfism.

Q.55 What are the important features of dwarfism?

Stunted growth is the prominent feature of dwarfism. The different parts of the body are almost proportionate. Only the head becomes slightly longer. All other functions including mental activity are normal.

Q.56 What is Laron dwarfism?

The stunted growth in children because of the lack of somatomedin is known as Laron dwarfism. The secretion of growth hormone is normal.

Q.57 What is psychogenic dwarfism?

Dwarfism due to exposure to extreme emotional deprivation or stress is called psychogenic dwarfism.

Q.58 What are the importants features of acromicria?

- Atrophy and thinning of hands and feet
- Hypothyroidism
- Hyposecretion of adrenocortical hormones
- Lethargy and obesity
- Loss of sexual functions.

Q.59 What is Simmond's disease or pituitary cachexia?

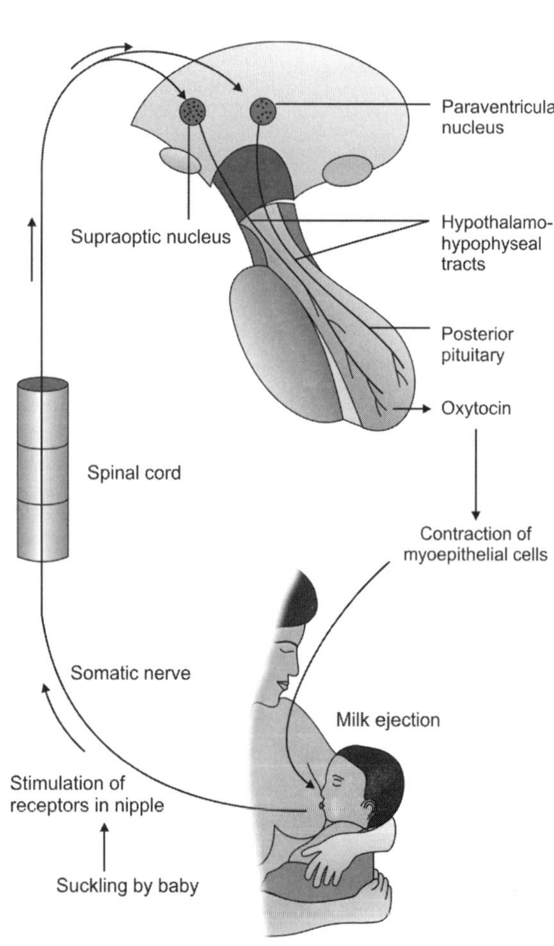

Fig. 13.5: Milk ejection reflex

It is a pituitary disease that occurs mostly in panhypopituitarism (hyposecretion of all the anterior pituitary hormones due to atrophy or degeneration of the gland).

Q.60 What are the features of Simmond's disease?

- Rapid development of senile decay and appearance of old age
- Loss of hair and teeth
- The skin over the face becomes dry and wrinkled.

Q.61 What is Laurence-Moon-Biedl syndrome?

It has following characteristics:

- Physical and mental retardation in growth
- Subnormal intelligence.
- Infantile gonads.
- Obesity with polydactylism
- Retinitis pigmentosa

All these are due to hypofunction of pituitary gland as a result of tumor of chromophobe cells or lesions in hypothalamus in the young.

Q.62 Name the nuclei secreting ADH and oxytocin?

- Supraoptic nuclei—ADH
- Paraventricular nuclei—Oxytocin.

Q.63 What is syndrome of inappropriate hypersecretion of antidiuretic hormone (SIADH)?

SIADH is the disease due to the excessive secretion of ADH.

Q.64 What are the features of SIADH?

- Decrease in urine output
- Increased water retention and ECF volume
- Secondary increase in urine output with more sodium ions
- Decreased sodium concentration in ECF
- Convulsions and coma in severe condition.

Q.65 Name the hormones secreted by thyroid gland.

- Triiodothyronine (T_3)
- Tetraiodothyronine (T_4 or thyroxine)
- Calcitonin.

Q.66 Which is more potent amongst T_3 and T_4? Why it is so?

T_3 is more potent than T_4 because T_3 is found freely in the plasma and can act immediately. But T_4 is bound with plasma proteins, so it takes time for it to be released and then to act.

Q.67 What are the substances necessary for the synthesis of thyroid hormones?

- Amino acid tyrosine
- Inorganic ion iodine.

Q.68 How much of iodine is required for the synthesis of normal quantity of thyroid hormones?

One mg of iodine per week or 50 mg per year.

Q.69 Name the stages in the synthesis of thyroid hormones.

The following are the stages in the synthesis of thyroid hormones (Fig. 13.6)

- Thyroglobulin synthesis
- Iodide trapping and iodide pump
- Oxidation of iodide into elemental iodine
- Iodination of tyrosine
- Coupling reactions.

Q.70 What are the enzymes involved in the synthesis of thyroid hormones?

- Peroxidase that converts iodide into elemental iodine
- Iodinase that accelerates the iodination of tyrosine.

Q.71 What is thyroglobulin?

Thyroglobulin is a large glycoprotein secreted by the endoplasmic reticulum and Golgi apparatus of follicular cells and stored in the follicles of thyroid gland.

Q.72 How are thyroid hormones released from thyroglobulin?

The follicular cells form pinocytic vesicles around thyroglobulin—hormone complex. Then the digestive enzymes like proteinase present in lysosomes of the follicular cells digest the thyroglobulin and release the hormones.

Q.73 How are the thyroid hormones transported in the blood?

Thyroid hormones are transported in the blood in combination with plasma proteins called thyroxine binding globulin (TBG), thyroxine binding prealbumin (TBPA) and albumin.

Q.74 What is the normal plasma level of T_3 and T_4?

T_3 =0.12 μg/dl

T_4 =8 μg/dl.

Q.75 What are the actions of thyroxine on protein metabolism?

Thyroxine increases:

- Translation of RNA
- Transcription of DNA into RNA
- Activity of cellular enzymes
- Mitochondrial activity.

Q.76 What are the actions of thyroxine on carbohydrate metabolism?

Thyroxine is a diabetogenic hormone. It increases:

- Glucose absorption from gastrointestinal tract
- Transport of glucose into the cells
- Breakdown of glycogen (glycogenolysis) into glucose
- Gluconeogenesis.

Q.77 What are the actions of thyroxine on fat metabolism?

- Mobilizes fat from fat depots and increases free fatty acids in the blood
- Increases deposition of fat in liver causing fatty liver
- Decreases the level of cholesterol, phospholipids and triglycerides in plasma.

Q.78 What are the actions of thyroxine on cardiovascular system?

Thyroxine increases overall activity of cardiovascular system. It:

- Increases the heart rate
- Increases force of contractions of the heart
- Causes vasodilatation and increases blood flow
- Increases systolic blood pressure and decreases diastolic pressure leading to increase in pulse pressure.

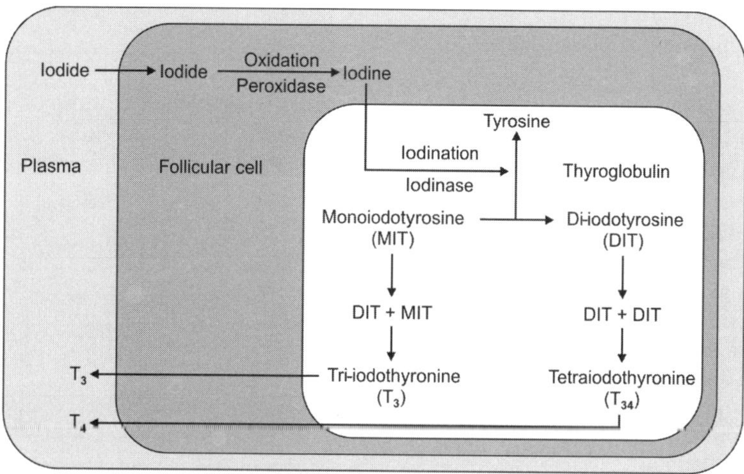

Fig. 13.6: Synthesis of thyroid hormones

Q.79 What is the action of thyroxine on respiratory system?
Thyroxine increases the rate and force of respiration.

Q.80 What are the actions of thyroxine on GI tract?
Thyroxine increases the secretions and movements of GI tract. It also increases appetite and intake of food.

Q.81 What are the actions of thyroxine on central nervous system (CNS)?
Thyroxine is necessary for the development of CNS during fetal life. In adult life, it stimulates and maintains the normal function of CNS.

Q.82 Name the factors increasing the secretion of thyroid hormones.
- Low basal metabolic rate
- Leptin
- Alpha melanocyte stimulating hormone

Q.83 Name the factors decreasing the secretion of thyroid hormones.
- Excess iodide intake
- Stress
- Somatostatin.

Q.84 What are the actions of thyroid stimulating hormone (TSH)?
TSH increases:
- Number and size of thyroid cells
- Secretory activity of thyroid cells
- Iodide pump and iodide trapping in thyroid cells
- Thyroglobulin secretion
- Iodination of tyrosine and coupling to form thyroid hormones
- Proteolysis of the thyroglobulin by which the thyroid hormones are released into the blood.

Figure 13.7 illustrates the regulation of secretion of thyroid hormones.

Q.85 What are the causes for hyperthyroidism?
- Presence of TSH like substances in the blood
- Thyroid adenoma
- Grave's disease.

Q.86 What are the important features of hyperthyroidism?
- Intolerance to heat
- Increased sweating
- Loss of weight
- Diarrhea
- Muscular weakness
- Nervousness
- Toxic goiter
- Oligomenorrhea or amenorrhea
- Exophthalmos
- Polycythemia
- Tachycardia and atrial fibrillation
- Systolic hypertension
- Cardiac failure.

Q.87 What is exophthalmos?
Protrusion of eyeballs is known as exophthalmos.

Q.88 What are the causes for exophthalmos in hyperthyroidism?
In hyperthyroidism, there is edema of the retro-orbital tissues and degenerative changes in the extraocular muscles. These two changes are responsible for protrusion of eyeballs.

Q.89 What are the effects of hypothyroidism?
Hypothyroidism leads to myxedema in adults and cretinism in children.

Q.90 What are the features of myxedema?
- Swelling of the face
- Bagginess under the eyes
- Nonpitting edema
- Atherosclerosis leading to arteriosclerosis and hypertension
- Anemia
- Fatigue and muscular sluggishness
- Somnolence
- Menorrhagia and polymenorrhea in females
- Decreased cardiovascular functions
- Increased body weight
- Constipation
- Mental sluggishness
- Depressed hair growth
- Scaliness of the skin
- Frog like husky voice
- Cold intolerance.

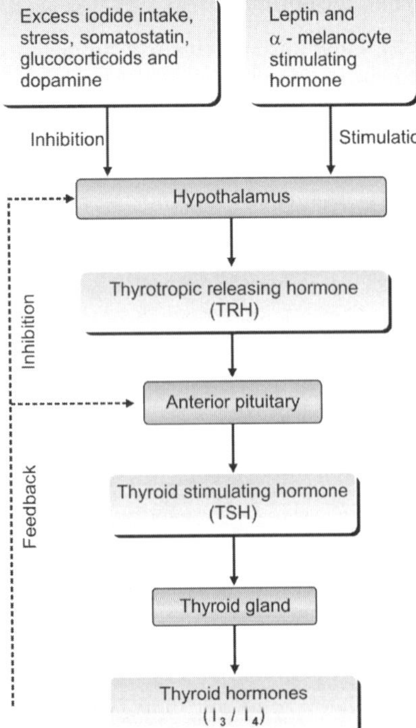

Fig. 13.7: Regulation of secretion of thyroid hormones

Q.91 What are the features of cretinism?
- Sluggish movements
- Croaking sound while crying
- Mental retardation
- Stunted growth
- Bloated body
- Protrusion of tongue with dripping of saliva
- Pot belly.

All these symptoms give idiotic look to the baby.

Q.92 What are the major differences between cretinism and pituitary dwarfism?
- In cretinism, there is mental retardation and in dwarfism, the development and functions of nervous system are normal
- The different parts of the body are disproportionate in cretinism but, in dwarfism, the different parts of the body are proportionate
- In cretinism, the reproductive function is abnormal whereas, it may be normal in dwarfism.

Q.93 What is goiter?
Enlargement of thyroid gland is known as goiter.

Q.94 How do you classify goiter?
Goiter can be classified as in Figure 13.8.

Q.95 What is toxic goiter?
Enlargement of thyroid gland with hypersecretion of hormones is known as toxic goiter.

Q.96 What is nontoxic goiter and what are the types of nontoxic goiter?
Enlargement of the thyroid gland with hyposecretion of hormones is known as nontoxic goiter.
It is of two types:
- Endemic colloidal goiter that is due to lack of iodine
- Idiopathic nontoxic goiter that is due to thyroiditis or presence of goiterogenic factors in foodstuffs.

Q.97 Name some antithyroid substances.
- Thiocyanate
- Thioureylenes
- Inorganic iodides in high concentration.

Q.98 Name the thyroid function tests.
- Measurement of T_3 and T_4 in blood
- Measurement of basal metabolic rate
- Measurement of TRH and TSH in blood.

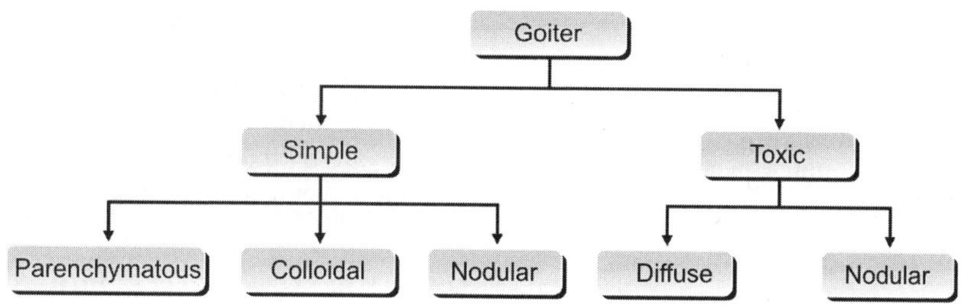

Fig. 13.8: Classification of goiter

Q.99 What is the important function of parathyroid glands in the body?
Parathyroid glands secrete parathormone that is very essential to maintain the blood calcium level.

Q.100 Why should the blood calcium level be maintained?
Because, calcium is very essential for many important activities in the body such as:
- Neuronal activity
- Muscular activity
- Cardiac function
- Secretory activities of the glands
- Coagulation of blood.

Q.101 What is the normal daily requirement of Ca^{++} and P and what is their normal blood level?

Substance	*Daily requirement*	*Blood level*
Ca^{++}	0.8 –1 gm	9–11 mg%
P	1–1.4 gm	2.5 –4 mg%

Q.102 Name the hormones which regulate Ca^{++} metabolism and their source.
- *Vitamin D*—Diet mainly and also skin by UV radiation.
- *PTH*—Chief cells of parathyroid gland.
- *Calcitonin*—Parafollicular cells (C-cells) of thyroid gland.

Q.103 What are the main signs and symptoms of hypoparathyroidism?
Hypocalcemia, hyperphosphatemia, increase in blood pH, neuromuscular hyperirritability causing tetany.

Q.104 What are the common signs present in tetany? Explain each of them.
- *Trousseau's sign or carpopedal spasm*—It is manifested in the upper limb as flexion at the wrist and thumb with hyperextension of remaining fingers called obstetric hand/Accoucheur's hand or carpopedal spasm. If this is demonstrated by occluding the blood supply to a limb through sphygmomanometer cuff, it is known as Trousseau's sign.
- *Chvostek's sign*—If skin in front of the ear is tapped, there is contraction or spasm of facial muscle.
- *Erb's sign*—It is depicted by the enhanced motor excitability of galvanic current.

Q.105 Name the hormones involved in the regulation of blood calcium level.
The hormones involved in the regulation of blood calcium level (Fig. 13.9) are:
- Parathormone secreted from parathyroid glands
- 1, 25 dihydroxycholecalciferol synthesized in kidney from vitamin D that is released from the liver
- Calcitonin secreted from parafollicular cells of thyroid gland.

Q.106 What are the actions of parathormone?
Parathormone increases the blood calcium level by increasing:
- Resorption of calcium from bones
- Reabsorption of calcium from renal tubules
- Absorption of calcium from intestine by activating vitamin D.

Q.107 How is 1, 25 dihydroxycholecalciferol (active form of vitamin D) formed?
1, 25 dihydroxycholecalciferol (active form of vitamin D—Fig. 13.10) is formed from vitamin D_3 (inactive form of vitamin D), which is also called cholecalciferol. Vitamin D_3 is converted into 25 hydroxycholecalciferol in liver and this is converted into 1, 25 dihydroxycholecalciferol in kidney in the presence of parathormone.

Q.108 What are the actions of 1, 25 dihydroxycholecalciferol?
It increases:
- Absorption of calcium from intestine
- Synthesis of ATPase in intestinal epithelium
- Alkaline phosphatase in intestinal epithelium.

Q.109 What are the actions of calcitonin?
Calcitonin decreases blood calcium level by:
- Increasing deposition of calcium in bones
- Increasing excretion of calcium through urine
- Decreasing the absorption of calcium from intestine.

Q.110 What is tetany?
Repeated convulsive muscular contractions, which occur due to hypoparathyroidism and hypocalcemia is known as tetany.

Q.111 What are the important features of tetany?

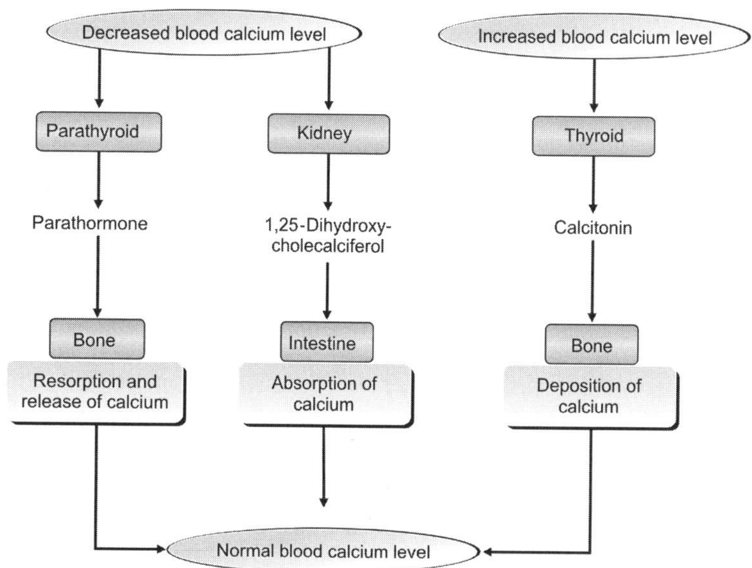

Fig. 13.9: Regulation of blood calcium level

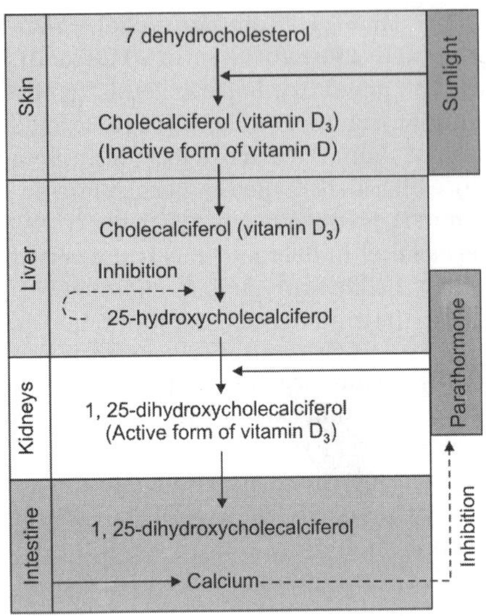

Fig. 13.10: Activation of vitamin D

- Carpopedal spasm
- Laryngeal stridor
- Cardiovascular changes like dilatation of heart, prolonged duration of ST segment and QT interval in ECG, arrhythmias, hypotension and heart failure.

Q.112 What are the important features of hypercalcemia?

- Depression of neuronal activities
- Sluggishness of reflex activities
- Reduction in the duration of ST segment and QT interval in ECG
- Lack of appetite
- Constipation.

Q.113 What are the major types of cells in bone? Mention their functions.

- Osteoblasts—concerned with bone formation that involves the formation of bone matrix and deposition of calcium (osteoblastic activity)
- Osteocytes—concerned with maintenance of bone
- Osteoclasts—concerned with bone resorption that involves the destruction of bone matrix followed by removal of calcium (osteoclastic activity).

Q.114 What do you mean by osteomalacia, osteoporosis and osteosclerosis?

Osteomalacia is the adult ricket characterized by:

- Decrease in mineral in bone/unit of bone matrix
- Generally limited to females usually after multiple pregnancy and lactation

Causes:

- Dietary deficiency of vit-D
- Malabsorption of vit -D
- Chronic renal failure
- Inadequate exposure to sun

Characteristic features:

- Bone pain and tenderness
- Fracture may occur
- Proximal myopathy
- Deformed bone with bowing legs
- Retarded growth — In children
- Thickening of wrists and ankle.

Osteoporosis: It is the clinical condition characterized by;

- Increase in all constituents of bone due to increase in bone resorption and decrease in bone formation.
- Causes
  - Postmenopausal women (due to low estrogen level resulted from increase in sensitivity of PTH to bone)
  - Hyperparathyroidism
  - Hyperthyroidism
  - Calcium deficiency

Osteosclerosis: Increased calcified bone in patient with metastatic tumor, lead poisoning and hypothyroidism.

Q.115 What is rickets? What is its cause?

It is a bone disease in children characterized by collapse of chest wall and curvature of spine.

It is due to the inadequate mineralization of bone matrix.

Q.116 What is endocrine part of pancreas?

Islets of Langerhans form the endocrine part of pancreas.

Q.117 Name the types of cells in islets of Langerhans.

- A or alpha cells which secrete glucagon
- B or beta cells which secrete insulin
- D or delta cells which secrete somatostatin
- F or PP cells which secrete pancreatic polypeptide.

Q.118 What are the actions of insulin?

- Insulin is the antidiabetogenic hormone, i.e. it decreases the blood sugar level by acting on carbohydrate metabolism
- It increases synthesis and storage of proteins
- It increases the synthesis and storage of fat
- It promotes growth of the body along with growth hormone.

Q.119 What are the actions of insulin on carbohydrate metabolism?

Insulin:

- Facilitates the transport of glucose into the cells
- Increases peripheral utilization of glucose
- Increases the conversion of glucose into glycogen in liver and muscle
- Inhibits glucogenolysis
- Inhibits gluconeogenesis.

By all these actions, insulin acts as an antidiabetogenic hormone, i.e. it decreases blood sugar level.

Q.120 What is the effect of insulin on growth?

Insulin promotes growth of the body by its anabolic effects on proteins and by its protein sparing effects.

Q.121 What is Houssay animal? What is its importance?

Houssay animal is the one in which both pancreas and anterior pituitary are removed.

This preparation proves the importance of insulin in growth of the animal along with growth hormone. When growth hormone alone or when insulin alone is administered to a Houssay animal, growth is not accelerated. But, when both growth hormone and insulin are given together, growth is accelerated very much.

Q.122 How is insulin secretion regulated?

Insulin secretion is regulated mainly by blood glucose level. When blood sugar level is more, insulin secretion increases. And, when blood glucose level is less, insulin secretion decreases.

Q.123 Name the stimuli for insulin secretion.

- Increase in blood sugar level
- Increase in amino acid level in blood
- The β ketoacids in blood
- Gastrointestinal hormones like gastrin, secretin, cholecystokinin and GIP
- Other endocrine hormones like glucagon, growth hormone and cortisol
- Stimulation of parasympathetic nerve fibers (right vagus) to pancreas.

Q.124 What are the actions of glucagon?

Glucagon:

- Increases the blood sugar level
- Increases the transport of amino acids into the liver cells leading to gluconeogenesis
- Shows lipolytic and ketogenic actions
- Inhibits gastric secretion and increases bile secretion.

Q.125 How does glucagon increase the blood sugar level?

Glucagon increases the blood sugar level by increasing glycogenolysis and gluconeogenesis.

Q.126 Name the factors which increase secretion of glucagon.
- Reduction in blood glucose level
- Increase in amino acid level
- Exercise
- Stress
- Some hormones such as gastrin, cholecystokinin and cortisol.

Q.127 Name the factors which inhibit secretion of glucagon.
- Increase in blood glucose level
- Somatostatin
- Insulin
- Free fatty acids
- Ketone bodies.

Q.128 What are the sources of secretion of somatostatin?
- Hypothalamus
- D cells present in islets of Langerhans
- D cells present in stomach and upper part of small intestine.

Q.129 What are the actions of somatostatin?
Somatostatin:
- Inhibits the secretion of insulin and glucagon
- Decreases the motility of stomach and small intestine
- Decreases the secretion of CCK, GIP and VIP
- Decreases the secretion of growth hormone (hypothalamic somatostatin).

Q.130 What is the action of pancreatic polypeptide?
Pancreatic polypeptide is believed to increase the secretion of glucagon.

Q.131 What is the necessity for regulation of blood sugar level?
Glucose is the only nutrient that can be utilized by the tissues like brain, retina and germinal epithelium of gonads. So, the blood sugar level has to be regulated within normal limits.

Q.132 What is the normal blood sugar level?
Fasting blood sugar = 80 to 90 mg%
Postprandial blood sugar = 120 to 140 mg%

Q.133 How is the blood sugar level maintained?
Blood sugar level is maintained by a regulating mechanism that is operated through liver and muscle under the influence of insulin and many other hormones like thyroxine, cortisol, glucagon and adrenaline.

Q.134 What are the hormones that regulate blood glucose level? Which one is the most important?
Insulin, glucagon, epinephrine, hydrocortisone, ACTH, growth hormone and thyroxin, out of which insulin is most important.

Q.135 Name the hormones which are antagonistic to the insulin?
GH, thyrotrophic hormone, ACTH and glucagon.

Q.136 What is the role of liver in the maintenance of blood sugar level?
Liver acts as an important glucose buffer system. When blood sugar level increases after meals, the excess glucose is converted into glycogen and stored in liver. Afterwards, when blood sugar level decreases, liver glycogen is broken into glucose that is released into blood. These actions are brought about under the influence of insulin and glucagon.

Q.137 What are the types of diabetes mellitus?
- Type I diabetes mellitus or insulin dependent diabetes mellitus (IDDM)—due to deficiency of insulin
- Type II diabetes mellitus or non-insulin dependent diabetes mellitus (NIDDM)—due to the absence or reduced number of insulin receptors in the cells of the body.

Q.138 What are the causes for Type I diabetes mellitus?
- Degeneration of beta cells in islets of Langerhans
- Destruction of beta cells by viral infection
- Congenital disorder of beta cells
- Autoimmunity against beta cells.

Q.139 What are the causes for Type II diabetes mellitus?
- Hereditary disorders
- Endocrine disorders.

Q.140 Name the endocrine disorders in which diabetes mellitus is common.
Gigantism, acromegaly and Cushing's syndrome.

Q.141 What are the features of diabetes mellitus?
- Glucosuria
- Osmotic diuresis
- Polyuria
- Polydipsia
- Polyphagia
- Asthenia
- Acidosis
- Acetone breathing
- Kussmaul breathing
- Circulatory shock
- Coma.

Q.142 What is the cause of hyperinsulinism?
Tumor of beta cells of islets of Langerhans.

Q.143 What do you understand by glucose tolerance test (GTT)?
It is a common clinical laboratory method to investigate the cases of diabetes mellitus and certain other conditions. The patient is kept on about 300 gm carbohydrate diet daily for 3 days. Fasting sample is collected in the morning after which the patient is administered glucose by oral route (1 gm/kg of b.w). The blood and urine samples are collected ½, 1, 1 ½ and 2 hours interval. The blood glucose values are estimated and urine is tested for presence of glucose. The values of glucose are plotted in a graph paper to obtain a characteristic graph.

Q.144 What are types of GTT graph curves?
Three types:
- Normal curve,
- Lag curve and
- Diabetic curve.

The lag curve is seen in early diabetic patients.

Q.145 Differentiate between hyperglycemic and hypoglycemic coma?

S. No.	*Parameter*	*Hyperglycemic coma*	*Hypoglycemic coma*
1.	***Cause***	Due to increase in blood glucose level (>400 mg%)	Due to fall of blood glucose level (< 40 mg%) and more severe.
2.	***Rate of onset***	Slow	Rapid
3.	***Signs and symptoms***		
	(i) *Breathing*	Deep and rapid breathing	Labored breathing called air hunger or Kussmaul breathing.
	(ii) *Sweating*	Absent	Usually marked.
	(iii) *Hydration*	Marked dehydration	Normal
	(iv) *Urine exam.*	Marked glycosuria	Not specific and ketonuria

Q.146 What are the features of hyperinsulinism?

- Hypoglycemia
- Manifestations of CNS like nervousness, tremor and excessive sweating. If not treated immediately, hyperinsulinism leads to clonic convulsions and unconsciousness leading to coma.

Q.147 Name the hormones secreted by adrenal cortex.

Adrenal cortex secretes three groups of hormone:

- Mineralocorticoids—aldosterone and 11 deoxycorticosterone
- Glucocorticoids—cortisol and corticosterone
- Sex hormones (androgens)—dehydroepiandrosterone, androstenedione and testosterone and small quantity of estrogen and progesterone.

Q.148 What are the hormones secreted by different parts of adrenal cortex?

Mineralocorticoids are secreted by zona glomerulosa. Glucocorticoids are secreted mostly by zona fasciculata and a small quantity is secreted by zona reticularis. Sex hormones are secreted mostly by zona reticularis and a small quantity by zona fasciculata (Fig. 13.11).

Q.149 What is secondary aldosteronism?

When aldosterone secretion is increased not due to adrenal cortical change but due to other factors like severe hemorrhage, diarrhea, dehydration, sweating, nephrosis, congestive heart failure it is known as secondary aldosteronism.

Q.150 What is the glucose fever?

The patient with adrenal cortex insufficiency if suffers from circulatory collapse, glucose infusion may cause high fever known as glucose fever.

Q.151 In Cushing's syndrome why the patient appears a moon like face and buffalo like hump?

In Cushing syndrome there is increased secretion of glucocorticoids which promote deposition of fat in unusual sites on the body to result ***moon like face*** and ***buffalo hump***.

Q.152 What is Conn's syndrome?

Primary aldosteronism or *Conn's syndrome* is the clinical condition due to excess aldosterone secretion due to tumor or hyperplasia of Z. Glomerulosa of adrenal cortex which is characterized by:

- Muscular weakness (due to prolonged hypokalemia)
- Hypokalemic nephropathy
- Increase in plasma Na^+,
- Increase in plasma aldosterone level without edema due to aldosterone escape
- Increase in urinary aldosterone level
- Decrease in plasma K^+
- Decrease in plasma renin
- Albuminuria

Q.153 What are the actions of aldosterone (mineralocorticoids)?

Aldosterone increases:

- Reabsorption of sodium ions
- ECF volume
- Blood pressure
- Excretion of potassium
- Excretion of hydrogen ions
- Reabsorption of sodium from sweat and salivary glands
- Absorption of sodium from intestine.

Q.154 What are the stimuli for the secretion of aldosterone?

- Increase in potassium ion concentration in ECF
- Decrease in sodium ion concentration in ECF
- Decrease in ECF volume
- ACTH.

Q.155 How is aldosterone secretion regulated?

Increase in potassium ion concentration in ECF directly acts on zona glomerulosa of adrenal cortex and increases the secretion of aldosterone. Reduction in sodium concentration and volume of ECF causes release of renin from juxtaglomerular apparatus of kidney. Renin converts angiotensinogen into angiotensin I. Angiotensin I is converted into angiotensin II by converting enzyme. Angiotensin II stimulates zona glomerulosa of adrenal cortex and increases the secretion of aldosterone (Fig. 13.12).

Q.156 What are the actions of cortisol on carbohydrate metabolism?

Cortisol is a diabetogenic hormone and it increases the blood sugar level by:

- Increasing gluconeogenesis.
- Decreasing glucose uptake and utilization by peripheral cells (anti-insulin action).

Q.157 What are the actions of cortisol on protein metabolism?

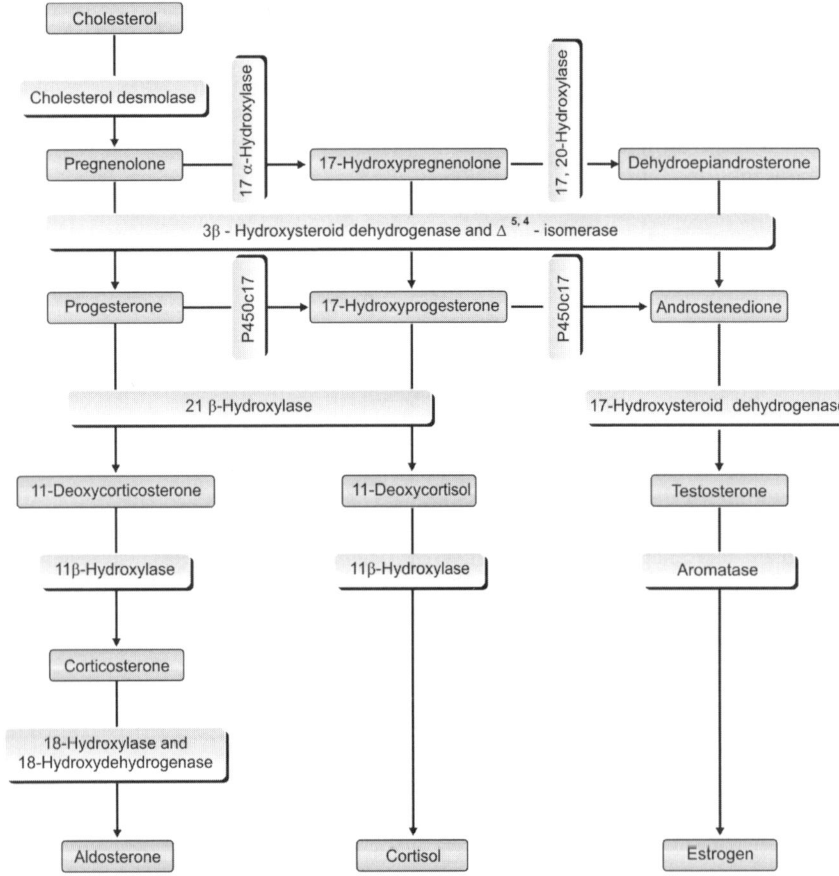

Fig. 13.11: Synthesis of hormones in adrenal cortex

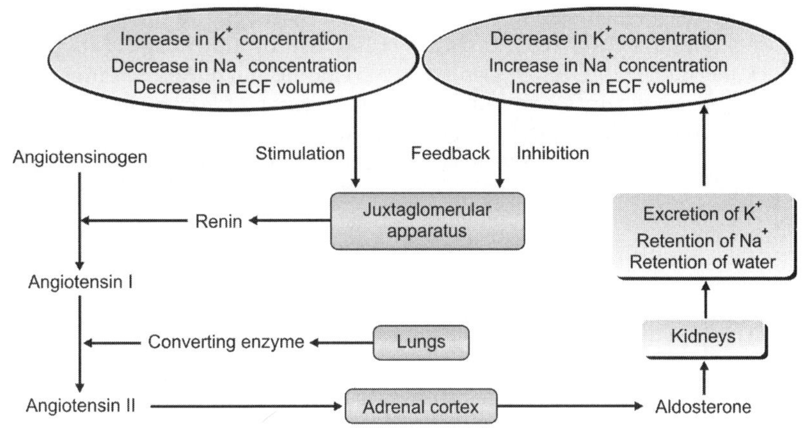

Fig. 13.12: Regulation of aldosterone secretion

Cortisol causes catabolism of proteins by:
- Increasing the breakdown of proteins
- Decreasing the synthesis of proteins.

Q.158 What are the actions of cortisol on fat metabolism?

Cortisol increases:
- Mobilization and redistribution of fats
- Fatty acids in the blood
- Utilization of fat for energy.

Q.159 What are the actions of cortisol on mineral metabolism?

Cortisol increases retention of sodium and water and excretion of potassium.

Q.160 What are the non-metabolic actions of cortisol?

- On blood cells—cortisol decreases the circulating eosinophils, basophils and lymphocytes and increases neutrophils, red blood cells and platelets
- On vascular system—cortisol is essential for vasoconstrictor action of adrenaline and noradrenaline
- On nervous system—it is essential for normal functioning of nervous system
- Permissive action—it is essential for execution of actions of some hormones
- Antistressor effects—it increases the resistance to stress
- Antiinflammatory effects—cortisol prevents inflammatory changes in cells caused by injury or infection
- Antiallergic actions—it prevents reactions in allergic conditions
- Immuosuppressive effect—cortisol suppresses immune system.

Q.161 How is the secretion of cortisol regulated?

By negative feedback mechanism through ACTH secreted by anterior pituitary and corticotropin releasing hormone by hypothalamus (Fig. 13.13).

Q.162 What are the actions of adrenocorticotropic hormone (ACTH)?

- Adrenal actions (on adrenal cortex):
- Maintains the structural integrity and vascularization of gland
- Converts cholesterol into pregnenolone from which the glucocorticoids are synthesized
- Causes release of glucocorticoids
- Prolongs the glucocorticoid action.
- Non-adrenal actions:
- Mobilizes the fats from fat tissues
- Melanocyte stimulating effect.

Q.163 What are the effects of hyperactivity of adrenal cortex?

- Cushing's syndrome
- Hyperaldosteronism and adrenogenital syndrome.

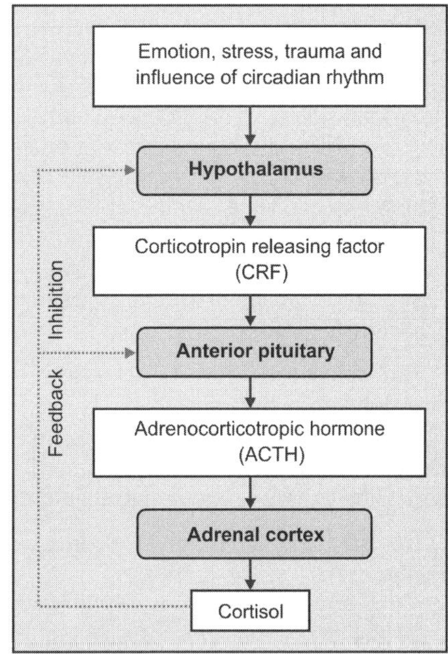

Fig. 13.13: Regulation of cortisol secretion

Q.164 List the features of Cushing's syndrome.

- Abnormal distribution of body fat resulting in moon face, torso, buffalo hump and pot belly
- Purple striae
- Thinning of extremities
- Thinning of skin and subcutaneous tissues
- Darkening of skin in neck
- Hyperpigmentation
- Facial redness
- Facial hair growth
- Muscular weakness
- Bone resorption and osteoporosis
- Hyperglycemia
- Hypertension
- Immunosuppression
- Poor wound healing.

Q.165 What is hyperaldosteronism?

Excessive secretion of aldosterone is known as hyperaldosteronism.

Q.166 Name the types and causes of hyperaldosteronism.

- Primary hyperaldosteronism (Conn's syndrome)—due to tumor in zona glomerulosa
- Secondary hyperaldosteronism—extra-adrenal causes like congestive cardiac failure, nephrosis, toxemia of pregnancy and cirrhosis of liver.

Q.167 What are the features of hyperaldosteronism?

- Increase in ECF volume and blood volume
- Hypertension
- Polyuria
- Polydipsia
- Muscular weakness
- Metabolic alkalosis.

Q.168 What is escape phenomenon in primary hyperaldosteronism?

In primary hyperaldosteronism, there is retention of sodium and water leading to increase in ECF volume. When ECF volume increases to certain level, atrial natriuretic peptide (ANP) is released from atrial muscles. ANP causes excretion of sodium and water from the kidney in spite of increased aldosterone secretion. This is known as escape phenomenon. Because of this, edema is not developed in primary hyperaldosteronism.

Q.169 What is adrenogenital syndrome?

Adrenogenital syndrome is the condition with increased activity of sex organs due to excessive secretion of sex hormones from adrenal cortex.

Q.170 What is virilism? What are its features?

Virilism is the development of male secondary sexual characters in females due to increased secretion of androgens.

Features:
- Increase in muscle bulk
- Deepening of voice
- Amenorrhea
- Enlargement of clitoris
- Male type of hair growth.

Q.171 What are the features of adrenogenital syndrome in males?
- Feminization
- Gynecomastia (enlargement of breast)
- Atrophy of testes
- Loss of interest in women.

Q.172 What are the effects of hypoactivity of adrenal cortex?
- Chronic adrenal insufficiency or Addison's disease
- Acute adrenal insufficiency or Addisonian crisis or adrenal crisis
- Congenital adrenal hyperplasia.

Q.173 What is the cause of Addison's disease?

Failure of adrenal cortex to secrete corticosteroids.

Q.174 What are the features of Addison's disease?
- Hyperpigmentation of skin and mucus membrane
- Muscular weakness
- Dehydration
- Hypotension
- Decreased cardiac output
- Hypoglycemia
- Nausea, vomiting and diarrhea
- Susceptibility to infections
- Inability to withstand stress.

Q.175 What is Addisonian crisis? When does it occur?

Sudden collapse of the person due to severe and acute need for large quantity of glucocorticoids is known as Addisonian crisis.
- It occurs in conditions like exposure to even mild stress.
- Hypoglycemia due to fasting
- Trauma
- Surgical operation
- Sudden withdrawal of glucocorticoid treatment.

Q.176 What is congenital adrenal hyperplasia?

It is the disease that develops due to the congenital absence of enzymes necessary for the synthesis of cortisol, particularly 21-hydroxylase. Since cortisol secretion is decreased, secretion of ACTH increases by feedback mechanism. ACTH acts on the adrenal glands and increases the number of cells leading to hyperplasia. Since, cortisol cannot be synthesized due to lack of enzymes, the synthesis of androgens increases leading to sexual abnormalities.

Q.177 What are the hormones secreted by adrenal medulla?

Adrenal medullary hormones are collectively known as catecholamines. Catecholamines are adrenaline or epinephrine, noradrenaline or norepinephrine and dopamine.

Q.178 What is general adaptation syndrome? What is it's role in combating stress?

The general manifestation of stress are called the ***general adaptation syndrome which is*** contributed by sympathoadrenal medullary system in which adrenal medullary hormones contribute to the ***fight or flight response*** by following ways:
- Allows more light to enter into eyes by relaxing accommodation and producing pupillary dilatation.
- Provides better perfusion of vital organs and muscles.
- Shortens the bleeding time (if wounded).
- Reinforcing the alert and arousal state by decreasing the threshold in reticular formation.
- Increasing glycogenolysis in liver and lipolysis in adipose tissue to increase energy supply.

Q.179 What is general adaptive syndrome?

General manifestation of stress is called general adaptive syndrome. It occurs in three stages:
1. Stage of alarm—No adaptation takes place.
2. Stage of resistance—Optimum adaptation occurs due to the interaction of adrenal cortex and adrenal medulla.
3. Stage of exhaustion—Due to continued stress.

Q.180 What is the mode of action of catecholamines?

The actions of catecholamines are exerted through some receptors present in the target organs called adrenergic receptors.

Q.181 What are the types of adrenergic receptors?
- Alpha adrenergic receptors, which are divided into $alpha_1$ and $alpha_2$ receptors
- Beta adrenergic receptors, which are divided into $beta_1$ and $beta_2$ receptors.

Q.182 What is the difference in the response of adrenergic receptors to adrenaline and noradrenaline?

Alpha receptors give more response to noradrenaline than for adrenaline. $Beta_1$ receptors have equal response to both adrenaline and noradrenaline. $Beta_2$ receptors give more response to adrenaline than to noradrenaline.

Q.183 What are the actions of adrenaline and noradrenaline on metabolism?

Adrenaline has metabolic actions but noradrenaline does not have metabolic effects. Adrenaline is a calorigenic hormone and it increases the basal metabolic rate. It increases blood glucose level by increasing glycogenolysis. On fats, it causes mobilization of fatty acids from adipose tissues.

Q.184 What is the action of adrenaline and noradrenaline on blood?

Adrenaline increases the red blood cell count and hemoglobin content of the blood by causing contraction of spleen. Noradrenaline does not show this action.

Q.185 What is the action of adrenaline and noradrenaline on heart?

Adrenaline increases overall activity of the heart, i.e. it increases the rate and force of contraction and excitability of the cardiac muscle. Noradrenaline has mild effect on heart.

Q.186 What are the actions of adrenaline and noradrenaline on blood vessels?

Noradrenaline has got stronger action on blood vessels. It causes vasoconstriction throughout the body thus, increasing the total peripheral resistance. So noradrenaline is called general vasoconstrictor. Adrenaline also causes vasoconstriction. But it causes the dilatation in some areas like skeletal muscle, liver and heart. So, adrenaline decreases the total peripheral resistance.

Q.187 What are the actions of adrenaline and noradrenaline on blood pressure?

Adrenaline increases systolic blood pressure by increasing the rate and force of contraction of heart and cardiac output. But, it decreases diastolic blood pressure by reducing the total peripheral resistance. Noradrenaline increases diastolic blood pressure to a greater extent because of its general vasoconstrictor action that increases the total peripheral resistance. It increases systolic pressure to a lesser extent.

Q.188 What are the actions of adrenaline on respiratory system?

Adrenaline increases the rate and force of respiration. When injected, it produces adrenaline apnea. It also causes bronchodilatation.

Q.189 What are the stimuli for the secretion of catecholamines?

Exposure to stress, cold and hypoglycemia are the stimuli for secretion of catecholamines.

Q.190 What is pheochromocytoma?

Pheochromocytoma is a condition characterized by hypersecretion of catecholamines. It is caused by tumor of chromophil cells in adrenal medulla. Hypertension, hyperglycemia and glucosuria are the important features of this condition.

Q.191 What is the function of pineal gland?

Pineal gland secretes the hormonal substance melatonin. In some animals, melatonin stimulates gonads and in some animals it inhibits the gonads. In humans, it inhibits the onset of puberty by inhibiting the gonads.

Q.192 What are the functions of thymus gland?

- It plays an important role in cellular immunity by processing the T lymphocytes
- It secretes thymosin (that helps in the proliferation of T lymphocytes) and thymin (that suppresses neuromuscular activity by inhibiting release of acetylcholine).

Q.193 Mention the hormones secreted by kidney.

- Erythropoietin
- Thrombopoietin
- Renin
- 1, 25 dihydroxy cholecalciferol
- Prostaglandins.

Q.194 Name the hormones secreted by heart. What is their action?

Atrial natriuretic peptide (ANP) and brain natriuretic peptide (BNP) are the hormones secreted by heart.

These hormones:

- Increase sodium excretion through urine (escape phenomenon)
- Decrease blood pressure.

Q.195 What are local hormones? What are the types of local hormones?

Local hormones are the hormonal substances, which execute their actions in the same area of secretion or in immediate neighborhood.

Types:

- Hormones synthesized in tissues
- Hormones synthesized in the blood.

Q.196 Name the local hormones synthesized in the tissues.

- Prostaglandins and related substances like thromboxanes, prostacyclin, leukotrienes and lipoxins.
- Other local hormones like acetylcholine, serotonin, histamine, substance P, heparin and GI hormones.

Q.197 Name the local hormones synthesized in the blood.

Serotonin, angiotensin and kinins.

Q.198 What is APUD cells?

GIT contains some cells which can take up amine precursors and decarboxylate them to convert it as amines. Therefore these cells are known as amine precursor uptake and decarboxylation (APUD) cells. Similar types of cell are also present in brain normally and also in some cases of lung cancer. G cell is one type of APUD cells.

Chapter 14

Reproductive System

Q.1 Name the sex organs in males.
- The primary sex organs—testes
- The accessory sex organs—seminal vesicles, prostate gland, urethra and external genitalia such as penis and scrotum (Fig. 14.1).

Q.2 Which chromosome determines the type of sex? What is H-Y antigen?
Y sex chromosome determines the type of sex. The testis determining gene product is known as H-Y antigen.

Q.3 What do you mean by SRY chromosome?
The gene present in the tip of the short arm of the human Y chromosome causes differentiation of indifferent or bipotential gonad to embryonic testis in the 7th–8th weeks after gestation. The region of the Y chromosome that contains the testis determining gene is called as SRY chromosome.

Q.4 What is sex chromatin or Barr body?
Soon after cell division has started during embryonic development one of the two X chromosomes of the somatic cell in normal female becomes functionally inactive. The inactive X chromosome is known as sex-chromatin or Barr body.

Q.5 What is the name of sex chromatin in male?
It is known as F body.

Q.6 To identify sex genotype certain cells are used for the cytological test. What are these cells?
These are: The epithelial cells of epidermal spinous layer, buccal mucosa epithelial cells, vaginal epithelial cell, leukocytes.

Q.7 Name the abnormalities of sexual differentiation due to nondisjunction of sex chromosome?
These are superfemale (44X XX), Klinefelter's syndrome (44XXY), Turner's syndrome (44X0).

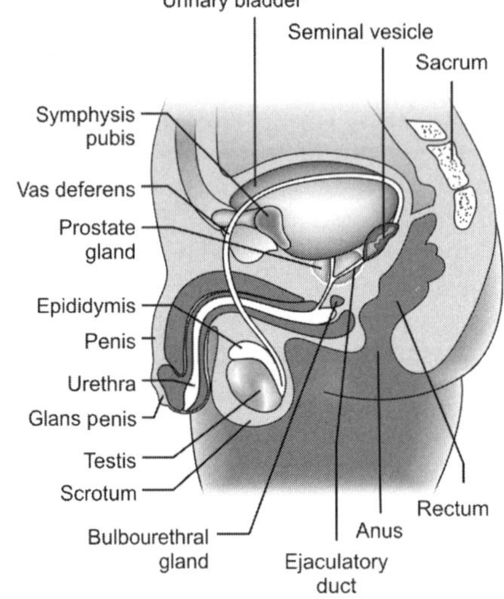

Fig. 14.1: Male reproductive system and other organs of pelvis

Q.8 What are the phenotypic features of Klinefelter's syndrome?
- *Characteristic features:*
  - Genetic sex is female
  - Chromosomal configuration 44XXY
  - Atrophied testis (Gonadal sex)
- *Phenotypic features:*
  - Male like appearance with feminine stigma
  - Bilateral gynecomastia
  - Sterile and impotent
  - Low or normal plasma testosterone level
  - High serum LH but normal FSH level
  - Small penis, testis, seminal vesicles, etc.
  - Secondary sex characters present

Q.9 Name the abnormality of sexual differentiation due to nondisjunction of autosome.
It is Down's syndrome or mongolism.

Q.10 What do you mean by male pseudohermaphroditism?
If the female internal genital organs develop in genital male due to less secretion of androgen by defective testis, it is known as pseudohermaphroditism.

Q.11 What are seminiferous tubules?
Seminiferous tubules are coiled tubular structures in the testes containing two types of cells, the spermatogenic cells and Sertoli cells.

Q.12 What are the spermatogenic cells?
Spermatogenic cells are the cells producing sperms in the testes. In children, only one type of spermatogenic cells is present called spermatogonia. After puberty, different stages of spermatogenic cells (spermatogonia, primary spermatocytes, secondary spermatocytes and spermatids) are found in the testes.

Q.13 What are Sertoli cells?
Sertoli cells are the supporting cells present in seminiferous tubules of testes.

Q.14 What are the functions of Sertoli cells?
Sertoli cells:
- Support and nourish the germ cells
- Provide necessary substances like hormones for spermatogenesis
- Convert androgens into estrogen.
- Secrete androgen binding protein, inhibin and Müllerian regression factor.

Q.15 What are the functions of testes?
- Gametogenic function—production of sperms
- Endocrine function—secretion of male sex hormones.

Q.16 What is spermatogenesis?
The production of sperms is known as spermatogenesis (Fig. 14.2).

Q.17 Name the stages of spermatogenesis.
- Stage of proliferation
- Stage of growth
- Stage of maturation
- Stage of transformation.

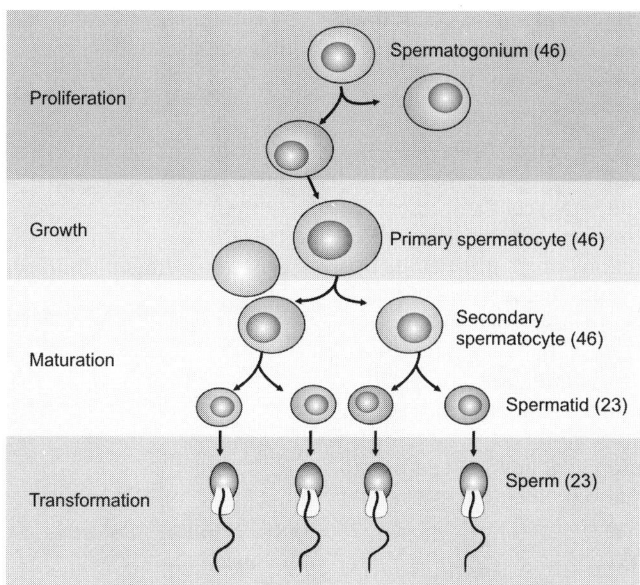

Fig. 14.2: Spermatogenesis. Number in parenthesis indicate chromosomal number

Q.18 At what stage of spermatogenesis the number of chromosomes becomes haploid?

At the stage of maturation, i.e. in the spermatids.

Q.19 Name the hormones necessary for spermatogenesis.

The following hormones are necessary for spermatogenesis (Fig. 14.3):

- Testosterone
- FSH
- LH
- Estrogen
- Growth hormone.

Q.20 Name some factors which inhibit spermatogenesis.

- Increase in temperature
- Infectious diseases such as mumps.

Q.21 What are the hormones secreted by testes?

The androgens or male sex hormones—testosterone, dihydrotestosterone and androstenedione.

Q.22 What is the source of secretion of testosterone?

Testosterone is secreted by the interstitial cells of Leydig present in testes. It is also secreted in small quantity in the adrenal cortex.

Q.23 What is the period of life during which testosterone is not secreted?

During the period between birth and puberty.

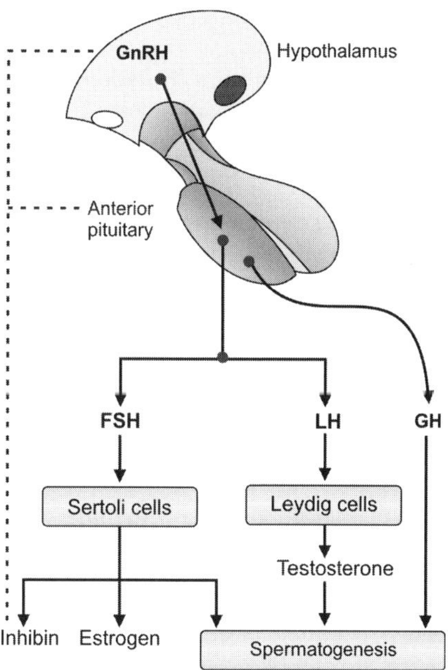

Fig. 14.3: Role of hormones in spermatogenesis → Stimulation → Inhibition

Q.24 What are the functions of testosterone in fetal life?

Testosterone helps in:

- Sex differentiation
- Development of sex organs
- Descent of testes.

Q.25 What are the functions of testosterone in adult life?

Testosterone:

- Increases the size of sex organs
- Causes development of secondary sexual characters such as muscular growth, bone growth, changes in skin, hair distribution, voice, BMR, blood, electrolyte concentration and water content in the body.

Q.26 How is testosterone secretion regulated?

In fetus, testosterone secretion is stimulated by human chorionic gonadotropin secreted from placenta. After puberty, testosterone secretion is stimulated by interstitial cell stimulating hormone (ICSH) secreted by anterior pituitary. The regulation is by negative feedback mechanism that involves ICSH and LH releasing hormone.

Q.27 What are the effects of extirpation of testes before puberty?

- The infantile sexual characters remain throughout life (eunuchism)
- Height is slightly more
- Bones are weak and thin
- Muscles are weak
- Sex organs do not increase in size and male secondary sexual characters do not develop
- Feminine distribution of fat occurs.

Q.28 What are the effects of extirpation of testes immediately after puberty?

- Functions of sex organs are depressed
- Seminal vesicles and prostate gland undergo atrophy
- Penis remains smaller in size
- Many of the secondary sexual characters such as male distribution of hair, musculature and thickness of bones are lost
- There is loss of sexual desire and sexual activities.

Q.29 What are the effects of extirpation of testes in adults?

- Accessory sex organs such as seminal vesicle and prostate gland degenerate
- Penile erection may occur but there is no ejaculation
- The secondary sexual characters and sexual desire may not be affected much.

Q.30 What is hypergonadism? What is its cause in males?

The condition characterized by hypersecretion of sex hormones from gonads is known as hypergonadism.

In males, it is due to the tumor of Leydig cells.

Q.31 What are the effects of hypergonadism in males?

- Rapid growth of muscles, bones, sex organs and secondary sexual characters

- Height of the person is less because of early closure of epiphysis
- Development of gynecomastia.

Q.32 What is hypogonadism? What are its causes in males?

The condition characterized by reduction in the functional activity of gonads is known as hypogonadism.

Causes in males:

- Congenital non functioning testes
- Underdeveloped testes
- Cryptorchidism
- Castration
- Absence of androgen receptors in testis
- Disorder of gonadotropes
- Hypothalamic disorder.

Q.33 What are the effects of hypogonadism in males?

Effects are similar to the effects of removal of testes before puberty. Refer Question No. 27 of this section.

Q.34 What are the functions of fluid secreted from seminal vesicles?

- It provides nutrition to the sperms
- The fibrinogen present in this fluid causes coagulation of semen
- Prostaglandin of the fluid enhances the fertilization of ovum by increasing the receptivity of cervical mucosa for the sperms and by increasing the rate of transport of sperms through reverse peristaltic movements of uterus and fallopian tube.

Q.35 What are the functions of prostatic fluid?

- Prostatic fluid provides optimum pH for motility of sperms
- The clotting enzymes in this fluid cause coagulation of semen
- Fibrinolysin present in this fluid causes lysis of coagulum.

Q.36 What is the nature of semen?

At the time of ejaculation, semen is liquid in nature. Immediately it is coagulated and the coagulated semen is known as coagulum. Finally it undergoes a secondary liquefaction.

Q.37 What are the properties of semen?

Specific gravity : 1.028
Volume : 4 to 6 mL/ejaculation
Reaction : Alkaline with a pH of 7.5

Q.38 What is the composition of semen?

See Figure 14.4.

Q.39 What is the normal sperm count?

100 to 150 millions/mL of semen.

Q.40 At what level of sperm count does the sterility occur in males?

When the sperm count is below 20 million/mL of semen.

Q.41 What is blood testis barrier? What is its function?

In between the Sertoli cells and other cells lining the seminiferous tubular wall there are tight junctions which prevent the free movement of substances across it. This is known as blood testis barrier. Its functions are:

- Helps in maintaining the composition of the fluid in the lumen of seminiferous tubule.
- It helps to prevent entry of sperm into blood and also protects the sperm from blood-borne noxious agents.

Q.42 What is the survival time of sperms after ejaculation?

About 24 to 48 hours at a temperature equivalent to body temperature.

Q.43 Why is scrotal temperature less than the body temperature?

A slightly lower scrotal temperature than the body temperature is essential for normal production of spermatozoa.

Q.44 Explain condition of cryptorchidism.

If the tested remains undescended, the seminiferous tubules do not develop due to higher temperature in the abdominal cavity and subsequently degenerate. Thus, there is no spermatogenesis resulting in sterility. However, the Leydig cells are unaffected and secrete testosterone at puberty, so that all the male secondary sex characters are normally present.

Q.45 Name the sex organs in females.

- The primary sex organs—ovaries
- The accessory sex organs—fallopian tubes, uterus, cervix, vagina and external genitalia such as labia majora, labia minora and clitoris.

Q.46 Name the hormones secreted by ovaries.

- Female sex hormones—estrogen and progesterone
- Inhibin
- Relaxin
- Small quantities of androgens.

Q.47 What are the sources of estrogen?

In a nonpregnant female: Follicles of ovaries

During pregnancy: Corpus luteum and placenta

A small quantity of estrogen is secreted from adrenal cortex throughout life.

Q.48 What are the actions of estrogen on uterus?

Estrogen causes:

- Enlargement of uterus
- Increase in blood supply to uterus
- Deposition of glycogen and fats in endometrium
- Proliferation and dilatation of endometrial blood vessels
- Proliferation and dilatation of endometrial glands
- Increase in spontaneous activity of uterine muscles and sensitivity to oxytocin
- Increase in the contractility of uterine muscles.

Q.49 What are the actions of estrogen on fallopian tubes?

Estrogen:

- Increases the number and size of ciliated epithelial cells lining the fallopian tubes
- Increases the activity of cilia that facilitates the movement of ovum through the fallopian tube
- Enhances the proliferation of glandular tissues in fallopian tubes.

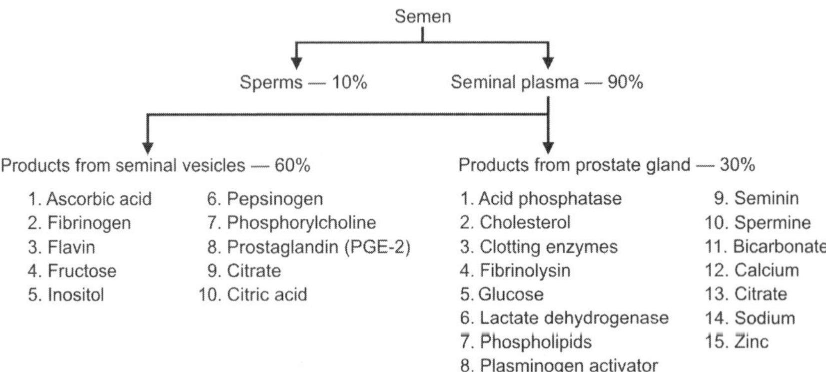

Fig. 14.4: Composition of semen

Q.50 What are the actions of estrogen on vagina?

Estrogen:

- Changes the cuboidal epithelium of vagina into stratified epithelium, which has more resistance to trauma and infection
- Increases the number of layers of vaginal epithelium by proliferation
- Reduces the pH of vagina causing more acidity.

Q.51 What are the actions of estrogen on mammary glands?

Estrogen increases the size of mammary glands by causing:

- Development of stromal tissues
- Extensive growth of ductile system
- Deposition of fat in the ductile system.

Q.52 What are the female secondary sexual characters influenced by estrogen?

- Hair growth in pubic region and axilla and profuse hair growth in scalp
- Softness, smoothness and increased vascularity of the skin
- Narrow shoulders, broad hip, converged thighs and diverged arms and deposition of fat in breasts and buttocks
- Retention of prepubertal voice with high pitch.

Q.53 What are the actions of estrogen on bones?

Estrogen increases osteoblastic activity that accelerates the height at the time of puberty. At the same time, it causes early closure of epiphysis.

Q.54 What are the actions of estrogen on metabolism?

Estrogen increases the protein synthesis and causes deposition of fat in the subcutaneous tissues, breasts, buttocks and thighs.

Q.55 How is the secretion of estrogen regulated?

The secretion of estrogen is regulated by FSH secreted from anterior pituitary through negative feedback mechanism. The secretion of FSH, in turn, is under the control of gonadotropic releasing hormone secreted from hypothalamus.

Q.56 What are the sources of progesterone?

In a nonpregnant female: Small quantity of progesterone is secreted from theca cells of ovary during follicular phase and large quantity is secreted from corpus luteum of ovary during the luteal phase of menstrual cycle.

In first trimester of pregnancy: Corpus luteum and placenta secrete a large quantity of progesterone.

A small quantity is secreted from adrenal cortex throughout life.

Q.57 What are the actions of progesterone on uterus?

Progesterone increases:

- Thickness of endometrium
- Size of the uterine glands
- Secretory activities of glandular epithelial cells
- Deposition of lipid and glycogen in the stromal cells
- Blood supply to endometrium. It decreases the frequency of uterine contractions, which favor the implantation and continuation of pregnancy.

Q.58 What is the action of progesterone on fallopian tubes?

Progesterone increases the secretion from mucosa of fallopian tube that is essential for the nutrition of fertilized ovum.

Q.59 What are the actions of progesterone on mammary glands?

Progesterone:

- Promotes the development of lobules and alveoli of mammary glands
- Makes the mammary glands secretory in nature.
- Increases the size of mammary glands by increasing the secretory activity and fluid accumulation in the subcutaneous tissue.

Q.60 Define menstrual cycle.

The cyclic events which take place in a rhythmic fashion during the reproductive period of a women's life is called menstrual cycle.

Q.61 What is the normal duration of menstrual cycle?

Normal duration of menstrual cycle is 28 days. Under normal conditions it ranges between 20 and 40 days.

Q.62 What is menarche? At what age does it occur?

The commencement of menstrual cycle is known as menarche. It occurs at the age of 12–15 years that marks the onset of puberty.

Q.63 Enumerate the changes taking place during menstrual cycle.

- Ovarian changes
- Uterine changes
- Vaginal changes
- Changes in cervix uteri.

Q.64 Name the phases of ovarian changes during menstrual cycle.

- Follicular phase—there is development of Graafian follicle and secretion of large amount of estrogen
- Luteal phase—there is development of corpus luteum and secretion of large amount of progesterone.

Q.65 Name the different ovarian follicles.

The different ovarian follicles are (Fig. 14.5):

- Primordial follicle
- Primary follicle
- Vesicular follicle
- Graafian follicle.

Q.66 When does ovulation occur?

It occurs on 14th day of menstrual cycle in a normal 28 days cycle.

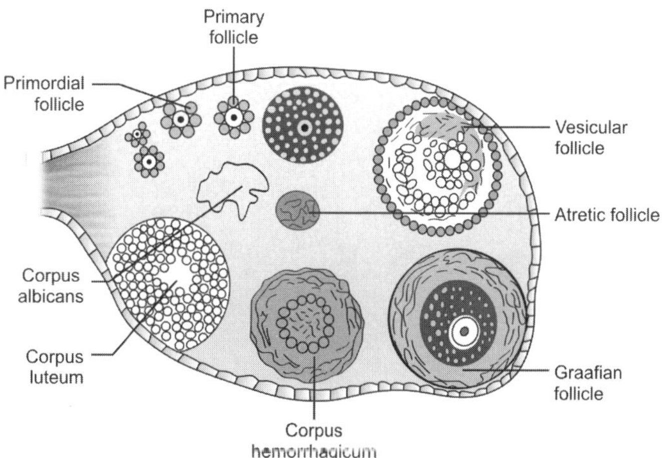

Fig. 14.5: Ovarian follicles and corpus luteum

Q.67 How does ovulation occur?

Ovulation occurs because of rupture of stigma which is a protrusion developed on the surface of the Graafian follicle. Rupture of Graafian follicle releases ovum into the abdominal cavity.

Q.68 What are the different phases of menstrual cycle and what is its cause?

There are four phases:

- *Menstrual phase:* It is due to withdrawal of progesterone secretion.
- *Proliferative phase:* It is due to estrogen secretion.
- *Ovulatory phase:* It is due to LH surge.
- *Secretory or luteal phase:* It is due to increase in secretion of progesterone.

Q.69 What is estrogen surge, FSH surge and LH surge?

- In the preovulatory phase of menstrual cycle rise of FSH concentration increases the serum concentration of estradiol to reach a peak at 12–13 days (in case of 28 days cycle), called **estrogen surge** (Fig. 14.6A).
- Within 24 hours of estrogen surge, the increased level of estrogen augments the responsiveness of the pituitary to GnRH which induces a burst of LH secretion. This peak rise of LH in serum just prior to ovulation is known as **LH surge** (Fig. 14.6B).
- At the same time when LH peak occurs serum concentration of FSH also increases suddenly to a peak level called **FSH surge** (Fig. 14.6B).

Q.70 What do you mean by withdrawal bleeding?

If no fertilization takes place, corpus luteum regresses by the process known as luteolysis resulting in sharp fall of estrogen and progesterone secretion. This inturn causes spasm in spiral arteries and thereby ischemia of superficial layer of endometrium. This ultimately leads to the shedding of superficial layer of endometrium and thereby release of blood and mucous through vagina known as withdrawal bleeding.

Q.71 How is ovulation time determined?

- By determining basal body temperature
- By determining the hormonal excretion in urine
- By determining hormonal level in plasma
- By ultrasound scanning.

Q.72 What is the physiological basis of BBT as indicator of ovulation?

At the time of ovulation body temperature rises by 0.3–0.5°C than the temperature at preovulatory phase. This increase in temperature is due to the increase of progesterone level in blood which is thermogenic.

Q.73 What is the importance of knowing ovulation time?

Determination of ovulation time is necessary to adopt rhythm method (safe period) of family planning.

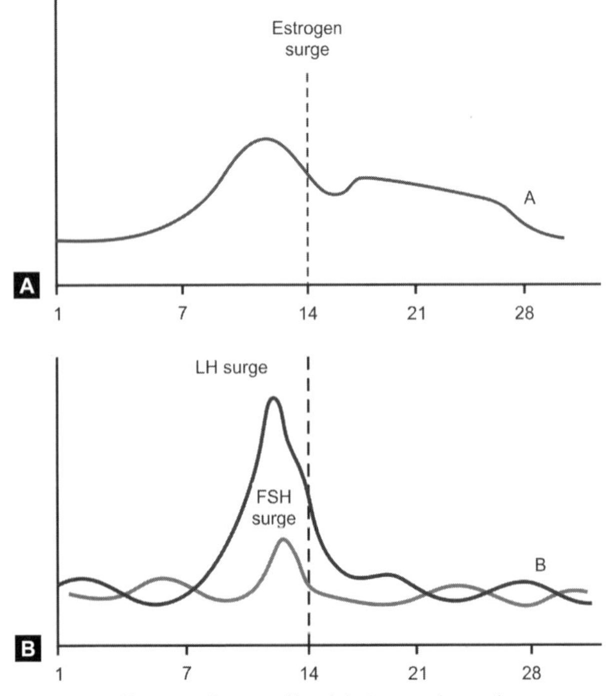

Fig. 14.6: Hormonal level during ovarian cycle

Q.74 What is corpus luteum?

The glandular yellow body that develops from the remaining cells of Graafian follicle after the release of ovum (after ovulation) is known as corpus luteum.

Q.75 Name the types of cells present in corpus luteum.

- Lutein cells derived from granulosa cells
- Cells of theca interna. The lutein cells are surrounded by cells of theca externa.

Q.76 What is the function of corpus luteum?

Corpus luteum:

- Functions as temporary endocrine gland and secretes large amount of progesterone and small amount of estrogen.
- Helps to maintain the pregnancy in the first trimester (till the placenta starts secreting the hormones).

Q.77 What is the fate of corpus luteum?

Fate of corpus luteum depends whether pregnancy occurs or not.

If pregnancy does not occur: It involutes and degenerates into corpus luteum menstrualis or spurium. The corpus luteum menstrualis is transformed into a whitish scar called corpus albicans.

If pregnancy occurs: It increases in size and remains for 3 to 4 months. During this period, it secretes large amount of progesterone and small amount of estrogen, which are essential to maintain pregnancy.

Q.78 Name the phases of uterine changes during menstrual cycle.

- Menstrual phase
- Proliferative phase
- Secretory phase.

Q.79 What are the uterine changes during menstrual phase?

The endometrium becomes involuted and desquamated. It is followed by vasoconstriction and hypoxia leading to necrosis and bleeding.

Q.80 What are the causes for uterine changes during menstrual phase?

At the end of menstrual cycle, there is sudden decrease in the level of estrogen and progesterone. This leads to sudden involution of endometrium at the beginning of next cycle. Since estrogen and progesterone are vasodilators, lack of these hormones causes severe vasoconstriction.

Prostaglandin secreted by the involuted endometrium also causes vasoconstriction. Due to severe vasoconstriction, hypoxia and necrosis occur in the endometrium. Necrosis causes rupture of blood vessels leading to bleeding.

Q.81 What is the composition of menstrual fluid?

- Blood (about 35 mL)
- Serous fluid (about 35 mL)
- Desquamated endometrial tissues.

Q.82 How much of blood is lost during menstrual phase?

About 35 mL

Q.83 Why the menstrual blood does not clot?

During menstruation, blood clots as soon as it oozes into the uterine cavity. Fibrinolysin released from the endothelium of damaged blood vessels causes lysis of the clot in the uterine cavity itself so that the menstrual blood does not clot.

Q.84 What are the uterine changes during proliferative phase?

- Endometrial cells proliferate
- Epithelium reappears on the surface of endometrium
- Uterine glands start developing
- Blood vessels also appear in stroma
- Endometrium reaches the thickness of 3–4 mm.

Q.85 What are the uterine changes during secretory phase?

- The uterine glands increase in size and become more tortuous
- Cytoplasm of stromal cells increases due to deposition of glycogen and lipids
- New blood vessels appear in endometrium
- Blood supply to the endometrium increases
- Thickness of endometrium increases to about 5–6 mm.

Q.86 What are vaginal changes during menstrual cycle?

During proliferative phase, the vaginal epithelium is cornified because of the influence of estrogen.

During secretory phase, there is proliferation of vaginal epithelium because of the action of progesterone. There is infiltration of leukocytes in the vaginal epithelium during this phase.

Q.87 What are the changes, which occur in cervix during menstrual cycle?

During menstrual phase, under the influence of estrogen, the mucus membrane of cervix becomes thin and alkaline. This helps for the survival and motility of sperms.

During secretory phase, because of the action of progesterone, mucus membrane of cervix becomes thick and adhesive.

Q.88 Name the hormones, which influence the ovarian changes during menstrual cycle.

During follicular phase	:	FSH, LH and estrogen
During ovulation	:	LH
During luteal phase	:	FSH and LH.

Q.89 Name the hormones, which influence the uterine changes during menstrual cycle.

During proliferative phase	:	Estrogen
During secretory phase	:	Progesterone
During menstrual phase	:	Sudden withdrawal of estrogen and progesterone.

Q.90 What are the abnormal types of menstruation?

- Amenorrhea—absence of menstruation during reproductive period of females
- Hypomenorrhea—decreased menstrual fluid
- Menorrhagia—excessive menstrual bleeding
- Oligomenorrhea—decreased frequency of menstrual bleeding
- Polymenorrhea—increased frequency of menstrual bleeding
- Dysmenorrhea—menstruation with pain
- Metrorrhagia—uterine bleeding in between menstruations.

Q.91 What is anovulatory cycle?

The menstrual cycle without ovulation is called anovulatory cycle.

Q.92 What is menopause?

In females, the permanent stoppage of menstruation in old age is known as menopause.

Q.93 What is the cause for menopause?

Throughout life there is degeneration of primordial follicles in the ovary. At the age of 45 years and above, the number of primordial follicles reduces leading to decrease in the secretion of estrogen by the ovary. When all the primordial follicles are atrophied estrogen secretion stops completely. This period is called menopause.

Q.94 What is postmenopausal syndrome?

After the onset of menopause, the woman develops certain physical, physiological and psychological changes, which are collectively known as postmenopausal syndrome.

Q.95 How is postmenopausal syndrome treated?

Postmenopausal syndrome can be treated by psychotherapy and hormone therapy. In hormone therapy, estrogen and progesterone are administrated with careful adjustment of dose.

Q.96 What are the causes for male infertility?

- Decrease in sperm count to about 20 millions/mL
- Presence of abnormal sperms like tailless sperms, two headed sperms and nonmotile sperms
- Obstruction of reproductive ducts like vas deferens.

Q.97 Where does fertilization of ovum occur?

Fertilization of ovum occurs in the fallopian tube.

Q.98 When does the zygote get implanted in the uterus?

After fertilization, the zygote takes 3 to 5 days to reach the uterus. In the uterus, the zygote remains freely in the uterine cavity for 2 to 4 days and then gets implanted. So, it takes about one week for the zygote to get implanted.

Q.99 What is the duration of pregnancy (gestation period)?

280 days (40 weeks) from the date of last menstrual period.

Q.100 What are the changes taking place in ovary during pregnancy?

When pregnancy occurs, follicular growth does not occur in ovary because of lack of FSH and LH. Corpus luteum grows in size and remains for three months and secretes large amount of progesterone and small amount of estrogen. After third month of pregnancy, when placenta starts secreting the hormones, corpus luteum degenerates.

Q.101 What are the changes taking place in uterus during pregnancy?

- Increase in the volume (from 0 to 5–7 liters), size and weight (from 30–50 gm to 1000–1200 gm) of the uterus
- Shape of the uterus changes from pyriform to globular
- Histological changes also occur with the development of decidua.

Q.102 What are the changes taking place in vagina during pregnancy?
- Size increases
- Violet coloration due to increase in blood supply
- Epithelial cells become less cornified
- Glycogen deposition increases in epithelial cells
- pH decreases to less than 3.5.

Q.103 What are the changes taking place in cervix during pregnancy?
- Increase in number of cervical glands
- Hypertrophy of endocervix which gives honeycomb appearance
- Increase in blood supply
- Increase in mucus secretion
- Softening of cervix
- Formation of mucus plug, which closes cervical canal.

Q.104 What are the changes taking place in mammary glands during pregnancy?
- Development of new ducts
- Formation of new alveoli
- Deposition of fat
- Increase in size
- Increase in vascularization
- The pigmentation of nipple and areola.

Q.105 How much is the weight gain of the body during pregnancy?

The average weight gain of the body during pregnancy is about 12 kg

Fetal weight	:	3.5 kg
Amniotic fluid weight	:	2.0 kg
Placental weight	:	1.5 kg
Increase in maternal body weight	:	5.0 kg

Q.106 What are the metabolic changes during pregnancy?
- Increase in BMR
- Increase in protein synthesis
- Increase in blood glucose level that may lead to diabetes in pregnancy
- Deposition of fat in maternal body with increased blood cholesterol level and ketosis
- Retention of water, sodium, calcium and phosphorus.

Q.107 What are the changes taking place in blood during pregnancy?
- Blood volume increases by about 20% (1 liter) mainly because of the increase in plasma volume
- Hemodilution occurs
- Anemia may develop.

Q.108 What are the cardiovascular changes during pregnancy?
- Cardiac output increases
- Blood pressure decreases slightly in second trimester
- Hypertension may develop later if proper prenatal care is not taken.

Q.109 What are the changes taking place in excretory system during pregnancy?
- Increase in renal blood flow, glomerular filtration rate and urine formation
- Formation of dilute urine
- Increase in frequency of micturition.

Q.110 What are the changes taking place in digestive system during pregnancy?
- Morning sickness involving nausea, vomiting and giddiness occurs during initial stage of pregnancy
- Movement of gastrointestinal tract decreases resulting in constipation
- Indigestion and hypochlorhydria may occur.

Q.111 What are the changes taking place in endocrine glands during pregnancy?

Generally, all the endocrine glands increase in size with increased hormonal secretion.

Q.112 What are the changes taking place in nervous system during pregnancy?

During early stages of pregnancy, there is excitement of nervous system leading to psychological imbalance such as change in the moods, excitement and depression.

Q.113 What is pre-eclampsia?

Toxemia of blood characterized by elevated blood pressure is known as pre-eclampsia.

Q.114 What is parturition?

Expulsion or delivery of the fetus from the mother's body at the end of pregnancy is known as parturition.

Q.115 Enumerate the hormones involved in the process of parturition.
- Maternal hormones—oxytocin, prostaglandins, cortisol, catecholamines and relaxin
- Fetal hormones—oxytocin, cortisol and prostaglandins
- Placental hormones—estrogen, progesterone and prostaglandins.

Q.116 What is the role of estrogen in parturition?

Estrogen increases the force of uterine contractions and the number of oxytocin receptors in the uterine wall. It also accelerates the synthesis of prostaglandins.

Q.117 What is the role of progesterone in parturition?

Progesterone does not play any role in parturition. But, it is responsible for the suppression of uterine contractions throughout the period of gestation. So, it is essential for the maintenance of pregnancy. At the end of gestation period, progesterone secretion decreases suddenly and parturition is induced.

Q.118 What is the role of oxytocin in parturition?

Oxytocin causes contraction of uterus and enhances labor through positive feedback mechanism and neuroendocrine reflex.

Q.119 What is double Bohr's effect?

Reduction in the affinity of hemoglobin for oxygen due to increased carbon dioxide tension is known as Bohr's effect. On the other hand, when the carbon dioxide tension decreases, the affinity for oxygen is increased. In fetus, along with metabolic end products, carbon dioxide is completely excreted from fetal blood into mother's blood. This develops low partial pressure of carbon dioxide in the fetal blood. So, the affinity of fetal hemoglobin for oxygen increases resulting in diffusion of more amount of oxygen from mother's blood into fetal blood.

Simultaneously, the partial pressure of carbon dioxide increases in mother's blood. This reduces the affinity of hemoglobin in mother's blood for oxygen resulting in diffusion of more amount of oxygen from mother's blood into fetal blood. This type of operation of Bohr's effect in both fetal blood and mother's blood is known as double Bohr's effect.

Q.120 What are the hormones secreted by placenta?
- Human chorionic gonadotropin (hCG)
- Estrogen
- Progesterone
- Human chorionic somatomammotropin (HCS)
- Relaxin.

Q.121 What are the actions of hCG?
- hCG is responsible for the preservation and maintenance of secretory activity of corpus luteum
- In male fetus, it stimulates the interstitial cells of Leydig and causes secretion of testosterone.

Q.122 What are the actions of human chorionic somatomammotropin (HCS)?
HCS:
- Causes enlargement of mammary glands in animals. But, in human beings, its action on mammary glands is not known
- Causes synthesis of proteins
- Reduces peripheral utilization of glucose in mother resulting in availability of more glucose for fetus
- Causes mobilization of fat from fat depots, thus making the availability of large quantity of free fatty acids for energy production in mother's body.

Q.123 What is fetoplacental unit?
Fetus and placenta are together called fetoplacental unit (Fig. 14.6) because of their interaction during the synthesis of steroid hormones.

Q.124 Explain the function of feto-placental unit briefly.
Cholesterol, the precursor for steroid hormones enters placenta from mother's blood. From cholesterol, placenta synthesizes pregnenolone. From pregnenolone, progesterone is synthesized. Some amount of pregnenolone enters fetus from placenta. Fetal liver also produces small amount of pregnenolone. From pregnenolone, dehydroepiandrosterone sulfate (DHEAS) and 16-hydroxy dehydroepiandrosterone sulfate (16 DHEAS) are formed. DHEAS and 16-DHEAS enter the placenta from fetus to form estrogen. Some amount of progesterone enters the fetus from placenta to form cortisol and corticosterone in fetal adrenal gland.

Q.125 What is the basis for pregnancy tests?
Determination of presence or absence of the hormone called human chorionic gonadotropin (hCG) in the urine of woman suspected for pregnancy.

Q.126 What is the principle of immunological test for pregnancy?
The principle of immunological test is to determine the presence or absence of agglutination of sheep's red blood cells or latex particles coated with hCG. Presence of agglutination indicates that the woman is not pregnant. And absence of agglutination indicates that the woman is pregnant.

Q.127 What are the advantages of immunological test for pregnancy?
- Immunological test is accurate
- The result is obtained within few minutes
- Procedure of the tests is easy to perform
- Test can be performed within first few days of conception.

Q.128 Name the hormones involved in the growth of mammary glands.
Estrogen, progesterone, prolactin, growth hormone, thyroxine, cortisol and placental hormones.

Q.129 What are the processes involved in lactation?
- Milk secretion
- Milk ejection.

Q.130 What are the phases of milk secretion?
- Initiation of milk secretion or lactogenesis
- Maintenance of milk secretion or galactopoiesis.

Q.131 What are the hormones involved in milk secretion?
Prolactin is necessary for initiation of milk secretion. Growth hormone, thyroxine and cortisol are necessary for maintenance of milk secretion.

Q.132 What are the contraceptive methods in females?
- Rhythm method
- By using mechanical barriers like cervical cap or diaphragm
- Pill method (oral contraceptives)
- By using intrauterine contraceptive devices (IUCD)
- Tubectomy.

Q.133 What is safe period? When does it exist?
The period of menstrual cycle during which there is no danger of pregnancy after sexual intercourse is known as safe period. It is 4 to 5 days after menstrual bleeding and 5 to 6 days before the onset of next menstrual cycle.

Q.134 What is the disadvantage of rhythm method of conception?
The knowledge of determining the time of ovulation is difficult for uneducated or less educated women. So, it is not a successful method among such women. Also, there must be understanding between the couples regarding this and self restrain is essential. Otherwise, it cannot be practiced.

Q.135 What are oral contraceptives?
The oral pills containing synthetic estrogen and progesterone are known as oral contraceptives.

Q.136 What is the mechanism of action of oral contraceptive pills?
Oral contraceptive pills prevent maturation of follicles and ovulation by suppressing the secretion of gonadotropins from pituitary. Thus, menstrual cycle becomes anovulatory in nature under the influence of these pills.

Q.137 Name the types of oral contraceptives.
- Classical pills
- Sequential pills
- Mini pills.

Q.138 What are the disadvantages of using oral contraceptive pills?
- Regular intake of pills without fail is difficult
- Long term use of these pills results in inhibition of synthesis of anticoagulants and clotting factors and endometrial carcinoma.

Q.139 What is the mechanism of action of intrauterine contraceptive device (IUCD)?
The IUCD prevents fertilization and implantation of ovum. The IUCD with copper content has got spermicidal action also.

Q.140 Name the commonly used IUCD.
Lippe's loop and copper T.

Q.141 What are the disadvantages of using IUCD?
- It causes heavy bleeding in some women
- It has the tendency to cause infection
- It may come out of uterus accidentally.

Q.142 What is medical termination of pregnancy? How is it done?
Abortion during first few months of pregnancy is called medical termination of pregnancy (MTP). There are three ways of doing MTP:
- Dilatation and curettage (D and C)
- Vacuum aspiration
- Administration of prostaglandin.

Q.143 What is the permanent method of sterilization in females?
Permanent method of sterilization in females is tubectomy. In this, the fallopian tubes are cut and the cut ends are ligated so that, the entry of ovum into uterus is prevented. Though this can cause permanent sterility, if necessary, recanalization of fallopian tube can be done using plastic tube.

Q.144 Name the contraceptive methods in males.
- Using condoms
- Vasectomy.

- Coitus interruptus
- Drugs which inhibit spermatogenesis (under research)

Q.145 What is the permanent method of sterilization in males?

Permanent method of sterilization in males is vasectomy. In this, the vas deferens is cut and the cut ends are ligated so that, the entry of sperms into ejaculatory duct and into semen is prevented. Though vasectomy causes permanent sterility, if necessary recanalization of vas deferens can be done.

Q.146 What are the differences between human's milk and cow's milk?

See Table 14.1.

Table 14.1: Differences in human's and cow's milk

S. No.	*Human's milk*	*Cow's milk*
1.	It contains less protein, less salts and more carbohydrates.	It contains more protein, more salts and less carbohydrates.
2.	Caseinogen is present more in amount.	Comparatively in less amount.
3.	It contains less fatty acids.	It contains more fatty acids.

Q.147 When does heart beat begins in fetus?

It begins by 4th week of pregnancy.

Q.148 When does GIT develop in the fetus?

It starts to develop by 4th month and by 7th month it grows almost upto normal stage.

Q.149 When do the kidney develops in the fetus?

These develop mostly by 3rd trimester of pregnancy but normal functioning becomes complete only few months before birth.

Q.150 What is the main source of energy in fetal metabolism?

Glucose is the main source of energy for fetus.

Chapter 15

Cardiovascular System

Q.1 What structural characteristics of cardiac muscle enable its continuous rhythmic contractions?

These are: Presence of pacemaker cell that initiates autorhythmicity, presence of special conductive tissue and presence of free branchings between the muscle fibers (syncytium) ensure the quick passage of impulse from pacemaker cell to all parts of heart to initiate continuous rhythmic contractions.

Q.2 Name the special conducting tissues of heart.

SA node, AV node, bundle of His and Purkinje fibers (Fig. 15.1).

Q.3 What is law of heart muscle?

It states that the size of muscle fibers, glycogen content and rate of conduction increases from nodal to Purkinje's fiber whereas length of systole, duration of refractory period and rhythmicity increases in the reverse direction.

Q.4 What is intercalated disc and what is its importance?

At the point of contact of two cardiac muscle fibers, extensive folding of cell membrane occurs which is known as intercalated discs. They provide a strong union between fibers so that the pull of one contractile unit can be transmitted to the next, thereby helps in increasing force of contraction.

Q.5 What is the role of gap junction in cardiac muscle?

Gap junction is present in the intercalated disc of cardiac muscle fibers and helps in rapid transferring of electrical currents, ions, etc. from one cell to another without coming in contact with ECF. Thus they provide low resistance bridge for the rapid spread out of electrical impulse, thereby helps the cardiac muscle to act as syncytium (functional).

Q.6 Name the valves and their location (Fig. 15.2).

There are four valves—two in between the atria and ventricles known as atrioventricular valves (A-V valves) and two are at the opening of the blood vessels arising from the ventricles (semilunar valves).

- *A-V valves:* These are present in between the atria and ventricles. The valve presents in between right atria and right ventricle is known as *tricuspid valve* and the valve presents in between left atria and left ventricle is known as *bicuspid valve.*
- *Semilunar valves:* There are two semilunar valves namely *pulmonary valve* and *aortic valve.* The pulmonary valve is present at pulmonary orifice which leads from RV to pulmonary artery and the aortic valve is present at aortic orifice which leads from LV to the aorta.

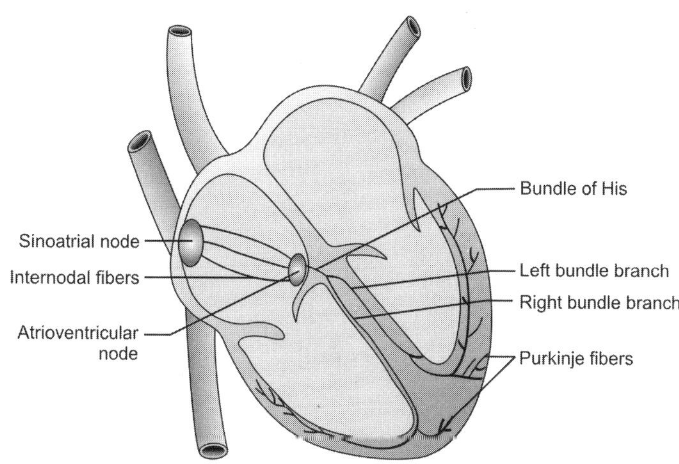

Fig. 15.1: Sinoatrial node and conductive system of the heart

Q.7 Name the special junctional tissues and their conduction rate.

The special junctional tissues and their rate of impulse generating capacity are:

Special junctional tissues	*Impulse generating capacity*
S A Node	75 ± 5 times/min
A V Node	60 times/min
Bundle of His	40 times/min
Purkinje's fiber	20 times/min

Q.8 What do you mean by pacemaker potential or diastolic depolarization?

The pacemaker tissue is characterized by unstable RMP due to slow depolarization resulting from leakage of Na^+ from outside to inside through Na^+ leak channels. This show leakage of Na^+ inside the cell causes increase in electropositively inside the cell which ultimately enables to induce another action potential easily. This slow polarization in between action potential is known as prepotential or pacemaker potential or diastolic depolarization.

Q.9 Why SA node is called as cardiac pacemaker?

SA node acts as a **pacemaker of heart** because the rate of impulse generation in normal heart is determined by this node because of its highest rate of impulse generating capacity (75 ± 5 times/min) than other junctional tissues. This is why it is known as **cardiac pacemaker**.

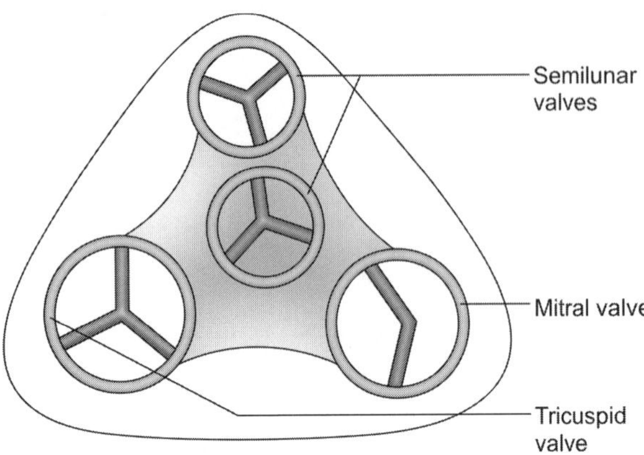

Fig. 15.2: Valves of heart

Q.10 What is ectopic pacemaker?

When the pacemaker is other than SA Node (e.g. AV node, etc.) it is called *ectopic pacemaker.*

Q.11 What is the duration of refractory period in cardiac muscle?

Refractory period is very long in cardiac muscles. It is about 0.53 seconds. In this, the absolute refractory period is 0.27 seconds and relative refractory period is 0.26 seconds.

Q.12 What is the significance of long refractory period in cardiac muscles?

Due to the long refractory period, the complete summation of contractions, fatigue and tetanus do not occur in cardiac muscle.

Q.13 What do you mean by nodal and idioventricular rhythm?

The AV node takes the charge of generating impulse rhythmically when SA node does not work. In this condition, atria and ventricles beat almost simultaneously at the rate of 60 times per min. This rhythm of heart is known as *nodal rhythm.* Whereas 2nd Stannius ligature applied over the A-V groove makes the atria to continue beating with its own rhythm whereas the ventricle stops beating due to blockade of impulse from atria to ventricles. After sometime ventricle generates its own impulse and starts beating at much slower rate. This rhythm of heart beat in which atria and ventricular beating do not follow any specific pattern is known as *idioventricular rhythm.*

Q.14 What is AV delay? What is its significance?

When the impulse reaches to AV node, there is a delay of about 0.1 sec to pass the impulse to bundle of His. This time gap is known as AV delay. It allows the atria to contract just ahead of ventricular contraction thereby atria is emptied before ventricular ejection.

Q.15 What is the ionic basis of plateau phase of cardiac action potential?

Immediately after depolarization voltage gated Na^+ channel's used to close resulting stoppage of entry of Na^+ ions and voltage gated K^+ channel start opening resulting exit of K^+. These result in rapid fall of electropositivity initially known as **rapid repolarization.** Afterwards, the rate of repolarization becomes slower due to prolonged opening of voltage gated Ca^{+2} channel through which Ca^{+2} enters inside. Thus the exit of K^+ is almost counter-balanced by entry of Ca^{+2} resulting in sustained depolarization known as **plateau phase**.

Q.16 Enumerate the properties of cardiac muscle.

- Excitability
- Rhythmicity
- Conductivity
- Contractility
  Contractility includes:
  - All or none law
  - Staircase phenomenon
  - Summation of subliminal stimuli
  - Refractory period.

Q.17 Is all or none law applicable in heart?

All or none law which states that if a stimulus is applied, whatever may be the strength of stimulus, the cardiac muscle responds maximally or it does not give any response at all (Fig. 15.3). Of course, it is **applicable only** in whole atrial muscle (i.e. atrial syncytium) or in whole ventricular muscle (i.e. ventricular syncytium) not **to a single cardiac muscle fiber.**

Q.18 Define staircase phenomenon. Why does it occur?

If stimuli are applied repeatedly, with an interval of 2 seconds to the cardiac muscles, the force of contraction increases gradually for the first few contractions. Later the force remains the same. The gradual increase in the force of contraction is known as staircase phenomenon or treppe response. It occurs because of the short interval of 2 seconds in between the stimuli. During this period, the beneficial effect is produced and this facilitates the force of successive contraction (Fig. 15.3).

Q.19 Why left ventricular subendocardial region is more prone to myocardial infarction (MI)?

The blood supply to the cardiac muscle in different areas of heart is not same. On the surface of the cardiac muscle there are large epicardial arteries supplying more blood to those areas whereas in the subendocardial region blood supply is less because it is supplied by smaller intramuscular arteries and plexus of subendocardial artery the diameter of

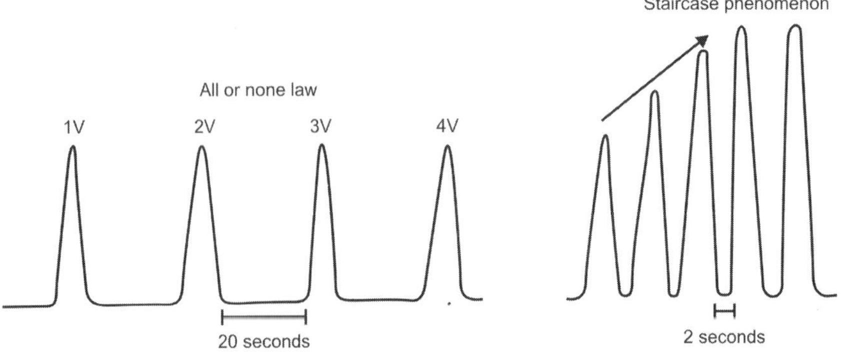

Fig. 15.3: All or none law and staircase phenomenon in cardiac muscle

which are less. This blood supply to the subendocardial plexus is further reduced during systole. Therefore the **subendocardial region is more prone to myocardial infarction**. Again as the left ventricular thickness is much more than that of right ventricle the occlusion is more severe in left ventricle. For this region **LV subendocardial region is more prone to MI**.

Q.20 What are the importance of anastomotic channels in heart muscle?

In the normal heart there are some collaterals among the smaller arteries which become active under abnormal conditions like myocardial ischemia. They open up within a few seconds after the sudden occlusion of larger artery and become double in number by the end of 2nd or 3rd day and reach to normal by one month. When atherosclerosis causes constriction of coronary arteries slowly over a period of many years, collateral vessels develop restoring normal blood and thus the patient **never experiences acute episode of cardiac dysfunction**.

Q.21 What is the importance of autoregulation in blood supply in heart muscle?

Like some other organs the heart has the capacity to regulate its own blood flow up to a certain limit in order to maintain an almost constant blood flow to the cardiac musculature in spite of any alteration of systemic blood flow. This is known as autoregulation of coronary blood supply.

Q.22 Why cardiac muscle cannot be tetanized?

It is because of its long absolute refractory period and thus summation of contractile response is not possible which is essential for tetanization of heart muscle.

Q.23 What is cardiogram?

The record of the mechanical activity of the heart is known as cardiogram.

Q.24 Mention the maximum and minimum pressure in heart during systole and diastole?

Chamber	*Peak pressure in systole*		*Min. pressure in diastole*	
Left ventricle	120	mm Hg	5–12	mm Hg
Right ventricle	25	mm Hg	2–6	mm Hg
Left atrium	15	mm Hg	5–8	mm Hg
Right atrium	6	mm Hg	1–5	mm Hg
Aorta	120	mm Hg	80	mm Hg
Pulmonary artery	25	mm Hg	5–12	mm Hg

Q.25 Define and give normal values of end diastolic volume, stroke volume and end systolic volume.

During ventricular diastole the intraventricular volume is increased which results filling of the ventricles. At the end of diastole the amount of blood filled by the ventricle is known as end diastole volume (EDV). It is about 120–130 mL.

During ventricular systole intraventricular volume decreases which results increase in pressure thus ejection of blood out of ventricles. During each systole the amount of blood pumped out by each ventricle is known as stroke volume (SV). **Normal value**: 70 mL/beat.

At the end of systole however some amount of blood is remained in each ventricle which is known end systolic volume (ESV). The **normal volume**: 50–60 mL/ beat.

Q.26 What do you mean by vagal escape? What is its cause?

If strong vagal stimulation to heart is continued then after a pause the ventricles resume to beat at a slow rhythm which is called vagal escape represented by Figure 15.4.

During prolonged vagal stimulation right auricle stops beating and distends due to blood overflow which leads to fall of BP → afferent impulse from carotid sinus to cardiac centers → stimulate ventricles to start its beat.

Q.27 What is the action of sympathetic nerves on heart?

Sympathetic nerves increase the rate and force of contraction of heart by secreting noradrenaline.

Q.28 What is sympathetic tone?

Continuous stream of accelerator impulses that arises from cardioaccelerator center and reaches the heart via sympathetic nerves is known as sympathetic tone or cardio accelerator tone. However, under resting conditions, the vagal tone is more dominant over the sympathetic tone.

Q.29 Define apex beat.

Apex beat is the impulse or throb which is felt and seen on the chest wall normally in the left 5th intercostal space just medial to left nipple.

Q.30 Name different phases of cardiac cycle. Mention the duration of each phase.

Phases of cardiac cycle	*Duration in sec*
• Atrial systole	0.1
• Atrial diastole	0.7
• Ventricular systole	0.3 Total
– Isovolumetric contraction	0.05
– Rapid ejection phase	0.1
– Slow ejection phase	0.15
• Ventricular diastole	0.5 Total
– Protodiastole	0.04
– Isovolumetric relaxation	0.08
– Filling phase	0.38 Total
– Ist rapid filling phase	0.1–0.12
– Slow filling phase	0.18–0.20
– Last rapid filling phase	0.06–0.10

Q.31 What is protodiastole? Is it part of systole or diastole?

Protodiastole is the very brief phase before diastole in which ventricular systole has ceased but relaxation yet to start.

It cannot be well defined whether the protodiastole is a part of systole or diastole as some workers include it in diastole as muscle contraction is stopped at this phase whereas some others believe that it is a part of systole as muscle relaxation has not yet started.

Q.32 Define cardiac cycle.

The sequence of events (mechanical, electrical, etc.) associated with consecutive heart beat is repeated cyclically which is known as cardiac cycle (Fig. 15.4). Normal duration is 0.8 sec if heart rate is 75 beats/ min.

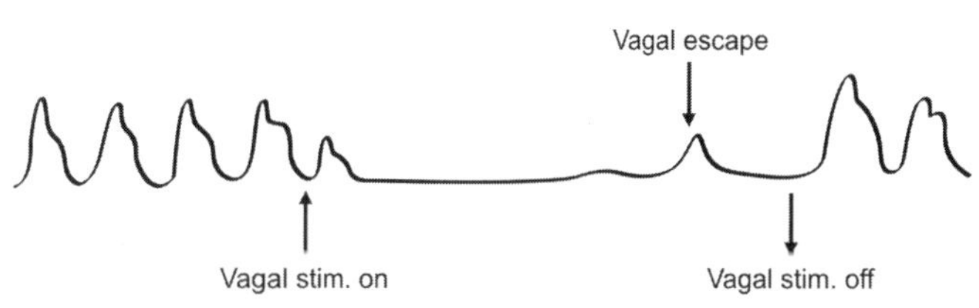

Fig. 15.4: Demonstration of vagal escape on heart muscle

Q.33 What are the causes of 1st heart sound?

These are:

- Closure and vibrations of AV valves at the beginning of ventricular systole.
- Vibrations of blood surrounding the AV valves.
- Vibrations of major blood vessels around the heart.
- Vibrations of walls of heart.

Q.34 What are the characteristics of 1st heart sound (Fig. 15.5)?

It is:

- Soft, prolonged with low pitch.
- Duration is 0.12 sec and occurs in peak or downstroke of R wave in ECG and just before onset of 'c' wave in jugular pulse tracing.
- Best heard at apex beat area and is associated with onset of ventricular systole.

Q.35 What is the significance of 1st heart sound?

It indicates force of contraction, condition of myocardium and competence of AV valves.

Q.36 What are the causes of 2nd heart sound?

These are:

- Closure and vibration of semilunar valves at the end of ventricular systole.
- Vibrations of blood surrounding these valves.
- Vibrations of walls of aorta and pulmonary artery.
- Vibrations of the wall of ventricles to a little extent.

Q.37 What are the characteristics of 2nd heart sound?

It is:

- Sharp, short and high pitched.
- Duration is 0.08 sec and follows T wave in ECG and coincides with 'v' wave in jugular venous pulse tracing.
- Best heard at 2nd right costal cartilage for aortic component and 2nd intercostal space at left sternal border for pulmonary component.
- Associated with onset of ventricular diastole.

Q.38 What is the significance of 2nd heart sound?

It indicates the competence of semilunar valves.

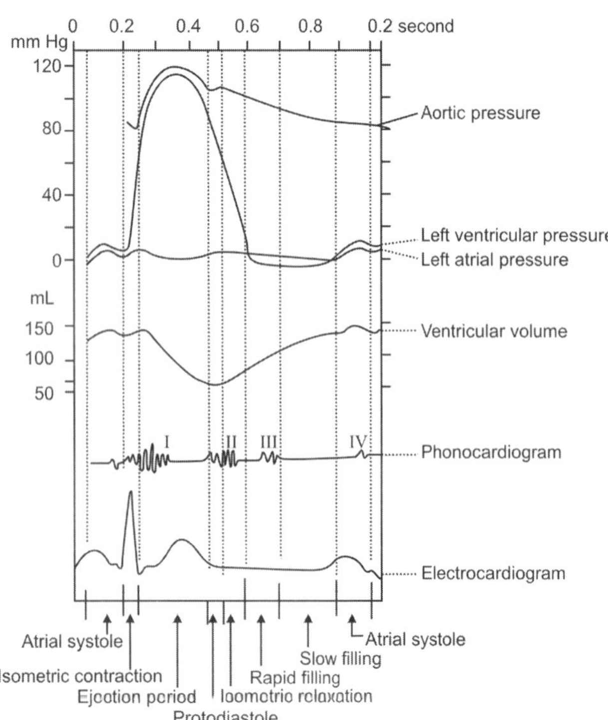

Fig. 15.5: Comprehensive diagram showing ECG, phonocardiogram, pressure changes and volume changes during cardiac cycle

Q.39 When and how 3rd heart sound is produced?

3rd heart sound is produced during the first 1/3 of ventricular diastole. It occurs due to the vibrations set up by the rushing of the blood during the rapid filling phase of ventricular diastole.

Q.40 Differentiate 1st and 2nd heart sound.

1st heart sound	*2nd heart sound*
• It is prolonged, low pitched and soft.	It is sharper, abrupt, clear and high pitched
• Coincides with carotid pulse	Does not coincide
• Coincides with R wave of ECG	May precede, coincide or follow the T wave of ECG.
• Best heard over the mitral area	Best heard over aortic and pulmonary area.
• Time interval between 1st and 2nd is shorter	Time interval between 2nd and next 1st is comparatively longer

Q.41 What is murmur?

It is the sound produced by turbulence produced in the blood by a forward flow through a stenosed (narrowed) valve or back flow (regurgitation) through a deformed or incompetent valve.

Q.42 How do you classify murmur?

It will be classified on the basis of their relationship with main heart sounds like presystolic, systolic, diastolic and also to and fro murmurs.

Q.43 What are the maximum and minimum pressure in heart?

- Maximum pressure in left ventricle is above 120 mm Hg.
- Max pressure in right ventricle is above 25 mm Hg.
- Minimum pressure in left ventricle is 80 mm Hg.
- Minimum pressure in right ventricle is few mm Hg.

Q.44 What is the normal heart rate? What are the factors affecting heart rate (HR)?

Normal value of HR is 72 beat/min with the normal range 60–90 beat/min.

The factors are: age, sex, body temperature, hypoxia, emotion, exercise, etc. and drugs like epinephrine and norepinephrine.

Q.45 Why HR is slightly higher in females than males?

It is because of two reasons:

- Lower systemic BP
- More resting sympathetic tone.

Q.46 What is Cushing reflex?

It is represented by following sequential events:

Increased intracranial pressure → decreases blood supply to medullary hypoxia and hypercapnia → stimulation of medullary vasomotor center → increase of systemic BP → stimulation of baroreceptors → stimulation of vagus nerve → decrease of HR and respiration. This reflex mechanism by which increased intracranial pressure results bradycardia is known as Cushing reflex.

Q.47 What do you mean by sinus arrhythmia?

Heart rate increases with inspiration and decreases during expiration. This phenomenon is known as sinus arrhythmia.

Q.48 State Marey's law.

If the other conditions remain constant then the HR is inversely related with systemic BP (Fig. 15.6).

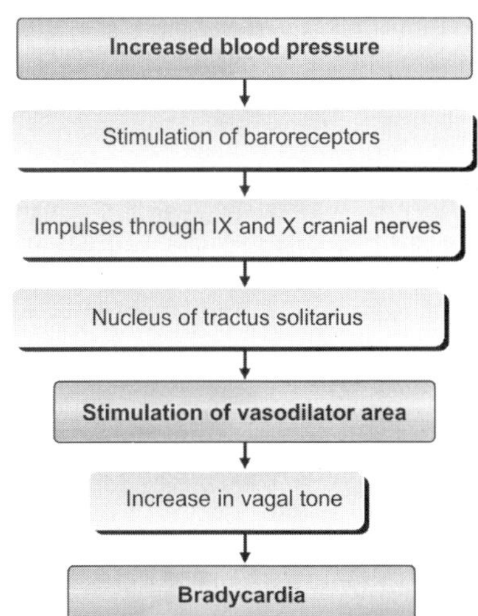

Fig. 15.6: Marey's (cardioinhibitory) reflex

Q.49 Define cardiac output and cardiac index.

Cardiac output: The amount of blood pumped out by each ventricle per min is called as cardiac output. The normal value is 5 lit/min/ventricle.

Cardiac index: It is the cardiac output per square meter of body surface area. The normal value is 3.2 L/m^2/min.

Q.50 What do you mean by extrinsic and intrinsic autoregulation of cardiac output?

If cardiac output is controlled by controlling only heart rate (as CO = HR × SV) it is known as extrinsic autoregulation of cardiac output whereas if it is regulated by regulating only stroke volume, it is known as intrinsic autoregulation.

Q.51 What is the difference between heterometric and homometric regulation of cardiac output?

To control cardiac output when ventricular contraction is regulated by controlling initial length of the muscle fiber, i.e. EDFL, then it is called heterometric regulation which is independent of cardiac nerves. Whereas when cardiac nerves regulate the myocardial contractility to control the cardiac output, it is known as homometric regulation of cardiac output.

Q.52 What is Frank-Starling's law of heart? What is its relation with venous return?

It states that within the physiological limit, the force of ventricular contraction is directly proportional to the initial length of muscle fibers (EDFL).

If venous return is increased the EDFL of the ventricular muscle is also increased resulting in more force of ventricular contraction thereby more cardiac output.

Q.53 What do you mean by Vis A Tergo and Vis A Fronte in relation to cardiac pump?

Vis A Tergo is the force which drives the blood forward from behind, e.g. the contraction of the heart drives the blood in forward direction, whereas Vis A Fronte is the force acting from front that attracts blood in the veins towards the heart, e.g. ventricular systolic and diastolic suction pressure.

Q.54 Enumerate the factors affecting venous return.

The factors are: Thoracic or respiratory pump, cardiac pump, muscle pump, total blood volume and increased sympathetic activities on veins.

Q.55 Name two methods by which cardiac output is measured.

These are:

- Direct Fick method and
- Indirect dye dilution method.

Q.56 Enumerate Fick's principle.

It states that the amount of a substance taken up by an organ or by whole body per unit time is equal to the arterial level of that substances minus the venous level (i.e. A-V difference) times the blood flow, i.e. amount of substance taken/min = A-V difference of the substance × blood flow/min.

Q.57 What are the disadvantages of Fick's method?

These are:

- As it is the invasive method the subject is exposed to all risk of hemorrhage, infection, etc.
- As the subject is conscious of the whole technique cardiac output may be higher than normal.

Q.58 Which dye is generally used in dye dilution method and why?

It is generally Evans blue or radioactive isotopes. Criteria for selection are as follows:

- These stay in the circulation during the test.
- These are not harmful and not toxic.
- Do not alter the hemodynamics of blood flow.
- Concentration of these substances can be easily measured.
- Excreted totally and neither reabsorbed nor secreted by the body.

Q.59 What is ballistocardiogram?

It is a record of the to and fro movements of the body in the headward to footward direction when the subject lies on a suitably suspended table. This is the another method of measuring cardiac output though it is now absolute.

Q.60 What is bundle of KENT?

In the individuals with WPW syndrome, there is one additional nodal connecting tissue in between atria and ventricles besides AV node which conducts the impulse more rapidly than AV node. This additional conducting pathway is known as bundle of KENT.

Q.61 Define blood pressure (BP).

It is the lateral pressure exerted by the moving column of blood on the wall of blood vessels during its flow.

Q.62 Define systolic, diastolic, mean and pulse pressure with each of their normal average values.

Systolic pressure (SP): It is the maximum pressure exerted during systole of the heart. Normal value = 120 mm Hg (Normal range: 110–140 mm Hg).

Diastolic pressure (DP): It is the minimum pressure during diastole of the heart. Normal value = 80 mm Hg (Normal range: 60–90 mm Hg).

Pulse pressure (PP): Pulse pressure is the difference between systolic and diastolic pressure. Normal value = 40 mm Hg.

Mean pressure (MP): It is average pressure during each cardiac cycle. Normal value = 93.3 mm Hg.

Q.63 Enumerate the significance of SP, DP, PP and MP.

- Systolic pressure (SP) indicates the extent of work done by the heart and also the force with which the heart is working. It also indicates the degree of pressure the arterial wall have to withstand.
- Diastolic pressure (DP) is the measure of the total peripheral resistance and it indicates the constant load against which heart has to work.
- Pulse pressure (PP) determines the pulse volume. Whereas mean pressure indicates the perfusion pressure head which causes the flow of blood through the arteries, arterioles, capillaries, veins and venules.

Q.64 Why does systolic pressure increase after meal?

After meal pressure over heart increases due to distended abdomen which in turn increases heart rate and also there is a release of epinephrine which also increases systolic blood pressure.

Q.65 What do you mean by baro-receptors? Where are they located?

Baroreceptors are the pressure receptors stimulated in response to change of pressure around them.

These are located in the wall of blood vessels (e.g. arterial baroreceptor-present in carotid sinus, aortic arch, root of right subclavian artery, junction of thyroid artery with common carotid artery, also pulmonary trunk) and also in the walls of the heart (e.g. atriocaval receptors, atrial receptors).

Q.66 What do you mean by buffer nerves? Why they are so called?

Carotid sinus nerve originated from carotid sinus and aortic nerve arised from arch of aorta are collectively known as buffer nerves as they prevent any change in systemic BP and thus help the BP to keep normal.

Q.67 What is Bain-bridge reflex?

Rapid injection of blood or saline in anesthetized animals produces a rise in heart rate if the initial heart rate is low. This is called Bain-bridge reflex (Fig. 15.7). This is due to the stimulation of stretch receptors in the wall of right atrium.

Q.68 Name different chemoreceptors responsible for BP regulation. What are their stimulants?

These are carotid bodies and aortic bodies. They get stimulated by hypoxia, hypercapnia, asphyxia and also acidemia.

Q.69 What is the effect of chemoreceptors on heart rate?

In conditions like hypoxia, hypercapnia and increased hydrogen ion concentration, the chemoreceptors send inhibitory impulses to vasodilator area (cardioinhibitory center). Now, the vagal tone is reduced and heart rate is increased.

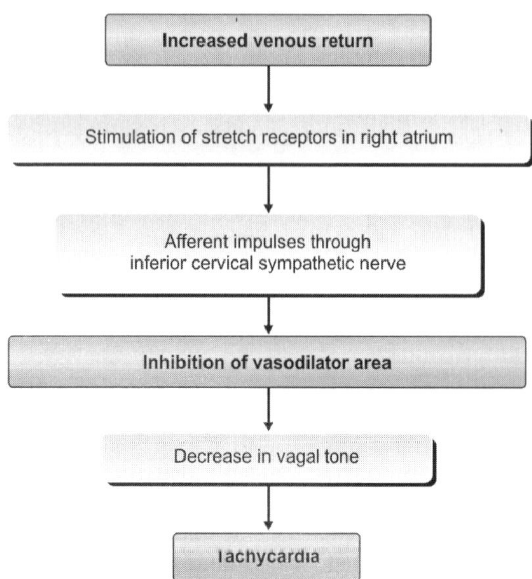

Fig. 15.7: Bain-bridge (cardioaccelerator) reflex

Q.70 Sudden standing increases diastolic BP—explain how?

On standing there is peripheral pooling of blood in lower parts of body → lowering of venous return to the heart → decrease cardiac output → thereby decrease systolic BP → leads to decrease baro-receptor discharge → thereby increases sympathetic activity → results increase of the total peripheral resistance due to vaso-constriction → ultimately leads to increase of diastolic pressure.

Q.71 If mean BP is decreased to 60 mm Hg then what compensatory mechanism will operate to bring it to normal?

Both baroreceptor mechanism (which operates in between 60 and 200 mm Hg mean blood pressure) and chemoreceptor mechanism which operates between 40 and 100 mm Hg of mean BP.

Q.72 If BP is decreased to 40 mm Hg then which compensatory mechanism will start into action?

Both chemoreceptor mechanism and CNS ischemic response.

Q.73 If mean BP is increased to 140 mm Hg then what compensatory mechanism will be operated?

Only baroreceptor mechanism.

Q.74 What do you mean by stress relaxation and reverse stress relaxation mechanism in relation to BP regulation?

Rise in arterial BP due to intravenous transfusion of blood increases perfusion pressure in blood storage organs that causes relaxation of blood vessels, thereby decreases venous return and thereby decreases cardiac output. This leads to decrease BP to normal level. This mechanism is known as stress relaxation.

The opposite phenomenon is known as reverse stress relaxation mechanism which is as follows:

Prolonged bleeding causes decrease of BP → thereby decreases perfusion pressure → leads to vasoconstriction of blood storage organs → results in increase of venous return and thus increases cardiac output → which in turn increases BP to normal level.

Q.75 What is hypertension? What do you mean by systolic hypertension and white coat hypertension?

Chronic elevation of blood pressure beyond 140/90 is generally labelled as hypertension. In advanced age, due to loss of elasticity of blood vessels, stretching of the wall of blood vessels decreases which results in increment of pressure during systole with normal diastolic pressure. This condition is known as systolic hypertension which is characterized by high pulse pressure.

Some hypertensive patients because of nervousness, have higher BP in the clinician's chamber than during their normal day time activity. This condition is known as white coat hypertension.

Q.76 What do you mean by malignant hypertension?

In some patients the blood pressure especially the diastolic pressure is increased to very high level (>120 mm Hg) within a short period. This condition is known as malignant hypertension.

Q.77 Which pressure is considered better to judge the hypertension–SP or DP? Justify your answer.

Clinically diastolic pressure is more useful to characterize the state of hypertension because diastolic pressure is comparatively constant and does not fluctuate like SP in response to day-to-day activity.

Q.78 What do you mean by labile hypertension?

In early stages of essential hypertension, systolic BP fluctuates. This is why it is referred to as labile hypertension.

Q.79 What is hypotension?

Chronic low BP specially the diastolic pressure below 60 mm Hg is called hypotension.

Q.80 What do you mean by postural hypotension?

In some hypotensive patients, sudden standing causes further fall of systemic BP that may result in dizziness, dimness of vision and even fainting. This is known as postural hypotension.

Q.81 What is the difference between pulse pressure and pressure pulse?

Pulse pressure is the difference of systolic and diastolic pressure whereas the pressure pulse or pulse is the wave transmitted to the arteries like radial arteries due to stretching and relaxation of wall of aorta in response to ventricular ejection of blood and ventricular filling respectively during cardiac cycle.

Q.82 What is the purpose of doing exercise tolerance test?

It is for determining the efficiency of the heart as a pumping organ.

Q.83 What is isometric (isovolumetric) contraction of the heart?

The period during which the ventricles of the heart contract as closed cavities (because all the valves are closed) without any change in the volume of ventricular chambers or in the length of muscle fibers is known as isometric (isovolumetric) contraction. During this period, the pressure increases very much.

Q.84 What is the significance of isometric contraction of the heart?

During isometric contraction, the pressure in the ventricles is greatly increased. When the ventricular pressure increases more than the pressure in aorta and pulmonary artery the semilunar valves open. Thus, the high pressure developed during isometric contraction is responsible for the opening of semilunar valves leading to ejection of blood from the ventricles.

Q.85 What is isometric or isovolumetric relaxation of the heart?

The period during which the ventricles of the heart relax as closed cavities (because all the valves are closed) without any change in the volume of ventricular chambers or in the length of muscle fibers is known as isometric or isovolumetric relaxation. The pressure decreases very much during this period.

Q.86 What is cardiac reserve?

It is the difference between the basal cardiac output of an individual and the maximum cardiac output that can be achieved in that person. It is also expressed as cardiac reserve percent.

Q.87 By observing HR can you predict the intensity of exercise or work done by a person?

Yes, - If HR is <100; it will be light exercise.
- If HR is 100–125; it will be moderate exercise.
- If HR is 126–150; it will be heavy exercise.
- If HR is >150; it will be severe exercise.

Q.88 What is apex-pulse deficit?

Normally the pulse rate and heart rate are identical but in some cases like extrasystoles and atrial fibrillations, some of the heart beats are too weak to be felt at the radial artery resulting in missing of that particular pulse. This causes higher heart rate than pulse rate. This condition is known as apex-pulse deficit or pulse deficit.

Q.89 Name the waves of normal arterial pulse tracing. What are their physiological basis?

In the normal arterial pulse recording, there are one steep upstroke called anacrotic limb and one rather slow down stroke called catacrotic limb. The end of anacrotic limb and beginning of catacrotic limb is designated as percussion wave (p). In the catacrotic limb there is also a negative wave called dicrotic notch (n) followed by a positive wave called dicrotic wave. Besides this, sometimes after the peak of the tracing there is another small wave called tidal wave (t). The waves are represented by Figure 15.8.

- *Percussion wave:* It is due to expansion of the artery for ventricular ejection during ventricular systole.
- *Catacrotic limb:* It is due to normalization of artery due to slow passing of blood towards periphery.
- *Dicrotic notch:* It is due to backflow of the blood from aorta towards heart due to pressure difference during ventricular diastole.
- *Dicrotic wave:* It is due to increase pressure again in the aorta due to prevention of back flow of blood towards heart by closure of aortic valve.

Q.90 Can you indicate the systolic and diastolic phases of the ventricle on the arterial pulse tracing?

Yes, the maximum ejection phase lasts from the start of the upstroke to peak of '*p*' wave while the reduced ejection phase lasts from peak of '*p*' wave to peak of dicrotic notch. The rest time period represents diastole.

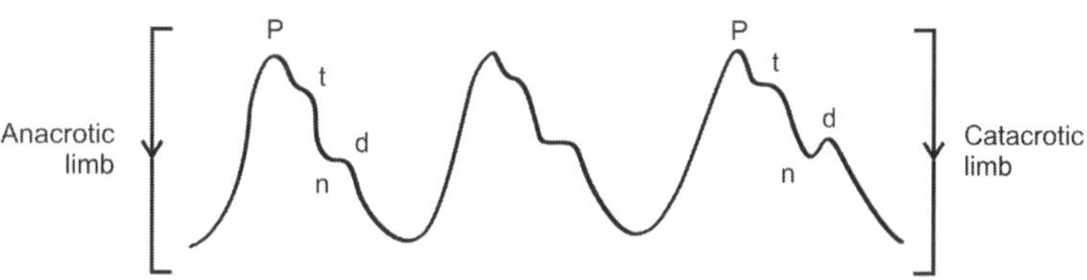

Fig. 15.8: Normal arterial pulse tracing

Fig. 15.9: Abnormal arterial pulse tracing (plateau pulse)

Q.91 What is dicrotic pulse?

There are two palpable waves—one in systole and another in diastole in congestive cardiomyopathy patients where stroke volume is low. This type of pulse is known as dicrotic pulse.

Q.92 What is plateau pulse?

During some pathological conditions like aortic stenosis the pulse wave rises slowly, followed by delayed and sustained peak and then the pulse faded slowly. Such type of pulse is known as plateau pulse as represented by Figure 15.9.

Q.93 What is anacrotic pulse?

Slow rising and slow fall of pulse wave due to prolonged ventricular ejection as occurs in aortic stenosis is known as anacrotic pulse.

Q.94 What do you mean by pulsus alterans and paradoxus?

Pulsus alterans is alternative weak and strong beating of pulse whereas the phenomenon when pulse disappears or becomes feeble during inspiration and becomes maximum during expiration is known as pulsus paradoxus.

Q.95 What is water hammer pulse?

In some conditions like aortic regurgitation there is sharp and steep rise followed by sleep fall of pulse which is known as water hammer pulse.

Q.96 How does jugular venous pulse record give the idea about right atrial pressure?

Jugular vein is connected directly with right atrium and as there is no valve at the junction of superior vena cava and right atrium, any change of right atrial pressure is directly transmitted to the jugular vein. That is why jugular venous pressure record gives the idea about right atrial pressure.

Q.97 Name the waves of jugular venous pulse and the causes of their onset.

The waves and their causes are as follows:

- 'a' wave—It is due to increase in pressure within atrium due to atrial systole.
- 'c' wave—It is due to increased pressure within atrium due to bulging of the tricuspid valve into the right atrium during isovolumic ventricular contractile phase.
- 'v' wave—It is due to the rise in atrial pressure due to atrial filling before the tricuspid valve opens during diastole.
- X descends—It is due to fall of intra-atrial pressure due to descend of the tricuspid valves.
- Y descends—It is due to the fall of intra-atrial pressure due to the opening of tricuspid valves to result ventricular filling.

Q.98 Define ECG.

It is the record of electrical activities of heart by electrocardiograph during different periods of cardiac cycle (Fig. 15.10).

Q.99 Enumerate the clinical significance of ECG.

Any abnormalities of the heart like ischemic heart disease, myocardial infarction, extrasystole, heart block, ventricular fibrillation and flutter, sinus arrhythmias, etc. are detected by the ECG record of the person.

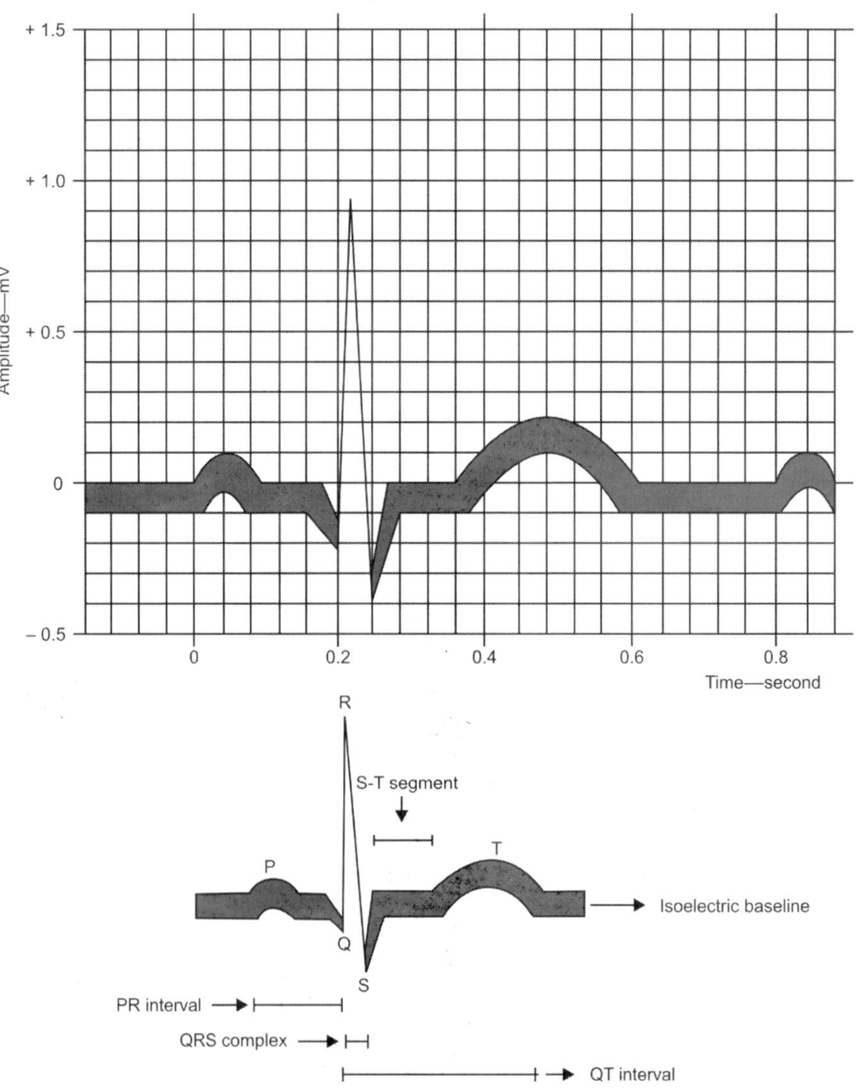

Fig. 15.10: Waves of normal ECG

Q.100 What does 'P' wave represent? What does it signify?

'P' wave represents the atrial depolarization. Any abnormalities of the 'P' wave means abnormality in the atria like larger 'P' wave denotes the atrial hypertrophy.

Q.101 What do QRST and QRS represent? What is the duration of ventricular complex?

QRST represents ventricular complex, i.e. ventricular depolarization and ventricular repolarization. Normal duration is 0.48 sec.

QRS complex represents ventricular depolarization only.

Q.102 What do Q and RS waves indicate?

'Q' wave indicates the ventricular septal activity whereas 'RS' wave indicates the excitation of ventricle proper with duration of 0.08–0.1 sec.

Q.103 What is the significance of T wave?

It is due to repolarization of ventricles and its normal duration is 0.27 sec. It indicates the functional activity of base of the heart. Clinically it signifies the myocardial damage in case of any abnormality in T wave.

Q.104 What does PR interval represent? What is its significance?

It represents atrial depolarization and conduction through bundle of His. Normal duration is 0.13–0.16 sec.

It is the interval from beginning of P wave to the beginning of Q or R wave. Prolonged PR interval signifies the conduction block.

Q.105 What is TP interval and what is its significance?

It is the period from the end of T wave to the beginning of P wave of next cardiac cycle. It represents the diastole or polarized state of whole heart. Normal duration is 0.2 sec at a HR of 75/min.

Q.106 What is QT interval and what does it represent?

It is the interval from the beginning of Q wave to the end of T wave (normal duration 0.40–0.43 sec). It represents ventricular events.

Q.107 What is ST interval? What does it represent?

End of S wave to the end of T wave is known as ST interval. The normal duration of which is 0.32 sec. It represents ventricular repolarization only.

Q.108 What is ST segment? What is its significance?

Following the QRS there is a long isoelectric period which extends from the end of S wave to the beginning of T wave called ST segment. Any change of the position of ST segment from the isoelectric line indicates the functional abnormalities of the heart. Deviation of ST segment more than 2 mm up from the isoelectric line is called elevated ST segment which is the clinical feature of MI. Similarly deviation of the same more than 2 mm downward from the isoelectric line is called depressed ST segment as seen in angina pectoris.

Q.109 Define lead.

The electrocardiographic connections, i.e. wires along with the electrodes to record ECG is known as lead.

Q.110 Classify leads.

Leads are classified as unipolar and bipolar leads which are again divided as follows:

- Unipolar lead
- Unipolar augmented limb lead
  - aVR
  - aVL
  - aVF
- Chest lead (V_1-V_6)
- Bipolar lead
  - Standard limb lead—I
  - Standard limb lead—II
  - Standard limb lead—III

Q.111 Why unipolar lead is so called?

In this type of leads, one electrode becomes inactive (indifferent electrode) whereas other one is active (exploring electrode). That is why it is known as unipolar lead.

Q.112 What do you mean by rule of thumb?

It is the general observation in the ECG record obtained from chest leads as follows:

- As we pass across the chest leads (V_1 - V_6) 'R' wave increases gradually in size and 'S' wave becomes smaller gradually. In lead V_3 both are equal.
- R wave in V_6 and S wave in V_1 represent left ventricular activity whereas R wave in V_1 and S wave in V_6 represent right ventricular activity.

Q.113 What is augmented limb lead? Why is it so called?

Augmented limb leads are unipolar type limb leads with slight modification in the recording technique where one electrode (active) is connected to the positive terminal of ECG machine and other two are connected through electrical resistant to the negative terminal of the ECG machine.

It is so called because the magnitude of different waves become larger by 50 percent than the same obtained from standard limb leads without any change of its normal pattern. These are classified as aVR, aVL and aVF.

Q.114 What do unipolar chest leads represent?

V_1 and V_2 are associated with right atrial and ventricular activity respectively whereas V_4, V_5 and V_6 represent left ventricular activity. V_3 is regarded as transitional zone.

Q.115 What do you mean by dextrocardiogram?

In case of damage of left branch of bundle of His, the impulse travels through right branch to the right ventricle resulting in predominant activity of right ventricle. Such a record is called dextrocardiogram.

Q.116 What is levocardiogram?

When right branch of bundle of His is damaged there is predominance of left ventricular activity. This type of record is called levocardiogram.

Q.117 What do you mean by Einthoven's triangle?

The equilateral triangle obtained by connecting the right arm, left arm and right leg, by means of electrical wires with current source as the heart at its center is known as Einthoven's triangle (Fig. 15.11).

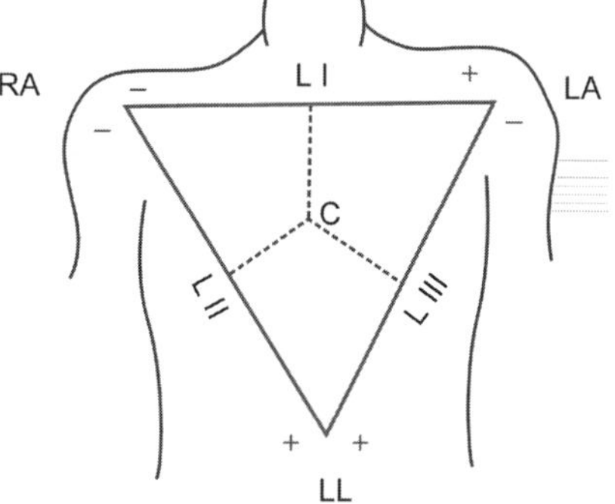

Fig. 15.11: Einthoven's triangle.
C = Center of electrical activity. RA = Right arm. LA = Left arm. LL = Left leg. LI, LII and LIII – Standard limb leads

Q.118 What is Einthoven's law?

It states that if the electrical potentials of any two of the three bipolar leads are known at any given instant, the 3rd one can be determined mathematically from the 1st two by simply summing the 1st two by considering the positive and negative signs of the different leads.

Q.119 What is J point? What is its significance?

J point is the end point of S wave and beginning of ST segment where no electrical activity of the heart exists. Normally the J point locates on the isoelectric line. Upward or downward deviation of this point indicates the heart diseases like MI and cardiac ischemia.

Q.120 What is vector?

It is an arrow that points the direction of the electrical potential generated by the current flow with the arrowhead in the positive direction.

Q.121 Mention the characteristics of vector.

These are:

- Direction of current flow is represented by the arrowhead and
- Length of the arrow is drawn proportionate to the voltage of the potential.

Q.122 What do you mean by vector cardiogram?

Vector of current flow through the heart changes rapidly as the impulse spreads through the heart muscle. These changes are:

- The vector increases and decreases in length because of the increasing and decreasing voltage of the vector.
- It also changes direction accordingly with the changes in the average direction of the electrical potential of the heart. The record that shows these changes in the vectors at different times during the cardiac cycle is called as vector cardiogram.

Q.123 What do you mean by electrical axis of the heart or cardiac vector?

Since the standard limb leads I, II, III are records of the potential difference between two points, therefore, deflection in each lead at any point indicates the magnitude and direction in the axis of the electromotive force generated in the heart. This is called as electrical axis of the heart.

Q.124 What is the effect of change in the blood sodium concentration on the heart?

Increased sodium concentration in blood decreases the rate and force of contraction. Very high sodium concentration can stop the heart in diastole. Very low level of sodium produces low voltage waves in ECG.

Q.125 What is the effect of hyperkalemia on the heart?

Normal potassium concentration in serum is about 3.5 to 5 mEq/L. When it increases above 6 mEq/L (hyperkalemia) the resting membrane potential in cardiac muscle is decreased leading to hyperpolarization. It reduces the excitability of the muscle. ECG shows a tall T wave.

The increased potassium concentration above 8 mEq/L affects the conductive system also. And in ECG, P-R interval and the duration of QRS complex are prolonged.

During severe hyperkalemia (above 9 mEq/L), atrial muscle becomes unexcitable. So, in ECG, P wave is absent and QRS complex merges with T wave.

In experimental animals, increased potassium concentration stops the heart in diastole immediately.

Q.126 What is the effect of hypokalemia on the heart?

Hypokalemia (decrease in potassium concentration) reduces the sensitivity of heart muscle. In ECG, S-T segment is depressed. Amplitude of T wave is reduced. In severe hypokalemia, T wave is inverted. U wave appears. P-R interval is prolonged.

Q.127 What is the effect of hypercalcemia on human heart?

Normal serum calcium level is 9–11 mg%. In hypercalcemia, there is reduction in duration of S–T segment and Q–T interval, with slight increase in excitability and contractility.

Q.128 What is calcium rigor?

The stoppage of the heart in systole when a large quantity of calcium ion is infused in experimental animals is known as calcium rigor. It is a reversible phenomenon. When the calcium ions are washed, the heart starts functioning normally.

Q.129 What is the effect of hypocalcemia on heart?

Hypocalcemia (reduction in serum calcium level) reduces the excitability of the cardiac muscle. In ECG, the duration of S–T segment and Q–T interval is prolonged.

Q.130 What do you know about U wave in ECG?

It is rarely seen as a small positive round wave after the T wave. It is due to slow repolarization of papillary muscles. It is more commonly seen in children.

Q.131 What do you mean by left and right axis deviation? From the ECG record how can you assess whether any person is having left or right axis deviation?

If the normal direction of mean QRS vector falls in between –30° and +30°, it is called left axis deviation which represents the horizontal position of heart. Similarly, if it falls in between +75° and +110°, it is known as right axis deviation which also represents vertical position of heart. Clinically axis deviations are made by finding the amplitude of R wave in the bipolar leads as follows:

- If R wave is the tallest in lead II, it is normal electrical axis of heart (+59°).
- If R wave is the tallest in lead I, it is left axis deviation.
- If R wave is the tallest in lead III it is called as right axis deviation.

Q.132 What are the physiological left or right axis deviation? What is the clinical significance of electrical axis of heart?

Physiological left axis deviation is seen:

- During expiration
- When a person lies down
- If the person is stocky and fatty.

Physiological right axis deviation is seen:

- During inspiration
- When a person stands up
- Normally in tall and lanky people.

Clinical significance: Hypertrophy of any ventricles and bundle branch block is indicated from the electrical axis of heart. In patients with hypertrophy of left ventricle and left bundle branch block, left axis deviation is seen whereas in hypertrophy of right ventricles and right bundle branch block patients, right axis deviation takes place.

Q.133 What is the extrasystole or premature contraction?

Sometimes, any part of the heart other than SA node can produce an impulse. This is called an ectopic focus. The ectopic focus produces an extra beat of the heart, which is called extrasystole or premature contraction.

Q.134 What is compensatory pause? What is its cause?

Extrasystole is always followed by a long pause where the heart stops. This temporary stoppage of heart, immediately after extrasystole is known as a compensatory pause. It occurs because the heart has to wait for the arrival of next natural impulse from the pacemaker.

Q.135 What is the difference between 1st degree and 2nd degree heart block?

When all atrial impulses reach the ventricles therefore atrial rate: ventricular rate becomes 1:1 but PR interval becomes longer than 0.2 sec, it is called 1st degree incomplete heart block.

Whereas when all atrial impulses are not conducted to the ventricles producing atrial and ventricular contraction at a rate of either 2:1 or 3:1 ratio with gradual lengthening of PR interval till one ventricular beat is missed, this type of heart block is known as 2nd degree incomplete heart block.

Q.136 What do you mean by Wenckebach phenomenon?

In case of 2nd degree heart block, there is a gradual increase of PR interval until one ventricular beat is missed. This is known as Wenckebach phenomenon.

Q.137 What is 3rd degree heart block?

Complete blockade of conduction of impulse from atria to ventricle is known as third degree or complete heart block.

Q.138 What is the difference between flutter and fibrillation?

S. No.	*Flutter*	*Fibrillation*
1.	This is due to spreading of regular circus movement of impulse through the heart	This is due to spreading of irregular circus movement in many areas of the heart
2.	In this case there is a coordinated contraction of heart	There is an incoordinated contraction of heart
3.	Heart rates are within 200 to 300 beats/min.	Heart rates are more than 300 beats/min.

Q.139 What are the clinical findings of ECG during MI?

- Elevation of ST segments in the leads overlying the area of infarct and
- Depression of ST segment in the reciprocal leads.

Q.140 What do you mean by Stokes-Adams syndrome?

In case of complete heart block, there is some delay before ventricles start beating at their own rate. During this period the systemic blood pressure falls to a very low level and blood supply to brain becomes inadequate. If ventricles do not beat for more than few seconds it causes dizziness and fainting called as Stokes-Adams syndrome.

Q.141 What are the ECG changes during bundle branch block? What changes take place in heart sound production during its bundle branch block?

The ECG changes are as follows:

- Prolonged QRS complex (>0.12 sec)
- Abnormal ST segment and T wave.
- The second heart sound is splited.

Q.142 What types of ECG changes take place in atrial flutter and atrial fibrillation?

In case of atrial flutter following changes are seen:

- Shortening of all time intervals, e.g. PR, TP intervals
- Merger of T wave with P wave of next cardiac cycle
- 2nd degree type (2:1) of heart block.

In case of atrial fibrillation following changes are seen:

- Absence of P wave.
- Appearance of fibrillation (f) waves
- Absence of T wave
- Irregular QRS complex.

Q.143 How does the ECG record change with time after MI?

- Within a few hours after MI: Elevation of ST segment.
- After some days of MI: Elevation of ST segment along with inversion of T wave.
- After several weeks of MI: ST segments return to normal but inversion of T wave is still present along with appearance of Q wave.
- After months and years of MI: T wave becomes normal and Q wave becomes deep.

Q.144 What do you mean by mean circulatory filling pressure and mean systemic filling pressure?

If the heart beat is stopped, the flow of blood every where in the circulation ceases after few seconds resulting in equal pressure within the whole circulation which is known as mean circulatory filling pressure.

Whereas the mean systemic filling pressure is the pressure measured everywhere in the systemic circulation after blood flow is stopped by the clamping of the large blood vessels at the heart. Normally the amount of both are almost equal.

Q.145 Name different types of blood vessels in vascular system with examples of each.

These are as follows:

- Distensible (Windkessel) vessels—aorta, pulmonary artery and their large branches.
- Resistance vessels—arterioles, meta-arterioles
- Exchange vessels—capillaries
- Capacitance vessels—venules and venous compartments
- Shunt vessels—AV anastomoses.

Q.146 What is windkessel effect?

The blood flow through aorta is pulsatile in nature, i.e. it increases during systole and decreases during diastole of the heart. However, the blood flow through other blood vessels becomes uniform and continuous. This is because, during systole, aorta (and to some extent the other larger blood vessels) dilates and later it recoils. This elastic recoiling of aorta causes the continuous blood flow through other blood vessels. Thus, the pulsatile blood flow is converted into continuous flow. This recoiling effect is known as windkessel effect and the blood vessels exerting this effect are called the windkessel vessels.

Q.147 What are the components of vasomotor system?

- Vasomotor center
- Vasoconstrictor fibers
- Vasodilator fibers.

Q.148 Where is vasomotor center situated?

Vasomotor center is situated in the reticular formation of medulla oblongata.

Q.149 What are the components of vasomotor center?

- Vasoconstrictor or pressor area
- Vasodilator or depressor area.
- Sensory area.

Q.150 Name the vasoconstrictor and vasodilator nerve fibers.

Vasoconstrictor fibers are the sympathetic vasoconstrictor fibers.

Vasodilator fibers are:

- Parasympathetic fibers
- Sympathetic cholinergic fibers
- Antidromic nerve fibers.

Q.151 What is the mode of action of sympathetic adrenergic fibers on blood vessels?

Sympathetic adrenergic fibers cause constriction of blood vessels (vasoconstriction) by secreting noradrenaline.

Q.152 What is vasomotor tone?

Vasomotor tone is the continuous discharge of impulses from vasoconstrictor center to arterioles through vasoconstrictor nerve fibers. Vasomotor tone maintains arterial blood pressure by producing constant partial constriction of blood vessels (peripheral resistance). The arterial blood pressure is directly proportional to vasomotor tone.

Q.153 Blood flow to the different body organs can be so effectively regulated by only small changes in the caliber of the arteries. How is it possible?

As resistance to blood flow is inversely proportional to the 4th power of the radius (r) of arterioles, the small changes of radius can cause greater changes of resistance to blood flow and thereby flow to the different body organ.

Q.154 What do you mean by critical closing pressure?

Extravascular tissues exert a small but definite pressure on vessels and when the intraluminal pressure falls below this extravascular pressure the vessel collapses. The pressure at which the flow ceases is called critical closing pressure.

Q.155 State the law of Laplace. What is its functional significance?

It states that the distending pressure (P) in a distensible hollow object is equal at equilibrium to the tension in the wall (T) divided by two principal radii of curvature of object (R_1 and R_2), i.e. $P = T(1/R_1 + 1/R_2)$.

Significance: (i) smaller the radius of the blood vessels lesser the tension in the wall necessary to balance the distending pressure. This is why (i) thin and delicate capillaries are less prone to rupture, (ii) dilated heart has to do more work than normal heart.

Q.156 What is axon reflex?

In response to a firm stroke in the skin the afferent impulses are relayed to the endings near cutaneous arterioles down the branches of sensory nerve to result cutaneous arteriolar dilatation. This neural pathway which does not involve CNS is known as axon reflex.

Q.157 What do you mean by cold blue skin and warm red skin?

Cold blue skin is the skin in which the arterioles are constricted and the capillaries are dilated whereas in warm red skin both arterioles and capillaries are dilated.

Q.158 What is triple response?

A firm and strong stroke on the skin by a blunt object evokes a series of responses which are

- Red reaction
- Flare and
- Wheal.

These responses to the injury are collectively known as triple response.

Q.159 What is the physiological basis of red reaction, flare and wheal?

Red reaction: It is due to the dilatation of precapillary sphincter due to release of histamine and/or bradykinin like vasodilator substances.

Flare: It is due to dilatation of arterioles, terminal arterioles and precapillary sphincter which causes increase in blood flow and thereby irregular erythematous area surrounding the red line.

Wheal: It is due to increased capillary permeability and rise of capillary pressure which ultimately causes local diffuse swelling at and near the site.

Q.160 What is white reaction?

When a pointed object is drawn lightly over the skin the stroke line becomes pale due to draining out of blood from the capillaries and small vein due to contraction of precapillary sphincters.

Q.161 What is the average total peripheral resistance of rest?

It is 1 PRU.

Q.162 On what factors the peripheral resistance does depend.

It depends on the elasticity of vessel wall, diameter of arterioles (inversely), viscosity and velocity of blood directly.

Q.163 Define Poiseuille's law.

It states that resistances to blood flow in a blood vessel proportionately varies with length of blood vessels and viscosity of blood and inversely with 4th power of radius of lumen of vessels.

Q.164 What is circulation time? Give the value of total circulation time.

It is time taken by blood to flow from one site to any other specific site. Normal total circulation time is 12–16 sec.

Q.165 Coronary blood flow fluctuates with each phases of cardiac cycle, explain.

During systole the coronary blood flow is reduced because of compression of coronary vessels due to contraction of cardiac muscle whereas during diastole as cardiac muscle relaxes, there is distention of coronary vessels to its original diameter and thus blood flow through it to heart muscle is increased.

Q.166 Why does the subendocardial portion of left ventricle is more prone for MI?

It is for two reasons as follows:

- No blood flows to this portion during systole because of poor blood supply in this region and also compression of blood vessels during systole.
- Anaerobic respiration goes on in inner layer which increases further under stress.

Q.167 What is the normal time taken for coronary circulation?

It is about 8 sec.

Q.168 What are the factors on which coronary blood flow depends?

These are mainly lumen of coronary vessels, mean aortic pressure and also by cardiac output, HR, body temperature, CO_2 concentration in blood and cardiac sympathetic stimulation.

Q.169 What is normal pulmonary blood flow rate?

It is about 3–5 L/min.

Q.170 What is the normal blood flow rate in liver?

It is about 1500 mL/min.

Q.171 What is the normal coronary blood flow?

It is about 225 mL/min.

Q.172 Give the normal value of cerebral blood flow.

It is approximately 750 mL/min.

Q.173 Define shock. Classify it.

Shock is a syndrome characterized by low cardiac output which is inadequate to maintain normal tissue perfusion. It is of four types—hypovolemic, vasogenic, cardiogenic and obstructive shock.

Q.174 What do you mean by cold shock?

When the amount of fluid in the vascular system is inadequate to fill it, resulting in decrease in circulatory blood volume it is known as hypovolemic or cold shock.

Q.175 What is warm shock?

When the diameter of capacitance vessels is increased by vasodilatation, there is a decrease of cardiac output in spite of normal blood volume. This type of shock is vasogenic shock and in this type of shock as skin becomes warm it is also called warm shock.

Q.176 What is congested shock?

The cardiogenic shock causes congestion of lungs, viscera and that is why it is called congested shock.

Q.177 What is Bezold-Jarish reflex?

The ventricular receptors are sensitive to chemicals or partial occlusion of aorta or coronary artery which are responsible for profound bradycardia, hypotension and apnea. This response is known as 'Coronary chemoreflex' or Bezold Jarish reflex which is clinically associated with **myocardial infarction** or **vasovagal syncope.**

Q.178 What is bradycardia? Where can you see the physiological bradycardia?

Decrease of heart rate below 60 beat/min is known as bradycardia which is physiologically seen in following conditions.

- Athelets
- Males
- Emotional stimuli like shock, grief, depression, etc.
- During expiration.

Q.179 What do you mean by laminar and turbulent flow? How does turbulence produce?

- *Laminar or stream line flow:* It is fixed layer wise, i.e. each layer of blood remains at the same distance from the wall blood vessels flowing through a long vessel and the velocity of blood is maximum in the core of the blood vessels and minimum in its periphery or surface. This type of steady rate of blood flow is known as laminar blood flow.
- *Turbulent blood flow:* When the blood flows crosswise in the vessels by forming whorls in the blood is called **eddy current.** This type of blood flow is known as turbulent blood flow. It is produced by obstruction of vessels or when it takes sharp U turn.

Q.180 What are the signs and symptoms of shock?

Different signs and symptoms manifested during shock are as follows:

- Reduction in arterial blood pressure.
- Reflex tachycardia and reduced stroke volume.
- Decrease in pulse pressure and appearance of thready pulse.
- Reduction in velocity of blood flow producing stagnant hypoxia and cyanosis.
- Pale and cold skin due to reflex vasoconstriction.
- Decreased urinary output due to reduced renal blood flow and GFR.
- Fainting due to reduced blood flow to the brain tissue.
- Feeling of intense thirst if the patient is conscious.
- Rapid and shallow breathing.
- Metabolic acidosis due to excessive production of lactic acid by myocardium.
- Death due to cerebral or cardiac failure.

Q.181 What are the symptoms of left ventricular failure?

These are:

- Difficulty in breathing on exertion.
- Dyspneic attack at night.
- Dyspnea in supine position.

Q.182 What are signs and symptoms of right ventricular failure?

These are:

- Engorgement of right atrium
- Increased venous pressure
- Swelling of liver
- Peritoneal and pleural effusion
- Cyanosis and dyspnea
- Edema.

Q.183 What are the common causes of left ventricular failure?

These are: essential hypertension, coronary insufficiency, myocardial fibrosis and mitral valve incompetence.

Q.184 What are the salient features of left ventricular failure?

These are:

- Decrease in cardiac output with vasoconstriction of peripheral vessels
- Pulmonary edema, dyspnea and anoxia (cardiac asthma).

Q.185 How do you differentiate left and right cardiac failure broadly on the basis of edema?

In left heart failure, pulmonary edema is seen whereas in right heart failure, edema is systemic in nature.

Q.186 Enumerate some of the effects of severe hemorrhage.

These are decreased blood volume, state of shock, increased heart rate, decreased systolic BP, vasoconstriction, hemodilution, rapid and shallow breathing, blurred vision, fainting, etc.

Chapter 16

Respiratory System and Environmental Physiology

Q.1 What is the normal respiratory rate?
12 to 16 per minute.

Q.2 What are the types of respiration?
- External respiration that involves exchange of respiratory gases, i.e. oxygen and carbon dioxide between the alveoli of the lungs and blood
- Internal respiration that involves exchange of respiratory gases between blood and tissues.

Q.3 Define respiratory unit.
Respiratory unit is the terminal portion of respiratory tract where the exchange of gases occurs.

Q.4 Name the structures of respiratory unit.
- Respiratory bronchiole
- Alveolar ducts
- Antrum
- Alveolar sacs
- Alveoli.

Q.5 List the non-respiratory functions of respiratory tract.
- Olfaction
- Vocalization
- Prevention of dust particles
- Defense mechanism
- Maintenance of water balance
- Regulation of body temperature
- Regulation of acid base balance
- Anticoagulant function
- Secretion of angiotensin converting enzyme (ACE)
- Synthesis of hormonal substances.

Q.6 What is the role of lungs in defense mechanism?
- *Lung's own defense:* Secretion of immune factors—defensins and cathelicidins
- *Leukocytes:* Neutrophils and lymphocytes kill the bacteria and virus
- *Macrophages:* Engulf dust particles and pathogens, act as antigen presenting cells; secrete interleukins, tumor necrosis factors and chemokines
- *Mast cell:* Produces hypersensitivity reactions
- *Natural killer cell:* First line of defense against virus
- *Dendritic cells:* Function as antigen presenting cells.

Q.7 What are the characteristic features of pulmonary circulation?
- The wall of pulmonary blood vessels is thin
- These blood vessels are more elastic
- Smooth muscle coat is not well developed in these blood vessels
- True arterioles have less smooth muscle fibers
- Pulmonary capillaries are larger than systemic capillaries.

Q.8 What is the normal pulmonary blood pressure?

Systolic pressure	:	25 mm Hg
Diastolic pressure	:	10 mm Hg
Mean arterial pressure	:	15 mm Hg
Capillary pressure	:	7 mm Hg.

Q.9 Enumerate the factors regulating pulmonary circulation.
- Cardiac output
- Pulmonary vascular resistance
- Nervous factors
- Chemical factors.

Q.10 Name the primary inspiratory and primary expiratory muscles with the nerve supply.
Primary inspiratory muscles:
- Diaphragm—innervated by phrenic nerve
- External intercostal muscles—innervated by intercostal nerves.

Primary expiratory muscles:
Internal intercostal muscles—innervated by intercostal nerves.

Q.11 Name the accessory respiratory muscles.
The accessory inspiratory muscles are sternomastoid, scalene, anterior serrati, elevators of scapulae and pectorals.

The accessory expiratory muscles are abdominal muscles.

Q.12 What are the movements of thoracic cage during inspiration?
Thoracic cage enlarges during inspiration and its size increases in all diameters. Increase in anteroposterior diameter is due to the elevation of upper costal series and the upward and forward movement of sternum. Increase in transverse diameter is due to the elevation of lower costal series. The increase in vertical diameter is due to descent of diaphragm.

Q.13 What is pump handle movement? What is its significance?
During inspiration the upper costal series (second to sixth pair of ribs) are elevated and the sternum moves upward and forward. This type of movement of ribs and sternum is called pump handle movement.
Significance: It increases the anteroposterior diameter of thoracic cage during inspiration.

Q.14 What is bucket handle movement? What is its significance?
During inspiration the central portions (arches) of upper costal series (second to sixth pair of ribs) and lower costal series (seventh to tenth pair of ribs) swing outward and upward. This is called bucket handle movement.
Significance: It increases the transverse diameter of thoracic cage during inspiration.

Q.15 What is the significance of contraction of diaphragm during inspiration?
When the diaphragm contracts, it is flattened. This increases the vertical diameter of thoracic cage during inspiration.

Q.16 What is Dalton's law?
It states that total pressure exerted by a mixture of gases is equal to the sum of the partial pressures of all the gases present within it.

Q.17 What is Henry's law?

It states that if temperature is kept constant, amount of gas dissolved in any solution is directly proportional to the partial pressure of that gas.

Q.18 Give the normal value of intrapulmonary or intra-alveolar pressure.

It is about 760 mm Hg.

Q.19 Why intra-alveolar pressure is equal to that of atmospheric pressure? How is it affected during inspiration and expiration?

It is equal to the atmospheric pressure as during quiet breathing, at the end of expiration and at the end of inspiration, no air is going in and out of the lungs.

During inspiration it decreases 3 mm Hg below its normal value, i.e. 757 mm Hg and during expiration it increases 3 mm Hg above its normal value, i.e. 763 mm Hg.

Q.20 What is Valsalva maneuver and Muller's maneuver?

Forced expiration against a closed glottis may produce positive intrapulmonary pressure of > 100 mm Hg above the atmospheric value. This voluntary act is known as Valsalva maneuver.

Forced inspiration against closed glottis can reduce the intrapulmonary pressure to < 80 mm Hg below the atmospheric value. This voluntary act to reduce the intrapulmonary pressure is known as Muller's maneuver.

Q.21 What is collapsing tendency of lungs?

The constant threat of compression of the lungs is called collapsing tendency of lungs.

Q.22 What are the factors causing collapsing tendency of lungs?

- Elastic property of lung tissues that induces the recoiling tendency of lungs
- Surface tension exerted by the alveolar fluid.

Q.23 What are the factors preventing the collapsing tendency of lungs?

- Intrapleural pressure that overcomes elastic recoiling tendency of lungs
- Surfactant that overcomes surface tension.

Q.24 What is surfactant? Name the cells secreting surfactant.

Surfactant is the lipoprotein substance that reduces the surface tension induced by the fluid lining in the alveoli.

It is secreted by type II alveolar epithelial cells of lungs and Clara cells situated in bronchioles.

Q.25 What is the function of surfactant?

Surfactant prevents collapsing tendency of lungs by reducing the surface tension in the alveoli.

Q.26 What is respiratory distress syndrome or hyaline membrane disease?

It is the condition in infants with collapse of lungs due to the absence of surfactant. In adults it is called adult respiratory distress syndrome (ARDS).

Q.27 Define and give normal values of intrapleural or intrathoracic pressure.

The intrapleural or intrathoracic pressure is the pressure existing in the pleural cavity.

It is always negative. During inspiration it is – 6 mm Hg and during expiration it is – 2 mm Hg.

Q.28 What is the cause for negative intrapleural pressure?

The intrapleural pressure is negative because of constant pumping of fluid (secreted by visceral layer of pleura) from the intrapleural space into lymphatic vessels.

Q.29 What is the significance of intrapleural pressure?

The intrapleural pressure prevents collapsing tendency of lungs. It is also responsible for respiratory pump that increases venous return.

Q.30 How is intrapleural pressure measured?

By using intraesophageal balloon.

Q.31 Define and give normal values of intraalveolar or intrapulmonary pressure.

The intraalveolar or intrapulmonary pressure is the pressure existing in the alveoli of lungs.

During inspiration it is – 4 mm Hg
During expiration it is + 4 mm Hg

Q.32 What is the significance of intraalveolar pressure?

- It causes flow of air into alveoli during inspiration and out of alveoli during expiration
- It helps in exchange of gases between alveoli and blood.

Q.33 What is transpulmonary pressure?

Transpulmonary pressure is the difference between the intraalveolar pressure and intrapleural pressure.

Q.34 What is compliance?

The expansibility of lungs and thorax is known as compliance. It is defined as change in volume per unit change in pressure.

Q.35 Define compliance in relation to intraalveolar pressure and give normal value.

In relation to intraalveolar pressure, compliance is defined as the volume increase in lungs per unit increase in intraalveolar pressure.

Compliance of lungs and thorax = 130 mL/cm H_2O. Compliance of lungs alone = 220 mL/cm H_2O.

Q.36 Define compliance in relation to intrapleural pressure and give normal value.

In relation to intrapleural pressure, compliance is defined as the volume increase in lungs per unit decrease in the intrapleural pressure.

Compliance of lungs and thorax = 100 mL/cm H_2O. Compliance of lungs alone = 200 mL/cm H_2O.

Q.37 Define work of breathing.

The work done by respiratory muscles during breathing to overcome the resistance in thorax and respiratory tract is known as work of breathing.

Q.38 What are the types of resistance for which energy is utilized during work of breathing?

- Airway resistance—that is overcome by airway resistance work.
- Elastic resistance of lungs and thorax—that is overcome by compliance work.
- Nonelastic viscous resistance—that is overcome by tissue resistance work.

Q.39 Define and give normal values of lung volumes.

- *Tidal volume:* The volume of air breathed in and out of lungs in a single normal quiet breathing.
  Normal value: 500 mL.
- *Inspiratory reserve volume:* The additional amount of air that can be inspired forcefully beyond normal tidal volume.
  Normal value: 3300 mL.
- *Expiratory reserve volume:* The additional amount of air that can be expired forcefully after normal expiration.
  Normal value: 1000 mL.
- *Residual volume:* The amount of air remaining in the lungs even after forced expiration.
  Normal value: 1200 mL.
  Figure 16.1 illustrates spirogram showing lung volumes and capacities

Q.40 What is lung capacity? Define and give normal values of lung capacities.

Two or more lung volumes together are called lung capacity (Fig. 16.1).

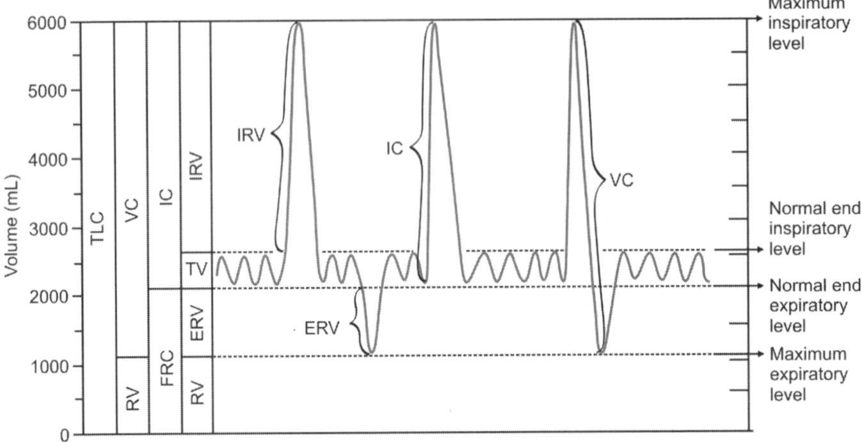

Fig. 16.1: Spirogram. TV = Tidal volume, IRV = Inspiratory reserve volume, ERV = Expiratory reserve volume, RV = Residual volume, IC = Inspiratory capacity, FRC = Functional residual capacity, VC = Vital capacity, TLC = Total lung capacity

Lung capacities:

- *Inspiratory capacity:* The maximum volume of air that can be inspired from the end expiratory position. It includes tidal volume and inspiratory reserve volume.
  Normal value: 3800 mL.
- *Vital capacity:* The maximum volume of air that can be expelled out forcefully after a maximal (deep) inspiration. It includes inspiratory volume, tidal volume and expiratory reserve volume.
  Normal value: 4800 mL.
- *Functional residual capacity:* The volume of air remaining in the lungs after normal expiration (after tidal expiration).
  It includes expiratory reserve volume and residual volume.
  Normal value: 2200 mL.
- *Total lung capacity:* The amount of air present in the lungs after a maximal (deep) inspiration. It includes all the four lung volumes i.e., inspiratory reserve volume, tidal volume, expiratory reserve volume and residual volume.
  Normal value: 6000 mL.

Q.41 Why the 'wheeze' sound is heard during expiration but not in inspiration of an asthma patient?

During inspiration the intrapleural and mediastinal negativity rises and as a result the bronchial diameter increases. Reverse occurs during expiration. Therefore, resistance to airflow is normally low in inspiration and high in expiration. This is why in bronchial asthma inspiration may not be difficult but expiration becomes difficult. This explains why the "wheeze" in bronchial asthma is heard during expiration but not in inspiration.

Q.42 What is the significance of residual volume?

- It helps in the exchange of gases in between breathing and during expiration
- It maintains the contour of the lungs.

Q.43 What are the instruments used to measure lung volumes and lung capacities?

- Spirometer
- Respirometer.

Q.44 Name the lung volumes and capacities, which can not be measured by spirometer.

- Residual volume
- Functional residual capacity
- Total lung capacity.

Q.45 How are residual volume and functional residual capacity measured?

- Helium dilution technique
- Nitrogen washout method.

Q.46 Define vital capacity. What is its importance?

It is the maximum volume of air which can be expired by forceful effort after a maximal inspiration.

It provides useful information about the strength of respiratory muscles and also provides useful information about other aspects of pulmonary function through FEV_1.

Q.47 In which posture VC is highest and why?

In standing posture it is the highest as in standing position diaphragm descends down thereby increasing intrathoracic volume. This increases intra-alveolar volume during inspiration.

Q.48 In whom the vital capacity is more?

- Heavily built persons
- Athletes
- People playing musical wind instruments like bugle.

Q.49 Name the pathological conditions when vital capacity is reduced.

- Asthma
- Emphysema
- Weakness or paralysis of respiratory muscle
- Congestion of lungs
- Pneumonia
- Pneumothorax
- Hemothorax
- Pyothorax
- Hydrothorax
- Pulmonary edema
- Pulmonary tuberculosis.

Q.50 Why does VC decrease during pregnancy?

During pregnancy diaphragm is pushed up by the growing fetus resulting in decrease of intrathoracic volume and thereby decrease of capacity to inspire air and there by VC is decreased.

Q.51 What is respiratory minute volume (RMV)? Give its normal value.

Respiratory minute volume is the amount of air that is breathed in and out of lungs during each minute. It is the product of tidal volume and respiratory rate.

Normal value: 6000 mL (500 mL × 12).

Q.52 What is maximum breathing capacity (MBC) or maximum ventilation volume (MVV)? What is its normal value?

It is the maximum amount of air that can be breathed in and out of lungs by forceful respiration (hyperventilation).

Normal value:

In healthy adult males – 150 to 170 liters/minute

In females – 80 to 100 liters/minute.

Q.53 What is forced expiratory volume (FEV) or timed vital capacity?

The amount of air that can be expired forcefully (after deep inspiration) in a given unit of time is called forced expiratory volume (FEV) or timed vital capacity (Fig. 16.2).

Q.54 What is FEV_1?

The amount of air that can be expired forcefully after deep inspiration in the first second is called FEV_1 (1 stands for 'first second').

Q.55 Give the normal values of FEV_1, FEV_2 and FEV_3.

FEV_1 = 83%; FEV_2 = 94%; FEV_3 = 97%.

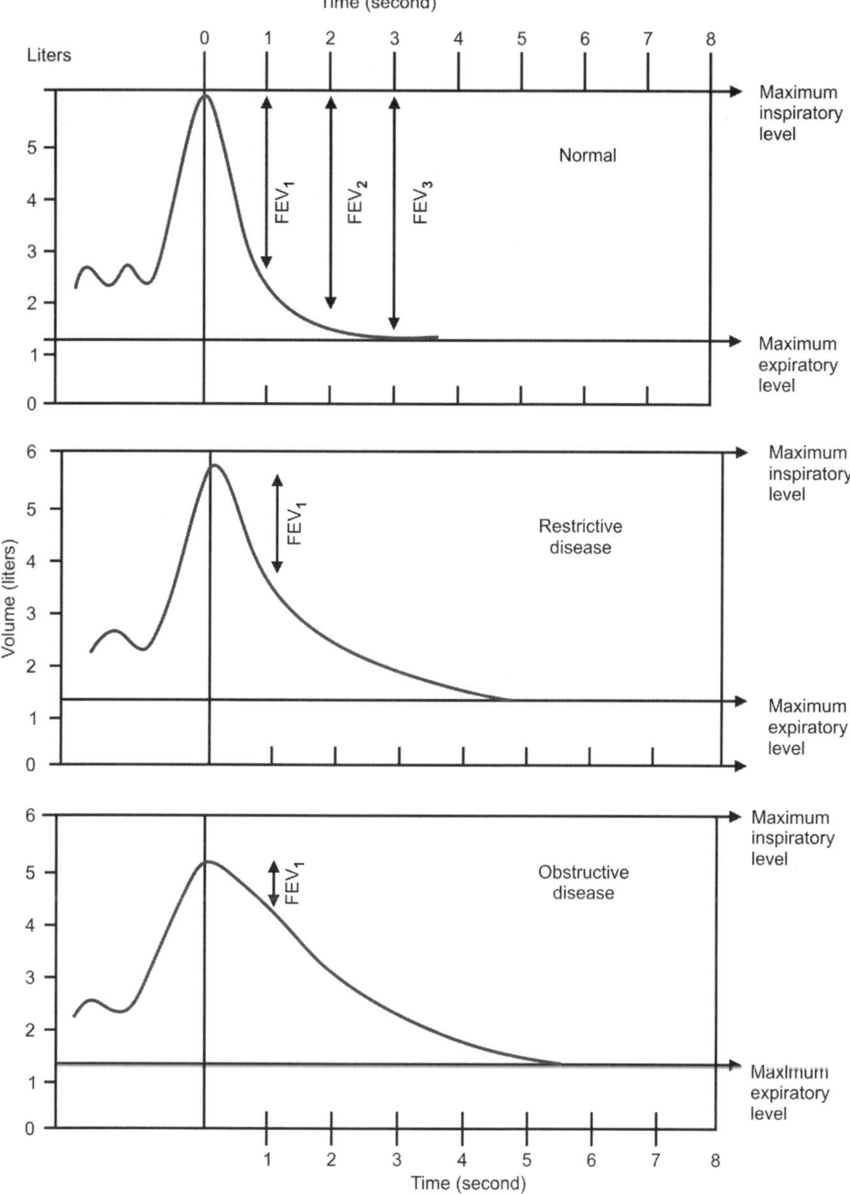

Fig. 16.2: Forced expiratory volume

Q.56 What is the significance of determining FEV?

Vital capacity may be almost normal in some of the respiratory diseases. However determination of FEV has greater diagnostic value, as it is decreased significantly in some respiratory disorders, particularly in obstructive diseases like asthma and emphysema.

Q.57 Define and give normal value of peak expiratory flow rate (PEFR).

The maximum rate at which air can be expired after deep inspiration is known as peak expiratory flow rate (PEFR).

Normal value: About 400 liters/minute.

Q.58 How is PEFR measured?

By using Wright's peak flow meter or mini peak flow meter.

Q.59 What is the significance of measuring PEFR?

Measurement of PEFR is useful in assessing the respiratory diseases, especially to differentiate the obstructive and restrictive diseases. It is about 200 liters/minute in restrictive diseases and it is only 100 liters/minute in obstructive diseases. It is valuable when measured serially to establish the pattern of airway obstructive disease and to monitor its responses in treatments, especially asthma.

Q.60 What is pulmonary ventilation? Give its normal value.

Pulmonary ventilation is the cyclic process by which fresh air enters the lungs and an equal volume of air is expired. It is defined as the amount of air breathed in and out of lungs in one minute. It is the product of tidal volume and respiratory rate. It is otherwise known as respiratory minute volume.

Normal value: 6000 mL/minute.

Q.61 What is alveolar ventilation? Give its normal value.

Alveolar ventilation is the amount of air utilized for gaseous exchange every minute.

Alveolar ventilation = (Tidal volume – Dead space volume) × Respiratory rate.

Normal value: 4200 mL.

Q.62 What is dead space? Give normal value.

The part of respiratory tract where the gaseous exchange does not occur is known as dead space. The air present in the dead space is called dead space air.

Normal value: 150 mL.

Q.63 What are the types of dead space?

- Anatomical dead space, which includes the volume of respiratory tract from nose up to terminal bronchiole.
- Physiological dead space which includes anatomical dead space and two additional volumes:
  - The volume of air in those alveoli, which are not functioning
  - The amount of air in those alveoli, which do not receive adequate blood flow.

Q.64 Why the physiological dead space is equal to anatomical dead space in normal conditions?

Because all the alveoli of both lungs are functioning and all the alveoli receive adequate blood supply in normal conditions.

Q.65 How is dead space measured?

By single breath nitrogen washout method.

Q.66 What is ventilation perfusion ratio? Give its normal value.

It is the ratio of alveolar ventilation (VA) and the amount of blood (Q) flowing through the lungs.

Ventilation perfusion ratio = VA/Q = 4200/5000.

Normal value: About 0.84.

Q.67 What are the differences between inspired air and alveolar air?

- Oxygen content is more in inspired air than in alveolar air
- Carbon dioxide is less in inspired air than in alveolar air
- Inspired air is dry whereas alveolar air is humid.

Table 16.1: Composition of inspired air, alveolar air and expired air

	Inspired air	*Alveolar air*	*Expired air*
Oxygen	20.84 mL% (159 mm Hg)	13.60 mL% (104 mm Hg)	15.70 mL% (120 mm Hg)
Carbon dioxide	0.04 mL% (0.30 mm Hg)	5.30 mL% (40 mm Hg)	3.60 mL% (27 mm Hg)
Nitrogen	78.62 mL% (596.90 mm Hg)	74.90 mL% (596 mm Hg)	74.50 mL% (566 mm Hg)
Water vapor	0.50 mL% (3.80 mm Hg)	6.20 mL% (47 mm Hg)	6.20 mL% (47 mm Hg)

Value in parenthesis is the partial pressure.

Q.68 What is the composition of inspired air (atmospheric air), alveolar air and expired air?

The composition of inspired air (atmospheric air), alveolar air and expired air is tabulated in Table 16.1.

Q.69 How is alveolar air collected?

By using Haldane-Priestley tube.

Q.70 How is inspired air collected?

Since the inspired air is the atmospheric air, it can be drawn from the atmosphere through the syringe.

Q.71 How is expired air collected?

By using Douglas bag.

Q.72 What is respiratory membrane?

The alveolar membrane and the capillary membrane in the lungs through which diffusion of gases takes place are together called respiratory membrane (Fig. 16.3).

Q.73 What are the layers of respiratory membrane?

From within outside:

- Surfactant
- Fluid lining the alveoli
- Alveolar epithelial cells
- Interstitial layer
- Basement membrane
- Capillary endothelial cells.

Q.74 What is diffusing capacity?

Diffusing capacity is the volume of gas that diffuses through respiratory membrane each minute for a pressure gradient of 1 mm Hg.

Q.75 Mention the diffusing capacity for oxygen and carbon dioxide.

Diffusing capacity for oxygen is 21 mL/minute/mm Hg and for carbon dioxide it is 400 mL/minute/mm Hg. Thus, the diffusing capacity for carbon dioxide is about 20 times more than that of oxygen.

Q.76 What are the factors affecting the diffusing capacity?

Diffusing capacity is directly proportional to

- Pressure gradient of gases between alveoli and blood in pulmonary capillary
- Solubility of gas in fluid medium
- Total surface areas of respiratory membrane.

Diffusing capacity is inversely proportional to:

- Molecular weight of the gas
- Thickness of respiratory membrane.

Q.77 State Hook's law in relation to lung.

Length is directly proportional to force within a physiological limit.

Q.78 Define lung compliance. What is 'hysteresis' curve of lung compliance?

The change of lung volume per unit change in airway pressure is called lung compliance.

In compliance curve, at identical intrapleural pressure, the volume of lung is less in inspiratory phase than in the expiratory phase. This different pressure volume relationship curve during inspiration and expiration is known as 'hysteresis 'curve as represented by Figure 16.4.

Q.79 What is specific compliance? What is its advantage to use?

The compliance when expressed as a function of FRC is known as specific compliance.

In individuals with one lung only, lung compliance is approximately half of the normal even if the normal distensibility of normal lung is present. Similarly in children compliance is lower than normal though the distensibility of lung remains normal. This fallacy is removed with specific compliance since FRC is proportionately reduced and specific compliance remains essentially constant.

Q.80 What is the oxygen content and partial pressure of oxygen (PO_2) in the blood?

Arterial blood:

Oxygen content = 19 mL%

PO_2 = 95 mm Hg

Venous blood:

Oxygen content = 14 mL%

PO_2 = 40 mm Hg.

Q.81 What is the carbon dioxide content and partial pressure of carbon dioxide (PCO_2) in the blood?

Arterial blood : Carbon dioxide content = 48 mL%

PCO_2 = 40 mm Hg

Venous blood : Carbon dioxide content = 52 mL%

PCO_2 = 45 mm Hg.

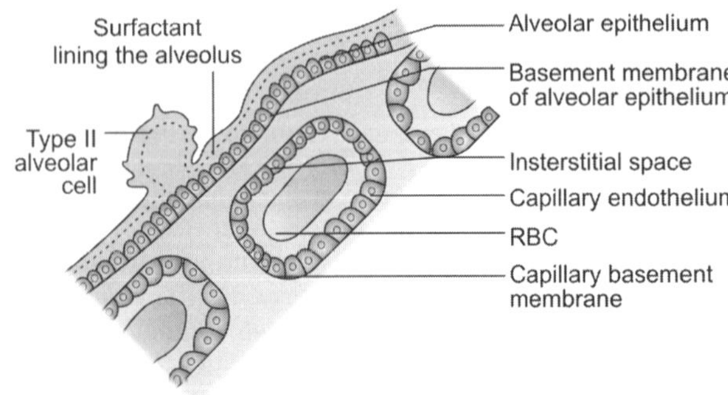

Fig. 16.3: Respiratory membrane

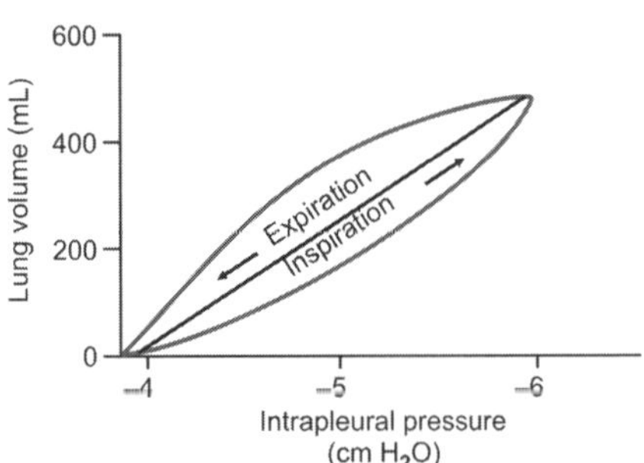

Fig. 16.4: Compliance curve of lungs

Q.82 What is coefficient of utilization?
The percent of blood that gives up its O_2 as it passes through the tissue capillaries is called as the coefficient of utilization. At rest it is about 25 percent and during heavy exercise it increases up to 75 percent.

Q.83 In which form CO_2 transported in blood?
Mainly in three forms:
- In dissolved form in plasma and RBC—0.3 mL%
- As bicarbonate form of Na^+ and K^+—3 mL%
- As carbamino compound form—0.7 mL%

Q.84 What is the CO_2 content and partial pressure of CO_2 in arterial and venous blood?

CO_2 content	PCO_2
Arterial blood—48 mL%	40 mm Hg
Venous blood—52 mL%	46 mm Hg

Q.85 In which form the venous CO_2 is mostly found?
In bicarbonate form.

Q.86 What are the effects of CO_2 addition to blood?
It causes increase in plasma bicarbonate ion, decrease in plasma chlorides and increase in RBC chlorides.

Q.87 What do you mean by maximum venous point and arterial point?
In deoxygenated blood with maximum PCO_2, 60-67 mm Hg, CO_2 content is 65 mL% called as the maximum venous point as represented by Figure 16.5. In oxygenated blood at PCO_2 40 mm Hg, CO_2 content is 48mL% called as the arterial point as represented by Figure 16.5.

Q.88 What do you mean by physiological CO_2 dissociation curve?
If we join maximum 'venous point' and 'arterial point' which corresponds to extreme CO_2 level in the body respectively, it will roughly reflect changes between PCO_2 and CO_2 content in the blood and called the physiological CO_2 dissociation as represented by curve C of Figure 16.5.

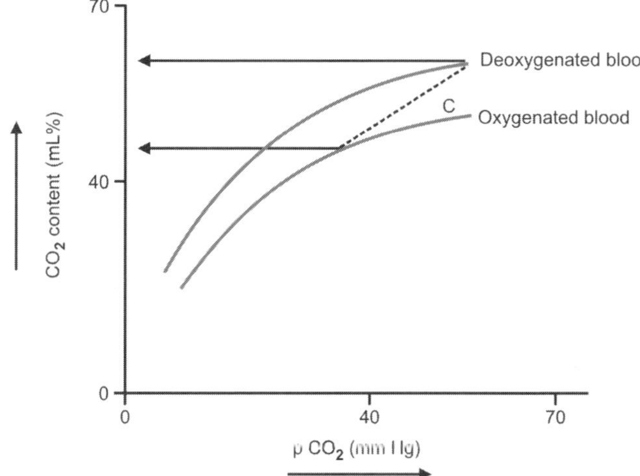

Fig. 16.5: CO_2 dissociation curve in whole blood

Q.89 What are the factors affecting CO_2 dissociation curve?
These are:
- Increase in body temperature shifts the curve to the left, i.e. at increased body temperature larger amount of CO_2 can be taken by the blood at a given PCO_2.
- Decrease in PO_2 shifts the curve to the left and there by helps in loading of CO_2 in blood.

Q.90 What is respiratory exchange ratio? Give its normal value.
It is the ratio between the amount of oxygen consumed (uptake) and the amount of carbon dioxide given out by the tissues.

It is 1.00 if only carbohydrate is utilized, 0.70 if only fat is utilized and 0.8 if only protein is utilized.

Q.91 How is oxygen transported by blood?
- As physical solution
- In combination with hemoglobin.

Q.92 What is the oxygen carrying capacity of hemoglobin and blood?
Oxygen carrying capacity of hemoglobin is 1.34 mL/g of hemoglobin. The oxygen carrying capacity of blood is 19 mL/100 mL of blood when the hemoglobin content in blood is 15 g%.

The oxygen carrying capacity of blood is only 19 mL% because the hemoglobin in the blood is saturated with oxygen only for about 95%.

Q.93 What is oxygen hemoglobin dissociation curve? What is its normal shape?
It is the curve that demonstrates the relationship between the partial pressure of oxygen and percentage saturation of hemoglobin with oxygen.

Normally, it is 'S' shaped or sigmoid-shaped (Fig. 16.6).

Q.94 Why this curve is sigmoid?
A Hb molecule contains 4 atoms of Fe^{++} each of which combines with O_2 in varied affinity. The combination of 1st heme in the hemoglobin molecule with O_2 increases the affinity of the 2nd heme for O_2 and oxygenation of 2nd heme increases the affinity of the 3rd and so on. This shifting of affinity of Hb for O_2 produces sigmoid shape.

Q.95 What is the significance of the sigmoid shape of O_2 dissociation curve?
- O_2 dissociation curve has the plateau above 60 mm Hg. This flat upper part indicates that even if the PO_2 increases from 60 mm Hg to 300 mm Hg, the O_2 content of the blood will not vary significantly. Similarly the effect of O_2 lack on the body will not be manifested until the PO_2 goes down below 60 mm Hg.
- The steep slope of the curve indicates that the slight decrease of PO_2 will cause greater release of O_2 from hemoglobin.

Q.96 What is the O_2 content in arterial and venous blood?
Arterial blood—19 mL%; venous blood—14 mL%.

Q.97 What is the partial pressure of O_2 in arterial and venous blood?
Arterial blood—100 mm Hg; Venous blood- 40 mm Hg.

Q.98 In which form O_2 is carried from lungs to tissues and in what amount?
- In dissolved form in plasma and RBC—0.3 mL%
- In oxyhemoglobin form—18.7 mL%

Q.99 What do you mean by O_2 carrying capacity of blood?
It is the O_2 carrying capacity of the total hemoglobin of blood. If the Hb content of a person is 16 gm% then his O_2 carrying capacity will be 16 × 1.34 mL (each gram Hb carry 1.34 mL O_2), i.e. 21 mL per deciliter of blood.

Q.100 What is the difference between O_2 content and O_2 capacity?
The O_2 content refers to the amount of O_2 actually present in a given sample of

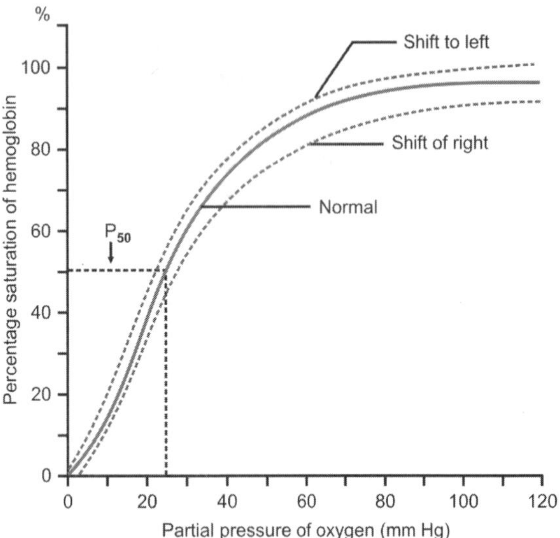

Fig. 16.6: Oxygen hemoglobin dissociation curve

blood where as O_2 capacity refers to the total amount of O_2 that can be carried by blood when the hemoglobin is fully saturated with O_2.

Q.101 What is the indication of shift to the right of oxygen dissociation curve? Name some factors causing it.

Shift to the right of oxygen dissociation curve indicates the dissociation or release of oxygen from hemoglobin.

It is caused by:

- Decrease in partial pressure of oxygen in blood
- Increase in partial pressure of carbon dioxide
- Increase in hydrogen ion concentration and decrease in pH (acidity)
- Increase in body temperature
- Excess of 2, 3 DPG (2,3, diphosphoglycerate).

Q.102 What is the indication of shift to the left in O_2 dissociation curve? When does it occur?

Shift to the left of oxygen dissociation curve indicates the acceptance (association or retention) of more amount of oxygen by hemoglobin.

It occurs:

- In fetal blood since fetal blood has more affinity for O_2 than the adult blood
- When hydrogen ion concentration in the blood decreases causing increase in pH (alkalinity).

Q.103 What is P_{50}?

The partial pressure of oxygen at which the hemoglobin saturation is 50% is called P_{50}. It is 25 mm Hg.

Q.104 What is Bohr's effect?

The presence of carbon dioxide decreases the affinity of hemoglobin for oxygen and enhances further release of oxygen to the tissues and oxygen dissociation curve is shifted to right. This is Bohr's effect.

Q.105 How is carbon dioxide transported in the blood?

- As physical solution
- As carbonic acid
- As bicarbonate
- As carbamino compounds.

Q.106 Name the method by which maximum amount of carbon dioxide is transported in the blood.

As bicarbonate (about 63%).

Q.107 What is chloride shift?

The negatively charged bicarbonate ions formed in the red blood cells diffuse out into the plasma. To maintain the electrolyte equilibrium, the negatively charged chloride ions move into the cells from plasma. This is known as chloride shift (Fig. 16.7).

Q.108 What is reverse chloride shift?

When the blood reaches the alveoli of lungs, the bicarbonate ions diffuse into the red blood cells from plasma. To maintain electrolyte equilibrium, chloride ions move out of the cells into the plasma. This is known as reverse chloride shift.

Q.109 What is carbon dioxide dissociation curve?

The curve that demonstrates the relationship between the partial pressure of carbon dioxide and the amount of carbon dioxide combined with blood is called the carbon dioxide dissociation curve.

Q.110 What is Haldane's effect? What is its cause?

Excess of oxygen content in the blood displaces carbon dioxide from hemoglobin and shifts the carbon dioxide dissociation curve to right. This is called Haldane's effect. This is because, when more amount of oxygen combines, the hemoglobin becomes acidic. The highly acidic hemoglobin causes the displacement of carbon dioxide from hemoglobin.

Q.111 Name the mechanisms involved in the regulation of respiration.

- Nervous mechanism
- Chemical mechanism.

Q.112 What are the respiratory centers?

- Two medullary centers situated in medulla oblongata:
  - Inspiratory center or dorsal group of neurons
  - Expiratory center or ventral group of neurons.

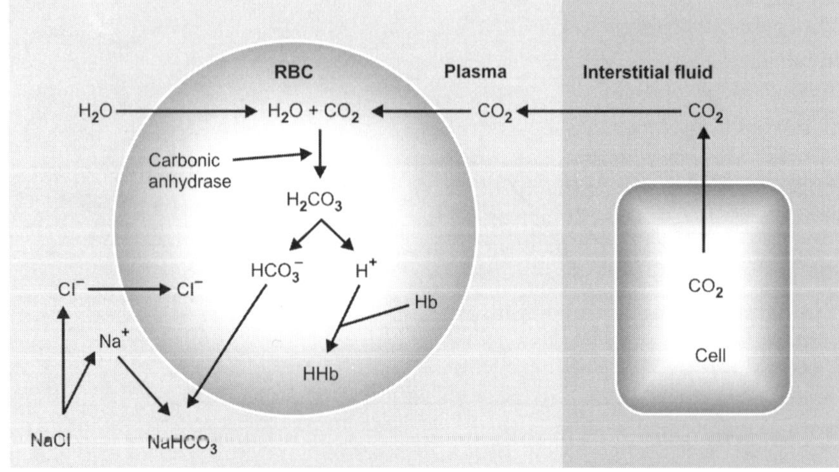

Fig. 16.7: Transport of carbon dioxide in blood in the form of bicarbonate and chloride shift

- Two pontine centers situated in pons:
  - Pneumotaxic center
  - Apneustic center.

Q.113 Mention the functions of each respiratory center.
Inspiratory center is concerned with inspiration. Expiratory center is concerned with expiration.

Expiratory center is inactive during quiet breathing and becomes active during forced breathing or when inspiratory center is inhibited.

Apneustic center increases inspiration by activating the inspiratory center.

Pneumotaxic center decreases inspiration by inhibiting apneustic center. By inhibiting the apneustic center, it reduces the duration of inspiration and thereby increases the rate of respiration.

Q.114 What is inspiratory ramp?
Normally, the discharge of impulses from inspiratory center is not uniform. To start with, the amplitude of action potential (impulse) is low because of activation of only few neurons. Later, when more and more neurons are activated, the amplitude increases gradually in a ramp fashion. The impulses are produced for 2 seconds during which inspiration occurs. This type of firing from inspiratory center is called inspiratory ramp.

Q.115 What is the significance of inspiratory ramp signals?
Significance of inspiratory ramp signals is that there is a slow and steady inspiration so that, the filling of lungs with air is also steady.

Q.116 What are the higher centers which alter the respiration by acting on the respiratory centers?
Anterior cingulated gyrus, genu of corpus callosum, olfactory tubercle and posterior orbital gyrus of cerebral cortex inhibit respiration. Motor area and Sylvian area of cerebral cortex facilitate breathing.

Q.117 What is apneusis? How it can be resulted?
It is the arrest of respiration in inspiratory phase. It can be experimentally resulted by transection in mid pons along with sectioning of vagus nerves.

Q.118 What are the various types of receptors in the lungs which alter the respiration?
- Stretch receptors present in the wall of bronchi and bronchioles of lungs
- 'J' receptors or juxta capillary receptors situated in the wall of alveoli near the capillaries
- Irritant receptors present in the wall of bronchi and bronchioles.

Q.119 What is the function of stretch receptors present in lungs?
Stretch receptors present in lungs prevent overstretching of lungs by producing Hering-Breuer reflex.

Q.120 What is Hering-Breuer reflex? What is its significance?
Stretching of lungs during inspiration stimulates the stretch receptors in the lungs. The stretch receptors in turn send inhibitory impulses to inspiratory center. So, inspiration stops and expiration starts. This is called Hering-Breuer reflex. This reflex is a protective reflex, because it restricts inspiration and prevents over stretching of the lungs.

Q.121 What is the function of 'J' receptors?
Role of 'J' receptors in physiological conditions is not known clearly. However, these receptors are responsible for hyperventilation in patients affected by pulmonary congestion and left heart failure.

Q.122 What is the function of irritant receptors?
When harmful chemical agents like ammonia and sulfur dioxide enter the lungs, the irritant receptors are stimulated. The stimulation of irritant receptors results in reflex hyperventilation and bronchospasm so that further entry of harmful agents into the lungs is prevented.

Q.123 What is the effect of stimulation of baroreceptors on respiration?
When arterial blood pressure increases, the baroreceptors are activated and send inhibitory impulses to respiratory centers. So, the respiration is inhibited.

Q.124 What is the effect of stimulation of proprioceptors on respiration?
During exercise, the proprioceptors situated in muscles, tendons and joints are stimulated and send impulses to cerebral cortex. Cerebral cortex in turn, activates the respiratory centers causing hyperventilation.

Q.125 What is the effect of stimulation of cold receptors (thermoreceptors) on respiration?
When body is exposed to cold, the cold receptors are activated and send impulses to cerebral cortex. Cerebral cortex in turn, activates the respiratory centers causing hyperventilation.

Q.126 What is the effect of stimulation of pain receptors on respiration?
Whenever pain receptors are stimulated, the impulses from them are sent to cerebral cortex via somatic afferent fibers. Cerebral cortex in turn, activates the respiratory centers causing hyperventilation.

Q.127 What are chemoreceptors?
Chemoreceptors are the receptors, which give response to change in chemical constituents of blood such as O_2, CO_2 and H^+.

Q.128 Classify chemoreceptors.
Depending upon the situation, the chemoreceptors are classified into two types:
- Central chemoreceptors situated in medulla oblongata near the inspiratory center and having close contact with blood and cerebrospinal fluid
- Peripheral chemoreceptors present in the carotid body and aortic body.

Q.129 Explain the function of central chemoreceptors briefly.
The activation of central chemoreceptors causes stimulation of inspiratory center resulting in increased rate and force of respiration. The main stimulant for central chemoreceptors is the increased hydrogen ion concentration. However, if the hydrogen ion concentration increases in blood, it cannot stimulate the central chemoreceptors because, the hydrogen ions cannot cross the blood-brain barrier. But, if the carbon dioxide increases in blood, it can cross the blood-brain barrier and enter interstitial fluid of brain or the cerebrospinal fluid. There, it combines with water forming carbonic acid that immediately dissociates into hydrogen ion and bicarbonate ion. Now, the hydrogen ions stimulate the central chemoreceptors causing increase in rate and force of respiration.

Q.130 Explain the function of peripheral chemoreceptors.
The main stimulant for peripheral chemoreceptors is reduction in partial pressure of oxygen (hypoxia). When partial pressure of oxygen decreases, the peripheral chemo-receptors are stimulated and send stimulatory impulses to inspiratory center. This causes increase in rate and force of respiration (Fig. 16.8).

Q.131 What is pulmonary chemoreflex?
Injection of veratridine or nicotine like alkaloid substances into pulmonary capillaries stimulate chemoreceptors present in pulmonary vessels producing bradycardia, hypotension and apnea

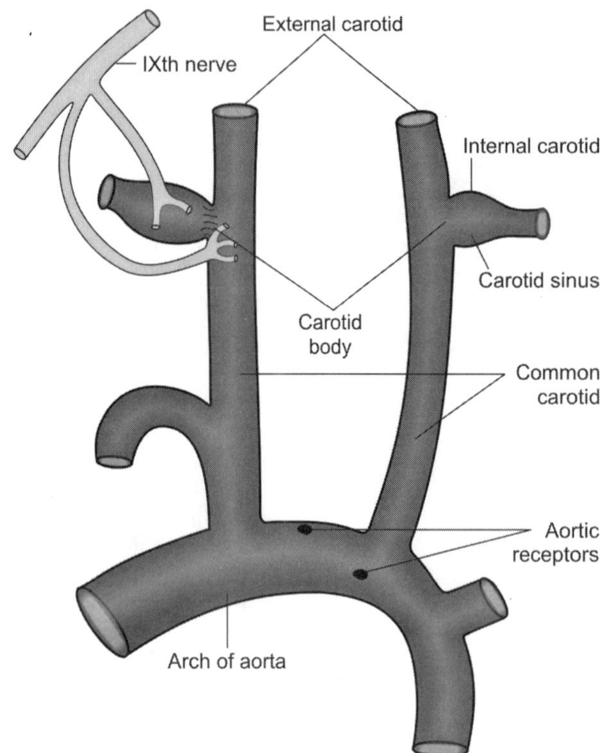

Fig. 16.8: Peripheral chemoreceptors (diagrammatic)

followed by tachycardia. This response is called pulmonary chemoreflex.

Q.132 What do you mean by CO_2 narcosis?

The accumulation of CO_2 in the body depresses the CNS, including respiratory centers and also produces headache, confusion, dizziness, apnea and eventually coma. This ill effect of excess CO_2 in the body is referred as CO_2 narcosis.

Q.133 What are the types of respiratory diseases?

- Obstructive diseases like asthma and emphysema
- Restrictive diseases like pneumothorax and pneumonia.

Q.134 What is the difference between obstructive and restrictive diseases of the lungs?

The difference between obstructive and restrictive lung diseases is tabulated in Table 16.2.

Q.135 Define the following.

Eupnea: The normal respiration

Tachypnea: Increase in the rate of respiration.

Bradypnea: Reduction in the rate of respiration.

Polypnea: Rapid shallow breathing resembling panting in dogs; the rate of respiration is increased significantly but the force is not increased significantly.

Hyperpnea: Highly significant increase in pulmonary ventilation due to increase in rate and force of respiration with more increase in rate.

Q.136 Define apnea. Name the conditions when apnea occurs.

Apnea is defined as temporary cessation of breathing.

It occurs:

- By voluntary effort—voluntary apnea or breath holding
- After hyperventilation
- During pharyngeal stage of deglutition—deglutition apnea
- During vagal stimulation—vagal apnea
- After adrenaline injection—adrenaline apnea.

Q.137 What is apnea time or breath holding time?

The time during which a person can voluntarily stop breathing is known as apnea time or breath holding time. It is about 40 to 60 seconds in a normal person.

Q.138 What is breaking point? What is its cause?

At the end of voluntary apnea, the person is forced to breathe. The time when the person is forced to breathe is called breaking point.

It is due to accumulation of CO_2.

Q.139 What is hyperventilation? When does it occur?

The increased pulmonary ventilation is known as hyperventilation. During this, both rate and force of respiration are increased.

Hyperventilation occurs in conditions like exercise when the partial pressure of carbon dioxide increases. It can also be produced voluntarily—voluntary hyperventilation.

Q.140 What are the effects of hyperventilation?

Carbon dioxide is washed out during hyperventilation leading to reduction in the partial pressure of carbon dioxide in blood. This causes suppression of respiratory centers resulting in apnea. Apnea is followed by Cheyne-Stokes breathing. After a period of Cheyne-Stokes breathing, normal respiration is restored.

Q.141 What is hypoventilation? When does it occur?

The decreased pulmonary ventilation due to reduction in rate and force of respiration is called hypoventilation.

It occurs in the following conditions:

- During the suppression of respiratory centers.
- After administration of some drugs.
- Due to partial paralysis of respiratory muscles.

Table 16.2: Differences in obstructive and restrictive lung diseases

Parameter	*Restrictive lung diseases*	*Obstructive lung diseases*
Effect on peak expiratory Flow rate	Reduction in peak expiratory flow rate is less in comparison to the obstructive diseases	Reduction peak expiratory flow rate is more in restrictive diseases
Effect on FEV_1 reduced	FEV_1 is only slightly	FEV_1 is very much reduced
Disease example	Polio, pneumonia, pleural effusion	Asthma, chronic bronchitis, emphysema

Q.142 What are the effects of hypoventilation?

Hypoventilation causes hypoxia and hypercapnea. So, there is increase in rate and force of respiration leading to dyspnea. Severe hypoventilation leads to lethargy, coma and death.

Q.143 Define hypoxia.

Hypoxia is defined as reduced availability of oxygen to the tissues of the body.

Q.144 Why the term hypoxia is preferred than anoxia?

Anoxia means the absence of oxygen. Since, there is no possibility for total absence of oxygen in living conditions, the term hypoxia is preferred.

Q.145 Classify hypoxia.

- Hypoxic hypoxia
- Anemic hypoxia
- Stagnant hypoxia
- Histotoxic hypoxia.

Characteristic features of different types of hypoxia are shown in Table 16.3.

Q.146 Explain hypoxic hypoxia briefly. Name some important causes for it.

Hypoxic hypoxia or arterial hypoxia means the decreased oxygen content in the blood and it is characterized by reduced partial pressure of oxygen. Oxygen carrying capacity of blood, rate of blood flow and utilization of oxygen are normal.

It is caused by:

- Low oxygen tension in inspired air (in atmosphere)
- Respiratory disorders
- Cardiac disorder.

Q.147 Explain anemic hypoxia briefly. Name some important causes for it.

Inability of the blood to carry enough amount of oxygen is known as anemic hypoxia. It is characterized by reduced oxygen carrying capacity of blood. Partial pressure of oxygen, rate of blood flow and utilization of oxygen are normal.

Any condition that leads to anemia will cause anemic hypoxia such as:

- Decreased red blood cell count
- Decreased hemoglobin content
- Presence of altered hemoglobin
- Combination of hemoglobin with gases other than oxygen and carbon dioxide (like carbon monoxide).

Q.148 Explain stagnant hypoxia briefly. Name some important causes for it.

Hypoxia due to decreased velocity of blood flow is known as stagnant hypoxia. It is characterized by reduced rate of blood flow. Partial pressure of oxygen, oxygen carrying capacity of blood, and utilization of oxygen are normal.

It is caused by:

- Congestive cardiac failure
- Hemorrhage
- Surgical shock
- Vasospasm
- Thrombosis
- Embolism.

Q.149 Explain histotoxic hypoxia briefly. Name some important causes for it.

Inability of tissue to utilize oxygen is called histotoxic hypoxia. It is characterized by reduced utilization of oxygen. Partial pressure of oxygen, oxygen carrying capacity of blood and rate of blood flow are normal.

It is caused by destruction of cellular oxidative enzymes and complete paralysis of cytochrome oxidase system due to cyanide or sulfate poisoning.

Q.150 What do you mean by O_2 poisoning.

Inhalation of O_2 in high O_2 pressure that occurs when O_2 is breathed at a very high alveolar oxygen pressure like in Caisson may result seizures followed by coma in most people. The other symptoms include nausea, muscle twitching, dizziness, disturbances of vision, irritability, etc. This phenomenon is called O_2 poisoning.

Q.151 What are the effects of severe acute hypoxia?

Severe acute hypoxia causes unconsciousness. If it is not treated immediately brain death occurs.

Q.152 What are the effects of chronic hypoxia?

- Red blood cell count increases due to release of erythropoietin from kidney
- Initially, the rate and force of contraction of heart are increased. Later, the rate and force of contraction of heart are decreased. There is reduction in cardiac output and blood pressure also
- Initially, the rate of respiratory rate is increased. Then, respiration becomes shallow and periodic. Finally, the rate and force of respiration are decreased
- Loss of appetite, nausea, vomiting and thirst occur.
- Urine becomes alkaline
- In mild hypoxia, symptoms of alcoholic intoxication like depression, apathy, and loss of self control occur
- The subject starts shouting, singing and crying. There is loss of orientation, discriminative ability, power of judgment and memory.

Q.153 How is hypoxia treated?

Hypoxia is treated by oxygen therapy.

Q.154 What is the efficacy of oxygen therapy in different type of hypoxia?

Oxygen therapy is not equally effective in all types of hypoxia.

Hypoxic hypoxia: Oxygen therapy is 100% useful

Anemic hypoxia: Oxygen therapy is moderately useful, i.e. about 70%

Stagnant hypoxia: Oxygen therapy is less than 50% useful

Histotoxic hypoxia: Oxygen therapy is of no use at all.

Table 16.3: Characteristic features of different types of hypoxia

S. No.	*Features*	*Hypoxic hypoxia*	*Anemic hypoxia*	*Stagnant hypoxia*	*Histotoxic hypoxia*
1.	PO_2 in arterial blood	Reduced	Normal	Normal	Normal
2.	O_2 carrying capacity of blood	Normal	Reduced	Normal	Normal
3.	Velocity of blood flow	Normal	Normal	Reduced	Normal
4.	Utilization of O_2 by tissues	Normal	Normal	Normal	Reduced
5.	Efficacy of O_2 therapy	100%	75%	> 50%	Not useful

Q.155 What is hypercapnea? When does it occur?

Increased carbon dioxide content in the blood is known as hypercapnea.

It occurs in conditions leading to asphyxia and breathing air containing more amount of carbon dioxide.

Q.156 What are the effects of hypercapnea?

- *Respiration:* Respiratory centers are stimulated leading to dyspnea
- *Blood:* pH of blood is reduced

- *Cardiovascular system:* Heart rate and blood pressure are increased. There is flushing of skin due to peripheral vasodilatation
- *Central nervous system:* Headache, depression, laziness, rigidity, fine tremors, generalized convulsions, giddiness and loss of consciousness occur.

Q.157 What is hypocapnea? When does it occur?

Decreased carbon dioxide content in the blood is known as hypocapnea.

It occurs in conditions associated with hypoventilation and prolonged hyperventilation.

Q.158 What are the effects of hypocapnea?

- *Respiration:* Respiratory centers are depressed. Respiratory alkalosis occurs
- *Blood:* pH of blood is increased
- *Central nervous system:* Dizziness, mental confusion, muscular twitching and loss of consciousness occur.

Q.159 What is asphyxia? When does it occur?

Asphyxia is the condition characterized by combination of hypoxia and hypercapnea due to obstruction of air passage.

It occurs in conditions like strangulation and drowning.

Q.160 What are the stages of asphyxia?

- Stage of hyperpnea
- Stage of convulsions
- Stage of collapse.

Q.161 What is dyspnea or air hunger?

The difficulty in breathing is called dyspnea or air hunger. It is defined as the consciousness of necessity for increased respiratory effort.

Q.162 What is dyspnea point?

Dyspnea point is the increased level of ventilation (increased rate and force of respiration) at which the difficulty in breathing becomes severe.

Q.163 Name the physiological and pathological conditions when dyspnea occurs.

Physiological condition: Severe muscular exercise.

Pathological conditions:

- Respiratory disorders like hindrance to respiratory movements and obstruction of respiratory tract
- Cardiac disorders like left ventricular failure and mitral stenosis
- Metabolic disorders like diabetic acidosis, uremia and increased hydrogen ion concentration.

Q.164 What is dyspneic index? What is the level of dyspneic index at which dyspnea occurs?

Dyspneic index is the index between breathing reserve and maximum breathing capacity. Breathing reserve is the difference between maximum breathing capacity (MBC) and respiratory minute volume (RMV).

$$\text{Dyspneic index} = \frac{\text{MBC} - \text{RMV}}{\text{MBC}} \times 100$$

Dyspnea occurs when the dyspneic index is reduced below 60%.

Q.165 Define periodic breathing. Mention the types of periodic breathing.

The abnormal or uneven respiratory rhythm is called periodic breathing (Fig. 16.9).

It is of two types:

- Cheyne-Stokes breathing
- Biot's breathing.

Q.166 What is Cheyne-Stokes breathing (Fig. 16.10)?

Cheyne-Stokes breathing is a type of periodic breathing characterized by two alternate periods namely, hyperpneic period and apneic period. During hyperpneic period, at the beginning, breathing is shallow. The force of respiration increases gradually and reaches the maximum. Then, it decreases gradually and reaches the minimum. This is called waxing and waning. When the force of respiration reaches the minimum apnea occurs. Then hyperpnea occurs and the cycle is repeated.

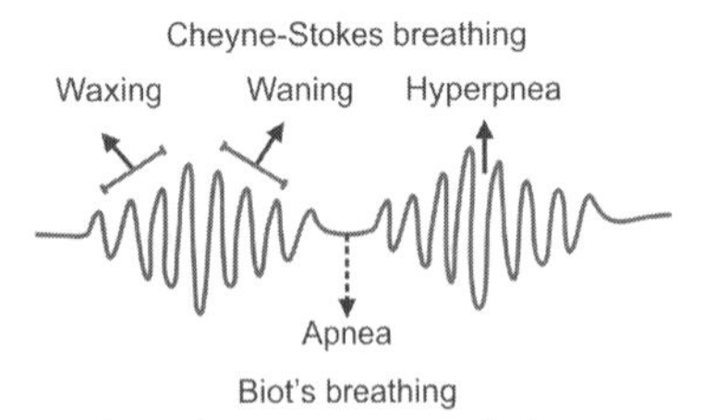

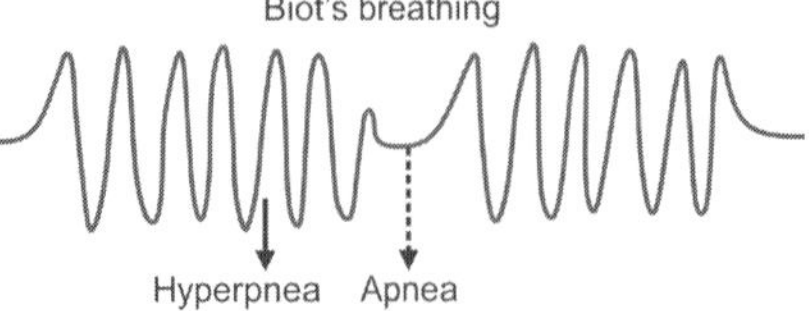

Fig. 16.9: Periodic breathing

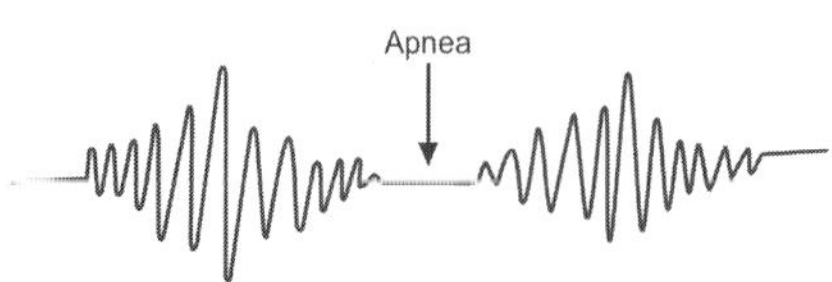

Fig. 16.10: Cheyne-Stokes breathing

Q.167 What are the causes for waxing and waning during Cheyne-Stokes breathing?

During forced breathing, excess of carbon dioxide is washed out of blood. When the carbon dioxide tension becomes very low, the respiratory centers become inactive and apnea occurs. During apnea, carbon dioxide is accumulated and oxygen tension is decreased. So, the respiratory centers are stimulated leading to gradual increase in force of breathing.

Q.168 What are the conditions when Cheyne-Stokes breathing occurs?

Physiological conditions:

- Sleep
- High altitude
- After prolonged hyperventilation
- During hibernation in animals
- In newborn babies
- After severe muscular exercise.

Pathological conditions:

- Increased intracranial pressure
- Cardiac failure
- Uremia
- Narcotic poisoning
- In premature infants.

Q.169 What is Biot's breathing?

Biot's breathing is a type of periodic breathing characterized by two alternate periods namely, period of apnea and period of hyperpnea. There is no waxing and waning. After apneic period hyperpnea occurs abruptly.

Q.170 What are the conditions when Biot's breathing occurs?

Biot's breathing occurs only in pathological conditions.

It occurs in nervous disorders due to lesion or injury to brain.

Q.171 Define cyanosis. What is its cause?

Cyanosis is defined as the diffused bluish discoloration of skin and mucus membrane.

It is due to the presence of large amount of reduced hemoglobin in blood. At least 5 gm% of reduced hemoglobin must be present to cause cyanosis.

Q.172 What are the areas of the body where cyanosis is seen markedly?

Though cyanosis is distributed all over the body, it is more marked in areas where the skin is thin like lips, cheeks, ear lobes, nose and fingertips above the base of nail.

Q.173 What are the conditions when cyanosis occurs?

Cyanosis occurs in:

- Arterial and stagnant hypoxia
- When altered hemoglobin is formed

- During sluggishness of blood flow as in the case of polycythemia.

Q.174 Why cyanosis does not occur in anemia?

Cyanosis usually occurs only when the amount of reduced hemoglobin is more than 5–7 gm% but in anemia the hemoglobin content itself is less. So, cyanosis cannot occur in anemia.

Q.175 Compare the central and peripheral cyanosis?

See Table 16.4

Table 16.4: Comparison between central and peripheral cyanosis

S. No.	*Central cyanosis*	*Peripheral cyanosis*
1.	It is due to hypoxic hypoxia.	It is due to stagnant hypoxia.
2.	Extremities become warmer due to increase blood flow to the tissue and hypertension which reflexly produce vasodilatation.	Extremities become cooler due to decrease tissueblood flow and hypotension.

Q.176 What do you mean by Caisson's disease?

It is the condition caused by sudden release of pressure if Caisson under water is suddenly brought out. This results in abdominal pain, disturbance of vital center in CNS and even sudden collapse of a person present within the Caisson.

Q.177 What do you mean by N_2 narcosis?

If the body is exposed to high atmospheric pressure, because of high N_2 pressure larger amount of N_2 will enter into lungs and thereby in body fluids ultimately causing euphoria, impairment of mental function and symptoms of alcoholic intoxication. These effects of N_2 in higher pressure is called N_2 narcosis.

Q.178 Define oxygen toxicity (poisoning). What is its cause?

Increased oxygen content in the tissues is called oxygen toxicity.

It is because of breathing pure oxygen at atmospheric pressure for long time or breathing pure oxygen at high pressure.

Q.179 What is the effect of breathing pure oxygen at atmospheric pressure for long time?

Breathing pure oxygen at atmospheric pressure for more than 12 hours leads to poisoning of lung tissues and pulmonary edema. The other tissues are not affected very much because of the hemoglobin oxygen buffer system.

Q.180 What is hyperbaric oxygen?

Pure oxygen at a high pressure of about 1500 mm Hg is known as hyperbaric oxygen.

Q.181 What are the effects of breathing hyperbaric oxygen?

Breathing hyperbaric oxygen poisons the lung tissues first and causes pulmonary edema. Afterwards, the failure of hemoglobin buffer system occurs. So, the other organs like brain are affected because of increased metabolic rate, production of excess of heat, destruction of cellular enzymes and damage of tissues. When brain is affected, hyperirritability occurs. It is followed by convulsions, coma and death.

Q.182 What is acute mountain sickness?

In some instances, the compensatory mechanism to high altitude breaks down and gives rise to serious symptoms known as Monge's disease or acute mountain sickness characterized by:

- Considerable increase of red cell mass and PCV.
- High pulmonary arterial pressure.
- Right heart failure in some cases.

Q.183 What is chronic mountain sickness?

It is the disease occurring in case of failure of long-term acclimatization process to the residents of high altitude. The signs and symptoms are:

- Extreme polycythemia
- Increase in viscosity of blood that results in fall of blood flow
- Increase in BP
- Cyanosis, fatigue, exercise intolerance
- Pulmonary edema.

Q.184 What are the acclimatization to the natives of high landers?

The acclimatization that occurs in the residents who are residing in the high altitude permanently for generations after generations, are as follows:

- Short body stature and large sized chest that results in high ratio of ventilatory capacity to body mass.
- Hypertrophy of right heart.
- Polycythemia
- Shifting of O_2 dissociation curve to right
- Increase in size of carotid bodies.

Q.185 What do you mean by CPR?

If the respiration fails along with stoppage of heart beat this procedure is followed till the person is not hospitalized for proper treatment. The procedures are:

- Cleaning of the airways
- Mouth-to-mouth breathing at the rate of 16–18/min
- External cardiac massage by pressing lower border of sternum by 4–5 cm at the rate of 80–90 times/min.
- After every 15 cardiac massage two mouth-to-mouth (15:2) breathing (if two subjects are present.

After every 5 cardiac massage 1 mouth-to-mouth (5:1) breathing (if one subject is available).

Q.186 What do you mean by Kussmaul breathing?

During some clinical conditions like diabetic coma there is rapid and deep breathing eliminating CO_2 and bicarbonate. This type of rapid and shallow breathing is known as *Kussmaul breathing*.

Q.187 What are the sources of carbon monoxide?

- Exhaust of gasoline engines
- Coal mines
- Gases from guns
- Deep wells
- Underground drainage system.

Q.188 What is the difference between the affinity of hemoglobin for carbon monoxide and oxygen?

Hemoglobin has got 200 times more affinity for carbon monoxide than for oxygen.

Q.189 What are the toxic effects of carbon monoxide?

- Carbon monoxide combines with hemoglobin and forms carboxy hemoglobin. This cannot take up oxygen so, anemic hypoxia occurs. The presence of carboxy hemoglobin decreases release of oxygen from hemoglobin and the oxygen dissociation curve shifts to left
- Carbon monoxide destroys the cytochrome system in the cells.

Q.190 What are the effects of carbon monoxide poisoning?

Breathing air with 1% carbon monoxide causes headache and nausea. Breathing air with more than 1% carbon monoxide causes loss of consciousness. When the percentage of carbon monoxide in the air is high, death occurs.

Q.191 What is atelectasis? What are its causes?

Atelectasis means collapse of a part or whole of lung.

Causes:

- Deficiency of surfactant
- Obstruction of bronchus or bronchiole
- Presence of air (pneumothorax), fluid (hydrothorax), blood (hemothorax) or pus (pyothorax) in pleural space.

Q.192 What are the effects of atelectasis?
- Decrease in the partial pressure of oxygen
- Dyspnea.

Q.193 Define pneumonia. What are its causes?
Pneumonia is the inflammation of lung tissues followed by accumulation of blood cells, fibrin and exudates in alveoli leading to consolidation of affected part of the lung.
Causes:
- Bacterial or viral infection
- Inhaling noxious chemical agents.

Q.194 What is delirium?
The extreme mental state due to cerebral hypoxia is called delirium.

Q.195 What are the features of delirium?
- Confused mental state
- Illusion
- Hallucination
- Disorientation
- Hyperexcitability
- Loss of memory.

Q.196 What is bronchial asthma?
Bronchial asthma is the respiratory disease characterized by difficulty in breathing with wheezing.

Q.197 What is wheezing? What is it due to?
Wheezing means whistling type of respiration noticed in bronchial asthma. It is marked during expiration.
It is due to obstruction of air passage by:
- Bronchiolar constriction
- Edema of mucus membrane in bronchioles
- Accumulation of mucus.

Q.198 What are the effects of bronchial asthma on respiratory system?
- Increase in residual volume and functional residual capacity
- Reduction in tidal volume, vital capacity, FEV_1, alveolar ventilation and partial pressure of oxygen in blood
- Acidosis
- Dyspnea
- Cyanosis.

Q.199 What is pleural effusion?
Accumulation of large amount of fluid in pleural cavity is called pleural effusion.

Q.200 What is emphysema?
Emphysema is an obstructive respiratory disease in which lung tissue especially alveolar membrane is damaged.

Q.201 What are the various factors affecting the body at high altitude?
- Hypoxia
- Expansion of gases
- Reduced atmospheric temperature
- Light rays.

Q.202 Why does hypoxia develop at high altitude?
Because of low atmospheric pressure in high altitude, the partial pressure of oxygen is reduced causing hypoxia.

Q.203 What are the effects of hypoxia at high altitude?
Refers Question Nos 151 and 152 of this Chapter for answer.

Q.204 What is mountain sickness?
Mountain sickness is the condition characterized by ill effects of hypoxia at high altitude. It is common in persons going to high altitude for the first time.

Q.205 What are the symptoms of mountain sickness?
- *Digestive system:* Loss of appetite, nausea and vomiting
- *Respiratory system:* Breathlessness caused by pulmonary edema due to hypoxia
- *Nervous system:* Headache, depression, disorientation, irritability, lack of sleep, weakness and fatigue.

Q.206 What is acclimatization?
The adaptation or the adjustment of the body to high altitude is known as acclimatization.

Q.207 What are the important changes in the body during acclimatization?
- *Blood:* Increase in red blood cell count, hemoglobin content and oxygen carrying capacity of blood
- *Cardiovascular system:* Increase in blood flow to vital organs like heart, brain and muscles due to increased heart rate and cardiac output
- *Respiration:* Increase in rate and force of respiration, pulmonary ventilation, pulmonary blood flow, diffusing capacity of gases in alveoli and uptake of oxygen in blood
- *Tissues:* Increase in the quantity of oxidative enzyme necessary for metabolism.

Q.208 Define decompression sickness. What are its other names?
Decompression sickness is the disorder that occurs when a person returns rapidly to normal surroundings (atmospheric pressure) after staying for a long time in a place with high atmospheric pressure like deep sea.
Other names of this disease:
- Compressed air sickness
- Caisson sickness
- Bends
- Divers palsy.

Q.209 Explain the cause for decompression sickness briefly.
High barometric pressure at deep sea compresses the gases causing reduction in the volume of the gases. Oxygen is utilized and carbon dioxide is expired. But, since nitrogen is an inert gas it is neither utilized nor expired. So, after compression it escapes from blood and gets dissolved in fat of the tissues and tissue fluid. When the person ascends rapidly to atmospheric pressure, nitrogen is decompressed and escapes from tissues in the form of bubbles. The bubbles obstruct the blood flow producing the embolism and decompression sickness.

Q.210 What are the symptoms of decompression sickness?
- Severe pain, numbness and itching
- Temporary paralysis and muscle cramps
- Occlusion of coronary artery and coronary ischemia
- Damage of brain tissue or spinal cord due to obstruction of blood flow
- Dizziness, shortness of breath and choking
- Unconsciousness and death.

Q.211 How is decompression sickness prevented?
While ascending from deep sea, ascent should be very slow with short stay at regular intervals. The person affected by decompression sickness is treated by recompression first and then he is brought slowly to atmospheric pressure.

Q.212 What is nitrogen narcosis? When does it occur?
Nitrogen narcosis is the unconsciousness or stupor (lethargy with suppression of sensations and feelings) produced by nitrogen. Its effects are similar to alcoholic intoxication.

It occurs in persons like deep sea divers or underwater tunnel workers who breathe pressurized air under high pressure.

Q.213 What is SCUBA?
SCUBA or self contained underwater breathing apparatus is the apparatus used

by deep sea divers and the underwater tunnel workers to prevent the ill effects of increased barometric pressure in deep sea or tunnels.

Q.214 What are the effects of sudden exposure of the body to cold?

When body is exposed to cold, large amount of heat is produced by increased metabolic activities and shivering. When the body is exposed to severe cold, the temperature regulating mechanism fails causing frostbite. And sleep or coma occurs.

Q.215 What is frostbite?

Freezing of surface of the body due to exposure to severe cold is known as frostbite. It is common in ear lobes and digits of hands and feet.

Q.216 What are the effects of exposure of the body to heat?

- Heat exhaustion
- Dehydration
- Heat cramps
- Heat stroke.

Q.217 What is heat stroke?

When body temperature increases above 41°C (106°F) during exposure to severe heat, some severe symptoms occur which are together called heat stroke.

Q.218 What are the effects of heat stroke?

The effects of heat stroke are dizziness, abdominal pain and unconsciousness. If not treated immediately, damage of brain tissue occurs resulting in death.

Q.219 What is sunstroke?

Prolonged exposure of the body to sun during summer in desert or tropical areas leads to a condition similar to heat stroke. This is called sunstroke.

Q.220 What are the conditions when artificial respiration is required?

Artificial respiration is required whenever there is arrest of breathing without cardiac failure. Arrest of breathing occurs during:

- Accidents
- Drowning
- Gas poisoning
- Electric shock
- Anesthesia.

Q.221 What are the methods of artificial respiration?

- Manual methods
- Mechanical methods.

Q.222 Name the manual methods of artificial respiration.

- Mouth-to-mouth breathing method
- Holger-Nielsen (back pressure arm lift) method.

Q.223 Name the mechanical methods of artificial respiration.

- Drinker's method
- Ventilator method.

Q.224 What are the effects of exercise on respiratory system?

- Increase in pulmonary ventilation
- Increase in diffusing capacity of oxygen
- Increase in the amount of oxygen consumption.

Q.225 What is VO_2 max? Give values.

Amount of oxygen consumed under maximal aerobic metabolism is called VO_2 max. It is the product of cardiac output and maximal amount of oxygen consumed by the muscles.

During resting condition VO_2 max is 35 to 40 mL/kg body weight/minute in males and 30 to 35 mL/kg body weight/minute in females. During exercise, it is increased by 50%.

Chapter 17

Nervous System

Q.1 What are the divisions of nervous system?

- Central nervous system (CNS) that includes brain and spinal cord.
- Peripheral nervous system (PNS) that includes:
  - Somatic nervous system that is concerned with movements
  - Autonomic nervous system (ANS) that is concerned with visceral functions.

Q.2 What are the parts of the brain?

- Prosencephalon (fore forebrain) that is divided into:
  - Telencephalon which includes two cerebral hemispheres
  - Diencephalon which includes thalamus, hypothalamus, metathalamus and subthalamus.
- Mesencephalon (midbrain).
- Rhombencephalon (hindbrain) that is divided into:
  - Metencephalon which includes pons and cerebellum
  - Myelencephalon or medulla oblongata.

Q.3 What are the parts of brainstem?

- Midbrain
- Pons
- Medulla oblongata.

Q.4 Define neuron or nerve cell.

Neuron or nerve cell is defined as the structural and functional unit of the nervous system.

Q.5 Classify the neurons.

Neurons are classified by three different methods:

- Depending upon number of poles:
  - Unipolar neurons
  - Bipolar neurons
  - Multipolar neurons.
- Depending upon the function:
  - Motor neurons
  - Sensory neurons.
- Depending upon length of axon:
  - Golgi type I neurons
  - Golgi type II neurons.

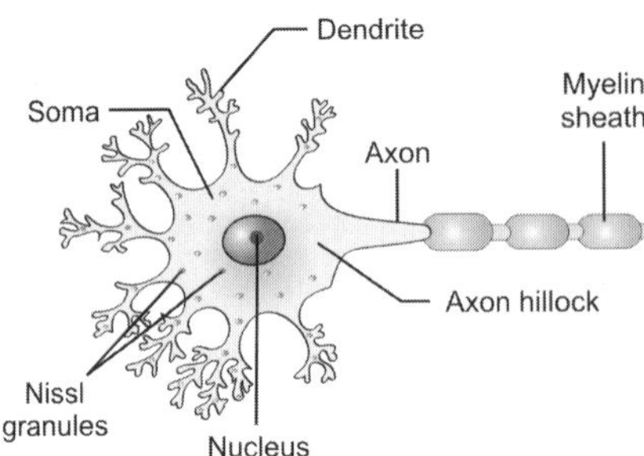

Fig. 17.1: Structure of neuron

Q.6 Name the parts of a neuron.

The parts of a neuron are (Fig. 17.1):

- Nerve cell body or soma
- Dendrite
- Axon.

Q.7 What are the important structures present in nerve cell body of the neuron?

Nucleus, Nissl bodies, neurofibrils, mitochondria and Golgi apparatus.

Q.8 What are Nissl bodies? What is their function?

Nissl bodies are the small granules present throughout the soma of neuron and dendrites but not in axon hillock and axon.

These bodies are responsible for the tigroid or spotted appearance of soma. Nissl bodies contain ribosomes and are concerned with synthesis of proteins in the neuron.

Q.9 What are the processes of neuron?

- Dendrite—the short process that carries the impulses towards the cell body.
- Axon—the long process that carries the impulses away from the cell body.

Q.10 Mention the number of axon and dendrite in each neuron.

Each neuron has only one axon. The dendrite may be absent or present. If present, it may be one or many in number.

Q.11 What is axis cylinder?

The axoplasm and the axolemma that covers the axon are together called axis cylinder.

Q.12 What is myelin sheath?

Myelin sheath is a thick tubular sheath covering the axis cylinder.

Q.13 What is node of Ranvier? And what is internode?

Myelin sheath is not continuous around the axon and it is absent at regular intervals. The area where the myelin sheath is absent is known as node of Ranvier. The segment of axon between the two nodes is called internode.

Q.14 What are the functions of myelin sheath?

- It is responsible for faster rate of conduction of impulses through nerve fiber. In myelinated nerve fiber, the impulses are conducted by means of saltatory conduction
- It has a high insulating capacity. Because of this it restricts the nerve impulse within the single nerve fiber and prevents stimulation of neighboring nerve fibers.

Q.15 What is myelinogenesis?

Formation of myelin sheath is called myelinogenesis.

Q.16 What are the Schwann cells? What is their function?

Schwann cells are a type of cells present in neurilemma close to axolemma. These cells are responsible for the development of the myelin sheath.

Q.17 What is neurilemma? What is its function?

Neurilemma (neurilemmal sheath or Schwann sheath) is the thin membrane that forms the outer most covering of the nerve fibers. It contains Schwann cells and so it is essential for myelinogenesis.

Q.18 Classify the nerve fibers.

Nerve fibers are classified by six different methods:

- Depending upon the structure:
  - Myelinated nerve fibers
  - Nonmyelinated nerve fibers.

- Depending upon distribution:
  - Somatic nerve fibers
  - Autonomic nerve fibers.
- Depending upon source of origin:
  - Cranial nerve fibers
  - Spinal nerve fibers.
- Depending upon the functions:
  - Motor nerve fibers
  - Sensory nerve fibers.
- Depending upon neurotransmitter secreted by them:
  - Adrenergic nerve fibers
  - Cholinergic nerve fibers.
- Depending upon the diameter and rate of conduction of impulse:
  - Type A fibers
  - Type B fibers
  - Type C fibers.

Type A fibers are again divided into A alpha, A beta, A gamma and A delta nerve fibers.

Q.19 Name the nerve fibers conducting the impulse with maximum and minimum velocity.

Type A alpha nerve fibers conduct the impulse with maximum velocity (70 to 120 meters/second).

Type C fibers conduct the impulse with minimum velocity (0.5 to 2 meters/second).

Q.20 Distinguish between Aδ and C fibers.

See Table 17.1.

Q.21 Name the properties of nerve fibers.

- Excitability
- Conductivity
- Refractory period
- Summation
- Adaptation
- Infatigability
- All or none law.

Q.22 What are the two types of potentials noticed in nerve fibers?

- Action potential (nerve impulse)—produced when the nerve is stimulated with adequate strength of stimulus (threshold or minimal stimulus). It is propagated and nongraded.
- Electrotonic potential or local response—is produced when the strength of stimulus is not adequate (subthreshold or subminimal stimulus). It is nonpropagated and graded.

Q.23 How much is the resting membrane potential in a nerve fiber?

About—70 mV.

Q.24 Differentiate between EPSP and AP.

See Table 17.2.

Q.25 What are the properties of generator potential?

The properties of GP are:

- It is non-propagatory in nature
- It is monophasic
- It does not obey all or none law.

Q.26 What do you mean by spatial and temporal summation?

- Simultaneous stimulation of two afferent nerves by a stimulus of subthreshold intensity can evoke action potential in motor neuron. This property is known as spatial summation.
- Whereas if subminimal stimuli are repeated at short intervals in a single nerve, reflex action can also be evoked which is known as temporal summation.

Q.27 What is the fractionation phenomenon?

Direct stimulation of motor nerve results in more response than reflex response or in other words the tension developed reflexly is always a fraction of response that is produced by direct motor nerve stimulation. This is known as fractionation phenomenon.

Q.28 What do you mean by law of forward conduction?

Synapse permits the conduction of impulse from presynaptic to postsynaptic neuron only, i.e. unidirectionally. This property is known as law of forward conduction.

Q.29 Name the receptors responsible for following sensations—touch, pressure, hot, cold and pain.

- *Touch*—Markel's disc or Meissner's corpuscle.
- *Pressure*—Pacinian corpuscle.
- *Hot*—Ruffini's end organs.
- *Cold*—Krause's end bulb.
- *Pain*—Free nerve endings.

Q.30 What is saltatory conduction?

In a myelinated nerve fiber, the action potential (nerve impulse) jumps from one node of Ranvier to another node of Ranvier, making the velocity of conduction faster. This type of conduction in a myelinated nerve fiber is called saltatory conduction.

Q.31 Explain the mechanism of saltatory conduction briefly.

Myelin sheath is not permeable to ions. So during the conduction of action potential, the entry of sodium ions from extracellular fluid into nerve fiber occurs only at the node of Ranvier, where the myelin sheath is absent. This causes depolarization only in successive node and not in internode. So, the action potential jumps from one node to another. Hence, it is called saltatory conduction (saltare = jumping).

Q.32 Why is the nerve fiber not fatigued?

Nerve fiber is not fatigued because it can conduct only one action potential

Table 17.1: Features of Aδ and C fibers

S. No.	Aδ fibers	C fibers
1.	Small myelinated, 2–5 μm diameter, 12–30 mm/sec. conduction velocity	Nonmyelinated, 0.4–1.2 μm diameter with conduction velocity 0.5–2 mm/sec
2.	Less in number	Relatively more
3.	Conduct impulse only to noxious stimulus	In response to thermal and mechanical stimulus
4.	Sensitive to electrical stimulus	Less sensitivity
5.	Most sensitive to pressure	Most sensitive to local anesthetics and chemical factors

Table 17.2: Features of EPSP and AP

S. No.	EPSP	AP
1.	Stimulus intensity to generate EPSP has no threshold	Has threshold level
2.	Does not obey all or none law	Obeys all or none law
3.	Absence of refractory period	Present
4.	Summation can occur	Never possible
5.	Nonpropagatory	Propagatory

at a time. At that time it is completely refractory and cannot conduct another action potential.

Q.33 What are the changes, which take place in nerve cell body during degeneration of nerve fiber?

- The Nissl granules disintegrate by chromatolysis
- Golgi apparatus also disintegrates
- Cell body swells due to accumulation of fluid and becomes round
- Neurofibrils disappear
- Nucleus is displaced towards the periphery. In extreme conditions, nucleus is extruded out of the cell.

Figure 17.2 illustrates the degeneration and regeneration of a nerve fiber.

Q.34 What is Wallerian degeneration?

Degenerative change in the distal cut end of the nerve fiber is called Wallerian degeneration.

Q.35 Explain the changes during Wallerian degeneration briefly.

- Axis cylinder swells and breaks up into small pieces. After few days, the debris is seen in the space that was occupied by the axis cylinder
- Myelin sheath disintegrates into fat droplets
- Neurilemmal sheath is not affected but the cells of Schwann multiply rapidly. The macrophages invade from outside and remove the debris of axis cylinder and fat droplets. So neurilemmal tube becomes empty and it is filled with cytoplasm of Schwann cell.

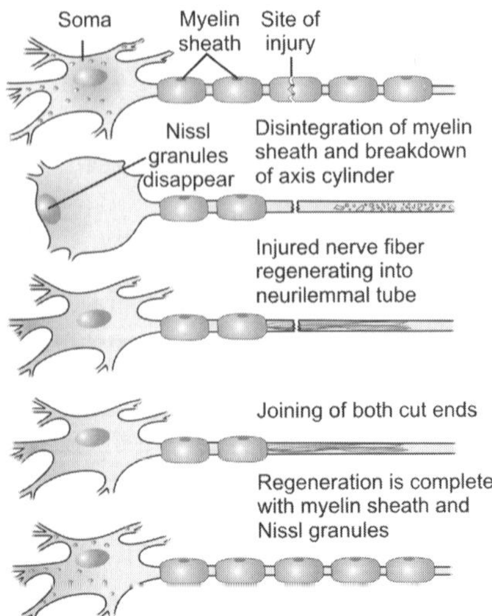

Fig. 17.2: Degeneration and regeneration of nerve fiber

Q.36 What is retrograde degeneration?

The degenerative change that occurs at the proximal cut end of the nerve fiber is called retrograde degeneration.

Q.37 What is transneuronal degeneration?

If an afferent nerve fiber is cut, the degeneration occurs in the neuron with which the afferent nerve fiber synapses. This is called transneuronal degeneration.

Q.38 What are the criteria for regeneration of nerve fiber?

- The gap between the cut ends of the nerve fiber should not exceed 3 mm
- Neurilemma should be present
- Nucleus must be intact
- The two cut ends should remain in the same line.

Q.39 Why regeneration does not occur in central nervous system?

Neurilemma is necessary for regeneration. But neurilemma is absent in central nervous system, so regeneration can not take place.

Q.40 Define neuroglial cell, neuroglia or glia.

Neuroglial cell, neuroglia or glia is the supporting cell of the nervous system (Fig. 17.3).

Q.41 Classify neuroglial cells.

- Central neuroglial cells in nervous system:
  - Astrocytes, which are divided into two subtypes, fibrous astrocytes and protoplasmic astrocytes
  - Microglia
  - Oligodendrocytes.
- Peripheral neuroglial cells:
  - Schwann cells
  - Satellite cells.

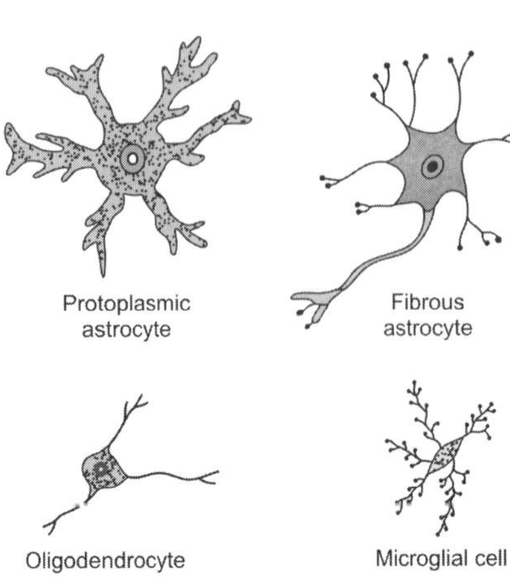

Fig. 17.3: Neuroglial cells in CNS

Q.42 What are the functions of neuroglia?

Astrocytes:

- Form the supporting network in brain and spinal cord
- Form the blood-brain barrier
- Maintain the chemical environment of ECF around CNS neurons
- Provide calcium and potassium ions and regulate neurotransmitter level in synapses
- Regulate recycling of neurotransmitter during synaptic transmission.

Microglia:

- Engulf and destroy the microorganisms and cellular debris by phagocytosis
- Migrate to the injured or infected area of CNS and act as miniature macrophages.

Oligodendrocytes:

- Provide myelination around the nerve fibers in CNS
- Provide support to the CNS neurons by forming a semi-stiff connective tissue between the neurons.

Schwann cells:

- Provide myelination (insulation) around the nerve fibers in PN
- Play important role in nerve regeneration
- Remove cellular debris during regeneration by phagocytosis.

Satellite cells:

- Provide physical support to the PNS neurons.
- Help in regulation of chemical environment of ECF around the PNS neurons.

Q.43 Define receptor.

Receptor is an afferent nerve terminal, which receives the stimulus. It is defined as the biological transducer that converts various forms of energy, i.e. stimulus into action potential in nerve fiber.

Q.44 Classify receptors.

- Exteroceptors:
  - Cutaneous receptors
  - Chemoreceptors
  - Telereceptors.
- Interoceptors:
  - Visceroreceptors
  - Proprioceptors.

Q.45 What are the cutaneous receptors or mechanical receptors?

Receptors situated in the skin are called cutaneous receptors. The different cutaneous receptors (Fig. 17.4):

- Touch receptors—Meissner's corpuscle and Merkel's disc
- Pressure receptors—Pacinian corpuscle
- Temperature or thermoreceptors—Krause's end organ for cold and Ruffini's end organ for warm
- Pain receptors or nociceptors—free (naked) nerve ending.

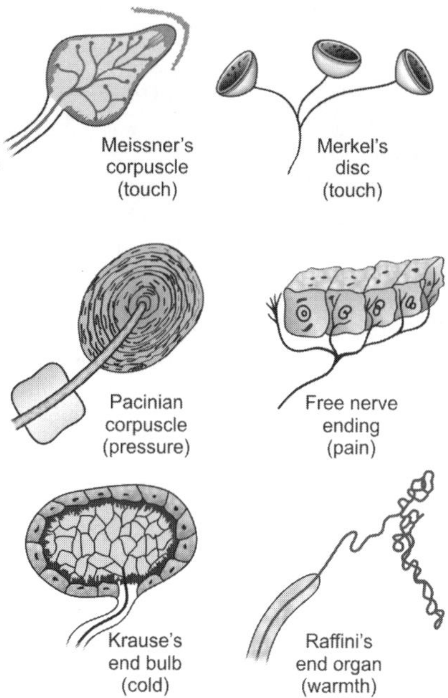

Fig. 17.4: Cutaneous receptors

Q.46 What are chemoreceptors, which belong to the group of exteroceptors?

Receptors giving response to chemical stimuli are known as chemoreceptors. The chemoreceptors, which belong to the group of exteroceptors are taste receptors in taste buds and olfactory receptors for smell in the nose.

Q.47 What are telereceptors?

The receptors, which give response to stimuli arising away from the body are called telereceptors. Telereceptors are:

- Hair cells of organ of Corti in the ear for hearing
- Rods and cones of retina in the eye for vision.

Q.48 What are visceroreceptors?

Receptors situated in the viscera are called visceroreceptors. Stretch receptors, baroreceptors, chemoreceptors and osmoreceptors are the visceroreceptors. Visceroreceptors are situated in heart, blood vessels, lungs, gastrointestinal tract, urinary bladder and brain.

Q.49 Define and classify the proprioceptors.

Proprioceptors are the receptors, which give response to change in the position of different parts of the body.

Proprioceptors are of two types:

- The receptors in labyrinthine apparatus
- Muscle spindle, Golgi tendon organ, Pacinian corpuscles and free nerve endings, which are situated in muscle, tendon, ligament, fascia and joints.

Q.50 Enumerate the properties of receptors.

- Specificity of response
- Adaptation
- Response to increase in strength of stimulus
- Electrical property—receptor potential.

Q.51 What is doctrine of specific nerve energies or specificity of response?

Each receptor gives response to a particular type of stimulus. For example, the pain receptors are stimulated by pain stimulus. This property of receptor is called doctrine of specific nerve energies or specificity of response.

Q.52 What is adaptation?

When a receptor is continuously stimulated with the same strength of stimulus, after some time receptor stops sending impulses through afferent nerve. This property is called adaptation.

Q.53 How receptors are classified based on adaptation? Give examples.

On the basis of adaptation, receptors are classified into two types:

- Phasic receptors which get adapted rapidly. Touch and pressure receptors are the phasic receptors
- Tonic receptors, which adapt slowly. Pain receptors and muscle spindle are tonic receptors.

Q.54 What is receptor potential?

When a receptor is stimulated, a nonpropagated depolarization occurs. This is called receptor potential or generator potential.

Q.55 Enumerate the properties of receptor potential.

- Nonpropagated
- Monophasic
- Does not obey all or none law.

Q.56 What is synapse?

The junction between two neurons is called synapse. It is only a physiological continuity between two nerve cells and not the anatomical continuation.

Q.57 How is synapse classified?

Synapse is classified by two methods:

- *Anatomical classification:* Synapse is divided into three types depending upon the axon ending:
  - Axosomatic synapse
  - Axodendritic synapse
  - Axoaxonic synapse.
- *Functional classification:* Synapse is divided into two types depending upon the transmission of impulses:
  - Electrical synapse
  - Chemical synapse.

Q.58 Explain the structure of axosomatic synapse briefly.

Axon of presynaptic neuron divides into many presynaptic terminals. This has a covering membrane called presynaptic membrane. The presynaptic terminal contains mitochondria and the synaptic vesicles. Synaptic vesicles contain neurotransmitter substance. The membrane of postsynaptic neuron is called postsynaptic membrane. It contains receptor proteins. The space between presynaptic and postsynaptic membrane is called synaptic cleft. Basal lamina of synaptic cleft contains cholinesterase.

Q.59 What is the function of synapse?

Main function of synapse is to transmit the impulses, i.e. action potential from one neuron to another. However, some of the synapses inhibit the transmission of impulses. Thus, synapses are of two types:

- Excitatory synapse that transmits the impulses—excitatory function
- Inhibitory synapse that inhibits the transmission of impulses—inhibitory function.

Q.60 Distinguish between electrical and chemical synapses.

See Table 17.3.

Q.61 Explain the synaptic transmission briefly.

When action potential reaches the presynaptic axon terminal, voltage gated calcium channels at the presynaptic membrane open and calcium ions enter the terminal. This causes release of acetylcholine from synaptic vesicles. Acetylcholine passes through presynaptic membrane and synaptic cleft and binds with receptor protein present on postsynaptic membrane. The acetylcholine receptor complex opens ligand gated sodium channels so that, sodium ions enter the synapse, i.e. soma. This produces excitatory postsynaptic potential (EPSP), which in turn causes development of action potential in the initial segment of axon of postsynaptic neuron.

Q.62 What is excitatory postsynaptic potential (EPSP)?

When action potential reaches presynaptic axon terminal, it causes the development of a nonpropagated electrical potential in the soma of postsynaptic neuron through acetylcholine. This potential

Table 17.3: Chemical and electrical synapses

S. No.	Chemical	Electrical
1.	Impulse is transmitted from preto postsynaptic site through release of neurotransmitter, i.e. chemical mediators	Impulse is transmitted through gap junction
2.	Most of synapses are chemical type	Present only in specific synaptic junction of brain
3.	Presence of synaptic cleft	Cleft is replaced by low resistance bridges
4.	Synaptic delay is present	Absent
5.	Sensitive to O_2 lack	Insensitive to O_2 lack

in postsynaptic neuron is known as excitatory postsynaptic potential (EPSP).

Q.63 What are the properties of EPSP?
- Nonpropagated
- Monophasic
- Does not obey all or none law.

Q.64 What is the significance of EPSP?
EPSP causes development of action potential in the initial segment of axon of postsynaptic neuron. Actually EPSP opens sodium channels in the initial segment of axon so that sodium ions enter the axon from ECF resulting in development of action potential.

Q.65 Name the types of synaptic inhibition.
- Postsynaptic or indirect inhibition
- Presynaptic or direct inhibition
- Renshaw cell inhibition.

Q.66 What is postsynaptic inhibition?
The failure of production of action potential in the postsynaptic membrane because of release of an inhibitory neurotransmitter from the presynaptic terminal is called postsynaptic inhibition.

Q.67 What is postsynaptic inhibitory potential (IPSP)?
The inhibitory neurotransmitter released from presynaptic axon terminal causes opening of potassium channels. This results in efflux of potassium ions from soma of postsynaptic neuron and development of hyperpolarization. This type of hyperpolarization is called postsynaptic inhibitory potential (IPSP).

Q.68 What is presynaptic inhibition?
In some synapses, the action potential reaching the presynaptic axon terminal fails to release neurotransmitter from the synaptic vesicles. So, the transmission of impulse is inhibited. This is called presynaptic or direct inhibition.

Q.69 What is Renshaw cell inhibition?
This occurs in spinal cord. Renshaw cell is a type of motor neuron situated near alpha motor neuron in anterior gray horn. When alpha motor neuron of spinal cord sends motor impulses via anterior nerve root fibers, some of the impulses reach the Renshaw cell by passing through collateral fibers. Renshaw cell in turn sends inhibitory impulses to alpha motor neuron so that, the discharge from motor neuron is reduced (Fig. 17.5).

Q.70 What is the significance of synaptic inhibition (Fig. 17.6)?
Synaptic inhibition offers restriction over the neurons and muscles so that the excess stimuli are inhibited and the various movements are performed properly and accurately.

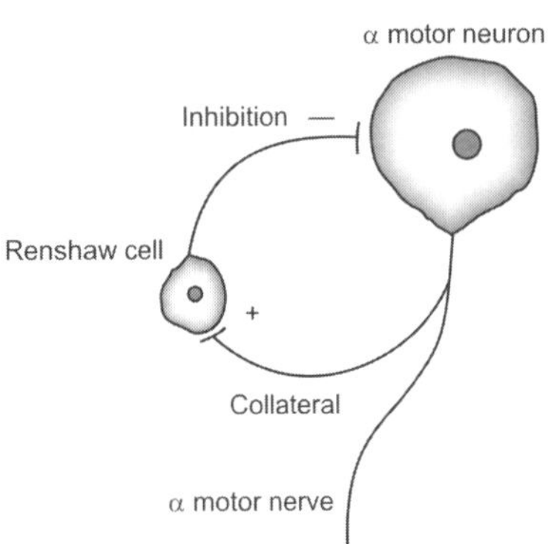

Fig. 17.5: Renshaw cell inhibition

Q.71 Name the properties of synapse.
- One way conduction
- Synaptic delay
- Fatigue
- Summation
- Electrical property—EPSP or IPSP.

Q.72 What is Bell-Magendie law?
The impulses are transmitted only in one direction in synapse, i.e. from presynaptic neuron to postsynaptic neuron. This is called Bell-Magendie law.

Q.73 What is synaptic delay? And, what is its cause?
The delay in the transmission of impulses through synapse is known as the synaptic delay.
It is due to the time taken for:
- Release of neurotransmitter
- Movement of neurotransmitter from axon terminal to postsynaptic membrane
- Opening of ionic channels in postsynaptic membrane by the neurotransmitter.

Q.74 What is the cause for fatigue in synapse?
Fatigue at synapse is due to the exhaustion of neurotransmitters.

Q.75 What is Weber Fechner law?
The frequency of action potential in a sensory nerve is directly proportional to the magnitude of generator potential which inturn is directly proportional to the intensity of stimulus. This relationship between intensity of stimulus, magnitude of GP and frequency of AP in the afferent nerve is known as Weber Fechner law.

Q.76 What is Muller's doctrine of specific nerve energy?
Sensation produced by impulses generated in a receptor depends on the specific

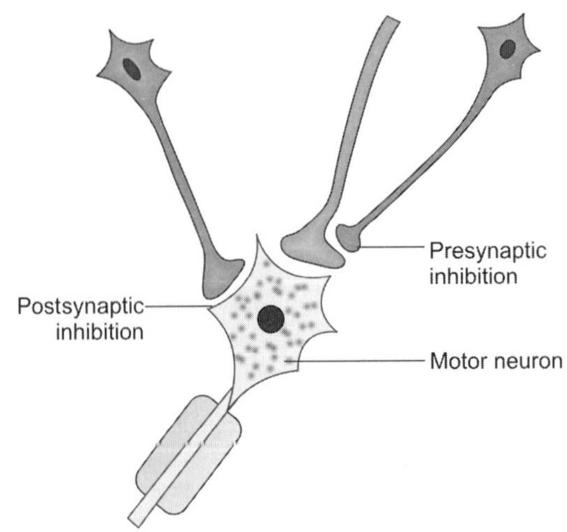

Fig. 17.6: Synaptic inhibition

part of brain, i.e. the specific pathways for specific sensation are separated from nerve organ to cerebral cortex. This is known as Muller's law.

Q.77 What do you mean by law of projection? What is phantom limb?

No matter where a particular sensory pathways is stimulated along its course to the cortex, the conscious sensation produced is referred to the location of the receptor. This principle is called law of projection.

A limb that has been lost by accident or amputation, the patient usually experiences intolerable pain and proprioceptive sensations in the absent limb and is called phantom limb.

Q.78 What do you mean by law of intensity discrimination?

The brain interpretes different intensities of sensations by varying the frequency of AP generated by receptor and/or by varying the number of receptors activated or both. This is known as law of intensity discrimination.

Q.79 Name some excitatory neurotransmitter substances.

Acetylcholine, noradrenaline and histamine.

Q.80 Name some inhibitory neurotransmitter substances.

Gamma amino butyric acid (GABA), glycine, dopamine and serotonin.

Q.81 Define reflex activity.

Response to a peripheral nervous stimulation that occurs without consciousness is known as reflex activity.

Q.82 What is reflex arc? Enumerate its components (parts).

The anatomical neural pathway for a reflex action is called reflex arc.

It has five components:

- Receptor
- Afferent or sensory nerve
- Center
- Efferent or motor nerve
- Effector organ.

Q.83 What are the methods of classification of reflexes?

Reflexes are classified by five different methods:

- Depending upon whether inborn or acquired
- Depending upon the situation of center
- Depending upon the purpose or functional significance
- Depending upon number of synapse
- Depending upon clinical basis (Refer next 5 questions for details).

Q.84 Classify the reflexes depending upon whether inborn or acquired.

- Unconditioned reflexes which are present at the time of birth. These reflexes do not require previous learning or training or conditioning but contact of a substance with the receptor is essential. Best example is the secretion of saliva when any object is kept in mouth
- Conditioned reflexes which are acquired after birth. These reflexes require previous learning or training or conditioning and contact of a substance with receptor is not necessary. The example is the secretion of saliva by the sight, smell, thought or hearing of a known edible substance.

Q.85 Classify the reflexes depending upon the situation of center.

- Cerebellar reflexes
- Cortical reflexes
- Midbrain reflexes
- Bulbar or medullary reflexes
- Spinal reflexes.

Q.86 Classify the reflexes depending upon the purpose or functional significance.

- Protective or flexor reflexes
- Antigravity or extensor reflexes.

Q.87 Classify the reflexes depending upon number of synapse.

- Monosynaptic reflexes (stretch reflexes)
- Polysynaptic reflexes.

Q.88 Classify the reflexes depending upon clinical basis.

- Superficial reflexes, which are elicited from surface of the body, i.e. from skin (superficial cutaneous reflexes) and mucus membrane (superficial mucus membrane reflexes)
- Deep reflexes which arise from structure beneath the skin
- Visceral reflexes which are elicited from organs in viscera
- Pathological reflexes which are the abnormal reflexes and can be elicited only in diseased conditions.

Q.89 Enumerate the properties of reflexes.

- One way conduction
- Reaction time
- Summation
- Occlusion
- Subliminal fringe
- Recruitment
- After discharge
- Rebound phenomenon
- Fatigue.

Q.90 What are the types of summation in reflex activity?

- *Spatial summation:* When two afferent nerve fibers supplying a skeletal muscle are stimulated separately with subliminal stimulus, there is no response. But, if both nerve fibers are stimulated together with same strength of stimulus, the muscle contracts. This is called spatial summation.
- *Temporal summation:* When one nerve is stimulated repeatedly with subliminal stimuli, these stimuli are summed up and cause contraction of the muscle. This is called temporal summation.

Q.91 What is occlusion?

In a muscle which is innervated by two motor nerves called A and B, when both nerves are stimulated simultaneously, the tension developed by the muscle is less than the sum of the tension developed when each nerve is stimulated separately. This type of response is called occlusion (Fig. 17.7).

For example, if nerve A is stimulated alone, the arbitrary unit of tension developed is 9. If nerve B is stimulated the tension developed is 9 units. So, the sum of tension developed when nerves A and B are separately stimulated = 9 + 9 = 18 units. But when, both A and B are stimulated together, the tension produced is (A+B) = 12 units only. This phenomenon is called occlusion and it is due to the overlapping of the nerve fibers during the distribution.

Q.92 What is subliminal fringe?

In some reflexes which involve the muscle with two nerve fibers called A and B, the tension developed by simultaneous stimulation of two nerve is greater than the sum of tension produced by the stimulation of these nerves separately.

For example, if nerve A or B is stimulated alone, the arbitrary unit of tension developed by muscle = 3 units. So, the sum of tension developed if nerves A and B are stimulated separately is 3 + 3 = 6 units. But, when both the nerves are stimulated together, the tension developed = (A + B) = 12 units. So, the tension here is greater than the sum of tension produced if A and B are separately stimulated. This phenomenon is called subliminal fringe (Fig. 17.8) and it is due to the effect of spatial summation.

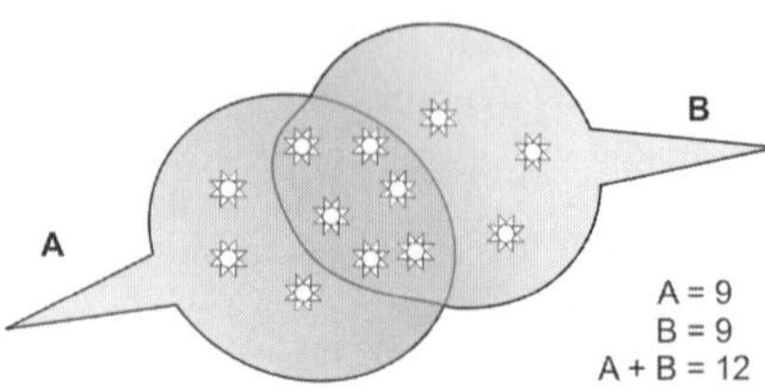

Fig. 17.7: Occlusion

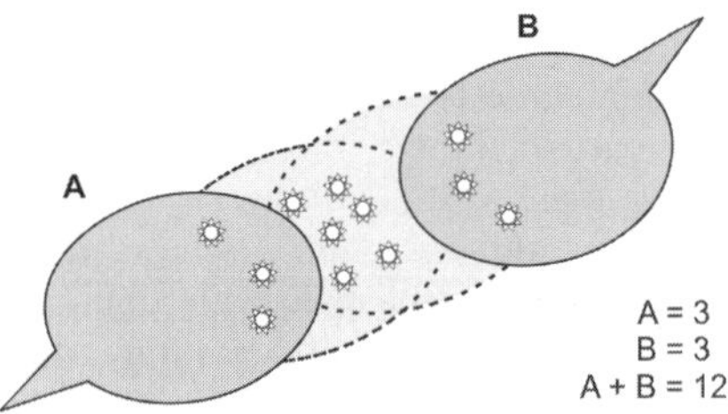

Fig. 17.8: Subliminal fringe

Q.93 What is recruitment?

When an excitatory nerve is stimulated with stimuli of constant strength for a long time, there is a progressive increase in the number of motor neurons activated. This phenomenon is called recruitment. It is similar to the effect of temporal summation.

Q.94 What is after discharge?

If a reflex action is elicited continuously for sometime, and then the stimulation is stopped, the reflex activity, i.e. contraction may continue for some time even after the stoppage of stimulus. This is called after discharge. The center discharges impulses even after stoppage of stimulus. This is because of internuncial neurons, which continue to transmit afferent impulses even after stoppage of stimulus.

Q.95 Which is the first seat of fatigue in reflex arc?

The center or the synapse of the reflex arc is the first seat of fatigue.

Q.96 What is crossed extensor reflex?

When a flexor reflex is elicited in one limb, the flexor muscles of that limb are stimulated and the extensor muscles are inhibited. But on the opposite limb, the flexors are inhibited and extensors are excited. This is called crossed extensor reflex. It is due to reciprocal innervation.

Q.97 What is Babinski's sign? When does it occur?

Babinski's sign is the abnormal plantar reflex. In normal plantar reflex, a gentle scratch over the outer edge of the sole of the foot causes plantar flexion and adduction of all toes and dorsiflexion and inversion of foot. But in Babinski's sign, there is dorsiflexion of big toe and fanning of other toes.

It is common in infants due to the nonmyelination of pyramidal tracts. In normal persons, it can be elicited during deep sleep. The pathological condition when it appears is upper motor lesion.

Q.98 What are pendular movements?

While eliciting a tendon jerk, some slow oscillatory movements are developed instead of brisk movements. These movements are called pendular movements and are common in cerebellar lesion.

Q.99 What are the effects of upper and lower motor neuron lesion on reflexes?

During upper motor neuron lesion, all the superficial reflexes are lost. The deep reflexes are exaggerated and the Babinski's sign is positive.

During lower motor neuron lesion, all the superficial and deep reflexes are lost.

Q.100 What are the segments of spinal cord?

Spinal cord is made up of 31 segments viz.

Cervical segments	=	8
Thoracic segments	=	12
Lumbar segments	=	5
Sacral segments	=	5
Coccygeal segment	=	1

Q.101 What are the neurons present in gray horn of spinal cord?

Anterior gray horn consists of motor neurons. Posterior gray horn consists of sensory neurons. The lateral gray horn contains intermediolateral horn cells, which give rise to sympathetic preganglionic fibers.

Q.102 Name the types of neurons present in the anterior gray horn.

- Alpha motor neurons
- Gamma motor neurons
- Renshaw cells.

Q.103 Name the types of neurons present in the posterior gray horn.

- Substantia gelatinosa of Rolando
- Marginal cells
- Chief sensory cells
- Clarke's column of cells.

Q.104 What are the white columns of spinal cord?

- Anterior white column—between the anterior median fissure on one side and anterior nerve root and anterior gray horn on the other side
- Lateral white column—between the anterior nerve root and anterior gray horn on one side and posterior nerve root and posterior gray horn on the other side
- Posterior gray column—in between posterior nerve root and posterior gray horn on one side and posterior median septum on the other side.

Q.105 Briefly classify tracts of spinal cord.

- Short tracts—connecting different parts of spinal cord itself:
  - Association or intrinsic tracts which connect the adjacent segments of spinal cord on the same side
  - Commissural tracts, which connect the opposite halves in the same segment of spinal cord.
- Long tracts or projection tracts connecting spinal cord with other parts of central nervous system:
  - Ascending tracts which carry sensory impulses from spinal cord to brain
  - Descending tracts, which carry motor impulses from brain to the spinal cord.

Q.106 Enumerate the ascending tracts in spinal cord.

- Anterior white funiculus:
  - Anterior spinothalamic tract.
- Lateral white funiculus:
  - Lateral spinothalamic tract
  - Ventral spinocerebellar tract
  - Dorsal spinothalamic tract
  - Spinotectal tract
  - Spinoreticular tract
  - Spino-olivary tract
  - Spinovestibular tract
- Posterior white funiculus:
- Tract of Goll
- Tract of Burdach
- Comma tract of Schultze.

Q.107 Name the type of fibers forming ascending tracts of spinal cord.

All ascending tracts of spinal cord are formed by the fibers of second order neurons (crossed fibers) except posterior column tracts.

The posterior column tracts are formed by the fibers of first order neurons (uncrossed fibers).

Q.108 Which sensations are carried by these nerves?

Fasciculus gracilis and	Fine touch, tactile localization, kinesthetic
Fasciculus cuneatus	movements, vibration, deep pressure
Lateral spinothalamic	Pain and temperature
Ventral spinothalamic	Crude touch

Q.109 What are the functions of spinothalamic tracts?

Anterior spinothalamic tract carries crude touch (protopathic) sensation and lateral spinothalamic tract carries pain and temperature sensations.

Q.110 What are nonsensory impulses?

The impulses of subconscious kinesthetic sensation are called nonsensory impulses.

Q.111 What are the functions of posterior column tracts?

Posterior column tracts carry the impulses of:

- Fine touch or epicritic tactile sensation
- Tactile localization
- Tactile discrimination
- Sensation of vibration
- Conscious kinesthetic sensation
- Stereognosis.

Q.112 Classify the descending tracts of spinal cord.

- Pyramidal tracts which give the appearance of a pyramid on the upper part of anterior surface of medulla oblongata while running from cerebral cortex towards spinal cord
- Extrapyramidal tract, which are the descending tracts other than pyramidal tracts.

Q.113 Name the pyramidal tracts.

- Anterior corticospinal tract
- Lateral corticospinal tract.

Q.114 Mention the origin of fibers of pyramidal tracts.

- Primary motor areas and supplementary motor areas in frontal lobe of cerebral cortex (30%)
- Premotor area in frontal lobe (30%)
- Parietal lobe particularly from somatosensory areas (40%).

Q.115 Briefly describe the course of pyramidal tracts.

After taking origin from cerebral cortex, the fibers of pyramidal tracts descend down through corona radiata, internal capsule, midbrain and pons and enter medulla. While running down through the upper part of anterior surface of medulla, these fibers give the appearance of a pyramid.

At the lower border of medulla, 80% of fibers from each side cross to opposite side forming pyramidal decussation or motor decussation. After crossing, these fibers descend through the lateral white column of spinal cord as lateral corticospinal tract. Remaining 20% of the fibers descend down in the same side through the anterior white column as anterior corticospinal tract.

Q.116 What are the functions of pyramidal tracts?

Pyramidal tracts are concerned with voluntary movements of the body and are responsible for fine and skilled movements.

Q.117 What are the effects of lesion of pyramidal tracts?

Lesion in pyramidal tracts is called upper motor neuron lesion. It causes:

- Loss of voluntary movements
- Increase in muscle tone leading to spastic paralysis of muscles
- Loss of all the superficial reflexes
- Exaggeration of deep reflexes
- Babinski's sign.

Q.118 Name the extrapyramidal tracts.

- Medial longitudinal fasciculus
- Anterior vestibulospinal tract
- Lateral vestibulospinal tract
- Reticulospinal tract
- Tectospinal tract
- Rubrospinal tract
- Olivospinal tract.

Q.119 What are the functions of medial longitudinal fasciculus?

Medial longitudinal fasciculus helps in the coordination of reflex ocular movements and the integration of ocular and neck movements.

Q.120 What is the function of vestibulospinal tracts?

Vestibulospinal tracts are concerned with adjustment of position of head and body during angular and linear acceleration.

Q.121 What are the functions of reticulospinal tract?

Reticulospinal tract is concerned with control of movements, maintenance of muscle tone, respiration and control of diameter of blood vessels.

Q.122 What are the effects of complete transection of spinal cord?

Complete transection of spinal cord causes immediate loss of sensation and voluntary movements below the level of lesion. The effects occur in three stages:

- Stage of spinal shock
- Stage of reflex activity
- Stage of reflex failure.

Q.123 What is paraplegia in flexion?

During the stage of reflex activity after complete transection of spinal cord, the tone returns to flexor muscles first. And the limbs in this condition tend to adopt a position of slight flexion. This type of paralysis is known as paraplegia in flexion.

Q.124 What are the effects of incomplete transection of spinal cord?

The effects of incomplete transection of spinal cord are similar to the effects of complete transection except that, during the stage of reflex activity, the tone returns to extensor muscles first.

Q.125 What is paraplegia in extension?

During the stage of reflex activity after incomplete transection of spinal cord, the tone returns to extensor muscles first. The limbs in this condition tend to adopt a position of slight extension. This is called paraplegia in extension.

Q.126 What is hemisection of spinal cord?

Injury to one lateral half of spinal cord is called hemisection.

Q.127 What is Brown-Séquard syndrome?

The signs and symptoms, which occur after hemisection of spinal cord are together called Brown-Séquard syndrome (Fig. 17.9).

Q.128 What are the effects of hemisection of spinal cord on the same side of the body below the lesion?

The sensations carried by uncrossed fibers of posterior column tracts namely, fine touch sensation, tactile localization, tactile discrimination, sensation of vibration, conscious kinesthetic sensation and stereognosis are lost.

The sensations carried by crossed spinothalamic tracts such as crude touch,

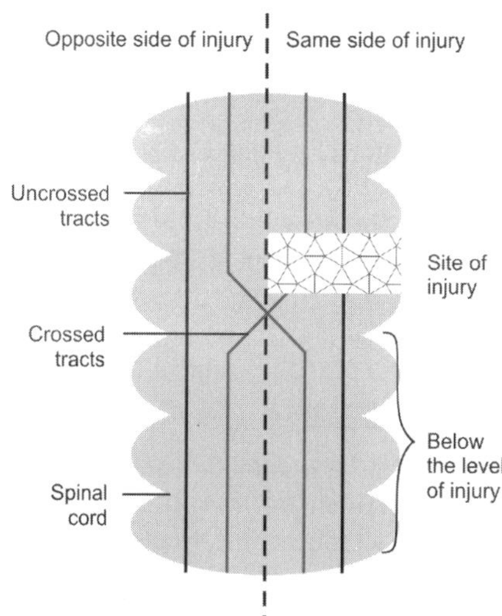

Fig. 17.9: Hemisection of spinal cord (Brown-Séquard syndrome)

pain and temperature sensations are not affected.

The motor changes resemble the effects of upper motor lesion.

Q.129 What are the effects of hemisection of spinal cord on the opposite side of the body below the lesion?

The sensations carried by crossed spinothalamic tracts such as crude touch, pain and temperature sensations are lost.

The sensations carried by uncrossed fibers of posterior column tracts namely, fine touch sensation, tactile localization, tactile discrimination, sensation of vibration, conscious kinesthetic sensation and stereognosis are not affected.

The motor functions are not affected. If affected, it would be mild and the effects resemble the effects of upper motor lesion.

Q.130 What are the effects of hemisection of spinal cord on the same side of the body at the level of the lesion?

There is complete anesthesia, i.e. all the sensations are lost. The motor changes resemble the effects of lower motor lesion.

Q.131 What are the effects of hemisection of spinal cord on the opposite side of the body at the level of the lesion?

The sensations carried by crossed spinothalamic tracts such as crude touch, pain and temperature sensations are lost.

The sensations carried by uncrossed fibers of posterior column tracts namely, fine touch sensation, tactile localization, tactile discrimination, sensation of vibration, conscious kinesthetic sensation and stereognosis are not affected.

The motor functions are not affected. If affected, it would be mild and the effects resemble the effects of lower motor lesion.

Q.132 What is syringomyelia? What is its cause?

Syringomyelia is a disease of spinal cord characterized by the presence of fluid filled cavities in the spinal cord. It occurs due to over growth of neuroglial cells in spinal cord accompanied by cavity formation and accumulation of fluid.

Q.133 What is dissociated anesthesia?

In case of syringomyelia there is loss of pain and temperature sensation whereas sense of touch is unaffected. This condition is known as dissociated anesthesia.

Q.134 What is tabes dorsalis? What is its cause?

Tabes dorsalis is a disease of spinal cord. It occurs due to degeneration of dorsal nerve roots. Degeneration of dorsal nerve roots is common in syphilis.

Q.135 What is the feature of tabes dorsalis?

The characteristic feature of tabes dorsalis is the slow progressive nervous disorder affecting the motor and sensory functions of spinal cord.

Q.136 Classify sensations.

- Somatic sensations:
  - Epicritic or light sensations
  - Protopathic or crude sensations
  - Deep sensations.
- Special sensations:
  - Visual sensation
  - Auditory sensation
  - Gustatory or taste sensation
  - Olfactory sensation or sensation of smell.

Q.137 Name the epicritic sensations.

- Fine touch or tactile sensation
- Tactile localization
- Tactile discrimination
- Temperature sensation with finer range.

Q.138 Name the protopathic sensations.

- Pressure sensation
- Pain sensation
- Temperature sensation with wider range.

Q.139 Name the deep sensations.

- Sensation of vibration or pallesthesia
- Kinesthetic sensation or kinesthesia
- Visceral pain sensation.

Q.140 How are the sensations from the face transmitted to the brain?

Through the ophthalmic, maxillary and mandibular divisions of trigeminal nerve.

Q.141 What is somatosensory system?

Somatosensory system is the sensory system involving the pathways, which convey the information from the sensory receptors present in skin, skeletal muscles and joints.

Q.142 Name the components of somatosensory pathways.

- Receptor
- First order neurons
- Second order neurons
- Third order neurons in some cases
- Center in the brain.

Q.143 Define the following.

- *Anesthesia:* Loss of all sensations.
- *Hyperesthesia:* Increased sensitivity to sensory stimuli.
- *Hypesthesia:* Reduction in the sensitivity to sensory stimuli.
- *General anesthesia:* Loss of all sensations with loss of consciousness produced by anesthetic agents.
- *Analgesia:* Loss of pain sensation.
- *Hyperalgesia:* Increased sensitivity to pain stimulus.
- *Illusion:* Mental depression due to misinterpretation of a sensory stimulus.
- *Hallucination:* Feeling of a sensation without any stimulus.

Q.144 What are lemnisci? Name the different lemnisci.

The prominent bundles of sensory nerve fibers in the brain are called lemnisci.

Different lemnisci:

- Spinal lemniscus formed by spinothalamic tracts
- Lateral lemniscus formed by fibers carrying sensation of hearing from cochlear nuclei to inferior colliculus and medial geniculate body
- Medial lemniscus formed by posterior column tracts
- Trigeminal lemniscus formed by fibers of sensory nuclei of trigeminal nerve.

Q.145 What is lateral motor system? Name its components.

Lateral motor system is the part of motor system formed by the motor nerve fibers, which terminate on motor neurons situated in lateral part of ventral gray horn in spinal cord and also on the corresponding motor neurons of cranial nerve nuclei in brainstem.

It includes:

- Lateral corticospinal tract
- Rubrospinal tract
- Part of corticobulbar tract.

Q.146 What are the functions of lateral motor system?

- Lateral corticospinal tract activates the muscles in the distal portions of the limbs and skilled voluntary movements
- Rubrospinal tract facilitates tone of flexor muscles
- Corticobulbar tracts are concerned with the movements of expression in lower part of face and movements of tongue.

Q.147 What is medial motor system? Name its components.

Medial motor system is the part of motor system formed by the motor nerve fibers, which terminate on the motor neurons situated in the medial part of ventral gray horn of spinal cord and on the corresponding motor neurons of cranial nerve nuclei in brainstem.

It includes:

- Anterior corticospinal tract
- Part of corticobulbar tracts not belonging to lateral motor system
- Lateral and medial vestibular tracts
- Reticulospinal tract
- Tectospinal tract.

Q.148 What are the functions of medial motor system?

- Maintenance of posture and equilibrium, chewing movements and eyebrow movements
- Movements of head in response to visual and auditory stimuli.

Q.149 What are upper motor neurons? Name them.

The neurons in the higher center of brain, which control the lower motor neurons are called upper motor neurons.

Upper motor neurons are:

- Motor neurons in cerebral cortex
- Neurons in basal ganglia and brainstem
- Neurons in cerebellum.

Q.150 What are the lower motor neurons? Name them.

Lower neurons are the anterior horns cells in the spinal cord and motor neurons of the cranial nerve nuclei situated in brainstem, which innervate the skeletal muscles directly. These neurons constitute the "Final common pathway" of motor system.

The lower motor neurons are the alpha motor neurons in anterior horns of spinal cord and the cells of nuclei of III, IV, V, VI, VII, IX, X, XI and XII cranial nerve.

Q.151 What are the effects of upper motor neuron lesion?

- Hypertonia
- Spastic paralysis of muscles without wastage
- Loss of superficial reflex
- Appearance of Babinski's sign
- Exaggeration of deep reflexes
- Clonus.

Q.152 What are the effects of lower motor neuron lesion?

- Hypotonia
- Flaccid paralysis with wastage of muscles
- Loss of all reflexes.

Q.153 What are the main differences between upper and lower motor neuron lesion?

LMNL	*UMNL*
• Single individual muscle is affected	• Group of muscles are affected
• Flaccid type of muscle paralysis due to hypotonia	• Spastic type of muscle paralysis due to hypertonia
• Disuse atropy of muscle takes place	• Not severe
• All reflexes are absent as motor pathway is damaged.	• Deep reflexes are hyperactive due increased g motor activity and some superficial reflexes like abdominal, cremasteric reflexes are lost.
• Babinski's sign is negative	• It is positive.

Q.154 What are the components of pain sensation?

- *Fast pain:* Whenever pain stimulus is applied, a fast, bright, sharp and localized pain sensation is produced. This is called fast pain
- *Slow pain:* The fast pain sensation is followed by a dull, diffused and unpleasant pain called slow pain.

Q.155 Name the nerve fibers transmitting pain sensation.

Fast pain is transmitted by type A delta afferent fibers and slow pain is transmitted by type C fibers.

Q.156 What are the causes for visceral pain?

- Ischemia
- Chemical stimuli
- Spasm of hollow organs
- Over distention of hollow organs.

Q.157 What is referred pain?

The pain sensation, produced in some parts of the body is felt in other structures away from the place of development. This is called referred pain.

Q.158 Give some examples of referred pain.

- Cardiac pain referred to inner part of left arm and shoulder
- Pain in ovary referred to umbilicus
- Pain in testis referred to abdomen
- Pain in diaphragm referred to right shoulder
- Pain in gallbladder referred to epigastric region
- Renal pain referred to loin.

Q.159 Name the neurotransmitter involved in pain sensation.

Substance P

Q.160 What is analgesia system?

The pain control system of central nervous system is called analgesia system. It inhibits the impulses of pain sensation.

Q.161 What are the pain control systems in brain and spinal cord?

The pain control system in brain is present in gray matter surrounding aqueduct of Sylvius and raphe magnus nuclei in pons.

In the spinal cord, the pain control system is in posterior gray horn which is considered as gateway for pain impulses.

Q.162 What is gate theory of pain?

When pain sensation is produced in any part of the body, along with pain fibers, some of the other afferent fibers particularly the touch fibers reaching the posterior column of spinal cord are also activated. The posterior column of spinal cord sends collaterals to cells of substantia gelatinosa in the posterior gray horn. Thus, some of the impulses ascending via posterior column fibers pass through the collaterals and reach substantia gelatinosa. Here, the impulses inhibit the release of substance P by pain fibers and pain sensation is suppressed. Thus, there is gating of pain in posterior gray horn level.

Q.163 Name important centers or nuclei present in medulla oblongata.

- Respiratory centers
- Vasomotor center
- Deglutition center
- Vomiting center
- Superior and inferior salivatory nuclei
- Nuclei of 12th, 11th, 8th and 5th cranial nerves
- Vestibular nuclei.

Q.164 What are the important structures present in pons?

- Pyramidal tract fibers
- Medial lemniscus
- Nuclei of 8th, 7th and 5th cranial nerves
- Pneumotaxic and apneustic centers for regulation of respiration
- Vestibular nuclei.

Q.165 What are the important structures present in midbrain?

- Tectum, which includes superior colliculus and inferior colliculus.
- Cerebral peduncles which include basis pedunculus, substantia nigra, tectum and red nucleus.

Q.166 What is red nucleus? What is its function?

Red nucleus is a large oval or round mass of gray matter between superior colliculus and hypothalamus.

It controls:

- Muscle tone
- Complex muscular movements
- Righting reflexes
- Eyeball movements
- Skilled movements.

Q.167 Name the different groups of thalamic nuclei.

- Midline nuclei
- Infralaminar nuclei
- Medial mass nuclei
- Lateral mass nuclei
- Posterior group nuclei.

Q.168 What is thalamic syndrome? What are its features?

The signs and symptoms that occur during thalamic lesion are together called thalamic syndrome.

The features are:

- Anesthesia—loss of sensations
- Astereognosis—inability to recognize a known object by touch with closed eyes
- Sensory ataxia—incoordination of voluntary movements due to loss of sensations
- Thalamic phantom limb—inability to locate the position of a limb with closed eyes
- Amelognosia—illusion felt by the patient that his limb is absent.
- Spontaneous pain.
- Involuntary movements like athetosis, chorea and intention tremor.
- Thalamic hand or athetoid hand - abnormal attitude of hand characterized by moderate flexion at wrist and hyperextension of all fingers.

Q.169 What is the cause for thalamic lesion?

Thalamic lesion occurs mostly because of blockage of thalamogeniculate branch of posterior cerebral artery by thrombosis.

Q.170 What is tremor? Which type of tremor occurs in thalamic syndrome?

Rapid alternate rhythmic and involuntary movement of flexion and extension in the joints of fingers and wrist or elbow is called tremor.

In thalamic syndrome, intension tremor (tremor while attempting to do any voluntary act) occurs.

Q.171 What is internal capsule? Where is it situated?

Internal capsule is the compact band of afferent and efferent fibers connecting cerebral cortex with brainstem and spinal cord.

It is situated in between thalamus and caudate nucleus on the medial side and lenticular nucleus on the lateral side.

Q.172 What are the nuclei of hypothalamus?

- Anterior or preoptic group—preoptic nucleus, paraventral nucleus, anterior nucleus and supraoptic nucleus
- Middle or tuberal group—dorsomedial nucleus, ventromedial nucleus, lateral nucleus and arcuate (tuberal) nucleus
- Posterior or mammillary group—posterior nucleus and mammillary body.

Q.173 What is the role of hypothalamic centers for regulation of body temperature?

- *Heat loss center:* When body temperature increases, heat loss center is stimulated. This causes cutaneous vasodilation and secretion of large amount of sweat so that heat is lost directly from skin or through sweat.
- *Heat gain center:* When body temperature decreases, heat gain center is activated. This causes increased production of heat by shivering and prevents loss of heat by causing cutaneous vasoconstriction.

Q.174 What is the role of hypothalamus in regulation of food intake?

Hypothalamus has two centers to regulate the food intake, feeding center and satiety center. Normally, feeding center is active and it is controlled by satiety center.

Q.175 Name the mechanisms involved in regulation of appetite and food intake.

- Glucostatic mechanism
- Lypostatic mechanism
- Peptide mechanism
- Hormonal mechanism
- Thermostatic mechanism.

Q.176 What is the role of hypothalamus in regulation of water balance?

Hypothalamus regulates water balance by two mechanisms:

- By thirst mechanism—when body water reduces the thirst center in hypothalamus is stimulated leading to water intake
- When body water reduces, osmolarity of body fluids increases. This, in turn stimulates ADH secretion from hypothalamus. ADH increases reabsorption of water from renal tubules.

Q.177 Name the hypothalamic centers concerned with behavior and emotional changes.

- Reward center
- Punishment center.

Q.178 What is rage? What is sham rage?

When punishment center is stimulated in animals, a violent aggressive emotional state is exhibited. This is called rage. It includes the reactions like development of defense posture, extension of limbs, lifting of tail, hissing, spitting, and severe savage attack even by mild provocation.

Normally the punishment center in hypothalamus is kept inhibited by cortical centers. So mild irritations are overcome or ignored. However, in animals or human beings with brain lesions, even a very mild stimulus can evoke violent and angry reactions of rage. This type of rage is called sham rage. It is due to release of hypothalamus from cortical control.

Q.179 Name the disorders caused by hypothalamic lesion.

- Diabetes insipidus.
- Dystrophia adiposogenitalis.
- Laurence-Biedl-Moon syndrome
- Narcolepsy
- Cataplexy.

Q.180 What is diabetes insipidus? What is its cause?

Diabetes insipidus is the disease characterized by excretion of large quantity of dilute urine. It is due to the failure of water reabsorption from renal tubules. It occurs due to deficiency or absence of ADH because of tumor of hypothalamus.

Q.181 Name the parts of cerebellum.

- Vermis
- Two cerebellar hemispheres.

Q.182 Name the functional divisions of cerebellum.

- Vestibulocerebellum
- Spinocerebellum
- Corticocerebellum.

Q.183 What are the histological structures of cerebellum?

- Gray matter or cerebellar cortex—made up of nervous structures arranged in three layers:
  - Molecular or plexiform layer
  - Purkinje layer
  - Granular layer.
- White matter—formed by nerve fibers and gray masses called cerebellar nuclei.

Q.184 Name the afferent nerve fibers to cerebellar cortex.

- Climbing fibers
- Mossy fibers.

Q.185 Name the cerebellar nuclei.

- Fastigial nucleus
- Globosus nucleus
- Emboliform nucleus
- Dentate nucleus.

Q.186 What are the nerve fibers of white matter of cerebellum?

- Projection fibers, which connect cerebellum with other parts of central nervous system
- Association fibers, which connect different regions of same cerebellar hemisphere
- Commissural fibers, which connect the areas of both halves of cerebellar cortex.

Q.187 How are the projection fibers of cerebellum arranged?

Projection fibers of cerebellum are arranged in three bundles:

- Inferior peduncles between cerebellum and medulla oblongata
- Middle peduncles between cerebellum and pons
- Superior peduncles between cerebellum and midbrain.

Q.188 What are the components of vestibulocerebellum?

Vestibulocerebellum includes flocculonodular lobe which is formed by nodulus of vermis and the lateral extension on either side called flocculus.

Q.189 What are the functions of vestibulocerebellum?

Vestibulocerebellum regulates tone, posture and equilibrium because of its connections with vestibular apparatus, vestibular nuclei and spinal motor neurons.

Q.190 Name the components of spinocerebellum.

Lingula, central lobe, culmen, lobulus simplex, declive, tuber, pyramid, uvula, paraflocculi and medial portions of cerebellar hemispheres.

Q.191 What are the functions of spinocerebellum?

- Spinocerebellum forms the receiving area for tactile, proprioceptive, auditory and visual impulses. It also receives the cortical impulses via pons.
- Spinocerebellum regulates the postural reflexes by modifying muscle tone
- Spinocerebellum also receives impulses from optic and auditory pathway and helps in adjustment of posture and equilibrium in response to visual and auditory impulses.

Q.192 What are the components of corticocerebellum?

Lateral portions of cerebellar hemispheres.

Q.193 What are the functions of corticocerebellum?

Corticocerebellum is concerned with integration and regulation of muscular activities because of its afferent and efferent connections with cerebral cortex through cerebro-cerebello-cerebral circuit. Cerebellum also receives feedback signals from the muscles during muscular activity.

Q.194 What is Charcot's triad?

It is a syndrome characterized by nystagmus, intention tremor and scanning speech due to disturbances of cerebellar connection with brainstem which generally occurs during disseminated sclerosis.

Q.195 Name the mechanisms of action of corticocerebellum.

Corticocerebellum acts by:

- Damping action
- Controlling ballistic movements
- Timing and programming the movements
- Servomechanism
- Comparator function.

Q.196 What are the effects of cerebellar lesion?

- Disturbances in tone and posture
- Disturbances in equilibrium
- Disturbances in movements.

Q.197 What are the disturbances in tone and posture during cerebellar lesion?

- Atonia
- Change in attitude
- Deviation movement
- Change in the response of deep reflexes.

Q.198 What are the disturbances in equilibrium during cerebellar lesion?

- *While standing:* The legs are spread to provide a broad base and the body sways from side to side with oscillation of head.
- *While moving:* Staggering, reeling and drunken like gait is observed.

Q.199 What are the disturbances in movements during cerebellar lesion?

- Ataxia
- Asynergia
- Asthenia
- Dysmetria
- Intention tremor
- Astasia
- Nystagmus
- Rebound phenomenon
- Dysarthria
- Diadochokinesis.

Q.200 What are basal ganglia?

Basal ganglia are the scattered masses of gray matter submerged in subcortical substances of cerebral hemisphere. Basal ganglia form the part of extrapyramidal system, which is concerned with integration and regulation of motor activities.

Q.201 What are the primary components of basal ganglia?

- Corpus striatum
- Substantia nigra
- Subthalamic nucleus of Luys.

Q.202 Give an account of hyperkinetic syndrome of basal ganglia.

Hyperkinetic syndrome may be due to:

- *Lesion of caudate N and putamen*: It is characterized by purposeless involuntary jerky movements which do not follow definite pattern known as chorea.
- *Lesion of globus pallidus*: It is characterized by slow confluent writhing or worm-like movement called atheosis.

Q.203 What are the parts of corpus striatum?

Corpus striatum includes:

- Caudate nucleus
- Lenticular nucleus which is divided into outer putamen and inner globus pallidus.

Q.204 What are the functions of basal ganglia?

- Control of voluntary motor activity
- Control of muscle tone
- Control of reflex muscular activity
- Control of automatic associated movements
- Role in arousal mechanism.

Q.205 Name the disorders of basal ganglia.

- Parkinson's disease
- Wilson's disease
- Chorea
- Athetosis
- Choreoathetosis
- Huntington's chorea
- Hemiballismus
- Kernicterus.

Q.206 What is the cause for Parkinson's disease?

Parkinson's disease or Parkinsonism is due to the damage of basal ganglia. It is mostly because of deficiency or lack of dopamine that is secreted by dopaminergic fibers of nigrostriatal pathway.

Q.207 What are the symptoms of Parkinson's disease?

- Rigidity
- Poverty of movements
- Static (drum beating) tremor
- Akinesia or hypokinesia
- Festinant gait
- Emotional changes.

Q.208 Name the lobes of cerebral cortex.

- Frontal lobe
- Parietal lobe
- Temporal lobe
- Occipital lobe.

Q.209 What are the functional divisions of frontal lobe?

Frontal lobe is divided into two parts on the basis of functions:

- Precentral cortex situated anteriorly
- Prefrontal cortex situated posteriorly.

Q.210 What are the parts of precentral cortex or excitomotor cortex?

- Primary motor area, which has area 4 and area 4S
- Premotor area, which has areas 6, 8, 44 and 45
- Supplementary motor area.

Q.211 What are the functions of precentral cortex?

The primary motor area of precentral cortex is concerned with initiation of voluntary movements and speech. Premotor area is responsible for movements of tongue, lips and larynx, which are involved in speech. Supplementary motor area is concerned with skilled movements.

Q.212 How is the localization or homunculus in motor area designed?

Muscles of various parts of the body are represented in area 4 in an inverted way from medial to lateral surface. The lower parts of the body are represented in medial surface and upper parts of the body are represented in lateral surface. The order of representation from medial to lateral surface is toe, ankle, knee, hip, trunk, shoulder, arm, elbow, wrist, hand, fingers and face. However, the face is not represented in inverted manner.

Q.213 What are the areas present in prefrontal cortex or orbitofrontal cortex?

Areas: 9, 10, 11, 12, 13, 14, 23, 24, 29, and 32.

Q.214 What are the functions of prefrontal cortex?

- It forms the center for higher functions like emotion, learning, memory and social behavior
- It is the center for planned actions
- It is the seat of intelligence and hence, it is called organ of mind
- It is responsible for personality of the individuals
- It is responsible for various autonomic changes during emotional conditions.

Q.215 What is frontal lobe syndrome? What are its important features?

The signs and symptoms, which occur due to injury or ablation of prefrontal cortex are together called frontal lobe syndrome.

Important features:

- Emotional instability
- Lack of concentration and fixing attention
- Lack of initiation and planning any action
- Loss of recent memory
- Loss of moral and social sense
- Failure to realize the seriousness of condition
- Functional abnormalities.

Q.216 What do you mean by amnesia, retrograde amnesia and anterograde amnesia?

- *Amnesia:* It is defined as the inability to recall the memories of recent events.
- *Retrograde amnesia:* It is inability to recall memories from the distant past.
- *Anterograde amnesia:* It is the inability of the person to establish new long-term memories of those types of information that are basis of intelligence. It occurs due to lesion in hippocampus.

Q.217 Name the functional areas of parietal lobe.

- Somesthetic area I or primary somesthetic or primary sensory area which has areas 3, 1 and 2
- Somesthetic area II
- Somesthetic association area which has areas 5 and 7.

Q.218 What are the functions of somesthetic areas of parietal lobe?

Somesthetic area I is responsible for perception and integration of cutaneous and kinesthetic sensations. Area 1 is concerned with sensory perception. Areas 2 and 3 are involved in the integration of these sensations.

Somesthetic area II is also concerned with perception of sensation.

Somesthetic association area is concerned with combined sensations like stereognosis.

Q.219 How is the localization or homunculus in primary sensory area designed?

The sensory areas (Fig. 17.10) of the body are represented in primary sensory area in an inverted manner. The toes are represented in lower part of medial surface, legs at the upper border, then from above downwards—knee, thigh, hip, trunk, upper limb, neck and face. The representation of face is not inverted.

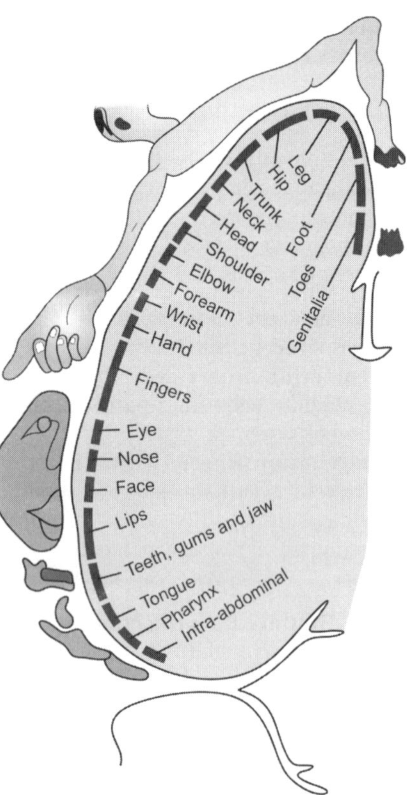

Fig. 17.10: Topographical arrangement (homunculus) of sensory areas in cerebral cortex

Q.220 What is sensory motor area?

The sensory area in postcentral gyrus extends anteriorly into precentral gyrus of frontal lobe, i.e. the motor area. Similarly, the motor area is extended from precentral gyrus posteriorly into postcentral gyrus. So, the precentral and postcentral gyri are knit together by association neurons and are interrelated functionally. This area where both motor and sensory neurons are present is called sensory motor area.

Q.221 What is the function of sensory motor area?

The sequential movements, which are timed and programmed by corticocerebellum are stored in the sensory motor area.

Q.222 Name the areas of temporal lobe.

- Primary auditory area, which includes areas 41, 42 and Wernicke's area
- Auditopsychic area, which includes area 22
- Area for equilibrium.

Q.223 What are the functions of primary auditory area?
Primary auditory area is concerned with perception and interpretation of auditory impulses. Areas 41 and 42 are concerned with perception of auditory impulses.

Wernicke's area (along with auditopsychic area—area 22) is responsible for the interpretation of sound.

Q.224 What is temporal lobe syndrome or Kluver-Bucy syndrome?
This is the condition that occurs in animals particularly in monkeys after the bilateral ablation of temporal lobe along with amygdaloid and uncus. It occurs in human beings during bilateral lesion of these structures.

Q.225 What are the manifestations of temporal lobe syndrome?
- Aphasia
- Auditory disturbances like tinnitus and auditory hallucinations
- Disturbances in smell and taste sensations
- Dreamy states
- Visual hallucinations.

Q.226 What are the areas of visual cortex?
- Primary visual area—area 17
- Visual association area—area 18
- Occipital eye field—area 19.

Q.227 What are the functions of areas of visual cortex?
Primary visual area (17) is concerned with perception of visual impulses. Visual association area (18) is concerned with interpretation of visual impulses. Occipital eye field (19) is concerned with movement of eyeballs.

Q.228 Define limbic system.
Limbic system is a group of cortical and subcortical structures, which form a limbus or ring around the hilus of cerebral hemisphere.

Q.229 What are the structures of limbic system?
- Paleocortical structures:
  - Hippocampus
  - Pyriform cortex
  - Olfactory lobe with olfactory tubercle.
- Juxtallocortical structures:
  - Cingulate gyrus
  - Orbitoinsulotemporal cortex
- Subcortical structures:
  - Amygdaloid
  - Septal nuclei
  - Thalamic nuclei
  - Hypothalamic nuclei
  - Caudate nucleus
  - Reticular formation of midbrain.

Q.230 What is Papez circuit?
The interconnections between the various structures of limbic system form a complex closed circuit called Papez circuit.
It includes: Hippocampus → mammillary bodies → thalamus → cingulate gyrus of cortex → hippocampus.

Q.231 What are the functions of limbic system?
- Role in olfaction
- Regulation of endocrine glands
- Regulation of ANS
- Regulation of food intake
- Control of circadian rhythm
- Regulation of sexual function
- Role in emotional state
- Role in memory
- Role in motivation.

Q.232 Define reticular formation.
Reticular formation is a diffused mass of neurons and nerve fibers forming an ill-defined meshwork of reticulum in central portion of brainstem.

Q.233 Name the divisions of reticular formation.
- Ascending reticular activating system
- Descending reticular system which includes:
  - Descending inhibitory reticular formation
  - Descending facilitatory reticular formation.

Q.234 What are the functions of ascending reticular activating system (ARAS)?
- It is concerned with arousal phenomenon, alertness, maintenance of attention and wakefulness.
- It causes emotional reactions
- It plays an important role in regulating the learning processes and development of conditioned reflexes.

Q.235 What are the classical or specific sensory pathways?
Classical or specific sensory pathways are the pathways which transmit the sensory impulses from receptors to cerebral cortex via thalamus.

Q.236 What are the functions of descending inhibitory reticular formation?
- Control of somatomotor system—it is responsible for smoothness and accuracy of movements by regulating the muscle tone
- Control of vegetative functions like cardiac function, blood pressure, respiration, gastrointestinal function and body temperature.

Q.237 What are the functions of descending facilitatory reticular formation?
- Facilitation of movements of the body and maintenance of muscle tone
- Facilitation of autonomic functions
- Role in wakefulness and alertness by activating the ARAS.

Q.238. What is decorticate preparation?
The animal in which all the connections of cerebral cortex are cut is called decorticate preparation. The basal ganglia and brainstem are kept intact.

Q.239 What are the features of decorticate animal?
There is extension of lower limbs and flexion of upper limbs at elbow joint across the chest. The wrists and fingers are also flexed. When neck is turned to one side, there is flexion of lower and upper limbs on the opposite side.

Q.240 What is decerebrate preparation?
In this, all the connections of cerebral hemispheres are removed at the level of midbrain by sectioning between superior and inferior colliculi.

Q.241 What are the features of decerebrate preparation?
It is characterized by a state of stiffness called decerebrate rigidity resembling the effects of upper motor neuron lesion. There is extension of all the limbs, extension of tail and arching of the back. This type of attitude is called opisthotonus.

Q.242 Name the intrafusal fibers which form the muscle spindle.
- Nuclear bag fiber
- Nuclear chain fiber.

Q.243 What is the unique feature of muscle spindle?
Muscle spindle is the only receptor in the body, which is innervated both by sensory nerve fibers and motor nerve fibers.

Q.244 Brief the innervation of muscle spindle.
The following are nerve supply of muscle spindle (Fig. 17.11):
Sensory nerve supply: By two types of sensory nerve ending:
- Primary sensory nerve ending belonging to type A alpha fiber.
- Secondary sensory nerve ending belonging to type A beta fiber.

Motor nerve supply: By gamma motor neuron belonging to type A gamma fibers.

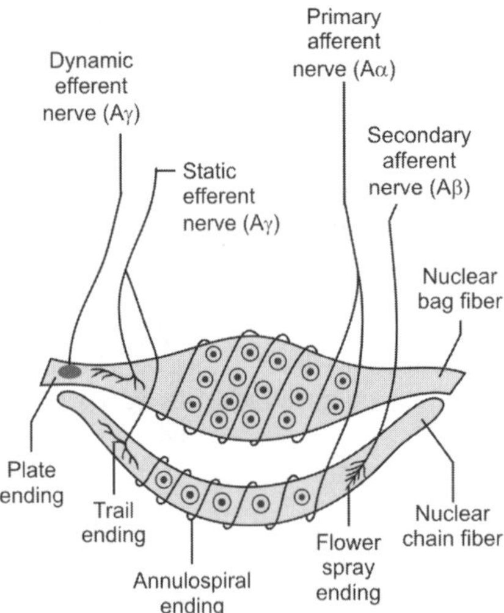

Fig. 17.11: Nerve supply to muscle spindle. Red = Afferent (sensory) nerve fibers. Blue = Efferent (motor) nerve fibers. Letters in parenthesis indicate the type of nerve fibers

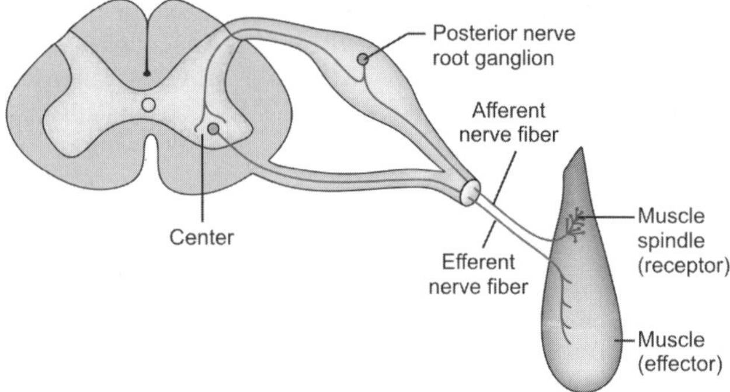

Fig. 17.12: Stretch reflex

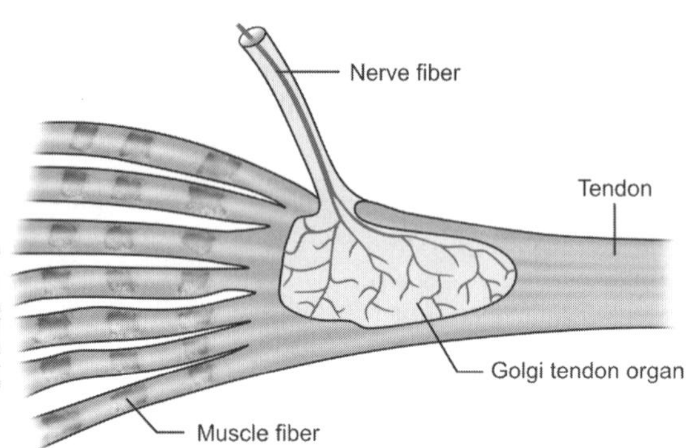

Fig. 17.13: Golgi tendon apparatus

Q.245 What are the functions of muscle spindle?

Muscle spindle gives response to change in the length of the muscle fiber. It has two functions:

- It is the receptor organ for stretch reflex.
- It plays an important role in the maintenance of muscle tone.

Q.246 What is stretch reflex?

When a muscle is stretched, it contracts reflexly. This is called stretch reflex or myotatic reflex (Fig. 17.12). It is a monosynaptic reflex and the quickest of all the reflexes.

Q.247 What is Golgi tendon organ? What is its nerve supply?

It is a proprioceptor situated in the tendon of the muscle (Fig. 17.13). It is supplied by sensory nerve fiber belonging to type A beta group.

Q.248 What are the functions of Golgi tendon organ?

Golgi tendon organ is concerned with:

- Forceful contraction
- Inverse stretch reflex
- Lengthening reaction.

Q.249 What is inverse stretch reflex?

The relaxation of the muscle due to powerful stretch is called the inverse stretch reflex. It is the inhibition of contraction of extrafusal fibers due to excessive stretching. So, it is called autogenic inhibition.

Q.250 What do you mean α–γ linkage?

For the maintenance of muscle tone, the impulses from the γ motor neuron causes contraction of end portion of intrafusal fibers resulting in stretching of muscle spindle. This leads to the discharge of impulses from the primary sensory nerve endings. These impulses stimulate the α motor neurons of spinal cord which in turn send impulses to extrafusal fibers and cause contraction of extrafusal muscle. This is known as α–γ linkage.

Q.251 What is reciprocal inhibition?

When a stretch reflex is induced, activity of afferent fibers from muscle spindle excites the motor neurons supplying the muscle from which the impulses come and inhibits those supplying its antagonist muscle. This phenomenon is called as reciprocal inhibition.

Q.252 Compare monosynaptic and polysynaptic reflex.

Parameter	*Monosynaptic*	*Polysynaptic*
• *No of synapse*	Only one	Many
• *Latent period*	Shorter	Comparatively longer
• *Important feature*	Do not have phenomenon of after discharge	Present
• *Example*	Stretch reflex	Withdrawal and superfacial reflex.

Q.253 What is lengthening reaction?

In a decerebrate animal, some resistance is offered when the arm is flexed at elbow joint passively. This resistance is offered because of stretch reflex developed in the triceps muscle. However, if the forearm is flexed forcefully, the resistance to flexion is abolished suddenly, leading to quick flexion of arm. This is called lengthening reaction. And it is due to activation of Golgi tendon organ.

Q.254 Define posture.

Posture is defined as the subconscious adjustment of tone in different groups of muscles in accordance to every movement of the body.

Q.255 What is the significance of posture?

Posture helps to have smooth and accurate movements and also to maintain the body in equilibrium with line of gravity.

Q.256 Name the basic phenomena of posture.

Muscle tone and stretch reflex.

Q.257 What are the types of postural reflexes?

- Static reflexes which occur at rest
- Statokinetic reflexes which occur during movement.

Q.258 Name the types of static postural reflexes.

- General static or righting reflexes
- Local static or supporting reflexes
- Segmental static reflexes
- Statotonic or attitudinal reflexes (Table 17.4).

Q.259 Define general static or righting reflexes.

General static or righting reflexes are the postural reflexes, which help to maintain the upright position of the body.

Q.260 Name the righting reflexes.

- Labyrinthine righting reflexes acting upon the neck muscles
- Neck righting reflexes acting upon the body
- Body righting reflexes acting upon the head
- Body righting reflexes acting upon the body
- Optical righting reflexes.

Q.261 Where are the centers for righting reflexes situated?

Centers for all the righting reflexes except optic righting reflexes are in midbrain. The center for optical righting reflexes is in the cerebral cortex.

Q.262 What is the significance of local static or supporting reflexes?

Local static or supporting reflexes support the body against the gravity in different positions and also protect the limbs against hyperextension or hyperflexion.

Q.263 Name the supporting reflexes. Where is their center situated?

- Positive supporting reflexes
- Negative supporting reflexes.

The center is in spinal cord.

Q.264 Name the segmental static reflex.

Crossed extensor reflex is the segmental static reflex. (Refer Question 96 of this section for details).

Q.265 Define statotonic or attitudinal reflexes.

Statotonic or attitudinal reflexes are the postural reflexes developed according to the attitude of the body.

Q.266 What are the types of statotonic reflexes? Where is their center situated?

- Tonic labyrinthine and neck reflexes acting upon limbs
- Tonic labyrinthine and neck reflexes acting upon eyes.

  The center is in the medulla oblongata.

Q.267 What are the statokinetic reflexes?

Postural reflexes concerned with angular (rotatory) and linear (progressive) movements are known as statokinetic reflexes.

Q.268 What are the parts of labyrinth or inner ear?

- Vestibular apparatus
- Cochlea.

Q.269 What is vestibular apparatus? What are its parts?

Vestibular apparatus is a part of inner ear concerned with maintenance of posture and equilibrium.

It consists of semicircular canals and otolith organ.

Q.270 What is otolith organ?

Otolith organ is the part of vestibular apparatus. It is formed by utricle and saccule.

Q.271 Name the semicircular canals of vestibular apparatus.

- Anterior or superior canal
- Posterior canal
- Lateral or horizontal canal.

Q.272 How are the semicircular canals situated?

Anterior and posterior semicircular canals are situated vertically. The lateral semicircular canals are situated horizontally.

Q.273 What is ampulla?

The enlarged portion of each semicircular canal is known as ampulla.

Q.274 What is crista ampullaris?

Crista ampullaris is the receptor organ in the ampulla of semicircular canal.

Q.275 Name the receptor cells of crista ampullaris?

Type I and type II hair cells.

Q.276 What are stereocilia?

Stereocilia are the cilia arising from cuticular plate in the apex of hair cells of crista ampullaris.

Q.277 What is kinocilium?

Kinocilium is the largest cilium of hair cells of crista ampullaris.

Q.278 What is macula?

Macula is the receptor organ of otolith organ.

Q.279 How is macula of otolith organ situated?

Macula of utricle is situated in horizontal plane so that, the hair cells are in vertical position.

The macula of saccule is situated in vertical plane so that, the hair cells are in horizontal position.

Q.280 Name the nerve supplying the vestibular apparatus.

Vestibular division of vestibulocochlear nerve.

Q.281 What are the functions of vestibular apparatus?

- It is responsible for detecting the position of head during different movements
- It causes reflex adjustments in the position of eyeballs, head and body during postural changes.

Table 17.4: Static position reflexes

Reflex	*Center*	*Animal preparation to demonstrate*
General static reflexes (Righting reflexes) • Labyrinthine righting reflexes acting upon the neck muscles • Neck righting reflex acting upon the body • Body righting reflexes acting upon the head • Body righting reflexes acting upon body • Optical righting reflexes	Red nucleus situated in midbrain Occipital lobe	Thalamic or normal blind folded animal Labyrinthectomized animal
Local static reflexes • Positive supporting reflexes • Negative supporting reflexes	Spinal cord	Decorticate animal
Segmental static reflexes • Crossed extensor reflex	Spinal cord	Spinal animals
Statotonic or attitudinal reflexes • Tonic labyrinthine and neck reflexes acting on the limbs • Labyrinthine and neck reflexes acting upon the eyes	Medulla oblongata	Decerebrate animal

Q.282 What is the function of semicircular canals?
Semicircular canals are responsible for the maintenance of posture and equilibrium during rotatory movements or angular acceleration of the head in vertical, anteroposterior or transverse axis.

Q.283 What is the function of otolith organ?
Otolith organ is responsible for the maintenance of posture and equilibrium during linear acceleration or progressive movements. Utricle is concerned with horizontal acceleration and saccule is concerned with vertical acceleration.

Q.284 What is nystagmus?
The characteristic to and fro movements of eyeball, which occur during rotation in horizontal plane is known as nystagmus.

Q.285 What are the components of nystagmus?
- *Slow component:* At the beginning of the rotation of head, the eyeballs rotate in the opposite direction of head slowly.
- *Quick component:* When slow movements of eyeballs stop, the eyeballs move quickly to the new fixation point in the direction of rotation of head.

Q.286 What are the causes for the slow and quick components of nystagmus?
Slow component of nystagmus is due to vestibuloocular reflex which is produced when the labyrinthine impulses reach the ocular muscles.

Quick component is due to the activation of some centers in brainstem.

Q.287 What are the effects of labyrinthectomy?
Removal of labyrinthine apparatus on both sides (bilateral) causes loss of equilibrium and loss of hearing sensation. Removal of labyrinthine apparatus on one side (unilateral) causes less effect on postural reflexes. Some autonomic symptoms like nausea, vomiting and diarrhea occur.

Q.288 Define electroencephalogram or EEG.
Record or graphical registration of electrical activities of the brain is known as electroencephalogram or EEG.

Q.289 What is the significance of EEG?
EEG is useful in the diagnosis of neurological disorders and sleep disorders.

Q.290 Name the waves of EEG.
- Alpha rhythm
- Beta rhythm
- Delta rhythm.

Q.291 What is alpha block?
While recording EEG with closed eyes, alpha rhythm appears. When the eyes are opened, the alpha rhythm disappears and fast, irregular and low voltage waves appear. Disappearance of alpha rhythm is known as alpha block.

Q.292 Name the physiological conditions when delta waves appear in EEG.
Delta waves are common in early childhood during waking hours. In adults, deep sleep is the only physiological condition when delta waves appear in EEG.

Q.293 Name the pathological conditions when delta waves appear in EEG.
- Brain tumor
- Epilepsy
- Increased intracranial pressure
- Mental deficiency or depression.

Q.294 What are theta waves of EEG?
Theta waves are the low frequency and low voltage waves appearing in EEG in children below five years of age.

Q.295 Define sleep.
Sleep is the mental and physical relaxation either superficially or deeply with closed eyes.

Q.296 What are the important physiological changes during sleep?
- Decreased plasma volume
- Reduced heart rate and blood pressure
- Reduced rate and force of respiration and appearance of Cheyne-Stoke's breathing
- Decreased salivary secretion
- Decreased urine formation and increased specific gravity of urine
- Increased sweat secretion
- Reduced lacrimal secretion
- Reduced muscle tone
- Absence of some reflexes like knee jerk and appearance of Babinski's sign
- Constriction of pupil and movement of eyeballs up and down.

Q.297 Name the types of sleep.
- Nonrapid eye movement sleep or NREM sleep (non-REM sleep) (Table 17.5)
- Rapid eye movement sleep or REM sleep or paradoxical sleep.

Table 17.5: Characteristics of REM sleep and non-REM sleep

S. No.	*Characteristics*	*REM sleep*	*Non-REM sleep*
1.	Rapid eye movement	Present	Absent
2.	Dreams	Present	Absent
3.	Muscle twitching	Present	Absent
4.	Heart rate	Fluctuating	Stable
5.	Blood pressure	Fluctuating	Stable
6.	Respiration	Fluctuating	Stable
7.	Body temperature	Fluctuating	Stable
8.	Neurotransmitter	Noradrenaline	Serotonin

Q.298 What is REM sleep?
This is a type of deep sleep during which the eyeballs move frequently and dreams may appear. This occupies 20 to 30% of total sleeping period.

Q.299 What are the changes noticed in EEG during REM sleep?
EEG shows irregular waves (desynchronized waves) with high frequency and low amplitude.

Q.300 What is NREM sleep?
This is the type of sleep during which the eyeballs do not move. This occupies 70–80% of total sleeping period.

Q.301 What are the stages of NREM sleep?
NREM sleep is divided into four stages based on EEG pattern:
- Stage of drowsiness
- Stage of light sleep
- Stage of medium sleep
- Stage of deep sleep

Q.302 What are the changes noticed in EEG during different stages of NREM sleep?
- *Stage of drowsiness:* Alpha waves appear.
- *Stage of light sleep:* Alpha waves are diminished and abolished. Low voltage fluctuations and infrequent delta waves appear.
- *Stage of medium sleep:* Spindle bursts superimposed by low voltage delta waves appear.
- *Stage of deep sleep:* More prominent delta waves appear.

Q.303 What is the mechanism of sleep?
Sleep occurs due to activation of sleep inducing centers (raphe nucleus and locus ceruleus) and inhibition of ascending reticular activating system (ARAS).

Q.304 What are the neurotransmitters causing sleep?
Serotonin secreted by nerve fibers from raphe nucleus of pons—induces non-REM sleep.

Noradrenaline secreted by nerve fibers from locus cereleus—induces REM sleep.

Q.305 Define epilepsy.
Epilepsy is the brain disorder characterized by convulsive seizure or loss of consciousness or both.

Q.306 What is the cause for epilepsy?
Epilepsy is due to excessive discharge of impulses from some parts of brain particularly cerebral cortex.

Q.307 What are convulsive seizures?
Uncontrolled involuntary muscular contractions are called convulsive seizures.

Q.308 What are the types of epilepsy?
- Generalized epilepsy or general onset epilepsy or general onset seizure due to excessive discharge of impulses from all parts of brain
- Localized epilepsy or local epilepsy or focal epilepsy or local seizure due to discharge of impulses from one part of brain.

Q.309 Name the types of generalized epilepsy.
- Grand mal epilepsy
- Petit mal epilepsy
- Psychomotor epilepsy.

Q.310 Define learning.
Learning is defined as process by which new information is acquired.

Q.311 Name the types of learning.
- Associative learning
- Nonassociative learning.

Q.312 What is habituation?
Habituation means getting used with something to which a person is constantly exposed. When a person is exposed to a stimulus repeatedly he starts ignoring the stimulus slowly.

Q.313 What is sensitization?
When a stimulus is applied repeatedly, habituation occurs. But if the same stimulus is combined with another type of stimulus, which may be pleasant or unpleasant, the person becomes more sensitive to the original stimulus. It is called the amplification of response or sensitization.

Q.314 Define memory.
Memory is defined as the ability to recall the past experience. It is also defined as retention of learned materials.

Q.315 Classify memory on physiological basis.
- Explicit memory that involves conscious recollection of past experience
- Implicit memory in which the past experience is utilized without conscious awareness.

Q.316 Define short-term and long-term memories.
Short-term memory is the recalling of events of hours or days. Long-term memory is the recalling of events of weeks, months or years.

Q.317 What is the basic mechanism of short-term memory?
Basic mechanism of short-term memory is the development of new neural circuits by the formation of new synapses.

Q.318 What is the basic mechanism of long-term memory?
Basic mechanism of long-term memory is reinforcement of newly formed neuronal circuits by using it often leading to consolidation and encoding of memory in different areas of brain.

Q.319 Name the sites of encoding of memory.
Hippocampus, Papez circuit, frontal areas and parietal areas.

Q.320 What is memory engram or memory tracing?
It is the process by which the memory is facilitated and stored in brain by means of structural and biochemical changes.

Q.321 What is consolidation of memory?
It is the process by which a short-term memory is crystallized into a long-term memory.

Q.322 Name the drugs, which facilitate memory.
Caffeine, physostigmine, amphetamine, nicotine, strychnine and metrazol.

Q.323 What is amnesia?
Loss of memory is known as amnesia.

Q.324 What is dementia?
Progressive deterioration of intellect, emotional control, social behavior and motivation associated with loss of memory is known as dementia.

Q.325 What is Alzheimer's disease?
It is a progressive neurodegenerative disease due to degeneration, loss of function and death of neurons in many parts of brain particularly cerebral cortex. Dementia is the common feature of this disease.

Q.326 Define conditioned reflex.
Conditioned reflex is a reflex response acquired or learnt by experience.

Q.327 Classify the conditioned reflexes.
- Classical conditioned reflexes which are established by a conditioned stimulus followed by an unconditioned stimulus
- Instrumental or operant conditioned reflexes which are established by conditioned stimulus followed by reward or punishment (behavior of the person is instrumental).

Q.328 How are the properties of classical conditioned reflexes demonstrated?
By the classical salivary secretion experiments devised by Ivan Pavlov.

Q.329 Classify the classical conditioned reflexes.
- Positive conditioned reflex:
  - Primary conditioned reflex—with one conditioned stimulus
  - Secondary conditioned reflex—with two conditioned stimuli
  - Tertiary conditioned reflex—with three conditioned stimuli.
- Negative conditioned reflex:
- External or indirect inhibition
- Internal or direct inhibition.

Q.330 What is the significance of instrumental conditioned reflexes?
Instrumental conditioned reflexes play an important role in the development of behavior pattern in an individual particularly during learning process in childhood.

Q.331 What is the physiological basis of conditioned reflexes?
Learning and memory form the physiological basis of conditioned reflexes.

Q.332 Define speech.
Speech is the expression of thought by production of articulate sound, bearing a definite meaning.

Q.333 What is the mechanism of speech?
Mechanism of speech is by the coordinated activities of central speech apparatus and peripheral speech apparatus.

Central speech apparatus consists of higher centers, i.e. the cortical and subcortical centers. Peripheral speech apparatus includes larynx or sound box, pharynx, mouth, nasal cavities, tongue, and lips. All the structures of peripheral speech apparatus work in coordination with respiratory system.

Q.334 What are the cortical areas concerned with speech?

- Motor areas:
  - Broca's area (area 44) or speech center or motor speech area—it controls the movements of vocalization
  - Upper frontal area—it controls the coordinated movements concerned with writing.
- Sensory areas:
  - Auditopsychic area—it is concerned with memories of the spoken words
  - Visuopsychic area—it is concerned with storage of memories of visual symbols.

Q.335 What is Wernicke's area, Dejerine area and sensory speech center?

- Wernicke's area is auditory speech center located in the region at the posterior end of the superior temporal gyrus in the dominant hemisphere and is concerned with comprehension, i.e. interpretation and understanding of auditory information.
- Dejerine area is visual speech center located in the angular gyrus behind Wernicke's area. These two areas are collectively known as sensory speech area.

Q.336 What are the motor speech areas?

It includes Broca's area (area 44) and Exner's area (motor writing center). Broca's area is located in the inferior frontal gyrus in the region of the anterior and ascending rami of the lateral sulcus in the dominant hemisphere whereas Exner's area is located in the middle frontal gyrus in the dominant hemisphere.

Q.337 Define aphasia? What is its cause?

Aphasia is defined as loss or impairment of speech due to brain damage. It is due to damage of speech centers.

Q.338 Name the types of aphasia.

- Broca's aphasia
- Wernicke's aphasia
- Global aphasia
- Anomic aphasia
- Other types of aphasia—motor aphasia, sensory aphasia and agraphia.

Q.339 What is dysarthria or anarthria? What is it due to?

Difficulty or inability to speak is known as anarthria or dysarthria. It is due to the paralysis or ataxia (lack of coordination) of muscles involved in speech.

Q.340 What is pure word deafness?

It is failure to recognize spoken speech with no other defect of speech or intelligence.

Q.341 What is cerebrospinal fluid or CSF?

The fluid present in the central canal of spinal cord, subarachnoid space and cerebral ventricles is known as cerebrospinal fluid or CSF. It is a part of ECF.

Q.342 Which is the site of formation of CSF?

CSF is formed by the choroid plexus, which is situated in the ventricles of cerebral hemispheres. Major portion of CSF is formed in lateral ventricles.

Q.343 What is the mechanism of formation of CSF?

CSF is formed by process of secretion by the choroid plexus. It involves active transport with expenditure of energy.

Q.344 What are the properties of CSF?

Volume	: 150 mL
Rate of formation	: 0.3 mL/minute
Specific gravity	: 1.005
Reaction	: Alkaline.

Q.345 What is normal glucose level in CSF?

50–70 mg%.

Q.346 What is the normal cerebral blood flow?

750 mL/min.

Q.347 What is the composition of CSF?

CSF consists of 99.13% of water and 0.87% of solids. Solids:

- Organic substances such as proteins, amino acids, sugar, cholesterol, urea, uric acid, creatinine and lactic acid
- Inorganic substances such as sodium, potassium, calcium, magnesium, chlorides, phosphates, bicarbonates and sulfates
- Blood cells—lymphocytes - 5/cu mm.

Q.348 Describe the circulation of CSF briefly.

Major quantity of CSF (Fig. 17.14) is formed in lateral ventricles and flows to third ventricle through foramen of Monro. From here, it passes to fourth ventricle through aqueductus Sylvius. From fourth ventricle, CSF enters cisterna magna and cisterna lateralis through foramen of Magendie (central opening) and foramen of Luschka (lateral opening).

A portion of cisternal fluid circulates through spinal subarachnoid space. Larger part of fluid passes upwards over the brainstem through the surface of the cerebral hemispheres.

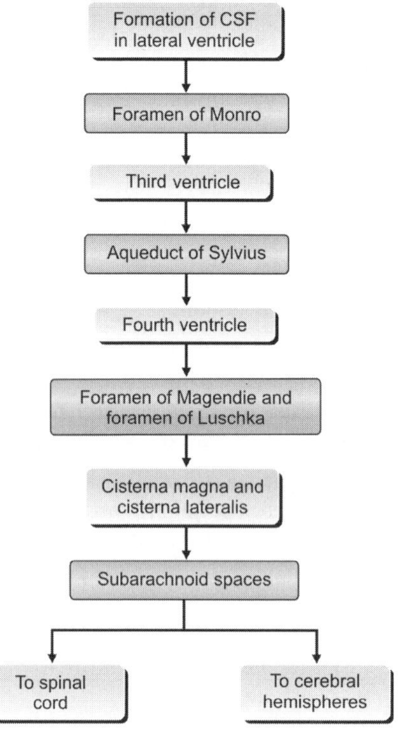

Fig. 17.14: Schematic diagram of CSF circulation

Q.349 How is CSF absorbed?

CSF is mostly absorbed by the arachnoid villi into dural sinuses and spinal veins. A small amount is absorbed along the perineural spaces into cervical lymphatics and into perivascular spaces.

Q.350 What is the mechanism of absorption of CSF?

CSF is absorbed by means of filtration due to the gradient between hydrostatic pressure in subarachnoid space and the pressure exerted by blood in dural sinus.

Q.351 What is the normal pressure exerted by CSF?

In lateral recumbent position: 10 to 18 cm H_2O.

In sitting posture: About 30 cm H_2O.

Q.352 What are the functions of CSF?

- Protection of brain against severe blow
- Regulation of cranial content volume
- Medium for exchange of nutritive substances, respiratory gases and waste products between the blood and brain tissues.

Q.353 What is contrecoup injury?

When head receives a severe blow, the brain moves forcefully and hits against the skull bone at a point opposite where the blow was applied. This leads to damage of brain tissues. This is called contrecoup injury.

Q.354 How is CSF collected?
CSF is mostly collected by lumbar puncture by passing a needle into the subarachnoid space between 3rd and 4th lumbar spines.

It is also collected by cisternal puncture by passing a needle into cisterna magna between occipital bone and atlas.

Q.355 What is hydrocephalus? What are its effects?
Abnormal accumulation of CSF in the skull associated with enlargement of head is called hydrocephalus.

It causes atrophy of brain tissues, mental weakness and convulsions.

Q.356 What is blood-brain barrier?
The barrier for passage of certain substances from blood into brain tissues is known as blood-brain barrier.

Q.357 How is blood-brain barrier developed?
Blood-brain barrier is developed by the formation of tight junctions between the endothelial cells of capillaries and development of foot processes of astrocytes (neuroglia) around the capillaries.

Q.358 Name some substances, which can pass through blood-brain barrier.
Oxygen, carbon dioxide, water, glucose, amino acids, electrolytes and some drugs like sulfonamides, tetracycline and many lipid soluble substances.

Q.359 Name some substances, which cannot pass through blood-brain barrier.
Catecholamines, penicillin and bile pigments.

Q.360 What are the functions of blood-brain barrier?
- It protects brain from damages caused by entry of injurious substances from blood into brain tissues
- It maintains the constant neuronal environment in central nervous system by preventing the escape of neurotransmitter into the blood.

Q.361 What is blood cerebrospinal fluid barrier?
The barrier between the blood and cerebrospinal fluid existing in choroid plexus is called blood cerebrospinal barrier fluid. It allows the movements of all the substances, which are allowed by blood-brain barrier.

Q.362 What is autonomic nervous system (ANS)?
ANS is the part of peripheral nervous system that is concerned with regulation of visceral or vegetative function of the body. It is also called vegetative or involuntary nervous system.

Q.363 What are the divisions of ANS?
- Sympathetic division or thoracolumbar outflow which includes the nerve fibers arising from lateral gray horns of all the 12 thoracic segments and the first two lumbar segments of spinal cord
- Parasympathetic division or craniosacral outflow which includes some cranial nerve fibers and fibers arising from sacral segments of spinal cord.

Q.364 Name the ganglia present in sympathetic division of ANS.
- Paravertebral ganglia or sympathetic ganglia
- Prevertebral or collateral ganglia
- Terminal or peripheral ganglia.

Q.365 Name the nerves, which constitute the parasympathetic division of ANS.
- Cranial nerves–III, VII, IX and X nerves
- The pelvic nerve formed by sacral nerve fibers arising from I, II and III sacral segments of spinal cord.

Q.366 What are the functions of ANS?
ANS is concerned with regulation of vegetative functions in the body, which are beyond voluntary control. By regulating various vegetative functions, ANS plays an important role in homeostasis.

Q.367 What are the neurotransmitters secreted by sympathetic fibers?
Preganglionic sympathetic fibers: Acetylcholine. Postganglionic sympathetic adrenergic fibers: Noradrenaline.

Postganglionic sympathetic cholinergic fibers: Acetylcholine.

Q.368 Name the structures innervated by sympathetic cholinergic nerve fibers.
Blood vessels to:
- Heart
- Skeletal muscles
- Sweat glands.

Q.369 What is the neurotransmitter secreted by parasympathetic fibers?
Both preganglionic and postganglionic fibers of parasympathetic nerves secrete acetylcholine.

Q.370 What are sympathomimetic drugs? Give examples.
Drugs, which produce the effects similar to the effects of stimulation of sympathetic nerves are called sympathomimetic drugs. Examples are phenylephrine, isoproterenol, albuterol, ephedrine, tyramine and amphetamine.

Q.371 What are sympathetic blockers? Give examples.
The drugs, which prevent the actions of sympathetic neurotransmitters are known as sympathetic blockers.

Examples are reserpine, guanethidine, benzamine, phentolamine, metaprolol and hexamethonium.

Q.372 What are parasympathomimetic drugs? Give examples.
Drugs, which produce the effects similar to the effects of stimulation of parasympathetic nerves are called parasympathomimetic drugs.

Examples are pilocarpine, methacholine, neostigmine and physostigmine.

Q.373 What are parasympathetic blockers? Give examples.
Drugs, which prevent the actions of parasympathetic nerve fibers are called parasympathetic blockers.

Examples are hematopine and scopolamine.

Q.374 What are ganglionic blockers? Give examples.
Drugs, which prevent the transmission of impulses from preganglionic neurons to postganglionic neurons are called ganglionic blockers.

Examples are tetraethyl ammonium, hexamethonium and pentolinium.

Chapter 18

Special Senses

Q.1 Define special senses.
The complex sensations are called special sensations or special senses.

Q.2 Name the special senses.
- Visual sensation
- Auditory sensation
- Gustatory or taste sensation
- Olfactory sensation or sensation of smell.

Q.3 What is optic axis?
The line joining the anterior pole and posterior pole of the eyeball is called optic axis.

Q.4 What is visual axis?
The line joining a point in cornea little medial to anterior pole and fovea centralis (which is situated little lateral to posterior pole) is known as visual axis.

Q.5 What is the significance of visual axis?
The significance of visual axis is that the light rays from the object passes through this axis and reach the retina of eye.

Q.6 What is conjunctiva? What are its parts?
Conjunctiva is a thin mucous membrane that covers the exposed part of eye.
Its parts:
- Bulbar portion that covers the anterior surface of eyeball
- Palpebral portion that covers the inner surface of eyelids.

Q.7 What is lacrimal gland? Where is it situated?
Lacrimal gland is the glandular structure that secretes tear. It is situated in the bone that forms upper and outer border of eye socket.

Q.8 How is tear drained?
From lacrimal gland, tear flows over the surface of conjunctiva and drains into nose via lacrimal ducts, lacrimal sac and nasolacrimal duct.

Q.9 What are the layers of wall of eyeball?
- Outer layer or tunica externa or tunica fibrosa that includes cornea and sclera
- Middle layer or tunica media or tunica vasculosa that includes choroid, ciliary body and iris
- Inner layer or tunica interna or tunica nervosa or retina.

Q.10 What is cornea?
The transparent structure that forms the anterior 1/6th of outer layer of eyeball is called cornea.

Q.11 What is sclera?
Sclera is the posterior 5/6th of outer layer of eyeball.

Q.12 What is choroid? How is it formed?
Choroid is the posterior 1/6th of middle layer of eyeball. It is formed by capillary plexus, small arteries and veins.

Q.13 What is ciliary body?
Ciliary body is a ring-like structure formed by anterior part of middle layer of eyeball.

Q.14 Name the parts of ciliary body.
- Orbiculus ciliaris
- Ciliary body proper
- Ciliary processes.

Q.15 What is iris? And what is pupil?
Iris is a circular diaphragm formed by anterior most portion of middle layer of eyeball and it is placed in front of lens.

Pupil is the circular opening in the center or iris. The anterior and posterior chambers communicate through the pupil.

Q.16 What is retina? Name the layers of retina.
Retina is the layer of eyeball that forms the sensory part. It contains the receptors for vision.
Layers of retina (Fig. 18.1):
- Layer of pigment epithelium
- Layer of rods and cones
- External limiting membrane
- Outer nuclear layer
- Outer plexiform layer

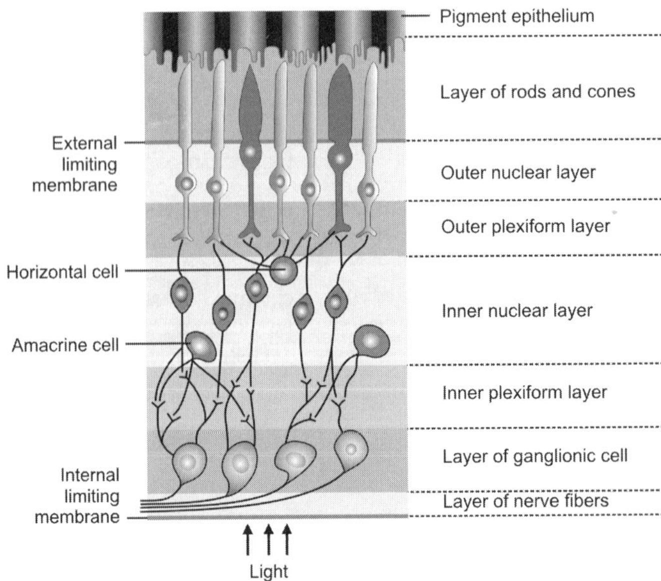

Fig. 18.1: Layers of retina

- Inner nuclear layer
- Inner plexiform layer
- Ganglion cell layer
- Layer of nerve fibers
- Internal limiting membrane.

Q.17 What is fundus oculi? How is it examined?

The posterior part of interior of eyeball is known as fundus oculi or fundus. It is examined by using ophthalmoscope.

Q.18 Name the important structures of fundus oculi.

- Optic disk
- Macula lutea.

Q.19 What is optic disk?

Optic disk is a pale disk-like structure situated near the center of the posterior wall of eyeball. It is formed by convergence of optic nerve fibers. It is also called blind spot because it is insensitive to light since there are no rods and cones here.

Q.20 What is macula lutea?

Macula lutea is a yellow spot situated little lateral to optic disk in the posterior wall of the eyeball. The yellow color is due to the presence of a yellow pigment. There is a small depression in the center of macula densa called fovea centralis.

Q.21 What is the importance of fovea centralis?

Fovea centralis is the region of acute vision because it contains only the cones.

Q.22 Name the intraocular fluids.

- Vitreous body
- Aqueous humor.

Q.23 What is vitreous body? How is it formed?

Vitreous body is a gelatinous substance present in the space between the lens and retina. It is formed by network of proteoglycan molecules.

Q.24 What is aqueous humor? How is it formed?

Aqueous humor is a thin fluid present in the space between lens and cornea. It is formed from plasma by diffusion, ultrafiltration and active transport of substances through epithelial cells lining the ciliary processes.

After formation, aqueous humor reaches the posterior chamber by passing through suspensory ligaments. From here it reaches the anterior chamber via pupil.

Q.25 How is aqueous humor drained?

Aqueous humor is drained from anterior chamber into extra ocular veins by passing through limbus (the angle between cornea and iris), the meshwork of trabeculae and canal of Schlemm.

Q.26 What is the physiological and clinical importance of canal of Schlemm?

Aqueous humor once formed from the ciliary process passes from the posterior chamber then via pupil enters into anterior chamber which then passes into the intrascleral venous plexus through canal of Schlemm, thereby this canal helps to drainage the aqueous humor continuously secreting from ciliary body. Blockade of the canal of Schlemm leads to increase intraocular pressure above 80 mm Hg resulting in pain and degeneration of blood vessels of retina and choroid. This condition is known as glaucoma.

Q.27 What are the functions of aqueous humor?

Aqueous humor:

- Maintains the shape of eyeball
- Maintains the intraocular pressure
- Provides nutrition, oxygen and electrolytes to avascular structures—lens and cornea.

Q.28 What is the normal intraocular pressure? How is it measured?

Normal intraocular pressure is 12–20 mm Hg.

It is measured by tonometer.

Q.29 What are the changes taking place in lens during old age?

After 40–45 years of age, the lens looses its elastic property and presbyopia occurs. After 55 to 60 years, lens becomes opaque resulting in cataract.

Q.30 What is cataract?

Cataract is the opacity or cloudiness in natural lens of the eye.

Q.31 Name the ocular muscles.

- Superior rectus
- Inferior rectus
- Medial rectus
- Lateral rectus
- Superior oblique
- Inferior oblique.

Q.32 What are the nerves supplying ocular muscles?

- Oculomotor (III) nerve that supplies superior rectus, inferior rectus, medial rectus and inferior oblique muscle
- Trochlear (IV) nerve that supplies superior oblique muscle
- Abducent (VI) nerve that supplies lateral rectus muscle.

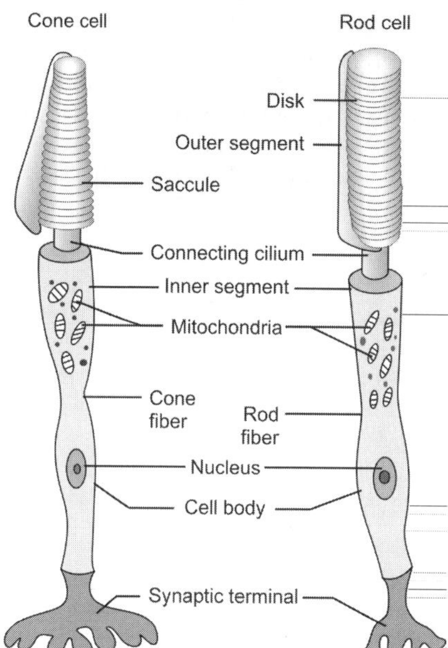

Fig. 18.2: Structure of visual receptors

Q.33 What is the refractory power of cornea and lens?

Refractory power of cornea is 42 D (Diopter) and refractory power of lens is 23 D.

Q.34 What are the visual receptors? Explain their distribution briefly.

Visual receptors (Fig. 18.2) are rods and cones present in the retina of eyeball. In fovea centralis, only the cones are present. While proceeding from fovea towards periphery of retina, rods increase and cones decrease in number. At the periphery of retina, only rods are present.

Q.35 What is the function of rods?

Rods have low threshold for light stimulus and are responsible for dim light vision or night vision or scotopic vision.

Q.36 What are the functions of cones?

Cones have high threshold for light stimulus and are responsible for bright vision or day light vision or photopic vision. Cones are also responsible for acuity of vision and color vision.

Q.37 What is rhodopsin?

Rhodopsin or visual purple is the photosensitive pigment present in the outer segment of rod cells.

Q.38 What is phototransduction?

Phototransduction or visual phototransduction is the process by which the light stimulus causes development of receptor potential in the visual receptors.

Q.39 What is the difference between the resting membrane potential in visual receptors and other cells of the body?
Resting membrane potential in visual receptors is very less and it is only about -40 mV whereas in other cells of the body it is - 70 to - 90 mV.

Q.40 What is the difference between the receptor potential in rod cells and other sensory receptors?
Usually, the receptor potential is in the form of depolarization in the sensory receptors. But, in rod cells, it is in the form of hyper-polarization, i.e. the negativity increases to about - 70 to - 80 mV.

Q.41 What are the photosensitive pigments present in cone cells?
Porphyropsin, iodopsin and cyanopsin.

Q.42 What is dark adaptation? What are its causes?
When a person enters a room with dim light after spending long time in a bright light area, he cannot see any object in the beginning. After about 20 minutes time, he starts seeing the object. This process is called dark adaptation.
Causes:
- Increase in the sensitivity of rods due to resynthesis of rhodopsin
- Dilatation of pupil.

Q.43 What is light adaptation? What are its causes?
When a person enters a bright light area from a dim light area, he feels discomfort for some time due to dazzling effect of bright light. After about 5 minutes, he is able to see the objects without discomfort. This process is called light adaptation.
Causes:
- Reduction in the sensitivity of rods due to breakdown of rhodopsin
- Constriction of pupil.

Q.44 Define electroretinogram (ERG).
Electroretinogram (ERG) is the record of electrical activity produced in retina when it is stimulated by the light rays.

Q.45 Define acuity of vision.
Ability of the eye to determine the precise shape and details of any object is called acuity of vision or visual acuity.

Q.46 Name the receptors responsible for acuity of vision.
Cones are responsible for acuity of vision.

Q.47 How is acuity of vision tested?

For distant vision	:	By using Snellen's chart
For near vision	:	By using Jaeger's chart.

Q.48 Define field of vision.
Part of external world seen by one eye when it is fixed in one direction is known as field of vision.

Q.49 What is binocular vision?
In man and some animals in whom the eyeballs are situated in front of head, the visual fields of both eyes overlap, i.e. a portion of external world is seen by both eyes. This is known as binocular vision.

Q.50 What is monocular vision?
In some animals like horse in whom the eyeballs are situated at the sides of head, the visual fields of both eyes overlap only to a very small extent, i.e. different portions of external world is seen by each eye. This is known as monocular vision.

Q.51 What are the divisions of visual field?
- Temporal field that extends to about 100° laterally
- Nasal field that extends to about 60° medially
- Upper field that extends to about 60° above
- Lower field that extends to about 75° below.

Q.52 What are corresponding retinal points?
While looking at an object, the points of retina of both eyes on which the light rays from the object fall are called corresponding retinal points.

Q.53 What is diplopia? How does it occur?
Diplopia means double vision. While looking at an object, if the eyeballs are directed in such a way that the light rays do not fall upon the corresponding point of retina of both eyes, a double vision or diplopia occurs i.e., one single object is seen as double.

Q.54 What are the causes for permanent and temporary diplopia?
Permanent diplopia is caused by paralysis or weakness of ocular muscles.

Temporary diplopia occurs due to imbalanced actions of ocular muscles in conditions like alcoholic intoxication.

Q.55 What is blind spot?
Optic disk is called blind spot. While looking at an object, if the image of the object falls upon optic disk, the object cannot be seen because the visual receptors are absent in optic disk. So, this part of retina is called blind spot.

Q.56 How is visual field determined?
By:
- Using perimeter
- Using Bjerrum's screen
- Confrontation test.

Q.57 What is photopic, scotopic and mesopic vision?
- *Photopic*—It is daylight vision due to cone receptor.
- *Scotopic*—It is dim light vision and a function of rods.
- *Mesopic*—It is a full moon night vision where reading becomes difficult.

Q.58 Trace the pathway for visual sensation.
Visual pathway (Fig. 18.3) includes:
- Receptors—rods and cones
- First order neurons—bipolar cells in retina
- Second order neurons—ganglionic cells in retina
- Optic nerve—formed by axons of ganglionic cells
- Optic chiasma—crossing of medial fibers of optic nerve
- Optic tract—formed by crossed and uncrossed fibers of optic nerve
- Third order neurons—lateral geniculate body
- Optic radiation
- Visual cortex.

Q.59 Where is the center for vision?
Center for vision is in visual cortex that is situated in calcarine fissure in medial surface of occipital lobe.

Q.60 Define the term anopia, homonymous hemianopia, heteronymous hemianopia, scotoma.
- *Anopia*—It is the complete loss of visual field in an eye (Fig. 18.4A)
- *Hemianopia*—Refers to the blindness of half of the visual field (Fig. 18.4B)
- *Homonymous hemianopia*—It refers to the loss of field of vision of same halves in two eyes (Fig. 18.4B)
- *Heteronymous hemianopia*—When different halves of field of vision in two eye are lost (Fig. 18.4C).
- *Scotoma*—Loss of vision in an eye which is confined to the center of the visual field.

Q.61 Name the effects of lesion at different levels of visual pathway.
- Lesion in optic nerve—total blindness
- Lesion in lateral fibers of optic chiasma on one side—nasal hemianopia

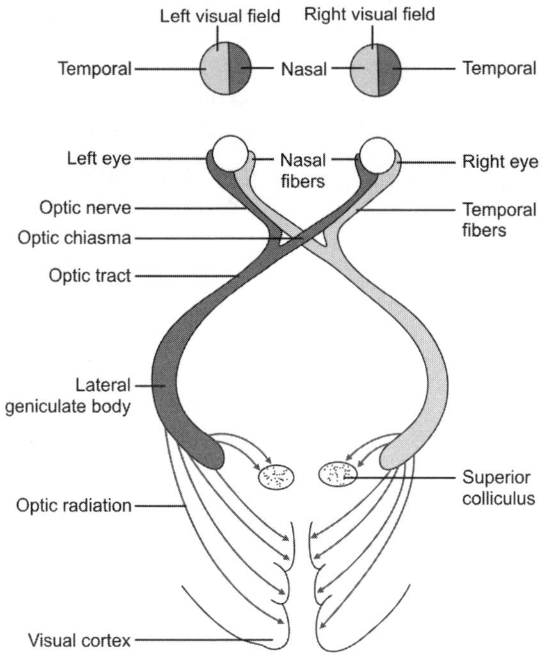

Fig. 18.3: Visual pathway

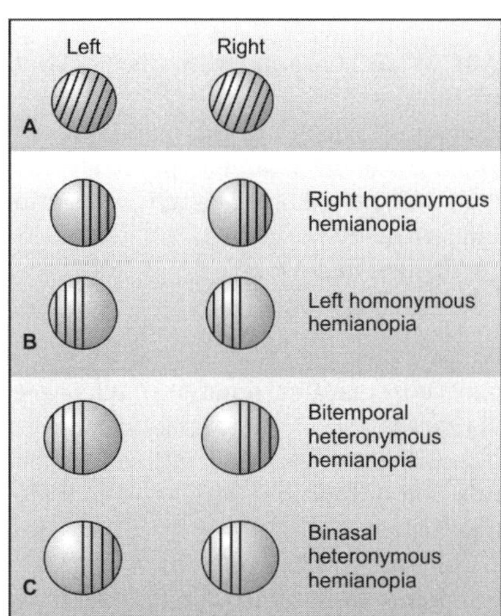

Fig. 18.4: Types of hemianopia

- Lesion in lateral fibers of both the sides of optic chiasma - binasal hemianopia
- Lesion in medial fibers of optic chiasma—bitemporal hemianopia
- Lesion in left optic tract, left lateral geniculate body, left optic radiation or left visual cortex—right homonymous hemianopia
- Lesion in right optic tract, right lateral geniculate body, right optic radiation or right visual cortex—left homonymous hemianopia.

Figure 18.5 illustrates the effect of lesion of visual pathway.

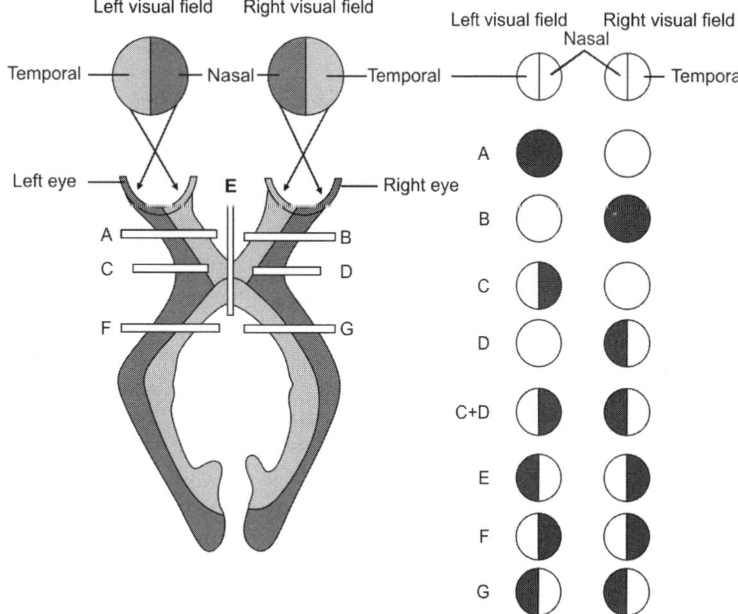

Fig. 18.5: Effects of lesions of optic pathway. Dark shade in circles indicates blindness
A. Lesion of left optic nerve—Total blindness of left eye
B. Lesion of right optic nerve—Total blindness of right eye
C. Lesion of lateral fibers in left side of optic chiasma—Left nasal hemianopia
D. Lesion of lateral fibers in right side of optic chiasma—Right nasal hemianopia
C + D. Lesion of lateral fibers in both sides of optic chiasma—Binasal hemianopia
E. Lesion of medial fibers in optic chiasma—Bitemporal hemianopia
F. Lesion of left optic radiation—Right homonymous hemianopia
G. Lesion of right optic radiation—Left homonymous hemianopia

Q.62 What is macula sparing?

In homonymous hemianopia, the macular vision is not affected in spite of lesion in visual cortex. This is called macula sparing. This is because the optic fibers from each eye are projected to visual cortex of both sides.

Q.63 Define pupillary reflexes. Name them.

Pupillary reflexes are the reflexes, which cause the alteration in the diameter of pupil.

Pupillary reflexes:
- Light reflex
- Ciliospinal reflex
- Accommodation reflex.

Q.64 Define and classify light reflex.

Light reflex is the reflex in which, the flash of light into the eye causes constriction of pupil.

Light reflex is classified into two types:
- Direct light reflex in which, the flash of light in one eye causes constriction of pupil in the same eye
- Indirect or consensual light reflex in which the flash of light in one eye causes constriction of pupil in the same eye as well as in the opposite eye.

Q.65 What is Wernicke pupillary reflex?
In case of partial damage of light reflex fibers, when light is focused on the blind part of retina light reflex is lost and if light is focused on the sound retinal part light reflex persists. This reflex is known as Wernicke pupillary reflex.

Q.66 Trace the pathway for light reflex.
Pathway for light reflex includes:
- Afferent fibers—fibers from optic pathway ending in pretectal nucleus of midbrain
- Center—pretectal nucleus
- Efferent fibers—fibers from pretectal nucleus reach Edinger-Westphal nucleus of III cranial nerve. The fibers from this go to ciliary ganglion. Short ciliary nerves arising from this supply constrictor pupillae muscles of iris.

Q.67 What is ciliospinal reflex?
Stimulation of skin over neck causes dilatation of pupil. This is known as ciliospinal reflex.

Q.68 What is the nerve supply to the muscles of iris?
Constrictor pupillae muscle of iris is supplied by parasympathetic nerve fibers from Edinger-Westphal nucleus of III cranial nerve.

Dilator pupillae muscle is supplied by sympathetic fibers.

Q.69 Define accommodation of eyeball.
Accommodation is the adjustments made in eyeballs while looking at near object.

Q.70 What is Argyll Robertson pupil and reverse Argyll Robertson pupil?
In case of lesion in aqueduct and superior colliculi, there is a loss of light reflex keeping the convergence accommodation reflex intact. This type of pupil is referred to as Argyll Robertson pupil. Where as due to the lesion in frontal lobe (bilaterally) or damage of its descending fibers to III nerve nucleus pupillary constriction in response to light is present but accommodation is lost. This is known as reverse Argyll Robertson pupil.

Q.71 What are the adjustments made in eyeballs during accommodation?
- Convergence of eyeballs—due to contraction of medial recti
- Constriction of pupil - due to contraction of constrictor pupillae
- Increase in the anterior curvature of lens—due to contraction of ciliary muscle.

Q.72 What is Young-Helmholtz theory of accommodation?
It describes how the curvature of the lens increases during accommodation.

Q.73 Explain briefly the mechanism of increase in the anterior curvature of lens during accommodation.
During distant vision, lens is flat due to the traction by suspensory ligaments. During near vision, ciliary muscle contracts and draws the choroid forward. So the ciliary processes are brought closer to lens and the suspensory ligaments are slackened. Now, the tension on the lens is released. Due to the elastic property, the lens bulges forward so that anterior curvature of lens increases.

Q.74 What are Purkinje-Sanson images?
Purkinje-Sanson images are the images of flame of a lighted candle held in front of eye. These images are used to demonstrate the increase in the anterior curvature of lens during accommodation.

Q.75 Trace the pathway for accommodation.
Pathway for accommodation includes:
- Afferent fibers—visual fibers from retina to visual cortex in occipital lobe and the association fibers from there to frontal eye field (area 8) in frontal lobe.
- Center—frontal eye field
- Efferent fibers—fibers from frontal eye field to Edinger-Westphal nucleus of III cranial nerve. Fibers from this nucleus reach ciliary ganglion. Nerve fibers from this pass through short ciliary nerves and supply constrictor pupillae. Some fibers from frontal eye field reach somatic motor nucleus of III cranial nerve and fibers from this supply the medial recti.

Q.76 What are the spectral colors? Name them.
Colors forming the spectrum are called spectral colors. Spectral colors are violet, indigo, blue, green, yellow, orange and red (VIBGYOR).

Q.77 What are the primary colors? Name them.
Primary colors are those, which can produce white when combined together. Primary colors are red, green and blue.

Q.78 What are the complementary colors? Give examples.
When two colors are mixed or combined in right proposition, white is produced. Such two colors are called complementary colors.
Examples:
- Red and greenish blue
- Orange and cyan blue
- Purple and green.

Q.79 Name the theories of color vision.
- Thomas Young's trichromatic theory
- Helmholtz trichromatic theory
- Granit's modulator and dominator theory
- Hartridge's polychromatic theory
- Hering's theory of opposite colors.

Q.80 Define and classify the color blindness.
Failure to appreciate one or more color is known as color blindness. Color blindness is classified into three types: (i) Monochromatism, (ii) Dichromatism, (iii) Trichromatism.

Q.81 What is monochromatism? What are its types?
Monochromatism is the condition in which the subject cannot appreciate any color and the whole spectrum is seen in different shades of gray.
It is divided into two types:
- Rod monochromatism
- Cone monochromatism.

Q.82 What is dichromatism? What are its types?
Dichromatism is the condition when only two of the three primary colors are appreciated.
It is of three types:
- Protanopia in which the first primary color, red cannot be appreciated
- Deuteranopia in which green cannot be appreciated
- Tritanopia in which blue cannot be appreciated.

Q.83 What is trichromatism? What are its types?
Trichromatism is the condition in which all the three primary colors are appreciated but the perception of one of the colors is very weak.

It is divided into three types:

- Protanomaly in which perception of red color is weak
- Deuteranomaly in which perception of green is less
- Tritanomaly in which perception of blue is less.

Q.84 How is color blindness determined?

By using: (i) Ishihara's color charts, (ii) Colored wool, (iii) Edridge-Green lantern.

Q.85 Name the errors of refraction.

The errors of refraction are (Fig. 18.6):

- Myopia
- Hypermetropia
- Anisometropia
- Astigmatism
- Presbyopia.

Q.86 Define emmetropia.

Emmetropia is the condition with normal refractory power of eye.

Q.87 Define ametropia. What are its types?

Any deviation in the refractory power of eye from normal condition is known as ametropia.

It is of two types:

- Myopia
- Hypermetropia.

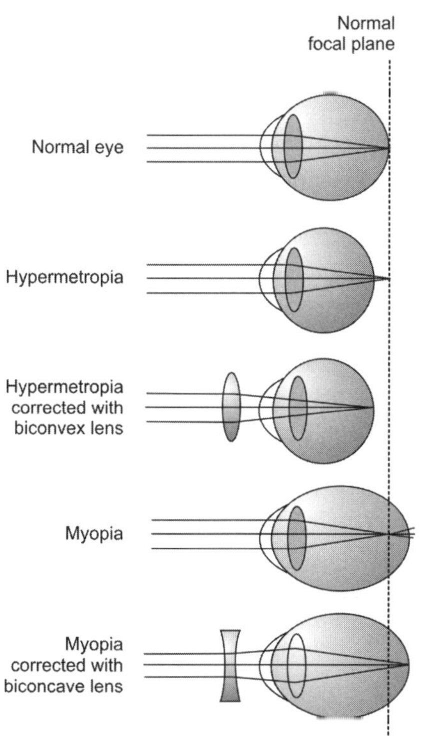

Fig. 18.6: Errors of refraction

Q.88 What is myopia or short sightedness? What is its cause?

Myopia or short sightedness is the condition in which the near vision is normal but the distant vision is defective.

It is caused by increase in anteroposterior diameter of the eyeball. So, the image from distant object is brought to a focus in front of retina.

Q.89 How is myopia corrected?

By using concave lens.

Q.90 What is hypermetropia or long sightedness? What is its cause?

Hypermetropia or long sightedness is the condition in which distant vision is normal but the near vision is affected.

It is caused by the decrease in the anteroposterior diameter of eyeball. So, the light rays are brought to a focus behind retina.

Q.91 How is hypermetropia corrected?

By using convex lens.

Q.92 What is anisometropia?

Anisometropia is the condition in which the refractory power of both the eyes are not the same.

Q.93 Define astigmatism. What is its cause?

Astigmatism is the defect in which the light rays are not brought to a sharp point upon retina.

It is caused by irregularity in the curvature of lens and unequal refractory power of lens in different meridians.

Q.94 What are the types of astigmatism?

- *Regular astigmatism:* In this, the refractory power is unequal in different meridians but, in one single meridian, it is uniform throughout
- *Irregular astigmatism:* In this, the refractory power is unequal not only in different meridians but also in different points of same meridian.

Q.95 How is astigmatism corrected?

By using cylindrical lens.

Q.96 What is presbyopia?

In old age, the amplitude of accommodation reduces and the near object cannot be seen clearly. Thus, the inability to see the near objects in old age is known as presbyopia.

Q.97 What are the causes for presbyopia?

- Decreased elasticity of lens that prevents the increase in the anterior curvature during near vision
- Decreased convergence of eyeballs due to weakness of ocular muscles.

Q.98 How is presbyopia corrected?

By using convex lens.

Q.99 Name the parts of ear (Fig. 18.7).

- External ear
- Middle ear
- Internal ear.

Q.100 What is tympanic membrane?

Tympanic membrane is a semitransparent structure that separates the middle ear from external auditory meatus.

Q.101 Define auditory ossicles. Name them.

Auditory ossicles are the miniature bones arranged in middle ear in the form of a chain from tympanic membrane to oval window.

The auditory ossicles are:

- Malleus
- Incus
- Stapes.

Q.102 What are the skeletal muscles attached to auditory ossicles?

Tensor tympani and stapedius.

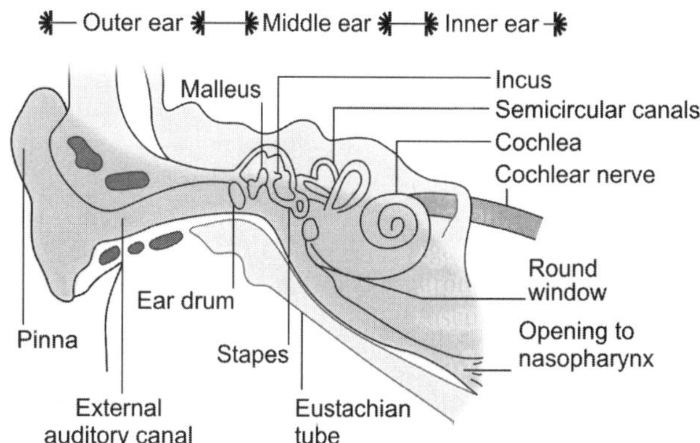

Fig. 18.7: Functional anatomy of human ear

Q.103 What is tympanic reflex? What is its significance?

Tympanic reflex is a reflex action in which loud sound causes contraction of muscles of middle ear, tensor tympani and stapedius.

Significance:
- Tympanic reflex prevents rupture of tympanic membrane by loud noise
- It also prevents fixation of footplate of stapes against oval window during exposure to loud noise
- It also protects cochlea from loud noise.

Q.104 Name the sense organs present in internal ear or labyrinth.
- Cochlea for hearing
- Vestibular apparatus for equilibrium.

Q.105 What are the compartments of cochlea?
- Scala vestibuli
- Scala media or cochlear duct
- Scala tympani.

Q.106 What are the membranes, which divide cochlea into three compartments?
- Vestibular membrane or Reissner's membrane that separates scala vestibuli and scala media
- Basilar membrane that separates scala media and scala tympani.

Figure 18.8 shows a cross-section of spiral canal of cochlea

Q.107 Name the fluids present in cochlea.
- Perilymph in scala vestibuli and scala tympani
- Endolymph in scala media.

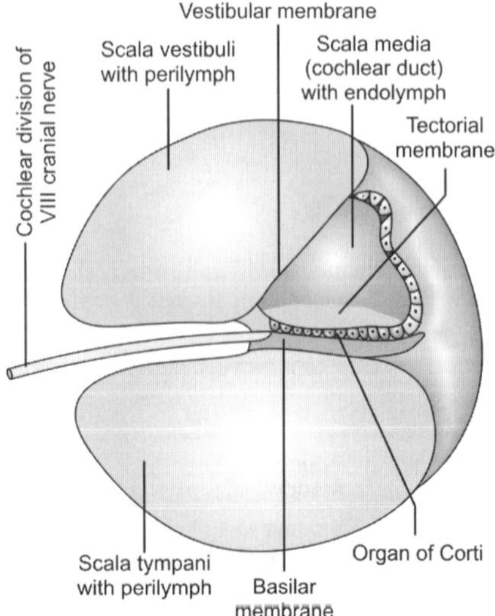

Fig. 18.8: Cross-section of spiral canal of cochlea

Q.108 What is helicotrema?

Helicotrema is small canal that connects scala vestibuli and scala tympani at the apex of cochlea.

Q.109 What is ductus reunions?

Ductus reunions is a slender canal that connects scala media with saccule of vestibular apparatus.

Q.110 What is organ of Corti (Fig. 18.9)?

Organ of Corti is the sensory part of cochlea situated in the upper surface of basilar membrane.

Q.111 What is eustachian tube and what is its importance?

It connects the middle ear cavity with the pharynx. Normally its pharyngeal opening is closed but opens during act of swallowing, chewing or yawning and thereby helps the air to enter into middle ear. Therefore, it serves to equalize the pressure on the two sides of tympanic membrane when atmospheric pressure changes.

Q.112 Why does there is pain in ear and even loss of hearing in sore throat?

Due to infection during sore throat there is an inflammation occurring in pharynx causing closure of pharyngotympanic tube. Thus middle ear cavity becomes a closed cavity due to inability to open eustachian tube. When the air within the middle ear gets absorbed, its pressure decreases resulting in inward bulging of tympanic membrane that causes pain sensation and in severe cases there may be rupture of tympanic membrane resulting in loss of hearing.

Q.113 Name the receptor cells of organ of Corti.

Inner and outer hair cells.

Q.114 What is tectorial membrane? What is its function?

Tectorial membrane is the membrane present at the roof of organ of Corti and it is in contact with processes of hair cells of organ of Corti.

When sound waves reach the inner ear, the endolymph in scala media vibrates. This causes movements of tectorial membrane. The movements of tectorial membrane stimulate the hair cells.

Q.115 What are the divisions of vestibulocochlear nerve (VIII cranial nerve)?
- Vestibular division that supplies the vestibular apparatus
- Cochlear division that supplies the cochlea.

Q.116 What are the cortical areas for auditory sensation?
- Primary auditory areas - areas 41 and 42
- Wernicke's area
- Auditopsychic area - area 22.

Q.117 What are the functions of cortical areas for auditory sensation?

Primary auditory areas (areas 41 and 42) are concerned with perception of auditory impulses.

Wernicke's area and auditopsychic area (area 22) are concerned with analysis and interpretation of auditory impulses.

Q.118 What is the role of external ear in hearing?

External ear directs the sound waves towards the tympanic membrane.

Q.119 What is role of middle ear in hearing?

The role of middle ear in hearing is to conduct the sound waves. When sound waves reach the tympanic membrane, it vibrates. The vibrations from tympanic membrane are transmitted by auditory ossicles in the middle ear to perilymph of internal ear through oval window.

Q.120 What is sound impedance? How is impedance offered in the ear?

Impedance means obstruction or opposition to the passage of sound waves. In the ear, impedance is offered by the perilymph present in cochlea.

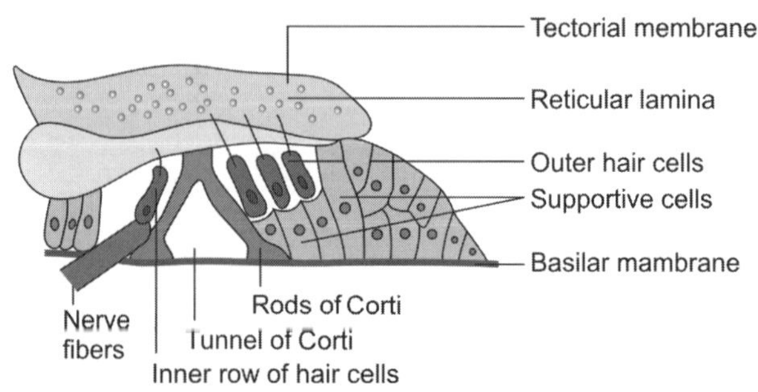

Fig. 18.9: Organ of Corti

Q.121 What is impedance matching?
Impedance matching is the mechanism by which the sound impedance offered by the perilymph in the cochlea is decreased. This is done by the tympanic membrane and the lever system of auditory ossicles. These two factors convert the sound energy into mechanical vibration in the fluid of internal ear, which helps to overcome the impedance.

Q.122 Name the types of conduction of sound waves in the ear.
- Ossicular conduction
- Bone conduction
- Air conduction.

Q.123 What is traveling wave?
The vibrations from tympanic membrane reach the oval window and cause movement of fluid in scala vestibuli, scala media and scala tympani. The movement of fluid in scala tympani initiates a wave in basilar membrane near round window. This wave travels through basilar membrane towards the apex of cochlea. This wave is called traveling wave.

Q.124 What is the significance of traveling wave?
It produces the vibration in basilar membrane, which in turn, causes stimulation of hair cells in organ of Corti.

Q.125 Name the electrical potentials involved during the process of hearing.
- Receptor potential or cochlear microphonic potential
- Endocochlear potential or endo lymphatic potential
- Action potential in auditory nerve fibers.

Q.126 What is cochlear microphonic potential or cochlear receptor potential?
The slight depolarization that develops in the hair cells of organ of Corti when stimulated is known as cochlear microphonic potential or cochlear receptor potential. The resting membrane potential in these cells is -60 mV. When stimulated, it comes down to about - 50 mV causing the slight depolarization.

Q.127 What are the properties of cochlear microphonic potential?
It is:
- Monophasic
- Non-propagative.

Q.128 What is the significance of cochlear microphonic potential?
Cochlear microphonic potential causes generation of action potential in auditory nerve fiber.

Q.129 What is endocochlear or endolymphatic potential? What is its significance?
An electrical potential exists between endolymph and perilymph with endolymph having a potential of +80 mV. This is known as endocochlear or endolymphatic potential. It increases the excitability and response of hair cells.

Q.130 What are the theories of hearing?
- Theories of first group, according to which the analysis of pitch of the sound is the function of cerebral cortex:
  - Telephone theory of Rutherford
  - Volley theory.
- Theories of second group, according to which the analysis of pitch of the sound is the function of cochlea:
  - Resonance theory of Helmholtz
  - Place theory
  - Traveling theory.

Q.131 Name the auditory defects or deafness.
- Conduction deafness
- Nervous deafness.

Q.132 What is conduction deafness? What are its causes?
Deafness due to the defect in conduction of sound in external ear and/or middle ear is known as conduction deafness.
Causes:
- Obstruction of external auditory meatus by wax
- Thickening of eardrum by repeated middle ear infection
- Perforation of eardrum by unequal pressure on either side
- Otosclerosis—the fixation of footplate of stapes against oval window.

Q.133 What is nervous deafness?
Deafness due to damage of any structure in cochlea or lesion in auditory pathway is known as nervous deafness.

Q.134 Name the tests for hearing.
- Rinne's test
- Weber's test
- Audiometry.

Q.135 What is the frequency of tuning fork that is used for hearing tests?
512 cycles/second.

Q.136 Which type of conduction is better in persons with normal hearing?
In persons with normal hearing, air conduction is better than bone conduction.

Q.137 Which type of conduction is better in conduction deafness?
In conduction deafness, bone conduction is better than air conduction.

Q.138 What does happen to conduction of sound in nerve deafness?
In nerve deafness, both air conduction and bone conduction are reduced or lost.

Q.139 What is audiometry?
Audiometry is a technique used to determine the nature and extent of auditory defects.

Q.140 Name the sense organs for taste or gustatory sensation.
Taste buds are the sense organs for taste sensation.

Q.141 What are the types of papillae on the tongue?
- Filiform papillae situated over the dorsum of tongue
- Fungiform papillae situated over the anterior surface of tongue near the tip
- Circumvallate papillae arranged in the shape of 'V' over the posterior part of tongue.

Q.142 Name the types of cells present in the taste buds. Which are the receptor cells?
The types of cells present in taste buds are (Fig. 18.10):
- Type I cells or sustentacular cells
- Type II cells
- Type III cells
- Type IV cells or border cells

Type III cells are the receptor cells in taste bud.

Q.143 Trace the pathway for taste sensation.
Pathway for taste sensation includes:
- Receptors—receptor cells in taste buds
- First order neurons—neurons in the nuclei of the cranial nerves namely, facial nerve, glossopharyngeal nerve and vagus nerve

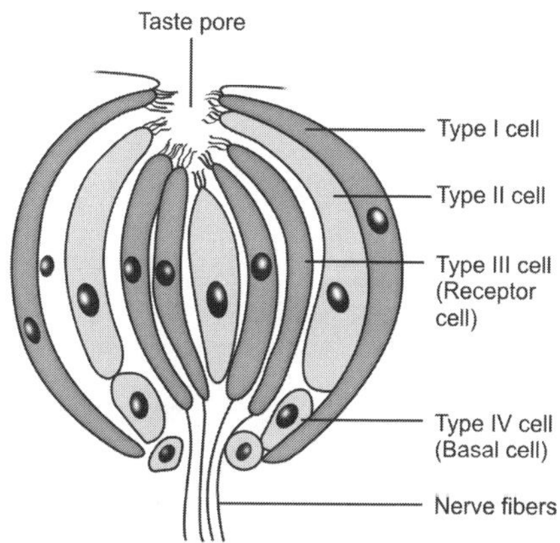

Fig. 18.10: Taste bud

- Second order neurons—neurons in the nucleus tractus solitarius
- Third order neurons—neurons in the posteroventral nucleus of thalamus
- Taste center—opercular area of cerebral cortex.

Q.144 Name the nerves carrying taste sensation.

- Chorda tympani branch of facial nerve carries taste sensation from anterior two thirds of tongue
- Glossopharyngeal nerve carries taste sensation from posterior two thirds of tongue
- Vagus nerve carries taste sensation from other areas.

Q.145 Name the primary taste sensations.

- Sweet
- Salt
- Sour
- Bitter.

Figure 18.11 shows the distribution of primary taste receptors on the dorsal surface of tongue.

Q.146 Name the chemical substances producing taste sensation.

Sweat taste: Organic substances like monosaccharides, polysaccharides, glycerol, alcohols, aldehydes, ketones and chloroform and inorganic substances like lead and beryllium.

Salt taste: Chlorides of sodium, potassium and ammonium, nitrates of sodium and potassium and some sulfates, bromides and iodides.

Sour taste: Hydrogen ions in acids and acid salts.

Bitter taste: Organic substances like quinine, strychnine, morphine, glucosides, picric acid and bile salts and inorganic substances like salts of calcium, magnesium and ammonium. Bitter taste is mainly due to cations.

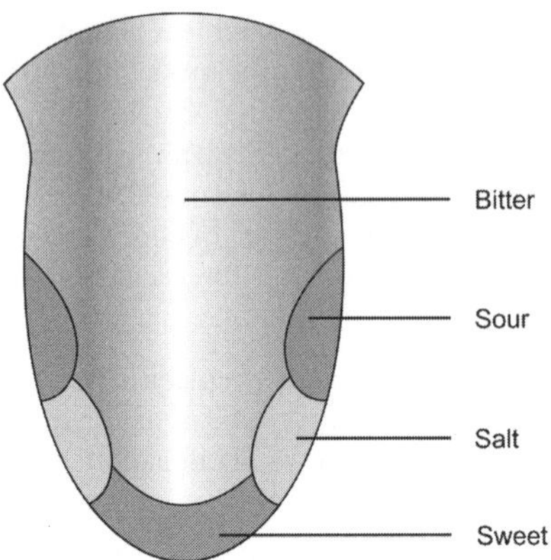

Fig. 18.11: Distribution of primary taste receptors on the dorsal surface of tongue

Q.147 Name the taste sensations having very low threshold value and very high threshold.

Bitter taste has a very low threshold (quinine in 1 in 2,000,000 dilution) value and sweet taste has a very high threshold (sugar in 1 in 200 dilution) value.

Q.148 What is ageusia? What are its causes?

Loss of taste sensation is known as ageusia. Ageusia in anterior two thirds of the tongue is caused by lesion in facial nerve, chorda tympani or mandibular division of trigeminal nerve. Ageusia in anterior one thirds of tongue is caused by lesion in glossopharyngeal nerve.

Q.149 What is hypogeusia?

Decrease in taste sensation is called hypogeusia.

Q.150 What are the receptors for sensation of smell or olfactory sensation?

Receptors for sensation of smell or olfactory sensation are the expanded end of dendrite of bipolar neurons situated in olfactory mucus membrane.

Q.151 Trace the pathway for olfactory sensation.

Receptors are the ending of dendrite of bipolar cells in olfactory mucus membrane. Axons of these cells synapse with dendrites of mitral cells, which form the olfactory glomeruli in olfactory bulb. Axons from olfactory bulb form olfactory tract that terminates in the center situated in olfactory cortex. Olfactory cortex includes the structures of limbic system namely, olfactory nucleus, prepyriform cortex, olfactory tubercle and amygdala.

Q.152 What are the different types of odor? Give examples.

- Aromatic or resinous odor—camphor, lavender, clove and bitter almond
- Ambrosia odor—musk.
- Burning odor—burning feathers, tobacco, roasted coffee and meat
- Ethereal odor—fruits, ethers and bees wax
- Fragrant or balsamic odor—flowers and perfumes
- Garlic odor—garlic, onion and sulfur
- Goat odor—caproic acid and sweet cheese
- Nauseating odor—decayed vegetables and feces.
- Repulsive odor—bed bug.

Q.153 What is anosmia?

Loss of sensation of smell is known as anosmia.

Q.154 What is hyposmia? What is its common cause?

Reduction in olfactory sensation is called hyposmia. Its common cause is the constant exposure to a particular odor like that of perfume that is often used in excess.

Q.155 What is hyperosmia?

Increased olfactory sensation is called hyperosmia.

Chapter 19

Skin and Body Temperature Regulation

Q.1 Name the functions of skin.

Skin has varied functions. The important functions are: Protection, regulation of body temperature, excretion, synthetic function, receptive function, secretory function, absorptive function, water balance and storage function (the dermis of the skin and subcutaneous tissue can store fats, water, salts and glucose).

Q.2 Classify the sweat gland and differentiate it.

Sweat glands are of two types: Eccrine and apocrine (Table 19.1).

Table 19.1: Eccrine and apocrine glands

Parameter	*Eccrine*	*Apocrine*
• Location	Found in all over the body	Found in axilla, mons pubis, scrotum, nipple, etc.
• Type of secretion	Clear, watery and thin	Milky, opalescent and having characteristic smell on decomposition.
• Stimulus	Increase of body temperature	Stress and sexual stimulation.

Q.3 What do you mean by homeothermic and poikilothermic? Give examples of each.

- The animals capable of maintaining constant body temperature inspite of wide variations in environmental temperature are known as homeothermic (warm blooded) animals, e.g. man, mammals, birds.
- Whereas the animals showing variation of body temperature in accordance with environmental temperature are called poikilothermic or cold blooded animals e.g. reptiles, fish, amphibians, etc.

Q.4 What is the normal body temperature in man? What do you mean by comfortable or neutral zone temperature?

- Normal body temperature (BT) of man is 98.4°F or 37°C.
- *Comfort zone:* It is the ambient temperature at which there is no active heat gain or heat loss mechanism operated by the body. It is 27 ± 2°C.

Q.5 What is the normal skin and oral temperature?

- *Normal oral temperature:* 36.3–37.1°C (97–98.8°F)
- *Normal skin temperature:* 29.5–33.9°C (85–93°F).

Q.6 What is basal temperature? Give the value of core temperature. What are the site for recording core temperature?

Basal temperature is the body temperature recorded under complete physical and mental rest which is recorded generally in morning after awaking.

Core temperature is 0.5 1°C more than oral temperature, i.e. its value is 37.5–38°C in an average.

Site of recording of core temperature includes rectum, vagina, esophagus and tympanic membrane.

Q.7 What will be the effect on body if core temperature is changed in following ways?

- *If it is decreased to 26°C or less:* It will lead to death of that person due to cardiac failure.
- *If it is increased to 43.5°C or more:* It will lead to death due to heat stroke.
- *If it is increased to 41°C for prolonged period:* There will be irreversible brain damage.

Q.8 Why regulation of body temperature is required?

It has following reasons:

- Speed of chemical reaction in the tissues varies with temperature.
- Enzyme system of our body has got narrow range of optimum temperature at which it functions properly. Thus the normal body function depends on a relatively constant body temperature.

Q.9 What is the average daily sweat secretion?

It is about 1 lit/day.

Q.10 What are the main functions of sweat?

It helps to regulate body temperature, maintains water electrolyte and acid-base balance, helps to excrete some excretory products and also keeps the skin moist.

Q.11 What do you mean by thermal sweating, non-thermal sweating, emotional sweating?

- *Thermal sweating:* It occurs in response to rise of environmental or body temperature and mediated by eccrine sweat glands.
- *Non-thermal sweating:* When sweating is stimulated by increased epinephrine level in the blood this type of sweating is called as non-thermal sweating. It is mainly mediated through apocrine type of gland.
- *Emotional sweating:* This is the type of sweating which takes place during emotion controlled by premotor area of cerebral cortex.

Q.12 How thermal sweating is controlled?

It is by hypothalamus.

Q.13 Name the main tissues where heat is produced in most.

It is in liver and muscle.

Q.14 Name the heat gain mechanisms.

These are: Shivering, increase in thyroid-stimulating hormone (TSH) and adrenaline secretion, continuous indirect vasoconstriction.

Q.15 What are the heat loss mechanisms?

These are: By radiation from the body to cooler object, by conduction and convection to surroundings, by evaporation through sweating, by excreta in urine and feces.

Q.16 Name the main calorigenic hormones in the body.

It is adrenaline and thyroxin.

Q.17 How much is the approximate daily heat loss through various channels?

Through skin—2200 cal.; through lungs—150 cal; through warming of air and food—100 cal; through urine and feces—50 cal.; total = 2500 cal.

Q.18 What is the role of brown fat in BT regulation?

Brown fat which plays a role in BT regulation mainly in infants, is present between the scapula, at the nape of neck, along the great vessels in the thorax and abdomen. These fat cells contain numerous mitochondria and thereby by increasing fatty acid oxidation it produces heat.

Q.19 What do you mean by insensible perspiration? What is its role in BT regulation?

Insensible perspiration is the passage of water by continuous diffusion through the epidermis which cannot be seen or felt. Its amount is 50 mL/hr.

It helps in loosing the heat from the body by 30 Kcal/ hour.

Q.20 Why one feels hotter in a humid day?

In a humid weather, the heat loss by evaporation becomes difficult as rate of evaporation depends on relative humidity. As humidity is high, the rate of evaporation becomes low and thereby heat loss becomes less.

Q.21 Name the heat gain and heat loss center.

Heat loss center is posterior hypothalamus whereas heat gain center is anterior hypothalamus.

Q.22 Can a person be made poikilothermic?

Yes, lesion in posterior hypothalamus causes body termperature to fall towards environmental temperature as both hot and cold regulating mechanisms are destroyed as anterior hypothalamic fiber passes via the posterior hypothalamus.

Q.23 What is critical temperature?

It is defined as the temperature at which a naked body needs the help of accessory chemical reactions to maintain the BT.

Q.24 What is pyrexia and hyperpyrexia?

Pyrexia is the state of the body when BT ranges from 37.2°C to 40.5°C (99°F to 105°F) where as hyperpyrexia is the state when BT rises above 40.5°C or 105°F.

Q.25 What is hypothermia and deep hypothermia?

Hypothermia is a state of body when BT falls to 30–32°C whereas deep hypothermia is a state when BT falls below 25°C.

Q.26 Explain briefly nervous control of body temperature.

The thermal centers for temperature regulation are situated in hypothalamus and are known as heat production center and heat loss center. These two centers have reciprocal action, i.e. stimulation of heat production center should simultaneously cause inhibition of heat loss center and *vice versa.*

Q.27 Name the hormonal glands associated with control of body temperature.

These are thyroid glands, adrenal medulla and adrenal cortex, secreting thyroxins, adrenaline, and adrenal corticoid hormones respectively. These hormones are calorigenic in action.

Q.28 What do you mean by heat stroke, heat cramp and heat exhaustion?

- *Heat stroke:* It is caused due to high environmental temperature, i.e. more than 41°C resulting in impairment of body temperature regulating mechanism.
- *Heat exhaustion:* It is caused by excessive sweating in response to heat which results in loss of water, sodium chloride through sweat and thereby reduction of blood volume.
- *Heat cramps:* Sometimes in people working in hot weather the muscles become hyperexcitable due to excessive loss of Na^+ and Cl^- from the body due to excessive sweating. This condition is called as heat cramps.

Chapter 20

Practical Viva in Hematology

Q.1 Which blood is generally used in hematological practical–capillary blood or venous blood?
Capillary blood.

Q.2 What is the difference between capillary blood and venous blood?
Capillary blood is obtained from punctured capillaries, smallest arterioles or venules by a skin puncture usually over a finger or ear lobe or the heel of the foot (in infants) and shows lower cell counts, lower hemoglobin concentration and PCV values as some tissue fluid always dilute the blood, whereas the venous blood is obtained from a superficial vein by venopuncture which shows comparatively higher cell counts, higher Hb percentage and PCV values as it is not contaminated with tissue fluid.

Q.3 Why the capillary blood is called peripheral blood?
Capillary blood is called as peripheral blood as it comes from the peripheral blood vessels like venules or smallest arterioles or capillaries in contrast to venous blood.

Q.4 Why the thumb or little finger is not pricked for collecting blood?
It is because the underlying palmar fascia from these digits extends up to the forearm. So in case of any infection at the site of injury, there is a chance of the infection to spread up to the forearm.

Q.5 In case of infants, from where is the capillary blood collected?
It is collected from either big toe or heel as the fingers are too small.

Q.6 What measures will you take to prevent the spreading of hepatitis infection following pricking of finger?
The needle used to prick the finger should be heated over flame.

Q.7 Why should the pricked finger not to be squeezed?
Squeezing the finger results in coming out of the tissue fluid that dilutes the capillary blood and thus giving lower values.

Q.8 What are the features of ideal blood film?
These are:

- It should be tongue shaped, uniformly thick, neither too thick nor too thin and should occupy the middle 2/3rd of the slide.
- Microscopically all the cells should be separate without any overcrowding and rouleaux formation.

Q.9 What is the composition of Leishman stain? What are the function of each constituent and why the stain should be acetone free?
The composition and function of each constituents is:

- ***Leishman powder:*** 0.15 gm
  - *Eosin:* An acidic dye stains basic part of cell, e.g. cytoplasm
  - *Methylene blue:* A basic dye stains acid part of cell, e.g. nucleus
- ***Methyl alcohol (acetone free):*** 100 mL (as a fixative and solvent)

Acetone free methyl alcohol is used because the acetone being a strong lipid solvent can even destroy the cell by lysing the cell membrane.

Q.10 Why buffer solution is used instead of distilled water in Leishman's staining?
The pH of buffer solution is adjusted at 6.8 and at this particular pH the ionization of stain is optimum, so the stain particles can easily penetrate the cell to stain it.

Q.11 Why is Leishman's stain diluted after 1–2 min?
During the initial 1–2 min staining does not take place, as the stain particles cannot enter the cell as long as they are not ionized by addition of water. During this period the absolute alcohol of Leishman's stain serves two functions:

- Fixes the blood cells on the glass by precipitating the plasma proteins, which act as glue.
- Preserves the normal shape and chemistry of cells.

Q.12 Can tap water be used for diluting the stain after fixation?
It should not be used as methylene blue of Leishman's stain may be unable to stain the cells because of improper pH.

Q.13 Name any other stain that can be used to stain the blood film.
Giemsa's stain.

Q.14 What do you mean by 'vital staining'?
It is the special staining method to stain the living cells.

Q.15 Why is cedar wood oil required to use oil immersion lens?
It is because the refractive index of this oil is similar to that of glass avoiding the refraction of the light. Otherwise the image will be faint and blurred.

Q.16 Which part of the blood film should be avoided for counting the cells?
"Head" and extreme "tail" part of the slide as the cells present in these area are few in number and also distorted.

Q.17 How do you differentiate between RBC pipette and WBC pipette?
RBC and WBC pipettes are differentiated from Figures 20.1A and B and Table 20.1.

Table 20.1: Difference between RBC and WBC pipettes

Parameter	*RBC pipette*	*WBC pipette*
Upper gradation	101	11
Diameter of bulb	More	Less
Color of mouthpiece	Red	White
Color of bead in the bulb	Red	White

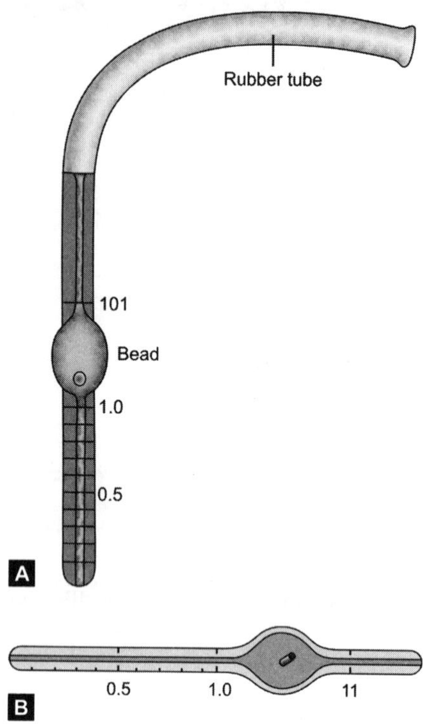

Figs 20.1A and B: (A) RBC pipette, (B) WBC pipette

Q.18 What are the functions of the bulb in a diluting pipette?

It helps the blood to be diluted and also to be mixed with the diluting fluid.

Q.19 What are the functions of the bead present inside the bulb of the diluting pipette?

It:

- helps to mix the blood with the diluting fluid.
- helps to identify the pipette by just glancing it.
- gives an idea whether the pipette is wet or dry. If it is dry the bead rolls freely inside the bulb.

Q.20 Why is it important to discard the first two drops of diluted blood from the pipette before charging the counting chamber?

The stem of the pipette contains only the cell free diluent which is to be discarded before charging the chamber; otherwise the count will be low and thus erroneous.

Q.21 Why any small excess of blood drawn into the pipette should not be removed by a piece of cotton?

If the cotton is used to remove the excess blood drawn in the pipette then it will absorb only the fluid not the cells. This will result in higher RBC/WBC count than the actual value.

Q.22 How will you clean the pipette in case of clotting of the blood inside the stem of the pipette?

It is to be kept in strong nitric acid or alkali or H_2O_2 for 24 hrs and then washed in the running tap water. A flexible suitably thick metal wire is now inserted to clean the capillary bore, finally rinse with alcohol or ether to dry it.

Q.23 How will you clean the pipette?

It is by sucking up and blowing out distilled water several times followed by sucking up and blowing out acetone for drying it.

Q.24 How will you clean the chamber and cover slip?

It is by washing it first with soap and water and then with alcohol.

Q.25 Can these pipettes be used for any other purpose?

The RBC pipette can be used for counting platelets, WBCs (when their count is very high as in leukemia) and also for counting the spermatozoa in the semen.

Q.26 What is a Neubauer's counting chamber?

Neubauer's double counting chamber (Figs 20.2A to D) is a thick rectangular glass with a polished transverse bar in the center, separated from the rest of the slide by two parallel grooves on either side. The polished bar is divided into two equal platforms by a groove in the middle resulting in 'H' shaped depression. The surface of the platforms is 1/10 mm below the surface of the rest of the slide. So if a cover glass is placed over the surface of the counting chamber, the under surface of the coverglass remains 1/10 mm above the polished surface of the platform. The counting area is in the form of a central ruled area on the polished surface of each platform. It is a square, with each side measuring 3 mm. This square is divided into 9 equal squares, each having a side of 1 mm. Of these, the four corner squares are used for WBC counting. The central 1 mm square is divided into 25 equal small squares of 1/5 mm side, by means of triple lines of which the 4 corner ones and the

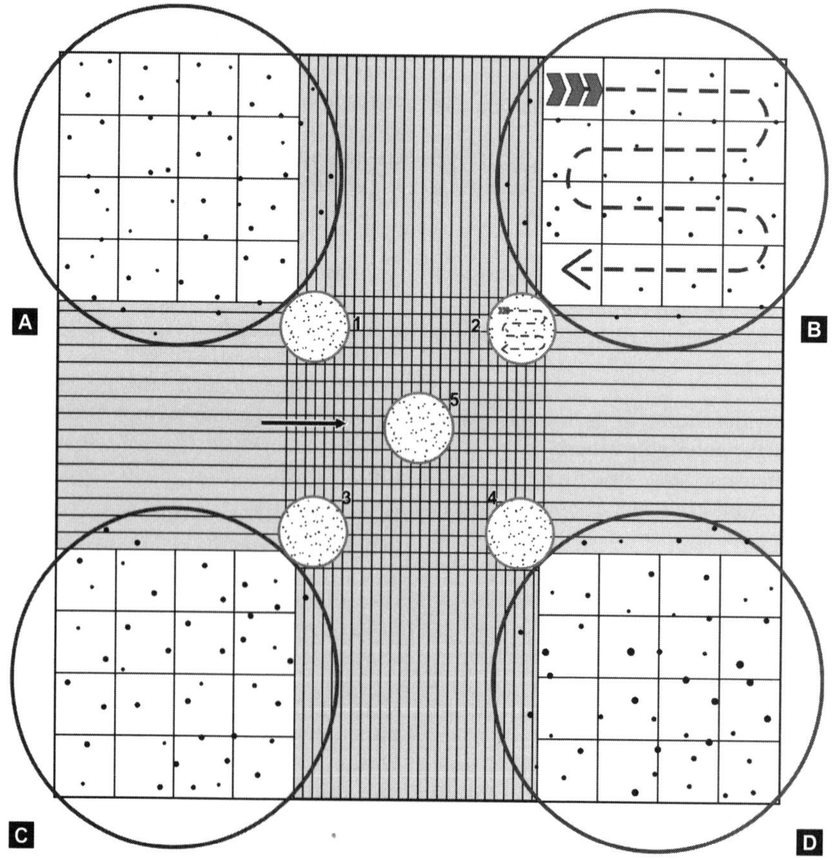

Figs 20.2A to D: Neubaur's counting chamber: A-B-C-D are fields used for doing the white blood cell count. 1-2-3-4-5 are fields used for doing the red blood cell count

central one are used for RBC counting. Each of these squares is subdivided into 16 smallest squares each of 1/20 mm side.

Q.27 Describe the procedure of filling the pipette for doing RBC/WBC count.

The right type of pipette must be chosen, WBC pipette or the RBC pipette. The correct type of diluting fluid must be then selected (Turk's fluid for WBC count and Hayem's fluid for the RBC count). The blood must be sucked up to the mark of 0.5 of the pipette and the dilution fluid must be filled up to the 101/11 mark depending on whether the RBC or WBC pipette is used. The contents must be mixed by gently rolling the pipette. This fluid is then used for charging the Neubaur's counting chamber.

Q.28 Describe the procedure of charging the chamber for doing RBC/ WBC count. Pipette filled with blood is provided to you.

Firstly the chamber and the cover slip are cleaned. The chamber is then mounted on the mechanical stage of the microscope. The initial few drops of the solution are discarded and the tip of the pipette is wiped. The pipette is then placed at 45° angle at the edge of cover slip and the chamber is charged.

The counting chamber must be kept undisturbed for 2–3 minutes, on the stage so that the cells settle down on the ruled area. The WBC count is done under the low power, whereas the RBC count is done under high power. A chart of the squares must be drawn on the paper and the number of cells in each square must be counted and written down.

Method of total leukocyte count

Let the number of cells counted in 64 squares be "N"

Therefore number of cells in one square = N/64

Side length of 1 square = 1/4 mm

Area of 1 square = 1/16 mm^2

Depth of fluid film = 1/10 mm

So, Volume of fluid in 1 square 1/16 × 1/10 = 1/160 mm^3

Number of cells in 1/160 cu.mm of diluted blood = N/64

Number of cells in 1 cu.mm of diluted blood = N × 160/64

Dilution factor = 1/20

Therefore, no. of cells in 1 cu.mm of undiluted blood = (N × l60 × 20)/64 = N × 50 cells/cu.mm

Method of RBC count

RBC count is done under the high power objective of the microscope after identifying the RBC counting area.

At least 5 squares, each having 16 smallest squares (preferably 4 corner and 1 central) should be counted to obtain a satisfactory value. While counting each small square, cells touching the top and left margin of each square should be omitted and cells touching bottom and right margin of each square should be counted. Draw a chart of the counting squares on the paper and record the number of cells in each square.

Calculation

Let the number of cells counted in (5 × 16) = 80 smallest squares be "N"

Number of cells in 1 smallest square is N/80

Side of 1 square = 1/20 mm

Area of 1 square = 1/400 mm^2

Depth of fluid film in counting chamber is 1/10 mm

Volume of diluted blood in 1 square = 1/400 × 1/10 = 1/4000 mm^3

Number of cells in 1/4000 mm^3 diluted blood = N/80

Number of cells in 1 mm^3 of diluted blood = (N/ 80 × 4000)

= (N × 4000)/ 80

(Total diluted volume in bulb of the pipette is 100 parts, out of which 0.5 is blood. So dilution is 0.5 in 100, i.e.1 in 200)

So number of cells in 1 mm^3 of undiluted blood = (N × 4000 × 200)/80 = N × l0000

Q.29 What are the dimensions of WBC and RBC squares?

- *Each smallest square for RBC counting:*
  - *Area:* 1/20 mm × 1/20 mm = 1/ 400 mm^2
  - *Volume:* 1/400 mm^2 × 1/10 = 1/4000 mm^3
- *Each smallest square for WBC counting*:
  - *Area:* 1/4 mm × 1/4 mm =1/16 mm^2
  - *Volume:* 1/16 mm^2 × 1/10 mm = 1/160 mm^3.

Q.30 What are the features of an ideally charged chamber?

These are:

- No flowing of blood into the trenches
- No air bubbles.

Q.31 When blood is taken to the mark 0.5 and the diluting fluid to mark 101, why is the dilution 1 in 200 and not 1 in 202?

The dilution of the blood occurs not in its stem but in the bulb of the pipette, the volume of which is 101–1 = 100. Hence half volume in hundred gives a dilution of 1 in 200.

Q.32 If Hayem's fluid is not available can you use any other?

0.9 gm% normal saline can be used but the cells have to be counted within an hour after filling the pipette, even though the red cells are likely to form rouleaux.

Q.33 How do you differentiate red cells from dust particles?

The red cells are appeared in round discs of uniform size and light pink in color whereas the dust particles are angular with varying size and colors.

Q.34 What is the composition of Turk's fluid? What is the function of each constituents?

The composition and function of each constituents is;

- *Glacial acetic acid:* 1.5 mL (hemolyses the red cells)
- *Gentian violet:* 1.5 mL (stains the nuclei of WBCs)
- *Distilled water:* Up to 100 mL (as a solvent).

Q.35 What is the composition of Hayem's fluid

The Hayem's fluid is composed of the following substances:

- Sodium chloride 0.5 gm
- Sodium sulphate 2.5 gm
- Mercuric perchloride 0.25 gm
- Distilled water 100 mL.

Sodium chloride and sodium sulphate together helps in maintaining the isotonicity of fluid. Sodium sulphate also prevents clumping of red cells. Mercuric perchloride fixes the cells and acts as a preservative.

Q.36 Name any other accurate method to do RBC count?

Use of electronic cell counter.

Q.37 Why is it necessary to follow the rules of counting?

It is to avoid the error of missing some cells and counting other more than once.

Q.38 What is the fate of leukocytes in this experiment?

The leukocytes in this experiment are as much diluted that its number are considered very negligible to consider because:

- The low count of WBC in comparison to RBC.
- Dilution of the WBC by 100 times.

Occasionally leukocyte may be seen but its concentration is very less (1 WBC for every 600–700 RBCs, i.e. one WBC in 80 squares). So even if it is counted along with the RBCs the RBC count will not vary too much (i.e. 10,000/cmm of blood).

Q.39 What do you mean by the term 'glacial'? Why it should be glacial acetic acid in Turk's fluid?

Glacial means pure. Only the pure form of acetic acid can give the refractivity around the WBCs that helps the WBC to be differentiated from the dust particles (which are opaque).

Q.40 What is the fate of the RBCs in this experiment (total count of leukocytes)?

RBCs are hemolyzed by the glacial acetic acid otherwise it would not be possible to count the WBCs.

Q.41 Can any other agent be used to hemolyze the RBCs?

No, any strong agent will also lyse the WBCs and any weak agent will take long time to lyse them completely.

Q.42 What is the difference between DLC and absolute leukocyte count?

In DLC the percentages of different leukocytes are determined whereas in absolute count the actual number of different leukocytes per cu mm of blood are calculated.

Q.43 How does the DLC of a child differ from that of adult?

In adults the granulocytes (mostly neutrophil) predominate whereas in children the lymphocytes predominate.

Q.44 Can you get rough idea of TLC by doing DLC?

Yes, if the cells appear more frequently amongst the RBCs the TLC will be high and *vice versa*.

Q.45 Enumerate the sources of error in hemocytometry?

These are:

- ***Pipette error,*** *i.e.* inaccuracy in calibration and in measurement.
- ***Field error,*** *i.e.* unequal distribution of cells over the counting chamber due to:
  - Overcharging or undercharging of the chamber.
  - Presence of grease or oil on slides or cover slip.
- ***Personal error,*** *i.e.* wrong counting of cells.
- ***Stastical error***: It is inversely proportional to the square root of the number counted.

Q.46 Can you use oxalate mixture in Westergren method and citrate solution in Wintrobe's method?

No, the anticoagulants used for each method can not be interchanged as both the methods are standardized with the specific anticoagulant.

Q.47 What are the advantages and disadvantages of Wintrobe's and Westergren method?

Wintrobe's method:

- *Advantage*: Same sample of oxalated blood can be used for ESR first and then after one hour for PCV by centrifuging it.
- *Disadvantage*: The method is less sensitive as the column of blood is not high.

Westergren method:

- *Advantage:* The method is more sensitive as the column of blood is high.
- *Disadvantage:* Citrate solution used in this case dilutes the red cells and thus tends to raise the ESR, however as the fibrinogens and globulins of plasma are also diluted there is also tendency of lowering the ESR.

Figures 20.3A and B show Westergren's tube and Wintrobe's tube respectively.

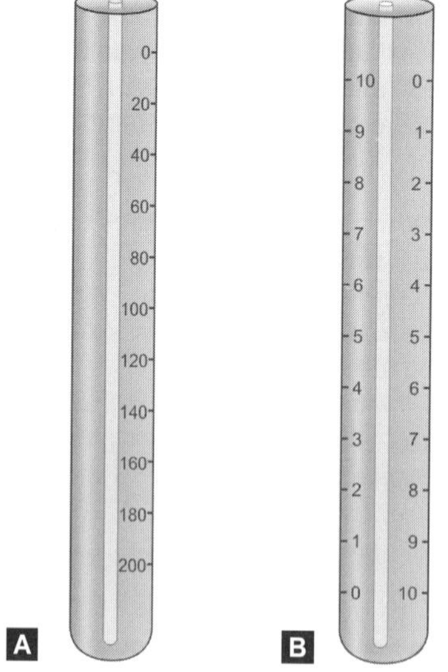

Figs 20.3A and B: (A) Westergren's tube, (B) Wintrobe's tube

Q.48 Why ESR reading is taken after one hour?

This is because more than 95–98% of RBCs settle down by the end of this time and after that the rate of sedimentation of RBCs do not affect the ESR significantly.

Q.49 Why the normal values of ESR are more in Wintrobe's method than that of Westergren method?

It is because of:

- Effect of atmospheric pressure over the blood column as the tube is kept open in its top.
- Nature of powdered mixture of oxalate solution used as an anticoagulant.

Q.50 What is the importance of determining hematocrit?

It is simple but more accurate test for determining the presence of anemia or polycythemia. It is also used for determining various absolute corpuscular values.

Q.51 Which cells make up the buffy layer? How thick it is? When its thickness increases?

The buffy layer consists of packed leukocytes and platelets. It is 1 mm thick. Its thickness increases in severe leukocytosis, leukemia and thrombocytosis.

Q.52 What is the difference between PCV of arterial blood and venous blood? What is the reason behind this difference?

The PCV of venous blood is higher than that of arterial blood. It is because in the venous blood the RBCs gain an extra weight due to the entry of water within it resulted due to chloride shift.

Q.53 Why color index is not an appropriate index of hemoglobin content of RBC?

It is because of wide range of normal value of RBC.

Q.54 Which absolute corpuscular value is most useful?

It is MCHC because:

- It expresses the actual Hb concentration in RBCs only, not in whole blood.
- It does not consider the RBC count for its calculation.

Q.55 Is it possible to know the sex of a person from the blood film?

Yes, in the blood film of females of the Barr body, i.e. chromatin of the sex chromosome is seen in some neutrophils.

Q.56 What are the features of a senile neutrophil?

These are less motile and least effective. These cells commonly break up during the spreading of the blood film.

Q.57 Which stage of neutrophil is most effective?

3-lobed neutrophil is the most motile and functionally most effective in killing the bacteria.

Q.58 Why Cook-Arneth count is not used as a routine investigating tool?

This is because:

- During some physiological conditions neutrophils used to enter the circulation from various storage pools, whereas during some other conditions there is shifting of neutrophils to the storage pools, resulting in shift to the left or right. This normal phenomenon may give the false indication about the status of bone marrow if we totally depend on this investigating tool.
- Besides this, the better method like bone marrow biopsy are now available for assessing bone marrow function.

Q.59 What are the indications of doing reticulocytes count?

It is to assess the red cell forming and releasing activity of the bone marrow.

Q.60 How does a reticulocyte differs from the RBC?

The reticulocytes are comparatively larger than RBCs and also contain dots, strands and filaments of bluish stained material.

Q.61 Why does the ABO incompatibility rarely produce hemolytic disease on the newborn?

This is because the anti-A and anti-B antibodies are IgM type of immunoglobulins that do not cross the placenta because of their large MW and thus there is no chance of antigen antibody reaction.

Q.62 What do you mean by *zone phenomenon*?

For agglutination to occur the concentration of antigen and antibody has to be same, otherwise there will be no antigen-antibody reaction. This is termed as *zone phenomenon*.

Q.63 What are the earliest effects of a mismatched transfusion?

These are: Severe pain anywhere in the body, sense of suffocation, feeling of tightness in chest, shivering and even fever.

Q.64 Which blood substitutes may be used to restore blood volume if suitable donor is not available?

Crystalloid solution (glucose saline) and colloid solutions like human albumin, dextrose with NaCl, etc.

Q.65 Why does calcium deficiency not cause a bleeding disorder though it is essential in blood coagulation?

It is because the calcium required for the blood clotting is in minute quantities.

Q.66 What is a thrombocytopenic purpura and thromboasthenia purpura?

Purpura with normal platelet count is called as athrombocytopenic purpura and purpura with normal count but abnormal circulating platelets is called thromboasthenia purpura.

Q.67 What do you mean by fragility and hemolysis?

Fragility means the susceptibility of red cells to being broken down by osmotic or mechanical stresses. Whereas the hemolysis means the breaking down of red cells resulting in release of hemoglobin into the surrounding fluid.

Q.68 What is the effect of 5% glucose, 10% glucose, urine and urea solution of any strength on red cells?

- ***5% glucose:*** It is isotonic with blood, so no change in size and shape of RBC.
- ***10% glucose:*** It is hypertonic, so there is shrinkage of red cells due to exoosmosis.
- ***Urine:*** Urine is hypotonic so the red cells will swell up due to entry of some water.
- ***Urea solution:*** Hemolysis of red cells due to entry of urea followed by water into the red cells.

Q.69 Describe the procedure of estimation of hemoglobin by Sahli's method.

Firstly N/10 HCl solution is taken and put until the 2 gm% mark in the hemoglobin tube. Next, the capillary pipette is selected and the patient's blood is sucked up to the mark of 20 in the pipette. After placing the pipette in the lower part of the hemoglobin tube, the blood is gently transferred into the HCl solution and the contents are gently stirred. The hemoglobin tube must be placed in the stand so that the scale is turned away from site and cannot be seen. Dilute the contents with the pure water until the color is same as that of the standard. The results must be read exactly three minutes after blood has been added to the HCl solution. Hemoglobin values are read at the meniscus of the brown solution.

Q.70 Can strong acids or alkalis be used instead of HCl to measure hemoglobin?

No. The strong acids will oxidize the hemoglobins and the strong alkalis will cause disruption of Hb. So in both the cases there will be no formation of acid hematin.

Q.71 What happens if more or less amount of N/10 HCl is taken instead of required amount?

If less amount of acid is taken all the hemoglobin will not be converted to acid hematin resulting in a low value. Besides this there may be clot formation due to improper mixing of blood with acid. On the other hand if more than the required amount of acid is taken the final color developed in case of severe anemia would be much lighter than the standard.

Q.72 Why it is necessary to convert hemoglobin in acid hematin?

If hemoglobin is not converted into acid hematin then the color of oxyhemoglobin, which has a wide spectrum of colors, cannot be standardized.

Q.73 Can tap water be used for diluting and color matching?

No, as the salt present in the tap water may cause turbidity which will interfere with color matching.

Q.74 Can N/10 HCl be used for diluting and color matching?

Yes.

Q.75 While matching the color why it is important to lift the stirrer above the solution but not take it out?

If the stirrer is kept in the solution it will lighten the color and thus matching will occur earlier resulting in low value. On the other hand if it is taken out every time during the color matching, some solution will go out of the tube and thus again giving a low value.

Q.76 Classify the severity of anemia as per the Hb concentration?

Depending on the Hb level the anemia may be graded as:

- Mild : Hb 10–2 gm%
- Moderate : Hb 5–8 gm%
- Severe : Hb below 5 gm%

Section 3

BIOCHEMISTRY

CHAPTERS

Chapter 21

Biophysics

Q.1 What is pH?
pH is defined as the negative logarithm of the hydrogen ion concentration.

$$pH = -\log [H^+]$$

or

$$pH = \frac{1}{\log [H^+]}$$

Q.2 What is the relationship between H^+ ions and pH?
There is an inverse relationship between the two. As H^+ ion concentration increases and vice versa.

Q.3 What are the methods by which pH can be determined?
pH can determined by:
1. Indicators
2. pH paper
3. Buffers
4. pH meter.

Q.4 What is the pH of distilled water, gastric juice, intestinal juice, pancreatic juice, blood and urine?

Distilled water pH	=	7.
Gastric juice pH	=	0.9–1.
Intestinal juice pH	=	7–8.
Pancreatic juice pH	=	7.5–8.
Blood pH	=	7.4.
Urine pH	=	5.5.– 6.5.

Q.5 What is the pH of each of the following solutions?
i. 10^{-3}N HCl
ii. 10^{-2} N NaOH
i. 3
ii. 12.

Q.6 What are buffers?
Buffers are solutions which resist changes in their pH when small amount of acids or alkalies are added to them. Buffers act like shock absorber against the sudden changes of pH.

Q.7 What is the composition of a buffer?
Buffer is pair of weak acid and its salt with a strong base.
Example: CH_3COOH/CH_3COONa

Q.8 Give few examples of commonly used buffers in the laboratory.
- Acetate buffer (sodium acetate/acetic acid)
- Phosphate citrate buffer (Na_2HPO_4 KH_2PO_4)
- Citrate buffer (sodium citrate/citric acid)
- Barbitone buffer (sodium diethyl barbiturate/diethyl barbituric acid).

Q.9 How the pH of a buffer solution can be calculated?
pH of a buffer solution can be calculated by Henderson Hasselbalch equation.

$$pH = pK + \log \frac{\text{Concentration of salt}}{\text{Concentration of acid}}$$

Q.10 What is pK?
pK is that pH at which the acid is half neutralized.

Q.11 Name the various body buffers.
Buffers operating in the body are:
1. $H_2CO_3/BHCO_3$
2. H.protein/B.protein
3. BH_2PO_4/B_2HPO_4
4. HHb/BHb
5. $HHbO_2/BHbO_2$
6. H. organic acid/B. organic acid.

Q.12 What is the importance of buffers in the body?
Buffers maintain the pH of the various fluids of the body compartments constant despite wide variation in the H^+ ion concentration which are produced in the normal course of the metabolism of the body as bye-product which otherwise could have lowered down the pH.
1. To regulate the pH of the body fluids.
2. To control the pH in chemical reactions catalysed by enzymes.

Q.13 Name some pH disorders.
If pH of body decreases, it is referred as acidosis and if increases it is referred as alkalosis. The cause of both may be metabolic or respiratory.

Q.14 What is diffusion?
It is movement of a particles from their higher concentration to lower concentration.

Q.15 What is osmosis?
Movement of solvent molecules from a pure solvent to dilute solution through a semipermeable membrane is called osmosis.

Q.16 What is a semipermeable membrane?
A membrane which allows the solvent molecule to pass but does not allow the passage of the solute molecule.

Q.17 What is osmotic pressure?
It is the pressure generated by osmosis. When solvent moves from concentrated solution to dilute solution through semipermeable membrane, then particles of solvent exert a pressure on semipermeable membrane, it is called osmotic pressure.

Q.18 What are the factors on which the osmotic pressure depends?
Osmotic pressure depends only on the number of dissolved particles and is independent of the size of the particles.

Q.19 Why an ionized solution has more osmotic pressure than an unionized solution?
- Ionized solution gives rise to more number of particles or ions on ionization and each ion will exert an osmotic pressure; hence, the osmotic pressure is more whereas in unionized solution, the number of molecules remains the same.

Q.20 Which will have greater osmotic pressure?
a. 1 molar solution of NaCl.
1 molar solution of $CaCl_2$.
1 molar solution of glucose.
b. What will be the order of osmotic pressure?
a. 1 molar solution of $CaCl_2$ will have the maximum osmotic pressure because $CaCl_2$ on ionization gives rise to 3 ions and each ion will exert an independent osmotic pressure. Whereas NaCl gives two ions on ionization and glucose does not ionize.

b. The order of osmotic pressure is: 1 molar solution of $CaCl_2$ > 1 molar solution of NaCl > 1 molar solution of glucose.

Q.21 What is Gibbs-Donnan equilibrium?
The unequal distribution of diffusible ions across the membrane when a non-diffusible ion is present on one side of a membrane leads to Gibbs-Donnan equilibrium.

Q.22 Explain the importance of Gibbs-Donnan equilibrium.
1. In the maintenance of differential concentrations between the various compartments of the body.
2. In the process of absorption.
3. In the process of secretion.

Q.23 Explain the following in terms of osmotic pressure.
1. Isotonic solution?
2. Hypertonic solution?
3. Hypotonic solution?
- *Isotonic solution:* They have the same osmotic pressure as within the cells.
- *Hypertonic solution:* They have the higher osmotic pressure than within the cells.
- *Hypotonic solution:* They have the lower osmotic pressure than within the cells.

Q.24 What is surface tension?
The force by which the surface molecules are held together and form a membrane over the surface of the liquid is called surface tension.

Q.25 Name the substance which lower down the surface tension.
Bile salts, organic substances, ammonia, strong mineral acids.

Q.26 What is adsorption?
The process of holding up of substances from the solution on surface is called adsorption.

Q.27 What is hydrotropy?
The process whereby water insoluble substances are made soluble without undergoing any chemical change.

Q.28 What are crystalloids?
Crystalloids are those substances which diffuse readily through membranes.

Q.29 What are colloids?
Colloids are those substances which do not diffuse through membranes.

Q.30 What are protective colloids?
Colloids which prevents other substances from being precipitated are called protective colloids.

Q.31 What are the roles played by protective colloids?
Protective colloids play an important role physiologically.
1. Proteins of milk serve as protective colloids to calcium phosphate present in the milk.
2. Blood proteins serve as protective colloids to the calcium phosphate of the blood.

Q.32 What are indicators?
Indicators are substances which change in color with change in pH of solutions in which they are present. They behave like weak acids or weak bases, the ionized and unionized forms of which differ in color.

Q.33 What is the nature of indicators?
Indicators are dyes which are weak organic acids or weak organic bases and have the property of dissociating.

Q.34 What are the common indicators generally used? Give their effective pH range and their acid and alkaline color.

Indicators	*pH range*	*Acid color*	*Alkaline color*
Thymol blue (acid range)	1.2–2.8	Red	Yellow
Methyl yellow	2.9–4.0	Red	Yellow
Methyl orange (Topfers indicators)	3.1–4.4	Red	Yellow orange
Methyl red	4.3–6.1	Red	Yellow
Phenol red	6.7–8.3	Yellow	Red
Thymol blue (alkaline range)	8.0–9.6	Yellow	Blue
Phenolphthalein	8.2–10	Colorless	Red

Q.35 What is milliequivalent?
Milliequivalent is one thousandth of a gram equivalent weight.

Q.36 What is normality?
Normality is number of equivalents of solute in one liter of solution.

Q.37 What is molarity?
Molarity is number of moles of solute in one liter of solution.

Q.38 What is osmolarity?
Osmolarity is same as molarity except the fact that in osmolarity only osmotically active particles are considered.

Q.39 What is molality?
Molality of a solution refers to number of moles of solute in 1000 gm of solvent.

Q.40 What is the relationship between milligram percent and milliequivalent?
milliequivalent/liter =

$$\frac{\text{milligram percent} \times 10 \times \text{valency}}{\text{molecular weight or atomic weight}}.$$

Q.41 What is the milligram percent of calcium when its concentrations is 5 mEq/L?

$$\frac{\text{mg\%} \times 10 \times 2}{40} = 10 \text{ mg\%}.$$

Q.42 What is the milliequivalent per liter of sodium when its concentration is 322 milligram percent.

$$\frac{322 \times 10 \times 1}{23} = 140 \text{ mEq/L}.$$

Q.43 What is chromatography?
Chromatography is defined as the analytical technique used for separating mixtures on the basis of difference in affinity for a stationary and a mobile phase.

Q.44 What are the various types chromatography?
1. Thin layer chromatography
2. Column chromatography
3. Paper chromatography
4. Gas chromatography.

Q.45 What kind of paper is required for paper chromatography?
We use chromatographic strip or Whatman filter paper 1 for paper chromatography.

Q.46 What are the units for expressing chromatography?
R_f

Q.47 What is R_f?
R_f is defined as the ratio of distance traveled by the compound to the distance traveled by the solvent.

Q.48 What is partition coefficient?
Partition coefficient is ratio of concentration of solute in phase-1 to that of other phases.

Q.49 How partition coefficient is designated?
K_d.

Q.50 What is the significance of partition coefficient?
The position to which a given solute moves up the paper is largely dependent of its partition coefficient at given temperature.

Q.51 What are the uses of chromatography?
1. Separation of amino acid in patients with inborn errors of metabolism.
2. Identification of reducing sugars in serum.
3. Separation of drug metabolites.
4. For detection of poisons.

Colorimetry

Q.1 What is Lambert's law?

Lambert's law states that the proportion of light absorbed by an absorbing substance is independent of the intensity of the incident light.

Q.2 What is Beer's law?

Beer's law states that the proportion of light absorbed depends only on the total number of absorbing molecules through which light passes.

Q.3 Define photometry.

Photometry is the most common analytical technique used in clinical biochemistry. Principle of photometry is based on physical laws of radiant energy. The intensity of absorbed transmitted or reflected light is measured and is related to concentration of test substrate.

Q.4 Define colorimetry.

This is a technique of measurement, i.e. quantitative analysis of substance in all biological fluids. Basis of doing this, is to convert the substance into colored product by performing various specific reaction. The intensity of color is directly proportional to the amount of substance present in the sample.

Q.5 What is optical density?

Optical density (OD) is the logarithmic ratio of the intensity of the incident light to that of the emergent light.

$$\text{OD} = \log_{10}\frac{I_0}{I}$$

where, I_0 is the intensity of incident light.
I is the intensity of emergent light.

Q.6 What is transmission?

Transmission is defined as the ratio of the intensity of the transmitted light to that of the incident light.

$$\text{T} = \frac{I}{I_0}$$

where, I_0 is the intensity of incident light.
I is the intensity of emergent light.

Q.7 What is Lambert-Beer's law?

Lambert-Beer's law states that when monochromatic light passes through a colored solution, the amount of light transmitted decreases exponentially.

a. With decrease in thickness of the layer of solution through which the light passes.
b. With increase in concentration of the colored substance.
   The relation is

$$\frac{I}{I_0} = e^{-kct}$$

where, I = intensity of emergent light.
I_0 = intensity of incident light.
k = a constant.
c = concentration of the colored substance.
t = thickness of the layer of the solution.

Q.8 What is the relation between the optical density and transmittance?

Optical density and transmittance are related by the formula.
Optical density = 2 – log (Transmittance).

Q.9 Define standard curve.

- When OD or absorbance on Y-axis is plotted against concentration on X-axis, a straight line passing through origin is obtained which means Beer's law is followed.
- With the help of standard curve, the concentration of unknown can be readily determined.

Q.10 What is the use of blank in colorimetric estimation?

The function of the blank is to eliminate the effect of light absorption by the reagents used, since otherwise they might lead to falsely high values for optical density and consequently to falsely high values for the concentrations of the substance it is desired to determine.

Q.11 What are complementary colors?

Colors of light	*Complementary colors*
Violet	Yellow-green
Blue-green	Orange
Blue	Yellow
Green-blue	Red
Green	Purple/red
Yellow	Blue

DO YOU KNOW?

- The best method for hemoglobin estimation is colorimetry.

Chapter 23

Carbohydrates

Q.1 What is the composition of Benedict's qualitative reagent?

Benedict's qualitative reagent contains:

1. *Copper sulfate:* This supplies cupric (Cu^{++}) ions.
2. *Sodium carbonate:* This makes the medium alkaline.
3. *Sodium citrate:* It prevents the precipitation of Cu^{++} ions as $Cu(OH)_2$ or $CuCO_3$ by, forming loosely bound complex with Cu^{++} ions, i.e. (Cu^{++} sodium citrate complex) which on dissociation gives a continuous supply of Cu^{++} ions.

Q.2 What is the composition of Benedict's quantitative reagent?

Benedict's quantitative reagent contains:

1. Copper sulfate
2. Sodium carbonate
3. Sodium citrate
4. *Potassium ferrocyanide:* It keeps cuprous oxide (Cu_2O) in the solution.
5. *Potassium thiocyanate:* It precipitates Cu^+ ions as cuprous thiocyanate (CuCNS).

Q.3 Why Benedict's quantitative reagent gives a white precipitate while Benedict's qualitative reagent gives a brick-red precipitate?

Benedict's quantitative reagent gives a white precipitate of CuCNS.

$$Cu^{++} \xrightarrow[\Delta]{\text{Red sugar}}$$

$$Cu^{+} \xrightarrow{\text{KCNS}} CuCNS\downarrow$$

whereas Benedict's qualitative reagent gives a red precipitate of Cu_2O.

$$Cu^{++} \xrightarrow[\Delta]{\text{Red sugar}}$$

$$OH^{+}Cu^{+} \longrightarrow 2\,Cu\,(OH)\,Cu_2O\downarrow$$

Q.4 What are the substances which give false Benedict's test?

Glucuronates, salicylates (aspirin), vitamin C, homogentisic acid, etc.

Q.5 What is the difference between Benedict's qualitative test and Barfoed's test?

Benedict's qualitative test is a reduction test carried out in alkaline medium, whereas Barfoed's test is a reduction test carried out in acidic medium.

Q.6 Define carbohydrates in chemical term.

Carbohydrates are defined chemically as aldehyde or ketone derivatives of the higher polyhydric alcohols or compounds which yield these derivatives on hydrolysis.

Q.7 How will you classify carbohydrates?

Carbohydrates are classified into four major groups:

- *Monosaccharide (simple sugars):* They cannot be hydrolyzed into simpler forms.
- *Disaccharides:* They yield two molecules of same or different monosaccharide units on hydrolysis.
- *Oligosaccharides:* They yield three to six molecules of monosaccharides on hydrolysis.
- *Polysaccharides (glycans):* They yield more than 6 molecules of monosaccharides on hydrolysis.

Q.8 How are monosaccharides further classified?

Monosaccharides are further classified into two groups depending on:

- The number of carbon atoms they possess, e.g. trioses, tetroses, pentoses, hexoses, etc.
- Whether aldehyde (-CHO) or ketone (-CO) group is present, e.g. aldoses, ketoses.

Q.9 Give an example of an aldohexose and a ketohexose which is of biological importance.

- Aldhexose: D-Glucose
- Ketohexose: D-Fructose.

Q.10 How will you classify polysaccharides?

Polysaccharides are classified into two main groups:

- Homopolysaccharides (hemoglycans): Polymer of same monosaccharide units, e.g. starch, glycogen, insulin, dextrins, cellulose, etc.
- Heteropolysaccharides (heteroglycans): Polymer of different monosaccharide units or their derivatives, e.g. mucopolysaccharides (MPS).

Q.11 What is the general test of carbohydrates?

Molisch reaction, anthrone test.

Q.12 What is the principle of Molisch test?

Carbohydrates on treatment with concentrated sulfuric acid undergoes dehydration to give furfural or furfural derivatives which on condensation with β-naphthol gives a characteristic purple or violet color ring at the junction.

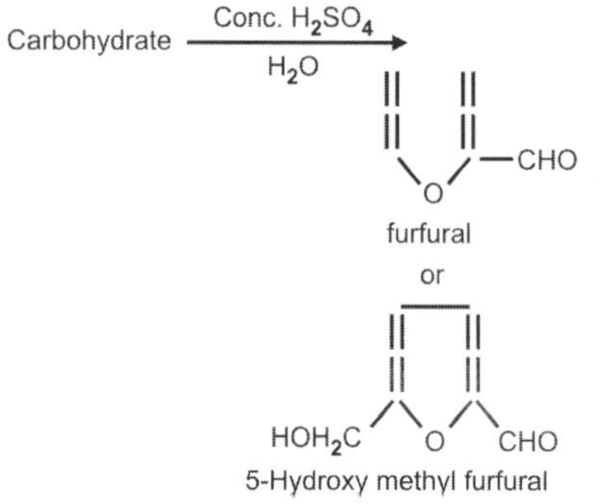

Q.13 Is iodine test for polysaccharides a physical or chemical reaction?
It is physical reaction in which iodine molecules get adsorbed on the surface of polysaccharides.

Q.14 What are reducing sugars?
Reducing sugars are those which possess free aldehydic or ketonic group in their structure.

OR

Sugars having free anomeric carbon atom in their structure are called reducing sugars.

Q.15 What are epimers? Give examples.
- Carbohydrates that differ in their configuration around a specific carbon atom other than the carbonyl carbon atom are called epimers.
- Glucose and galactose are epimers as they differ in their configuration around C-4 carbon atom. Similarly, glucose and mannose are epimers as they differ around C-2 carbon atom.

D-Galactose	D-Glucose	D-Mannose
CHO	^{1}CHO	CHO
H—C—OH	H—^{2}C—OH	HO—^{2}C—H
HO—C—H	HO—^{3}C—H	HO—C—H
HO—^{4}C—H	H—^{4}C—OH	HO—C—H
H—C—OH	H—^{5}C—OH	H—C—OH
CH_2OH	$^{6}CH_2OH$	CH_2OH

Q.16 What is an asymmetric carbon?
A carbon atom to which four different atoms or groups of atoms are attached is said to be an asymmetric carbon.

Q.17 What are the effects of presence of asymmetric carbon in a compound?
The presence of asymmetric carbon atoms in a compound produces the following effects:
- Gives rise to the formation of stereoisomers of that compound.
- Also confers optical activity to the compound.

Q.18 What are stereoisomers?
The compounds which are identical in composition and differs only in spatial configuration are called stereoisomers. Two such stereoisomers of glucose, D-glucose and L-glucose are mirror-images of each other.

Q.19 What are optical isomers?
When a beam of plane polarized light is passed through sugar solution exhibiting optical activity, it will be rotated to the right or left according to the type of the compound. Such compounds are called "optical isomers" (or "enantiomorphs").

Q.20 Draw the structure of L-glucose.

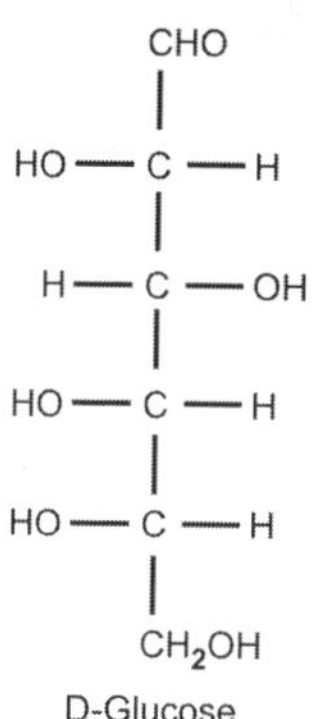

D-Glucose

Q.21 What are anomers? Give examples.
Carbohydrates that differ only in their configuration around the carbonyl carbon atom are called anomers. The carbonyl carbon atom is called the anomeric carbon atom.
α-D-glucose and β-D-glucose are the anomeric form of D-glucose.

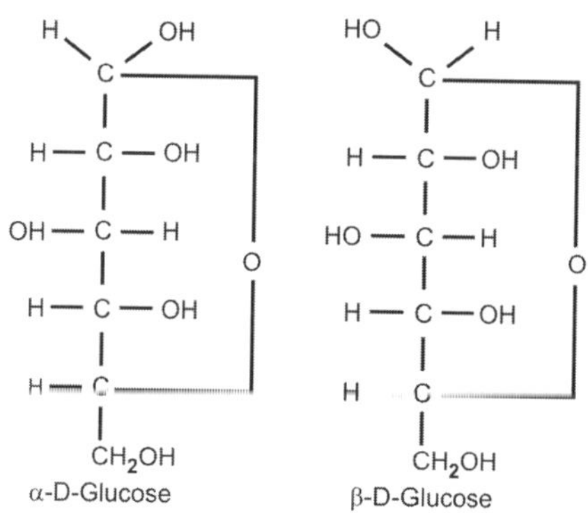

Q.22 How will you determine the D and L series of the sugars?
If the hydroxyl group on the highest asymmetric carbon atom is right oriented then the sugar belong to D-series and if the hydroxyl group is left oriented then it belongs to L-series.

Q.23 What is mutarotation?
The change in specific rotation of an optically active solution without any change in other properties is known as mutarotation.

Q.24 Give details of mutarotation.
Crystalline glucose is α-D-glucopyranose. The cyclic structure is retained in solution but isomerism takes place about position-I, i.e about anomeric carbon atom to give a mixture of α-glucopyranose (38%) and β-glucopyranose (62%). This equilibrium is accompanied by optical rotation, i.e. mutarotation, as hemiacetal ring opens and reforms with change of position of –H and –OH group on carbon-1.

Q.25 What are the optical rotations shown by glucose in solution?
The glucose solution shows rotation according to its form: a form shows +112° and b-form shows +19°. When the solution has an equilibrium mixture of a and b forms, it shows fixed rotation of +52.5°.

α-D-Glucose **Fixed**
+112° ⟶ +
rotation **β-D-Glucose**
52.5° ⟵ +19°

Q.26 What is meant by pyranose form?
The pyranose forms of the sugars are internal hemiacetals formed by combination of the aldehyde or ketone group of the sugar with the –OH group on the 5th carbon atom from the aldehyde or ketone group.

Q.27 What is meant by the furanose form of sugar?
The furanose forms of sugars are formed by reaction between the aldehyde or ketone group with the –OH group on the 4th carbon from the aldehyde or ketone group.

Q.28 Which property of reducing sugars best explains the ring or cyclic structure of the carbohydrates?
Mutarotation.

Q.29 Why glucose and fructose give the same osazones?
Glucose and fructose differ in their structure at first two carbon atoms, i.e. C_1 and C_2 only. In osazone formation these two carbon atoms takes part in the reaction and during osazone formation the structural dissimilarity at C_1 and C_2 disappears. Hence, they give the same osazone.

Q.30 What are the shapes of osazones of glucose, fructose, lactose and maltose?
Glucose: Needle shaped.
Fructose: Needle shaped.
Lactose: Cotton ball shaped.
Maltose: Sunflower shaped.

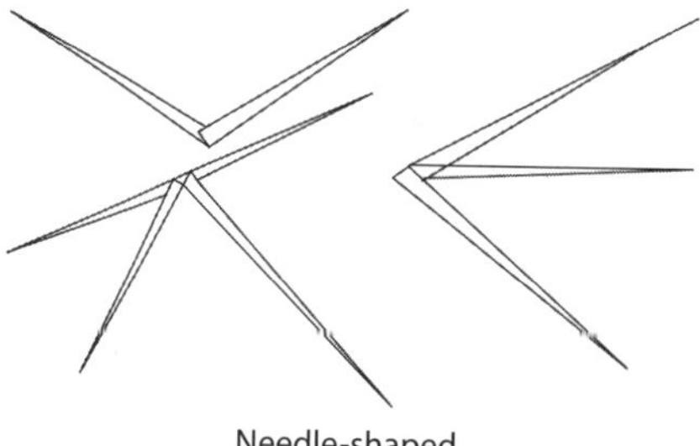
Needle-shaped

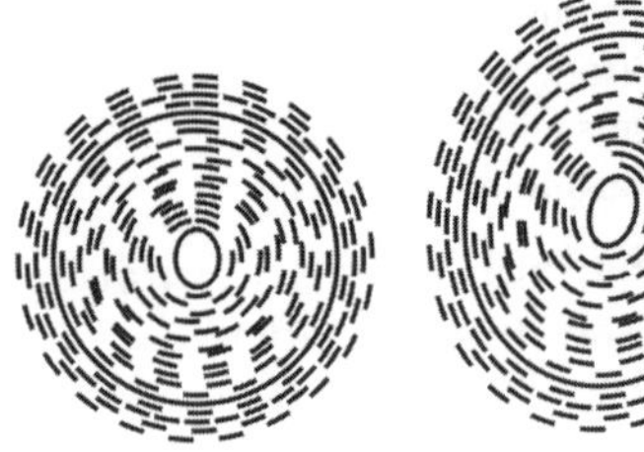

Lactose (Cotton ball-shaped)

Maltose (Sunflower-shaped)

Q.31 What are the reduction products of glucose, mannose, galactose and fructose?

- On reduction the monosaccharides produce sugar alcohols. Thus,
- D-Glucose→D-Sorbitol
- D-Galactose→D-Dulcitol
- D-Mannose→D-Mannitol
- D-Fructose→D-Mannitol+D-Sorbitol

Q.32 Why sucrose is a nonreducing sugar?

Sucrose consists of glucose and fructose which are linked through their reducing sugars, i.e. aldehyde group of glucose is linked to keto group of fructose. As a result of this linkage, both the reducing groups are blocked. Hence, sucrose is a nonreducing sugar.

Q.33 What is inversion?

The process by which dextrorotatory sucrose is converted to a levorotatory mixture of glucose and fructose is called inversion.

$$\underset{(+65.5)}{\text{Sucrose}} \xrightarrow{H^+} \underset{(+52.7)}{\text{Glucose}} + \underset{(-92)}{\text{Fructose}}$$

Q.34 Why sucrose is called an invert sugar?

Sucrose on hydrolysis gives glucose and fructose. Fructose has greater specific rotation than glucose. The resulting mixture is levorotatory. The mixture is known as invert sugar.

Q.35 Name one biological fluid which is rich in fructose. What is the source of fructose in this fluid and its importance?

- Seminal fluid is rich in fructose.
- ***Source:*** It is formed from glucose in the seminiferous tubular epithelial cells.
- ***Importance:*** Spermatozoa utilizes fructose for energy.

Q.36 What is the fate of disaccharide (i.e. sucrose) when injected into the blood?

Sucrose will be excreted as such in the urine as there is no enzyme sucrase present in the blood to hydrolyse it.

Q.37 What are the components of lactose?

Lactose contains glucose and galactose.

Q.38 What are the components of sucrose?

Sucrose contains glucose and fructose.

Q.39 What are the components of maltose?

Maltose contains two molecules of glucose only.

Q.40 What is the nature of linkage in lactose?

β(1, 4) linkage

β Galactose → C_4 glucose.

Q.41 What is the nature of linkage in maltose?

α(1,4) linkage.

Q.42 What is the nature of linkage in sucrose?

α(1,2).

Q.43 What is the nature of linkage in starch?

Starch is formed by amylose and amylopectin. Amylose is nonbranching. Amylopectin consists of 1,4 linkage but at branching point 1,6 linkage is present.

Q.44 What is the nature of linkage in glycogen?

Glycogen has α(1,4) linkage but at branching point α(1,6) linkage is present.

Q.45 What is the nature of linkage in cellulose?

β(1,4).

Q.46 Name the components of starch.

Starch contains two components:

1. Amylose
2. Amylopectin.

Q.47 What is difference between amylose and amylopectin?

Amylose	*Amylopectin*
• Linear molecule containing α-(1,4) linkages	• Branched molecule containing α-(1,4) linkages and α-(1,6) linkages at branching
• Blue color with iodine	• Violet color with iodine
• Water soluble	• Sparingly soluble

Q.48 What are hydrolytic products of starch and state their reaction with I_2 solution?

Hydrolysis of starch yields succession of polysaccharides of diminishing molecular size as follows:

Course of hydrolysis	Reaction with I_2 solution
Starch ↓	Blue
Soluble starch ↓	Blue
Amylodextrin ↓	Purple
Erythrodextrin ↓	Red
Achróódextrin ↓	Colorless
Maltose	

Q.49 What is aglycone?

The noncarbohydrate portion of a glycoside is called aglycone.

Q.50 What is the difference between starch and cellulose?

In starch, the glucose units are linked by α (1,4) glycosidic linkages whereas in the cellulose, the glucose units are linked by β (1,4) glycosidic linkages.

Q.51 Why cellulose is not utilized by human body?

The enzyme responsible for the cleavage of β (1,4) linkages in the cellulose is absent in the human system. Hence, it cannot be utilized.

Q.52 What are amino sugars?

Sugars containing an NH_2 group in their structure are called amino sugars. The alcoholic OH group on carbon 2 is usually replaced by $-NH_2$ group.

Examples: D-Glucosamine, D-Galactosamine.

Q.53 What are glycosides?

Glycosides are compounds containing a carbohydrate and a non-carbohydrate residue in the same molecule; carbon 1 of carbohydrate is attached to the non-carbohydrate residue by an acetyl linkage.

Q.54 What are dextrins?

Dextrins are the partially degraded breakdown products of starch.

Q.55 Differentiate dextrins and dextrans.

Dextrins	*Dextrans*
• Are hydrolytic products of starch • Used in infant feeding	• Synthetic polymer of D-glucose • Used as a plasma expander when given IV in cases of hemorrhage (blood loss), it increases the blood volume

Q.56 What is the difference between starch and glycogen?

Starch	Glycogen
• Plant origin.	• Animal origin.
• It is a branched molecule. Branching occurs after every 20–24 glucose units.	• Highly branched than starch. Branching occurs after every 8–10 glucose units.
• Blue color with iodine solution	• Red color with iodine solution.

Q.57 What are mucopolysaccharides?
Mucopolysaccharides are acidic substances containing uronic acids and N-acetylated amino sugars in combination with proteins.

Q.58 Give few examples of mucopolysaccharides.
Hyaluronic acid, heparin, chondroitan sulfate, dermatan sulfate.

Q.59 What are proteoglycans?
Proteoglycans are conjugated proteins (called "core" proteins) covalently linked to any of the glycosaminoglycans (GAGs). The amount of carbohydrates in proteoglycans is much greater (up to 95%) as compared to glycoproteins.

Q.60 State the type of linkages found in proteoglycans?
Three types of linkages with core protein and GAG is observed. They are:
- ***O-glycosidic linkage:*** between N-acetyl galactosamine (Gal NAc) and serine/threonine of core protein.
- ***N-glycosyl amine linkage:*** Formed between N-acetyl glucosamine (Glc NAc) and amide N of asparagine (ASn) of core protein.
- ***O-glycosidic linkage with xylose:*** formed between xylose and serine of core protein.

Q.61 What are the repeating units of hyaluronic acid?
Glucuronic acid: N-acetylglucosamine.

Q.62 What are the various pathways by which glucose is utilized?
1. Glycolysis followed by tricarboxylic acid cycle.
2. Hexose monophosphate shunt pathway.
3. Conversion to glycogen.
4. Conversion to galactose and then to lactose.
5. Give rise to nonessential amino acids.
6. Conversion to fat.

Q.63 What is glycolysis. What are rate-limiting enzymes of glycolysis?
Glycolysis can be defined as oxidation of glucose or glycogen to pyruvate and/or lactate. The rate-limiting enzymes of glycolysis are:
- Hexokinase
- Phosphofructokinase
- Pyruvate kinase.

  Glycolysis is sometimes also known as Embden-Meyerhof Pathway (EM pathway)

Q.64 What is the role of O_2 in glycolysis?
The glycolytic pathway is unique in the sense that it can occur in presence of O_2 if available ("aerobic" phase) and it can function also in absence of O_2 ("anaerobic" phase).

Q.65 In which part of the cell glycolysis occurs and where the enzymes are located?
Glycolysis occurs in cytosol and enzymes involved are cytosolic (extra mitochondrial).

Q.66 State the biomedical importance of glycolysis.
- Provides energy.
- Cardiac muscle has poor glycolytic activity and poor survival under conditions of ischemia.
- Rate of glycolysis is very high in fast growing cancer cells. Enhanced glycolysis produces more pyruvic acid than TCA cycle can handle. Hence pyruvic acid accumulates and forms lactic acid producing local lactic acidosis which is congenial for certain cancer therapy.
- Inherited enzyme deficiencies like *hexokinase* and *pyruvate kinase* produce hemolytic anemia.

Q.67 Name the steps of glycolysis where ATP is consumed.
ATP is utilized for phosphorylations:
- For conversion of glucose→glucose-6-P
- For conversion of fructose-6-P→ fructose 1,6-bi-P
- Body spends two ATP molecules (-2 ATP)

Q.68 Which enzymes are required for phosphorylations?
- For phosphorylation of glucose: *Hexokinase*/and/or *glucokinase* enzymes.
- For phosphorylation of fructose-6-P: *Phospho-fructokinase* enzyme.

Q.69 State five differences between hexokinase and glucokinase.

Hexokinase	Glucokinase
• Found in all tissues	• Found only in liver
• More stable	• More labile
• Non-specific, can phosphorylate any of the hexoses	• Specific only for glucose
• Km is low, hence high affinity for glucose	• Km is high, hence low affinity for glucose
• Main function to make available glucose to tissues for oxidation at lower blood glucose level	• Main function to clear glucose from blood after meals and at blood levels greater than 100 mg per dl

Q.70 Which is the most energy-yielding step in glycolytic pathway?
Oxidation of glyceraldehyde-3-P, in presence of O_2, by the enzyme *glyceraldehyde-3-P dehydrogenase* which is NAD^+ dependent. 2 NADH when oxidized in ETC gives 6 ATP.

Q.71 What is the end-product of glucose oxidation by glycolysis?
Pyruvic acid.

Q.72 Name the inhibitors of glycolysis.
- ***Iodoacetate and iodoacetic acid***: Inhibit *glyceraldehyde-3-P-dehydrogenase.*
- ***Arsenite***: Inhibits *phosphoglycerate kinase* indirectly and no ATP is formed at substrate level.
- ***Fluoride***: Inhibits the enzyme *enolase.*

Q.73 Name the tissues which solely depends for energy from glycolysis.
- Red blood cells and,
- Brain and nervous tissue.

Q.74 Enumerate the sources of pyruvic acid (PA) in the body.
- By glycolysis—principal source.
- Oxidation of LA→PA in presence of O_2.
- Deamination of alanine.
- Other pyruvic acid forming amino acids, e.g. glycine, serine, cysteine/cystine, threonine.
- Decarboxylation of oxaloacetic acid (OAA)
- From malic acid by malic enzyme.

Q.75 Enumerate the fate of pyruvic acid (PA) in the body.
- Oxidative decarboxylation of PA to form acetyl CoA in presence of O_2.
- Reduction of PA→LA in absence of O_2.
- Amination to form alanine.
- Conversion to glucose (gluconeogenesis).
- Conversion to malic acid

- Formation of oxaloacetate (OAA) by "CO_2-fixation reaction".

Q.76 State the irreversible steps in glycolysis.

- Glucose-6-P→Glucose.
- Fructose-1, 6-bi-P→Fructose-6-P
- Phosphoenol pyruvate→Enol-pyruvate.

Q.77 What is anaplerotic reaction or anaplerosis?

- A sudden influx of PA or acetyl CoA to the TCA cycle might seriously deplete the supplies of OAA required for the *citrate synthase* reaction.
- Two reactions that are auxiliary to TCA cycle operate to prevent this situation. These are called ***anaplerotic reactions*** (or "filling-up" reactions), and the phenomenon is called ***anaplerosis.***

Q.78 Name the two anaplerotic reactions.

- Conversion of PA to OAA by ***CO_2-fixation reaction*** by the enzyme *pyruvate carboxylase* which requires biotin, ATP, Mg^{++} and acetyl CoA. ***Acetyl CoA acts as a +ve modifier***; it helps the enzyme to maintain "active" conformation.
- Conversion of PA to OAA through malic acid formation (Fig. 23.1)

Q.79 Name the inhibitor of *lactate dehydrogenase* enzyme (LDH).

Oxamate. It competitively inhibits *lactate dehydrogenase* (LDH) and prevent reoxidation of NADH.

Q.80 How many ATPs are produced in glycolysis in presence of O_2 (Aerobic phase)? Explain.

- ***Produces 8 (Eight) ATP.***
- Details as follows:

(a) Loss:
- Phosphorylation of glucose = –1 ATP
- Phosphorylation of fructose = –1 ATP

–2 ATP

(b) Gain:
- Oxidation of glyceraldehyde-3-P = +6 ATP
- Phosphoglycerate kinase reaction (substrate level) = +2 ATP
- Pyruvate kinase reaction (substrate level) = +2 ATP

+10 ATP

* Net gain = 10 ATP – 2 ATP = 8 ATP

Q.81 How many ATPs are produced in glycolysis in absence of O_2 (anaerobic phase)?

- In absence of O_2, NADH + H^+ produced by oxidation of glyceraldehyde -3-P, cannot be oxidized in ETC NADH is converted to NAD^+ in reduction of pyruvate to lactate. Hence 6 ATP is not produced.
- In anaerobic phase, per molecule of glucose oxidized,

4 ATP – 2 ATP = 2 ATP will only be produced.

Q.82 What is the ATP yield under aerobic conditions?

38 ATP per molecule of glucose metabolised.

i.e. glycolysis 8 ATP

TCA cycle 30 ATP

Q.83 What is TCA cycle?

- A TCA cycle is the final common pathway for metabolism of carbohydrates, lipids and proteins (IIIrd phase of metabolism). It is a cyclic process and involves a sequence of compounds interrelated by oxidation-reduction and other reactions which finally produce CO_2 and H_2O.
- TCA cycle is also known as Krebs' cycle or citric acid cycle (Fig. 23.2).

Q.84 How many ATPs are formed in the TCA cycle?

i. From acety CoA

ii. From pyruvate.

i. 12 ATPs

ii. 15 ATPs.

Q.85 State the over-all bioenergetics in complete oxidation of glucose/glycogen in glycolysis-cum-TCA cycle in presence of O_2.

A. ***Glycolysis***	***ATP yield per hexose unit***
• Glycogen→F-1, 6-bi-P	-1 ATP
• Glucose→F-1, 6-bi-P	-2 ATP

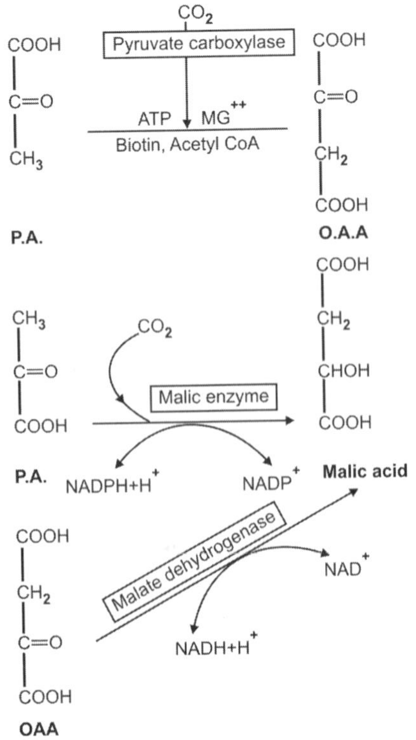

Fig. 23.1: Anaplerotic reactions

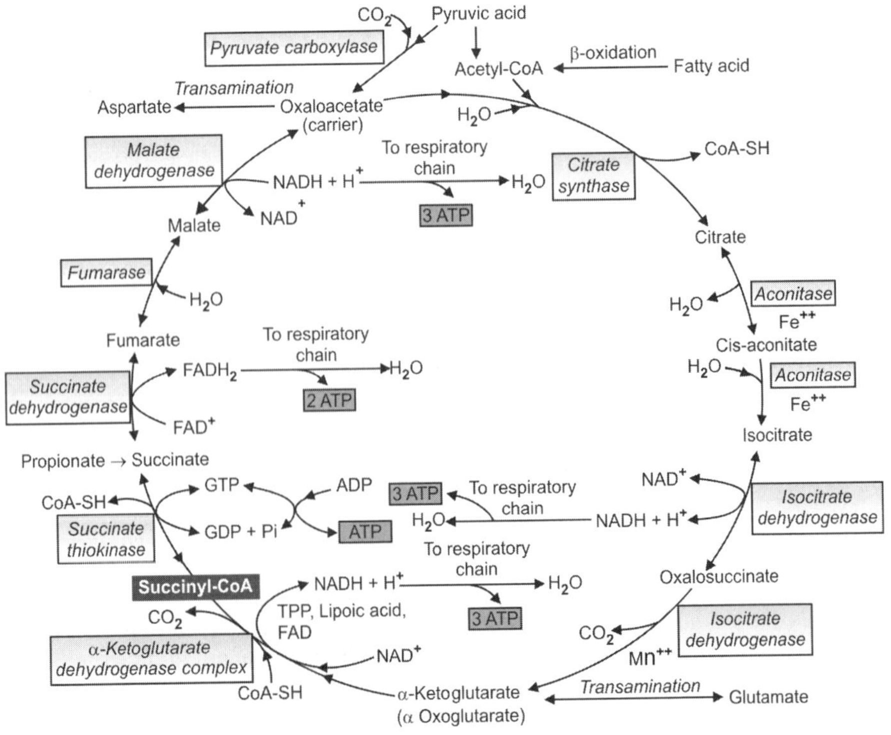

Fig. 23.2: Citric acid cycle

- Glyceraldehyde-3-P dehydrogenase (2 NADH→2 NAD^+) +6 ATP
- Substrate level phosphorylation:
- Phosphoglycerate kinase +2 ATP
- Pyruvate kinase +2 ATP
  Net gain in glycolysis:
  - For glycose = +8 ATP
  - For glycogen = +9 ATP

B. ***Oxidative decarboxylation of P.A.:***
- PDH complex (2 NADH → 2NAD^+) +6 ATP

C. ***TCA cycle:***
- Isocitrate dehydrogenase (2 NADH→ 2 NAD^+) +6 ATP
- α-oxoglutarate dehydrogenase (2 NADH → 2 NAD^+) +6 ATP
- Substrate level phosphorylation: Succinate thiokinase 2 (GTP or ITP)→ 2 ATP + 2 ATP
- Succinate dehydrogenase 2 FAD. H_2 → FAD + 4 ATP.
- Malate dehydrogenase 2 NADH → 2 NAD^+ +6 ATP
  Total = + 24 ATP

∴ Total energetics:
- Per mole of Glucose = 24 + 6 + 8 ATP = 38 ATPs
- Per mole of Glycogen = 25 + 6 + 9 ATP = 39 ATPs

Note: Under anaerobic conditions (in absence of O_2):
- Glucose = +2 ATPs
- Glycogen = +3 ATPs.

Q.86 State the efficiency of complete oxidation of glucose.
- One mole of glucose after complete oxidation produces = 38 ATPs
- Total energy captured in ATP per mole of glucose oxidized= 7600 × 38 = 2,88,800 calories.
- Oxidation of one molecule of glucose "*in vitro*" produces = 6,86,000 calories

∴ Hence efficiency

$$= \frac{2,88,800}{6,86,000} \times 100 = 42\%.$$

Q.87 How 2,3-DPG formation takes place?
When bisphosphoglycerate mutase acts upon 1,3 bisphosphoglycerate formation of 2,3 bisphosphoglycerate taken place in erythrocytes.

Q.88 What is Rapoport-Luebering cycle or shunt (RLC or RLS)?
- RLC or RLS is a diversion in glycolytic pathway in red blood cells. Conversion of 1, 3-BPG to 3 PG does not occur and ATP is not formed at substrate level. It forms 2, 3-BPG.
- It is calculated to deplete and waste the energy needed by the RB cells (Fig. 23.3).

Q.89 What is the function of 2,3 DPG?
2,3-DPG reduces the affinity of oxygen with hemoglobin so, is responsible for normal oxygen delivery to peripheral issues.

Q.90 What is the net energy change during formation of 2-3-DPG in glycolysis?
No net production of ATP takes place.

Q.91 Why citric acid cycle is considered the common pathway for carbohydrate, fat and protein metabolism?
Citric acid cycle is the common pathway for the metabolism of carbohydrates, fats and proteins since it provides the complete oxidation of acetyl CoA to carbon dioxide and water.

Acetyl CoA comes from all the three metabolism:

1. Carbohydrate metabolism: Glycolysis
2. Fat metabolism: β-oxidation
3. Protein metabolism: Transamination

Hence, citric acid cycle is the common pathway for the metabolisms of carbohydrate, fat and protein.

Q.92 What is the inhibitor of aconitase step in Krebs' cycle?
Aconitase, which converts citrate to isocitrate is inhibited by fluoroacetate.

Q.93 Which step of Krebs' cycle is inhibited by arsenite?
Arsenite inhibits the α-ketoglutarate dehydrogenase complex thus impending the conversion of α-ketoglutarate to succinyl-CoA.

Q.94 What is the inhibitor of succinate dehydrogenase?
Succinate dehydrogenase is competitively inhibited by malunate and oxaloacetate.

Q.95 What is the peculiarity of succinate dehydrogenase?
It is only enzyme of TCA cycle which is found to inner mitochondrial membrane unlike others which are present in matrix of mitochondria.

Q.96 Enumerate the vitamins which play an important role in TCA cycle?
1. Riboflavin
2. Niacin
3. Thiamin
4. Pantothenic acid.

Q.97 What is oxidative decarboxylation?
Oxidation accompanied by decarboxylation is called oxidative decarboxylation.

Q.98 What is the oxidative decarboxylation product of pyruvic acid?

$$\text{Pyruvic acid} \xrightarrow[\text{—2H,—CO}_2]{\text{Oxidative decarboxylation}} \text{Acetyl CoA.}$$

Q.99 What is substrate level phosphorylation?
When energy is liberated without entrance to electron transport system, it is termed substrate level phosphorylation.

Q.100 When substrate level phosphorylation occurs in TCA cycle?
When succinyl CoA is converted into succinate. One ATP is liberated by substrate level phosphorylation.

Q.101 Can TCA cycle function in absence of O_2?
TCA cycle cannot function in absence of O_2.

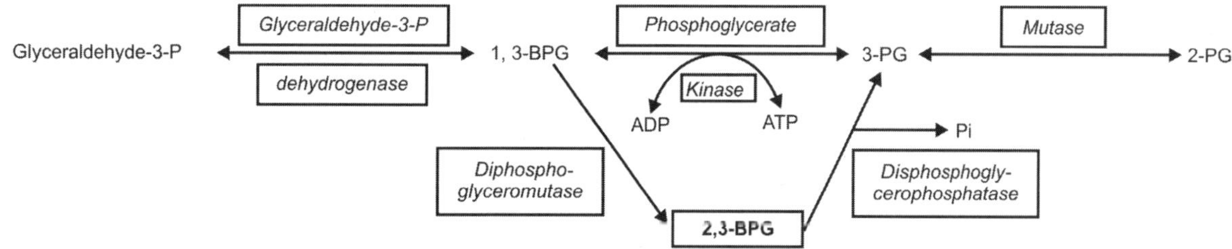

Fig. 23.3: Rapoport-Luebering shunt or cycle

Q.102 Where are the enzymes of TCA cycle located?

Enzymes of TCA cycle are located in mitochondrial matrix, either free or attached to the inner surface of the inner mitochondrial membrane, which facilitates the transfer of reducing equivalents to the adjacent enzymes of ETC.

Q.103 Succinyl CoA is an intermediate in TCA cycle. How it is formed?

Succinyl CoA is formed by oxidative decarboxylation of α-oxoglutarate by *"α-oxoglutarate dehydrogenase complex"* which requires TPP, lipoic acids, CoA-SH, FAD, NAD^+ and Mg^{++} ions as coenzymes/cofactors (Fig. 23.4).

Q.104 State the inhibitors of TCA cycle.

- ***Fluoroacetate***: Inhibitor of *"aconitase"* and allows citrate to accumulate.
- ***Arsentie:*** Inhibits *"α-oxoglutarate dehydrogenase"* enzyme complex and allows accumulation of α-oxoglutarate (α-ketoglutarate)
- ***Malonate/OAA:*** Inhibits *succinate dehydrogenase* by competitive inhibition and allows accumulation of succinate.

Q.105 What is the importance of OAA in TCA cycle?

- It is required to start the cycle
- A small quantity is necessary.
- At the end of the cycle, OAA is regenerated by oxidation of malate by *matate dehydrogenase.*
- Thus OAA acts ***catalytically*** to restart the cycle again.

Q.106 What will happen to TCA cycle if OAA is not available?

In absence of OAA, TCA cycle will not operate. Acetyl CoA will accumulate and will be diverted to form ketone bodies and biosynthesis of FA and cholesterol.

Q.107 Why TCA cycle is said to be amphibolic in nature?

- TCA cycle has dual role:
- ***Catalytic role:*** The acetyl CoA produced by metabolism of carbohydrates, lipids and proteins are completely oxidized to produce CO_2, H_2O and energy.
- ***Anabolic role*** (*synthetic role*): Intermediates of TCA cycle are utilized for synthesis of various biologically important compounds in the body, e.g.
  - Synthesis of non-essential amino acid.
  - Formation of glucose (gluconeogenesis)
  - FA synthesis
  - Synthesis of cholesterol and steroids.
  - Heme synthesis.

Q.108 What is the enzyme of the above oxidative decarboxylation reaction?

Enzyme involved is pyruvate dehydrogenase complex. This enzyme complex consists of three enzymes:

i. Pyruvate dehydrogenase.
ii. Dihydrolipoyl transacetylase.
iii. Dihydrolipoyl dehydrogenase.

Q.109 What are the cofactors of the above reaction?

1. Thiamine pyrophosphate (TPP).
2. Lipoic acid.
3. Coenzyme A (CoA-SH).
4. Flavin adenine dinucleotide (FAD).
5. Nicotinamide-adenine dinucleotide (NAD^+).
6. Mg^{++} ions.

Q.110 What is the active form of PDH?

"Active" forms of PDH is the dephosphorylated form. *Insulin stimulates phosphatase enzyme and converts* 'inactive' → to 'active' form by dephosphorylation.

Q.111 What is the inactive form of the PDH enzyme?

- "Inactive" form is the phosphorylated form catalyzed by the enzyme *"PDH kinase"*. Following favor the formation of "inactive" form:
- Rise in ATP/ADP ratio
- Rise in $NADH/NAD^+$ ratio
- Acetyl CoA/CoA. SH ratio
- Increased cyclic AMP level in cells.

Q.112 Pyruvic acid is formed in cytosol by glycolysis but oxidative decarboxylation takes place in mitochondrion. Pyruvic acid is impermeable to mitochondrial membrane. How it is done?

Pyruvic acid formed in cytosol is not permeable to mitochondrial membrane, it is transported to mitochondrion by a specific ***transport*** protein.

Q.113 What is the importance or metabolic significance of hexose monophosphate shunt pathway?

1. Provides NADPH which is used in fatty acid synthesis.
2. Provides pentose sugar: which is the building block of nucleic acid.

Q.114 Acetyl CoA is formed inside mitochondria but the fatty acid synthesis (*de novo*) from acetyl CoA occurs in cytosol. Acetyl CoA is not permeable to mitochondrial membrane. How acetyl CoA made available in cytosol?

- Acetyl CoA is transported out in the form of citrate, an intermediate of TCA cycle, to cytosol, as citrate is readily permeable to mitochondrial membrane.
- In the cytosol, citrate is cleaved by the enzyme *citrate cleavage enzyme* (ATP-citrate lyase) to acetyl CoA and OAA, so that acetyl CoA can be used for FA synthesis.

Q.115 In glycolysis, NADH is produced in cytosol, but it is oxidized in ETC in mitochondria to produce ATP. NADH is not permeable to mitochondrial membrane. Explain how it is achieved?

NADH produced in cytosol by glycolysis transfer the reducing equivalents through the mitochondrial membrane via substrate pairs linked by suitable dehydrogenases by "shuttle systems". Two such shuttle systems are:

- Glycerophosphate shuttle.
- Malate shuttle.

Q.116 Show schematically glycerophosphate shuttle.

See Figure 23.5

Note: *α-glycero-P-dehydrogenase in mitochondrion is Fp-dependent.*

Hence produces 2 ATP per mole of glucose oxidized. Hence in this 36 ATP is produced per mole of glucose oxidized.

Q.117 Show schematically malate shuttle.

See Figure 23.6.

Note: *Malate shuttle is commonly used by the body.*

Use of malate shuttle forms 38 ATP.

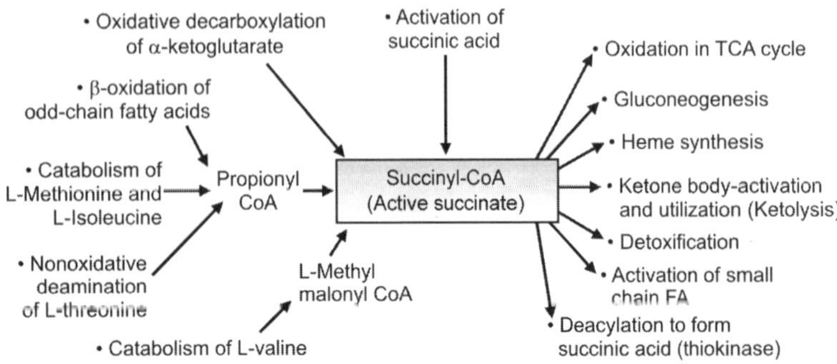

Fig. 23.4: Succinyl CoA: Formation and fate

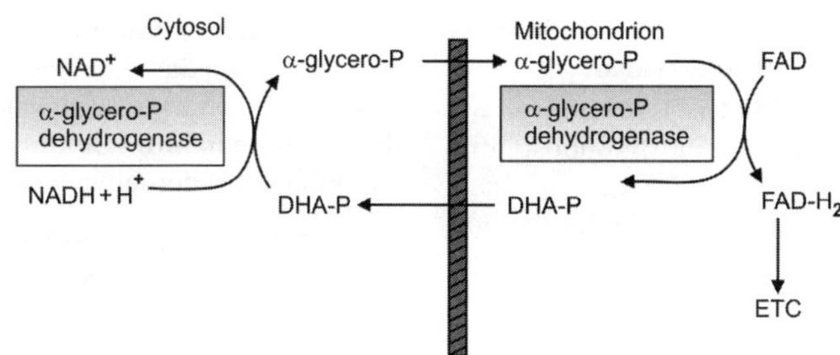

Fig. 23.5: Glycerophosphate shuttle

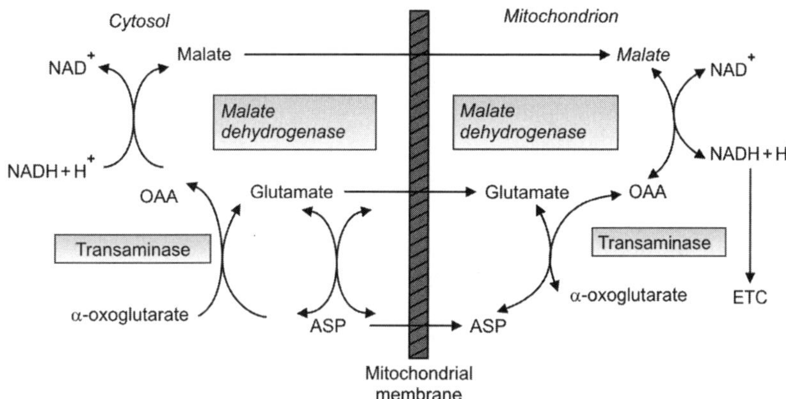

Fig. 23.6: Malate shuttle

Q.118 Name the tissues where hexose monophosphate shunt pathway is active?
HMP shunt pathway is active in liver, adipose tissue, mammary gland, adrenal cortex, testis, etc.

Q.119 What are the clinical problems do G-6-PD deficient persons have to face?
The patients with G-6-PD deficiency when given antimalarials like primaquine, they develop hemolysis.

Q.120 Why is it so?
G-6-PD is responsible for maintenance of glutathione in reduced state. Moreover, when primaquine is given it leads to generation of more free radicals. These two factors contribute to hemolysis.

Q.121 Name two keto acids involved in carbohydrate metabolism.
i. Pyruvic acid
ii. α-keto-glutaric acid.

Q.122 Name two reactions catalyzed by kinases.
Reactions catalyzed by kinases:
1. Glucose → Glucose -6-PO_4 .
2. Fructose-6-PO_4 → Fructose-1, 6-diPO_4.

Q.123 Name two reactions catalyzed by mutases.
Reactions catalyzed by mutases:
1. 3-Phosphoglycerate → 2-Phosphoglycerate.
2. Glucose-6-PO_4 → Glucose 1-PO_4.

Q.124 What is glycogenesis?
It is the formation of glycogen from glucose in the body.

Q.125 Show schematically the steps of glycogenesis.
See Figure 23.7

Q.126 What happens to liberated UDP by glycogen synthase action?
UDP is converted to UTP again by the enzyme *nucleoside diphosphokinase* and is reutilized again for UDPG formation.

Q.127 What is the primer? How the first primer formed?
- The first "primer" originally supposed to be synthesized on a protein backbone which is a process similar to synthesis of other glycoproteins.
- A pre-existing glycogen molecule or "primer" is a must, so that UDPG can add the glucose molecule to the outer end of a chain.

Q.128 How much ATP is spent by the body to add one glucose unit to outerchain?
- In each addition of glucose unit, 2 ATP molecules are expended by the body:
  - One ATP is utilized in phosphorylation of glucose→ Glucose-6-P.
  - Another ATP is used to convert UDP to UTP again.

Q.129 What is key and rate-limiting enzyme in glycogenesis?
Glycogen synthase (synthetase) enzyme.

Q.130 What are the active and inactive forms of the enzyme glycogen synthase?
- Glycogen synthase enzyme occurs as 'active' GS 'α' or 'inactive' GS 'β' forms and both are interconvertible.
- GS 'α'→GS 'β' by phosphorylation which is modulated by *cyclic-AMP dependent protein kinase* and glycogenesis is stopped.
- GS 'β'→ GS 'α' is formed by dephosphorylation catalyzed by the enzyme *"protein phosphatase-1"* when glycogenesis starts.

Q.131 What is UDPG? What are its functions?
- UDPG is uridine diphosphate glucose. It is an intermediate in glycogenesis. It is activated glucose which adds one glucose unit to a chain in 'primer' molecule, by α 1 → 4 glycosidic linkage.
- Other functions:
  - It is formed as an intermediate in uronic acid pathway required for formation of D-glucuronic acid.

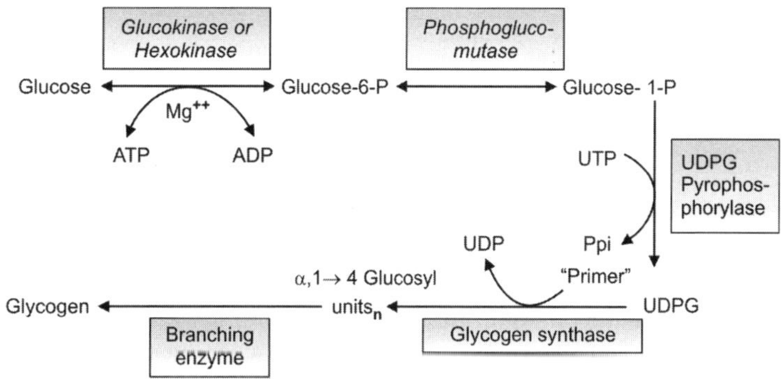

Fig. 23.7: Steps of glycogenesis

- It is also required for synthesis of lactose from glactose in lactating mammary gland.

Q.132 State the factors that bring about stimulation and inhibition of glycogenesis.

- ***Stimulators:***
  - Insulin
  - Glucocorticoids
  - High concentration of glucose.
- ***Inhibitors:***
  - Increase concentration of glycogen (by "feedback" inhibition).
  - Increased cyclic AMP in the cells, which can be brought about by hormones viz.,
    - Epinephrine
    - Norepinephrine
    - Glucagon
    - Thyroid hormones.

Q.133 How insulin increases glycogenesis?

Insulin directly stimulates the enzyme *protein phosphatase-1* thus brings about dephosphorylation of *glycogen synthetase* and forms "active" glycogen synthase, GS 'α', and increases glycogenesis.

Q.134 How does increased cyclic AMP level in cells inhibits glycogenesis?

- Increased cyclic AMP in the cells converts *inactive protein kinase* (C_2R_2) → *active protein kinase* (C_2).
- Active protein kinase (C_2) has following two effects:
  - Brings about phosphorylation of 'GS' enzyme with the help of ATP and thus GS 'α'→ GS 'β' inhibiting glycogenesis.
  - Also converts a protein factor *inhibitor 1* (inactive) and phosphorylates it to form "active inhibitor 1-P", which in turn inhibits *protein phosphatase-1*, so that conversion of inactive GS 'β'→to active GS 'α' does not occur thus inhibiting glycogenesis.

Q.135 What do you know about branching and debranching enzymes?

- Branching enzymes produce a branching in the molecule by establishing an α (1,6) glycosidic linkages.
- Debranching enzymes produce a cleavage of α(1,6) linkages.

Q.136 What is glycogenolysis?

Breakdown of glycogen to glucose is called glycogenolysis.

Q.137 Which is the key and rate-limiting enzyme in glycogenolysis?

Phosphorylase enzyme.

Q.138 What are the active and inactive forms of liver phosphorylase?

- Active phosphorylase is the phosphorylated form "active" phosphophosphorylase 'α'.
- Inactive phosphorylase is the dephosphorylated form "inactive" dephosphophosphorylase 'β'.

Q.139 What are the basic differences between liver and muscle phosphorylase?

- There is no cleavage of structure with liver phosphorylase as compared to muscle phosphorylase.
- Four molecules of pyridoxal-P is required for activity of muscle phosphorylase, not so with liver phosphorylase.
- Muscle phosphorylase is not affected by glucagon (no receptor on muscle).

Q.140 Name the hormones which bring about glycogenolysis through increased cyclic AMP levels in cells.

- Catecholamines-epinephrine and norepinephrine
- Glucagon, and
- Thyroid hormones.

Q.141 What is the product of phosphorylase activity on glycogen molecule?

Active phosphorylase in the presence of inorganic Pi brings about phosphorolytic cleavage of α 1→4 glycosidic bond from outermost chain of glycogen molecule and glucose is released as glucose-1-P and not free glucose.

Q.142 What is the fate of glucose-1-P released by phosphorylase activity.

- Glucose-1-P is converted to glucose-6-P by the enzyme *phosphoglucomutase.*
- In liver and kidney (not in muscle), glucose-6-P is acted upon by the enzyme *glucose-6-phosphatase* and free glucose is formed.

Q.143 Why glucose is not formed in muscle from glycogen break-down?

In muscle, *glucose-6-phosphatase* enzyme is *absent*, hence glucose-6-P enters the glycolytic cycle and forms pyruvates and lactates.

Q.144 How does glucagon action differ from catecholamines in glycogenolysis?

- Catecholamines cause breakdown of liver as well as muscle glycogen.
- But glucagon breaks down only liver glycogen and ***not muscle glycogen*** (as receptor for glucagon is not present in muscle).

Q.145 What is calmodulin?

Calmodulin is a Ca^{++} dependent regulatory protein which is specific for calcium. It is a flexible protein, with 4 binding sites for Ca distributed in 4 domains.

Q.146 What is gluconeogenesis. Name few gluconeogenic substances.

- The process of formation of glucose from non carbohydrate substances is called gluconeogenesis.
- The gluconeogenic substances are pyruvic acid, propionate, lactic acid, glycerol and amino acids (Fig. 23.8).

Q.147 Name the rate-limiting enzymes of gluconeogenesis.

Four rate-limiting enzymes are:

- Pyruvate carboxylase (mitochondrial).
- Phosphoenolpyruvate carboxykinase (cytosol).
- Fructose-1, 6-biphosphatase (cytosol).
- Glucose-6-phosphatase (cytosol).

Q.148 Name the irreversible steps in gluconeogenesis and enzymes used to circumvent the irreversible steps.

Irreversible steps (Energy barrier)	*Enzymes used to circumvent*
• Pyruvate → phosphoenol pyruvate.	• Pyruvate carboxylase (Mitochondrial)—conversion of PA to OAA by CO_2 fixation reaction.
	• Phosphoenol pyruvate carboxykinase (cytosol) converts OAA to phosphoenol pyruvate.
• Fructose-1-5-bi-P → fructose-6-P	• Fructose-1, 6-bi-phosphatase (cytosol).
• Glucose-6-P →glucose.	• Glucose-6-phosphatase (cytosol).

Q.149 Name the tissues where gluconeogenesis occur and name one disease and one condition in which gluconeogenesis is significantly enhanced.

- Principally occurs in liver (85%) and kidney (15%).
- Uncontrolled diabetes mellitus and prolonged starvation.

Q.150 State how glucose is formed from glycerol?

- Glycerol is phosphorylated in presence of the enzyme *"glycerol kinase"* and ATP to form α-Glycero P.
- α-Glycero-P is converted to Di-OH-acetone-P by dehydrogenase and NAD^+.
- Di-OH-acetone-P and glyceraldehyde-3-P forms fructose, 1-6, biphosphate

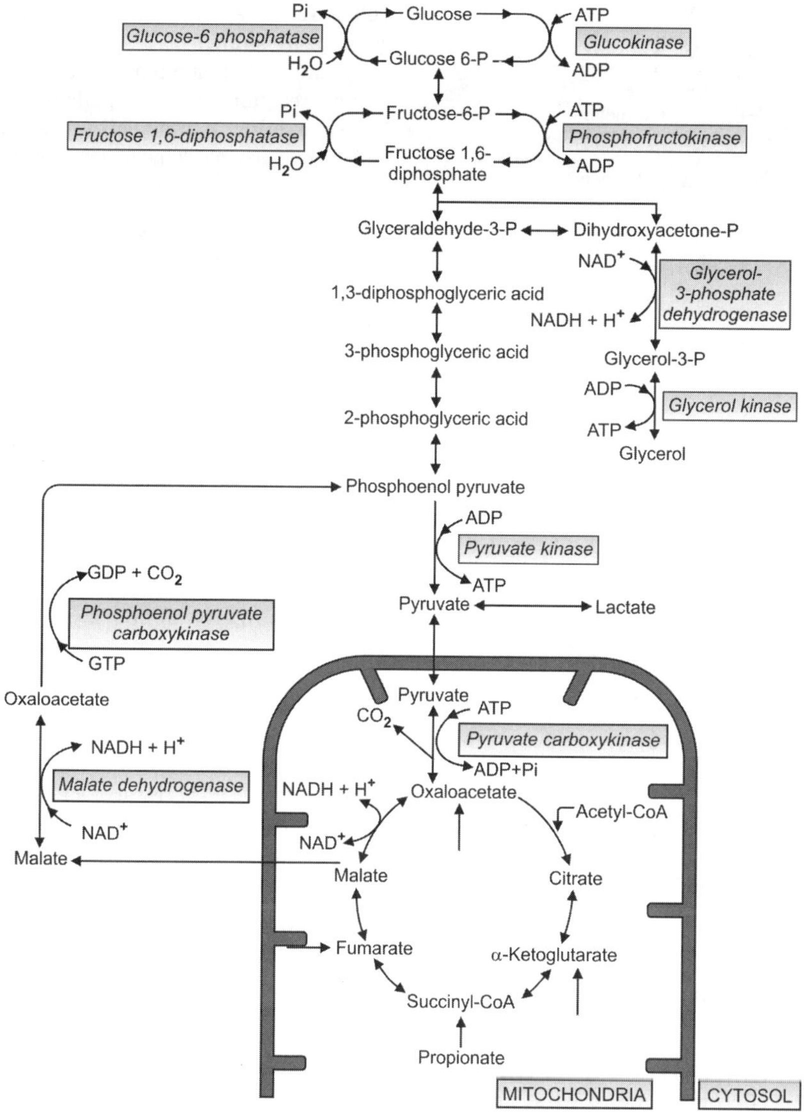

Fig. 23.8: Pathway of glyconeogenesis in the liver

which by reversal of glycolysis form glucose.

Q.151 Mention the sources of propionyl-CoA in humans.

- From catabolism of L-methionine
- Catabolism of isoleucine
- Oxidation of odd-chain FA
- Synthesis of bile acids
- Nonoxidative deamination of L-threonine.

Q.152 Show schematically how glucose is formed from propionic acid.

See Figure 23.9.

Q.153 What is the role of hormones in gluconeogenosis?

- *Glucagon:* Increases gluconeogenesis from lactic acid (LA) and amino acids.
- *Glucocorticoids:* Stimulate gluconeogenesis by increasing protein catabolism in the peripheral tissues and increasing hepatic uptake of amino acids. It increases the activity of transaminases and other key enzymes concerned in gluconeogenesis.

Q.154 What are the enzymes concerned with the reversal of glycolysis, i.e. gluconeogenesis?

1. Pyruvate carboxylase.
2. Phosphoenol pyruvate carboxy kinase.
3. Fructose 1,6 -diphosphatase.
4. Glucose-6-phosphatase.

Q.155 How pyruvate is converted to glucose?

Pyruvate is converted to glucose by gluconeogenesis. Reaction is given as:

Pyruvate → Oxaloacetate
Oxaloacetate → Phosphoenol pyruvate
Phosphoenopyruvate → Glucose.
(Fig. 23.10)

Q.156 What is Cori cycle?

Lactic acid produced in the muscle reaches the liver through blood where it is converted to glucose by gluconeogenesis, which again becomes source of energy for utilization. This process continues and is called Cori cycle (Fig. 23.11).

Q.157 Can muscle glycogen be the source of blood glucose?

Muscle glycogen cannot be a source of blood glucose, because it lacks an enzyme glucose 6-phosphatase.

Q.158 Give the breakdown products of glycogen.

$$\text{Glycogen} \xrightarrow{\text{Phosphorylase}} \text{Glucose-1-PO}_4$$

$$\text{Glucose-I-PO}_4 \xrightarrow{\text{Mutase}} \text{Glucose-6-PO}_4$$

$$\text{Glucose-6-PO}_4 \xrightarrow{\text{G-6 Pase}} \text{Glucose}$$

Q.159 Name the six classical types of GSDs and indicate the enzyme deficiencies.

Type/Name	*Enzyme deficiency*
• Type I-von Gierke's disease	• Glucose-6-Pase
• Type II-Pompe's disease	• Acid maltase (Limit dextrinosis)
• Type III-Forbe's disease	• Debranching enzyme
• Type IV-Andersen's disease (Amylopectinosis)	• Branching enzyme.
• Type V-McArdle's disease	• Muscle phosphorylase
• Type VI-Her's disease	• Liver phosphorylase

Q.160 Explain von Gierke's disease.

In von Gierke's disease there is deficiency of glucose-6-phosphatase.

Q.161 Explain Mc Ardle's disease.

In Mc-Ardle's disease there is deficiency of muscle phosphorylase.

Q.162 What is hexose monophosphate pathway or shunt?

It is an alternate pathway of glucose oxidation which takes place in certain special tissues to serve certain special functions.

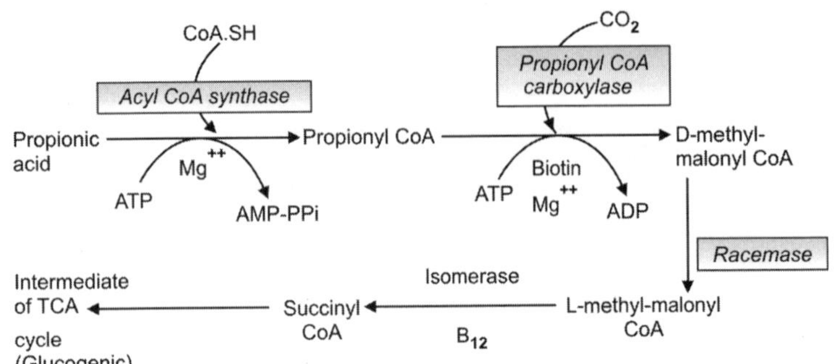

Fig. 23.9: Formation of glucose

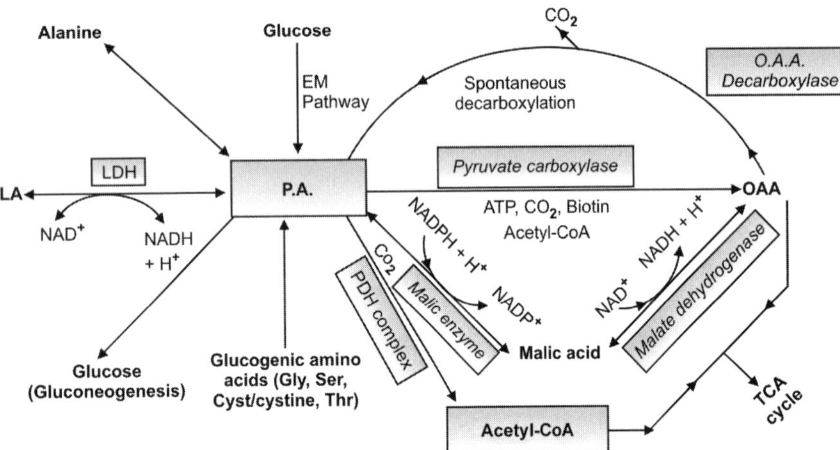

Fig. 23.10: Formation and fate of pyruvic acid

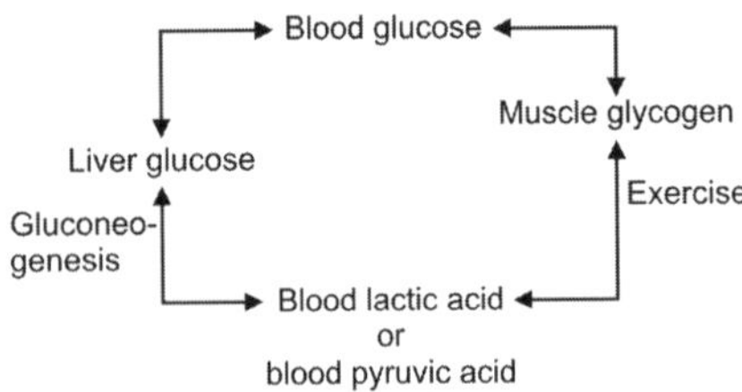

Fig. 23.11: Cori cycle

Q.163 How does HMP shunt differ from EM pathway?

Essential differentiating points are:

HMP shunt	*EM pathway*
• Occurs in certain special tissues only	• Occurs in all tissues
• A multicyclic process	• Not so
• $NADP^+$ acts as H-acceptor	• NAD^+ acts as H-acceptor
• ATP is not produced pathway Not meant for energy	• Energy producing ATP is produced
• CO_2 is formed	• CO_2 is never formed

Q.164 In which part of the cell HMP shunt operates?

In the cytosol.

Q.165 Why HMP shunt is called a multicyclic process and what are the products.

It is called a multicyclic process, as 3 mols of glucose-6-P enter the cycle, producing 3 mols of CO_2, 6 mols of NADHP and 3 mols of 5-C residues which rearrange to give 2 mols of glucose-6-P (re-enters the cycle) and one mol of glyceraldehyde-3-P.

Q.166 Which is the key and rate-limiting enzyme in HMP shunt?

Glucose-6-phosphate dehydrogenase (G-6-PD).

Q.167 What is the role of thiamine in HMP shunt?

TPP the active form of thiamine is required as a coenzyme with *"transketolase"* enzyme for two transketolation reactions.

Q.168 Name the tetrose sugar formed in HMP shunt.

Erythrose-4-P.

Q.169 How does $NADP^+$/NADPH ratio regulate HMP shunt?

- If the ratio is high, it enhances the rate-limiting reaction and shunt pathway.
- If the ratio is low, it inhibits G-6-PD and 6-phosphogluconate dehydrogenase enzymes and decreases shunt pathway.

Q.170 State the hormones that regulate HMP shunt.

- ***Insulin:*** It induces the synthesis of dehydrogenase enzymes and enhances the activity of the pathway.
- ***Thyroid hormones:*** It stimulates the activity of G-6-PD enzyme of the pathway.

Q.171 State the metabolic role of HMP shunt.

- Does not produce energy.
- ***Provides NADPH:*** required for various reductive synthesis.
- ***Provides pentoses:*** required for nucleic acid synthesis.
- ***Role in lens metabolism:*** Provides reduced glutathione (G-SH) necessary for maintenance of lens proteins.
- ***Role in RB Cells:*** Provides reduced glutathione (G-SH) necessary for RB cells membrane integrity.
- CO_2 is produced which is used in CO_2 fixation reaction.

Q.172 Mention some metabolic reactions where NADPH is used for reductive processes.

- NADPH is used in a number of metabolic processes where it provides hydrogen for reductive synthesis, e.g.
- Extramitochondrial *de novo* fatty acid synthesis.
- In synthesis of cholesterol.
- In synthesis of steroids.
- In conversion of oxidized glutathione to reduced glutathione.
- For conversion of phenylalanine to tyrosine.
- For conversion of methemoglobin→Hb.
- In synthesis of sphingolipids.
- In "microsomal" chain elongation of FA.
- In uronic acid pathway.

Q.173 Muscle tissues contain very small amounts of the dehydrogenases enzymes but still skeletal muscle is capable of synthesizing pentose sugars. How?

Probably this is achieved by ***reversal of shunt pathway*** utilizing fructose-6-P, glyceraldehyde-3-P and the enzymes *transketolase* and *transaldolase* (by non-oxdative pathway).

Q.174 What is Wernicke-Korsakoff syndrome?

- A genetically variant form of *transketolase* occurs which cannot bind TPP thus affecting transketolation reaction.
- The patient shows severe neuropsychiatric symptoms characterized by lesions and hemorrhages near IIIrd ventricle.
- The patient shows deranged mental function, loss of memory, depression, disorientation and mental confusions.

Q.175 What is galactosemia?

Excretion of galactose in urine due to the deficiency of enzyme galactose-1-PO_4 uridyl transferase leads to a condition known as galactosemia (Fig. 23.12).

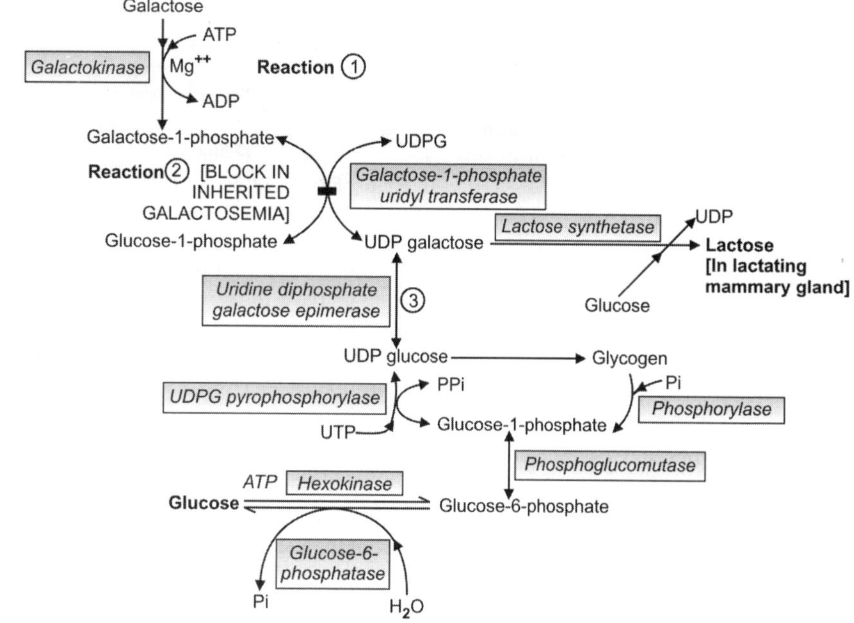

Fig. 23.12: Metabolism of galactose

Q.176 State the salient clinical features in galactosemia.

- ***Infants*** appear normal at birth but later:
  - fails to thrive
  - becomes lethargic and may vomit.
  - develops hypoglycemia and
  - may develop jaundice.
- ***After 2 to 3 months:***
  - develops cataracts in both eyes.
  - shows mental retardation due to accumulation of galactose and galactose-I-P in cerebral cortex.
  - liver may have fatty infiltration and produces cirrhosis liver.

Q.177 What is the cause of cataracts in galactosemia?

- Excess of galactose in lens of the eye is reduced to "galactitol" (Dulcitol), an alcohol by the enzyme *aldose reductase*. Galactitol cannot escape from lens cells. Osmotic effect of sugar alcohol contributes to injury to lens proteins producing cataracts.
- Excess of galactose inhibits the enzyme G-6-PD of HMP shunt leading to less NADPH which results to low reduced glutathione (G-SH).

Q.178 What are the enzymes deficient in galactosemia?

Enzymes deficient are:

a Galactose kinase.

b. Galactose-1-PO_4-uridyl transferase.

Q.179 What is uronic acid pathway?

It is an alternative pathway for glucose oxidation. This pathway also does not produce energy (Fig. 23.13).

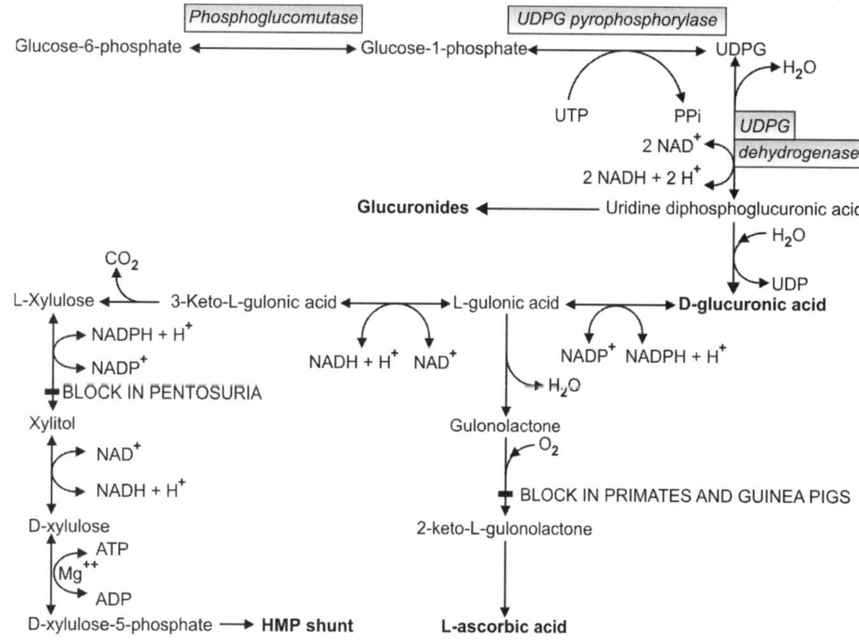

Fig. 23.13: Uronic acid pathway

Q.180 What is the metabolic significance of uronic acid pathway.

- Formation of D-glucuronic acid which is used for detoxication.
- Produces vitamin C in lower animals, but cannot synthesize vitamine C in man, and other primates including guinea pigs due to absence of enzymatic machinery.
- Inherited deficiency of the enzyme *"L-xylitol dehydrogenase"* produces essential pentosuria.
- Excessive xylitol or parenteral administration of xylitol may lead to oxalosis.

Q.181 What are the metabolic functions of D-glucuronic acid produced in uronic acid pathway?

Detoxication: Detoxicates drugs, chemicals, antibiotics, hormones, etc. convert them to corresponding soluble glucuronides which are excreted.

Examples:

- Aromatic acid like benzoic acid.
- Phenol and secondary/tertiary aliphatic alchohols.
- Drugs and other xenobiotics—they are first hydroxylated by mono-oxygenase cyt.P_{450} system and then conjugated with-D-glucuronic acid.

  Antibiotics like chloramphenicol.
- Steroid hormones and thyroid hormones.

- Bile pigments: unconjugated bilirubin is conjugated with UDP-glucuronic acid and converted to soluble mono and diglucuronides.
- Synthesis of heteroglycans containing D-glucuronic acid, e.g. heparin, hyaluronic acid, chondroitin SO_4.

Q.182 Name some drugs which increase the formation of D-glucuronic acid by uronic acid pathway.

- Barbiturates
- Aminopyrine and
- Antipyrine.

Q.183 What is true glucose?

True glucose means glucose only without taking into account the presence of any other reducing substances in the blood.

Q.184 Name some dietary sources of fructose.

- Principal source is sucrose (canesugar/table sugar), which on hydrolysis in intestine gives fructose.
- Other sources are fruit juices and honey.

Q.185 Name the specific enzyme that phosphorylates fructose and the product formed.

- Specific enzyme is *fructokinase.*
- Product is fructose-1-P.

Q.186 State some metabolic importance of fructose.

- Frustose is easily metabolized and a good source of energy.
- Seminal fluid is rich in fructose and spermatozoa utilizes fructose for energy.
- Excess dietary fructose is harmful, leads to increased synthesis of TG.
- In diabetes mellitus, fructose metabolism through ***sorbitol pathway*** may account for development of cataracts and neuropathy.
- Inherited deficiency of the enzyme *"aldolase-B"* produces inherited disorder ***hereditary fructose intolerance.***

Q.187 How is glucose converted to fructose in seminiferous tubular epithelial cells?

- It is achieved by sorbitol pathway.
- Sorbitol pathway for conversion of glucose to fructose is shown schematically (Fig. 23.14).

Q.188 How do you explain biochemically the formation of cataract and neuropathies in diabetes mellitus?

- Formation of sorbitol from glucose by sorbitol pathway proceeds rapidly in the lens of the eye and Schwann cells of the nervous system.

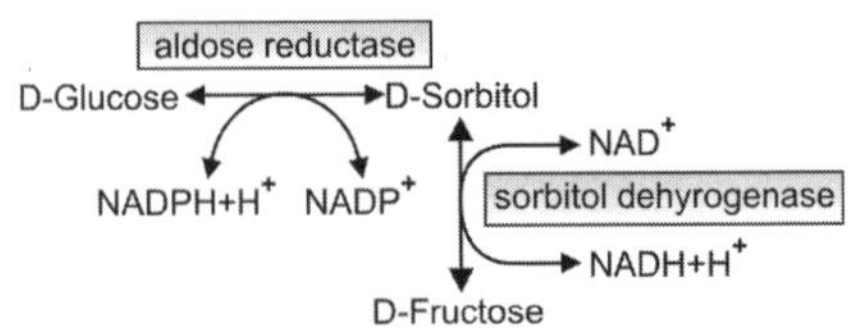

Fig. 23.14: Sorbitol pathway

- Elevated sorbitol concentration in these cells increase the osmotic pressure which may be responsible for the development of cataracts of lens of the eye and diabetic neuropathy.

Q.189 What is hereditary fructose intolerance?

- It is an inherited disorder, due to inherited deficiency of the enzyme *aldolase B.* It leads to excessive rises of fructose and fructose-1-P in blood.
- Blood glucose falls leading to hypoglycemia. The cause of hypoglycemia is probably due to:
  - excessive insulin secretion and
  - inhibition of *phosphoglucomutase* enzyme, by fructose-1-P
- Diets low in fructose and sucrose is beneficial.

Q.190 How will you estimate true glucose?

True glucose is estimated by glucose oxidase method.

Q.191 What is use of glucose vial?

Glucose vial is used to collect blood for esumation of glucose level.

Q.192 Glucose vial constitutes?

It contains sodium fluoride and potassium oxalate in 1:3 ratio. Sodium fluoride inhibits glycolysis and potassium oxalate inhibits coagulation of blood.

Q.193 What are different methods of measuring reducing sugars level?

1. Reduction method:
   - King and Asatoor method
   - Folin and Wil method
   - Ferricyanide method.
2. Orthotoluidine method.
3. Enzymatic method.
4. Dextrostick method.

Q.194 What is glucose postprandial level?

Glucose postprandial level is 100–140 mg%.

Q.195 What are different types of glucose transporters?

GLUT-1	Brain, kidney RBC
GLUT-2	Liver, pancreas
GLUT-3	Kidney, placenta
GLUT-4	Heart, skeletal muscle, adipose tissue
GLUT-5	Small intestine
SGLT-1	Small intestine and kidney

Q.196 What are the conditions in which blood sugar level is low?

Blood sugar level is low in:

1. Overdosage of insulin in the treatment of diabetes.
2. Hypothyroidism.
3. Addison's disease.

Q.197 What are the conditions in which blood sugar level is raised?

Conditions in which blood sugar level is raised are.

1. Diabetes mellitus.
2. Hyperthyroidism.
3. Hyperadrenalism.
4. Hyperpituitarism.
5. Thyrotoxicosis.

Q.198 What is the kidney threshold for glucose?

180 mg.

Q.199 What are the hormones which keep the blood sugar level high?

1. Glucagon.
2. Epinephrine.
3. Adrenal cortex hormones.
4. Growth hormone and ACTH.
5. Thyroid hormone.

Q.200 What is the hormone which regulates blood sugar level?

Insulin.

Q.201 Why insulin cannot be given orally?

Insulin is polypeptide and is digested by the enzymes of digestive system into amino acids before it reaches in the blood. Hence, it cannot be given orally.

Q.202 What is renal glycosuria?

As a result of low kidney threshold, glucose appears in the urine. Blood sugar level remains normal.

Q.203 What is diabetes mellitus?

Diabetes mellitus is a metabolic disorder due to the deficiency of insulin, resulting in high blood glucose level and glucose excretion in the urine.

Q.204 What is IDDM?

In this disease there is autoimmune destruction of b-cells of pancreas and it is also known as juvenile diabetes.

Q.205 What is NIDDM?

In this disease there is resistance to insulin so in spite of normal or elevated levels of insulin hyperglycemia occurs.

Q.206 What is the abnormality in the urine sample of diabetic patient and the urine sample of starving patient?

In diabetic patient, urine will show the presence of glucose and ketone bodies; Whereas in case of starving patient only ketone bodies will be present in the urine.

Q.207 What are the different reducing sugars that appears in urine and under what conditions?

1. *Glucose*: Appears in urine in renal glycosuria and diabetes mellitus.
2. *Lactose*: During later stages of pregnancy and lactation.
3. *Galactose*: In galactosemia due to the deficiency of the enzyme galactose-1-PO_4 uridyl transferase. This condition is encountered in infants.
4. *Fructose*: Due to the consumption of a lot of fruits containing fructose such as grapes, plums, cherry, etc.
5. *Pentoses*: Due to the consumption of a lot of fruits containing pentoses. Also in congenital abnormality characterized by inability to metabolize L-xylulose.

Q.208 How the collection of blood specimen is done for estimating blood glucose?

The blood sample is collected in potassium oxalate: sodium fluoride bottle.

Q.209 What is the function of each?

Sodium fluoride: It prevents glycolysis by inhibiting the enzyme enolase of the glycolytic pathway.

Potassium oxalate: It acts as an anti-coagulant.

Q.210 What is the role of isotonic $CuSO_4$ in blood sugar estimation?

Isotonic $CuSO_4$ prevents the hemolysis of red blood cells so that glutathione which is present in the red blood cells does not come out, otherwise will also reduce the alkaline $CuSO_4$ and give rise to high blood sugar values.

Q.211 What are the functions of insulin?

1. Insulin promotes the entry of glucose in all the tissues of the body except liver.
2. Insulin helps in glycogenesis.
3. Insulin prevents glycogenolysis.
4. Insulin inhibits gluconeogenic enzymes.

Q.212 How glucose is removed from the blood?

- Oxidation of glucose in tissues to provide energy.
- Hepatic glycogenesis.
- Glycogenesis in muscles and other tissues.
- Conversion to fat (lipogenesis) in adipose tissue mainly.
- For synthesis of lactose, fructose, ribose sugars, glycoproteins, glycolipids, MPs, etc.
- Excretion in urine when blood glucose exceeds renal threshold (abnormal condition).

Q.213 What is the role of insulin on carbohydrate metabolism?

- Net effect produces fall in blood glucose concentration.
- Diminished supply of glucose to blood due to:
  - decrease hepatic glycogenolysis
  - increased hepatic glycogenesis.
  - Decreased gluconeogenesis
- Increased in the rate of utilization of glucose by tissue cells:
  - increased uptake by tissues
  - increased oxidation for energy
  - increased lipogenesis

Q.214 Name the emergency hormones which increase blood glucose.

- Catecholamines viz. epinephrine and norepinephrine.
- Glucagon.

Q.215 What is hyperglycemia?

Increase in blood glucose level above normal is called hyperglycemia.

Q.216 Enumerate some causes of hyperglycemia.

- Diabetes mellitus.
- Hyperactivity of thyroids (thyrotoxicosis), anterior pituitary (acromegaly/gigantism) and adrenal cortex (Cushing's syndrome).
- In diffuse diseases of pancreas, e.g. in pancreatitis, carcinoma of pancreas.
- In sepsis and in some infectious diseases.
- In intracranial diseases, e.g. meningitis, encephalitis, intracranial tumors, and hemorrhage.
- In emotional 'stress.'

Q.217 What is hypoglycemia?

Decrease in blood glucose level below normal is called hypoglycemia.

Note: Hypoglycemia manifests clinically when the blood glucose is below < 40 mg % ("true" glucose).

Q.218 Enumerate some causes of hypoglycemia.

- Overdosage of insulin in treatment of DM—most common cause and clinically important.
- Hypoactivity of thyroids (myxedema, cretinism), anterior pituitary (Simmond's disease), and adrenal cortex (Addison's disease).
- Insulin secreting tumor (insulinoma)—rare cause.
- In severe liver diseases.
- Leucine sensitive hypoglycemia.
- Glycogen storage diseases (GSDs).
- Idiopathic hypoglycemia in children.
- Nonendocrine tumors like retroperitoneal fibrosarcoma can produce insulin-like hormones.

Q.219 State important biochemical changes that occur in an untreated uncontrolled DM.

- *Hyperglycemia:* Increase in blood glucose.
- *Glycosuria:* Excretion of glucose in urine.
- *Hypercholesterolemia:* Increase in blood cholesterol.
- *Ketonemia:* Increased ketone bodies in blood-acetoacetic acid, β OH-butyric acid and acetone.
- Increased urea and nonprotein nitrogen in blood.
- *Dehydration.*
- *Acidosis*: Lowered pH, hyperventilation (Metabolic) and Kussmaul breathing.
- Lowered HCO_3 and alkali reserve.
- Lowered sodium in blood—hyponatremia and disturbance in fluid and electrolyte balance.

Q.220 What is meant by glucose tolerance?

The ability of the body to utilize glucose is ascertained by measuring its glucose tolerance. It is indicated by the nature of blood glucose curve following the administration of a standard dose of glucose.

Q.221 Name some conditions where you find decreased glucose tolerance.

- Diabetes mellitus.
- Hyperthyroidism.
- Hyperactivity of adrenal cortex and anterior pituitary.

Q.222 Name some conditions where you find increased glucose tolerance.

- Hypothyroidism,
- Hypofunction of adrenal cortex (Addison's disease),
- Hypopituitarism,
- Hyperinsulinism,
- Decreased absorption of glucose like sprue/celiac disease, etc.

Q.223 How will you perform a standard glucose tolerance test (GTT)?

- A 'fasting' sample of venous blood is collected in fluoride bottle ("fasting" sample of blood).
- The bladder is completely emptied and urine is collected for qualitative test for glucose and ketone bodies ("fasting" urine).

- The individual is given 75 gm of glucose dissolved approximately in 250 mL of water to drink orally. Time of oral glucose administration is noted.
- A total of five specimens of venous blood and urine are collected every ½ hr, after the oral glucose (½ hr, 1 hr, 2 hr, and 2½ hr samples of blood and urine).
- Glucose content of all six (including fasting sample) samples of blood are estimated. Corresponding urine samples are tested qualitatively for glucose and ketone bodies.
- The results of blood glucose values are plotted as a graph against time. The curve thus obtained is called ***glucose tolerance curve***.

Q.224 Describe a typical normal glucose tolerance curve (GTC).

- Fasting glucose is within normal limits of 60 to 100 mg % ("true" glucose).
- The highest peak value is reached within one hour.
- The highest value does not exceed the renal threshold, i.e. 170 to 180 mg%.
- At 2 hr, there is a *hypoglycemic dip* (10 to 15 mg lower than fasting value).
- The fasting level is reached by 2½ hr.
- No glucose or ketone bodies are defected in any specimens of urine.

Q.225 Describe a diabetic type of GTC.

- Fasting blood glucose is definitely raised >110 mg% ("true" glucose).
- The highest vlaue is usually reached after 1 to 1½ hr.
- The highest value exceeds the normal renal threshold. The blood glucose does not return to the fasting level within 2½ hr, remains much higher than the fasting value. This is the most characteristic feature of DM.
- Urine samples show presence of glucose. Urine may or may not contain ketone bodies depending on the type of DM and its severity.
- According to severity, the GTC may be:
  - Mild diabetic curve.
  - Moderately severe diabetic curve.
  - Servere diabetic curve.

Q.226 What is primary lactose, Intolerance?

It is a hereditary deficiency of lactose, most commonly found in persons of Asian and African desert. Common symptoms of lactose intolerance include vomiting, bloating, cramps, watery diarrhea and dehydration. Diagnosis is based on positive hydrogen breath test after an oral lactose load.

Q.227 What is secondary lactose intolerance?

It is precipitated at any age by gastrointestinal disturbances such as celiac sprue, coli tis or viral-induced damage to intestinal mucosa.

Q.228 What is the treatment for lactose intolerance?

Dietary restriction of milk and milk products or using drug like lactose pills.

Q.229 What is galactokinase?

1. Galactosemia
2. Galactosuria
3. Cataract in early childhood.

Q.230 What is the treatment of galactokinase deficiency?

Eliminate the source of galactose from diet.

Avoid dairy products.

Q.231 What is essential fructosuria?

It is a genetic deficiency of fructokinase and is usually benign. It is a presence of fructose in urine.

Q.232 What is hereditary fructose intolerance?

It is one of the serious diseases resulting due to the accumulation of fructose I-phosphate in liver and in proximal tubule in kidney. Present with vomiting, apathy, liver damage can lead to jaundice. Renal demage can lead to Fanconi like syndrome and sever hypoglycemia.

Q.233 What is the treatment for hereditary fructose intolerance?

Eliminate fructose from diet.

Q.234 Explain Pompe's disease.

In this disease there is deficiency by lysosomal L-1,4 gluasidose. It is infantile onset and death usually occur by 2 years.

Q.235 Explain Cori's disease.

In this disease there is deficiency of glycogen debranching enzyme.

DO YOU KNOW?

- Muscle stores excess glucose as glycogen.
- Phosphoenolpyruvate (PEP) in glycolysis acts as a substrate level phosphoration and works irrespective of oxygen and mitochondria hence work when ATP needed in anaerobic phase in myocardial infarction.
- Thiamine deficiency patients along with alcoholic also suffer from hypoglycemia. So if you administer only glucose in these alcoholic patients it would not enter the cell so you need to administer thiamine alongwith glucose to get the best result.
- Congestive heart failure may be complication owing to insufficient ATP and accumulation by keto acid in the cardiac muscle.
- Vitamins are very important for the activity of PDH. Any deficiency of vitamine causes decrease function of PDH and decrease glycolysis or acarbohydrate metabolism.

Chapter 24

Lipids

Q.1 What are lipids?
Lipids are ester-like compounds of fatty acids which are insoluble in water but are soluble in fat solvents.

Q.2 Give the classification of lipids.
Lipids are classified into:
1. Simple lipids.
2. Compound lipids.
3. Derived lipids.

Q.3 What is biological importance of lipids in the body?
1. Lipids are the most concentrated source of energy.
   Their caloric value is 9 KCal/gm.
2. Lipids provides essential fatty acids.
3. Supplies fat-soluble vitamins.
4. As components of cell wall.
5. Lipids act as insulating material.

Q.4 What are fats?
Fats are the esters of fatty acids with glycerol.

Q.5 What is the difference between fat and an oil?
Fats are solid at ordinary temperature whereas oil are liquid, i.e. (liquid) fat at ordinary temperature.

Q.6 What are saturated fatty acids? Give examples.
Fatty acids which do not have any double bond in their structure are called saturated fatty acids.
Examples: Acetic acid (CH_3 COOH), propionic acid (C_2H_5COOH), butyric acid (C_3H_7COOH), etc. Higher homologues like palmitic acid ($C_{15}H_{31}COOH$), stearic acid ($C_{17}H_{35}COOH$), etc.

Q.7 What are unsaturated fatty acids? What are the types?
- Fatty acids which contain double bonds in their structure are called unsaturated fatty acids (UFA).
- ***Types:*** Depending on the degree of unsaturation they are divided into two groups:
  - Monounsaturated (monoethenoid): containing only one double bond.
  - Polyunsaturated (polyethenoids): containing more than one double bond in their structure.

Q.8 Give an example of monosaturated FA (monoethenoid) found in our body fat.
Oleic acid $C_{17}H_{33}COOH$ (Formula 18: 1; 9) is found in abundance in our body fat.

Q.9 Name the three polyunsaturated fatty acids (polyethenoids).
Three polyunsaturated fatty acids of biological importance are:
- ***Linoleic acid:*** Two double bonds between C_9 and C_{10} and another between C_{12} and C_{13} (Formula: 18 : 2; 9, 12).
- ***Linolenic acid:*** Contains three double bonds between carbons 9 and 10, 12 and 13, and 15 and 16. (Formula: 18 : 3; 9, 12, 15).
- ***Arachidonic acid:*** It is a 20 C fatty acid and contains four double bonds between 5 and 6, 8 and 9, 11 and 12, and 14 and 15. (Formula: 20 : 4; 5, 8, 11, 14).

Q.10 What are the saturated fatty acids most abundant in fats?
1. Palmitic acid.
2. Stearic acid.

Q.11 What are essential fatty acids?
Essential fatty acids are those which cannot be synthesized by the body and hence are supplied in the diet. They are also called 'polyunsaturated acids'. They are:

	No. of double bonds	*No. of carbon atoms*
1. Linoleic acid	2	18
2. Linolenic acid	3	18
3. Arachidonic acid	4	20

Q.12 What are the functions of essential fatty acids?
1. Essential fatty acids prevent deposition of cholesterol in atherosclerosis.
2. In the synthesis of prostaglandins.

Q.13 Which EFA is important?
- Linoleic acid is most important as arachidonic acid can be formed in the body from linoleic acid.
- Biologically arachidonic acid is very important as prostaglandins (PGs) and leukotrienes (LTs) are formed in the body from it.

Q.14 What is saponification?
Hydrolysis of fat by an alkali is called saponifications.

Q.15 What is saponification number?
Saponification number is defined as the mg of KOH required to saponify 1 gm of fat.

Q.16 What does the saponification number of a fat indicates?
- Saponification number of a fat gives an idea about the average chain length of the fatty acids present in the fat.
- Higher the saponification number, the shorter (or smaller) will be the chain length of the fatty acid and vice versa.

Q.17 What is iodine number?
Iodine number is defined as the grams of iodine absorbed by 100 gm of the fat.

Q.18 What does the iodine number indicate?
Iodine number indicates the degree of unsaturation of the fat. Higher the iodine number, the more is the degree of unsaturation of the fat.

Q.19 What is an acid number?
The amount of KOH in mg required to neutralize free fatty acids presents in 1 gm of fat.

Q.20 Which will have a higher acid number, a pure fat or rancid fat?
A rancid fat will have a higher acid number.

Q.21 What is glycerol?
Glycerol is commonly called as glycerine. Chemically it is a trihydric alcohol containing three -OH groups in the molecule.

Q.22 What are the sources of glycerol in the body?
Sources of glycerol in the body:
- ***Exogenous:*** From dietary fats, approximately 22% of glycerol formed in the gut by lipolysis of dietary TG is absorbed directly to portal blood.
- ***Endogenous:*** From lipolysis of fats (TG) in adipose tissue.

Q.23 What is acrolein test?
The presence of glycerol in a compound is detected by acrolein test. Glycerol, when heated with $KHSO_4$, two molecules of water are removed and it produces ***acryl aldehyde*** which has characteristic pungent or acrid odor.

Q.24 Is glycerol produced in the body by lipolysis a waste product or useful?
Glycerol produced in the body is not a waste product. It is useful in:
- Re-esterified to form TG again
- It has nutritive value. It can be converted to glucose/and glycogen by the process called gluconeogenesis.

Q.25 Can glycerol be used clinically in medicine?
Glycerol has been used orally or IV in cases of cerebrovascular diseases. It is nontoxic and it reduces cerebral edema with improvement in CS fluid. There is no rebound increase in intracranial pressure on discontinuation of therapy.

Q.26 What is rancidity?
Rancidity is a chemical resulting in the unpleasant odor and taste in the fat on storage when they are exposed to light, heat, and moisture.

Q.27 What are antioxidants?
Antioxidants are substances which prevents rancidity.

Q.28 Give classification of antioxidants?
1. *Primary antioxidants:* Prevent formation of new free radicals, e.g. superoxide dismutase, glutathione peroxidase, ceruloplasmin.
2. *Secondary antioxidants:* Remove preformed free radicals before they can initiate chain reaction, e.g. vitamin E and C, β carotene, uric acid, bilirubin.
3. *Tertiary antioxidants:* Repair cell structures damaged by free radical attack, e.g. DNA repair enzymes.

Q.29 Which vitamin prevents rancidity?
Vitamin E.

Q.30 What are phospholipids?
Phospholipids are compound lipids that contain in addition to fatty acids, and glycerol/or other alcohol, a phosphoric acid residue, nitrogen contaning base and other constituets.

Q.31 State how phospholipids are classified?
As per Celmer and Carter's classification phospholipids are classified into three groups:
1. ***Glycerophosphatides:*** Containing glycerol as alcohol group, e.g. phosphatidyl choline (lecithin), phosphatidyl ethanolamine (cephalin), phosphatidylserine, plasmalogens, etc.
2. ***Phosphoinositides:*** Containing inositols as alcohol, e.g. phosphatidyl inositol (lipositol).
3. ***Phosphosphingosine:*** Containing an unsaturated 18 carbon aminoalcohol called sphingosine (sphingol), e.g. sphingomyelin.

Q.32 What are phospholipases? What are the types? Mention their site of action.
Phospholipases are enzymes which hydrolyze phospholipid in a characteristic way. Five types described. Their specific site of action on lecithin are as follows:
1. ***Phospholipase A_2:*** Attacks β position and forms lysolecithin + one mole of FA.
2. ***Phospholipase A_1:*** Attacks ester bond in position 1 of lecithin.
3. ***Phospholipase B:*** Substrate is lysolecithin (lysophospholipase). It hydrolyzes ester bond in a position and forms glyceryl phosphoryl choline + one mole of FA.
4. ***Phospholipase C:*** Hydrolyzes phosphate ester bond and forms α, β-diacyl glycerol + phosphoryl choline.
5. ***Phospholipase D:*** Splits off choline and forms phosphatidic acid.

Q.33 What is sphingomyelin?
- It is a phospholipid. It does not contain glycerol, instead contains an 18 carbon unsaturated aminoalcohol called sphingosine (sphingol).
- ***On hydrolysis,*** sphingomyelin yields one molecule FA + phosphoric acid + nitrogenous base choline + sphingosine (alcohol).

Q.34 What are plasmalogens? How does it differ from lecithin/cephalin?
- Plasmalogens are phospholipids. They are predominantly found in brain and nervous tissue.
- ***On hydrolysis***, they yield one molecule of FA + glycerol + phosphoric acid + a nitrogenous base usually ethanolamine (sometimes choline) + one molecule of long chain aliphatic aldehyde. Thus it differs from lecithin cephalin in having a long chain aliphatic aldehyde.

Q.35 What is cardiolipin?
It is a phospholipid found in mitochondrial inner membrane. Chemically, it is diphosphatidyl-glycerol and is formed from phosphatidyl-choline.

Q.36 What is dipalmitoyl lecithin (DPL)? What is its clinical importance?
- Dipalmitoyl lecithin (DPL), is secreted in lung alveoli by type II granular pneumocytes lining the alveolar wall. It acts as a surfactant and lowers the surface tension in lung alveoli and prevents collapse of lung alveoli.
- Absence of DPL, in premature infants, may produce collapse of lung alveoli producing difficulties in respiration and death. It is called as ***respiratory distress syndrome (RDS) or hyaline membrane disease.***

Q.37 What are glycolipids? Give examples.
- Glycolipids are lipids that contain carbohydrate moiety in their molecule.
- They are mainly of two types:
  - Cerebrosides, and
  - Gangliosides.

Q.38 What are cerebrosides?
Cerebrosides are glycolipids. A cerebroside is built as follows:

Sphingosine (alcohol) <
- FA of high molecular weight
- Usually galactose (may contain glucose sometimes)

Q.39 What are the different types of cerebrosides? How do they differ from each other.
Cerebrosides are mainly four types. They differ from each other by the fatty acid content. They are:
- ***Kerasin:*** Contains FA lignoceric acid.
- ***Cerebron:*** Contains hydroxy lignoceric acid called "cerebronic (phrenosin) acid".
- ***Nervon:*** Contains an unsaturated homologue of lignoceric acid called "nervonic acid".
- ***Oxynervon:*** Contains hydroxy derivative of nervonic acid.

Q.40 What is a ceramide?

Ceramide is formed by esterification of sphingosine with FA of high molecular weight.

Principally found in white matter of brain, in myelin sheaths and medullated nerves.

Q.41 What is the composition of lecithins?

Lecithins contain glycerol, fatty acids, phosphoric acid and choline.

Q.42 What is the composition of cephalins?

Cephalins contain glycerol, fatty acids, phosphoric acid and ethanolamine.

Q.43 What is the sugar component present in cerebrosides?

Galactose.

Q.44 What are sphingolipidoses? Describe some important sphingolipidoses.

Sphingolipidoses are a class of heterogeneous group of inherited disorders relating to sphingolipids and they primarily affect the central nervous system (Table 24.1).

Table 24.1: Some examples of sphingolipidoses

Disease	*Enzyme deficiency*	*Lipid accumulating (see key below)*	*Clinical symptoms*
• **Niemann-Pick disease**	***Sphingomyelinase***	Cer + p-cholines-phingomyelin	Enlarged liver and spleen, mental retardation, fatal in early life
• **Gaucher's disease**	***b-glucosidase***	Cer + Glc glycosylceramide	Enlarged liver, massive splenomegaly, erosion of long bones, mental retardation in infants
• **Tay-Sachs disease**	***Hexosaminidase A***	Cer – Glc – Gal (NeuAc) + Gal NAC GM_2 ganglioside	Mental retardation, muscle weakness, blindness
• **Metachromatic leukodystrophy**	***Arylsulfatase A***	Cer – Gal – Gal + O SO_3 3 sulfogalactosyl ceramide	Mental retardation, demyelination, psychologic disturbances in adults
• **Fabry's disease**	***α-Galactosidase***	Cer – Glc – Gal + Gal Globotriaosyl ceramide	Skin rash, renal failure (full symptoms only in males X-linked recessive)
• **Krabbe's disease**	***β-Galactosidase***	Cer + Gal galactosyl ceramide	Mental retardation myelin almost absent

Key: Cer – Ceramide
Glc – Glucose
Gal – Galactase
Neu-Ac – N-acetyl neuraminic acid
+ Denotes site of deficient enzyme reaction

Q.45 What are steroids?

Substances possessing cyloperhydrophenanthrene nucleus are called steroids. (Fig. 24.1)

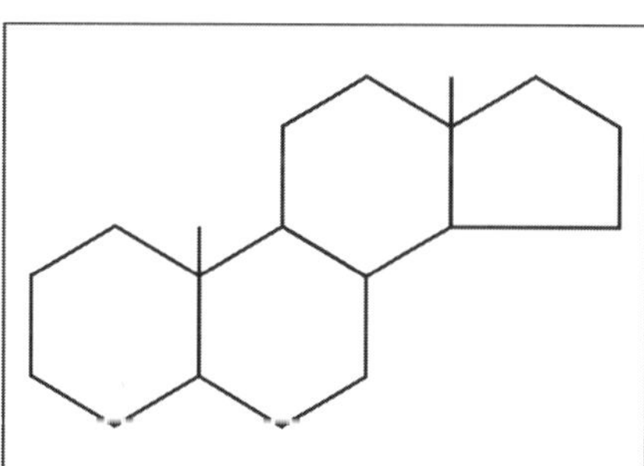

Fig. 24.1: Structure of steroids

Q.46 Give the structure of cholesterol.

See Figure 24.2.

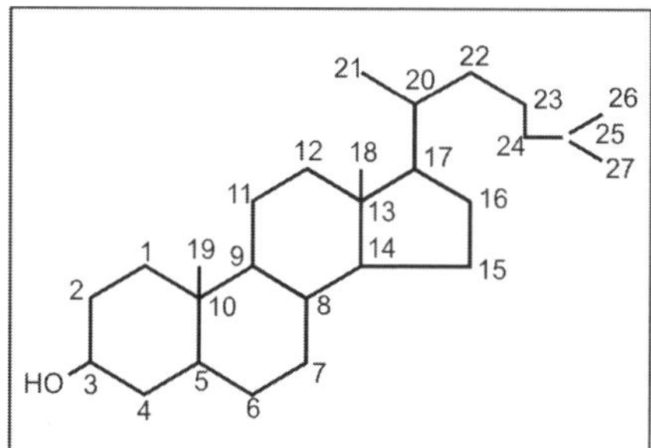

Fig. 24.2: Structure of cholesterol

Q.47 State the characteristic features of cholesterol structure.

Characteristic features of cholesterol structure are:

- Possesses "cyclopentanoperhydrophenanthrene" nucleus.
- α-OH group at C_3.
- It contains an unsaturated double bond between C_5 and C_6.
- It has two-CH_3 groups at C_{10} and C_{13}
- It has an eight carbon side chain attached to C_{17}.

Q.48 What are the sources of cholesterol in the body?

Two sources:

- ***Exogenous:*** Dietary cholesterol approximately 0.3 gm/day. Diet rich in cholesterol are butter, egg yolk, milk, cream, meat, etc.
- ***Endogenous:*** Synthesized in the body from two carbon units, acetyl CoA

  $(CH_3.\overset{\overset{O}{||}}{C}\sim S-CoA)$. Approximately 1.0 gm/day.

Q.49 What are the forms in which cholesterol exist in blood and tissues?

Cholesterol occurs both in ***free*** form in which -OH group on C_3 is free and in ***ester*** form in which -OH group is esterified with fatty acids.

Q.50 How does cholesterol esterified?

Two ways:

1. Some cholesterol esters are formed in tissues by the transfer of "acyl" groups from acyl-CoA to cholesterol by the enzyme *acyl transferase.*
2. By interaction of lecithin and cholesterol: Plasma cholesterol esters are produced in plasma by transfer of an acyl group, usually unsaturated, from β-position of lecithin to cholesterol by the enzyme *lecithin-cholesterol acyl transferase* (LCAT).

$$\text{Lecithin} + \text{Cholesterol} \xrightarrow{\text{LCAT}} \text{Cholesterol ester} + \text{Lysolecithin}$$

Q.51 What are bad effects of cholesterol?

Excessive cholesterol is harmful to body in that it gets deposited in arterial walls producting atherorsclerosis. This can narrow the lumen of the blood vessel impeding blood flow which can cause thrombosis.

Q.52 Is cholesterol good for the body? What are the good effects?

- In normal quantities cholesterol has number of good effects. They are:
- Bile acids are produced from cholesterol by its oxidation in liver.
- Cholesterol is converted to steroid hormones in adrenal cortex and to sex hormones in gonads.

- Cholesterol forms 7-dehydrocholesterol and in skin it is converted to vit D_3 by UV rays.
- Cholesterol is poor conductor of heat and hence acts as an insulator.
- It is also a poor conductor of electricity and has a high dielectric constant.
- Cholesterol is in abundance in brain and nervous tissues where it functions as an insulating covering of structures which generate and transmit electrical impulses.

Q.53 What are the tests by which cholesterol is detected?

1. Liebermann-Burchard reaction.
2. Salkowski test.

Q.54 What is Liebermann-Burchard reaction?

A chloroform solution of cholesterol when treated with acetic anhydride and conc. H_2SO_4 gives a grass green color (cholesta polyene sylphonic acid is formed). This reaction is utilized in colorimetric estimation of cholesterol in blood by Sackett's method.

Q.55 What is Zak's reaction?

When solution of cholesterol is treated with glacial acetic acid, ferric chloride and conc H_2SO_4, it gives purple red color (cholestapolyene carbonium ion). This reaction forms the basis for the colorimetric estimation of cholesterol by Zak's method.

Q.56 What is 7-dehydrocholesterol? Why it is called pro-vitamin D_3?

7-dehydrocholesterol is produced in the body from cholesterol and it is present in skin epidermis. Ultravoilet ray of sunlight changes 7-dehydrocholesterol to vitamin D_3 (cholecalciferol). Hence it is called ***pro-vitamin D_3***.

Q.57 What is pro-vitamin D_2?

Ergosterol is a plant sterol. When it is irradiated with UV rays [long wave 265 μm (millimicron)] it changes to vitamin D_2. Over irradiation may produce toxic substances viz. Toxisterols and suprasterols.

Q.58 What are chylomicrons?

Chylomicrons are the central core of triglycerides, phospholipids and cholesterol combined with small amount of proteins.

Q.59 What are lipoproteins?

The combination of lipid with proteins is called lipoproteins.

Normal lipid profile in humans shown in Table 24.2.

Table 24.2: Normal values of lipoproteins (Normal lipid profile)

Lipid fraction	*Normal values*
• **Total cholesterol**	150 to 240 mg/dL
• **Serum HDL—cholesterol**	Males—35 to 60 mg/dL Female—40 to 70 gm/dL
• **Serum TG (Triacylglycerol)**	Males—60 to 165 mg/dL Females—40 to 140 mg/dL
• **Serum chylomicrons**	Up to 28 mg/dL (14 hrs Post-absorptive state)
• **Serum pre-β lipoproteins (VLDL)**	Males—up to 240 mg/dL Females—up to 210 mg/dL
• **Serum β-lipoproteins (LDL)**	Up to 550 mg/dL
• ***Serum LDL—cholesterol**	Up to 190 mg/dL

****Serum LDL—cholesterol can be calculated by the Friedewald formula:***

- LDL—cholesterol in mg/dL =

$$\text{Total cholesterol—HDL cholesterol—}\frac{TG}{5}$$

- LDL—cholesterol in m.mol/L =

$$\text{Total cholesterol—HDL cholesterol—}\frac{TG}{2.2}$$

Note: The formula is not much reliable at TG concentration > 4.5 m.mol/L (> 400 mg/dL)

Q.60 What are the various functions of plasma lipoproteins?

1. High density lipoproteins, i.e. α-lipoproteins.
2. Low density lipoproteins, i.e. β-lipoproteins.
3. Very low density lipoproteins, i.e. pre-α-lipoproteins.
4. Chylomicrons.

Q.61 Which fraction of the lipoprotein contains maximum cholesterol content?

Low density lipoprotein fraction (LDL).

Q.62 LCAT is present on which lipoprotein?

LCAT is present on HDL molecule. It converts cholesterol and lecithin to cholesteryl ester and lysolecithin respectively.

Q.63 What is the other name for VLDL remnant?

Intermediate density lipoprotein (IDL).

Q.64 LDL receptor can be expressed by another which method?

APO B-100, E receptor.

Q.65 What are apolipoproteins?

The protein component of plasma lipoproteins are called apolipoproteins.

Q.66 What are bile salts?

They are sodium and potassium salts of glycocholates and taurocholates.

Q.67 What is the role of bile salts?

1. As powerful emulsification agents.
2. Bile salt lowers the surface tension of the media and thus aid in the absorption of fats.

Q.68 At which step bile acid synthesis is regulated?

7-α hydroxylase step.

Q.69 Name the substrates required in adipose tissue for TG synthesis.

Two substances are required:

1. α-glycerol-P and
2. Acyl CoA-activated FFA (Fig. 24.3).

Q.70 Name the enzymes required for breakdown of TG (lipolysis).

Three enzymes are required:

1. *Triacyl glycerol lipase*—hormone sensitive and key rate limiting enzyme. The enzyme can exist in "active" or "inactive" state.
2. *Diacyl glycerol lipase*
3. *Monoacyl glycerol lipase*

(2 and 3:) Both are non-hormone sensitive.

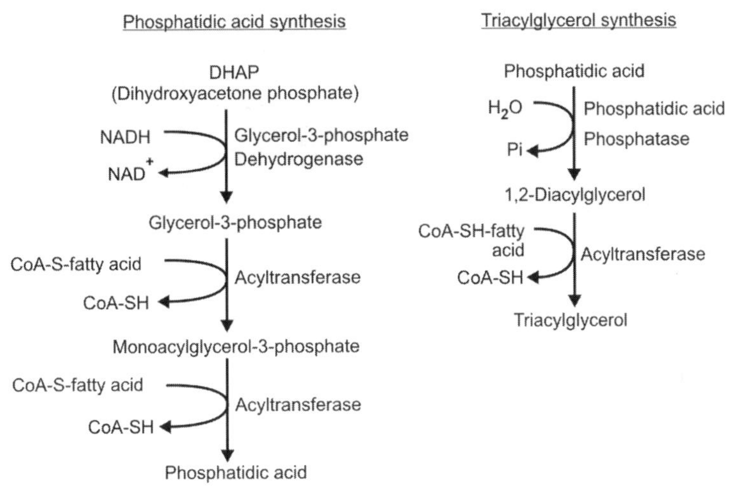

Fig. 24.3: Synthesis of triglycerides

Q.71 How the action of *TG lipase* is regulated?

- *TG lipase* can exist in active "a" and inactive "b" forms.
- Increased cyclic AMP↑ level in the cells converts "inactive" cyclic AMP dependent protein kinase (C_2R_2) to "active" protein kinase (C_2), which in turn phosphorylates "inactive" hormone-sensitive TG lipase "b" to "active" TG lipase "a" which breaks down TG to form DG + FFA.
- Active TG lipase "a" is converted to "inactive" TG lipase "b" by dephosphorylation.

Q.72 What is the action of insulin on adipose tissue?

- Insulin increases esterification (TG formation) as it enhances the glucose uptake by adipose tissue cells. Also increases glucose oxidation which provides α-glycero-P through di-OH-acetone-P.
- Insulin inhibits lipolysis. This is brought about by decreasing the levels of cyclic AMP in the cells. This is achieved by:
  - Inhibiting *adenyl cyclase* enzyme and
  - Increasing the *phosphodiesterase* activity. Lowered cyclic AMP brings about dephosphorylation of TG lipase "a"® to form TG lipase "b" (inactive) through protein kinase.

Q.73 What is the net effect of insulin on plasma FFA level?

Net result of insulin on adipose tissue is to inhibit the release of FFA from adipose tissue which results in fall of circulating plasma FFA level.

Q.74 List the hormones that increase the rate of breakdown of TG (lipolysis) in adipose tissue.

- Catecholamines:
  - Epinephrine and norepine phrine are the principal lipolytic hormones.
- Other hormones are:
  - Glucagon,
  - Growth harmones (GH),
  - Glucocorticoids (GC),
  - Thyroid hormones,
  - ACTH, a and b MSH TSH and vasopressin.

Q.75 What is lipoprotein lipase? Where it is found? What is the action of this enzyme?

- The enzyme *lipoprotein lipase* is located in the walls of the blood capillaries in various organs.
- TG of circulating chylomicrons and VLDL is acted upon by *lipoprotein lipase* which brings about progressive delipidation. The enzyme requires PL and apo-C II as cofactors.

$$\text{TG of chylomicrons and VLDL} \xrightarrow[\textit{Lipoprotein lipase}]{\text{PL, apo-C II}} \text{FFA + Glycerol}$$

Q.76 What is brown adipose tissue? Where is it located?

- Brown adipose tissue is involved in metabolism particularly when heat generation is necessary.
- The tissue is extremely active:
  - in arousal from hibernation
  - in animals exposed to cold and
  - in heat production in new borns.
- Location site: It is present particularly in the thoracic region.

Q.77 What is β-oxidation?

Oxidation takes place at the β-carbon atom and the β-carbon is oxidized to carboxyl group (Fig. 24.4).

Q.78 Where does β-oxidation take place?

Mitochondria.

Q.79 What is the site of β-oxidation of FA in a cell? Mention the enzyme responsible.

- ***Site:*** In mitochondrion of the cell.
- ***Enzyme:*** Serveral enzymes known collectively as *FA oxidase enzyme system* are found in the mitochondrial matrix, adjacent to the respiratory chain (ETC). These enzymes catalyze β-oxidation of a long chain FA to acetyl CoA.

Q.80 Activation of long chain FA occurs in cytosol, but β-oxidation occurs in mitochondrial matrix. Activated long chain FA (acyl CoA) is impermeable to mitochondrial membrane. Explain how acyl CoA reach mitochondria?

Acyl CoA penetrate to the inner mitochondrial matrix only in combination with a substance called ***carnitine*** present in mitochondrial membrane.

Q.81 What is carnitine?

- Carnitine is chemically "β-OH-γ-trimethyl ammonium butyrate". It is widely distributed in liver, and other tissues, particularly in large quantities in muscles.
- Carnitine level of tissues is considerably influenced by dietary methionine and choline levels.
- It is synthesized from lysine and methionine principally in liver and in kidney.

Q.82 What are the functions of carnitine?

- Carnitine does not penetrate the mitochondrial membrane. It is considered as a *carrier molecule* and acts as a *ferry boat* to transport long chain Acyl CoA across mitochondrial membrane for its oxidation in mitochondrial matrix.
- Also facilitates exit of acetyl-CoA and acetoacetyl CoA from mitochondria to cytosol where FA synthesis occurs.

Q.83 State the enzymes involved in transportation of acyl CoA to mitochondrial matrix by carnitine.

- Carnitine-palmitoyl-transferase I
- Carnitine-acyl-carnitine-translocase
- Carnitine-palmitoyl-transferase II.

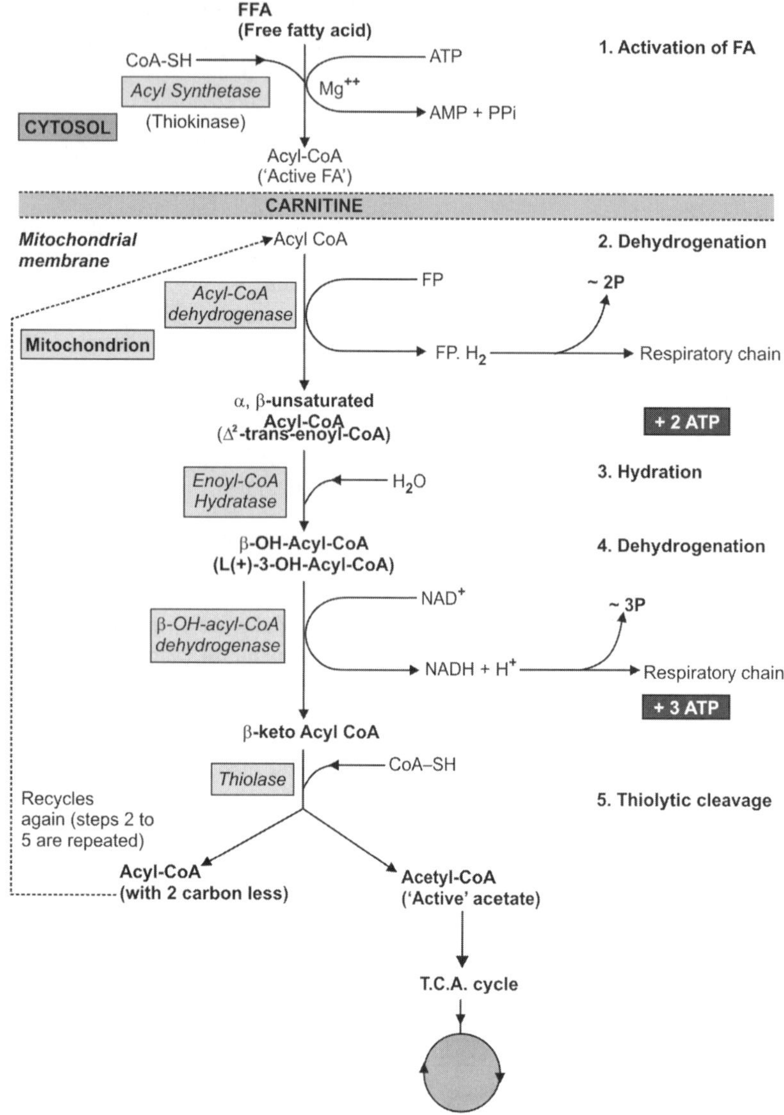

Fig. 24.4: Beta-oxidation of fatty acids

Q.84 Name the steps of β-oxidation.

Once acyl CoA is transported by carnitine in the mitochondrial matrix, it undergoes β-oxidation by the enzyme complex *fatty acid oxidase system*. The steps of β-oxidation are:

- ***Dehydrogenation*** by the enzyme *acyl CoA dehydrogenase*. H-acceptor is FAD.
- ***Hydration*** by the enzyme *enoyl hydrolase*, addition of one molecule of H_2O.
- ***Dehydrogenation*** by the enzyme *3-OH acyl CoA dehydrogenase*. H-acceptor is NAD^+.
- ***Thiolytic cleavage*** by the enzyme *thiolase* and CoA-SH.

Q.85 How many ATPs are formed by one turn of β-oxidation of FA?

Five ATPs:

- Two from oxidation of reduced FAD in ETC.
- Three from oxidation of reduced NAD in ETC.

Q.86 How many acetyl CoA are produced from β-oxidation of one molecule of palmitic acid?

- Palmitic acid is $C_{15}H_{31}COOH$.
- For complete oxidation by β-oxidation, it will undergo 7 cycles producing 7 acetyl CoA in each turn + one acetyl CoA extra will be produced in last cycle.

Therefore, β-oxidation of one molecule of palmitic acid will form 8 acetyl CoA.

Q.87 How much energy will be produced by β-oxidation of palmitic acid? What is the efficiency?

Each cycle produces 5 ATP, hence

7 cycles will produce	7 × 5 = 35~P
Total 8 molecules of acetyl CoA when oxidized in TCA cycle will form (12 × 8)	96~P
	Total = 131~P
In initial activation of palmitic acid ~ P bonds utilized.	= –2~P
	Total = 129~P

∴ Energy production = 129 × 7.6 = 980 KC

Caloric value of palmitic = 2340 KC/mole acid

$$\therefore \text{Efficiency} = \frac{980}{2340} \times 100 = 41\%$$ of total energy of combustion of palmitic acid.

Q.88 How propionyl CoA enters the citric acid cycle?

$$\text{Propionyl CoA} \xrightarrow[\text{CO}_2\text{, Biotin}]{\text{Propinoyl CoA Carboxylase}} \text{D-Methyl malonyl CoA.}$$

$$\text{D-Methylmalonyl CoA} \xrightarrow{\text{Recemase}} \text{L-Methylmalonyl CoA.}$$

$$\text{L-Methylmalonyl CoA} \xrightarrow[B_{12}]{\text{Isomerase}} \text{Succinyl CoA.}$$

Q.89 What is α-oxidation?

- α-oxidation is another alternative pathway for oxidation of FA which involves decarboxylation of the -COOH group after hydroxylation and the formation of a FA containing an "odd-number" of carbon atoms which subsequently undergoes repeated β-oxidation producing $(\text{acetyl CoA})_n$ + propionyl CoA.
- It occurs in microsomes of brain and liver, an aerobic process requires molecular O_2.

Q.90 What is ω-oxidation?

In ω-oxidation, fatty acids undergo oxidation at ω-carbon, producing a dicarboxylic acid, which is then subjected to β-oxidation and cleavage to form successively smaller dicarboxylic acids.

Q.91 What is precursor of cholesterol biosynthesis?

Acetyl CoA.

Q.92 Where the enzyme system for cholesterol biosynthesis located?

Enzyme system of cholesterol biosynthesis are associated with:

- Cytoplasmic particles (microsomes)
- Soluble fraction of cytosol.

Q.93 Show schematically the steps of cholesterol biosynthesis.

See Figure 24.5.

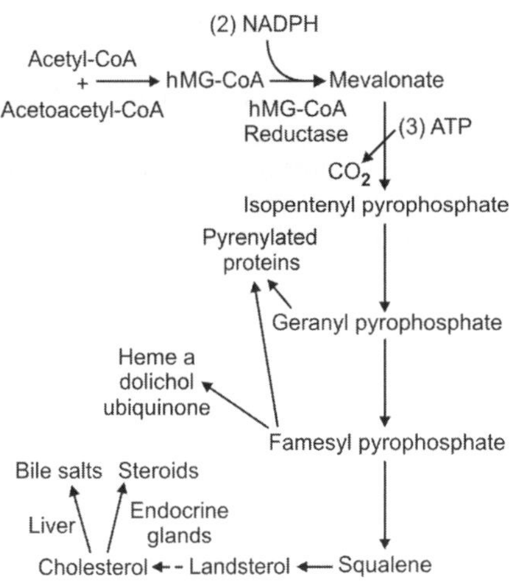

Fig. 24.5: Biosynthesis of cholesterol

Q.94 What is the key and rate-limiting enzyme in cholesterol biosynthesis?

hMG-CoA reductase which converts hMG-CoA to mevalonate.

Q.95 How cholesterol biosynthesis is regulated by hMG-CoA reductase?

- Cholesterol itself inhibits the enzyme *hMG-CoA reductase* by "feed-back" inhibition.
- Fasting/starvation inhibits the enzyme and decreases cholesterol synthesis.
- Increased dietary intake reduces the endogenous hepatic biosynthesis of cholesterol by reducing the activity of *hMG-CoA reductase.*
- ***Hormones:***
  - Insulin and thyroid hormones increases hMG-CoA reductase activity
  - Glucagon and corticosteroids decreases the activity of the enzyme.

Q.96 What are the function of cholesterol?

1. Cholesterol is an important tissue component. Because of its conductivity, cholesterol plays a role in insulating nerves and brain structure.
2. Cholesterol through the formation of ester of fatty acids appears to a play a role in the transport of fatty acids in the body.
3. Cholesterol neutralizes the hemolytic action of a number of agent such as shake venom, bacterial toxins, etc.
4. Cholesterol gives rise to provitamin D.
5. Cholesterol is a precursor of cholic acids in the body.
6. Cholesterol gives rise to sex hormones.

Q.97 What is the relationship between polyunsaturated acid and cholesterol?

Polyunsaturated acids tends to lower the plasma cholesterol level.

Q.98 What are the source of acetyl CoA?

1. Carbohydrate metabolism, i.e. glycolysis.
2. Fat metabolism, i.e. β-oxidation of fatty acids.
3. Protein metabolism, i.e. transamination.

Q.99 What are the various pathways by which acetyl CoA is utilized?

1. Acetyl CoA gives rise to CO_2 and H_2O by citric acid cycle.
2. Acetyl CoA gives rise to fatty acid synthesis via malonyl CoA.
3. Acetyl CoA gives rise to cholesterol synthesis.
4. Acetyl CoA gives rise to acetoacetic acid, i.e. ketone bodies.
5. Acetyl CoA undergoes acetylation reactions.

Q.100 What is Refsum's disease?

An inherited disorder due to deficiency of the enzyme *phytase a-oxidase,* as a result phytanic acid cannot be converted to pristanic acid which accumulates in blood and tissues. Principal manifestation is neurological-polyneuropathy with muscle atrophy.

Q.101 What is Zellweger's syndrome (hepatorenal syndrome)?

A rare inherited disorder in which there is inherited absence of peroxisomes in all tissues. Due to absence of peroxisomes and its enzymes fail to oxidize long chain FA in peroxisomes, resulting to accumulation of long-chain FA (C_{26} to C_{38}) in brain and other tissues like liver/kidney.

Q.102 What is the enzyme system present in higher animals including mammals for "*de novo*" synthesis?

In higher animals, the synthesis is carried out by a multienzyme complex called *fatty acid synthase (or synthetase),* which also incorporates ACP. This multienzyme complex is a dimer, i.e. made up of two identical monomeric units (I and II) aligned "head to foot" on either side. One end has the 4-phosphopantetheinyl group of ACP (Pan-SH), while the other end has cysteinyl SH (cys-SH). Each monomeric unit has seven enzymes required for total synthesis of fatty acid palmitic acid from acetyl CoA (Fig. 24.6).

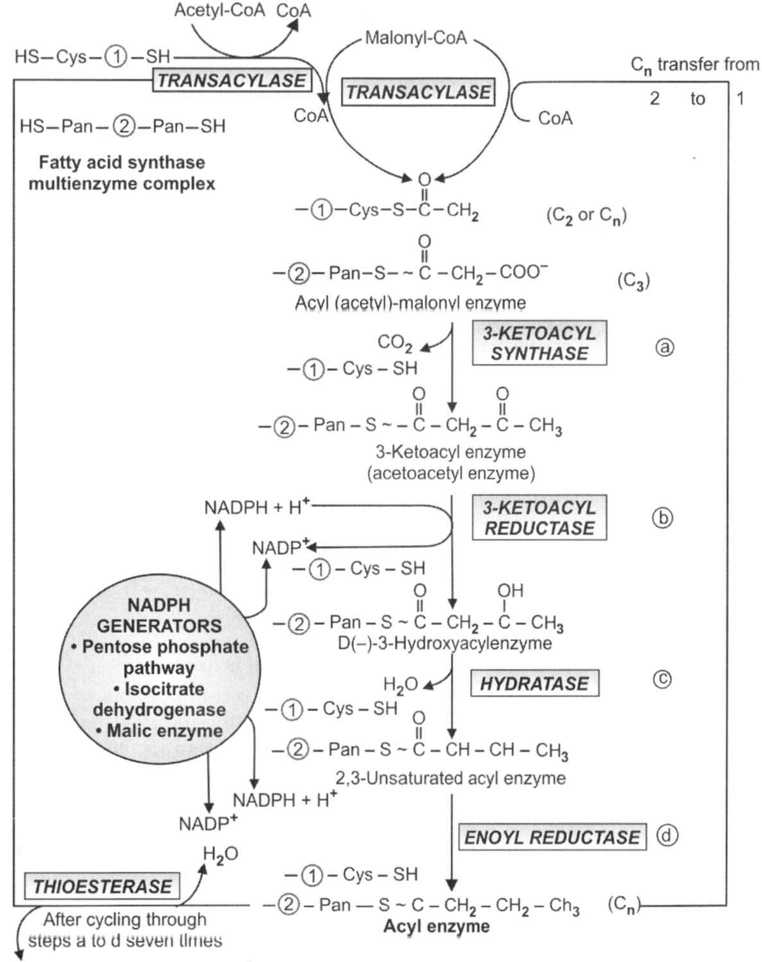

Fig. 24.6: *De novo* biosynthesis of fatty acids

Q.103 Name the tissues in which "*de novo*" synthesis of FA take place.

FA synthesis occurs in adipose tissue, liver, brain kidney, mammary gland, and lungs in which the multienzyme complex have been found in soluble cytosolic fraction of these tissues.

Q.104 What are the sources of NADPH in FA synthesis?

- HMP shunt is the principal source of NADPH
- Other alternative sources are:
  - Isocitrate dehydrogenase
  - Malic enzyme.

Q.105 State the role of hormones in lipogenesis.

- ***Glucagon:*** Inhibits FA synthesis by inhibiting the key enzyme *acetyl CoA carboxylase.*
- ***Insulin:*** Increases FA synthesis—in several ways:
  - By decreasing lipolysis
  - Activation of protein phosphatase
  - Stimulating synthesis and activation of *citrate lyase.*
  - Enhancing formation of acetyl CoA by stimulating glycolysis.

Q.106 Differentiate mitochondrial and microsomal chain elongation system. Mention the salient points.

Microsomal system	*Mitochondrial system*
• Usual common pathway	• Not a common pathway
• Operates in endoplasmic reticulum (ER) of microsomes	• Operates in mitochondria
• Acyl CoA of saturated C_{10} to C_{16} FA are the starting material	• Palmityl CoA is usually the starting material
• End product is next higher homologue	• Stearic acid is produced
• Requires O_2 (aerobic)	• Operates under relative anaerobiosis. High NADH/NAD^+ ratio favors
• Acetyl CoA is added through malonyl CoA	• Acetyl CoA is directly incorporated in palmityl CoA
• Pyridoxal-P is not required	• Pyridoxal-P is required
• NADPH is required	• NADPH is required

Q.107 How short chain fatty acids are transported through mitochondrial membrane?

Short chain fatty acids directly enter into the mitochondrial membrane.

Q.108 How long chain fatty acids are transported through mitochondria membrane?

Long chain fatty acids are transported through mitochondrial membrane by continue palmitoyl-transferase and carnitine acylcarnitine translocase.

Q.109 What is a fatty liver?

Significant accumulation of triglycerides in the liver give rise to fatty liver.

Q.110 What are lipotropic factors?

Substances which prevent the deposition of triglycerides in the liver are called lipotropic factors.

Q.111 Name lipotropic substances?

Liptropic substances are choline, methionine.

Q.112 What are ketone bodies?

Ketone bodies are:

1. Acetoacetic acid.
2. β-hydroxybutyric acid.
3. Acetone.

Q.113 What is the nature of ketone bodies?

Ketone bodies are acidic.

Q.114 Who is the net producer of ketone bodies?

Liver.

Q.115 Can liver utilize the ketone bodies? If not why?

Liver cannot utilize the ketone bodies because the activating enzyme required for ketone bodies utilization is absent in the liver.

Q.116 What are the tissues which prefer ketone bodies for utilization?

Extrahepatic tissue prefer ketone bodies for utilization because they possess the activating enzymes.

Q.117 What is ketosis?

Significant accumulation of ketone bodies in blood (ketonemia) and their excretion in the blood (ketonuria) leads to a condition known as ketosis.

Q.118 What is the normal excretion of ketone bodies?

1 mg per day.

Q.119 How ketone bodies are formed?

Ketone bodies are formed as intermediatory breakdown products of fat metabolism. If carbohydrate metabolism is defective, more fat breaks for energy purpose. Hence, more of ketone bodies are formed.

Q.120 Name the ketone bodies.

- Acetoacetic acid
- β-OH butyric acid
- Acetone.

Q.121 Show schematically the inter-relationship of three ketone bodies.

Figure 24.7.

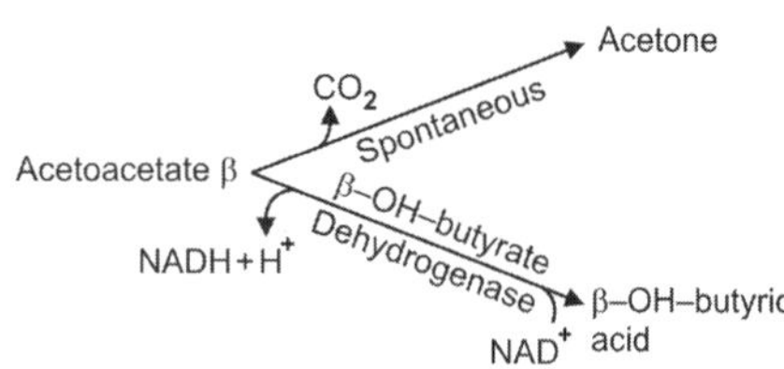

Fig. 24.7: Inter-relationship of three ketone bodies

Q.122 How ketone bodies are formed in the liver?

- ***Principal pathway*** is by hMG-CoA formation (Fig. 24.8)
- Acetyl CoA + Acetyl CoA → Acetoacetyl CoA
- Acetoacetyl CoA + Acetyl CoA.

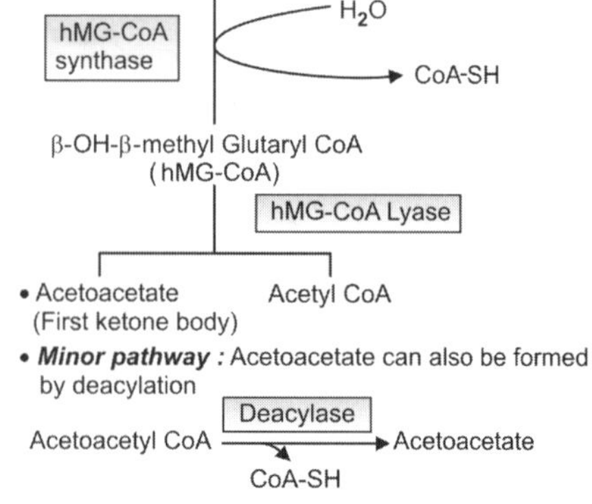

Fig. 24.8: Principal pathways

Q.123 How are ketone bodies utilized by extrahepatic tissues?

Ketone bodies once framed in the liver flow to the blood from where ketone bodies are taken up by extrahepatic tissues, and utilized as "fuel".

- ***Main pathway:*** Uses succinyl CoA

  Acetoacetate + Succinyl CoA

  Thiophorase

  Acetoacetyl CoA + Succinic acid

Note: *Thiophorase* enzyme also called *CoA transferase.*

- ***Minor pathway (Fig. 24.9)***

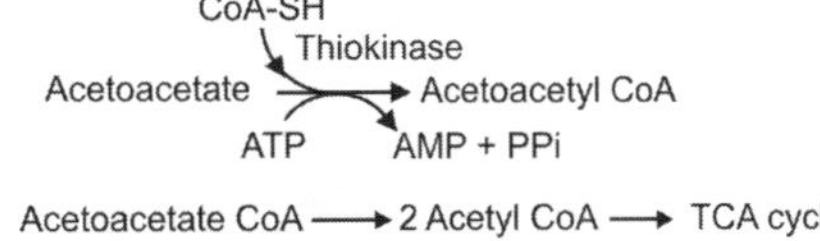

Fig. 24.9: Minor pathway

Q.124 What is Tay-Sachs disease?

Deposition of Ganglioside Gm_2 in brain and eye (macula) gives rise to Tay-Sachs disease. Death usually occur in <2 yrs.

Q.125 What is I cell disease?

Accumulation of inclusion bodies in the cell due to defect in phosphorylation of mannose.

Q.126 What is sphingolipids?

They are the important constituent of cell membrane. They contain no glycerol but are similar in structure to glycerophospholid.

Q.127 What is MCAD (Medium chain acyl CoA dehydrogenase) deficiency?.

MCAD deficiency include profound fasting hypoglycemia and lethargy, coma and even death if not heated.

Q.128 What is the Rx (treatment) of MCAD deficiency?.

IV glucose.

DO YOU KNOW?

- Excreation of methylmalonic acid indicate α_1 Vitamin + B_{12} deficiency rather than folate deficiency.
- In alcoholics 3-hydroxybutyrate levels are always high and that also helpful diagnosis to find out whether a person is alcoholic or not.
- Sphingosine is the precursor of all sphingolipids.
- Acetone has fruits odor which is a helpful diagnosis in ketoacidosis.

Chapter 25

Amino Acids and Proteins

Q.1 What is an α-amino acid?
Amino acid containing amino group and carboxyl group on the α-carbon atom.

$$R-\overset{\overset{NH_2}{|}}{\underset{\underset{H}{|}}{C^{\alpha}}}-COOH$$

α-amino acid

Q.2 What are glycogenic amino acids?
Those amino acids which give rise to the intermediates of carbohydrate metabolism are called glycogenic amino acids.

Q.3 Name the glycogenic amino acids?
Alanine, arginine, aspartic acid, cysteine, glutamic acid, glycine, histidine, proline, methionine, serine, leucine, threonine.

Q.4 What are ketogenic amino acids?
Amino acids which give rise to acetyl CoA during its course of metabolism are called ketogenic amino acids.

Q.5 Name some amino acids which are both glycogenic and ketogenic.
1. Phenylalanine.
2. Tryptophan.
3. Tyrosine.

Q.6 What are essential amino acids?
Essential amino acids are those amino acids which cannot be synthesized by the body and hence have to be supplied in the diet.

Q.7 Name ketogenic amino acids?
Leucine.

Q.8 Name the essential amino acids?
Isoleucine, Leucine, Tryptophan, Threonine, Phenylalanine, Valine, Methionine, Lysine.
The essential amino acids can be remembered using the formula "MTTVILPL".

Q.9 What are semi-essential amino acids?
Those amino acids which are partly synthesized by the body are called semi-essential amino acids, e.g. arginine and Histidine.

Q.10 What are non-essential amino acids?
Those amino acids which are synthesized by the body.

Q.11 How are amino acids classified?
Amino acids are classified mainly into three groups depending on their reaction in solution as follows:
- Neutral amino acids
- Acidic amino acids
- Basic amino acids

Q.12 What are neutral amino acids? Give examples.
Neutral amino acids form the largest group and they can be further classified as follows:
- ***Aliphatic amino acids:*** Having an aliphatic side chain and are called simple monoamino monocarboxylic acids, e.g. glycine (gly), alanine (ala), valline (val), leucine (leu) and isoleucine (Ile).
- ***Hydroxyamino acids:*** Having an—OH gr, e.g. serine (ser), threonine (thr).
- ***Aromatic amino acids:*** Contain a benzene ring, e.g. phenylalanine (phe), tyrosine (tyr).
- ***Heterocyclic amino acids:*** Containing heterocyclic group like indole ring or imidazole ring, e.g. tryptophan (trp), histidine (his).
- ***Sulfur containing amino acids:*** Containing, Sulfur, e.g. methionine (met), cysteine (cys) and cystine.
- ***Imino acids:*** They do not have a free —NH_2 group but have a basic, pyrrolidone ring, e.g. proline (pro), hydroxyproline (hyp).

Q.13 What are acidic amino acids? Give examples.
Acidic amino acids have two —COOH groups and one NH_2 group and they are monoamino dicarboxylic acids, e.g. aspartic acid (asp) glutamic acid (glu).

Q.14 What are basic amino acids? Give examples.
Basic amino acids have one —COOH group and two —NH_2 groups and they are diaminomonocarboxylic acids, e.g. arginine (arg), lysine (lys) and hydroxylysine (hyl).
Note: *Histidine (his) can also be classified as basic amino acid due to presence of imidazole ring.*

Q.15 Which is the simplest of the amino acids?
Glycine is the simplest of the amino acids.

Q.16 What is the chemical name of glycine?
Chemically glycine is called aminoacetic acid.

Q.17 Which amino acids do not have an asymmetric carbon and do not show optical activity?
Glycine is the only amino acid which does not have an asymmetric carbon and does not exhibit optical activity.

Q.18 Name the amino acids that do not occur in proteins. Give five examples.
There are compounds similar to basic skeletal structure of amino acid but do not occur in proteins. *Examples:*
- β-alanine found in coenzyme A
- α-amino butyrate—a neurotransmitter
- Taurine—present in bile salt
- Ornithine and citrulline—metabolites of urea cycle
- DOPA—precursor of pigment melanin.

Q.19 Give examples of three amino acids which produce specific biologic compounds.
- ***Tyrosine:*** produces
  - Thyroid hormones,
  - Catecholamines, and
  - Melanin pigments
- ***Glycine:*** Required for—
  - Heme synthesis
  - Formation of creatine
  - Formation of glutathione

- ***Tryptophan:*** Necessary for
  - Formation of serotonin
  - Forms vitamin nicotinic acid.

Q.20 How does an amino acid behave in an acidic and basic media?

- Every amino acid has at least two ionizable groups:
  - The —NH_2 group and
  - The —COOH group.
- ***In acidic medium:*** The —NH_2 group behaves as a base and accepts a proton and becomes positively charged (cationic form).

$$\begin{array}{c} R-CH-COO^- \\ | \\ NH_2^+ \end{array}$$

- ***In basic medium:*** The —COOH group acts as proton donor and the amino acid becomes negatively charged (anionic form).

$$\begin{array}{c} R-CH-COO^- \\ | \\ NH_2^- \end{array}$$

Q.21 What is isoelectric pH and what are zwitterion (or hybrid ion)?

At a specific pH the amino acids carry both the charges in equal number and thus exists as ***dipolar ion*** or ***zwitterion***. At this point the net charge on the amino acids is zero, i.e. the positive and negative charges on the amino acids equalize. *The pH at which the amino acid or protein is in zwitterion form is called isoelectric pH (pI).*

Q.22 What is the isoelectric pH (pI) for albumin, hemoglobin and casein?

- Isoelectric pH (pI) of:
  - Albumin is 4.7
  - Hemoglobin is 6.7
  - Casein is 4.6.

Q.23 Name the sulfur containing amino acids.

1. Methionine.
2. Cysteine.
3. Cystine.

Q.24 Which sulfur containing amino acid is an essential amino acid?

Methionine.

Q.25 Deficiency of essential amino acid causes what?

Deficiency of essential amino acids leads to negative nitrogen balance.

Q.26 What are the sources of amino acids in the blood?

1. Tissue protein breakdown.
2. Dietary proteins.
3. Intracellular synthesis.

Q.27 What is the role of amino acids?

1. As building block of proteins.
2. As precursors of:
   a. Hormones
   b. Purines
   c. Pyrimidines
   d. Porphyrins
   e. Certain vitamins such as pantothenic acid, folic acid, etc.

Q.28 What are the amino acids containing aromatic ring?

1. Phenylalanine.
2. Tyrosine.
3. Tryptophan.

Q.29 Guanidino group is present in which amino acid?

Arginine.

Q.30 Imidazole group is present in which amino acid?

Histidine.

Q.31 Indole group is present in which amino acid?

Tryptophan.

Q.32 Name an imino acid.

Proline.

Q.33 How amino acids are detected?

1. Aromatic amino acids containing benzene ring by xanthoproteic reaction.
2. Hydroxy aromatic amino acids by Millon's reaction.
3. Tryptophan by Hopkin's Cole reaction.
4. Arginine by Sakaguchi reaction.

Q.34 What is the role of methionine?

As methylgroup donar.

Q.35 Which amino acid gives rise to nicotinic acid during its metabolism?

Tryptophan.

Q.36 What are proteins?

Proteins are polypeptides, formed by the linking of amino acids through peptide bonds.

Q.37 Give the classification of proteins.

Proteins are classified into:

1. Simple proteins.
2. Conjugated proteins.

Q.38 What is the configuration of amino acids present in the naturally occurring proteins?

L-configuration.

Q.39 What are the general tests for proteins?

1. Biuret reaction.
2. Ninhydrin test.

Q.40 Give one common reaction for all amino acids.

All amino acids on heating with a solution of ninhydrin in acetone to about 100°C give a purple-blue colored compound (ninhydrin reaction).

Q.41 Which amino acids do not give purple-blue color in ninhydrin reaction?

Proline and hydroxyproline are exceptions and they give yellow color.

Q.42 What is biuret reaction?

Protein solution in water, two or three drops of dilute copper sulphate and excess of NaOH (about one mL) produces a pink or purple-violet color.

Q.43 What is the mechanism of biuret reaction?

The color depends upon the presence of two or more peptide bonds. It is due to coordination of cupric ions with the unshared electron pairs of peptide nitrogen and the oxygen of water to form the colored co-ordinate complex.

Note: *Thus di-peptides and free amino acids do not give the biuret test.*

Q.44 Name one amino acid which can give +ve biuret reaction.

Only histidine can give a +ve biuret reaction.

Q.45 What is xanthoproteic reaction?

The aromatic amino acids like phenylalanine, tyrosine and tryptophan present in protein molecule give yellow precipitate when heated with concentrated HNO_3. The reaction is due to nitration of aromatic ring.

Note: *Collagen and gelatin do not give a positive reaction.*

Q.46 Give a specific color test for tyrosine in protein.

Millon's test: It is a specific color test for tyrosine of protein. Proteins having tyrosine give a white precipitate with Millon's reagent (10% mercurous chloride in H_2SO_4) on heating. On addition of $NaNO_2$ the precipitate turns pink-red.

Q.47 Mention a specific color test for arginine in protein.

Sakaguchi test: It is a specific color test for arginine of protein. Sakaguchi reagent consists of alcoholic α-naphthol and a drop of sodium hypobromite. Guanidine group of arginine reacts to give the red color.

Q.48 Mention a specific color test for sulfur containing amino acids.

Lead acetate test: It is specific for sulfur containing amino acids. The proteins having S-containing amino acids when boiled with strong alkali split out sulfur as sodium sulfide which reacts with lead acetate to give black precipitate of lead sulfide (Pbs).

Q.49 What is biuret?

Biuret is a compound formed by heating urea. It contains two peptide linkages.

$$2\,C(NH_2)_2{=}O \xrightarrow{\text{heating}} NH_2\text{—CO—NH—CO—}NH_2 + NH_3$$

NH$_2$
|
2C—O —heating→ NH$_2$—CO—NH—CO—NH$_2$
| +
NH$_2$ NH$_2$

Q.50 What are the methods by which proteins can be estimated?

1. Biuret method.
2. Lowry method.
3. Micro-Kjeldahl method.
4. Spectroscopy.

Q.51 What is the factor used in the conversion of nitrogen to protein?

6.25.

Q.52 How does the factor 6.25 come?

The average nitrogen content of proteins is 16 gm%, i.e. 16 gm of nitrogen is present in 100 gm protein. 1 gm of nitrogen is present in 6.25 gm protein.

Q.53 Name two dipeptides found in our body.

The two dipeptides present in muscle tissues are:

- ***Carnosine***: β-alanine + histidine
- ***Anserine***: β-alanine + methyl-histidine

Q.54 Name tripeptide of immense biological importance.

- Glutathione is a tripeptide consisting of three amino acids—Glutamic acid + Cysteine + and Glycine.
- It functions in the body in oxidation-reduction system.

Q.55 What are fibrous proteins? Give two examples.

When the axial ratio of length: Width of a protein molecule is more than 10, it is called a fibrous protein.

Examples:

- α-keratin from hair
- Collagen of connective tissues.

Q.56 What are globular proteins? Give two examples.

When the axial ratio of length: width of a protein molecule is less than 10, it is called a globular protein.

Examples:

- Hemoglobin
- Myoglobin

Q.57 What are chromoproteins? Give suitable examples.

Chromoproteins are conjugated proteins that contain a simple protein with colored substance as the prosthetic group.

Examples:

- All hemoproteins are chromoproteins which contain "heme" as prosthetic group, e.g. hemoglobin, cytochromes, catalase, peroxidase.
- Flavoproteins contain riboflavin, a yellow colored substance.
- Visual purple (rhodopsin) contains protein 'opsin' + prosthetic group 11-cisretinal.

Q.58 What are phosphoproteins? Give two examples which have dietary importance.

Phosphoproteins are conjugated proteins containing phosphoric acid as the prosthetic group. The phosphoric acid is esterified through the -OH groups of serine and threpnine.

Examples:

- Caseinogen of milk
- Vitellin of egg yolk

Q.59 What are metalloproteins? Give three examples.

Metalloproteins are conjugated proteins which contain a metal ion as their prosthetic groups.

Examples:

- *Carbonic anhydrase* contains zinc
- Ceruloplasmin contains copper
- Ferritin contains Fe

Q.60 What is the essential difference between glycoproteins and mucoproteins?

- Both are conjugated proteins and contain carbohydrate moiety as prosthetic group.
- They differ by carbohydrate content in that glycoproteins contain less than 4% of carbohydrates whereas mucoproteins contain more than 4% of carbohydrates.

Q.61 How does primary derived proteins differ from secondary derived proteins? Give examples of each group.

Primary derived proteins are synonymous with denatured proteins. In this, primary structure is not disturbed and peptide bonds remain intact.

Examples: Proteins, metaproteins-acid/ alkaline, coagulated proteins.

Secondary derived proteins are formed by progressive hydrolysis of the peptide bonds. Protein is gradually hydrolyzed to produce smaller molecules.

Examples: Proteoses (albumoses), peptones, peptides.

Q.62 What are the various levels of protein structure?

1. Primary structure.
2. Secondary structure.
3. Tertiary structure.
4. Quaternary structure.

Q.63 What is primary structure?

Primary structure indicates the sequence of amino acids linked through peptide bonds. It has a N-terminus and for the other end C-terminus.

Q.64 What is an α-helix?

The coiling of polypeptide chain into a right hand helix.

Q.65 How will you determine the N-terminal amino acid of a protein?

1. Sanger's method.
2. Edman's method.

Q.66 Which reagent is used in Sanger's method?

2,4 dinitrofluorobenzene (2,4 DNFB).

Q.67 Which reagent is used in Edman's method?

Phenyl isothiocyanate.

Q.68 Which method is superior, Sanger's or Edman's?

Edman's method is superior over Sanger's method because Edman's method involves the removal of one amino acid at a time from the N-amino acid of a peptide, leaving the remaining peptide chain intact. The process can be repeated and sequence of amino acid is obtained from N-terminal end. Where as in Sanger's method only the N-terminal amino acid is identified because after the removal of N-terminal amino acid with DNFB, the remaining peptide chain breaks into amino acid mixture.

Q.69 What is secondary structure?

- Secondary structure refers to the twisting of the peptide chain into a helical form. Hydrogen bonds are responsible for the secondary structure.
- Hydrogen bonds in secondary structure may form either:
  - α-helix or
  - β-pleated sheet.

Q.70 Name different secondary structures of proteins.

1. α-helix.
2. β-pleated sheath.
3. β-bend.
4. Super secondary (motifs).
5. Non-repetitive secondary structures.

Q.71 Which amino acid is conspicuously absent in α-helix?
Proline is never found in α-helix and is conspicuously absent.

Q.72 Which amino acids tend to destabilize α-helix?
More polar residues such as arginine, glutamic acid and serine may repel and destablize α-helix.

Q.73 Which protein tends to form triple helix?
Collagen is rich in proline and hydroxyproline hence it cannot form α-helix or β-pleated sheet; it forms a triple helix. The triple helix is stabilized by both non-covalent as well as covalent bonds.

Q.74 What is tertiary structure? Which bonds are found in tertiary structure?
- Tertiary structure arises due to the folding of the helical structure into globular, ellipsoidal or any other conformations.
- The folding is brought about by:
  - Disulfide bonds (s-s)
  - Polar or/salt linkages between atoms with positive and negative charges.
  - Hydrophobic bonds, and
  - Van der Waal forces.

Q.75 What is quaternary structure?
When proteins consist of two or more peptide chains held together by non-covalent interactions or by covalent cross-links, it is referred to as the quaternary structure.
Examples:
- ***Hemoglobin***: A tetramer having four polypeptide chains
- ***Lactate dehydrogenase (LDH)***: Four polypeptide chains.

Q.76 What is meant by +ve ion precipitation?
The +ve ions most commonly used are those of heavy metals like Pb^{2+}, Zn^{++}, Ca^{++}, Hg^{++}, etc. The metals precipitate proteins at the pH alkaline to its isoelectric pH (pI). At this pH, proteins behave as anions and +ve metal ions combine with -COO group to give insoluble precipitate of metal proteinate.

Q.77 What is meant by –ve ions precipitation?
Negative ions like tungstic acid, trichloroacetic acid, picric acid, tannic acid, etc. combine with proteins when the pH of the medium is on acidic side of its isoelectric pH (pI). In acidic pH, proteins exist as Pr^+ and forms precipite with -ve ions.

Q.78 What is electrophoresis?
Migration of charged molecules under the influence of electricity is called electrophoresis.

Q.79 What are the fractions obtained in paper electrophoresis of serum?
Albumin being lighter moves fast. It is followed by α_1 globulin, α_2 globulin, β-globulin and last γ-globulin near the origin.

Q.80 What is RF value?
The ratio of the distance moved by a compound/amino acid in paper chromatography to the distance moved by the solvent front is known as Rf value.

Q.81 Name six important plasma proteins.
- Six plasma proteins are:
  Albumin, α_1-globulin, α_2-globulin, β-globulin, fibrinogen and γ-globulin.

Q.82 Which fraction will be absent in serum proteins and why?
Fibrinogen is absent in serum protein because fibrinogen is used up in clot formation in separation of serum.
Serum = Plasma – Fibrinogen.

Q.83 Where is albumin synthesized in the body?
Albumin is mainly synthesized in the liver. Rate of synthesis is approximately 14.0 gm/day.

Q.84 State the important functions of albumin.
- Nutritive function.
- Exerts low viscosity
- Contributes 70–80% of osmotic pressure.
- Plays important role in exchange of fluids between blood and tissue.
- Helps in transport of several substances viz. FFA, unconjugated bilirubin, Ca^{++} and steroid hormones.
- Binding of certain drugs: Sulphonamides, aspirin, penicillin are bound to albumin and carried in the blood.

Q.85 What is haptoglobin? What is its clinical importance?
Haptoglobin (Hp) is an α_2-globulin synthesized in liver. Haptoglobin binds "free" Hb if present in blood. Average binding capacity of Hp irrespective of phenotype is approximately 100 mg/dL. Hp-Hb complex circulates in the blood and ultimately destroyed by RE cells.

Q.86 What is methemalbumin? What does it signify?
Normally blood does not contain methemalbumin. When the degree of intravascular hemolysis is rapid and severe, as may happen in incompatible blood transfusion, free Hb released exceeds the binding capacity of Hp. The free Hb combines with albumin to form methemalbumin, which can be detected by a sensitive test called Schumm's test. *Detection of methemalbumin points to IV hemolysis.*

Q.87 What is Bence Jones protein? What is the clinical significance?
Bence Jones protein is an abnormal protein which occurs in blood and urine of people suffering from a disease called multiple myeloma (a plasma cell tumor). It is monoclonal light chain either "κ" or "λ" and excreted in urine of multiple myeloma patient.

Q.88 How Bence Jones protein can be detected in urine in clinical laboratory?
Presence of Bence Jones protein in urine can be identified easily by a simple "heat test". Take urine in a clean test-tube and heat upto 50–60°C, when Bence Jones protein if present, is precipitated. But when heated further the precipitate dissolves again. Reverse occurs on cooling.

Q.89 What are the normal values of total proteins and differential proteins?
- Total proteins: 7.0 to 7.5 gm%
- Differential protein:

Gm% (by precipitation)	*% of total proteins (by electrophoresis)*
• Albumin—3.7 to 5.3	50 to 70%
• Globulins—1.8 to 3.6	29.5 to 54%
• α_1-globulins—0.1 to 0.4	2.0 to 6.0%
• α_2-globulins—0.4 to 0.8	5.0 to 11.0%
• β-globulins—0.5 to 1.3	7.0 to 16%
• γ-globulins—0.6 to 1.5	11.0 to 22.0%

Q.90 What is the normal level of fibrinogen in plasma?
0.2 to 0.4 gm% (200 to 400 mg%).

Q.91 What is the normal A:G ratio? In which condition the A:G ratio is reversed?
- Normal A:G ratio is 2.5 to 1.0 (usually 2:1)
- It is reversed characteristically in cirrhosis liver.

Q.92 What are the types of hypergammaglobulinemias? Give a few examples.
- Polyclonal gammopathies:
  - Chronic infections, e.g. TB, kala-azar
  - Chronic liver disease—cirrhosis liver
  - Sarcoidosis

- Autoimmune diseases, e.g. rheumatoid arthritis, systemic lupus erythematosus (SLE)
- Monoclonal gammopathies:
  - Multiple myeloma
  - Macroglobulinemia, etc.

Q.93 What is hypoproteinemia?
Decrease of total proteins below normal is called hypoproteinemia.

Q.94 State some causes of hypoproteinemia.
Hypoproteinemia can occur as follows:
- ***Hemodilution:*** Water intoxication (overload), IV infusion of fluids. In this both albumin and globulins are decreased and A:G ratio remains unaltered.
- ***Hypoalbuminemia:*** Conditions resulting in low albumin level, accompanied either by no increase in globulin or by an increase which is less than the fall in albumin. A:G ratio is decreased.
  - ***Loss through kidney:*** Nephrotic syndrome
  - ***Loss through GI Tract:*** Protein losing enteropathy, etc.

Examples:
- **Hypogammaglobulinemia:**
  - Conditions due to decrease in γ-globulin.

Q.95 Name four transport binding proteins.
- ***Transferrin (siderophilin):*** Binding and transport of Fe.
- ***Thyroxine binding globulin (TBG):*** Binds and transport thyroxine (T_4)
- ***Transcortin:*** Binding and transport of cortisol.
- ***Retinol binding protein (RBP):*** Binding and transport of retinol (Vit A).

Q.96 Which carbamoyl phosphate synthase is there in urea cycle?
Carbamoyl phosphate synthase I.

Q.97 Name the hormones which favor tissue uptake of amino acids.
- Insulin
- Growth hormone (GH) and
- Testosterone favor the uptake of amino acids by tissues. Hence these hormones are called "anabolic" hormones.
- Estradiol stimulates selectively the uptake of amino acids by uterus.
- Epinephrine and glucocorticoids (GC) stimulate uptake of amino acids by the liver only.

Q.98 What is meant by nitrogen balance or nitrogen equilibrium?
- In an adult healthy individual maintaining a constant weight, the amount of intake of 'N' in food (as dietary proteins mainly) will be balanced by an excretion of an equal amount of 'N' in urine (in the form of urea mainly, uric acid, creatine and to a small extent by amino acids) and
- In faces (mainly as unabsorbed N)
- Such an individual is then said to be in nitrogen balance or in nitrogen equilibrium.

Q.99 What is meant by positive nitrogen balance?
- If the amount of nitrogen consumed exceeds the amount of nitrogen lost, the person is said to be in *positive nitrogen balance*. This happens in:
  - Growing children
  - Pregnancy
  - Convalescence after illness/surgery, and
  - After administration of anabolic hormones.
- In positive nitrogen balance, the body puts on weight and N-intake will be greater than N-output since new cells are formed and N is retained as tissue proteins.

Q.100 What is meant by negative nitrogen balance?
A subject whose intake of N is less than the output of N is said to have a negative nitrogen balance. The person loses weight and the tissue proteins are catabolized to a greater extent than they are formed. Such a situation occurs in:
- Old age
- Starvation
- Wasting diseases like tuberculosis
- Cancer
- Prolonged illness
- Postoperative conditions and burns.

Q.101 What non-protein nitrogenous substances are synthesized from amino acids? List them.
- *Creatine* from glycine, arginine and methionine.
- *Heme* from glycine and succinyl CoA.
- *Melanins* from tyrosine.
- *Choline* from serine through ethanolamine.
- *Purine nucleotides* from glycine with 5-Phosphoribosylamine.
- *Pyrimidine nucleotides* from aspartate and carbamoyl phosphate.

Q.102 What happens to the N-part of an amino acid?
In mammalian tissues, α-NH_2 group of amino acids, derived either from the diet or breakdown of tissue proteins, is first converted to NH_3 and then to urea and excreted in urine.

$$-NH_2 \text{ group} \rightarrow NH_3 \rightarrow \text{urea} \downarrow \text{excreted in urine}$$

Humans are therefore called ureotelic.

Q.103 Why proteins are amphoteric in nature?
Proteins contain some free amino group and carboxyl group. It will give the tests of both acid and basic groups. Hence, they are amphoteric in nature.

Q.104 Why proteins behave as buffers?
On either side of pI, proteins behave as cations or anions depending upon the pH of the media and hence they behave as buffers.

Q.105 What is denaturation of proteins?
Denaturation is defined as loss in secondary, tertiary and quaternary (if present) structure of proteins on heating. Denaturation results in loss in biological activity of the proteins.

Q.106 What is a nitrogen balance?
If the ratio of N (protein) intake to N out take is equal to 1, the person is in a nitrogen balance.

$$\frac{\text{N intake}}{\text{N out take}} = 1.$$

Q.107 What is transamination?
Transamination is an intermolecular transfer of amino group of an amino acid to a keto group of keto acid resulting in the formation of corresponding keto acid and amino acid.

Transaminases

$$R_1 \text{ amino acid} + R_2 \text{ keto acid} \xrightleftharpoons{\text{Transaminases}} R_1 \text{ Keto acid} + R_2 \text{ amino acid.}$$

Q.108 What is the co-factor required in transamination?
Pyridoxal phosphate.

Q.109 Name the tissues in which transamination reaction occur.
Transamination takes place principally in
- Liver
- Kidney
- Heart
- Brain, but to a small extent transamination can take place in all tissues.

Q.110 What are the amino acids which participate most in transamination?
Alanine, aspartic acid, glutamic acid.

Q.111 What are the amino acids which does not participate in transamination?
Lysine, threonine.

Q.112 Name two specific transaminases which are of clinical importance. What are the substrates and end-products?

- *Aspartate transaminase* (or aspartate aminotransferase): Previously used to be called "S-GOT"— serum glutamate oxaloacetate transaminase.

Aspartate + α-oxoglutarate ⇌ OAA + Glutamate

(Donor (amino acid)	(recipient keto acid)	(New keto acid)	(New (amino acid)

- *Alanine transaminase* (or Alanine aminotransferase): Previously used to be called as "S-GPT"— serum glutamate "pyruvate transaminase".

Alanine + α–oxoglutarate = Pyruvic Acid + Glutamate

(Donor amino acid)	(recipient keto acid)	(New keto acid)	(New amino acid)

Note: In both the cases, glutamic acid is produced as new amino acid.

Q.113 What are normal levels of S-GOT and S-GPT?

- **S-GOT** (aspartate transaminase): Normal serum activity is 4 to 7 i.u/L (7 to 35 Karmen units/mL).
- **S-GPT** (alanine transaminase): Normal serum activity is 3 to 15 i.u/L (6 to 32 Karmen units/mL).

Q.114 What is deamination? What are the types?
Deamination is the process by which N of amino acid is removed as NH_3.
Two types of demination:

- Oxidative deamination
- Non-oxidative deamination.

Q.115 What is the site of oxidative deamination?

- Liver and
- Kidney are the main organs where oxidative deamination takes place.

Q.116 What is the nature of enzyme and coenzyme that take part in oxidative deamination?

- ***Enzyme***: *L-amino acid oxidase* which acts specifically on L-amino acids and oxidatively deaminate to form NH_3.
- ***Coenzyme***: is flavoprotein (FMN) which acts as H-acceptor.
- ***Nature of the enzyme***: The *L-amino acid oxidases* are auto-oxidizable flavoproteins. The reduced flavoproteins FMN. H_2 are re-oxidized at substrate level directly by molecular O_2 forming H_2O_2 which is converted to H_2O by the enzyme *catalase*.

Q.117 What are the toxic substances that can form in oxidative deamination of L-amino acids by "L-amino acid oxidase?

- *Superoxide anion* O_2^-: This is scavenged by the enzyme *superoxide dismutase*.
- H_2O_2: Scavenged by the enzyme *catalase*. Both superoxide anion O_2^- and H_2O_2 are toxic and they can oxidize membrane proteins, lipids, -SH groups and methionine sulfur and produce harmful effects.

Q.118 What is the function of presence of *D-amino acid oxidase* in cells, when it is known there is absence of D-amino acids in cells?
There is no D-amino acids in the body tissues, hence it is not clear about the presence of *D-amino acid oxidase* in cells and its functions. Probably the dietary deamino acids may be oxidized by *D-amino acid oxidase* and be isomerized through their ketoacids to L-amino acids.

Q.119 Does oxidative deamination fulfill a major role in formation of NH_3?
L-amino acid oxidase does not fulfill a major role in mammalian amino acid catabolism and formation of NH_3 as:

- It is restricted only to liver and kidney.
- Activity of the enzyme in these tissues is low.
- It does not have any effect on glycine or the L-isomers of the dicarboxylic acid or b-OH-α-amino acids.
- It does contribute to formation of NH_3 to some extent.

Q.120 What is non-oxidative deamination? Give two examples.

- Certain amino acids like OH-amino acids, histidine and S-containing amino acids can be deaminated non-oxidatively by specific enzymes to form NH_3. They also do not fulfill major role in NH_3 formation.
- Two examples (Fig. 25.1).

Q.121 State the principal method by which NH_3 is formed from L-amino acids.

- ***By transdeamination:*** It is combination of transamination + Deamination.

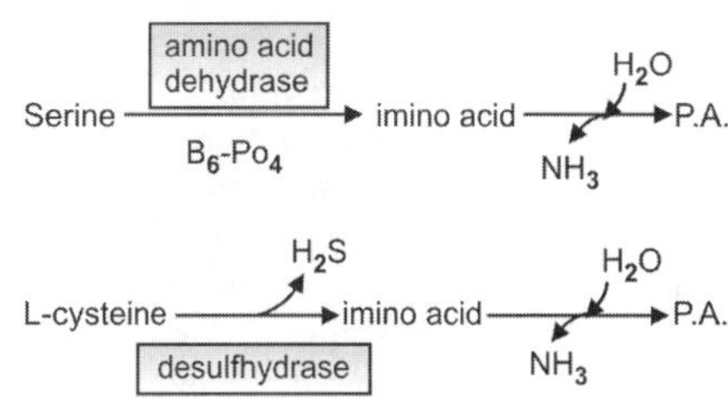

Fig. 25.1: Examples of nonoxidative deamination

- The-NH_2 group of most L-amino acids are transferred to α-oxoglutarate by specific transaminases by the process of transamination forming L-glutamate as end-product. L-glutamate is oxidatively deaminated by a specific Zn-containing enzyme *glutamate dehydrogenase,* of high activity and wide distrubution, which requires NAD^+ or $NADP^+$ as coenzyme, to form NH_3.
- As the process involves first transamination and coupled with oxidative deamination, it is called ***transdeamination***.

Q.122 List the sources of NH_3 in the body.

- Transdeamination—the major pathway.
- Oxidative deamination by *L-amino acid oxidase*
- Non-oxidative deamination
- Absorption of NH_3 from gut produced by intestinal bacterial flora:
  - From dietary proteins
  - From urea present in fluids secreted into the gut
  - From uric acid secreted in bile (This assumes importance in intestinal obstruction and portacaval shunt)
- Pyrimidine catabolism (small fraction).

Q.123 How albumin and globulin are separated?

- Albumin can be separated by full saturation of ammonium sulphate. Globulin can be separated by half saturation of ammonium sulphate.

Q.124 Name some biologically active peptides.

1. Glutathione.
2. Oxytocin.
3. Vasopressin.
4. Gramicidin-S.

Q.125 Which amino acid gives rise to thyroxine?
Phenylalanine.

Q.126 What is the fate of ammonia?

1. Glutamine synthesis.
2. Synthesis of urea.
3. Reanimation process.

Q.127 Describe the urea cycle. What is another name for urea cycle?

This is a cycle through which the toxic NH_3 in the body is detoxicated to form non-toxic urea in the liver. An alternative name for urea cycle is Krebs' Henseleit cycle. Urea cycle is shown in Figure 25.2. In this process, one molecule of NH_3 and the molecule of CO_2 are converted into urea through five sequential enzymatic reactions in each turn of cycle. Ornithine is regenerated in the last reaction which acts as a catalytic agent and repeats the cycle again.

Q.128 What are non-protein nitrogenous substances?

Nitrogen containing substances other than amino acids or proteins are called non-protein nitrogenous substances. They are urea, uric acid, creatine, purine, pyrimidine, amino acids, ammonia, etc.

Q.129 Which organ is associated with urea formation?

Liver is the only organ which can form urea and all the enzymes involved have been isolated from the liver tissue.

Q.130 Can other organs like kidney and brain synthesize urea?

- ***Kidney:*** Can form upto arginine but cannot form urea as enzyme *"arginase"* is absent in kidney tissues.
- ***Brain:*** Can synthesize urea if citrulline is available but lacks the enzyme *ornithine transcarbamylase* which forms citrulline from ornithine.

Q.131 State the five steps of urea cycle indicating the enzymes involved and location of the enzyme.

		Enzyme	*Location*
• Reaction 1:	Synthesis of carbamoyl-P	Carbamoyl synthetase I	Mitochondria
• Reaction 2:	Synthesis of citrulline	Ornithine transcarbamylase	Mitochondria
• Reaction 3:	Synthesis of argininosuccinate	Argininosuccinate synthetase	Cytosol
• Reaction 4:	Cleavage of argininosuccinate	Argininosuccinase	Cytosol
• Reaction 5:	Cleavage of arginine to form urea and ornithine	Arginase	Cytosol

Q.132 What is the role of N-acetyl glutamate (AGA) in the first reaction in formation of carbamoyl-P?

N-acetyl glutamate (AGA) acts as an allosteric activator of the enzyme *carbamoyl synthetase I*, it brings about conformational changes in the enzyme so that the second molecule of ATP can bind.

Q.133 Show schematically the first reaction, synthesis of carbamoyl-P. What are the starting materials? Is biotin necessary?

- Starting materials: CO_2 and NH_3
- Biotin is not required.

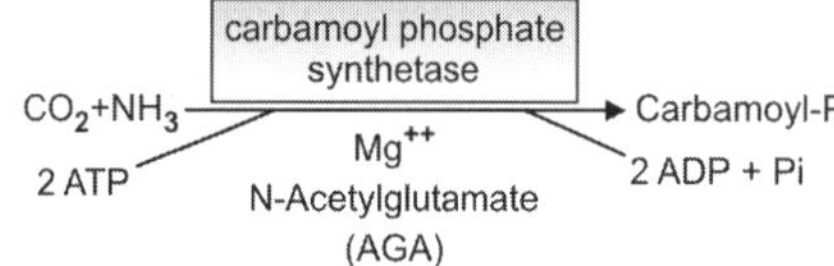

Q.134 How many amino acids take part in the urea cycle?

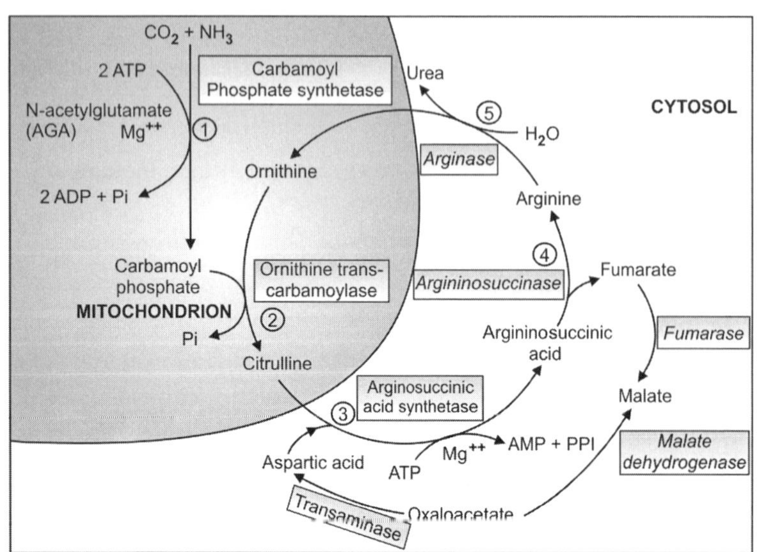

Fig. 25.2: Biosynthesis of urea or ornithine (urea cycle)

Five amino acids participate in urea cycle. They are:

1. Ornithine
2. Citrulline
3. Glutamic acid
4. Aspartic acid
5. Arginine.

Q.135 What is the energy requirement in formation of one molecule of urea in urea cycle?

Three molecules of ATP are utilized for the formation of one molcule of urea. As one ATP is converted to AMP, it is equivalent to utilizing four high energy bonds.

Q.136 Urea contains two nitrogen (NH_2 groups), what are the sources of the two nitrogens?

- One nitrogen of NH_2 group is derived from the NH^+_4 ion (Reaction 1)
- Other nitrogen of NH_2 group is provided by aspartic acid (Reaction 3).

Q.137 What is the normal blood level of urea in the body?

- Normal level ranges from 20 to 40 mg%.
- Indians take less proteins in diet, hence in Indians normal level is taken as 15 to 40 mg%.

Q.138 In what conditions the blood level of urea increase:

Causes may be:

- Prerenal
- Renal
- Postrenal.

Q.139 Mention some prerenal causes.

Conditions in which plasma volume/body fluid are reduced, e.g.:

- Salt and water depletion
- Severe and protracted vomiting as in pyloric/intestinal obstruction,
- Severe and prolonged diarrhea,
- Hemorrhage and shock,
- In burns.

Q.140 State some renal causes:

- In acute *glomerulonephritis.*
- In type II nephritis (nephrotic syndrome)—in later stages.
- Chronic pyelonephritis.
- Malignant nephrosclerosis.
- Mercurial poisoning.

Q.141 Mention some post-renal causes.

- Enlargement of prostate—benign and malignant
- Stone in urinary tract (urinary lithiasis)
- Stricture of urethra
- Tumors of the bladder affecting urinary flow.

Q.142 What is glutamine?

- Glutamine is chemically δ-amide of α-amino glutaric acid. Glutamine formation helps in removal of NH_3 which is toxic (detoxication of NH_3).
- Glutamine serves as an important reservoir of nitrogen in tissues which is readily available and can be drawn upon for various synthetic processes, when the body needs.

Q.143 How glutamine is synthesized in the body? Name the tissues involved in synthesis of glutamine.

Glutamine is synthesized in tissues like:

- Liver
- Kidney
- Brain and retina. It is formed from:
  - Glutamic acid and
  - NH_3 by the action of a mitochondrial enzyme *"glutamine synthetase"* in presence of ATP and Mg^{++}.
- The reaction is "irreversible", so that glutamine cannot reform glutamic acid and NH_3 by the same enzyme.

Q.144 How NH_3 can be obtained from glutamine?

- Glutamine is reservoir of NH_3 and readily available source of NH_3.
- Glutamine is hydrolyzed readily by a specific enzyme *glutaminase*. The reaction is 'irreversible' and can occur in various tissues like liver, kidney, brain and retina.

Q.145 State some important functions of glutamine in the body.

- ***Transamidation:*** The amide N of glutamine can be transferred to keto-group of fructose-6-P to form glucosamine-P.
- Formation of guanylate (GMP).
- ***Role of glutamine in kidney:*** in conservation of Na^+ (NH_4 - Na^+ exchange)
- ***Role of glutamine in brain:*** Brain tissues cannot detoxicate NH_3 to form urea. Excess NH_3 entering brain, e.g. in hepatic failure is detoxicated by glutamine formation.
- Role of glutamine in conjugation reaction (detoxication).
- ***Role in cancer:*** Certain tumors exihibit abnormally high requirement of glutamine/asparagine for their growth.

Q.146 What are the clinical manifestations of ammonia intoxication?

The symptoms of NH_3 intoxication include:

- A peculiar flapping tremor
- Slurring of speech
- Blurring of vision
- In severe cases lead to coma and death.

Q.147 How much urea is excreted daily in urine?

20–30 gm in 24 hours urine.

Q.148 Name the inherited disorders associated with urea cycle.

- Hyperammonemia type I—due to deficiency of enzyme carbamoyl-P-synthetase I
- Hyperammonemia type II, (also called ornithinemia)—due to the deficiency of enzyme ornithine transcarbamylase
- Citrullinemia—due to deficiency of the enzyme argininosuccinate synthetase
- Argininosuccinic aciduria—due to deficiency of enzyme argininosuccinase
- Hyperargininemia—due to deficiency of enzyme arginase.

Q.149 What are the amino acids involved in creatine synthesis?

Glycine, arginine, methionine (as S-Adenosyl methionine).

Q.150 What is the role of creatine phosphate?

As as source of energy in the contraction of muscles.

Q.151 What are the uses of glycine?

Glycine is involved in the synthesis of

1. Hemoglobin.
2. Creatine phosphate.
3. Purines.
4. Glutathione.
5. Proteins.
6. Phospholipids through ethanolamine, choline and serine.
7. In detoxification reactions.
8. Bile acids.

Q.152 Which is the major non-protein nitrogenous constituent of the urine?

Urea.

Q.153 Where carbamoyl phosphate synthase II is required?

In pyrimidine *de novo* biosynthesis.

Q.154 What is the role of N-acetyl glutamic acid in the first step of urea cycle?

N-acetyl glutamic acid is the modifier of the enzyme carbamoyl phosphate synthetase. It keeps the enzyme in the correct information.

Q.155 How much protein is potentially glucogenic?

It has been shown by experimental studies that 60 gm (approximately 58 gm) of glucose are formed and excreted in urine for 100 gm of proteins metabolized. Thus 60% of proteins (amino acids) is potentially glucogenic.

Q.156 Name at least ten amino acids which are glucogenic.

Glycine, alanine, serine, cysteine/cystine, threonine methionine, proline, valine, arginine, glutamate.

Q.157 Name the only amino acid which is ketogenic.

L-leucine.

Q.158 Name at least three amino acids which are both glucogenic and ketogenic.

Phenylalanine/tyrosine, tryptophan, isoleucine.

Q.159 1. What is GABA?
2. How it is formed?
3. What is the coenzyme of this reaction?

1. GABA is γ-aminobutyric acid, important in brain metabolism.
2. It is formed by the decarboxylation of glutamic acid.

$$NH_2-\underset{}{CH}(COOH)-CH_2-CH_2-COOH$$

Glutamic acid

$\downarrow -CO_2$

$$NH_2-CH_2-CH_2-CH_2-COOH$$

GABA

3. Coenzyme of the reaction is pyridoxal phosphate.

Q.160 Define albinism.

Deficiency of enzyme tyrosinase in the tyrosine metabolism leads to an inborn error of metabolism called albinism.

Q.161 What is tyrosinosis?

Deficiency of enzyme p-hydroxyphenyl pyruvate hydroxylase in tyrosine metabolism leads to tyrosinosis.

Q.162 What is alkaptonuria?

Deficiency of enzyme homogentisic acid oxidase in tyrosine metabolism leads to alkaptonuria.

Q.163 What is phenylketonuria?

Deficiency of the enzyme phenylalanine-4 hydroxylase in tyrosine metabolism leads to phenylketonuria.

Q.164 Name some biological important amines.

1. Histamine.
2. γ-aminobutyric acid (GABA).
3. Epinephrine (adrenaline).
4. Norepinephrine (noradrenaline).

Q.165 What are catecholamines?

Catecholamines are

1. Adrenaline.
2. Noradrenaline.
3. Dopamine.

Q.166 Name at least five biogenic amines of clinical importance. State the amino acid from which it is derived and mention biological importance in the body.

Biogenic amine	*Name of the amino acid*	*Biologic importance*
• Histamine	Histidine	• Vasodilator, BP↓ • HCl↑ • Pepsin, • Liberated in anaphylactic reaction
• γ-amino butyric acid (GABA)	Glutamic acid	• Presynaptic inhibitor in brain • Forms a by-pass in TCA cycle (Gaba shunt)
• Taurine	Cysteic acid (derived from cysteine)	• Constituent of bile acid (taurocholic acid)
• Tryptamine	Tryptophan	• Tissue hormone, a derivative of 5-OH tryptamine (serotonin) • Vasoconstriction, BP↑
• Ethanolamine	Serine	• Forms choline • Constituent of PL like cephalin.

Q.167 What is GABA shunt? Show schematically.

GABA by its conversion to succinic acid can form a "by-pass" in TCA cycle and this is called as GABA shunt (Fig. 25.3).

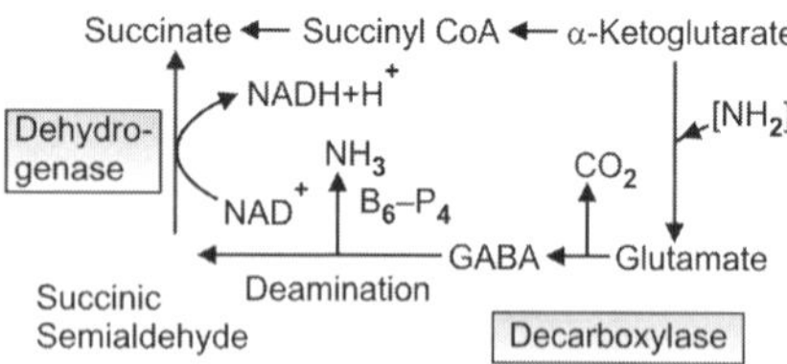

Fig. 25.3: GABA shunt

Q.168 Name the polyamines.

- Spermidine
- Spermine and
- Their precursor putrescine.

Q.169 What is carbamoyl-P synthetase II? What is its function?

Carbamoyl-P-synthetase II is an enzyme present in cytosol. It can form carbamoyl-P from CO_2 and amide-N of glutamine, which is used in pyrimidine synthesis.

Q.170 What are the aromatic amino acids? Name them, which is essential and which is not.

- Aromatic amino acids are:
  - Phenylalanine and
  - Tyrosine
- Phenylalanine is essential and must be provided in the diet
- Tyrosine is not essential as it can be formed from phenylalanine in the liver.

Q.171 State how phenylalanine is converted to tyrosine in the body.

Reactions occur in two stages:

- ***Stage 1:*** Reduction of molecular O_2 to H_2O and conversion of phenylalanine → to tyrosine. Reduced from of folic acid FH_4 acts as H-donor to molecular O_2 and is converted to FH_2. In this stage, there is incorporation of one atom of molecular O_2 to p-position of phenylalanine to form tyrosine, while other atom of O_2 is reduced to H_2O.
- ***Stage 2:*** Reduction of FH_2 to reform FH_4 takes place by NADPH, acting as a H-donor, catalyzed by the enzyme *dihydrobiopterin reductase*.

Note: Both the reactions are irreversible.

Q.172 What is meant by sparing action?

The feeding of tyrosine/or cysteine decreases the need of phenylalanine/or methionine in the diet respectively. This is called "sparing" action. Phenylalanine/ or methionine is spared from synthesis of tyrosine/or cysteine respectively.

Q.173 Tyrosine is a non-essential amino acid. In which condition it becomes an essential amino acid?

In phenylketonureas, where phenylalanine cannot be converted to tyrosine in the body due to inherited absence of the enzyme *phenylalanine hydroxylase*, tyrosine becomes an essential amino acid to the patient.

Q.174 What is albinism?

It is an inherited disorder of tyrosine metabolism in which the enzyme *tyrosinase* is lacking resulting in non-formation of melanin pigment for which the hair, skin and retina of eye are not black.

Q.175 What is the relationship between cysteine and cystine?

- Two molecules of L-cysteine oxidized to give one molecule of cystine which is joined by S-S-bond.
- One molecule of cystine is reduced to form two molecules of cysteine.

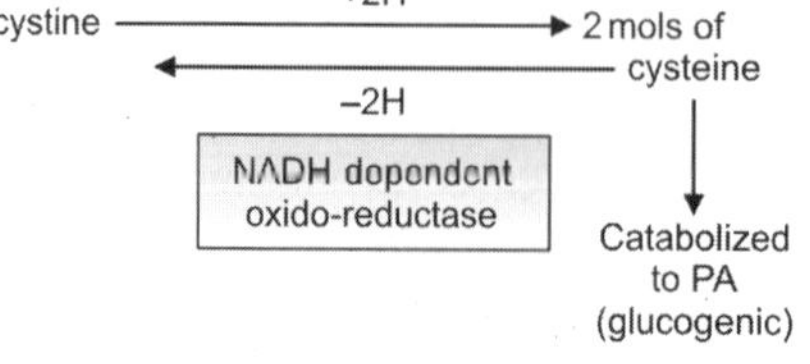

Q.176 What is active methionine? What is it known as chemically?

- "Active methionine" is an active compound in which CH_3 group is biologically lable and it acts as a methyl (CH_3) group donor to a suitable "acceptor" so that it can form biologically important compounds.
- Chemically "active methionine" is S-adenosyl methionine.

Q.177 What is meant by transmethylation reaction?

Certain compounds of the body, with structures containing CH_3 group attached to an atom other than carbon can take part in enzymic reactions, whereby these CH_3 groups are transferred to a suitable "acceptor" which have no methyl group. Such reactions are called as transmethylation reaction.

Q.178 Name the compounds which possess biologically labile methyl group.

- Active methionine (S-adenosyl methionine): containing -S ~ CH_3 group.
- Choline: containing -CH_3 groups attached to N as follows:

$$N^+ \begin{matrix} CH_5 \\ —CH_5 \\ CH_5 \end{matrix}$$

- Betaine: an oxidation product of choline.

Q.179 Give five examples of transmethylation reactions, where active methionine acts as methyl donor.

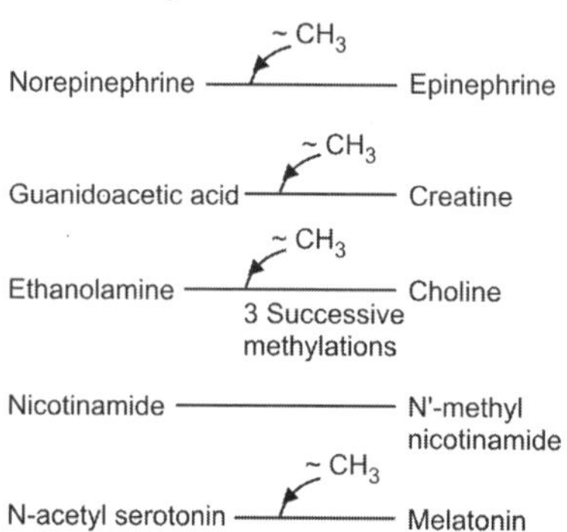

Q.180 Enumerate the metabolic role or biomedical importance of L-methionine.

- Methionine is glucogenic.
- Methionine is converted to L-cysteine in the body.
- ***Lipotropic function:*** provides choline, a lipotropic agent, by methylations of ethanolamine.
- Transmethylation reactions: by forming active methionine.
- In polyamine synthesis.
- Formation of methyl mercaptan in liver diseases, which accounts for foul odor in breath (fetor hepaticus).

Q.181 Show schematically how L-cysteine is formed from L-methionine.

See Figure 25.4.

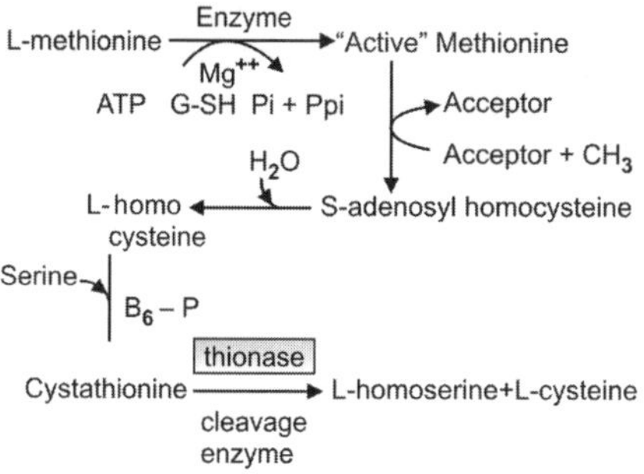

Fig. 25.4: Formation of L-cysteine

Q.182 What is homocystinuria? What is the enzyme deficiency?

- An inherited disorder of metabolism of L-methionine or more specifically its metabolic intermediates homocysteine/ or homocystine.
- ***Enzyme deficiency:*** *Cystathionine synthase* enzyme deficiency leads to accumulation of homocystine which is excreted in urine.

Q.183 State five metabolic role/or biomedical importance of glutathione in the body.

- Reduced glutathione (G-SH) is necessary for maintaining integrity of red cell membranes and lens protein.
- It helps to destory H_2O_2 and other peroxides in cells. The reaction is catalyzed by selenium containing enzyme *glutathione peroxidase.*
- Acts as coenzyme with liver enzyme *glutathione-insulin transhydrogenase* which helps in catabolism of insulin.
- Glutathione takes part in "gamma-glutamyl cycle" for absorption of amino acids from gut.
- Reduced glutathione (G-SH) is required as coenzyme/cofactor for *PG synthetase system* in PG synthesis and also for formation of "active methionine".

Q.184 What is serotonin?

Serotonin is chemically 5-OH tryptamine (5-HT), and it is formed from amino acid tryptophan. It is present in blood and is produced in tissues like gastric mucosa, intestine, brain, mastcells and platelets. (Probably stored in platelets?)

Q.185 How serotonin is synthesized by the serotonin-producing cells in the body?

- Tryptophan is first hydroxylated to form 5-OH tryptophan in liver by *hydroxylase* enzyme in presence of O_2 and NADPH.
- 5-OH tryptophan is next decarboxylated by the enzyme *decarboxylase* to form 5-OH tryptamine (5-HT), called serotonin, the reaction requires B_6P as coenzyme.

Q.186 Name the enzyme which degrades serotonin. What is the degradation product?

- Enzyme is *monoamine oxidase* (MAO)
- Degradation product is 5-OH-indole acetic acid (5-HIAA) which is excreted in urine.

Q.187 In which condition excessive amount of serotonin is produced in the body?

- Excessive amount of serotonin is produced in carcinoids a malignant tumor of serotonin-producing cells. It is also called as argentaffinoma.
- The clinical features associated with it is called as carcinoid syndrome.
- In carcinoids, 60% of tryptophan is utilized for serotonin formation as compared to 1% in normals.

Q.188 What is melatonin?

- Chemically melatonin is "N-acetyl-5-methoxy serotonin". It is a hormone produced by the pineal gland and peripheral nerves of man and some other higher animals.
- The hormone lightens the color of melanocytes in the skin of frog and blocks the action of MSH (melanocyte stimulating hormone) and ACTH.

Q.189 What is Hartnup's disease?

- It is an inherited disorder associated with the metabolism of tryptophan. The clinical symptoms are mental retardation, intermittent cerebellar ataxia and skin rash.
- Urine of patients with Hartnup's disease contains increased amounts of indoleacetic acid and tryptophan.

Q.190 What is the metabolic defect in Hartnup's disease?

- An inherited deficiency of the enzyme *tryptophan dioxygenase* in liver.
- Could also be due to a defective indole transport and consequent impairment of tryptophan metabolism.

Q.191 What is the relationship between creatine-P and creatinine?

- Both are nonprotein nitrogenous compounds.
- Creatinine-P is chemically methyl guanidinoacetic acid".
- Creatinine is anhydride of creatine, the reaction is irreversible and non-enzymatic. Creatinine has a ring structure.

Q.192 Show schematically the biosynthesis of creatine.

See Figure 25.5.

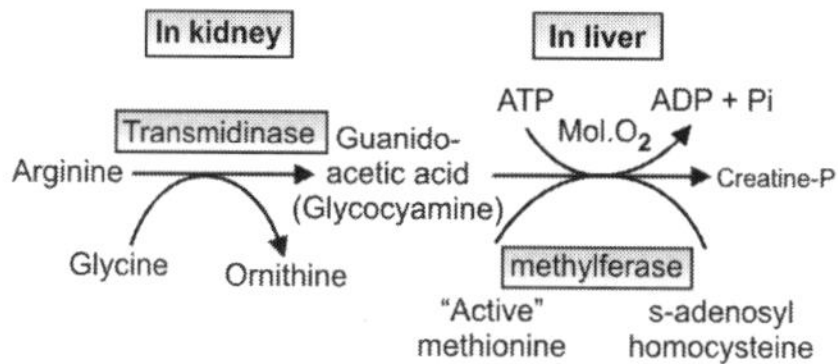

Fig. 25.5: Synthesis of creatinine

Q.193 Where does the formation of creatinine occur?

In liver and in kidney.

Q.194 What is meant by creatinine clearance test.

Endogenous creatinine clearance is used as a renal function test. At normal levels of creatinine in the blood, this metabolite is filtered at the glomerulus but neither secreted nor reabsorbed by the tubules. Hence its clearance measures the glomerular filtration rate (GFR).

Q.195 What is the normal endogenous clearance?

Normal values for creatinine clearance varies from 95 to 105 mL/min.

Q.196 What are the biological importance and metabolic role of arginine?

- Semiessential amino acid required in growing children, in pregnancy and lactation
- Required for creatine synthesis
- Required in urea formation (an intermediate metabolite)
- Tissue protein formation
- Glucogenic amino acid.

DO YOU KNOW?

- Most excess nitrogen is converted to urea in the liver and goes through the blood to the kidney, where it is eliminated in urine.
- Methotrexate inhibits DHF reductase (Dihydrofolate reductase) making it a very useful antineoplastic drug.

Chapter 26

Nucleoproteins

Q.1 What is a nucleoside?
Sugar-base combination is called nucleoside.

Q.2 What is a nucleotide?
Sugar-base-phosphate combination is called nucleotide.

Q.3 What are nucleic acids?
Nucleic acid are polynucleotides.

Q.4 What is the difference between adenine, adenosine and adenylic acid?

- Adenine is a purine base.
- Adenosine is a nucleoside. It is a adenine-ribose combination. Adenylic acid is a nucleotide. It is a adenine-ribose-phosphoric acid.

Q.5 What are the types of nucleic acid?
Two types of nucleic acid are:
1. Deoxyribose acid (DNA).
2. Ribo nucleic acid (RNA).

Q.6 What are the bases present in DNA?
Purine bases: Adenine, guanine.
Pyrimidine bases: Cytosine, thymine.

Q.7 What are the bases present in RNA?
Purine bases: Adenine, guanine.
Pyrimidine bases : Cytosine, uracil.

Q.8 How are the bases united in nucleic acid?
Bases are united through hydrogen bonding.

Q.9 What is the number of hydrogen bonds present between?
i. Adenine—Thymine
ii. Cytosine—Guanine
Adenine—Thymine........Two hydrogen bonds
Guanine—Cytosine...........Three hydrogen bonds.

Q.10 What is the difference between RNA and DNA?

DNA	*RNA*
1. Bases present are adenine, guanine, cytosine, thymine	1. Bases present are adenine, guanine, cytosine, uracil
2. Sugar present is deoxyribose.	2. Sugar present is ribose.
3. It is double strand molecule.	3. It is a single strand molecule.

Q.11 What is the function of transfer RNA?
Transfer RNA carries a specific amino acid to the site of protein synthesis on the m-RNA template.

Q.12 What are functions of mRNA?
It carries message from DNA to cytoplasm and is having codons on it in sequential manner.

Q.13 What are functions of rRNA?
It consists of 80% of total RNA and helps in protein synthesis by finding to ribosomes.

Q.14 What is replication?
Copying of DNA to from identical daughter molecules is called the replication.

Q.15 What is transcription?
Transcription is the process by which the genetic message in DNA is transcribed into the form of messenger RNA to be carried to the ribosomes.

Q.16 What is translation?
Translation is the process by which the genetic message is decoded on the ribosomes.

Q.17 What is genetic code?
Genetic code refers to the language used by the genetic material, i.e. the DNA to transfer genetic information to the site of new product formation. Genetic code is universal.

Q.18 What is a triplet or codon?
The sequence of three non-overlapping nucleotides is called a triplet.

Q.19 What is the chain initiating codon in protein synthesis?
Chains initiating codon is AUG.

Q.20 Chain initiating codon AUG codes for which amino acid?
AUG codes for:
i. N-formylmethionine in prokaryotes.
ii. Methionine eukaryotes.

Q.21 Which is the other initiator of protein synthesis in addition to AUG?
In some lower animals GUG serves the function of initiation.

Q.22 What are the chain terminating codons?
UAA, UAG, UGA.

Q.23 What are the sources of C and N of purine skeleton (Fig. 26.1)?
N_1-aspartic acid
C_2, C_8—formate.
C_4, C_5, N_7—glycine.
C_6—respiratory CO_2.

Q.24 What is the fate of purine bases?
Purine base are converted to uric acid.

Q.25 What is the normal uric acid level?
2-6 mg%.

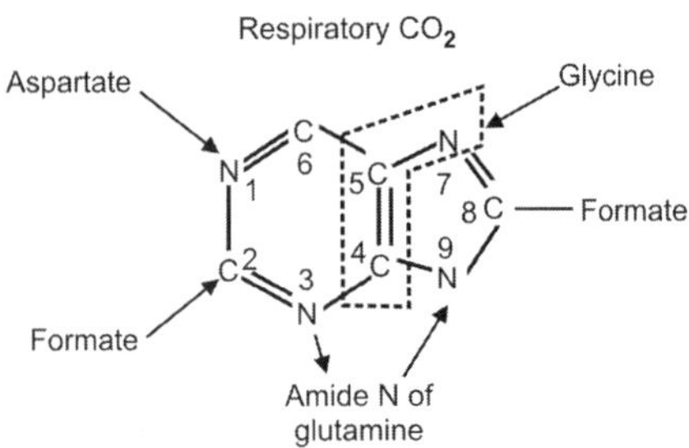

Fig. 26.1: Purine skeleton

Q.26 What is a gout?
Increased uric acid level in the blood given rise to a condition known as gout.

Q.27 How are pyrimidines catabolized?
Pyrimidines are catabolized to β-alanine and amino-isobutyric acid.

Q.28 What are the unusual bases present in nucleic acid?
1. 5-methyl cytosine.
2. N^6-methyl adenine.
3. N^6, N^6-dimethyl adenine
4. N^6,N^7-dimethyl guanine
5. 5-hydroxy methyl cytosine.

Q.29 Enumerate some plant derived methylated purines.
- Caffeine (1,3,7-trimethyl xanthine)
- Theophylline (1,3-dimethyl xanthine)
- Theobromine (3,7, dimethyl xanthine).

Q.30 Enumerate some adenosine derivatives.
- ATP, cAMP, adenosine-3-phosphate-5-phosphosulfate, S-adenosyl, methionine.

Q.31 Name some synthetic nucleotide analogs.
- 5-fluorouracol
- 5-iododeoxyuridine
- 6-thioguanine
- 6-mercaptopurine
- 6-azauridine
- 8-azaguanine
- Allopurinol
- Azathioprine.

Q.32 What is the main difference between *de novo* and salvage pathways of purine production?
Salvage reactions require for less energy than *de novo* synthesis of purines.

Q.33 What is mode of action of methotrexate?
Methotrexate, an anticancer drug, inhibits the enzyme dihydrofolate reductase.

Q.34 Name some medicinally important nucleotides with uses.
1. Allopurinol G or gout and hyperuricemia.
2. Azathioprine-immunosuppressive in organ transplantation.
3. 5-iododeoxyuridine for herpetic keratitis.

Q.35 Explain Lesch Nyhan syndrome.
An overproduction of uric acid with arthritis and bizzare syndrome of self mutilation is due to defect in HGPRTase, an enzyme of purine salvage, this causes rise in intracellular PRPP, results in purine overproduction.

Q.36 Who demonstrated the DNA transformation experiment?
DNA-transformation experiment was conducted by Griffith. But the demonstration that DNA contains genetic information was first made by Avery Macleod Mc-Carty.

Q.37 What is hyperchromicity of denaturation?
By denaturation of DNA strands there is increase in optical absorbance of purine and pyromidine bases.

Q.38 What is the cap of mRNA?
Cap is present on 5' end of mRNA and consists of 7-methyl guanosine triphosphate.

Q.39 Which histones are present in nucleosome?
Histone is octameric molecule consisting of H_1, H_2A, H_2B, H_3, H_4.

Q.40 Which diseases are related with defective DNA repair?
1. Xeroderma pigmentosum.
2. Ataxia telangiectasia.
3. Fanconi anemia.
4. Bloom's disease.

Q.41 Name some antibiotics that inhibits protein synthesis.
1. Puromycin.
2. Cyclohexinide.
3. Diphtheria toxin.

Q.42 Who proposed operon model?
Jacob and Monod in year 1961.

Q.43 What are restriction endonuclease?
Restriction endonuclease are enzymes that cut the DNA segment in specific sequence, e.g. Eco RI cut the DNA strand at $\frac{\text{GAATTC}}{\text{CTTAAG}}$

Q.44 Define intron.
Intron is a sequence of gene that is transcribed out excised before translation.

Q.45 What is southern blotting?
A method for transferring DNA from an agarose gel to nitrocellulose filter, on which DNA can be detected by a suitable probe (e.g. cDNA).

Q.46 Explain polymerase chain reaction (PCR).
An enzymatic method for the repeated copying of two strands of DNA that make up a particular gene sequence.

Q.47 Define probe.
A molecule used to detect the presence of a specific fragment of DNA/RNA.

Chapter 27

Enzymes

Q.1 What are enzymes?
Enzymes are biological catalysts.

Q.2 What is the nature of enzymes?
Enzymes are protein in nature.

Q.3 Name the fastest acting enzyme?
Carbonic anhydrase (CA).

Q.4 What is Km (Michaelis constant)?
Km is defined as that substrate concentration which produces half of the maximum velocity.

Q.5 Km is defined in terms of what?
It is defined in terms of substrate concentration.

Q.6 Which Km value is preferred, low or high?
Low Km value is preferred because at low substrate we can achieve the maximum velocity.

Q.7 Give Michaelis Menten equation (Fig. 27.1)?

$$V_1 = \frac{V_{max}[s]}{Km + [s]}$$

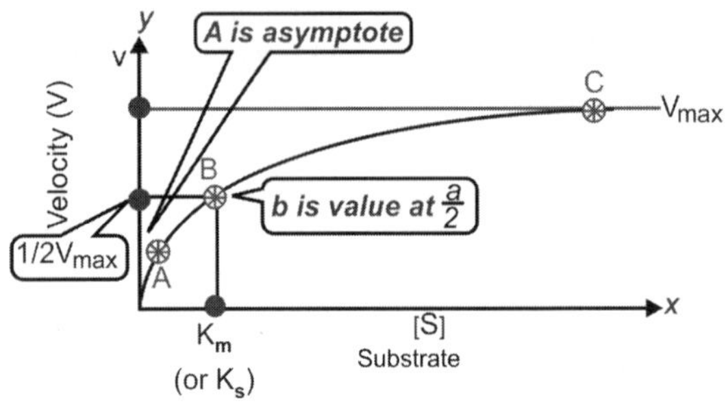

Fig. 27.1: Michaelis-Menten plot (saturation plot of enzyme catalyzed reaction)

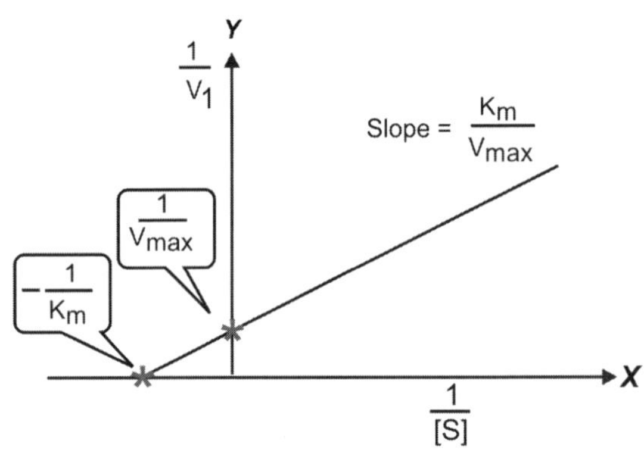

Fig. 27.2: Lineweaver-Burk plot

Q.8 Some enzymes do not follow Michaelis Menten equation then they follow which reaction?
Hill equation.

Q.9 Give example which follows Hill equation?
Hemoglobin.

Q.10 Why hemoglobin follows Hill equation not Michaelis Menten equation?
Due to property of heme-heme interaction, hemoglobin does not follow Michaelis Menten equation as no straight line is followed.

Q.11 Why Lineweaver-Burk curve or equation is preferred (Fig. 27.2)?
Lineweaver-Burk equation is in the form of a straight line. For a straight line, we need two points, i.e. by selecting few different concentrations, we can plot the curve and hence find out the Km.

Q.12 What are the factors which influence enzyme activity?
Factors which influence enzyme activity are:

Table 27.1: Classification of enzymes

S. No.	*Classification*	*Action*
1.	Oxidoreductases	Act on many chemical groupings to add or remove hydrogen atoms.
2.	Transferases	Transfer functional groups between donor and acceptor molecules. Kinases are specialized transferases that regulate metabolism by transferring phosphate from ATP to other molecules.
3.	Hydrolases	Hydrolytically cleaves molecules by adding water across a bond.
4.	Lyases	Add water, ammonia or carbon dioxide across double bonds, or remove these elements to form double bonds.
5.	Isomerases	Carry out many kinds of isomerization: L to D isomerization, mutase reaction (shifts of chemical groups) and others.
6.	Ligases	An ATP dependent reaction that catalyze reactions in which two chemical groups are joined (or ligated) with the use of energy from ATP.

1. Substrate concentration.
2. Enzyme concentration.
3. pH.
4. Temperature.
5. Effect of activators and coenzymes.

Q.13 What are competitive inhibitors?
Competitive inhibitors are those which have structural similarity with the substrate molecules for the active site of the enzyme.

Q.14 What is the characteristics of line weaver Berk plot in competitive inhibition?
In competitive inhibition V_{max} remains the same but Km value increases.

Q.15 What is the characteristics of Lineweaver-Burk plots in non-competitive inhibition?
Reverse of above is followed in non-competitive inhibition.

Q.16 What are non-competitive inhibitors?
Non-competitive inhibitors are those which are attached not to the active site but to some other site of the enzyme.

Q.17 Give the example of competitive inhibitors.
Inhibition of succinate dehydrogenase by malonate. L-lactate dehydrogenase (substrate lactate) by oxamate. Cis-aconitase by fluorocitrate.

Q.18 Give the example of non-competitive inhibition.
1. Inhibitions of acetyl choline esterase, trypsin, chymotrypsin by diisopropyl-fluorophosphate.
2. Inhibition by heavy metal ions like Ag^+ and Hg^{++} ions.

Q.19 What is the difference between prosthetic group and co-enzyme?
- Prosthetic group is tightly bound to the enzyme.
- Whereas co-enzymes exist in the solution in free state and contact the enzyme only at the site of reaction.

Q.20 What is an allosteric site?
Site other than the active site is called the allosteric site.

Q.21 What are isoenzymes? Give two examples.
- Isoenzymes are the multiple forms of the same enzymes
- Lactate dehydrogenase,
- Alkaline phosphatase.

Q.22 Name the isoenzymes of LDH?
- LDH has 5 iso enzymes
  - HHHH, HHHM, HHMM, HMMM, MMMM.

Q.23 Name the diagnostic enzymes in MI?
SGOT (AST), LDH.

Q.24 Name the diagnostic enzyme in viral hepatitis?
SGPT (ALT).

Q.25 Which enzyme increases in serum in acute pancreatitis?
α Amylase.

Q.26 In Wilson disease, which enzyme plays diagnostic role?
Ceruloplasmin.

Q.27 Name the diagnostic enzyme for muscle disorders.
Creatine kinase.

Q.28 Acid phosphatase is diagnostic enzyme for which disease?
Carcinoma of prostate.

Q.29 Enzyme that is diagnostic tool for bone and liver disease.
Alkaline phosphatase.

Q.30 What are anti-enzymes?
Anti-enzymes are the substances produced as a result of repeated injection of certain enzymes in the serum, which prevents the normal action of the enzyme injected.

Q.31 How will you differentiate whether the given reaction is enzyme catalyzed or not?
We can differentiate the reaction by two test.
1. Heat sensitive test.
2. Acid test.

Q.32 What are constitutive enzymes?
Constitutive enzymes are those enzymes which are present in constant concentration during the life of the cell.

Q.33 What are inductive enzymes?
Inductive enzymes are those enzymes whose amount present in the cell is variable and depending upon the requirement.

Q.34 What is feedback inhibition?
In multienzyme reaction, the end-product of the reaction sequence may act as a specific inhibitor of an enzyme at or near the beginning of the sequence with the result the rate of entire sequence of reactions is determined by the steady state concentration of the end-product. This type of inhibition is called feedback inhibition.

Q.35 Give an example of feedback inhibition.
Cholesterol inhibits cholesterol synthesis in the liver by inhibition of HMG-CoA reductase.

Chapter 28

Biological Oxidation

Q.1 What is oxidation?
Oxidation is defined as loss of electrons

$Fe^{++} \rightarrow Fe^{+++}$

or

removal of hydrogen

$C_2H_5OH \rightarrow CH_3CHO$

or

addition of oxygen

$CH_3CHO \rightarrow CH_3COOH$.

Q.2 What do you mean by biological oxidation?
The stepwise degradation of metabolite for the liberation of energy carried out in the system.

Q.3 Give Gibbs free energy equation.
$\Delta G = \Delta H - T\Delta S$
ΔH = enthalpy of reaction
T = temp (K)
ΔS = entropy change.

Q.4 What inference is drawn from this reaction?
If ΔG is +ve then the reaction is endothermic while -ve implies, reaction to be exogonic.

Q.5 Who gave the concept of high energy phosphate bond?
Lippmann introduced the concept of high energy phosphate bond.

Q.6 Name the components of respiratory chain.
$NAD^+ \rightarrow Fp \rightarrow CoQ \rightarrow$ Cyt b $\rightarrow$ Cyt $C_1 \rightarrow$ Cyt c $\rightarrow$ Cyt a $\rightarrow$ Cyt a_3.

Q.7 In which order the components of respiratory chain are arranged?
The components of respiratory chain are arranged in order of increasing redox potential. NAD^+ has the lowest redox potential where as Cyt a_3 has the highest redox potential in the respiratory chain.

Q.8 Name the various ATP sites in the respiratory chain.
1. NAD^+ to FAD.
2. Cyt b to Cyt C_1.
3. Cyt a to Cyt a_3.

Q.9 What should be the minimum redox potential difference for the generation of ATP?
0.2 volts.

Q.10 What is the ATP yield in respiratory chain?
32 molecules of ATP (28 + 2 + 2).

Q.11 Which part of the NAD^+ accepts electrons?
Nicotinamide.

Q.12 Which part of the FAD accepts electrons?
Flavin.

Q.13 What is oxidative phosphorylation?
Oxidative phosphorylation is the process in which ATP is formed as electrons are transferred from NADH or $FADH_2$ to oxygen by a series of electron carries.

Q.14 Name the inhibitors of various energy sites in respiratory chain?
Inhibitors of site I = rotenone, amobarbital, piericidin.
Inhibitors of site II = Antimycin, BAL.
Inhibitors of site III = H_2S, CO_2, CN^-.

Q.15 Give the examples of substrate level phosphorylation.
D-Glyceraldehyde-$3PO_4 \rightarrow$
3 Phosphoglyceric acid.
Phosphoenol pyruvic acid $\rightarrow$ Pyruvate
Succinyl CoA $\rightarrow$ Succinic acid.

Q.16 What is P/O ratio?
P/O ratio is defined as the number of inorganic phosphate taken to phosphorylate ADP per atom of oxygen consumed.

$$\frac{P}{Q} = \frac{\text{Number of inorganic phosphate to phosphorylate ADP}}{\text{No. of atom of oxygen consumed}}$$

Q.17 Name few reactions where P/O ratio is 2.
1. Succinate $\longrightarrow$ Malate (TCA cycle)
2. Acyl CoA $\longrightarrow$ β unsat. Acyl CoA (β-oxidation of fatty acids).

Q.18 Name few high energy compounds.
Adenosine triphosphate, adenosine diphosphate, Inosine triphosphate, etc.

Q.19 What are uncouplers?
Uncoupler are those substances which separates oxidation from phosphorylation, e.g. dinitrophenol.

Q.20 Name the different classes of oxidoreductases.
1. Oxidases.
2. Dehydrogenases.
3. Hydroperoxidases.
4. Oxygenases.

Q.21 Give some examples of oxidases.
1. Cytochrome oxidase.
2. FMN, FAD.
3. Xanthine oxidase.
4. Glucose oxidase.
5. L-amino acid oxidase.

Q.22 Which enzyme is responsible for lowering of superoxide ion?
Superoxide dismutase (SOD).

Q.23 Superoxide dismutase (SOD) contains which ions?
- Superoxide dismutase enzymes is of two types:
  1. Up to plasmic SOD.
  2. Mitochondrial SOD.
- Up to plasmic SOD contains copper and uric ions where as mitochondrial SOD contains manganese.

Q.24 Who proposed the chemiosmotic theory?
Peter Mitchelle was the person to propose chemiosmotic theory.

Q.25 Name the inhibitor of adenine nucleotide transporter.
- Atractyioside is inhibitor of adenine, nucleotide transporter.

Chapter 29

Vitamins

Q.1 What are vitamins?
Vitamins are defined as organic substances present in natural occurring food and are required in small amounts for maintenance, growth and reproduction.

Q.2 What are the types of vitamins?
Vitamins are two types:
1. Fat-soluble vitamins.
2. Water-soluble vitamins.

Q.3 What are fat-soluble vitamins?
Fat-soluble vitamins are:
Vitamin A, D, E and K.

Q.4 What are water soluble vitamins?
Water-soluble vitamins are:
i. Vitamin B complex.
ii. Vitamin C.

Q.5 What are the members of vitamin B complex?
1. Thiamine (B_1)
2. Riboflavin (B_2)
3. Pyridoxine (B_6)
4. Cyanocobalamin (B_{12})
5. Niacin (B_5)
6. Pantothenic acid (B_7)
7. Biotin (B_9)
8. Folic acid
9. Lipoic acid
10. Para-amino-benzoic acid (PABA)
11. Choline
12. Inositol.

Q.6 Name the vitamins which are the components of electron transport chain.
1. Niacinamide
2. Riboflavin
3. Ubiquinone.

Q.7 Name the vitamins which are partly synthesized in the intestinal flora.
1. Vitamin K
2. Thiamine
3. Riboflavin
4. Pyridoxine
5. Niacin.

Q.8 What are provitamins?
Provitamins are substances which as such do not possess vitamin activity but on conversion give rise to vitamins. Provitamins of vitamin A are, a, b, g, carotenes, etc.

Q.9 What are the sources of vitamin A?
Butter, milk, cod liver oil, halibut liver oil, egg yolk, carrots.

Q.10 What is provitamin form of vitamins A?
Carotenes.

Q.11 How many molecules of vitamins A are formed from?
a. α-carotenes
b. β-carotenes
c. γ-carotenes.
a. 1 molecule
b. 2 molecule
c. 1 molecule.

Q.12 What are the functions of vitamin A?
1. Visual function.
2. Growth and reproduction.
3. Proper healing of epithelium.
4. Bones and teeth.
5. Necessary for synthesis of mucopolysaccharides.
6. It is also involved in nucleic acid metabolism.
7. It is also involved in the electron transport chain and in oxidative phosphorylation.

Q.13 What is the daily requirement of vitamin A?
5000 IU.

Q.14 What is the deficiency disease associated with vitamin A?
Night blindness.

Q.15 What are the active forms of vitamin D?
The active forms are:
1. 25-hydroxycholecalciferol (25 HCC).
2. 1,25 dihydroxycholecalciferol (1,25-DHCC).

Q.16 What are the organs which synthesize the active forms of vitamin D?
Formation of 25-HCC takes place in liver whereas formation of 1, 25 DHCC takes place in kidney.

Q.17 What are the precursors of vitamin A?
1. *Ergosterol:* It gives rise to calciferol (vitamin D_2).
2. *7-dehydrocholesterol:* It gives rise to ergocalciferol (vitamin D_3).

Q.18 What are functions of vitamin D?
1. In the absorption of calcium and phosphorus from the intestines.
2. Growth.
3. Mineralization of bones.
4. Normal functioning of parathormone.

Q.19 What is the daily requirement of vitamin D?
400 IU.

Q.20 What are the deficiency symptoms of vitamin D?
- Rickets in children
- Osteomalacia in adults.

Q.21 What is hypervitaminosis?
Excessive intake of vitamin A or D gives rise to condition known as hypervitaminosis.

Q.22 What are the functions of vitamin E?
1. As an antioxidant.
   a. Prevents the autoxidation of vitamin A and carotenes.
   b. Prevents the formation of fatty acid peroxidases in tissues due to the autooxidation of unsaturated fatty acids with oxygen.
   c. Protects the lipids of biological membranes against oxygen by acting as antioxidants.
2. Prevents rancidity.

Q.23 What are the functions of vitamin K?
1. In the synthesis of prothrombin.

2. In the oxidative process taking place in the photosynthesis in plant kingdoms.

Q.24 What is the structure of vitamin C?

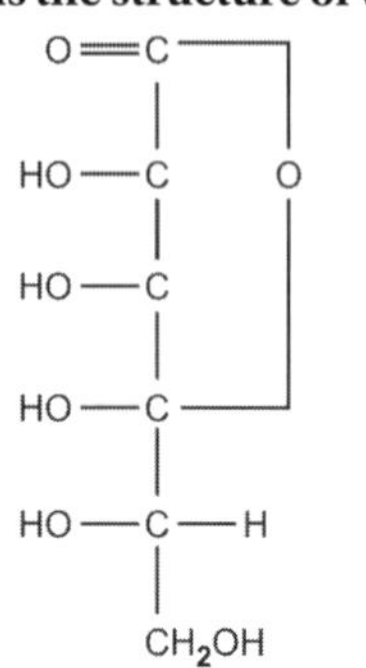

Q.25 What are the functions of vitamin C?

1. In the synthesis of collagen.
2. In the synthesis of steroid hormones both in adrenal cortex and corpus luteum.
3. As cofactor in the following reactions:
   a. In phenylalanine metabolism p-hydroxy phenyl pyruvic acid → homogentisic acid.
   b. Dopamine → Norepinephrine.
   c. Folic acid → Folinic acid.
4. In the synthesis of carnitine in the liver.
5. Necessary for the absorption of iron by reducing ferric form to ferrous form.
6. In tissue respiration.

Q.26 What are the sources of vitamin C?

Citrus fruits such as lemon, orange, pineapple, etc. Indian gooseberry, green-pepper, cauliflower, tomatoes, spinach, potatoes.

Q.27 What is the daily requirement of vitamin C?

60 mg.

Q.28 What is the normal level of vitamin C in blood?

0.6–1.5 mg per 100 mL of blood.

Q.29 What is the deficiency disease of vitamin C?

Scurvy.

Q.30 Which animal can synthesize vitamin C?

Rat, rabbit, dog, and birds.

Q.31 What is biological active form of vitamin B_1?

Thiamine pyrophosphate (TPP) (Fig. 29.1).

Q.32 What are the functions of vitamin B_1?

Thiamine pyrophosphate participates as coenzymes

1. In oxidative decarboxylation of α-keto acids.

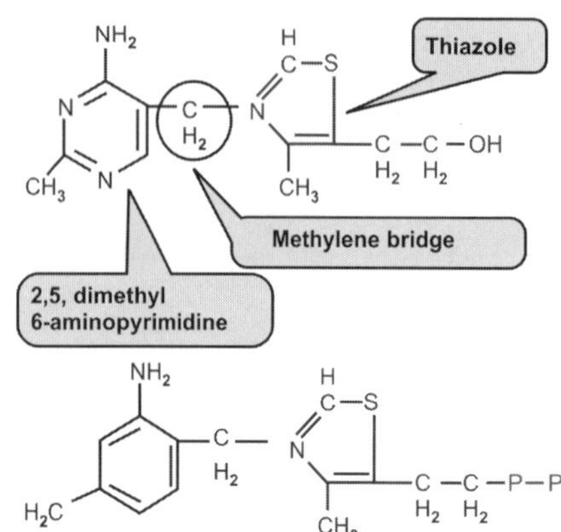

Fig. 29.1: Thiamine pyrophosphate (TPP)

   a. Pyruvic acid → Acetyl CoA.
   b. α-ketoglutaric acid → Succinyl CoA.
2. In trasketolation reaction of hexose monophosphate shunt pathway.

D-xylulose 5-PO_4 + D-ribose 5PO_4 ⟶ D-sedopheptulose 7-PO_4 + D-glyceraldehyde-3-PO_4

Q.33 What is the deficiency symptom of vitamin B_1?

Human: Beriberi
Rats: Bradycardia.

Q.34 What are the sources of vitamin B_1?

Germinating seeds, legumes, wheat, pork, eggs.

Q.35 What is the daily requirement of vitamin B_1?

1.4 gm.

Q.36 Why vitamin B_2 is also called lactoflavin?

Vitamin B_2 was first of all isolated from milk, hence it is called lactoflavin.

Q.37 What are the coenzymes forms of vitamins B_2?

1. Flavin mononucleotide (FMN).
2. Flavin adenine dinucleotide (FAD).

Q.38 What are the various active forms of vitamin B_6?

1. Pyridoxine
2. Pyridoxal
3. Pyridoxamine.

Q.39 What is the biologically active form of vitamin B_6?

Pyridoxal phosphate and pyridoxamine phosphate.

Q.40 What are the reactions mediated by vitamin B_6?

1. In transamination reaction
2. In decarboxylation reaction
   i. Histidine → Histamine
   ii. Tyrosine → Tyramine
   iii. Glutamic acid → γ-amino butyric acid (GABA)
   iv. α-amino-β-keto adipic β-keto adipic acid → δ-amino levulinic acid.
3. In dehydrases reaction
   i. Serine → Pyruvic acid
   ii. Threonine → α-ketobutyric acid
4. In transulfurase reaction
   Homocysteine → Serine.
5. In desulfuration reaction
   Cystine pyruvic acid
   Homocysteine → α-Keto butyric acid.
6. In tryptophan metabolism
   Kynurenine → Anthranilic acid
7. In the transport of amino acids and in the absorption of amino acid.
8. In essential fatty acid synthesis.

Q.41 What are the functions of biotin?

Biotin is required as cofactor in carbon dioxide fixation reaction.

Carbon dioxide fixation reactions are:

1. Acetyl CoA → Malonyl CoA.
2. Pyruvate → Oxaloacetate.
3. Propionyl CoA → Methyl malonyl CoA.
4. Formation of carbamoyl phosphate in urea cycle.
5. In purine skeleton, i.e. C_6.

Q.42 What are carbon dioxide fixing or carboxylation reactions?

Carboxylation reactions are those in which a molecule of carbon dioxide is added to produce a carboxyl group.

Q.43 What is the active form of folic acid?

Tetrahydrofolic acid.

Q.44 What are the functions of folic acid?

As a carrier of one carbon moiety.

Q.45 Name one carbon moiety.

1. Methyl group (—CH_3).
2. Hydroxy methyl group (—CH_2OH).
3. Formyl group (—CHO).
4. Formimino group (—CH=NH).

Q.46 Name the reactions mediated by one carbon moiety.

1. Ethanolamine → Choline
2. Glycine → Serine
3. Norepinephrine → Epinephrine
4. Guanidoacetic acid → Creatine
5. Uracil → Thymine
6. Ribonucleotides → Deoxyribonucleotides
7. Formation of N-formylmethionine transfer RNA.

8. In purine synthesis (i.e. C-2 and C-8 positions in purine skeleton comes from one carbon moiety).

Q.47 What are the sources of folic acid?
Green leafy vegetables, cauliflower, liver, kidney, etc.

Q.48 What are the 3 D's of niacin deficiency?
1. Diarrhea.
2. Dermatitis.
3. Dementia.

Q.49 What are the functions of vitamin B_{12}?
1. Glutamic acid → β Methyl aspartic acid.
2. L-Methyl malonyl CoA → succinyl CoA.
3. Ribonucleotides → Deoxy ribonucleotides.

Q.50 Name the components of coenzyme A.

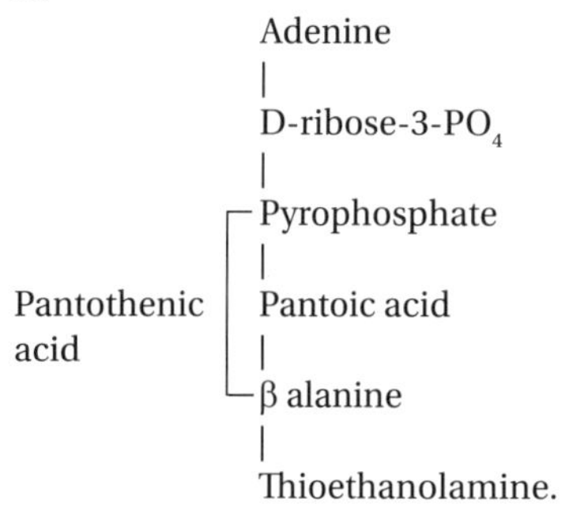

Q.51 The increased excretion of the following in the urine is a measure of the deficiency of which vitamins?
a. Formiminoglutamic acid
b. Methyl malonic acid
c. Xanthurenic acid
d. Homogentisic acid
e. Pyruvic acid.
a. Folic acid.
b. Vitamin B_{12}.
c. Vitamin B_6.
d. Ascorbic acid.
e. Thiamine.

Q.52 What is the deficiency disease of vitamin B_{12}?
Pernicious anemia.

Q.53 What are the sources of vitamin B_{12}?
Liver, kidney, meat, milk, cheese.

Q.54 Which vitamin is present in coenzyme A?
Pantothenic acid.

Q.55 What is scurvy?
It is poor wound healing, easy bruising, bleeding gums, bleeding time and painful glossitis. It can ultimately lead to anemia.

Q.56 What are the deficiencies of folic acid?
Megloblastic anemia.
Hemocystinemia
Deficiency in early pregnancy causes neural tube defects in fetus.

Q.57 Who are the people more affected by folic acid deficiency?
Alcoholics
Pregnant women in their early pregnancy.

Q.58 Name the proteins undergoing vitamin K-dependent carboxylation?
Coagulation factors II, VII, IX, X and protein C and S.

Q.59 What are the deficiencies of vitamin K?
- Fat malabsorption which can lead to bile duct occlusion
- Prolong treatment with broad spectrum antibiotics supply of vitamin K by killing intestinal bacteria
- Breastfed newborns as mother milk is very low in vitamin K
- Home births where babies were not given vitamin K injections
- Infants whose mothers were treated with anticonvulsants during pregnancy.

Q.60 Name some anticoagulants?
Warfarine, dicumarol, heparine.

Q.61 Sources of vitamin D?
Sunlight is a very good source of vitamin D. Vitamin D_3 is found in salmon (saltwater fish) and egg yolks.

VITAMINS
Fat-soluble Vitamins

Vitamins	*Functions*	*Sources*	*Deficiency diseases*
A	1. Visual cycle. 2. Maintenance of proper health of epithelium tissues 3. Stability of cellular and subcellular membranes 4. Synthesis of mucopolysaccharides. 5. Growth and reproduction 6. Bones and teeth. 7. Nucleic acid metabolism. 8. Electron transport chain and in oxidative phosphorylation	1. *Ready made sources*: Fish liver oils such as shark, cod, halibut fish liver oils, milk products, egg yolk. 2. *Provitamin sources*: Carrot, papaya, tomatoes	Night blindness
D	1. In the absorption of calcium and phosphorous from the intestines. 2. Growth. 3. Mineralization of bones.	Fish liver oils, egg yolk, milk products	Rickets
E.	1. As powerful antioxidants a. Prevent autooxidation of vitamin A and carotenes b. Prevent formation of fatty acids peroxidases in tissues due to autooxidation of unsaturated fatty acids with oxygen. c. Protect lipids of biological membranes against oxygen 2. Prevent rancidity	Wheat germ oil, corn oil, peanut oil, soyaben oil, sunflower oil: egg yolk, spinach, alfalfa	Sterility in rats
K.	1. In the synthesis of prothrombin. 2. In oxidative process taking place in photosynthesis in plant kingdom. 3. In electron transport chain and in oxidative phosphorylation.	Alfalfa, spinach, cabbage, cauliflower, egg yolk, liver.	1. Hemorrhage conditions 2. Prolongation clotting time

Water-soluble Vitamins			
Vitamins	*Functions*	*Sources*	*Deficiency diseases*
Thiamine (B_1)	1. Oxidative decarboxylation of α-Ketoacids, i.e. i. Pyruvic acid-acetyl CoA ii. α-keto gluta-succinic acid CoA 2. Transketolation reaction	Yeast, outercoating of seeds, cereals, legumes, wheat, pork, egg.	Human: Beriberi Rats: Bradycardia Beriberi
Riboflavin (B_2)	Forms the part of FMN and FAD which functions as prosthetic group of various enzymes I. FMN i. Warburg yellow enzyme ii. L-Amino acid oxidase iii. Cytochrome C reductase. II. FAD i. Acyl CoA dehydrogenase ii. D-Amino acid oxidase.	Yeast, milk, eggs, meat, fish liver, kidney, green leafy vegetables.	
Pantothenic acid (B_3)	Forms the part of coenzyme A which serves as a carrier of acyl group in enzymatic reactions. They are: 1. Fatty acid oxidation. 2. Fatty acid synthesis 3. Pyruvic acid oxidation. 4. Biological acetylations 5. Cholesterol biosynthesis. 6. In acyl carrier proteins.	Yeast, liver, kidney, egg yolk, molasses.	
Niacin (B_5)	As components of NAD and NADP+ which acts as coenzymes for many anaerobic dehydrogenases reactions. Enzymes requiring NAD+ or NADH coenzymes are: i. glyceraldehyde-3-PO_4 dehydrogenase. ii. Lactate dehydrogenase. iii. Malic dehydrogenase. Enzymes requiring NADP+ or NADPH as coenzymes are: i. Isocitrate dehydrogenase. ii. Glucose-6-PO_4 dehydrogenase. iii. Aldolase reductase.	Yeast, meat, liver, kidney, eggs, legumes.	*Man*: Pellagra tongue
Pyridoxal (B_6)	1. Transamination reaction. 2. Decarboxylation reaction. a. Histidine → Histamine b Tyrosine → Tyramine c. 5 hydroxy → 5 hydroxy Tryptophan tryptamine d. Glutamic acid → γ amino butyric acid GABA e. α-amino β-δ-amino Keto adipic acid → levulinic acid 3. In dehydrase reaction a. Serine → Pyruvic acid b. Threonine → α keto butyric acid 4. Transulfurase reaction Homocysteine → Serine 5. Desulphuration reaction a. Cystine → Pyruvic acid b. Homocystine → α-keto butyric acid 6. In tryptophan metabolism 7. In transport of amino acids and in absorption of amino acids. 8. In essential fatty acid synthesis.	Yeast, liver, egg yolk, rice polishings	
Biotin (B_7)	In carboxylation reactions, i.e. carbon dioxide fixing reactions: a. Acetyl CoA → Malonyl CoA b. Pyruvic acid → Oxaloacetic acid c. Propionyl CoA → D Methyl malonyl CoA d. $CO_2 + NH_3$ → Carbamoyl phosphate urea cycle e. In purine ring synthesis	Yeast, egg yolk, milk, molasses, Peanuts.	

Contd...

Contd...

Vitamins	*Functions*	*Sources*	*Deficiency diseases*
Folic acid (B_9)	As carrier of one carbon moiety. 1. Ethanolamine → Choline. 2. Glycine → Serine. 3. Norepinephrine → Epinephrine 4. Guanido → Creatine 5. Uracil → Thymine 6. Ribonucleotides → Deoxynucleotides 7. In purine synthesis 8. In formation of N-formylmethionine transfer RNA.	Yeast, liver, kidney, green vegetables	
Cyanoco-balamin (B_{12})	1. Glutamic acid → β Methyl asparatic acid 2. L-Methylmalonyl CoA → Succinyl CoA 3. Ribonucleotides → Deoxyribonucleotides.	Liver, kidney, meat, milk, cheese.	Pernicious anemia
Vitamin C	1. In collagen synthesis 2. In synthesis of steroid hormones both in adrenal cortex and corpus luteum 3. a. P-hydroxy phenyl pyruvic acid → Homogentisic acid. b. Dopamine → Norepinephrine c. Folic acid → Folinic acid. 4. In synthesis of carnitine 5. Absorption of iron 6. In tissue respiration.	Citrus fruits such as lemon, orange, pineapple, etc. Indian gossebery, green pepper, cauliflowers, tomatoes, spinach, potatoes.	Scurvy

DO YOU KNOW?

- Anticonvulsant drugs interfere with vitamin K absorption.
- Vitamin A is highly teratogenic hence should not be given to the pregnant mothers as it can cross the blood-brain barrier and causes teratogenic effects.
- Patients with end-stage renal disease develop renal osteodystrophy. IV/or oral 1,25 DHCC may be given.
- Isotretinoin is a form of retinoic acid and is used in treatment of acne. It is also highly teratogenic and should not be given to pregnant women.

Chapter 30

Blood

Q.1 What are the main functions of blood?

1. As a carrier of oxygen.
2. As a carrier of metabolic wastes of the body of kidney, lungs, skin and intestine for removal.
3. In the maintenance of acid-base balance.
4. In the maintenance of body temperature.
5. In transporting food materials to the tissues.
6. In regulating water balance.

Q.2 What is the compositions of blood?

1. Cellular fraction. This fraction contains erythrocytes, leukocytes and platelets.
2. Plasma fraction.

Q.3 What is the pH of blood?

7.4.

Q.4 What is the average volume of blood in the body?

5–7% of the body weight.

Q.5 What is the specific gravity of blood?

1.054–1.060.

Q.6 Name the steps involved in blood clotting.

$$\text{Prothrombin} \xrightarrow{\text{Thromboplastin, } Ca^{++}} \text{Thrombin}$$

$$\text{Fibrinogen} \rightarrow \text{Fibrin}$$

Q.7 Name the pathways of blood coagulation.

1. Intrinsic pathway.
2. Extrinsic pathway.

Q.8 Name the substance of blood clot.

Fibrin.

Q.9 Name the inhibitors of blood coagulation.

Oxalates, citrates, heparin.

Q.10 What are the factors which delay blood coagulation?

1. Low level of prothrombin.
2. Deficiency of vitamin C.
3. Obstruction of bile duct.

Q.11 Name blood buffers.

Blood buffers are:

$$\frac{BHCO_3}{H_2CO_3} \quad \frac{B_2HPO_4}{BH_2PO_4} \quad \frac{\text{B Proteinate}}{\text{H Protein}} \quad \frac{B.Hb}{H.Hb} \quad \frac{B.HbO_2}{H.HbO_2}$$

Q.12 What is plasma?

Plasma is obtained from blood which is prevented from clotting.

Q.13 What is the difference between serum and plasma?

Serum differs from plasma, in that it contains no fibrinogen.

Plasma–Fibrinogen = Serum.

Q.14 What is normal hemoglobin level in blood?

10–15 gm%.

Q.15 Hemoglobin is measured by which method?

Sahli method.

Q.16 Where is hemoglobin present?

Hemoglobin is present in RBC.

Q.17 Hemoglobin belong to which class of proteins?

It belongs to a class of conjugated protein.

Hemoglobin = Heme + globin.

Q.18 Give the structure of hemoglobin.

See Figure 30.1

Q.19 What is the prosthetic group of hemoglobin?

Heme.

Q.20 What is heme?

Heme is ferrous protoporphyrin.

Q.21 What is the state of iron in hemoglobin?

Iron is present in ferrous state (i.e. + 2 state).

Q.22 What is the state of iron in oxyhemoglobin?

Ferrous (+ 2 state).

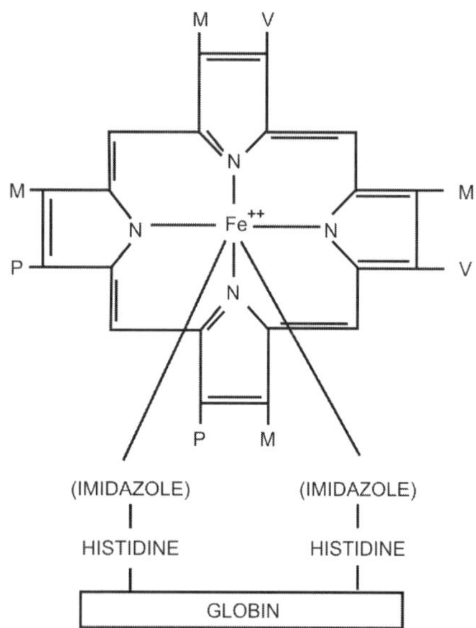

Fig. 30.1: Structure of hemoglobin

Q.23 Which amino acid is responsible for the buffering action of hemoglobin?

Histidine.

Q.24 What are the starting materials of hemoglobin synthesis?

Succinyl CoA and glycine.

Q.25 What is porphyria?

Accumulation of coproporphyrin and uroporphyrin in the blood with the excessive elimination in urine and feces is called porphyria.

Q.26 What is methemoglobin?

In methemoglobin, iron is in the ferric state.

Q.27 Persons with sickle cell anemia are having which advantage?

Plasmodium falciparum cannot invade sickled RBC so, the person heterogenous for sickle cell anemia escape from the deadly disease.

Table 30.1: Primary factors of blood coagulation

Factor	*Trivial name*	*Pathway*	*Characteristic*
Prekallikrein	Fletcher factor	Intrinsic	——
High-molecular-weight kininogen (HMWK) factor	Contact activation cofactor; Fitzgerald, Flaujeac William's	Intrinsic	——
I	Fibrinogen	Both	——
II	Prothrombin	Both	Contains N-term *gla* segment
III	Tissue factor	Extrinsic	——
IV	Calcium	Both	——
V	Proaccelerin, labile factor, accelerator (Ac-) globulin	Both	Protein cofactor
VI (Va)	Accelerin	——	This is Va, redundant to factor V
VII	Proconvertin, serum prothrombin conversion accelerator (SPCA), cothromboplastin	Extrinsic	Endopeptidase with *gla* residues
VIII	Antihemophilic factor A, antihemophilic globulin (AHG)	Intrinsic	Protein cofactor
IX	Christmas factor, antihemophilic factor B, plasma thromboplastin component (PTC)	Intrinsic	Endopeptidase with *gla* residues
X	Stuart-Prower factor	Both	Endopeptidase with *gla* residues
XI	Plasma thromboplastin antecedent (PTA)	Intrinsic	Endopeptidase
XII	Hageman's factor	Intrinsic	Endopeptidase
XIII	Protransglutaminase, fibrin stabilizing factor (FSF) fibrinoligase	Both	Transpeptidase

Q.28 What are the various types of hemoglobin? Give important difference between them.

Chains

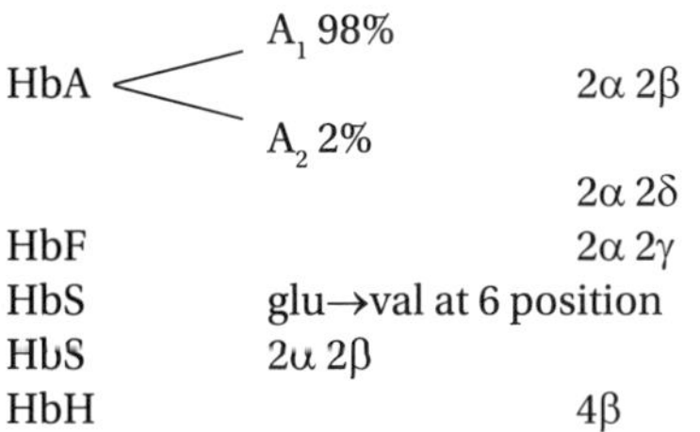

Q.29 What are thalassemias? What are its other names?

Mutations of regulator genes may sometimes represses the synthesis of one type of polypeptide chain of globin, with a compensatory increase in the synthesis of other types of polypeptide chains of globin. The latter may replace the repressed polypeptide chains in the globin molecules thus producing abnormal hemoglobin. Such genetic mutations are called thalassemias. It is also called Cooley's anemia or Mediterranean anemia.

Q.30 Which mutation is present in thalassemia?

Frame shift mutation.

Q.31 Give inhibitor of vitamin K which acts as anticoagulant.

Dicumarol, i.e.warfarin.

Q.32 What is the shape of O_2-Hb-dissociation curve?

Sigmoidal

Q.33 What is the shape of oxygen myoglobin dissolution curve?

Rectangular hyperbola.

Q.34 In lead poisoning RBCs are termed as:

Cabot's ring Howell-Jolly bodies.

Q.35 Which problem arises by massive blood transfusion?

Hemosiderosis.

Q.36 Which problem arises by long standing blood transfusion?

Since 2, 3-DPG level decreases in long time stored blood. So, affinity of O_2 towards Hb increase. So, person receiving blood cannot get O_2 adequately.

Q.37 Which Hb is having highest affinity for O_2 and why?

HbF (fetal hemoglobin) because of low 2,3-DPG level.

Q.38 If all four chains of Hb are b type, then condition is known as?

Hydrops fetalis.

Q.39 Name a Rh incompatibility disease.

Erythroblastosis fetalis.

Q.40 This condition arises in which circumstances?

This condition arises when Rh +ve male is married to Rh –ve female.

Q.41 Define Rh factor.

Presence or absence of protein is referred to as Rh factor.

Q.42 What are different type of Rh factor?

There are two types:

1. Rh + ve
2. Rh –ve.

In Rh +ve if your blood contains the protein then it is called Rh +ve. and If your blood does not contain the protein your blood is said to be Rh –ve.

Chapter 31

Liver Function Tests

Q.1 What are the various liver function tests commonly performed?

1. Serum total proteins, A/G ratio.
2. Prothrombin time.
3. Serum cholesterol.
4. Serum bilirubin.
5. SGOT.
6. SGPT.
7. S. alkaline phosphatase.

Q.2 What is the importance of liver function tests?

1. To assess severity of liver damage.
2. To differentiate different type of jaundice.
3. To find out the presence of latent liver disease.

Q.3 What is the normal serum protein level?

6-8 gm%.

Q.4 What is the level of albumin and globulin?

Albumin 4.0–5.5 gm%.
Globulin 1.5–3.0 gm%.

Q.5 What is the normal A/G ratio?

1.2:1

Q.6 What are the conditions in which total proteins are low?

1. Malnutrition.
2. Liver disease.
3. Nephrotic syndrome.

Q.7 Where is albumin synthesized?

Liver.

Q.8 What happens to A/G ratio, if liver is damage?

If liver is damaged, albumin synthesis is decreased and hence A/G ratio lowers down.

Q.9 If total proteins is low, what happens to A/G ratio.

A/G ratio lower decreases.

Q.10 How will you separate albumin and globulin?

1. By precipitation reactions using 28% sodium sulphite.
2. By electrophoresis.

Q.11 In case of nephrotic syndrome, what will be the abnormality in urine sample?

Urine will show the presence of albumin.

Q.12 Why albumin comes in the urine first?

Albumin is low molecular weight and smaller in size as compared to globulin and hence it is filtered by the glomeruli first.

Q.13 How is albumin detected in urine?

1. By heat coagulation test.
2. By Heller's test.

Q.14 What are the various fractions of plasma protein separated by electrophoresis?

Albumin
α_1-globulin
α_2-globulin
β-glubulin
γ-globulin
Fibrinogen

Q.15 What is the normal serum cholesterol level?

150–250 mg%.

Q.16 What is the percentage of esterified and non-esterified form of cholesterol?

Nonesterified form 30%
Esterified form 70%

Q.17 What are the conditions in which cholesterol level in the blood is increased?

1. Diabetes mellitus.
2. Atherosclerosis.
3. Hypothyroidism.
4. Obstructive jaundice.
5. Nephrotic syndrome.
6. Myxedema.
7. Xanthomatosis.

Q.18 What are the conditions in which cholesterol level in the blood is decreased?

1. Hyperthyroidism.
2. Pernicious anemia.
3. Hemolytic jaundice.
4. Malabsorption syndrome.

Q.19 By which method cholesterol is estimated.

Cholesterol is estimated colorimetrically by Liebermann Burchard reaction.

Q.20 What are the foods rich in cholesterol?

Egg yolk, liver, brain, etc.

Q.21 What is the site of cholesterol synthesis?

Liver, adrenal cortex, skin, intestine, testes, aorta, etc.

Q.22 What is the starting material for cholesterol synthesis?

Acetyl CoA.

Q.23 What are the steps in cholesterol synthesis?

Acetyl CoA → Acetoacetyl CoA → HMG CoA → Mevalonate → Squalene → Cholesterol.

Q.24 How cholesterol biosynthesis is regulated?

Cholesterol biosynthesis is regulated by negative feedback mechanism. The rate-limiting enzyme is HMG-CoA reductase. Dietary cholesterol inhibits the biosynthesis of cholesterol in liver by depressing the synthesis of HMG-CoA reductase in liver. Whereas fasting inhibits cholesterol biosynthesis by diverting HMG-CoA to ketone bodies formation. While high fat diet accelerate the cholesterol production.

Q.25 What are the metabolic products of cholesterol breakdown?

1. Bile acids.
2. Steroid hormones.
3. Fecal steroids (Coprostanol and cholesterol).

Q.26 What does SGOT/SGPT stand for?

- SGOT stands for serum glutamate oxaloacetate transaminase.
- SGPT stands for serum glutamate pyruvate transaminase.

Q.27 What is atherosclerosis?

Deposition of cholesterol and cholesterol esters in the arterial walls leads to a condition known as atherosclerosis.

Q.28 What is the normal serum bilirubin level?

0.2–0.8 mg%.

Q.29 Name the reaction by which bilirubin in estimated colorimetrically.

van den Bergh reaction.

Q.30 What is thymol turbidity test?

When thymol barbitone is added in serum:

a. Faint opalescent color when g globulin is normal, i.e. in prehepatic and posthepatic jaundice.
b. Turbid solution when g globin level increases, i.e. in hepatic jaundice.

Q.31 Give the results of van den Bergh test in different jaundice.

Prehepatic → Indirect +ve.
Hepatic → Biphasic.
Posthepatic → Direct +ve.

Q.32 What is direct bilirubin?

Bilirubin diglucuronide is called direct bilirubin. It is water soluble.

Q.33 What is the indirect bilirubin?

Albumin bound bilirubin is called indirect bilirubin. It is water insoluble.

Q.34 How bilirubin is formed?

Bilirubin is formed as a breakdown product of hemoglobin.

Q.35 What is jaundice?

Jaundice is due to the accumulation of bile pigments (i.e, bilirubin and biliverdin) in the blood giving rise to increased level of these in blood which imparts yellow color in the eyes and the skin.

Q.36 When does bilirubin appear in urine?

Bilirubin appear in urine in obstructive jaundice.

Q.37 What does the presence of bilirubin in urine suggest?

The presence of bilirubin in urine suggests the increased level of direct bilirubin (water soluble) in the blood.

Q.38 What is obstructive jaundice?

Obstructive jaundice is due to the obstruction or blockage of bile duct.

Q.39 What is hemolytic jaundice?

Hemolytic jaundice is due to increased rate of break down of red cells (hemolysis) leading to increased rate of bile pigment formation at a rate exceeding the capacity of the liver to remove the pigment.

Q.40 What is hepatocellular jaundice?

This type of jaundice is due to the impaired capacity of the liver to conjugate bilirubin and secrete the conjugate into the bile.

Q.41 Name some enzymes whose levels are elevated in liver disease.

1. Serum alkaline phosphatase.
2. SGPT.

Q.42 Name the bile pigments.

Bilirubin, biliverdin.

Q.43 What is the normal value of prothrombin time?

14–17 seconds.

Q.44 What are the conditions in which prothrombin time is affected?

1. In liver damage.
2. In obstructive jaundice.

Q.45 What is the vitamin required in formation of prothrombin?

Vitamin K.

Q.46 What is the harm of increased prothrombin time?

Increased prothrombin time signifies a low level of prothrombin, which mean blood will take more to clot and hence the loss of blood will be more.

Q.47 What is neonatal jaundice?

During first 10 days of life a child usually develops jaundice called physiological or neonatal jaundice.

Q.48 What is the reason of developing neonatal jaundice?

1. Excessive destruction of RBC.
2. Hepatic immaturity.

Q.49 What is the treatment of neonatal jaundice?

Phototherapy.

Q.50 Give the procedure of phototherapy?

Skin is illuminated with white light. So that albumin is converted into bilirubin which has shorter life span.

Chapter 32

Detoxification

Q.1 What is detoxification?
Detoxification is a biochemical changes taking place in the body whereby foreign molecules (toxic) are converted to harmless compounds which are more readily excretable.

Q.2 What are xenobiotics?
Xenobiotics refer to all foreign pollutants, food additives, chemicals, drugs, carcinogens, etc.

Q.3 What are the major phases involved in detoxification?
Phase I—hydroxylation (mainly)
Phase II—conjugation.

Q.4 What are the various processes involved in detoxification?

i. Oxidation
ii. Reduction
iii. Hydrolysis
iv. Conjugation

Q.5 What is responsible for hydroxylation?
Cytochrome P_{450}.

Q.6 What is the chemical nature of cytochrome P_{450}?
Cytochrome P_{450} is hemoproteins.

Q.7 What are the main sites of detoxification?
Liver—main seat of detoxification.
Kidney and other organs also participate to some extend.

Q.8 Give the examples of detoxifications?

- **Oxidation:**
  i. Methyl alcohol → Formic acid.
  ii. Indole → Indoxyl
- **Reduction:**
  i. Picric acid → Picramic acid.
  ii. p-Nitrobenzene → p-Nitrophenol
- **Hydrolysis:**
  i. Phenylacetic acid → Phenol
  ii. Aromatic amides → Corresponding acids.
- **Conjugation**
  i. Glycine + benzoic acid → Hippuric acid
  ii. Bilirubin + glucuronic acid → Bilirubin diglucuronide.

Q.9 What factors influence xenobiotics metabolizing enzymes?

1. Age.
2. Sex.
3. Some genetic factors.

Q.10 What is hydroxylation?
It is any chemical process that introduces one or more hydroxyl group (-OH) into a compound thereby oxidizing it.

Q.11 Describe reduction.
It is described as uptake of an electron by a molecule or atom.

Chapter 33

Urine

Q.1 What is the pH of normal urine?
6–6.5.

Q.2 Why does the pH of the urine increases (i.e. becomes alkaline) on standing?
Due to the bacterial conversion of urea to ammonia, which raises the pH of the urine.

Q.3 What is the normal output of urine per day?
1200–1500 mL.

Q.4 What are the conditions in which urine volume is increased?

- *Physiological conditions:*
  1. Excessive intake of water.
  2. Excitement.
  3. Cold climate
  4. High protein diet.
- *Pathological conditions:*
  1. Diabetes mellitus.
  2. Diabetes insipidus.
  3. In certain types of kidney disease.

Q.5 What are the conditions in which urine volume is decreased?

- *Physiological conditions:*
  i. In summer or hot weather due to increase loss of water by perspiration.
- *Pathological conditions:*
  i. Acute nephritis.
  ii. Fever.
  iii. Diseases of heart and lung.
  iv. Diarrhea and vomiting.

Q.6 What is the specific gravity of normal urine?
1.010–1.025.

Q.7 What does the specific gravity of urine indicates?

1. Concentrating power of kidney.
2. State of hydration of body.
3. Presence of solutes in the urine.
4. Effect of ADH.

Q.8 What are the normal constituents of urine?
See Table 33.1.

Table 33.1: Constituents of urine

Organic constituents	*Inorganic constituents*
Urea	Chloride as NaCl
Uric acid	Sodium
Creatinine	Potassium
Hippuric acid	Calcium
Amino acid nitrogen	Phosphorus as phosphates sulphur as sulphates

Q.9 Give the amount of various normal constituents of urine.

Urea	20–30 gm.
Uric acid	0.7 gm.
Amino acids	0.5–1 gm.
Ammonia	0.7 gm.
Creatinine	1.4 gm.
Ascorbic acid	15–20 mg.
Chloride (as NaCl)	10–15 gm.
Inorganic sulphate (as sulphur)	60–120 mg.
Neutral sulphate (as sulphur)	80–160 mg.
Sodium	3–5 gm.
Calcium	0.1–0.3 gm.
Phosphates (as inorganic phosphorous)	1.0–1.5 gm.

Q.10 How will you calculate the total solids excreted in the urine?
The total solids excreted in the urine in gm/liter can be calculated by Long's coefficient (2.66). The figures of second and third decimal places of specific gravity of urine are multiplied with 2.66.

Q.11 What are the abnormal constituents of urine?

i. Reducing sugars.
ii. Proteins (mainly albumin).
iii. Ketone bodies.
iv. Blood
v. Bile pigments (mainly bilirubin).

Q.12 What are the conditions in which abnormal constituents appear in urine?

1. Glucose—diabetes mellitus and renal glycosuria.
2. Proteins—Nephrotic syndrome.
3. Ketone bodies—Starvation and severe diabetes mellitus.
4. Blood—Hematuria.
5. Bilirubin—Obstructive jaundice
6. Urobilinogen—Hemolytic jaundice.

Q.13 What is the hormone which regulates chloride excretion?
Aldosterone.

Q.14 Which hormone is associated with Addison's disease?
Aldosterone.

Q.15 Why it is called aldosterone?
Aldosterone contains an aldehyde group at position C_{18}.

Q.16 What is the normal urinary excretion of chloride?
10–15 gm of chloride as sodium chloride.

Q.17 What are the conditions in which urinary excretion of chloride is increased?
Urinary chloride is increased in:

1. High intake of chloride.
2. Upon resolution of exudate.
3. Addison's disease.

Q.18 What are the conditions in which urinary excretion of chloride is decreased?

1. Low salt intake.
2. Diarrhea, vomiting.
3. Diabetes insipidus.
4. Chronic nephritis.
5. Cushing syndrome.

Q.19 How urine is formed?
Urine is formed by the following four processes.

1. Filtration of blood plasma by the glomeruli.
2. Selective reabsorption by the tubules of materials.

3. Secretion by tubules of certain substances.
4. Exchange of H^+ ions and production of NH_3.

Q.20 Name the hormones involved in urine formation.

i. Aldosterone.
ii. Antidiuretic hormone (ADH).

Q.21 What is the end-product of protein metabolism?

Urea.

Q.22 What is the end product of purine metabolism?

Uric acid.

Q.23 What is the daily excretion of urea?

25–30 gm.

Q.24 What is the level of creatinine in blood?

1–2 mg%.

Q.25 What is the method by which creatinine is estimated?

Jaffe's method.

Q.26 What is creatinine coefficient?

Creatinine coefficient is defined as the number of milligrams of creatinine plus creatinine nitrogen excreted per kilogram of body weight daily.

Q.27 Where it is present?

Creatinine is present in muscle, brain, blood, etc.

Q.28 In which form it is present?

Creatinine is present in free as well as in phosphorylated form.

Q.29 What are the precursors of creatine?

Glycine, arginine, methionine.

Q.30 What is the normal excretion of creatinine?

0.4–1.8 g/day.

Q.31 What are the conditions in which creatinine excretion is decreased?

1. Starvation.
2. Later stages of muscular dystrophy.
3. Muscular weakness.

Q.32 What are the conditions in which creatinine excretion is increased?

i. Early stages of muscular dystrophy when muscle destruction is occurring rapidly.
ii. In any wasting disease involving increased tissue catabolism.
iii. Hyperthyroidism.

Q.33 What tests will you do to assess the function of kidney?

Urine analysis is the biggest kidney function test.

Other kidney function tests are
1. Urea clearance test.
2. Insulin clearance test.
3. Creatinine clearance test.

Q.34 What is urea clearance?

Urea clearance is defined as the number of ml of blood which contain urea when excreted by kidneys in a minute.

$$\text{Urea Clearance} = \frac{\text{mg of urea excreted per minute}}{\text{mg urea per mL of blood}}$$

Q.35 What is maximum urea clearance?

If the rate of excretion of urine is 2 mL or more per minute.

Then maximum urea clearance is defined as:

$$\frac{\text{Observed urea clearance}}{\text{Average normal maximum urea clearance}} \times 100$$

Q.36 What is standard urea clearance?

If the rate of excretion of urea is less than 2 mL.

Standard urea clearance is defined as:

$$\frac{\text{Observed urea clearance}}{\text{Average normal standard urea clearance} \times \text{V}} \times 100$$

Q.37 What is maple syrup urine disease (MSUD)?

It is an extremely rare inherited disease characterized by sweat odor of the urine and sweat.

Q.38 What is urine incontinence?

It is an inability to control the passage of urine. This can range from an occasional leakage of urine to a complete inability to hold any time.

Q.39 What are the types of urinary incontinence?

Two main type of urinary incontinence are:
1. Stress incontinence—Occurs while coughing, sneezing, laughing or while exercise.
2. Urge continence—Involves a strong sudden need to urinate followed by instant bladder contraction and involuntary loss of urine.

Q.40 What is UTI (Urinary tract infection)?

An infection that can happen anywhere along the urinary tract, i.e. the kidney, the ureters, the bladder.

Q.41 What are the risk factors of increased chances of getting UTI?

1. Pregnancy and menopause
2. Kidney stones
3. Sexual intercourse, espically with multiple partners
4. Prostate inflammation
5. Decreased drinking fluids
6. Bowel incontinence
7. Catheterization.

Q.42 What are the symptoms of UTI?

1. Pressure in lower pelvis
2. Pain or burning with urination
3. Frequent or urgent need to urinate
4. Cloudy urine
5. Foul or strong urine odor

Q.43 What is discoloration of urine?

Urine of an abnormal color appears different from the usual straw-yellow color

Q.44 Characteristic color of urine indicating the disease?

1. Cloudy—Urinary tract infection.
2. Dark brown or clear urine—acute viral hepatitis or cirrhosis.
3. Pink, red or smoky brown color urine—kidney cancer, bladder stones, Wilms' tumor, hemolytic anemia, trauma to kidney.
4. Dark yellow or orange urine—laxative or B complex vitamins.
5. Green or blue urine—artificial color in food or drug. Drugs like amitriptyline, indomethacin.

DO YOU KNOW?

- The acient Romans used urine as a bleaching agent for cleaning cloths.
- Darker yellow or brown urine is often observed in the morning after the night's drinking of large quantity of alcohol.
- Women, elderly people and people with diabetes are more prone to urinary tract infections.

Chapter 34

Water and Mineral Metabolism

Q.1 What is the average body water content?

60-70% of the body weight.

Q.2 What are the biological functions of water?

1. Solvent power.
2. Catalytic action.
3. Lubricating action.
4. High latent heat vaporization.
5. High dielectric constant.

Q.3 What is the distribution of water in the body?

1. Intracellular fluid—50% of the body weight
2. Extracellular fluid—20% of the body weight.
   a. Plasma—4.5% of body weight.
   b. Interstitial fluid and lymph fluids—8% of body weight.
   c. Dense connective tissues—6% of body weight
   d. Transcellular fluids—1.5% of body weight.

Q.4 What are the effects of dehydration?

1. Dehydration leading to electrolyte imbalance due to loss of fluid with the electrolyte.
2. Fall in circulating fluid volume leading to shock.

Q.5 What is the principal cation of extracellular fluid?

Sodium.

Q.6 What is the principal cation of intracellular fluid?

Potassium.

Q.7 What is the normal Na^+ and K^+ levels in the serum?

Na^+=137–148 mEq/L.
K^+=3.9–5.0 mEq/L.

Q.8 What are the principal minerals required by the body?

Sodium, potassium, magnesium, phosphorus, sulphur, chloride and calcium.

Q.9 What are the trace elements required by the body?

Iron, iodine, copper, zinc, manganese, cobalt, molybdenum, selenium, chromium, fluoride.

Q.10 What are the general functions of minerals?

1. As structural components of body tissues.
2. In the regulation of body fluids.
3. In acid-base balance.
4. In the transport of gases.
5. In muscular contractions.

Q.11 What are the functions of potassium?

1. Intracellular cation in acid-base balance.
2. In muscle contraction.
3. Conduction of nerve impulse.
4. Cell membrane function.
5. Enzyme action.

Q.12 What are the conditions in which serum potassium level is increased?

Serum K^+ level is increased in:
1. Addison's disease.
2. Advanced chronic renal disease.
3. Diabetic acidosis.
4. Shock.

Q.13 What are the conditions in which serum potassium is decreased?

Serum K^+ level is decreased in:
1. Diarrhea.
2. Metabolic alkalosis
3. Familial periodic paralysis.

Q.14 What are the functions of sodium?

1. In the regulation of acid-base balance.
2. In the maintenance of osmotic pressure of body fluids.
3. In the preservation of normal irritability of muscles and permeability of the cells.

Q.15 What are the conditions in which serum sodium level is increased?

Serum Na^+ is increased in:
1. Cushing disease.
2. Excessive sweating.

Q.16 What are the conditions in which serum sodium level is decreased?

Serum Na^+ is decreased in
1. Acute addison's disease.
2. Vomiting, diarrhea.
3. Intestinal obstruction.
4. Nephrosis.
5. Severe burns.

Q.17 What is the daily requirement of calcium?

800 mg.

Q.18 What are the foods rich in calcium?

Milk, cheese, egg yolk and nuts.

Q.19 What is the normal serum calcium?

9–11 mg%

Q.20 What are the conditions in which serum calcium level is increased?

Serum Ca^{++} is increased in:
1. Hyperparathyroidism.
2. Hypervitaminosis-D.

Q.21 What are the conditions in which serum calcium is decreased?

Serum Ca^{++} is decreased in:
1. Hypothyroidism.
2. Decreased dietary intake.
3. Decreased absorption from the intestines.
4. Increased loss of calcium due to kidney disease.

Q.22 What is the difference between rickets and osteomalacia?

Deficiency of vitamin D gives rise to rickets in children and osteomalacia in adults.

Q.23 What are the functions of calcium?

1. In bones and teeth formation
2. In nerve impulse transmission
3. In muscle contraction
4. In the clotting of blood
5. In the coagulation of milk.
6. Activates certain enzyme systems.
7. In neuromuscular excitability.

Q.24 What should be ideal calcium: phosphorous ratio in the diet?
Calcium: phosphorous ratio should be 1:1 in the diet for ideal absorption.

Q.25 What are the factors which affect calcium absorption?
1. Vitamin D promotes the absorption of calcium.
2. High protein diet promotes calcium absorption.
3. Absorption of calcium requires acidic pH.
4. Phylates, oxalates and phosphates inhibit calcium absorption.

Q.26 What is the normal serum inorganic phosphorus level?
2.5–4.5 mg%.

Q.27 When is the phosphorus level lowered?
1. Rickets.
2. Hyperparathyroidism.
3. Diabetic coma.

Q.28 What are the functions of phosphorus?
1. Formation of bones and teeth
2. Formation of phospholipids.
3. Formation of high energy compounds.
4. Formation of co-enzymes.
5. Formation of organic phosphates.

Q.29 What are the sources of iron?
Liver, kidney, egg yolk, nuts, dates and spinach.

Q.30 How iron is absorbed in the body?
Iron is present in the foodstuffs as Fe^{+++} form. Gastric HCl separates Fe^{+++} from the other combination. Fe^{+++} form is reduced to Fe^{++} form by the reducing substances of the food. Now Fe^{++} form is more soluble and easily absorbed (Fig. 34.1).

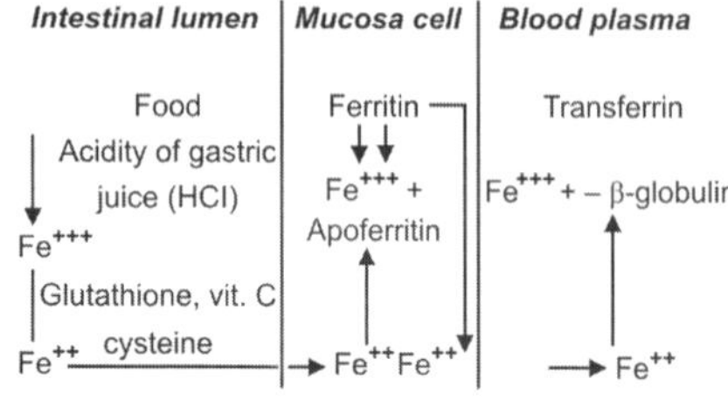

Fig. 34.1: Absorption of iron in the body

Q.31 What is the transported form of iron?
Transferrin.

Q.32 What is the state of iron in transferrin?
Ferric form.

Q.33 Name few iron containing compounds.
Hemoglobin, myoglobin, ferritin, transferrin. Enzymes: catalases, peroxidases, cytochromes b, C_1, c, a.

Q.34 What is ferritin?
It is the stored form of iron in the body.

Q.35 What is the state of iron in ferritin?
Ferric form.

Q.36 What is the daily requirement of iron?
15–20 mg.

Q.37 What is hemosiderosis?
Excessive deposition of iron in the tissues leads to hemosiderosis.

Q.38 What is ceruloplasmin?
It is a transported form of copper in the plasma in combination with proteins.

Q.39 What are the functions of copper?
1. Constituent of certain enzymes such as:
   a. Cytochromes.
   b. Cytochrome oxidase.
   c. Catalase.
   d. Peroxidase.
   e. Tyrosinase.
2. In hemoglobin synthesis.
3. Necessary for growth and bone formation.
4. Helps in the absorption of iron from gastrointestinal tract.

Q.40 What is Wilson's disease?
In Wilson's disease, copper is deposited in liver and brain causing hepatolenticular degeneration.

Q.41 What are the functions of zinc?
1. Necessary for certain enzymes such as:
   a. Carbonic anhydrase.
   b. Carboxy peptidase.
   c. Lactate dehydrogenase.
   d. Alkaline phosphatase.
2. Necessary for protein synthesis and protein digestion.
3. Necessary for optimum insulin action.

Q.42 What is fluorosis?
Excessive intake of fluorine causing mottling and discoloration of the enamel of the teeth.

Q.43 What is the deficiency of magnesium?
Magnesium deficiency can lead to:
1. Mitral valve prolapse
2. Migraine
3. Attention deficient disorder
4. Fibromyalgia
5. Asthma
6. Allergies.

Q.44 What is the distribution of magnesium in human body?
The adult human body contains 25 gm of magnesium. Over 60% of all magnesium is found in skeleton, 27% in muscle, 6.7% is found in other cells and less than 1% is found outside the cell.

Q.45 What are the functions of magnesium?
1. Energy production—Magnesium is required by the adenosine triphosphate (ATP) synthesizing protein in mitochondria.
2. Synthesis of essential molecule—Required at number of steps during synthesis of nucleic acid (DNP and RNA), enzymes participating in the synthesis of carbohydrate and lipids.
3. Structure roles—Play important role in bone, cell membranes and in chromosomes.
4. Ion transport across cell membrane.
5. Cell signaling.
6. Cell migration.

Q.46 What are iron disorders?
1. Hemochromatosis.
2. Acquired iron overload.
3. Sickle cell anemia.
4. Juvenile hemochromatosis.
5. Thalassemia.
6. Porphyria cutanea tarda.
7. Sideroblastic anemia.
8. Iron deficiency anemia.

DO YOU KNOW?

- Soybean protein may lower blood pressure.
- Magnesium is involved in more than 300 essential metabolic reactions.

Nutrition and Energy Requirement

Q.1 What is the purpose of food in the body?
The purpose of food is to provide energy.

Q.2 What is a balanced diet?
A balanced diet is one which contains all the constituents of food in proper proportions so as to provide energy and nutritional requirements for growth and reproduction.

Q.3 What are the components of food?
1. Carbohydrates
2. Fats.
3. Proteins.
4. Minerals.
5. Vitamins.
6. Water.

Q.4 What is the average calorie requirement of a normal person?
1. Sedentary (2500 Kcal)
2. Light work (3000 Kcal)
3. Moderate work (35000 Kcal)
4. Heavy work (4000 Kcal)

Q.5 What is the normal requirement of carbohydrates, fats and proteins?

Carbohydrates	400 gm
Fats	100 gm
Proteins	100 gm

Q.6 What is the protein of milk?
Casein.

Q.7 What does the biological value of protein suggests?
1. The presence and amounts of various essential amino acids.
2. Digestibility.
3. Availability of digested products.

Q.8 What is the biological value of a protein?
The biological value of a protein is a measure of the degree to which nitrogen can be used for growth or maintenance of total body function.
The biological value is expressed as:

$$\text{N intake} = \frac{\text{N loss in feces, sweat, urine}}{\text{N intake} - \text{N feces}}$$

$$\frac{\text{N retained}}{\text{N absorbed}} \times 100$$

Q.9 What do you mean by the quality of protein?
Quality of protein indicates the amino acids make up of the proteins.

Q.10 Why animals proteins are preferred?
Animals proteins are rich in essential amino acids and hence are preferred.

Q.11 What is kwashiorkor disease?
Deficiency of proteins in children leads to a condition called kwashiorkor disease.

Q.12 Which amino acid is deficient in pulses?
Lysine.

Q.13 Why corn or maize diet gives rise to niacin deficiency?
Corn or maize are distinctly deficient in an essential amino acid tryptophan. It is tryptophan which gives rise to the synthesis of niacin during its course of metabolism. Hence corn or maise diet gives rise to niacin deficiency.

Q.14 What are first class and second class proteins?
- First class proteins are those which contains all the essential amino acids in adequate amounts and support good growth. Animal proteins are first class proteins.
- Second class proteins are those which are partially lacking in one or more essential amino acids but cannot support growth. Vegetable proteins are second class proteins.

Q.15 How will you measure the caloric value of foodstuffs?
In bomb calorimeter.

Q.16 What is the unit of measuring energy?
Kilocalorie.

Q.17 What is the caloric value of carbohydrates, fats and proteins?

Carbohydrates	4.1 kcal/gm
Fats	9.3 kcal/gm
Proteins	4.1 kcal/gm

Q.18 What is respiratory quotient (RQ)?
RQ is the ratio of volume of carbon dioxide eliminated to the volume of oxygen utilized in oxidation.

$$RQ = \frac{\text{Volume of } CO_2 \text{ produced}}{\text{Volume of } O_2 \text{ consumed}}$$

Q.19 What is the RQ for carbohydrates?
RQ for carbohydrates is 1.

Q.20 What is basal metabolic rate (BMR)?
The amount of energy required just to maintain the body processes when a person is at complete rest.

Q.21 What are the conditions for the measurement of BMR?

1. Postabsorptive state.
2. Free from mental and physical strain.
3. Patients should be awake.
4. Environment temperature around 20 to 25°C.
5. Recumbent position during the test.

Q.22 What are the conditions in which BMR is

i. Low

ii. High?

BMR is low in

i. Hypoparathyroidism

ii. Starvation.

BMR is high in

i. Hyperparathyroidism

ii. Fever

iii. Polycythemia.

Q.23 What is specific dynamic action (SDA)?

SDA is defined as the extra-amount of heat produced over and above the calorie value of foodstuffs when burnt inside the body.

Q.24 Which has the highest SDA?

Proteins. The SDA of proteins is 30 percent.

Q.25 What is the protein sparing action of carbohydrates and fats?

Carbohydrates and fats make themselves available for energy purpose and thus spare proteins, for body building purposes.

Hormones

Q.1 What are hormones?

Hormones are defined as chemical substances formed in one part of the body, carried in the bloodstream to the other organs where they exert the action.

Q.2 How does hormones differ from enzymes?

Hormones	*Enzymes*
1. The site of origin is far from the site of action	Enzymes are utilized at the site where they are produced.
2. Hormones have to be discharged in the blood before carrying the reaction.	Enzymes show there action in the cell
3. Hormones can be proteins, small peptides, amino-acids or steroids.	Enzymes are only protein in nature.

Q.3 What is a second messenger?

c-AMP

Q.4 How cyclic AMP is formed?

Cyclic AMP is formed the ATP by an enzyme adenyl cyclase.

$$\text{ATP} \xrightarrow{\text{Adenyl cyclase}} \text{c-AMP}$$

Q.5 What are the hormones secreted by pancreas?

1. Insulin.
2. Glucagon.

Q.6 What is the chemical nature of insulin?

It is hypoglycemic hormone.

Q.7 What is the metabolic effects of insulin?

Metabolic effects	*Insulin*
1. Secretion stimulated	Hyperglycemia
2. Blood glucose	Decrease
3. Liver glycogen	Increase
4. Glucose utilization	Increase
5. Gluconeogenesis	Decrease
6. Protein synthesis	Increase
7. Fatty acid synthesis	Increase
8. Triglyceride breakdown	Decrease.

Q.8 What is the metal present in insulin?

Zinc.

Q.9 Which enzyme is responsible for activation of proinsulin?

1. Trypsin like enzyme
2. Carboxy peptidase like enzyme.

Q.10 What is the nature of glucagon?

Glucagon is hyperglycemic hormone.

Q.11 What are the metabolic effects of glucagon?

Metabolic effects	*Glucagon*
1. Secretion stimulated	Hypoglycemia
2. Blood glucose	Increase
3. Liver glucogen	Decrease
4. Gluconeogenesis	Increase
5. Protein synthesis	Decrease
6. Fatty and synthesis	Decrease
7. Triglyceride breakdown	Increase

Q.12 What are the hormones of thyroid gland?

Thyroxine and triiodotyronine.

Q.13 What are T_4 and T_3?

T_4 is thyroxine or tetraiodothyronine.
T_3 is triiodotyronine.

Q.14 Which is more active T_4 or T_3?

T_3.

Q.15 What is the cause of goiter?

Thyroxine deficiency.

Q.16 What are the functions of thyroid hormones?

1. BMR in hyperfunctioning of thyroid gland, BMR increase and vice versa.
2. Intestinal absorption of glucose is increased.

Q.17 Name some thyroid poisons?

Chlorate, perchlorate, pertechnitate, thiocyanate, periodate, nitrate, cyanide, dinitrate, thiouracol, carbimazole, sulphanamide.

Q.18 What are the hormones of parathyroid glands?

1. Parathormone.
2. Calcitonin.

Q.19 Calcitonin is secreted by which cells of thyroid?

C cells or parafollicular cells.

Q.20 How the secretion of parathormone is regulated?

Secretion of parathormone is regulated by the concentration of serum calcium. If the blood calcium is low, parathormone secretion is increased and vice versa.

Q.21 What are the hormones of adrenal medulla?

1. Epinephrine.
2. Norepinephrine.

Q.22 Name the hormones of adrenal cortex.

1. Mineralocorticoides.
2. Glucocorticoides.
3. Sex corticoides

Q.23 What is the difference between epinephrine and norepinephrine?

1. Structural difference:
   Epinephrine is a methylated form whereas norepinephrine is nonmethylated form.
2. Epinephrine is mainly secreted from adrenal medulla whereas norepinephrine is released mainly at nerve ending.
3. Epinephrine is released in blood and from there acts at various organ tissue whereas norephinephrine is acted only at the site of release because of nerve ending.

Q.24 What are the hormones of anterior pituitary?

1. Thyrotropin or thyroid stimulating hormone (TSH).
2. Corticotropin or adrenocorticotropic hormone (ACTH).
3. Growth hormone (GH) or somatotropin (STH).

4. Prolactin.
5. Follicle stimulating hormone (FSH).
6. Luteinising hormone (LH).

Q.25 What are male hormones?
Testosterone.

Q.26 What are female hormones?
1. Estrogen
2. Progesterone.

Q.27 What are functions of testosterone?
1. Stimulates growth maintenance and functions of secondary male sex organs like prostate, seminal vesicles, penis, etc.
2. Develops and maintains the accessory male sex character like beards, moustache, etc.
3. Promotes body growth
4. Growth of seminiferous tubules and spermatogenesis
5. Stimulates development of male external genitalia in the fetus.

Q.28 What are the hormones of gastrointestinal tract?
1. Secretin.
2. Cholecystokinin.
3. Gastrin.
4. Pancreozymin.
5. Enterogastrone

Q.29 What is the target of secretion CCK-PZ?
Gallbladder and pancreas.

Q.30 What are the hormones of posterior pituitary?
1. Vasopressin.
2. Oxytocin.

Q.31 Give another name for vasopressin?
Antidiuretic hormone (ADH).

Q.32 If ADH is less in secretion then which disorder is developed?
Diabetes insipidus.

Q.33 What are the functions of vasopressin and oxytocin?
Vasopressin:
1. It is an antidiuretic hormone (ADH). The absence of ADH causes diabetes insipidus.
2. Raises blood pressure.

Oxytocin:
1. Uterine contraction.
2. Milk secretion.

Chapter 37

Prostaglandins

Q.1 What are prostaglandins?
Prostaglandins are the cyclic derivates of unsaturated fatty acids having twenty carbon atoms.

Q.2 What is the precursor of prostaglandins?
Linoleic acid and linolenic acid.

Q.3 What are the various series of prostaglandins?
Prostaglandins are A, B, E, F, and G or H series.

Q.4 What are the functions of prostaglandins?
1. They lower blood pressure.
2. They are used in the induction of labor, termination of pregnancy and prevention of labor.
3. They are used in the treatment of gastric ulcer.
4. They are used to prevent inflammation.
5. They are used in asthma.

Q.5 What are eicosanoids? How will you classify?
Eicosanoids are compounds derived from "eicosa (20C) polyenoic acid". They consist mainly of two groups (Fig. 37.1):
- Prostanoids (PGs) and
- Leukotrienes (LTs)
- Prostanoids are further subdivided into three main types:
  - Prostaglandins (PGs)
  - Prostacyclins (PGI)
  - Thromboxanes (Tx)

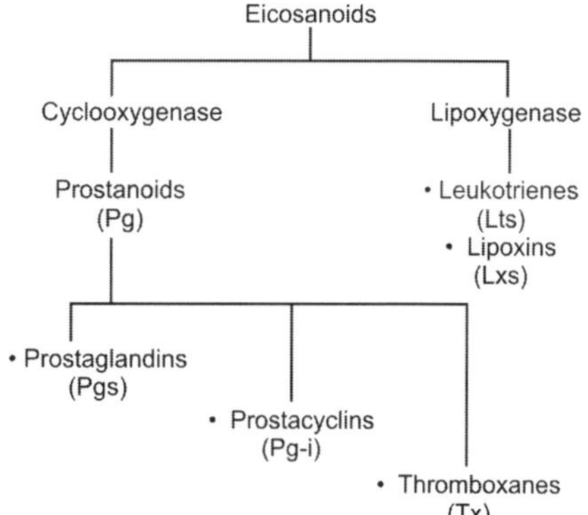

Fig. 37.1: Classification of eicosanoids

Q.6 Name the enzyme and its nature alongwith the cofactors required for PG synthesis.
- PGs are synthesized aerobically from polyunsaturated 20C FA, arachidonic acid with the help of a multienzyme complex, now called *prostaglandin-endoperoxide synthase* which consists of 2 components (Fig. 37.2).
- Cyclo-oxygenase system and
- peroxidase system

The following cofactors are required for the enzyme action:
- Presence of molecular O_2
- Reduced glutathione (G-SH)
- Tetrahydrofolate ($F.H_4$) and
- Heme.

Q.7 Enumerate briefly the functions of PGs.
- Antihypertensive—Lowers BP
- Inhibits platelets aggregation
- *Inhibits gastric secretion:* thus useful in alleviating gastric ulcers.
- *Stimulates GI musculature:* Thus have "purgative action".
- *Produces bronchodilatation:* Thus useful in the treatment of bronchial asthma.

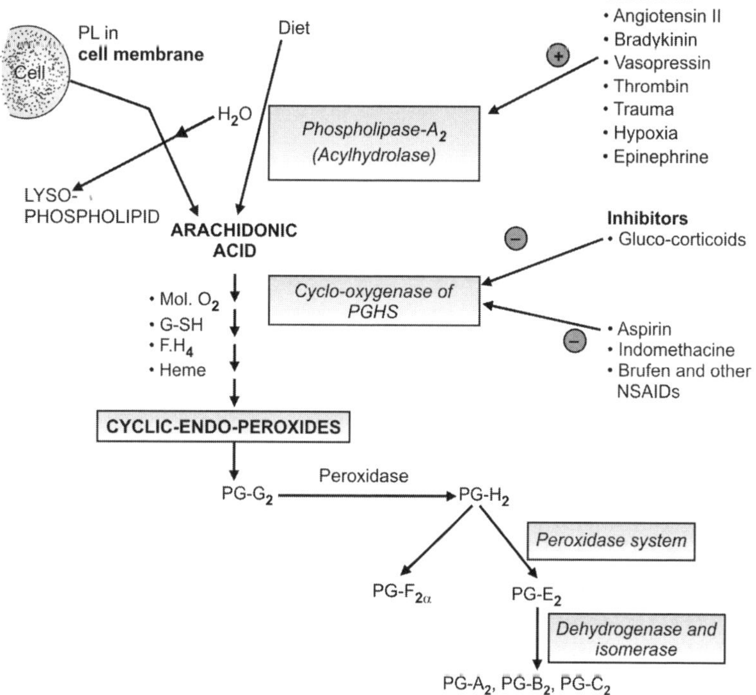

Fig. 37.2: Biosynthetic pathway and regulation of PGs

- *Increases uterine contraction:* Thus can be used as "abortifacient" and "induction of labor".
- *Renal action:* increases RPF, increases GFR, diuresis, matriuresis and kaliuresis.
- Stimulates renin secretion from JG cells.
- *Metabolic effects:*
  - Decreases lipolysis
  - Insulin-like effects
  - PTH-like action produces hypercalcemia
  - TSH-like activity
  - Steroidogenesis.
  - Produces luteolysis

Q.8 What is prostacyclin (PG-I_2) How is it synthesized?

Prostacyclin (PG-I_2) is one of the prostanoids formed in vascular endothelium. It is formed from PG-H_2, cyclic endoperoxide by the action of the microsomal enzyme *prostacyclin synthase.*

Q.9 Mention two most important functions of PG-I_2.

- PG-I_2 inhibits platelet aggregation.
- Produces vasodilatation and decreases BP.

Q.10 What is thromboxane (Tx)? How is it formed in the body?

- Thromboxane is prostanoids. It differs structurally from PGs and PG-I_2 in that it contains an **oxane ring** instead of cyclopentane ring. It is formed principally in platelets, but also in lungs, brain, polymorphs, etc. It is of two types: TxA_1 and TxA_2. In humans TxA_2 is the main type.
- It is synthesized from cyclic endoperoxide PG-H_2 by the action of the enzyme *thromboxane synthetase.*

Q.11 How aspirin prevents thrombus formation?

Aspirin has been found to be a most affective drug which prevents platelet aggregation and thrombus formation. A small amount 30 mg of aspirin can act as antiplatelet aggregator.

Mechanism:

- Aspirin irreversibly acetylates the platelets cyclo-oxygenase system and inhibits TxA_2 formation.
- At the same time, it also inhibits PG-I_2 synthesis in vascular endothelial cells.
- But unlike platelets, the vascular endothelial cells regenerate cyclo-oxygenase very soon.
- Thus over all balance shifts towards the formation of PG-I_2 which prevents platelets aggregation and thrombus formation.

Q.12 What are leukotrienes (LTs)?

Leukotrienes are conjugated trienes formed from eicosanoic acids in leukocytes, mast cells and macrophages by the *lipo-oxygenase* enzyme system in response to immunologic and non-inflammatory stimuli. LTs do not have any ring in its structure but have three characteristic conjugated double bonds.

Q.13 How LTs are synthesized in the body?

LTs are synthesized from arachidonic acid by the addition of hydroperoxy groups to arachidonic acid and produces "hydroperoxyeicosatetranoates" (HPETE) by lipo-oxygenase system. Leukotrienes are formed in our body mainly from 5 HPETE.

Q.14 State the clinical significance of leukotrienes (LTs).

- LTs in general appear to act as mediators in inflammation and anaphylaxis.
- Produce capillary dilatation and increase vascular permeability.
- Inhalation of LTs (C_4, D_4 or E_4) causes bronchospasm. They are about 30 times (E_4) to 1000 times (C_4 and D_4) more potent bronchoconstrictors in normal subjects as compared to histamine.
- Leukotrienes C_4 and D_4 are found to be potent stimulators of mucus secretion from human air-way tissues.
- SRS-A (Slow reacting substance of anaphylaxis) produced by mast cells during anaphylactic reaction has now been shown to be mixtures of LTs like LT-C_4, D_4 and E_4.

Q.15 What are the diseases caused by prostaglandin deficiency?

An absolute Q releative deficiency of prostaglandin can cause:

1. Natural disorder—such as peptic ulcer and diabetes mellitus.
2. Acquired as iatrogenic conditions—cyclosporine nephrotoxicity and gastropathy induced by nonsteroidal anti-inflammatory drugs.

B. Pre-eclampsia.

DO YOU KNOW?

- NSAIDs inhibit cyclooxygenase and reduce prostaglandin synthesis.

Chapter 38

Important Lab Values to Remember

Blood

1. **Hemoglobin**
   - *At birth* : 22–24 gm% (Avg-22 gm%)
   - *In adults*
     - Male : 13.5–18 gm% (Avg-15.5 gm%)
     - Female : 11–16 gm% (Avg-14 gm%)
2. **RBC count**
   - *At birth* : 6–7 million/cumm of blood
   - *In adults*
     - Male : 5–6 million/cumm of blood
     - Female : 4.5–5.5 million/cumm of blood
3. **WBC count (Total)**
   - *At birth* : 20000/cumm of blood
   - *In adults* : 4000–11000/cumm of blood
4. **WBC count (Differential)**
   - Neutrophil : 50–70%
   - Eosinophil : 1–4%
   - Basophil : < 1%
   - Lymphocyte : 20–40%
   - Monocyte : 2–8%
5. **WBC count (Absolute)**
   - Neutrophil : 3000–6000/cumm of blood
   - Eosinophil : 150–300/cumm of blood
   - Basophil : 10–100/cumm of blood
   - Lymphocyte : 1500–2700/cumm of blood
   - Monocyte : 300–600/cumm of blood
6. **Reticulocyte count** : 24000–84000/cumm of blood
7. **Platelet count** : 2–5 lac/cumm of blood
8. **ESR**
   - *Westergren*
     - Male : 3– 5 mm in 1st hour
     - Female : 5–7 mm in 1st hour
   - *Wintrobe's*
     - Male : 0–10 mm in 1st hour
     - Female : 0–20 mm in 1st hour
9. **RBC indices**
   - PCV (Hct)
     - Male : 40–50%
     - Female : 36–47%
   - MCV : 78–94 μm^3
   - MCH : 27–32 pg
   - MCHC : 30–38%
   - CI : 0.85–1.15
10. **Fragility of red cells**
    - Hemolysis begins : 0.5% NaCl
    - Hemolysis completes : 0.36% NaCl
11. **Bleeding time**
    - Duke's method : 2–5 min
    - Ivy's method : 3–9 min
12. **Coagulation time**
    - Capillary blood (Capillary glass tube method) : 3–8 min
    - Venous blood (Lee and White method) : 5–15 min
13. **Prothrombin time** : 14–18 sec
14. **Total blood volume** : 5–6 L
15. **Specific gravity**
    - Of blood : 1.048–1.066
    - Of RBC : 1.092–1.095
    - Of plasma : 1.026–1.035
16. **Frequency distribution of blood group in India**
    - A blood group : 21%
    - B blood group : 39%
    - AB blood group : 9%
    - O blood group : 31%
    - Rh positive : 95%
    - Rh negative : 5%

Plasma/Serum

1. **Plasma volume**
   - Male : 39 mL/ kgbw
   - Female : 40 mL/kgbw
2. **Osmolality**:
   - Plasma : 285–295 mosm/kg of serum water
   - ECF : 3000 mosm/L
   - ICF : 300 mosm/L
3. **NPN Substances** : 20–30 mg/dL
4. **Proteins (Total)** : 5.5–8 gm/dL
   - Albumin : 3.5–5 gm/dL
   - Globulin : 2–3.5 gm/dL
   - A/G ratio : 1.2:1
   - Fibrinogen : 0.2–0.5 gm/dL
5. **Amino acids** : 6–8 gm/dL
6. **Total lipid** : 450–1000 mg/dL
7. **Cholesterol** : 120–220 mg/dL
8. **Fatty acids** : 160–270 mg/dL
9. **Triglycerides** : 25–150 mg/dL
10. **Glucose**
    - Fasting : 80–110 mg/dL
    - Postprandial
      - Normal : 140 mg/dL
      - Diabetic : > 200 mg/dL

11. **Sodium** : 135–145 mEq/L
12. **Potassium** : 3.5–5 mEq/L
13. **Chloride** : 95–108 mEq/L
14. **Bicarbonate** : 26–28 mmol/L
15. **Calcium** : 2.12- 2.62 mEq/l
    : 8.5–10.5 mg/dL
16. **Magnesium** : 2–3 mg/dL
    : 1.5–2.0 mEq/L
17. **Phosphate** : 2.5–4.5 mg/dL
18. **Amylase** : 13–53 nmol/L
    : 60–180 Somogyi unit/dL
19. **SGOT** : 100–300 µmol/L
20. **SGPT** : 50–430 µmol/L
21. **Alkaline phosphatase** : 0.4–1.5 µmol/L
22. **Bilirubin**
    - Total (indirect) : Up to 1 mg/dL
    - Conjugated (direct) : Up to 0.4 mg/dL
23. **Creatinine** : 0.6–1.5 mg/dL
24. **Urea** : 20–40 mg/dL
25. **Uric acid**
    - *Male* : 3.6–8.5 mg/dL
    - *Femal* : 2.3–6.6 mg/dL
26. **Protein bound iodide** : 3.5–8 µg/dL
27. **Thyroid binding globulin** : 7–17 mg/L
28. **Thyroxin** : 70–140 nmol/L
29. **Tri-iodothyronine** : 1.2–3.0 nmol/L
30. **Serum iron** : 65–175 µg%
31. **Serum iron binding capacity** : 250–410 µg%

Cardiovascular System

1. **Pressure inside the heart (in mm Hg)**
    - Right atrium : 0–8
    - Right ventricle (Systolic) : 15–30
    - Right ventricle (Diastolic) : 0–8
    - Left ventricle (Systolic/aortic) : >150
    - Left ventricle (End diastolic) : 3–12
    - Pulmonary artery (Systolic) : 15–30
    - Pulmonary artery (End diastolic) : 3–12
    - Pulmonary capillary : 8
2. **Capillary pressure (Systemic)**
    Arterial end : 32 mm Hg
    Venous end : 12 mm Hg
    Mean : 20 mm Hg
3. **Heart rate (Resting)**
    - In adult : 60–90 beat/min
    - In infant : 120–140 beat/min
4. **Arterial BP (Resting)**
    Systolic : 110–140 mm Hg
    Diastolic : 60–90 mm Hg
    Mean : 100 mm Hg
5. **Cardiac output** : 5–6 L/min
6. **Cardiac index** : 2.5–3.6 L/min/m^2 BSA
7. **Stroke volume index** : 40 mL/sq.m /beat
8. **Stroke volume** : 70–90 mL
9. **End systolic volume** : 50 mL
10. **End diastolic volume** : 130 mL
11. **Coronary blood flow** : 225 mL/min
12. **Cerebral blood flow** : 750 mL/min
13. **Renal blood flow** : 1300 mL/min
14. **Circulation time:**
    Arm to tongue (saccharin) : 14.2 ± 2.4 sec
    Arm to lung (ether) : 4–8 sec
    Arm to foot (radiosodium) : 15–105 sec.

Respiratory System

	Gas	Percentage	Partial pressure (mm Hg)
1. **Inspired air (Atmospheric air)**			
	O_2	20.96	160
	CO_2	0.04	0.15
	N_2	79	600
2. **Alveolar air**			
	O_2	13.2	101
	CO_2	5.3	40
	N_2	75.3	572
	Water	6.2	47
3. **Expired air**			
	O_2	15.5	116.5
	CO_2	3.6	27.5
	N_2	74	569
	Water	6.2	47
4. **Arterial blood**			
	O_2	19.3	100
	CO_2	48	40
5. **Venous blood**			
	O_2	14.1	40
	CO_2	52	46

6. **Respiratory rate** : 12–18 breaths/min
7. **Respiratory quotient** : 0.8
8. **O_2 consumption at rest** : 250 mL/min
9. **O_2 diffusion capacity**
    In male : 27–42 mL/min/mm Hg
    In female : 28–30 mL/min/mm Hg
10. **Tidal volume** : 500–800 mL
11. **Inspiratory reserve volume** : 2000–3300 mL
12. **Expiratory reserve volume** : 1100 mL
13. **Residual volume** : 1200 mL
14. **Inspiratory capacity** : 3500 mL
15. **Expiratory capacity** : 1600 mL
16. **Functional residual capacity** : 2200 mL
17. **Vital capacity**
    In male : 4500–5500 mL
    In female : 3500–4500 mL
18. **Total lung capacity** : 5700 mL
19. **Pulmonary ventilation (RMV)** : 6–7 L/min
20. **Alveolar ventilation** : 4.9 L/min
21. **Maximum voluntary ventilation:**
    - *In male* : 150–170 L/min
    - *In female* : 80–100 L/min
22. **FEV_1** : 75–85%
23. **Dsypnoeic index** : 60–70 %
24. **Respiratory dead space volume** : 150 mL
25. **Lung compliance:**
    - *Lungs alone* : 220 mL/cm of water
    - *Lungs and Thorax* : 130 mL/cm of water
26. **O_2 transported through plasma** : 0.3 mL/100 mL of blood/100 mm Hg PO_2
27. **O_2 carrying capacity of blood** : 20 mL of O_2

28. CO_2 transport

Through plasma: 0.3 mL/100 mL of blood (7%)

- Through bicarbonate: 3 mL/100 mL of blood (70%)
- Through carbamino compound: 0.7 mL/100 mL of blood (0.7%)

Excretory System

1. **No. of nephron/kidney** : 1 million
2. **GFR** : 125 mL/min (180 L/day)
3. **Renal fraction** : 12–30%
4. **TmG of glucose:**
   In male : 300–450 mg/min
   In female : 250–350 mg/min
5. **Inulin clearance:**
   In male : 124 ± 25.8 mL/min
   In female : 119 ± 12.8 mL/min
6. **Creatinine clearance** : 91–130 mL/min
7. **PAH clearance (Renal plasma flow):**
   In male : 560 ± 630 mL/min
   In female : 490 ± 700 mL/min
8. **Urea clearance:**
   Maximal clearance : 60–90 mL/min
   Standard clearance : 40–65 mL/min
9. **Normal urinalysis**

Total 24 hours volume	1.5 liters (average)
Color	Straw or yellowish
Transparency	Clear
Sediments	Nil
Odor	Urinous (aromatic)
Reaction	Acidic (Average pH 6.0)
Specific gravity	1.010–1.025
Albumin	Nil
Sugar	Nil
Acetone bodies	Nil
Bile pigments	Nil
Bile salts	Nil
Urobilinogen	Not increased
Crystals	May or may not be present
Pus cells	Nil
RBCs	Nil
Casts	Nil
Bacteria	Nil

10. **Composition of urine:**

Various constituents of the urine, with 1.5 liters average volume, are as follows:

Inorganic constituents:

NaCl	15 gm
Phosphates	3 gm
Potassium as KO_2	3 gm
Calcium	0.3 gm
Magnesium	0.2 gm
Sulfates	2.5 gm
Iron	0.003 gm

Nitrogenous constituents:

Urea	20–30 gm
Uric acid	0.7–1 gm
Ammonia	0.6–0.7 gm
Creatinine	1.5 gm

Other constituents

Hippuric acid	0.6 gm
Indian	0.01 gm
Urinary	17-ketosteroids
Males:	6–18 mgm/24 hrs urine
Females:	4–14 mgm/24 hrs urine.

Reproductive System

1. **Volume of semen** : 3.1 mL
2. **Fructose content of semen** : 40–400 mg/dL
3. **Duration of spermatogenesis** : Approx. 74 days
4. **Testosterone production** : 10 mg/day
5. **Duration of menstrual cycle** : 28 ± 4 days
6. **Duration of menstrual blood flow** : 3–5 days
7. **Menstrual blood loss** : 10–55 mL
8. **Menarche** : 9–16 years
9. **Menopause** : 45–50 years
10. **Gestational period** : 280 days

Composition of CSF

Total volume	100–110 mL
Specific gravity	1004–1006
Pressure	100–110 mm water
Cell count	0–3 per c. mm
Sodium	334 mgm%
Potassium	10–12 mgm%
Calcium	5.3 mgm%
Chloride	700 mgm%
Bicarbonates	40–60 mgm%
Phosphates	1.8 mgm%
Sulfates	0.6 mgm%
Glucose	50–70 mgm%
Proteins	20–40 mgm%
Cholesterol	0.06–0.2 mgm%
Urea	10–30 mgm%

DAILY INTAKE OF VITAMINS

Fat-soluble vitamins

1. Vitamin A — 5000 IU
2. Vitamin D — 400–800 IU
3. Vitamin E — 30 mg
4. Vitamin K — Endogenous synthesis of vitamin K is adequate to meet the requirement.

WATER-SOLUBLE VITAMINS

Vitamin C — 60 mg

Vitamin B complex

1. Thiamine (Vitamin B_1) — 1.4 mg
2. Riboflavin (Vitamin B_2) — 0.6–1.8 mg
3. Pantothenic acid (Vitamin B_3) — 5–15 mg
4. Niacin (Vitamin B_5) — 15–20 mg
5. Pyridoxine (Vitamin B_6) — 2 mg
6. Biotin (Vitamin B_7) — 100–300 μg
7. Folic acid (Vitamin B_9) — 300–500 μg
8. Cyanocobalamin (Vitamin B_{12}) — 1.5 μg

BIOCHEMICAL CHANGES IN ACID-BACE BALANCE

	Normal	Respiratory		Metabolic	
		Acidosis	Alkalosis	Acidosis	Alkalosis
Plasma					
pH	7.4	↓	↑	↓	↑
H^+	40 nano Eq/L	↑	↓	↑	↓
PCO_2	40 mm Hg	↑	↓	↓	↓
HCO_3	24 mEq/L	↑	↓	↓	↓
H_2CO_3	1.2 nano mEq/L	↑	↓	↓	↓
Na^+	142 mEq/L	↑	↑	↓	↓
K^+	4 mEq/L	↓	↑	↓	↓
Cl^-	100 mEq/L	↑	↑	↓	↓
Urine					
pH	6 ± 1	5 ± 1	7 ± 1	5 ± 1	7 ± 1
NH_4	35 mEq/L	↑	↓	↑	↑
Titrability acid	35 mEq/L	↑	↓	↑	↓
Na^+	4 g/day	N	N	N	N
K^+	2.8 g/day	↑	↑	↓	↑
Cl^-	10 g/day (as NaCl)	↑	↓	↑	↓

DAILY INTAKE OF MINERALS

Sodium	4–6 gm Na^+
Potassium	2–3 gm
Calcium	800–1000 mg
Magnesium	150–450 mg
Iron	10–15 mg
Copper	2.5 mg
Chloride	10–15 gm of Cl^- as NaCl
Phosphorus	800–1500 mg
Iodine	120–130 µg
Zinc	15–20 mg

Appendix

Preparation of Standard Solutions

1. **Percent solution**
   a. **Weight per weight (W/w):**

   $$\text{Percent solution} = \frac{w}{W} \times 100$$

   Where,
   w = Weight of solute
   W = Weight of solvent

   Example: 5 percent NaCl — 5 gm NaCl + 95 gm H_2O.

   b. **Weight per volume (W/V):**

   $$\text{Percent solution} = \frac{W}{V} \times 100$$

   Where,
   W = Weight of solute.
   V = Volume of solvent.

   Example: 5 percent NaCl—5 gm of NaCl dissolved in H_2O and make up the volume up to 100 mL.

   c. **Volume per volume (v/V):**

   $$\text{Percent solution} = \frac{v}{V} \times 100$$

   Where,
   v = Volume of solute
   V = Volume of solvent

   Example:10 percent acetic acid—10 mL of glacial acetic acid is diluted with waer and make up the volume up to 100 mL.

2. **Normal solutions:** The number of gram equivalent of solute dissolved in 1 liter is called normal solution.

   $$\text{Normality} = \frac{\text{Gram equivalent of solute dissolved in 1 liter}}{\text{Equivalent weight}}$$

   $$\text{Equivalent weight} = \frac{\text{Molecular weight}}{\text{Valency}}$$

   a. For monovalent compounds normality = molarity
   b. If valence is more than 1 then
   Normality = Molarity × valency

   Example:

   i. *For solids*: To prepare 1 N NaOH solution. As NaOH is monovalent, therefore equivalent weight is 40 (atomic weight of Na = 23, O = 16, H = 1).
   ∴ 40 gm NaOH dissolved in 1 liter of water = 1 N.

   ii. *For liquids*: To prepare 10 N H_2SO_4 solution $N_1V_1 = N_2V_2$
   Where, N_1 = Normality to be prepared
   V_1 = Volume to be prepared in mL
   N_2 = Known normality (H_2SO_4 = 36)
   V_2 = To calculate (in mL)

   $\therefore\ 10 \times 1000 = 36 \times V_2$

   $$V_2 = \frac{10 \times 1000}{36}$$

   $\therefore\ V_2 = 277$ mL

   ∴ 277 mL of H_2SO_4 is diluted to 1 liter = 10 N

3. **Molar solution:** The number of gram moles or molecules dissolved in 1 liter of water is called molar solution.

   Example: Molecular weight + (atoms per mole) of NaOH = 40 (atomic weight of Na = 23, O = 16, H = 1).
   i.e., 40 gm of NaOH dissolved in 1 liter of water is 1 M NaOH.

4. **To convert mg/100 mL to mmol/L**
   mg/100 mL × 10 = mg/L

   $$\frac{\text{mg/100 mL} \times 10}{\text{mol. weight}} \text{ or } \frac{\text{mg/L}}{\text{mol. weight}} = \text{mmol/L}$$

   $$\frac{\text{mmol/L} \times \text{mol. weight}}{10} = \text{mg/100 mL}$$

S. No.	Element		Atomic number	Relative atomic mass	Valency
1.	Aluminum	(Al)	13	26.982	3
2.	Arsenic	(As)	33	74.992	3, 5
3.	Bismuth	(Bi)	83	208.98	3, 5
4.	Boron	(B)	5	10.811	3
5.	Bromine	(Br)	35	79.904	1, 3, 5, 7
6.	Calcium	(Ca)	20	40.078	2
7.	Carbon	(C)	6	12.011	3, 4
8.	Cobalt	(Co)	27	58.933	2, 3
9.	Copper	(Cu)	29	63.546	1, 2
10.	Chlorine	(Cl)	17	35.453	1, 3, 5, 7
11.	Fluorine	(F)	9	18.998	1
12.	Hydrogen	(H)	1	1.0079	1
13.	Iron	(Fe)	26	55.845	1, 3, 6
14.	Iodine	(I)	53	126.90	1, 3, 5, 7
15.	Lead	(Pb)	82	207.2	2, 4
16.	Lithium	(Li)	3	6.941	1
17.	Magnesium	(Mg)	12	24.305	2
18.	Manganese	(Mn)	25	54.938	2, 3, 4, 6, 7
19.	Mercury	(Hg)	80	200.59	1, 2
20.	Molybdenum	(Mo)	42	95.94	2, 3, 4, 5, 6
21.	Nitrogen	(N)	7	14.007	2, 3, 4, 5
22.	Oxygen	(O)	8	15.999	2, 1

S. No.	Element		Atomic number	Relative atomic mass	Valency
23.	Phosphorus	(P)	15	30.974	3, 5
24.	Potassium	(K)	19	39.098	1
25.	Selenium	(Se)	34	78.96	4, 6
26.	Silver	(Ag)	47	107.87	1, 2
27.	Sodium	(Na)	11	22.990	1
28.	Sulphur	(S)	16	32.066	2, 4, 6
29.	Zinc	(Zn)	30	65.39	2

Abbreviations

1. AMP : Adenosine Monophosphate
2. ADP : Adenosine Diphosphate
3. ATP : Adenosine Triphosphate
4. ALT : Alanine Transaminase
5. AST : Aspartate Transaminase
6. A : G ratio : Albumin Globulin ratio
7. AIDS : Acquired Immuno Deficiency Syndrome
8. ALA Syn. : Aminolevulinic Acid Synthetase
9. BSLF : Blood Sugar Level Fasting
10. BSL PP : Blood Sugar Level Postprandial
11. BMR : Basal Metabolic Rate
12. BCG : Bromocresol Green
13. B_6PO_4 : Pyridoxal phosphate
14. cAMP : Cyclic Adenosine Monophosphate
15. CSF : Cerebrospinal Fluid
16. DPL : Dipalmitoyl Lecithin
17. DNA : Deoxyribonucleic Acid
18. DM : Diabetes Mellitus
19. ETC : Electron Transport Chain
20. FMN : Flavin Mononucleotide
21. FAD : Flavin Adenine Dinucleotide
22. GTP : Guanosine Triphosphate
23. GFR : Glomerular Filtration Rate
24. GDP : Guanosine Diphosphate
25. GTT : Glucose Tolerance Test
26. GOD : Glucose Oxidase
27. GIT : Gastrointestinal Tract
28. HDL : High Density Lipoprotein
29. HMG CoA : Hydroxymethylglutaryl Coenzyme A
30. HbS : Sickle Hemoglobin
31. IU : International Unit
32. Kcal : Kilo calories
33. K_m : Michaelis-Menten dissociation constant
34. LDL : Low Density Lipoprotein
35. NAD : Nicotinamide Adenine Dinucleotide
36. NADP : Nicotinamide Adenine Dinucleotide Phosphate
37. OAA : Oxaloacetoacetate
38. OD : Optical Density
39. POD : Peroxidase
40. PLP : Pyridoxal phosphate
41. PFF : Protein Free Filtrate
42. PPt : Precipitation/Precipitate
43. PAPS : Phospho Adenosyl Phospho Sulphate
44. PEM : Protein Energy Malnutrition
45. PG : Prostaglandins
46. RDA : Recommended Daily Allowance
47. RNA : Ribonucleic Acid
48. RQ : Respiratory Quotient
49. RE System : Reticuloendothelial System
50. RBCs : Red Blood Cells
51. Rf : Relative front
52. SGOT : Serum Glutamate Oxaloacetate Transaminase
53. SGPT : Serum Glutamate Pyruvate Transaminase
54. SDA : Specific Dynamic Action
55. TCA : Trichloroacetic Acid
56. TCA cycle : Tricarboxylic Acid Cycle
57. TPP : Thiamine Pyrophosphate
58. VLDL : Very Low Density Lipoprotein